Metabolism and Medicine

Foundations of Biochemistry and Biophysics

This textbook series focuses on foundational principles and experimental approaches across all areas of biological physics, covering core subjects in a modern biophysics curriculum. Individual titles address such topics as molecular biophysics, statistical biophysics, molecular modeling, single-molecule biophysics, and chemical biophysics. It is aimed at advanced undergraduate- and graduate-level curricula at the intersection of biological and physical sciences. The goal of the series is to facilitate interdisciplinary research by training biologists and biochemists in quantitative aspects of modern biomedical research and to teach key biological principles to students in physical sciences and engineering.

Authors are also welcome to contact the publisher (Physics Editor, Carolina Antunes: carolina.antunes@tandf.co.uk) to discuss new title ideas.

Light Harvesting in Photosynthesis
Roberta Croce, Rienk van Grondelle, Herbert van Amerongen, Ivo van Stokkum (Eds.)

An Introduction to Single Molecule Biophysics
Yuri L. Lyubchenko (Ed.)

Biomolecular Kinetics: A Step-by-Step Guide
Clive R. Bagshaw

Biomolecular Thermodynamics: From Theory to Application
Douglas E. Barrick

Quantitative Understanding of Biosystems: An Introduction to Biophysics
Thomas M. Nordlund

Quantitative Understanding of Biosystems: An Introduction to Biophysics, Second Edition
Thomas M. Nordlund, Peter M. Hoffmann

Entropy and Free Energy in Structural Biology: Thermodynamics, Statistical Mechanics And Computer Simulation
Hagai Meirovitch

Metabolism and Medicine: Two Volume Set
Brian Fertig

https://www.crcpress.com/Foundations-of-Biochemistry-and-Biophysics/book-series/CRCFOUBIOPHY

Metabolism and Medicine

The Physics of Biological Engines

Volume 1

Brian J. Fertig, M.D., F.A.C.E.

Associate Professor, Robert Wood Johnson Medical School
Chairman, Department of Diabetes & Endocrinology,
Hackensack Meridian Health @ JFK University Medical Center
Founder Diabetes & Osteoporosis Center, Piscataway, N.J.

Editorial Assistance by Jack Tuszynski, Ph.D.

CRC Press
Taylor & Francis Group
Boca Raton London New York

CRC Press is an imprint of the
Taylor & Francis Group, an **informa** business

First edition published 2022
by CRC Press
6000 Broken Sound Parkway NW, Suite 300, Boca Raton, FL 33487-2742

and by CRC Press
2 Park Square, Milton Park, Abingdon, Oxon, OX14 4RN

Library of Congress Cataloging-in-Publication Data

Names: Fertig, Brian, author.
Title: Metabolism and medicine / Brian Fertig, M.D., Robert Wood Johnson
Medical School, Department of Diabetes & Endocrinology Hackensack
Meridian, Health @ JFK University Medical Center.
Description: First edition. | Boca Raton : CRC Press, 2022. | Series:
Foundation of biochemistry and biophysics | Includes bibliographical
references and index.
Identifiers: LCCN 2021027983 (print) | LCCN 2021027984 (ebook) | ISBN
9780367699918 (hardback) | ISBN 9780367712259 (paperback) | ISBN
9781003149873 (ebook)
Subjects: LCSH: Metabolism--Disorders. | Human body--Microbiology.
Classification: LCC RC627.5 .F47 2022 (print) | LCC RC627.5 (ebook) | DDC
616.3/9--dc23
LC record available at https://lccn.loc.gov/2021027983
LC ebook record available at https://lccn.loc.gov/2021027984

ISBN: 9780367699918 (hbk)
ISBN: 9780367712259 (pbk)
ISBN: 9781003149873 (ebk)

DOI: 10.1201/9781003149873

Typeset in Times
by Deanta Global Publishing Services, Chennai, India

Dedicated to my wife, Eileen, and my son, Matthew,

without whom none of this would have even mattered.

Contents

Preface to Volume One

Dear reader,

First of all, I wish to express my appreciation to you for deciding to invest your time and money to read my book. The potential audience to which you belong spans a wide range of backgrounds from clinical physicians and other health care providers, to medical students and experts from a diversity of basic scientific disciplines as well as laypersons interested in biomedical research. Being aware of this, I made a concerted effort to meet increased demands for clarity and wherever possible simplicity, without compromising precision. Due to the interdisciplinary nature of this work, it covers many fields of natural and life sciences that have a profound impact on our current and future understanding of health and disease. I do hope that each reader will learn something new and exciting in these two volumes covering both basic and applied aspects of metabolism.

As you embark on this intellectual journey with me, it is important that I explain to you the compelling motivation that inspired me to write this book series. The reason has to do with the question of what motivates one to become a physician in the first place. There are three major drives that affect our decision-making: autonomy, mastery, and purpose. Autonomy allows one to decide within meaningful limits what, when, how, and with whom we engage to do things together. Mastery is a drive that equates to the motivation to improve. Mastery of skills has an asymptotic learning curve based on the notion that perfection is only hypothetical but improvement is our constant objective. For example, with each patient experience, as an endocrinologist, I learn something new every day, no matter how subtle, about reading between lines to better understand both fears and expectations. This progressively enhances the fulfilling nature of patient care for both the patient and the provider. However, no matter how many patients one sees over a career, there is always more to learn. As eloquently stated by Daniel Pink, "autonomous people working toward mastery perform at very high levels".[1] Further, Pink beautifully goes on to describe purpose stating that, "those who do so in service of some greater objective can achieve even more. The most deeply motivated people—not to mention those who are most productive and satisfied—hitch their desires to a cause larger than themselves". The more of these elements of autonomy, mastery, and purpose a physician has, the more engaged they become, which is tantamount to performance. One final and crucial quote of Pink paints an accurate picture, "when it comes to motivation, there's a gap between what science knows and what business does. Our current operating system is built around external carrot-and-stick motivators that don't work, and often do harm. We need an upgrade". Lastly, Charles Black powerfully states "big vision and small daily steps = unexpected success".[2] These are the quotes that resonated with me as I made efforts toward the completion of this work.

The two books *Metabolism and Medicine: The Physics of Biological Engines* and *Metabolism and Medicine: The Metabolic Landscape of Health and Disease* can be viewed as a primer for a scientific foundation that is aimed to guide students, physicians, and other healthcare professionals (applied scientists), as well as the larger family of basic interdisciplinary and computer scientists, and to serve as a resource for the pursuit of these three basic drives.

Ten years ago, at the age of 50 and having practiced medicine for over 20 years, my feelings of fulfilment were ambivalent. While the kinship, beautiful memories, and real privilege to serve patients and their families over the years was intrinsically rewarding, there was also an unspoken void I experienced practicing medicine. The *modus operandi* of medicine can be briefly summarized as based on differential diagnosis through the use of pattern recognition, which is algorithmic, and to a large extent still intuitive. I felt disheartened about this state of affairs on two fronts. First, in order to truly help people live healthier and longer lives, we needed to know more about both the individual patient and the mechanism behind the disease. The prototypical process of patient care fundamentally is too generic, failing to invoke Big Data that includes metabolomics, proteomics, microbiomics, genetic uniqueness, and its epigenetic expression. Integrating Big Data and performing the necessary analytic tools is the next level of complexity that still awaits full implementation in medical practice. Perhaps most importantly, we need a quantitative model guiding our decision making; a model that integrates and maps out this vast amount of individualized data, and the patient's changing trajectory of the state of health over time. Additionally, essential in this process is to know more about our patients in order to understand the background to the disease they succumbed to. Included in the inadequacies is also the amount of time allocated to each patient, which is insufficient to delve into patient history beyond a few superficial questions. Next, in order to answer "the why", i.e. the root cause of the patient's state of health, we also need to know more both in terms of etiology and science. This is indeed the most important challenge facing medicine in the near future.

While the quest on each front is different, their solutions are intricately entwined and interdependent. Unlike any other area of human activity, health care is inherently complex and requires a multitude of fields of knowledge beyond any single individual's intellectual capabilities. Therefore, implementing new ideas in medical practice will require a much broader health care team of interdisciplinary contributors than is currently the case. A major inadequacy in the way healthcare is administered and delivered today is the fragmentation of medical disciplines. A multitude of specialists in each subfield of medicine practice their craft in what can metaphorically be described as "splendid isolation". This means that interactions between specialists are often limited to one-directional referrals and even proper patient data transfer, let alone data integration, is not commonplace. Instead, one would like to see a deeply interwoven network of multi-disciplinary interactions with other fields of medicine and interdisciplinary collaboration with fields outside medicine. In my opinion, this is one of the key aspects that can be successfully addressed by providing a comprehensive model of the patient's physiological state using a precision personalized scale of medicine. This faces the basic and redoubtable challenge of modeling the virtually infinite complexity of the fabric of biological systems such as the human body. This approach is still very nebulous and to progress it to the next stage we need major conceptual advances, which in my opinion are within our reach.

Therefore, I am very optimistic that the answers to these questions are available and can be implemented in medical practice within a reasonable time frame. In fact, the Physiological Fitness Landscape (PFL) model proposed in this book provides a tangible framework and vision on which to build a new horizon of transformative potential for the future of medicine. In this connection, deep interdisciplinary perspectives of both basic and applied sciences are integral to understanding the parameters of the PFL. There is a robust number of these model parameters and even more data points (hundreds of thousands to a million) that need to be introduced into basic "rules" to simplify the sheer complexity of the system and to be integrated into a PFL. Additional guidance gained from the use of artificial intelligence models such as IBM's Watson Oncology can trim down the numbers of essential parameters to a manageable few. To accomplish these goals, we'll require a very large basic interdisciplinary scientific team of physicists (specializing in biological thermodynamics, electromagnetism, complexity theory, biophysics including quantum biology, and mathematical and computer physics), biological chemists, molecular and cell biologists, physiologists, stress experts, and metabolic biologists. We will also require multidisciplinary sub-specialties of clinical medicine that include endocrinologists, cardiologists, etc. Finally, teams of scientific administrators will be necessary to collate, steer and attach all the information into each individual PFL. In order to maintain cost efficiency for delivering individualized care on a wide public health scale, it will be essential to organize a huge number of "data assembly lines" that can accommodate demand with minimal waste. All of this can be done because if there is a will, there is always a way.

I hope that this two-part series of books can provide the broad outlines of what needs to be integrated in the future of medicine as a more rigorous discipline of applied science with enormous importance to the knowledge- and information-driven society of tomorrow. Volume 1 introduces each of the major scientific disciplines in the context of medicine while Volume 2 is more focused on translational medicine in the context of metabolic physiology. The combined information of the two volumes attempts to answer the questions of etiology, "the why", and the sequence of physiology that allows the generation of optimally conducted PFLs. The latter can ultimately give the answers to which interventions, in a time-dependent fashion, can prevent and possibly even reverse disease, and accordingly allow maximum health span and longevity of each individual patient. For the sake of simplicity,

we illustrate the concept of the physiological fitness landscape where control parameters of health and disease are represented as horizontal axes and the vertical axis maps the physiological (or metabolic) fitness function measured in response to them. Valleys in the stress response plots correspond to stability regions while peaks and ridges delineate boundaries of these physiological stability zones. Since almost all chronic diseases of aging share the same control parameters, a common framework offered by the PFL model proposed here is more appropriate than a fragmented approach prevalent today. Additionally, three critical aspects on the periphery of modern medical interventions: chronophysiology, microbiota, and prolonged stress, are entering the mainstream of medical research thanks to advances in understanding how they affect our health. Their inclusion within the physiological fitness landscape, whose methodology is elaborated on in the second volume of this series, offers a practical approach to finding optimal solutions for healthy aging and precision-medicine therapies for age-related diseases.

I believe the solutions emerging from the integration of the new advances in science and technology into health care will have an enormously positive impact not only on chronic diseases but also on genetic diseases, developmental disorders such as cerebral palsy, and indeed all aspects of medical care and prevention. The trajectory of health and disease as well as the intricate contextual nature of pharmacologic and other therapeutic interventions on metabolism and physiology is part and parcel of the approach advocated here. We live at a time of an unprecedented pace of change in all aspects of our lives. What was unimaginable a decade ago is a new normal today. Using smartphones, almost everybody on the planet can store and access reams of information accumulated over centuries by humanity. This information revolution will undoubtedly transform health care sooner than we think. It is my belief that real-time, precise, and personalized monitoring and optimization of our state of health will be the new normal in our lifetime. Many of the building blocks of this transformation are discussed in these two volumes.

My hope is that this book series represents a small step toward opening the curtain to fill the two major voids stated above for future generations of professionals engaged in the practice of medicine. If this effort is not in vain, I will consider my mission to have been worthwhile. Indeed, there is no greater magnanimous way to dedicate motivation and purpose than the goal of helping humankind live without the burden of helplessness against premature diseases of physiology. Moreover, reaching out to present and future medical students is particularly important to me since they are most likely to embrace positive change and carry the torch of progress. However, medical education and educators must not stagnate but, in tandem with students, evolve, embrace new ideas, and be open to cross-fertilization with other fields of science and technology. As this is always a collective endeavor, feedback from readers like you will undoubtedly result in future improvements. Since this series of books is firmly grounded in several academic disciplines requiring some intellectual investment on your part, dear Reader, I'd like to encourage you to persevere on this journey by remembering the words of wisdom Theodore Roosevelt wrote:

Nothing in the world is worth having or worth doing unless it means effort, pain, difficulty … I have never in my life envied a human being who led an easy life. I have envied a great many people who led difficult lives and led them well.

For me personally, the process of research, collaboration, discussions with numerous scientists, students, colleagues, and lay people, learning, and writing has been very gratifying. It amounted to an investment of more than ten years of my life. For my wife, Eileen, and my son, Matthew, it was their ultimate sacrifice, which is the only part of this endeavor whose cost I can never reclaim and which will leave me with lasting regrets. The endless support and the beauty in the souls of my loved ones allowed me to persevere during the ups and downs of working on this *opus magnum*. Life choices are often made but not voluntarily, on both my part and theirs. With ten years gone like the blink of an eye, my conscience will always be challenged as to whether this achievement of enduring "effort, pain and difficulty" balances the real-life sacrifices of my two rocks, to whom these books must be dedicated. As a member of the noble profession of physicians, the guiding light of helping humanity through our work has allowed me to continue this task, overcoming challenges along the way.

I invite you to join me on this exciting and fulfilling exploration of the new horizons where multidisciplinary science meets the future of medicine. It's a bright future whose outlines we can already see and it makes the expectation of things to come so thrilling.

Respectfully yours,
Brian J. Fertig, M.D.
Piscataway, NJ

REFERENCES

1. Pink, Daniel. (2011). *Drive: The Surprising Truth About What Motivates Us.* Riverhead Books.
2. Black, C. (2020, May 21). *The Pursuit of Autonomy, Mastery, and Purpose in Medicine.* Op-Med. https://opmed.doximity.com/articles/the-pursuit-of-autonomy-mastery-and-purpose-in-medicine

Prologue

How can one explain the rich interconnectedness, which defines the amazing organizational complexity at many hierarchical scales of a living system, and especially the human being? Looking beyond the awe-inspiring structural architecture of biological systems, we see an even more impressive functional efficiency. To fully appreciate the latter, one must use the conceptual tools of physics because of the importance of concepts such as energy and force in analyzing metabolic processes. The second law of thermodynamics essentially states that energy dissipates as heat uniformly over time, with associated randomness or diffusion of particles. It is almost magical that the second law is abrogated in living systems, and that alone was the reason why the famous physicist Erwin Schrödinger devoted an entire book entitled *What Is Life* to the elucidation of this scientific puzzle.

One aspect of this explanation speaks to the manifestations of quantum physics in living systems that occur at many levels of biological organization. While the second law of thermodynamics is empirical and embodies the tendency of all closed physical systems toward maximum disorder, the first law of thermodynamics is an expression of the fundamental principle of the indestructibility of energy in all physical processes. In the context of physiology, it accounts for the source of a calorie, rooted in Einstein's famous equation $E = mc^2$, whose practical consequence is that solar nuclear fusion and fission processes release energy that is captured on Earth within the chemical bonds of nutrients of plants. This universal solar energy source is converted into the chemistry of food contained in plants, and then undergoes another quantum transformation in living cells, becoming the biological currency of ATP, efficiently produced in the process of oxidative phosphorylation along the electron transport chain of mitochondria.

Furthermore, the phenomenon of quantum metabolism shows that our metabolism is a quantum manifestation of biological energy production. Although it is a fundamental quantum phenomenon underpinning the deeply entangled healthy state of physiology and psychophysiology, it is likely not the only one. The interdisciplinary perspective which views "biology as really chemistry" and "chemistry as really physics" underscores the unavoidable recognition that the first and second laws of classical thermodynamics alone cannot explain the beautiful and exquisite design and potential that is intrinsic in human physiology. This entanglement of human physiology is rooted in metabolism and metabolic pathways that fundamentally distinguish living states from nonliving physical matter. The further any biological system can be moved away from the nonliving state—that is, separating it from death—the greater the metabolic entanglement which represents the interconnectedness defining complexity. The greater that complexity, the greater in parallel the metabolic health capable of providing the necessary bioenergetic resources with maximum efficiency. Conversely, the loss of ATP-producing capacity

achieved in the process of oxidative phosphorylation parallels mitochondrial dysfunction, and generates the excess heat from the chemistry of the rich organizational complexity. This latter effect is the manifestation of the breakdown of organization and often the beginning of metabolic disease and an associated pathology. Hence, mitochondrial dysfunction parallels senescence and age-related disease states. This is a deteriorating feed-forward, positive feedback destabilizing process that accelerates biological aging.

This forms the metaphorical basis for the application to the biology of physical ideas encapsulated in special theory of relativity, systems biology, chaos theory, and the theory of phase transitions. In addition to purely intellectual insights, this interdisciplinary cross-fertilization offers a promise of potential therapeutic advances with an ultimate goal of achieving phase transitions from diseased to healthy attractor states and prolonging longevity characterized by dilating time. This can be accomplished, in practical terms, by reducing the inflammatory and redox disturbance. An additional illustrative physical metaphor for a better understanding of biology is that aging and disease states are manifestations of an incinerating process, which degrades the quality of metabolic energy in the entropy-generating process conforming to the second law of thermodynamics. In particular, the energy that is locked in the chemistry underlying the organizationally complex biological structure can be viewed as representing negative entropy (or information). This state is highly functional, physiologically adaptive, and often coupled to deep levels of consciousness. However, it ultimately succumbs to the destructive force of nature that seeks the maximum entropic state that is characterized by loss of structure, function, and hence information.

It is hoped that the implications of this new way of thinking for the field of medicine will be transformative. First and foremost, it should be acknowledged that the field of physics is not only the most fundamental branch of science from which biological systems derive, but also a far more mature scientific discipline, rooted in mathematics as its language and logic as its *modus operandi*. Physics has a centuries-old history and a time-tested methodology that is a foundation of the scientific method in general. The very nature of the hypothesis in physics is consistent with the goal of asymptotically approximating the truth. Fundamentally, it involves searching for flaws and counter-examples to falsify the starting hypothesis, rather than validation—the latter having the aim of proving that the originator of the hypothesis was right while the former is attempting to bring one closer to the truth. The latter may be the way the legal profession or the political class operates, while the former is the first rule of scientific methodology. The objective of this method is to evolve the hypothesis, by gradually removing incorrect assumptions, and bringing our state of knowledge ever closer to objective truth.

The understanding of biological systems using insights from physics is, therefore, hoped to provide new models for finding solutions to healthcare problems. The goal, in general, is to slow the rate at which the aging process advances, i.e. causing an effective dilation of time invoking the metaphor of special relativity to medicine. The phenomenon of quantum metabolism exemplifies a mathematically validated process highlighting the correlated nature of energy production in a maximally efficient, organized, reduced-entropy state. It is worth stressing that all disease is rooted in dysfunction of the metabolic pathways. Expertise in the area of metabolism is central to clinical decision-making throughout the spectrum of susceptibility or chronic disease states.

There are many ways in which insights from physics can allow medicine to evolve. The dynamic model of a fitness landscape is a general framework for integrating insights from all disciplines—a Nobel Prize–worthy mathematical model that will be shown in this book to be applicable to physiology. We refer to it as the Physiological Fitness Landscape (PFL). However, to date, it has not yet been utilized for either biomedical research or clinical medicine. Physicians specialize in the application of science to the clinical setting of patient care, but this application typically involves an art that sits on top of the science, and uncertainty is inherent in the application of broad algorithms of care to a given patient. Improving the application of science to reduce uncertainty is the goal. Therefore, it must be recognized as a barrier that physicians—and certainly clinicians—speak a different language than the non-applied basic scientists. Bridging this divide is a very challenging but meritorious pursuit in the interest of public health. The science and the tools exist to shift the direction of clinical healthcare towards empowerment, but this goal requires an interdisciplinary assimilation from diverse areas of expertise, functionally related in such a way that the integration of the parts provides a greater whole.

This book describes the building blocks of understanding from a reasonable but not high-level technical language viewpoint, employing the perspective of a clinical physician. It attempts to assimilate concepts from five specific branches of physics relevant to biology and medicine, namely, biophysics, classical electromagnetism, thermodynamics, systems biology, and quantum mechanics. This framework includes the knowledge base afforded by molecular biology and biochemistry, integrated into an accessible model that invokes the dynamical fitness landscape. This endeavor is expected to provide a broader therapeutic perspective on a clinical scale, useful for biomedical research and ultimately helpful in the improved execution of patient care medicine. Hopefully, it does not lose the fidelity of meaning of the concepts described, so that it engages the interest of basic scientists seeking a perspective to clinical medicine. These physical concepts are presented in an accessible way in the first four chapters of the book. The final chapter is less technical and more clinically oriented, and highlights the fitness landscape concept applied for understanding susceptibility and disease states in the context of the stress response, allostatic and homeostatic parameters, and pathological states of disease. Importantly, insulin resistance/endogenous hyperinsulinemia appear to be inextricably linked to the stress response, underpinning a unifying model of chronic disease integrated into the framework of the fitness landscape. Identifying extrinsic and intrinsic control parameters to susceptibility states, as well as to disease states, provides the landscape to which therapeutic strategies can be applied. This is a highly dynamic construct whereby the strategy changes according to the stage and trajectory of the susceptibility or disease state, that is, the attractor state.

Also important is the concept of criticality and the threshold at which a disease becomes irreversible. Conversely, the Physiological Fitness Landscape approach also allows us to analyze, qualitatively and quantitatively, the existence of a potential phase transition toward a normal state, and the optimal path to achieve it. Crucially, irreversibility depends on the available intervention. For example, diabetes that was considered irreversible long-term, based on lifestyle and pharmacologically available interventions due to insulinopenia relative to the degree of peripheral insulin resistance, became reversible in the sense of achieving remission with the emergence of bariatric surgery. Application of the concepts of intrinsic and extrinsic control parameters, order parameters, susceptibility states, and the notion of criticality are relevant in the context of bariatric surgery for metabolic disease, providing insights that stem from the analysis of the associated fitness landscape.

A combination of lifestyle, existing polypharmacy, and metabolic surgery should be capable, including in refractory cases, of promoting phase transition of the manifestations of metabolic diseases such as central obesity, metabolic syndrome, diabetes, and even prediabetes states. Discussions of traditional pharmacologic therapy of metabolic disease are well-described in existing endocrinology and other literature and are promulgated by the American Association of Clinical Endocrinologists (AACE), the Endocrine Society, and the American Diabetes Association (ADA). A purpose of this book's content is to show a limited number of novel therapeutic strategies to fundamental targets of metabolic disease. These targets are control parameters to the order or fitness of human physiology or pathophysiology. They are discussed (and some are illustrated) in the multidimensional and dynamical (changing with time) fitness landscape model, which we are espousing as a clinical tool and even as a paradigm shift in medicine. These targets include the machinery of oxidative metabolism carried in mitochondria and insulin signaling. There is an emphasis on the stress response that orchestrates allostatic neuroendocrine and autonomic nervous system changes enlisted to preserve organismic health by maintaining biochemical parameters within tight homeostatic ranges. Higher-order control parameters, such as those that maintain or threaten the resilience of the stress response, are described, too. Also discussed are the limited number of truly extrinsic control parameters, as well as intrinsic control parameters (called secondary order parameters) defined with reference to a particular disease state. We also discuss in this book how the primary order parameters characterize the state of human health or disease, alternately stated as the state of biological aging compared to chronological aging. Within the framework of the stress response, allostatic swings of the CNS-mediated neuroendocrine and autonomic nervous system are acutely adaptive for calibrating bioenergetic priorities and immune function, and ultimately preserving homeostatic parameters and vital organ system function. There is an emphasis centrally on abnormal perceptions that drive the chronic and exaggerated stress response, and ultimately, pathophysiology. Peripherally, the importance of the gut microbiome and how it affects the gut endocrine system and bile acid metabolism in terms of metabolic health and chronic disease should be stressed. These are control parameters that also can be studied vigorously and dynamically in the model of the Physiological Fitness Landscape, on the basis of susceptibility to accelerated aging or a given chronic disease, as well as the reversibility of the susceptibility state or the chronic disease.

The Physiological Fitness Landscape model is highly applicable to precision-based medicine, and invokes available data from computational-based bioinformatics, which in turn utilizes genomic, proteomic, metabolomic, microbiomic data, and so on. The role of computers in medicine undoubtedly will shape the direction of the practice of medicine. The fitness landscape model is a mathematical one that will naturally emerge with the evolution of computers in medicine. Thus, it is historically timely to invoke the model of the Physiological Fitness Landscape, which demands an interdisciplinary coordination of computer scientists, mathematician physicists, biophysicists, and clinicians.

It is also critical that while connecting patient care to basic science, the intuitive non-algorithmic personal connection and the decision-making that recognizes patient fears, expectations, biases, and belief systems are not lost. The patient–physician relationship is powerful and must be prioritized to avoid losing this necessary therapeutic and traditional perspective of clinical healthcare with the Watson IBM computer already being utilized for diagnostic purposes with some success in the free market. Furthermore, medical training of physicians is increasingly channeling students towards either basic or clinical tracks, which reduces the number of years of training, and highlights the risk of the applied science of the medical profession being replaced by computers and essentially technicians. This risk is real and threatens the quality of healthcare delivery. There are advantages to computational medicine with the application of precision care, however, the high-level expertise and personal connection of physicians needs to be preserved as foundational to the expectations and therapeutic nature of healthcare. The Physiological Fitness Landscape provides a proactive model, and a framework capable of integrating valuable contributions in an interdisciplinary assimilation, while preserving the integrity of the skills and expertise of the clinical physician.

Content Overview

The first volume of this book is intended to popularize modern concepts in physics within the biomedical community. It contains the following technical chapters: 1) Biological Thermodynamics; 2) Biological Engines and the Molecular Machinery of Life; 3) From Quantum Biology to Quantum Medicine; 4) From Systems Biology to Systems Medicine; 5) Introduction to the Roadmap of Future Medicine; 6) Science Seen Through the Lessons of Life. Each chapter starts by posing some key questions relevant to each topic discussed and ends with a short summary of the contents. Below I elaborate on the significance of these chapters within an integrated approach to medicine, and the connections between them.

The laws of thermodynamics provide rules for the transformation of mass and energy from one form to another. The second law is unique in physics, because it explains the origin of the arrow of time, which is the basis for the gradual deterioration of all things including biological systems. Biological cycles of time define the reversibility of biological processes, but over time they, too, degrade and are superposed with the arrow of time that gives rise to aging. Biology has evolved specifically to protect the entropy reduction achieved by the biochemical processes developed by living systems.

Molecular motors are microscopic metaphors to machine engines. They are driven by ATP molecules using the first law, whereby the energy contained in their phosphate bonds is used to perform work, just as petroleum is converted into the mechanical work of an internal combustion engine. In both cases, mass motion is generated by motors with a remarkable design similarity between molecular motors and mechanical engines based on the same rotor and stator geometry.

The connection between electricity, magnetism, and energy generation is another parallel invoked in the electron transport chain of mitochondria, that may be based on quantum electromagnetism underpinning the phenomenon of quantum metabolism. Quantum biological processes are very likely the keys that can unlock the mysteries of cognition positioned at the center of quantum biological effects. Other areas of quantum biology include photosynthesis, magnetoreception in the retina of the eye of migratory birds and possibly humans, the mechanism of olfaction, and possibly the efficiency of protein folding processes.

Systems biology is a new, dynamically expanding approach to biological systems seen as quantifiable modules with interactions that can be envisaged similarly to process engineering. Various mathematical advances such as nonlinear dynamics, graph theory, bifurcations, and chaos theory are brought to bear on our improved understanding of the complex structure and function of hierarchically organized biological systems.

Traversing the various physical and mathematical concepts that allow us to understand biology with a more profound appreciation for its perfection, we must eventually confront the real-world applicability of these, sometimes, lofty ideas. In Chapter 5, we face these issues head-on. We discuss specific applications of physical concepts to human physiology and show how the Physiological Fitness Landscape and related methods can improve our clinical outcomes at the patient-specific level.

We complete this volume of the book with a selection of personal stories and anecdotes intended to amuse, illuminate and instruct by entertaining the reader. When writing this book and looking back at some of the events in the author's life, many times an unexpected insight was gained, a "Eureka!" moment took place where life imitated science and the art of medicine. This last chapter also provides a bridge to Volume 2 of this book, which transitions to medicine- and physiology-focused discussions with physics taking the back seat and clinical medicine clearly in the driver's seat.

Acknowledgments

There are numerous individuals to whom I owe a huge debt of gratitude for their assistance on this journey of discovery. I list and acknowledge their many positive effects on my development as a scientist and a writer, as well as their contributions to these two volumes.

First and foremost, my dynamic editorial team with multidisciplinary expertise did much of the heavy lifting needed to bring this project to completion. This amazing group of talented individuals was led by Gail Ferstandig Arnold, Ph.D. (Research Professor, Chemistry and Chemical Biology) who served as Associate Director, Graduate Studies and Academic Affairs as well as Associate Director, Institute for Quantitative Biomedicine (IQB), at Rutgers State University of New Jersey. Seeking editorial assistance through various Sections of the Science Division at Rutgers, I was referred to Gail in the IQB, and have had the pleasure of knowing her for about six or seven years. Gail should be credited for getting the ball moving towards the eventual amassment of an extraordinarily talented team of contributing editors. The process started with several graduate students who helped track down references for areas involving Volume 2. Moreover, Gail, upon her retirement from Rutgers, accepted my offer to personally apply her talented hand to the active editing of various important areas in Volume 2. Gail also has the essential skillset as an administrative coordinator. She has been vital for maintaining the coordinated activation of a robust team of editing and graphic design contributors. Gail's attitude is inspirational, and I couldn't thank her enough for this.

A major role in compiling material for Volume 1 was played by Dr. Shashidhar Rao, who has impressive experience in the pharma industry. A physical chemist with a penchant for biological chemistry, Shashidhar became the first senior scientist recruited to the project, at Gail's direction. Shashidhar Rao, a Rutgers scientist, was quite interested and capable of extending and assimilating his solid foundation of molecular chemistry to the realm of living systems. Shashi came on board about 6 years ago, helping with the revision of the Biological Thermodynamics Chapter in Volume 1. He taught me a lot about the dynamic chemistry of physical systems which is crucial for conveying analogies and relationships to living systems. Throughout the course of the ensuing years, his fastidious work ethic was valuable for proofreading and following up on missing elements. A couple of years later, Dr. Jack Tuszynski revisited the chapter to revise it to its current form. Shashidhar, always a gentleman, remained a loyal and stalwart partner, helping wherever he could, and has remained a strong contributor during the past six years.

Yasmin Zakiniaeiz (Yale Neuroscience postdoctoral fellow who has served as the Editor-in-Chief of the Yale Journal of Biology and Medicine) was a stalwart of reliability and solid content and presentation editor of both text and graphic design. Her major role focused on the comprehensive editing of the Stress Response Chapter. Yasmin, a natural self-starter who requires very little direction, continuously searched for voids and weaknesses anywhere in the book and took initiative to inform me as she was already taking steps to fill in these gaps, or asking the right questions. The more urgent the demand, the more she stepped into the role, sacrificing her personal time in the evenings, nights, and weekends.

Melissa Monsey (who served on the board of the Yale Journal of Biology and Medicine for several years and was an issue editor for journal issues on topics such as obesity, psychology, and psychiatry) served as an Organizer-in-Chief of sorts, helping to delegate editorial work where needed. Melissa was also my liaison for communications with outside experts, for whom she organized attachments and other work. Her organizing role helped prevent many important issues and pieces of work from slipping through the cracks. Melissa works at a high level and was my "right hand". She comprehensively edited the Calorie Restriction Chapter.

Bal Krishna Chaube, Abhishek Kumar Singh, and Sonal Shree are superb postdocs, who came to the US on J1 visas to do research at Yale University in various fields of metabolism. Abhishek's work focuses on endothelial metabolism and cardiovascular metabolic diseases. Sonal specializes in molecular biophysics and biochemistry, while Bal's interest predominantly is in the field of metabolic oncology. Abhishek, Sonal, and Bal were very helpful for technical content editing in various sections of the book. Bal was particularly generous in donating many hours of his time with phone discussions. During the course of these discussions, he would elucidate for my understanding complex but crucial components of metabolic pathways. The selfless work of these three talented postdocs is vastly appreciated.

Special thanks to Gerald Shulman and Philipp Scherer who directed me to Varman Samuel and Steve Mittelman, respectively. Steve Mittelman (M.D., Ph.D.) is a UCLA professor of pediatrics and endocrinologist, a world-renowned expert in the physiology of obesity and cancer. Varman Samuel (M.D., Ph.D.) is an associate professor and the section chief of endocrinology at the Yale Medical School, VA.

Varman and Steve were both vital to the enrichment and content editing of the material connecting metabolism to insulin resistance. Special thanks also must be given to Bart Staels who provided vital editorial revisioning of the Nuclear Hormone Receptor Chapter. Bart Staels (Ph.D.) is professor of pharmacy at the University of Lille, France, and director of the INSERM Unit UMR 1011. He is one of the highest-cited researchers in the world in the NHR area. Emily Manoogian, a postdoctoral fellow at the Salk Institute for Biological Studies was a huge help and crucial for the cohesion and content editing of the Circadian Biology of Time Chapter.

Michelle Hanlon, with excellent technical skills, greatly contributed to the graphical and editorial aspects of both volumes of the book. My deep appreciation also goes to Perrin

Beatty, Tiffany Louie, Jennifer (Kaplan) Goodell, Zhao Wang, Malcolm Watford, Cecile Martin, Olivia Tuma, Jacob Smith, Julie Meade, Chris Stewart, Luisa Torres, Nico Ekanem, Ryan Mischel and Rebecca Paszkiewicz, all of whom were vital to the project and could be counted on for their expertise and scientific rigor.

Beyond technical support, I'd like to acknowledge my intellectual mentors in the field of medicine, who helped shape my understanding of the complexities of the human body.

Dr. John Hogenesch is an Ohio Eminent Scholar and professor of Pediatrics at Cincinnati Children's Hospital Medical Center in the Divisions of Human Genetics and Immunobiology within the UC Department of Pediatrics. His lab studies genome biology with a focus on the molecular mechanisms of circadian rhythms in mammals. Dr. Hogenesch discovered several new bHLH-PAS transcription factors, including the hypoxia-inducible factors, as well as the core clock components, Npas2, Bmal2, Rora, and Bmal1, the only required component of the mammalian circadian clock. Dr. Hogenesch also pioneered the discovery of clock-regulated gene expression in plants, flies, mice, and man. As a genome biologist, Dr. Hogenesch's lab applies and integrates various disciplines including bioinformatics, genomics, molecular and cellular biology, and genetics. I attended and presented a poster at the Salk Institute 2017 Biology of Time Conference where I had a culturally enriching experience of widening perspectives in the field of circadian biology. The conference was dedicated to the 2017 Nobel Prize in Physiology or Medicine, awarded to Jeffrey C. Hall, Michael Rosbash, and Michael W. Young, for their discoveries of molecular mechanisms that control circadian rhythms. Additionally, the event was serendipitously timed to the writing of my Metabolism in Medicine book. There, I also had the fortunate and delightful experience of meeting John personally. He is one of the brilliant and pioneering scientists in this field, who discovered some of the core clock molecular components. John generously reviewed my work, still in need of content and enhancement editing. He offered the support of his top and right-hand postdoc Lauren Frances to juggle her day job responsibilities, with those required to marshal the clarity of the biology of time chapter. John emphatically endorsed the need for scientific writing, to the extent that is possible, at a tenth-grade level. He explained that it turns out that even advanced PhDs prefer to read at this level. He even referred me to two short paperback books, which he said in his lab have "rabbit ears". The goal of these books is to make the content attractive, clear, and readable. I am grateful for John's interest and help.

Don Mender (Lecturer in Psychiatry at Yale University, with expertise in integrative quantum physics) and Lloyd Demetrius (mathematician and theoretical biologist at the Max Planck Institute for Molecular Genetics at Berlin, Germany, and in the Department of Organismic and Evolutionary Biology at Harvard University) influenced my interest in studying quantum physics in the context of its possible profound role in living systems. The idea of using modern disciplines of physics to explain the exquisite and beautiful organizational perfection of biology and living systems struck me as being plausibly rooted in the concept of quantum metabolism. This term was coined by Lloyd Demetrius and promoted in several papers on the

topic including some co-authored with Jack Tuszynski. I had the fortunate opportunity to meet with Lloyd at a Princeton luncheonette, where he explained the phenomenon, which to me was cryptic based on just reading the papers. However, Lloyd's interest as a mathematician was to only accept the implications of this phenomenon as far as the math would take it. His interest in this area was at a fundamental science level, very detached from clinical medicine, where my interests lie. This experience epitomized, even galvanized my frustration that such brilliance can be so dispassionate toward extending these ideas to a broader world, where it can potentially have a massive and tangible impact. Nevertheless, by the end of our meeting, Lloyd provided a glimmer of hope when he endorsed Jack Tuszynski as an extraordinary talent with practical interests, and to whom he was grateful for including the concept of quantum metabolism in his medical student course series, The Future of Medicine.

Malcolm Watford, D.Phil (Professor of Biochemistry and Nutritional Sciences and Director, George H. Cook Scholars Program, Rutgers) is an unusual talent in the field of metabolism. Conveniently, he has been right in my backyard for almost thirty years, only a mile down the road from my office. It was only about a year ago that I introduced myself to Malcolm and supplicated his help. He was warm, humble, and welcoming, immediately giving a sense of family upon meeting him. After talking for about two hours, Malcolm's skill for good conversation was saliently apparent. A direct disciple of Sir Hans Krebs, having trained under him as a postdoctoral fellow at Oxford University in the UK, Malcolm is replete with the accomplishments, knowledge, and insights that one would expect from a researcher with such an impressive pedigree that also includes Otto Warburg (of whom Hans Krebs was a student). The concept of the Warburg effect, the role of Malcolm's contribution of glutamine metabolism to this effect, his explanation of the stages of starvation, and of course, Krebs's TCA cycle, the central hub of biochemical metabolism, all exemplify core relevance connecting this book's message to human health and disease. Malcolm's editorial augmentation of this book includes his elite level of content critique, in addition to his sharing valuable and engaging personal stories and anecdotes of Hans Krebs, a seminal pioneer of biological chemistry. I am profoundly grateful to Malcolm for allowing me to enlist his unusual expertise and historical gravitas for the purpose of perfecting this book's message regarding biochemical aspects of metabolism.

The late great Bruce McEwen (Alfred E. Mirsky Professor and head of the Harold and Margaret Milliken Hatch Laboratory of Neuroendocrinology at the university) was a major influence and inspiration for me. The terms allostatic load and overload were coined by Bruce, a stress biology endocrinologist Ph.D. from The Rockefeller Institute in NYC. Sadly, Bruce passed away January 2, 2020. I was privileged to be influenced by this true member of the scientific pantheon in the understanding of the stress response, who became a powerful collaborator and mentor. While these concepts apply to hormonal, autonomic, and immune system responses to psychogenic stress, they are also inextricably intertwined with the metabolism of mitochondrial bioenergetics and of insulin resistance. Moreover, the notion of stress is applicable to all hierarchical scales of living systems. In a general sense, stress should be considered the most fundamental driving parameter for the evolution of both health and disease, in humans and in fact all biological systems. This critical insight should further be invoked in diagnostic medicine, whereby the cardiac stress test can be viewed as a prototype for diagnostic stress testing. Accordingly, tissue and disease-specific metabolic stress testing that span the domain of all subspecialties of medicine should ultimately become employed as a new standard of care.

An epiphany, under the direction of another collaborating mentor, Jack Tuszynski, was to attach the concepts of allostatic load and overload to a metabolic/physiologic fitness landscape model, adapted from physics. This potentially transformative computerized mathematical model provides a precision personalized scale of medicine. My explorative journey into modern disciplines of physics, as well as of physical and biological chemistry, sought a deeper understanding of metabolism through the prism of a different lens. Too often people's health problems lie outside the fragmented and compartmentalized toolkits of applied science. The goal was to gain the insights needed, that lie beyond our currently available skillsets and knowledge, to truly help people. After seven years of self-learning and writing, the feeling of consternation and the inevitable failure to quench the thirst for this unrequited pursuit, help arrived to rekindle my optimism. I, therefore, wish to express my thanks to Dr. Jack Tuszynski (Allard Chair and Professor in Experimental Oncology in the Department of Oncology at the University of Alberta's Cross Cancer Institute, and a Professor in the Department of Physics) who has indeed reinforced my understanding of physics, biology and medicine as a rich and powerful playground with a real capacity for transforming medicine to levels currently unimaginable, gauged in the context of historical and present-day standards. Jack's expertise in biophysics, solid-state physics, as well as in the areas of electromagnetism and quantum physics, including the evolving and potentially explosive field of quantum biology, helped me bridge the enormous gap between physics and medicine. Even more special about Jack is his selfless dedication to advancing science, his humility, and an extraordinary generosity of his personal knowledge and insights. I have now had the amazing privilege of working with Jack over the span of four and a half years, including several hundreds of hours of stimulating conversations about physics and medicine. Taken together with his eloquent editorial revisions of this book, he made possible the metamorphosis of this project from failure to *magnum opus*. I am infinitely grateful for his taking me under his wing and believing that this work will be a valuable scientific contribution to medicine.

Last but not least, I wish to express my profound appreciation to Drs. Ralph Defronzo, C. Ronald Khan, Gerald Reaven, Alessio Fasano, Jeffrey Bland, Jeffrey Mechanic, Gerald Shulman, Phil Scherer, and Barbara Corkey, all of whom inspired and influenced me, and gave me numerous insights into the complexities of human physiology.

Having benefited enormously from the depth of knowledge of the numerous individuals mentioned and acknowledged above, it is nevertheless important for me to take full responsibility for the entirety of the book's contents, including any mistakes, faults, or omissions that the reader might find.

Author

Brian J. Fertig, M.D., F.A.C.E., is the Founder and President of the Diabetes & Osteoporosis Center in Piscataway, New Jersey (https://siomar2.wixsite.com/diabandosteocenter), established in 1994. Dr. Fertig's experience in diabetes, endocrinology and metabolism, including internship, residency, fellowship, and private practice, spans a period of 34 years. Dr. Fertig is also an Associate Professor at Robert Wood Johnson Medical School and the Chairman of the Department of Diabetes & Endocrinology at Hackensack Meridian Health—JFK University Medical Center. His passion for patient care and for finding the root problems of disease was his motivation and purpose for writing a two volume book series titled Metabolism and Medicine. On his exploratory journey, Dr. Fertig discovered modern conceptual tools to improve the problem-solving skill sets of great utility for him, his practitioner colleagues, and the next generation of healthcare providers. His genuine concern for the future of medical student education is one of the main factors that motivated him to write this book. Additionally, he is also concerned with the need for interdisciplinary expertise to augment the execution of patient care, but particularly in light of an expanding reliance on Nurse Practitioners and Physician Assistants. His philosophy is based on the concept that the greater the network of interdisciplinary scientific and clinical expertise, the greater the collective skill set and hence the benefit to the patient. The ever-increasing use of bioinformatics paves the way toward personalized-scale medicine. However, the way forward for medicine should also include skillful adaptation of methods developed in physics, which will allow quantification of therapeutic solutions with unprecedented precision. Since physics has a tendency to generalize empirical observations to formulate laws of nature while biology takes pains to tease out detailed specificities, the marriage of the two fields promises the most enlightened future direction for medicine as both an art and a science.

Personal Statements

"A comprehensive and original opus linking metabolism with disease and preventative directions for the future. A must-read for experts and students alike".

Michael Houghton, Ph.D.
2020 Nobel Laureate Medicine and Physiology
Professor at the Department of Medical Microbiology
and Immunology
Li Ka Shing Applied Virology Institute
University of Alberta
Edmonton, Canada

"Metabolism and Medicine by Dr. Brian Fertig is an unparalleled adventure into the fabric of physical law and medical impressionism, impelled by metaphysical questions and doubts to fashion new approaches to physiological questions in health and disease. This book is dense with science, synthesis, and interpretation, leaving the reader with just the right measure of resolve to learn even more. I strongly recommend Dr. Fertig's opus for those still interrogating metabolism and medicine for satisfying answers".

Jeffrey I. Mechanick, M.D., F.A.C.P., F.A.C.N., E.C.N.U., M.A.C.E.
Professor of Medicine
Medical Director, The Marie-Josee and Henry R.
Kravis Center for Cardiovascular Health
Director, Metabolic Support, Division of
Endocrinology, Diabetes and Bone Disease
Icahn School of Medicine at Mount Sinai
New York, USA
Past President of the American
Association of Clinical Endocrinologists

Prof. Mechanick served as: President of the American College of Endocrinology, President of the American Association of Clinical Endocrinologists, President of the American Board of Physician Nutrition Specialists, Member of the President's Council on Fitness, Sports and Nutrition – Science Board

He is Editor-in-Chief Emeritus, President's Council on Fitness, Sports and Nutrition Elevate Health, Chair, Board of Visitors, College of Computer, Mathematical, and Natural Sciences, University of Maryland, Chair, Physicians Engagement Committee, American Society for Parenteral and Enteral Nutrition

"An insightful and comprehensive summary of the state of the field! This should appeal to both a novice as well as to an expert. Well written and authoritative in all areas".

Philipp E. Scherer, Ph.D.
Professor, Department of Internal Medicine
Gifford O. Touchstone Jr. and Randolph G.
Touchstone Distinguished Chair in Diabetes Research
Director, Touchstone Diabetes Center
Interim Chair, Department of Cell Biology
The University of Texas Southwestern Medical Center
Texas, USA

Dr. Philipp Scherer is the first scientist to win what could be called the "Triple Crown" of diabetes research recognition – adding the top Asian award (the 2018 Manpei Suzuki. International Prize for Diabetes Research) to the American and European ones in recognition of his discovery of adiponectin, a hormone released by fat cells, and subsequent research into the hormone's role in fending off diabetes. His research has "deepened and widened our understanding of diabetes, obesity and energy homeostasis,"

"This transdisciplinary book beautifully tells the story of the dynamics of life in health and disease. In an unprecedented tour de force, Brian Fertig connects fundamental physicochemical mechanisms to the dynamics of metabolism. A mind-boggling book, for everyone to read!"

Bart Staels, Ph.D.
Professor of Pharmacology ('classe exceptionnelle')
University Lille, Inserm, CHU de Lille, U1011-EGID
Institut Pasteur de Lille
Lille, France

Prof. Staels has been awarded the Young Investigator Award of the European Atherosclerosis Society, the Bronze Medal of the CNRS and the Lifetime Achievement Award of the British Atherosclerosis Society, the pharmaceutical "Barré" 2007 prize from the Faculté de Pharmacie of Montreal, and the French "JP Binet" prize from the Fondation pour la Recherche Médicale.

"Dr. Fertig elegantly captures the complexity of metabolism, showing us how our physiology is intimately integrated from the quantum level to the whole organism. Insightful, comprehensive, and thoughtful, these books will enlighten the beginner and expert alike!"

Steven D. Mittelman, M.D., Ph.D.
Chief, Division of Pediatric Endocrinology
Interim Chief, Division of Pediatric Genetics
Professor of Pediatrics
Solomon A. and Maria M. Kaplan Chair of Pediatric
Endocrinology
UCLA Mattel Children's Hospital, David Geffen
School of Medicine
California, USA

Prof. Mittelman served as the Fellowship Director and Director of the Keck/Caltech Combined MD/PhD Program.

"Brian Fertig has melded a thought-provoking and eclectic blend of physics, physiology, psychology and philosophy into a highly original work of great scope and depth."

Frederick S. Kaplan, M.D.
Isaac and Rose Nassau Professor
Chief of the Division of Molecular Orthopaedic
Medicine
University of Pennsylvania School of Medicine
Pennsylvania, USA

"This is a masterpiece! A truly unique point of view that should make all physicians THINK more as they care for their patients.

I know you've been working hard for a long time on this and want to let you know—*you've succeeded.*"

Stan Schwartz, M.D.
Emeritus Associate Professor of Medicine
University of Pennsylvania
Pennsylvania, USA

"Before the Sars-COV-2 pandemic enveloped the world, many who study metabolism, dismayed by the increasing prevalence of obesity and related diseases, observed an epidemiological transition from infectious to metabolic diseases. At the time of this writing, nearly three million people around the world have died from COVID-19 and perhaps proving that aphorism false. But we now know that obesity, insulin resistance and diabetes are important risk factors for developing severe COVID-19 disease and, when the pandemic is contained, will remain important health concerns in nearly all countries. In these two volumes, Fertig has tackled the numerous complex pathways that comprise 'metabolism'. Most scientists spend careers studying one of these

pathways. Here, Fertig has compiled the work of many into a comprehensive work that is scholarly yet accessible. It is a love letter to the field, including important historical perspectives and key contributions of the physician scientists that have blazed the way. And, this work also provides an intricate framework with which to understand the coming tide of '-omic' data."

Varman Samuel, M.D., Ph.D.
Associate Professor of Medicine (Endocrinology)
Yale University
Section Chief
VA Connecticut Healthcare System
Connecticut, USA

"In *Metabolism and Medicine*, Brian Fertig provides an innovative, thought-provoking approach that unravels the pathophysiology of common metabolic diseases by relating them to basic physiochemical processes. A must read for everyone in the fields of endocrinology, diabetes, and metabolism."

Ralph A. DeFronzo, M.D.
Professor of Medicine
Chief, Diabetes Division
University of Texas Health Science Center
Texas, USA

"Dr. Brian Fertig demonstrates a great talent for combining knowledge in biology and physics with his own clinical expertise. He conceptualizes a mathematical model of a physiological fitness landscape based on energy economy, biological clock synchronization, and stress management. Thus, the model allows informed predictions for a precision personalized scale of medicine. In this book, the contours are shown of an upcoming revolution in the medical science comparable with that of information technology. The narrative captivates and is enriched with case studies, anecdotes, clear illustrations with summarizing captions, that all present a pleasant surprise to the innocent reader in a thought-provoking manner. Highly recommended!"

E. Ronald de Kloet, Ph.D.
Head Department of Medical Pharmacology
Leiden University
Leiden, the Netherlands

"Dr. Fertig's book is a must-read for anyone interested in gaining up-to-date knowledge about circadian optimization of human metabolism. In particular, the section on Circadian Biology, and the Biology of Time, captures the core of the chronophysiology perspective on human metabolism. These complex concepts are explained in a clear and cohesive fashion. The interdisciplinary metabolic perspective of biological time and aging is very engaging and provides an elucidating insight into this emerging field. This book has profound value to both researchers and clinicians."

Satchidananda Panda, Ph.D.
Professor
Salk Institute for Biological Studies
California, USA

"This is a wonderful book providing a comprehensive overview of metabolism from biological processes to pathophysiology. Undoubted, it will interest both basic scientists and clinicians. It provides a wealth of information in a clear and highly readable format. I especially enjoyed reading Chapter 8, which deals with the gut microbiota in health and disease. The chapter contains an enormous and highly readable account of this rapidly evolving area of research. The topic is of relevance to many disciplines including endocrinology, cardiology, neurology, psychiatry and immunology."

Ted Dinan, Ph.D.
Professor of Psychiatry
Microbiome Alimentary Pharmacobiotic Center
University College Cork
Cork, Ireland

Prof. Dinan was Chair of Clinical Neurosciences and Professor of Psychological Medicine at St. Bartholomew's Hospital, London. Prof. Dinan is a pioneer of gut microbiota research focusing on the influence of the brain function and development including the regulation of the hypothalamic-pituitary-adrenal axis in situations of stress. He was awarded the Melvin Ramsey Prize for this research into the biology of stress.

"It is both intriguing and gratifying to see that the quantum theory of consciousness has influenced modern thinking about metabolic processes in important ways, as shown in Brian Fertig's work. This book remarkably connects basic physiology to clinical medicine."

Stuart Hameroff, MD
Professor, University of Arizona

Pioneer in the science of consciousness, organizer of the annual TSC conferences and co-creator of the Penrose-Hameroff model of consciousness (Orch OR)

"A single author text is a rarity of late—especially one so comprehensive. These books combine text that is available to both novice and expert and provide insights to the history and personalities connected with discovery. Bravo to Dr. Fertig."

Mark M. Rasenick
Distinguished Professor of Physiology & Biophysics
and Psychiatry
Director, Biomedical Neuroscience Training
Program Research Career Scientist, Jesse Brown
VAMCU, Illinois Chicago College of Medicine

Dr. Rasenick's research, among many other contributions, was critical for the identification of G protein coupled receptors, the most pervasive and largest group of transmembrane receptors, important for hormonal and neurotransmitter signaling, physiology and behavior.

"This is a must-read for anyone interested in the intersection between physics and medicine, especially related to the understanding how applications of modern physical concepts may help achieve optimum health."

Deepak Chopra, MD
The Chopra Foundation
Chopra Global
DeepakChopra.com

"What a monumental work! It reads very well and is a wonderful history of the understanding and development of human metabolism and physiology in health and disease. How you put all this together astounds me, and I congratulate you.

I would recommend the 2 volumes for any student of the history of science and especially to those interested in the field of metabolism and it's development, as well as to people active in the field."

Seth Braunstein MD, PhD
Emeritus Associate Professor Medicine
University of Pennsylvania (Perleman)
School of Medicine
Past Director Diabetes Program University of
Pennsylvania (Perleman) School of Medicine

Biological Thermodynamics: On Energy, Information, and Its Evil Twin, Entropy

Why should a physician, who clearly is not an engineer or physicist, study such an old field of science as thermodynamics? What new concepts in thermodynamics are relevant to human health and disease? What is entropy and why should we think about it in the context of metabolic diseases? What is biological information? Is information related to energy? Is there both nutritional energy and nutritional information contained in the food we eat? Can the onset of a disease be viewed as a phase transition? If so, what are the implications for treating a disease?

well characterized by the use of quantities called order parameters and control parameters as well as the function, which depends on them that describes the state of the thermodynamic system, the free energy. In applications to living systems, we have borrowed generously from this vocabulary of physics but renamed free energy, a physiological fitness function, which is a mathematical formulation of the state of health of the living system. Much of the rest of this chapter describes in substantial detail key metabolic processes of energy generation and energy transformations that form the basis of the biochemistry background needed to understand metabolic processes and metabolic diseases.

Chapter Overview

The terms physicist and physician come from the same root word but the respective professions diverged over the centuries. Now may be a good time to shrink the gap between the two, especially with modern technology being such an integral part of health care delivery. Thermodynamics is a branch of physics that launched the first industrial revolution but it also directly links to the way our body works. In this chapter, we present an overview of the key concepts in thermodynamics, which is based on the static equilibrium of physical systems as well as on processes that take physical systems from one equilibrium state to the next. Energy conservation is the basis of the first law of thermodynamics while a tendency to reach a maximum entropy state is behind the second law. Both of them are important in the context of biology and medicine, although biological systems by definition are far from thermodynamic equilibrium, except at death. Thermodynamics teaches us about the irreversibility of processes that generate heat, and human metabolism is one such example. Paradoxically, living systems reduce entropy by utilizing energy coming from nutrients ingested by them. Nonetheless, with high but less than perfect efficiency, all living systems gradually succumb to the arrow of time and increase their entropy content bit by bit. Speaking of bits, information, which is of critical importance to all forms of life, is also referred to as negative entropy and it requires energy to be created or destroyed. Among the more recent applications of thermodynamics, the idea of phase transitions is of utmost importance to biology and medicine since a transition from health and disease can be understood as a phase transition between two states of living matter. Associated with it is another powerful concept, namely that of symmetry breaking and all life emerges as a broken symmetry, a new phase of matter, a living state. Phase transitions are

1.1 Introduction

As far as we know today, the laws of physics are universal both in space and time and are applicable equally to diverse physical phenomena whether they are in the realm of cosmology or medicine. In fact, the words "physician" and "physics" are similarly derived from the Latin word "*physicum*" and the French word "*physique*", originally referring to natural science, which encompassed all knowledge about the material world. In the 1300s, the word "physic" began to be used to describe natural science, which ultimately branched into the separate words of "physics" and "physician". The same laws of physics offer a framework for understanding the spontaneous nature of a chemical or a biochemical reaction just as easily as they help understand the trajectory of a spaceship launched toward Mars. Thus, understanding energy flow in biological systems would be greatly benefitted by an intimate knowledge of the fundamentals of physical laws of nature. To a physician like myself, specializing in endocrinology and metabolic disorders, the real world of clinical practice is compartmentalized into the labels such as "diabetes" or "thyroid condition". However, in the interest of a desegregated physiology and energy flow that is critically vitalizing to health, it is necessary to understand a connection to physics at its core.

This chapter interlocks the fundamental aspects of energy, entropy, and information with their relevance in the multidisciplinary context of the branches of endocrinology and metabolic health. It aims in part to blend an understanding of energy rooted in the laws of thermodynamics (commonly used to describe the thermal behavior of nonliving matter) with an application to organized living structures and complex biological organisms [1–4]. An emerging body of literature supports the extension of these disciplines to novel domains of biophysics including quantum biology and complex adaptive systems,

DOI: 10.1201/9781003149873-1

each contributing to an ignition key that turns on the biological engine and opens the door to improved insights into medical practice through research hypotheses and their tests. Further, extension of the thermodynamic concept of entropy to the domains of evolution and information theories holds promise as an organizing framework, which can assist in recognizing energy as information in biological systems. For example, it can even lead to a better understanding of cancer cells that survive by outcompeting ancestor cells for available resources. The latter scenario is an extension of a macro-evolutionary process. Concepts of robustness and flexibility are applied to various levels of drivers of the evolutionary process. The concepts of allostasis and homeostasis as strategies for physiological stability are described and explained below. Examples of allostatic overload are provided in cases when chronic stress parameters, or the mere chronic perception of stress, prevent successful homeostasis resulting in disease states and accelerated aging.

In this chapter, we discuss perhaps the most important aspect of living systems, namely their ability and necessity to absorb external energy stored in the form of chemical bonds of nutritional substances and to transduce this energy into useful and directed outputs such as work and heat production for the maintenance of the living state. This set of processes is termed metabolism and it clearly distinguishes living systems from inanimate matter. We examine the balance of energy involved and look into the molecular processes at the most fundamental level of metabolism. In the following, we will discuss various elements of thermodynamics that are fundamental to the understanding of the energy transformations associated with metabolism in biological systems. These include the first law of thermodynamics, the second law of thermodynamics, energy, enthalpy, entropy, and the Carnot cycle. While the mathematical formulations of these physical quantities can be found in more specialized texts in the literature, the intent of this chapter is to provide a qualitative understanding, through examples, of the impact of these physical elements of thermodynamics on the functioning of the biological engines in living cells. Their relationships to the ambient conditions in the physical world, i.e. temperature, pressure, and volume will be elucidated.

The word "doctor" originates from the Latin word of the same spelling which is the noun form of the Latin verb "*docere*" (to teach). Consequently, it alludes to being a teacher while the term "medical doctor" applies to healthcare practitioners for whom medical diagnosis is the primary form of execution of their profession. Medical diagnosis, a process of correlating disease type with symptoms and signs through the elimination of other possible causes via differential diagnosis, is derived from the physical examination and history of a person. This typically does not address the mechanism or etiology of a pathogenic process. It is, however, desirable that a medical diagnosis be accompanied by an understanding of the underlying events in human physiology that can be responsible for the diagnosed disease state. In that context, the diagnosis should serve as a way station toward such an understanding.

All biological systems require a sustained energy supply in the form of nutrients for their survival. This energy stored in the form of chemical bonds is then recycled through metabolic processes for use in specific physiological functions. Hence, this chapter frames a perspective on energy flow across biological structures, its origin, and its relationship to inanimate systems in which energy tends to be distributed evenly in a universal tendency to reach an equilibrium state of lowest energy. This is in contrast to the living systems, which are out of thermodynamic equilibrium but tend to follow homeostasis, which is a state of dynamical stability under external perturbations. We should stress here a major difference between two similar-sounding terms. Thermodynamic equilibrium is a state of the lowest free energy of a macroscopic physical system. *Homeostasis is a dynamical equilibrium of an open biological system, which has a tendency to follow the same trajectory (time course) in the space of biochemical reactions taking place in a living organism. Therefore, a physical condition of being at thermodynamic equilibrium is not what a living system seeks. A living system seeks dynamic stability in its far-from-thermodynamic steady state.* Nonetheless, taken together with its environment, a biological system must not violate laws of thermodynamics or any other physical laws. In particular, energy conservation expressed by the first law of thermodynamics must at all times be adhered to. While physical systems tend to spontaneously increase entropy, biological systems do the opposite, which is possible by the use of metabolic energy. In this chapter, we will closely examine how this seemingly paradoxical situation can be understood using the framework of physics and biochemistry.

The present chapter is organized in a stepwise fashion and it begins with defining thermodynamic quantities, which form a foundation for the understanding of thermodynamic devices such as heat engines and fuel cell engines. The concepts of information and information processing are discussed in the context of quantum mechanics. This evolves into a discussion of the metaphorical fuel cell and quantum mechanical-biological engines. A more detailed exposition of these concepts will be provided in separate chapters on quantum biology (Chapter 3) and on molecular engines (Chapter 2). Subsequent discussion of chaos and complexity theories will provide the basis for the metaphor of biological organisms as far-from-equilibrium engines. This will also frame the basis for understanding energy balance and metabolism.

Metabolic activity defines whether a biological system is alive or dead. Although in common parlance, inanimate objects such as furniture, automobiles, and communication media are thought of in terms of "life" associated with them, they are not characterized by metabolic activity and, obviously, are not alive. Intriguingly, the boundary between a material physical system and a living one is more blurred than one would imagine. As the Latin derivation of the words physician (homolog of "physics") and doctor (meaning "to teach") highlights, it has been my ambitious goal to collaborate with thought leaders in the basic science fields of physics to facilitate an understanding of biological systems from the perspective of their interface with physics. Accordingly, this opens the curtain to an explosive potential for growth in the field of medicine and therapeutic endeavors in the relatively near future.

Perhaps the most fundamental physics concept is energy because it organizes states of physical systems and determines their stability. Energy is, of course, of cardinal importance in the context of metabolism. However, what physics by and large ignores is information. Information is constantly absorbed, processed, interpreted, and acted upon by biological systems. Is there a relationship between energy and information? This question will be addressed later in the chapter but now we explore the basic concepts of matter from a physical perspective.

1.2 The Four Forces: Weak, Strong, Electromagnetic, and Gravitational; an Emphasis on the Weak Force

The building blocks of the universe are those that cannot be split up into smaller parts. The two major classes of building-block subatomic particles are the fermions and the bosons. The fermions consist of quarks and leptons. These particles are so classified because they all have an odd half-integer spin (for example 1/2, 3/2, 5/2) that cannot occupy the same state at the same location at the same time. This is an expression of the Pauli Exclusion Principle, which stated another way requires that two or more particles must not simultaneously occupy the same quantum state, for example, two electrons must not be in the same spin and orbital angular momentum state residing in the same atomic orbit. The electron shell structure of atoms is in fact a paragon example of this principle. *An electrically neutral atom contains electrons equal in number to the protons in the nucleus.* Electrons, a lepton subtype of fermion, avoid occupying the same quantum state of other electrons by having a different spin within any given electron orbital. For example, a full electron shell in the so-called s-state (1s, 2s, 3s, ...) having an orbital angular momentum $\ell = 0$, consists of two electrons, each with 1/2 spins and opposite spin projections $m_s = +1/2$ and $= -1/2$. This leads to anti-symmetric states required by Pauli. Bosons, particles with integer spin (for example 1, 2, 3) are the other class of subatomic particles. They characteristically are not subject to the Pauli Exclusion Principle. Hence, an arbitrary number of these particles can occupy the same state. These include the force-carrying particles or gauge bosons: Strong force (gluons); weak force ($W^{+/-}$, Z); electromagnetic force (photons); and gravitational force (gravitons, not yet discovered). The other boson particle most recently discovered is the Higgs boson, which is in a class of its own; it is neither a fermion nor a force carrier (i.e. gauge boson). It has a zero spin, no electric charge, and no color. It is the first elementary scalar particle found in nature. There are a host of remarkable implications to the Higgs boson, which will be discussed ahead.

The understanding of particle physics has much evolved since my days as an undergraduate student. In the late 1970s and early 1980s, we were taught about the nucleus of atoms containing protons and neutrons. These were as elemental as it got. However, these nuclear constituents are now known to be comprised of quarks in groups of three. Furthermore, protons represent the positive electric charge of the nucleus whereas neutrons are electrically neutral. *Quarks, the constituents of these nucleons, are responsible for the electrical net positive charge of protons as well as the net neutral charge of neutrons.* Additionally, quarks carry a nonelectrical type of charge referred to as color. Color in this model is a symbolic quantum variable, totally unrelated to the visual property of color. Another quantum variable ascribed to quarks is the notion of flavor, which again is totally unrelated to the taste property of flavor, but like color represents convenient models of description. Flavor, color and electrical charge quantities of quarks will be described ahead in this discussion. Color is a property of the strong force attributed to gluons and quarks introduced in the theory of quantum chromodynamics. Flavor is a quantum property of leptons and quarks, each of which comes in one of six flavors.

The weak force is one of four fundamental forces that govern the matter of the universe. The others are the strong force, electromagnetic force, and gravitational force. The weak force is the only one of the four that plays a greater role in things falling apart (that is decaying) in contrast to holding things together. The weak force, like the strong force, acts at the subatomic level and is important in creating elements as well as empowering the stars. It is responsible for radiative decay and hence natural radiation in the universe, ultimately transferred to Earth *via* electromagnetic rays empowering life. This radiation decay is referred to as β decay, referring to the loss of β particles, which have since been understood to be electrons. This is the process by which a neutron in a nucleus is transformed into a proton, or vice versa, and expels an electron to balance the charge and momentum. This process is a strategy for unstable atoms to become more stable.

Nuclear fusion occurs between colliding hydrogen nuclei in the Sun that occurs as a result of the force of gravity, which is much stronger than it is on Earth. The fused hydrogen nuclei result in heavier helium atoms. This occurs at the Sun's core where the temperature (at 15 million degrees Celsius) and density are greatest and most conducive for collisions leading to fusion, overcoming the natural repulsion between the positive charges of the nuclei. The heavier helium atom is slightly less than the sum of masses of the initial two hydrogen atoms. The difference in mass, Δm, called the mass defect, is released as energy and is represented by Einstein's famous formula: $E = mc^2$. This results in a remarkably high amount of energy despite only the tiny amount of mass lost in the process. *The energy that is released is transferred to Earth generating and sustaining life within its many life forms.* In contrast to nuclear fusion, the power generator source in the Sun, the energy necessary for life on Earth is propelled by the electrostatic force between atoms mediated by the electricity of shared electrons and their attraction to positively charged nuclei of the atoms.

Two hydrogen atoms bound together with a single nucleus consisting of two protons is an unstable form of helium. In contrast, a stable form of helium is composed of two protons and two neutrons within a single nucleus. Once the unstable form of helium is created, the weak force drives the ultimate formation of the stable forms of helium. The nucleus comprised of two protons represents an imbalance of protons and leads to the process of β decay whereby one of the protons spontaneously changes into a neutron. This produces deuterium, a heavy hydrogen atom consisting of one proton and one neutron in the atom's nucleus. Subsequent nuclear fusion between a deuterium atom and a proton merge to form a helium atom (He³) consisting of two protons and one neutron, in the process releasing a γ ray. This is the electromagnetic energy released in the process of nuclear fusion accompanying the release of a positron and a neutrino. The γ ray comes from the annihilation of a colliding positron with an electron in the vicinity. Finally, two He³ atoms merged to produce the stable helium isotope, He⁴, consisting of two neutrons and two protons with the immediate release of two protons in the process.

The weak force fundamentally involves various forms of nuclei responsible for natural and artificial forms of nuclear fusion as well as destabilizing atomic nuclei that change the identity of the atom to a different atom. In the latter case, the atomic mass of the element does not change; however, the atomic number (i.e. the number of protons in the nucleus) changes. The process in general involves transforming a neutron into a proton or vice versa.

Protons and neutrons are comprised of constituent parts called quarks. There are six varieties, or flavors, of quarks. The most common ones being the up and down quarks. The other larger four flavors of quarks are relatively scarce as they decay into the up and down quarks rapidly because they are inherently unstable. Protons consist of two up quarks, each with a charge of +2/3, along with one down quark with a charge of –1/3, hence with a net proton electrical charge of +1. Neutrons conversely consist of two down quarks, each with an electrical charge of –1/3 and a single up quark with a charge of +2/3, hence with a net neutral or 0 charge of the neutron. The weak force involves the exchange of particles called force carriers, as do the strong and electromagnetic forces, and likely the gravitational force as well. In the case of the weak force, it is mediated by three carriers, the two electrically charged W bosons (W⁺ and W⁻) and the electrically neutral Z boson. If a down quark of a neutron comes close enough to a neutrino, it is converted to an up quark by absorbing a W⁺ boson particle from the neutrino.

The neutrino having lost a positive boson (W⁺) becomes itself negatively charged turning into an electron (a lepton) and an anti-neutrino (an anti-lepton). The anti-neutrino is an anti-lepton particle that maintains symmetry. Additionally, it explains the energy released in turn from the loss of mass as a result of the decay process that is consistent with Einstein's famous $E = mc^2$. The energy of the electron is not enough to explain the total energy released; hence the neutrino particle provides that explanation. The weak force W⁺ boson emitted in this process represents the force released (i.e. it is the force carrier) or energy released as a manifestation of the mass lost in the transformation of the neutron to the smaller proton. The W⁺ boson interacts with the down quark turning it into an up quark. As a result, the two down and one up quarks of the neutron are replaced by two up and one down quark of a proton. This changes the atom to an entirely new element. For example, the weak force is responsible for changing the identity of the carbon-14 atom, comprised of six protons and eight neutrons, to that of nitrogen-14, comprised of seven protons and seven neutrons. Notably, in this case, whereby a neutron is converted to a proton, an electron is emitted with an anti-neutrino to maintain conservation of the lepton number. This process describes β⁻ decay.

Another example of the weak force mediating β⁻ decay transforming an element by changing the flavor of a down quark to that of an up quark and hence converting a neutron to the smaller sister nucleon proton involves the W boson being released or emitted from the down quark rather than being released or emitted by the neutrino. Alternatively, β⁺ decay describes a situation where a positron is emitted along with a neutrino accompanying the conversion of a proton to a neutron. In this case, the positron represents the anti-lepton while the neutrino is the lepton. In all cases that the weak force mediates the process of emitting a lepton, an accompanying antilepton is also emitted to maintain conservation or symmetry of lepton number as well as the property of spin. These are quantum properties that must obey the Pauli Exclusion Principle. The lepton number is conserved the same way as the electrical charge is although it does not represent electric charge per se. A positive lepton is, for example, an electron or a neutrino versus a negative lepton, which is an anti-electron (for example positron), or an anti-neutrino.

The application of weak force to nuclear β decay involving nuclear fusion as described above (the process above outlines only one but by far the most common path) provides both the energy and the source of heat of the Sun, without which neither the Sun itself nor living conditions on Earth could exist. The initial fusion of two protons leads to β⁺ decay that converts a proton to a neutron accompanying the release of a positron and a neutrino (lepton and anti-lepton). The deuterium hydrogen nucleus merges with another proton to create the He³ atom along the cascade that ultimately forms the isotope He⁴ as also described above. The γ ray produced as a result of the matter-antimatter annihilation process as a result of the positron colliding with an electron in the vicinity highlights the symmetry of the weak and electromagnetic forces, referred to collectively as the electroweak force. The electrically charged nature of weak boson carriers provided an initial hint to the speculation of the symmetry between these forces that led to the theoretical Standard Model that unifies the strong, weak, and electromagnetic forces. The weak force is responsible for all three forms of nuclear decay; α decay, β decay (discussed above), along with γ decay. Shortly ahead, more will be briefly highlighted about these three forms. In terms of the symmetry of the weak force and the electromagnetic force, it is worth pointing out that β decay of the weak force is ultimately responsible for γ decay, the process whereby γ electromagnetic rays are created, and that ultimately leads to the entire spectrum of solar electromagnetic radiation.

Another scenario whereby protons can turn into neutrons other than by β decay is a process called electron capture. This process occurs in atomic isotopes that have a relative abundance of protons to neutrons in the nucleus. However, the energy difference between the parent and daughter isotope is less than the amount of electron voltage yielded by the annihilation of an electron and a positron. That is, not enough energy would be available to allow positron (and neutrino) emission (positron emission as opposed to electron emission form of β decay being the other mechanism of transforming a proton to a neutron). An example of an isotope that will decay by electron capture is rubidium⁸³ (37 protons and 46 neutrons) to krypton⁸³ (36 protons and 47 neutrons). Like positron emission

β decay, the atomic mass does not change while the atomic number (i.e. proton number) reduces by 1. The atom remains neutral in charge but exists in an excited state following the electron being pulled into a proton quark from the innermost electron shell. Eventually, an electron from the outer shell will drop down to fill in the inner electron returning the atom to its unexcited ground state. During this process, an X-ray photon may be emitted. Additionally, the nucleus as a result of this process may be in an excited state due to a reconfiguration of nuclear components. A transition of an excited nuclear state to its ground state emits a gamma ray, another mechanism for producing γ rays aside from the positron emission β decay that annihilates with an electron in the vicinity.

The weak force interactions of nuclear decay involving α decay, β decay, and γ decay may be very simply summarized as follows. α decay refers to the emission of two neutrons and two protons, essentially a helium nucleus, from a larger atomic element. Since the helium nucleus has no electrons to neutralize the protons, α decay is a form of ionizing radiation because it is emitting two positively charged ions. The atomic elements from which the two neutrons and protons are emitted are not surprisingly reduced by two protons and two neutrons, hence producing a new element. *The purpose of α and β nuclear decay is to promote the nucleus of a given element into a more stable one.* β nuclear decay, as described above, involves turning a neutron into a proton. Accordingly, the neutron, which has a neutral charge, emits an electron, which is negatively charged in turn, which is responsible for assuming a positive charge. That is, a neutron is essentially a proton plus an electron. Once the electron is emitted, the neutron becomes the smaller-sized proton. By the same reasoning, another form of β nuclear decay transforms a proton into a neutron. As such, a proton emits a positron to become a neutral particle or neutron. This is called positron emission. Examples of these various forms of decay are as follows. U^{238} is converted into Th^{234} as an example of α decay. The numbers 238 and 234 represent the atomic mass of these elements, which is the total composite number of neutrons and protons. That is, two neutrons and two protons are released from uranium to produce thorium. The two neutrons and two protons again represent the release of a helium nucleus emitted from uranium responsible for transforming it into thorium. Be^7 converted to Li^7 is an example of positron emission nuclear decay. Seven is the atomic mass number of each element whereas beryllium has four protons and lithium three protons, their atomic numbers. A conversion of an up quark in one of the protons of beryllium to a down quark converting that proton to a neutron essentially changes the identity of the atom to a new element, i.e. lithium. β^- decay is exemplified by C^{14} changing to N^{14}. The nucleus of carbon^{14} consists of six protons and eight neutrons. Conversion of a neutron to a proton changes the identity of the carbon element to nitrogen. An interesting example of α decay involves the conversion of radon to polonium. Polonium is actually named after Poland because in the 1800s Poland did not exist as a separate country. It was partitioned between Prussia, Russia, and Austria in the late 1700s and remained occupied when Marie Curie discovered this element. She named polonium to stand for the country of her birth, Poland, to give support to its yearning for independence.

The weak force that is responsible for changing the flavor of quarks and subsequently allowing nuclear fusion or change in the identity of the atomic element is analogous to the strong force that is responsible for changing the color of quarks that promotes their adhesion to one another. Like flavor, color is also an essential property of the quark model. Analogous to electrically charged particles that interact by exchanging electromagnetic force carried by photons, color-charged particles (quarks) interact by exchanging the strong force carried by color-charged gluons. The attraction between the colors creating the very strong force field that serves as the quark adhesive is also analogous to the attraction of electrically charged protons and electrons in the Sun, for example, creating the electromagnetic force field that is ultimately the glue for all living matter.

The weak force in the Sun allows nuclear fusion to occur facilitating the strong force in cooperative fashion. In addition, it generates the electromagnetic force carried by visible light photons (derived from gamma rays) from the Sun to the chlorophyll of plants. This forms the food chain of our ecosystem.

Fusing lighter nuclei into heavier ones requires the strong force to hold them together. However, the strong force cannot hold two protons together very well, which in the Sun is the initial step in the fusion process. Therefore, the weak force provides the necessary help. It changes the quark flavor of the proton by changing an up quark to a down quark resulting in changing one of the protons to a neutron, hence changing the identity of the hydrogen atom isotope forming deuterium, a heavier hydrogen atom consisting of a proton and a neutron. The strong force is capable of holding a proton and neutron together *via* a poorly understood mechanism sometimes referred to as the residual strong force.

As a result of the weak force promoting nuclear fusion, it is responsible for the burning of stars, a process that creates heavier elements, which become the building blocks for planets and all of matter constituting the universe.

The Standard Model of physics developed in the early 1970s describes how three of the four forces guide the basic building blocks of reality from the 17 (six quarks, six leptons, four force-carrying particles or gauge bosons, and the Higgs boson) most elemental particles, particularly the up and down quarks and the most important lepton, the electron (Figure 1.1).

It has predicted remarkably well a wide variety of physical phenomena from how atoms behave to how stars burn. Gravity is not part of the Standard Model of particle physics, but because it is active over an infinite range of distances, it is far weaker than the other forces, particularly the strong and weak forces that act over a distance of less than the size of a proton. Thus, at the very microscopic range gravitational force is essentially irrelevant only becoming dominant at the scale of the human body or of the planets, for example. The gauge bosons of the weak force have significant mass and this has posed a major challenge to the Standard Model due to the notion that like the photon gauge boson that carries the

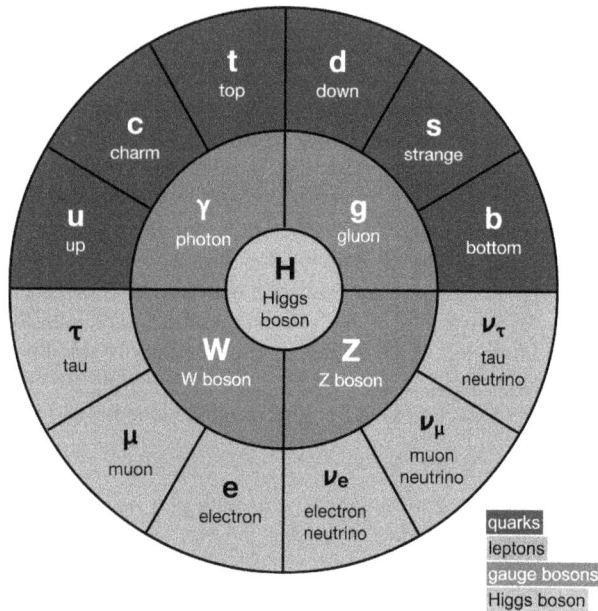

FIGURE 1.1 Schematic representation of the standard model of particle physics. Source: adapted from https://www.quantamagazine.org/a-new-map-of-the-standard-model-of-particle-physics-20201022/.

electromagnetic force, the W and Z bosons of the weak force need to also be massless in order for the two forces to represent a unified symmetry, that is derived from the same origin. The Higgs boson discovery, however, has provided a mechanism for explaining the fact that the weak force gauge bosons have a positive resting mass. The explanation that supports the Standard Model is based on the finding that *the Higgs boson field imparts mass to the W bosons and Z bosons*. Accordingly, at very high energies the two forces are similar whereas at low energies the forces become very different. Hence, when going from very high to low energy, the forces, initially very similar, break symmetry such that the electromagnetic force-carrying photon loses its mass. This occurred in the very early period of the universe expansion following the Big Bang when the force no longer was limited to a minuscule range less than the size of a proton. The Higgs boson was discovered in 2012 at CERN, Geneva, which merited the award of the Nobel Prize in 2013 to Peter Higgs and Francois Englert who proposed it in the 1960s. The weak and electromagnetic forces are, therefore, considered to be different manifestations of a collective origin and are referred to as the electroweak force. This shared origin, again, represents another symmetry of the physical world. Unified theories extend this system to include the strong nuclear force (*i.e.* the Standard Model) and even the gravitational force, albeit this unified theory of all fundamental forces remains an open line of research. Einstein coined the term Unified Field Theory and attempted to unify his theory of relativity (and gravity) with electromagnetism.

The weak force in nuclear fusion that powers the Sun has been technologically harnessed to create thermonuclear hydrogen bombs. *Nuclear reactions, both nuclear fusion and nuclear fission, that change the structure of atomic nuclei, are far stronger than the forces that hold together atoms of*

a molecule, i.e. molecular bonds. The former are the forces of the most powerful explosives humanity has ever developed. Hydrogen bombs, developed during the cold war in the 1950s, employing nuclear fusion are even more powerful than the bombs dropped on Japan by the U.S. toward the end of World War II, which harnessed the energy of nuclear fission. The initial bomb was dropped on Hiroshima on August 6, 1945, at 2:45 on a Monday morning. The process of nuclear fission involves splitting a heavy nucleus, for example, uranium, into two lighter ones, releasing energy in the process. Enriching uranium is the process of separating isotopes of U^{235} from the much more common U^{238} isotope (the latter representing roughly 99.3% by weight of natural uranium). Enriching uranium is the initial step in the process of nuclear fission. It is the weak force that causes the instability of uranium, which is responsible for nuclear fission. Consequently, extreme care must be taken in handling this element.

The amount of energy released by either process of nuclear fusion or fission is enormous and may be calculated by Einstein's $E = mc^2$ equation. The huge value of c^2 (c is the speed of light, 300 million meters per second) makes even a vanishingly small amount of mass equate to massive amounts of energy.

Energy is a very important physical concept that is vital for the functioning of various biological systems. It is the main driver for a network of biochemical reactions that form the basis of existence in living cells. We will further discuss a few key aspects of energy in the following section.

1.3 Energy in Its Various Forms

Our universe is made of matter and energy. In view of the 20th-century scientific theories on energy–mass equivalence formulated by Einstein, everything in the universe may be considered to be a form of energy. *Einstein demonstrated the energy-mass equivalence through his famous equation $E = mc^2$ where E, m, and c stand for energy, mass, and velocity of light*, respectively [5–7]. In physics, energy is a property of objects, a potential to perform work, and can only be converted between its various forms and can neither be created nor destroyed. Several common forms of energy, such as kinetic, potential, radiant, chemical, thermal and nuclear, are defined by their sources and capacity to be transformed into a specific form of work in the physical world. In addition, the concept of "informational energy" inspired from the kinetic energy expression in classical physics, is in vogue since the mid-1960s and was introduced [8, 9] as a measure of uncertainty or randomness of a probability system.

Every object has thermal energy at a temperature above absolute 0 defined as "0 Kelvin" (equal to –273.15° Celsius) and is created by atomic and molecular vibrations. That is, it has to be incredibly cold, i.e. –273.15° Celsius, for such vibrations to actually stop! Even in that state, according to quantum mechanics, so-called zero-point oscillations of the quantum ground state still persist. *Another form of energy is chemical energy defined as potential energy stored within the bonds between the atoms of a molecule and can be utilized in the course of a chemical reaction between a set of reactants to* form a set of products. By contrast, gravitational potential energy is stored gravitational energy due to a configuration of massive objects with respect to each other. For example, a ball placed at the top of a slide or a sloped road has greater potential energy than at the corresponding bottom and hence when released will spontaneously move towards the position corresponding to the lowest gravitational energy allowed by the constraints imposed on it (bottom of the valley). Electrical energy is the energy associated with the motion of charged particles (*e.g.* electrons and protons) under the influence of electric potential due to other charged objects interacting with them (including their own interactions). An additional form of energy is nuclear energy, which is locked up in the forces that bind the matter making up nucleons (protons and neutrons) of nuclei in atoms. The forces associated with this energy are the strong nuclear and weak nuclear forces (see section 1.2 for a detailed discussion). The former of these is involved in keeping the positively charged protons and neutrons together in the nucleus since the conventional electrostatic interactions would lead to the repulsion of positive charges between the protons of the nucleus. The weak nuclear forces are involved in the radioactive decay of the nucleus. An enormous amount of nuclear energy (which can potentially be destructive) is released in nuclear reactions involving nuclear fission where an atomic nucleus (*e.g.* Uranium-235) is split into lighter nuclei and neutrons.

1.4 Heat and Work

The importance of studying thermodynamics in the context of any life form anywhere in the universe is underscored by the fact that heat and work (manifestations of energy) are two most fundamentally integral components of life. In this context, one of Einstein's quotes, "Energy is everything", assumes particular significance. The root term "*dynamis*" in thermodynamics refers to power, the rate of doing work, equivalent to the amount of energy consumed per unit time. A commonly employed unit of mechanical power is "horsepower" (hp). An interesting side note concerns a historical perspective of this term coined by the engineer James Watt in 1782 while working on the performance of steam engines. More interesting than the definition of horsepower is the manner in which the definition came about into vogue. In the 18th century, horses were used to lift coal out of coal mines and Watt sought to define the power required by these animals to accomplish this task. By the horse using a pulley, he determined that the animal could lift 33,000 pounds of load over a distance of one foot per minute, corresponding to 33,000 foot-pounds of work coined as "horsepower" (abbreviated as hp). Although horses are far more powerful than humans, it is generally believed that Watt optimistically overestimated the capabilities of horses. In addition, one horsepower is equal to 746 watts (SI units of power) making a 20-watt light bulb equivalent to 0.027 hp or a 1 kW industrial light bulb equivalent to 1.3 hp. This implies that a Porsche that has a 500-horsepower engine is capable of pulling 16.5 million foot-pounds per minute. However, the horsepower of an engine is determined by multiplying the square of the cylinder diameter in inches by the number of cylinders

FIGURE 1.2 A powerful sports car—2021 Porsche 911. Source: https://www.guideautoweb.com/en/makes/porsche/911/2021/.

and then dividing that figure by 2.5. Using this method that grossly overestimates the actual horsepower of an engine, the actual horsepower of the motor is overestimated by a factor of 8. Therefore, a 500-horsepower sports car (Figure 1.2) realistically has only 62.5 horsepower relative to 1 horsepower of an actual horse. By comparison, a healthy human can sustain about 0.1 horsepower.

1.5 The Birth of Thermodynamics

The dawn of thermodynamics can be traced to 1823 when the book entitled *Reflection on the Motive Power of Fire* was published by one of Napoleon's officers, Sadi Carnot. In it, he proved the existence of a threshold for the transformation of heat into mechanical work. Next, in 1854, Rudolf Clausius, at the time professor at ETH in Zurich, extended the field of thermodynamics to all transformations occurring in nature including metabolic processes in living organisms. He introduced the fundamental concept of entropy, denoted S, whose meaning derives from the Greek words "*en*" (in) and "*tropé*" (transformation). Mathematically, entropy has been defined as $S = (E/T)$, *i.e.* as the ratio of the total energy of the system to its temperature. Another important scientist in the field of thermodynamics, Lord Kelvin (William Thomson), a British physicist and engineer introduced the absolute temperature T, which is employed in the definition of entropy. As will be discussed later, thermodynamics, in contrast to the previously developed fields of physics such as mechanics and electromagnetism, identified irreversible processes, which can take place only in one direction, and once the final state is reached it is impossible to return to the starting point traversing the same trajectory backward. This is associated with the existence of a so-called "arrow of time", which is indispensable for realistic descriptions of natural phenomena. The irreversibility effect is related to an increase in the entropy of a closed system as elaborated on in the second law of thermodynamics. This issue was addressed by Ludwig Boltzmann, a professor of physics at the University of Graz, who is considered today to be the founding father of statistical mechanics. He stressed the relevance of entropy to biological processes by stating that:

The general struggle for existence of animate beings is not a struggle for raw materials – these, for organisms, are air, water and soil, all abundantly available—nor for energy, which exists in plenty in any body in the form of heat, but of a struggle for entropy, which becomes available through the transition of energy from the hot Sun to the cold Earth [10].

Boltzmann was an admirer of Darwin, and hence he attempted to emulate the evolution of physical systems by living organisms. In 1872, in his most important paper, he described the evolution of gas molecules using an approach that combines statistics, used for the description of the distribution of velocities of these molecules with the dynamics of their collisions in mechanical terms. This very new type of reasoning at the time culminated in the so-called Boltzmann equation [11–12], which shows that *the spontaneous expansion of a gas is accompanied by an increase of entropy, in agreement with the second law of thermodynamics.* Importantly, in the Boltzmann equation, time is asymmetric, possessing a defined direction along the monotonic entropy increase trajectory. The greatest American physicist between the late nineteenth and early 20th century was Willard Gibbs (1839–1903), who was appointed the first chair in mathematical physics at Yale. He made fundamental contributions to thermodynamics by its extension to chemical systems, in which different phases (gas, liquid, and solid) can be present. In his monograph entitled *On the Equilibrium of Heterogeneous Substances*, he began with a quotation from Rudolf Clausius: "The energy of the world is constant. The entropy of the world tends towards maximum." Gibbs derived criteria for equilibrium and stability of thermodynamic systems [1–4, 13]. The energy of a thermodynamic system can be conveniently expressed by means of a function H, called enthalpy, that is, the sum of the total kinetic and of the interaction energies of the particles present in it, plus the product PV of the pressure, P, and volume V of the system itself. Consequently, in isobaric processes, *i.e.* transformations that occur under constant pressure, the change of enthalpy is equal to the heat transferred to the system plus the work done by changing its volume at constant pressure.

According to Clausius the change of the thermal part of the system's energy is given as the product of the absolute temperature, *T*, and the increase of entropy. Since it is not available to do purposeful work *W*, it is called "useless" energy. It must be subtracted from the work that could be performed by the system as a result of a change of its total energy, expressed by enthalpy. What remains then is called free energy because it can be used by the system to perform work through the slow variation of the macroscopic variables. This free energy has been traditionally denoted as *G* (to honor Gibbs) and can be simply computed as

$$\text{Available energy} = G = H - TS = \text{total energy} \\ - \text{useless energy} \tag{1.1}$$

At the molecular level, the useless part of the total energy mainly contributes to the kinetic energy of the molecules in the form of translational, vibrational, and rotational motions.

On the other hand, free energy can affect the electronic states of atoms and molecules and cause the breaking and making of intra-molecular bonds, i.e. it can lead to chemical transformations such as those taking place in the biochemical reactions of a living system. *Therefore, formation of new bonds can lead to molecular organization of a living system.* In order to evaluate the free energy cost required for molecular transformations, Gibbs introduced the concept of "chemical potential", which provided a critical link between thermodynamics and chemistry. In the context of living organisms, free energy is used to perform work for different subcellular activities. As explained elsewhere in this book, the transfer of free energy occurs typically by using the reaction of hydrolysis of ATP (adenosine triphosphate) and can result in the following types of transformations:

- Electric: transporting electric charges (ions, protons, and electrons) across the system as well as in and out of the system.
- Chemical: synthesis and biosynthesis of chemical compounds.
- Osmotic: active transport of molecules across membranes, partitions, and walls.
- Superficial: changing the surface area of a system as can be seen in morphogenesis, motility, and cell division.

The occurrence of a thermodynamic transformation implies the presence of nonzero differences (gradients) of temperature, pressure, or chemical potentials, across boundaries, which then become the driving forces for these transformations. At the final state of a transformation, in a system with constant internal energy, their values must be uniform everywhere, which is consistent with the maximum entropy principle. The question regarding the rates of these transformations is a crucial but difficult one and it was successfully addressed in 1930 by the Norwegian-born US scientist, Lars Onsager, who was awarded a Nobel Prize in chemistry for his work.

1.6 Microscopic Origin of Entropy

Entropy, as a physical quantity, is a very multifaceted entity including both microscopic and macroscopic domains. Although there are prosaic explanations of entropy, the idea was born from classical thermodynamics as a quantitative, not a qualitative, concept. Specifically, as mentioned above, its definition was first proposed by Rudolph Clausius (1867, "The Mechanical Theory of Heat") as *entropy (S) = the heat content of the system (Q) / the temperature of the system (T).*

Subsequently, William Thompson (Lord Kelvin) advanced the understanding of how mechanical work can be converted into heat and the concept of a gas being made up of tiny molecules whose average kinetic energy was defined as the temperature became understood. Furthermore, the equation $S = Q/T$ evolved to define the change of entropy according to the formula: $\Delta S = \Delta Q/T$ where ΔQ is the change in the heat content of the system (Q) and *T* is the temperature of the system (in

Kelvin). This equation recognizes that at a given temperature of a system in thermodynamic equilibrium a change in entropy moves in a parallel direction with the internal heat energy.

On a microscopic level, entropy describes the kinetics of the constituent atoms or molecules (*i.e.* particles) of any large sample of inanimate matter such as a gas, a liquid, or solid body. As stated above, entropy is related to the energy of a body or a system that cannot be used to do work. *Generally, entropy is greatest for gases, less so for liquids, and least so for solids because of the greater number of available energy states of a liquid compared to a solid and more so for a gas than a liquid. The amount of entropy is a function of the number of particles in the system and, specifically, the number of degrees of freedom available for their movement. Together, they represent the number of ways in which a system can exist or arrange itself in space in a given physical phase.* Each of these possibilities is an available energy state, which can be characterized in terms of the translational (linear), rotational, and vibrational (spring-like) energies of the constituent atoms and/or molecules. These terms correspond to degrees of freedom which are characterized in three general ways: 1) translational (allowing linear movements along the three dimensions of the Cartesian space, i.e. X, Y, and Z axes); 2) rotational (allowing for angular movement along the axes of three Euler angles ϕ, θ, and Ψ—for example, a diatomic molecule may spin about the center of a bond between its atoms or rotate along its bond axis [Figure 1.3]; an analogous example in the macroscopic world can be visualized by considering a baton being grasped at its center and twirled with the balls at the ends of the baton moving as a rigid body); and 3) vibrational (allowing a spring-like movement of stretching and contracting of bonds between atoms of molecules [Figure 1.4]).

Vibrational motion along a bond between two atoms represents a good example of conversion between different energy forms. In the compressed form of the spring-like motion, potential energy is stored while in the uncompressed state, that energy is released as kinetic energy. However, the spring constant of the bond forces the re-compression of the bond leading

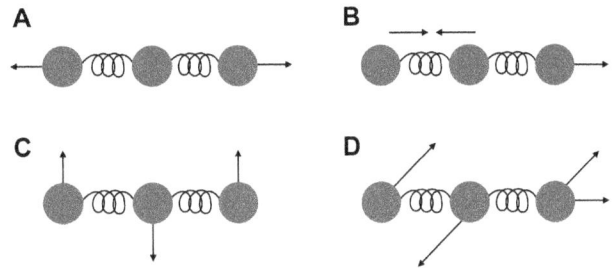

FIGURE 1.4 Various vibrational modes represented by arrows of a system of three atoms connected by two bonds represented as springs. Source: adapted from http://www.acamedia.info/sciences/J_G/envrad/microwaves/index.htm.

once again to the cycle of potential energy storage and release. For a diatomic molecule, rotational and vibrational movements occupy the space of two degrees of freedom and one degree of freedom, respectively. *The probability distribution of energy states is a way of representing the heterogeneity and disorder of the system in probabilistic terms and such a function is related to the entropy of the system.* It should be noted that the term "oscillation" is not equivalent to the vibration of atoms and molecules since molecular oscillations are generally relevant in the context of circadian clocks where levels of RNA and protein levels oscillate to regulate transcription of "period" and "timeless" genes.

The concept of the randomness of positions or of motions can illustrate the applicability of the notion of entropy to biological structures. Think about an unfolded protein (e.g. a denatured from its native biological form) flapping about in an aqueous solution as may occur in the natural conditions of biology. Compare this to the native conformation of a globular folded protein with a well-characterized biological function. In the case of the unfolded protein, at any given time, there would be several random ensembles of different conformations in solution. Moreover, the protein will change conformations randomly and rapidly within the solution.

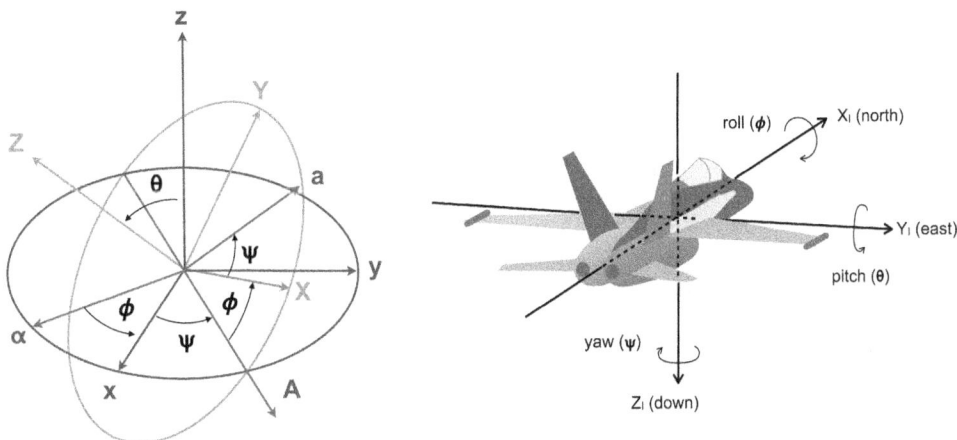

FIGURE 1.3 The Euler angles (ϕ, θ, Ψ) introduced by Leonhard Euler to describe the orientation of a rigid body. These angles can be thought of as twist, latitude, and longitude, which are the three parameters needed to describe the orientation of an object. In the world of aircraft flying, these angles are also defined as roll, pitch, and yaw, representing rotations of a plane around the X, Y, and Z axes.Source: adapted from a) original drawing by Timofei Shatrov and b) http://www.chrobotics.com/library/understanding-euler-angles.

Conversely, with the native molecule, the motions are far more restricted and the atoms of the molecule are all moving about equilibrium positions so that for practical purposes all the molecules have nearly the same average conformation. The average native conformation is well ordered due to strong intra-molecular interactions and hence it is in a lower entropy state than the more random unfolded form. However, the energy of the native folded conformation is considerably lower than that of the unfolded form as it is likely to be characterized by more stabilizing forces such as hydrogen bonds, hydrophobic packing, and aromatic stacking of side chains. Therefore, the loss of entropy (due to greater order) is more than compensated for by the gain in internal energy or enthalpy of the globular protein structure. In a reversible process, such a structure can potentially undergo a conformational transition to a higher energy state with higher entropy with a release of net heat energy to drive a coupled reaction (*e.g.* ATP hydrolysis).

When heat is added to a physical system (solid, liquid, or gaseous), the component responsible for increasing the system's internal energy will increase its temperature, created by the increase in the average kinetic energy of the particles within their allowed degrees of freedom. Kinetic energy (KE) of translational motion is defined as:

KE (translational) = ½(*mv²*), where m is mass and v is linear velocity. The kinetic energy of rotational motion is, on the other hand, defined as *KE (rotational) = 1/2 (I Ω^2)*, where I is the moment of inertia and Ω is the angular velocity. The moment of inertia (I) is the rotational analog of mass and is a measure of the distribution of an object's mass in space around a particular axis of rotation. Thus, for a point mass "m" located at a distance "r" from an axis of rotation, the moment of inertia (I) is defined as the product of the mass and the square of the distance to the axis of rotation (*I* = *mr²*). Extending this definition to an actual object requires the integration (or summation in simpler terms) of all such products throughout the extent of the substance. The key concept to be borne in mind is that moment of inertia is very much dependent on the axis of rotation. This implies that the moment of inertia depends additionally on the shape of the mass being considered. For example, a cylinder of a certain mass will have a different moment of inertia about an axis than a sphere or a rod of the same mass. Angular velocity (Ω) is the velocity of rotation, a function of the radius and the angle per unit time. To use a metaphor on a much more macroscopic level, a figure skater initially spins with her arms outstretched then moves her arms to her side bringing more of the mass closer to the axis of rotation, reducing the radius (*R*) and consequently reducing the moment of inertia (*I* = *MR²*). The reduction in the moment of inertia results in an increase in angular velocity (Ω) due to conservation of angular momentum (equal to the product of the moment of inertia and angular velocity) leading to an increase in her angular velocity of rotational motion. This in turn causes increased spinning of the athlete and transfer of energy to the surroundings in the form of heat, via the vibrations to the nearby air molecules, which in turn transfer such energy to their neighboring molecules. These energy transfers have the impact of raising the ambient temperature, however infinitesimally small. Consistent with the first and second laws of

thermodynamics, the energy allowing the skater's movement with the utilization of ATP is accompanied by the uncoupling of nutrient oxidation (*i.e.* the combustion of nutrient fuel) from ATP production (i.e. metabolic inefficiency) providing the dissipation of heat from the body in the form of sweat, which decreases the body temperature. It should be noted that such dissipation of heat is not limited to only rotational movements but applies to movements of any kind, of a body as a whole or of its internal components. The dissipation of such heat energy to the surroundings (not usable for any work) contributes to increasing the entropy of any system. The change in entropy may be in the form of the useless increased kinetic energy of the internal parts of a system that raises the body temperature of the individual. Because the body is physiologically isothermal, the increased movement or kinetic energy of the body as a whole and associated increased movement or kinetic energy of the component molecules of a person exercising will dissipate that heat to the surroundings leading to a change in its entropy.

The interrelationships between the usable and unusable energy of a system can be potentially understood through the definition of Gibbs free energy, which is useful in describing the spontaneity of biochemical processes where products end up with lower entropy than the reactants. In such processes, the lower enthalpy (or internal energy) of products enables the release of energy, which can be utilized in another simultaneously occurring process that consumes energy for the product formation (*e.g.* formation of ATP from ADP). *Thus, "available energy" from one biochemical reaction can be used to do "work" to drive product formation in a different but coupled reaction.* It may be noted that while the term "enthalpy" is commonly used in biochemical reactions, the equivalent "internal energy" is employed more often in inorganic chemical processes. In accordance with the first law of thermodynamics, work can be harnessed from any type of energy including nuclear, mechanical, electrical, light, or chemical. It is harnessed in industry in terms of pressure-volume work due to changes in volume at constant pressure (although the pressure may not remain constant during a specific process, its initial and final levels would be the same). For example, such work is derived from the expansion or contraction of a gas against external resistance as in the case of fuel combustion, which aids the locomotion of a vehicle or spinning of turbines (Figure 1.5) to produce electricity.

1.7 The Rule of Law in Physics: Energy Conservation

In accordance with the laws of conservation of energy, *the total energy in the universe is constant and can neither be created nor destroyed.* This is also true for any physical process so that energy can only change its form or be transferred from one system to another but cannot be created or destroyed. In processes and systems that involve heat, this concept is embodied in the unbreakable and unbendable First Law of Thermodynamics, which is an application of the law of the conservation of energy and it applies to a closed system like our universe or a suitable approximation of a closed system such as a thermally isolated container. Thermodynamics (derived as a combination of the

Generator

Stator

Rotor — Shaft

Turbine

Wicket gate

Water flow

Turbine blade

Hydroelectric Generator

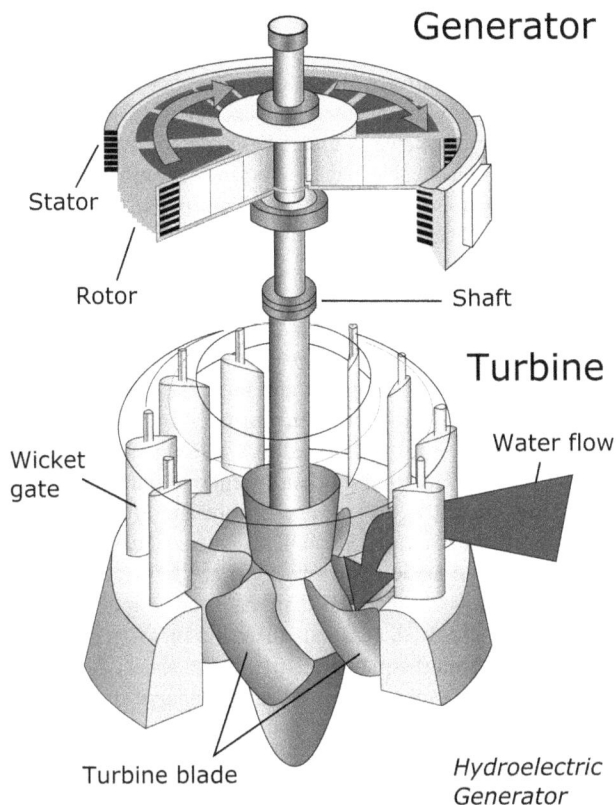

FIGURE 1.5 Use of turbines in the generation of hydroelectric power. Source: US Army Corps of Engineers (public domain).

two Greek words "*therme*" meaning heat and "*dynamis*" meaning power) is a branch of physics that studies heat, energy, and the ability of that energy to do work. Nobel Prize–winning American physicist Richard Feynman famously stated that: "It is important to know in physics today we have no knowledge of what energy is." [14]. Nonetheless, we define energy not for what it is, but for what it does. Accordingly, energy is defined as the capacity to do work or to produce heat. In contradistinction to work as typically indicating those things we "have to do" but do not want to do, in physics energy produces a force that acts on an object causing it to move or acquire kinetic energy at the expense of potential energy. *Force is defined as a rate of change (gradient) of energy per unit of distance.* The movement of an object driven by a physical force that takes place on a macroscopic scale should be distinguished from the microscopic vibrations within the object, associated with its thermal energy related to its temperature. In physics, the term "heat" is defined as an energy transfer between a system and its surroundings, other than that done by work (as seen in the transfer of energy by mechanical movements). Thus, *heat is deemed to be energy transfer by thermal interactions either in the form of radiation, convection or conduction.* Thermal radiation is the electromagnetic energy generated by the thermal motion of charged particles as in the case of solar energy, while thermal conduction is the kinetic energy of molecular motions. *Thermal convection involves mass movement of molecules that have a different kinetic energy (or temperature) than their surroundings and hence can carry that energy*

over distance. Atmospheric jet streams represent an example of thermal convection. In contrast to work, which is directed by the force applied to or by the system, heat is dissipated to the environment and hence such dissipation represents an irreversible process. While work can be recycled without any loss, heat cannot be recycled without loss.

1.8 The First and Second Laws of Thermodynamics

The industrial revolution that started in the latter part of the 18th century took place in societies, such as Britain, France, and Germany, which produced large-scale machinery and engaged in a competition to find the most energy-efficient machinery possible, pushing the limits of efficiency and productivity whose technological benefits are still felt today. This inventiveness frenzy prompted a growing legion of 19th-century physicists and engineers to analyze the flow of heat and associated chemical reactions in these machines (combustion) that resulted in the output of mechanical work. The industrial corporations emerging in Europe and America at that time were excited by the potential that this research could lead to machines that would produce and flawlessly convert heat into work with an increasing level of efficiency. However, taken to the extreme limit, the notion of 'perpetual motion machines' defined as devices that could in principle run continuously off its own exhaust, violated the most important and fundamental laws of natural science, *i.e.* the energy conservation law of thermodynamics, which define a relationship between heat, work, and energy with a constant energy balance over time.

The laws of thermodynamics provide the theoretical framework that enables one to describe the flow of energy through both inanimate and living systems. Accordingly, only reactions or processes that do not require energy input proceed spontaneously in nature unless coupled to a reaction that produces energy. The First Law of Thermodynamics is a version of the law of energy conservation, which states that the total energy of an isolated system is constant over time and that energy may be changed from one form to another, or transferred from one system to another, but can neither be destroyed nor created. As adapted to thermodynamic systems, this law states that the change in internal energy of a closed system is equal to the amount of heat supplied to the system, minus the amount of work done by the system on its surroundings. Mathematically, the law equates the change in internal energy of a system (U) to the heat added to the system (ΔQ) minus work done by the system (ΔW) ($\Delta U = \Delta Q - \Delta W$). An alternative form of this statement is $\Delta U = \Delta Q + \Delta W$, where ΔW is the work done "on the system". Thus, as a consequence of this law, if the internal energy of a system is constant ($\Delta U = 0$), then the amount of work done by the system has to be equal to the amount of heat added into the system. In other words, it would not be possible to do any work without expending some energy (either added or innate). Stated another way, perpetual motion machines are impossible to exist without the infusion of energy from an external source. Thus, the first law of thermodynamics allows energy to be borrowed (and stored) from one source for another to be used at a different location or at a different time. The generation of hydroelectric power represents a very common everyday example of such conservation. Here, potential energy associated with flowing water falling by the force of gravity and gaining kinetic energy as it accelerates flowing down can be used to turn turbines and generators that produce electricity representing the conversion of mechanical energy into electrical energy. Another common example, very relevant to all forms of life on Earth, is represented by the conversion of electromagnetic energy from the Sun to electrochemical energy resulting from the photosynthetic production of carbohydrates, proteins, phytonutrients, and vitamins in plants. These substances, in turn, are combusted or metabolized to high potential energy-containing molecules of ATP in our tissues, and the latter are utilized for biological work such as the mechanical activity of muscles for locomotion.

All things, living and inanimate, must obey the laws of physics including the laws of thermodynamics. Life forms absorb energy from the Sun and transform it into the structures and function of all the beauty that life displays. While energy utilization is manifested and perceived qualitatively in different forms, physicochemical measurements of such utilization are made using a number of metrics. For instance, mechanical work is measured in terms of joules, electrical power in terms of watts, nuclear energy in terms of MeV (million electron Volts), in chemistry Gibbs free energy, measured in kcal/mol, is the commonly used energy currency. It allows us to gain an understanding of available energy to do useful work ranging from the basic requirements of cell maintenance to the most energy-expensive physiological process of reproduction. Accordingly, it can be expected that biological thermodynamics should also be fundamental in the natural selection criteria for phenotypic traits in terms of the energy investment requirements. In this sense, the principles of physics must be followed and utilized by all living systems.

While the first law of thermodynamics deals with the quantity of energy in a system, it does not provide any insights into the direction of energy flow. This is elucidated *by the Second Law of Thermodynamics, which states that the total entropy of an isolated system always increases over time or remains constant in ideal cases where the system is in a steady state or undergoing a reversible process.* An appreciation of the second law requires a comprehensive understanding of the basic concept of entropy, which is generally deemed to be a measure of the number of microscopic configuration states of a thermodynamic system, which is typically defined by macroscopic parameters such as pressure, temperature, and volume. Stated another way, *entropy can be regarded to be a measure of randomness in a system and in accordance with the second law of thermodynamics, the randomness in the universe always increases.* Various aspects of entropy will be discussed in greater detail in a separate section in this chapter.

1.9 Energy Cannot Be Created but Can Be Transformed

Thermodynamics is important to the everyday life of people. It is the basis for our understanding of how matter and energy work. The first law of thermodynamics as described earlier deals with the conservation of energy. Broadly, it states that the total energy of an isolated system is constant. An isolated system (e.g. our universe) is one that neither allows matter nor energy to flow in or out. The basic relevance here is that one can manipulate energy to create change from one form to another such as using electricity to boil water or to burn petroleum to move a car. Additionally, in light of the energy-mass equivalence, matter can be converted to energy and energy to matter. The derivation of nuclear energy would be a very illustrative example of this principle. Thus, the first law of thermodynamics helps us appreciate the counterintuitive concept that we cannot in a real sense destroy a piece of paper. We can ignite the paper with a lighter and burn the paper to ashes. The paper has not, however, been destroyed from the perspective of physics, only transformed to a different form of matter and energy. The total amount of energy and mass in the closed system that is the universe before and after the paper was burned remains the same. This concept can be stated in terms of an equation whereby the mass of paper plus the mass of fire (*i.e.* the material that was used to make the fire such as firewood, fuel, etc.) plus the mass of oxygen (needed to promote burning) equals the mass of the ashes, the carbon dioxide and the smoke, generated as products of the burning. Similarly, the energy expended in burning the paper would be the same as the heat released through the smoke and any work that the escaping gases would have done on the surroundings. Thus, energy conservation is achieved in the process through inter-conversion.

Essential forms of energy include nuclear (potential energy required to bind protons and neutrons in the nucleus of atoms); light (energy possessed by oscillating electric and magnetic fields that make up the visible electromagnetic spectrum); chemical (the potential energy stored in the electrostatic bonding relationship among atoms in molecules); electrical (energy involved in initiating and maintaining electron flow); mechanical (the energy generated or stored by machines which induce or result from converted motion processes in a system; this includes gravitational potential energy); and heat energy (the kinetic energy associated with random motions of matter including vibration, rotation, and translation of molecules). The universe is replete with several examples of inter-conversion between these forms with resultant benefits to humanity (*e.g.* hydroelectric power stations to generate electricity without which modern human existence is impossible).

The second law of thermodynamics states that the entropy of an isolated system always evolves to a state of maximum entropy or thermodynamic equilibrium. The increase in entropy is associated with both the dissipation of energy and the dispersion of matter. In other words, localized energy will always disperse and spread out, and the total amount of entropy (disorder) in the universe will always increase. In the case of a burning piece of paper, the second law of thermodynamics is reflected in the spontaneous spread and dissemination of the fire to ultimately spread to the entire extent of the paper. Additionally, the smoke disperses rather than remaining confined to a spot. This process, like everything (asymptotically) in the universe, goes from order to disorder. Importantly, the second law applies to an isolated system. All living beings age and eventually die wherein the process of senescence and ultimately mortality is one that randomizes our cells and molecules, atoms, and subatomic particles to the universe in a process of increasing entropy in accordance with the second law of thermodynamics. In this context, the question of the complex development of organisms including human beings by the processes of mutation and natural selection underlying evolution deserves further investigation and attention and has been the subject of several theories that encompass the fields of information theory, quantum biology, and genetics. *While quantum biology (Chapter 3 in this volume) relates to the complexity of organisms and the processes of mutation and natural selection mechanistically, the framework of information theory allows information, considered as negative entropy, to explain some of the biological processes where complexity increases instead of decreasing as would be envisaged by the second law of thermodynamics (vide supra).* Driven by metabolic energy consumption, an increase in information is generally regarded to be associated with a decrease in entropy (*e.g.* embryonic development, cell differentiation, organ specialization, etc.).

A process spontaneously leading to the organized complexity of organisms including human beings would seem to violate the second law of thermodynamics as it is paradoxical and antithetical to the premise that everything in the universe goes from order to disorder but not the opposite. What allows us to evolve, i.e. to gain order and a greater level of organization, is that we are not a closed nor an adiabatic system but rather an open one, and a component of the isolated system of the universe constantly receiving a supply of energy from the Sun. Therefore, our evolution does not violate the second law of thermodynamics by the mere fact that this is an open system, in a far-from-equilibrium state sustained by an external energy supply. It should be recognized, however. that the Sun as a source of energy will eventually be transformed to a state where nuclear fusions would no longer be feasible causing the end of dispersion of life-sustaining energy. Thus, the end of all forms of life in the universe is an inescapable consequence of the second law of thermodynamics.

1.10 Heat, Entropy, and Energy Efficiency

It is recognized that not all of the input of heat added to a system is accountable by the free energy captured within the system or the work done by the system. Some energy escapes the system, lost to the surrounding environment as wasted energy with its associated entropy. This can be mathematically denoted as follows: the change in entropy (S) in a thermodynamic process is equal to the amount of heat added to the system divided by its temperature in Kelvin: $\Delta S = \Delta Q/T$. As with any thermodynamic property, the emphasis is on the change in entropy rather than its absolute value in a given state. If we consider an ideal gas or monoatomic gas, any heat added to a system will eventually be manifested in terms of increasing the number of possible states that such a collection of molecules will assume. This is also viewed in terms of the increase of randomness in the configuration of the atoms/molecules. In this light, for example, liquid water would have lower entropy than steam, as water molecules in the gaseous form have a

larger volume than in the liquid state and hence have access to molecular configurations with many more available energy states. *Thus, ΔS is a measure of dissipation of heat from the system as it attains a state of equilibrium following infusion of heat into it.* Following the definition of ΔS, for a given collection of molecules, the change in entropy for a given input of heat is higher at lower temperatures. Thus, entropy changes are not only sensitive to the phase of a substance, but also to the temperature at which heat is added. Entropy, like internal energy, is a "state function", that is, its value depends only on its current state and is independent of the path of achievement of such a state.

The concept of entropy carried the application of thermodynamics out of the industrial workplace to all of natural science. A natural outcome of the two laws of thermodynamics is a commentary on the efficiency of energy transformation. As will be shown later in this chapter, even in an idealized engine like the Carnot cycle, the efficiency of fuel conversion into work is always less than 1 (100%). This observation also extends to biochemical reactions, where some amount of heat is wasted to the surroundings.

Energy efficiency is described as the ratio of energy output (typically in the form of work done by the system) to the input energy needed to do work. In a biological engine, as a pertinent example, we can use glucose as the energy source. The combustion of one mole of glucose results in the production of 38 moles of ATP (the universal energy currency for living systems). Each mole of ATP upon hydrolysis to ADP (adenosine diphosphate) generates 7.3 kcal of Gibbs free energy. Hence, the combustion of glucose as the energy source leads to the generation of 38 moles of ATP, which corresponds to 277 kcal of stored energy available for subsequent work to be done by molecular motors that utilize ATP. The combustion of 1 mole of glucose (~180 grams) to six moles of carbon dioxide and six moles of water is an exothermic reaction that releases a total of 688 kcal/mole of Gibbs free energy. Thus, the combustion of glucose yields 3.82 kcal of free energy per gram. In this light, the efficiency of glucose combustion in terms of work done to produce ATP is obtained by the ratio of the total energy stored in the 38 moles of ATP to the total energy generated by one mole of glucose. Hence, the efficiency of the biological engine utilizing glucose as the energy source is ~ 40%.

1.11 Specific Heat

Translational kinetic energy on the microscopic level is the linear movement of particles, atoms, and molecules, analogous on the macroscopic level to the movement of an automobile. As heat is added to a liquid, its molecules experience increasing kinetic energy, which may be sufficient to break the weak intermolecular forces holding them together, e.g. the hydrogen bonds. With the increasing kinetic energy of individual molecules, the temperature of the liquid increases since the latter is proportional to the average of all kinetic energies. The relationship between the change in temperature in response to infusion of heat or thermal energy into any substance is defined by "specific heat", i.e. the amount of heat needed to increase the temperature of 1 kg of the substance by 1° Celsius.

TABLE 1.1

Specific Heats of Some Common Materials

Material	Specific Heat (joules/gram° C)
Liquid water	4.18
Solid water (ice)	2.11
Water vapor	2.00
Dry air	1.01
Basalt	0.84
Granite	0.79
Iron	0.45
Copper	0.38
Lead	0.13

The reference standard for specific heat is defined in terms of the value for water. In particular, it is the amount of added heat that is required to raise the temperature of 1 kg of water by 1° Celsius. For water, this value is 1 kilocalorie/kg°C (4,184 joule/kg°C) while it is smaller for ethanol (2,530 joule/kg°C), methanol (2,440 joule/kg°C), and propanol (2,390 joule/kg°C). As shown in Table 1.1, the specific heat values for metals such as iron, lead, and copper are much lower than that of water indicating that these materials conduct heat very easily.

The following question warrants further investigation: why is the specific heat of water higher than that of ethanol and much higher than that of iron? In the case of water, the individual molecules could be involved in up to four hydrogen bonding interactions with their neighbors. The oxygen atom with its lone pair electrons is capable of serving the role of an acceptor to the hydroxyl group (OH) from neighboring water (Figure 1.6a). Therefore, it can be potentially involved in two interactions as an acceptor since the oxygen atom in water has two lone pair electrons (Figure 1.6b).

In addition, the two hydroxyl groups of every water molecule can serve as hydrogen bond donors. These intermolecular hydrogen bonds dominate over the translational motion of the water molecules at lower temperatures and the competition between them is a decisive factor in the net kinetic energy of each molecule. However, as additional heat is added to the system, the kinetic energies of the individual molecules could

increase to a point where they could overcome the competition from the binding nature of the hydrogen bonds leading to a net increase in kinetic energy and hence the temperature. By contrast with the situation in water, ethanol (or methanol and propanol), which has only one hydroxyl (OH) group can at best be involved in two hydrogen bonds (one as a donor and the other as an acceptor). In this light, the intermolecular forces holding two ethanol molecules would be relatively weaker compared to those holding two water molecules together. Furthermore, the presence of the carbon atoms in the molecules (which represent the so-called hydrophobic component) contributes to the additional weakening of the hydrogen bonding interactions between the hydroxyl groups of two neighboring molecules. As a result, less heat would be needed in the case of ethanol, methanol, and propanol to increase the kinetic energy of individual molecules to a point where they can overcome intermolecular hydrogen bonding interactions. Furthermore, one may readily recall that ethanol is more "volatile" than water. Hence, the specific heat of a volatile liquid like ethanol tends to be lower than that of water, which is a better insulator of heat. In addition to translational motion, the molecules of liquids are endowed with rotational and vibrational degrees of freedom into which part of the heat coming in is transferred. The rotational motion of a molecule is the angular analog of the translational motion and is a rigid body motion. In the case of liquid water, for example, a water molecule could spin about an axis that bisects the bond angle H-O-H. The vibrational motions of a molecule (Figure 1.7) can result from combinations of bond stretching and angle bending and resemble the oscillating movements of the elastic coil of a slinky. Due to the rotational and vibrational motions of the molecules, any heat that flows into a liquid would be distributed in increasing the kinetic energy associated with these motions in addition to translational kinetic energy. It may be noted with interest that while a body of water in a glass may appear very passive and tranquil, the energies associated with the translational, vibrational, and rotational motions result in the movements of molecules at speeds of hundreds of meters per second.

In the case of metals, their specific heat is distinctly lower for the following reasons: 1) the atoms in a metallic object are very closely packed and their translational, vibrational, and

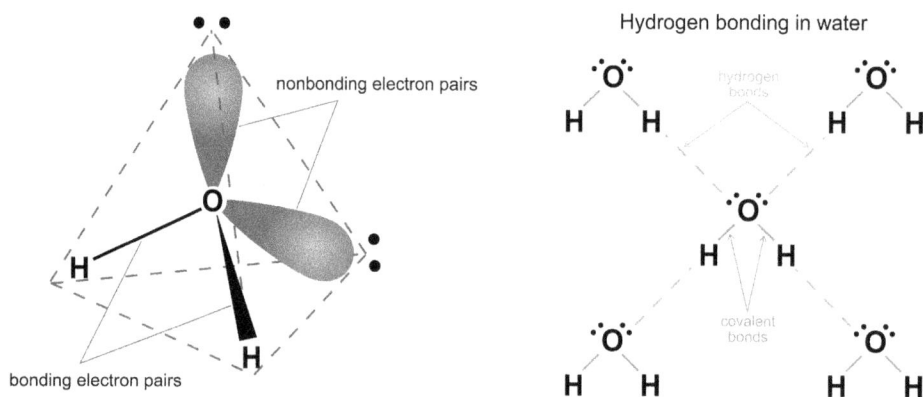

FIGURE 1.6 a) Structure of a water molecule (adapted courtesy of lectures at Penn State University). b) Hydrogen bonding patterns in a collection of water molecules. Source: adapted from "The Wonder of Water". (1997). In *Chemistry in Context*, ed. A. Truman Schwartz et al. Wm C Brown Publishers, Dubuque, Iowa. 2nd edition. A project of the American Chemical Society.

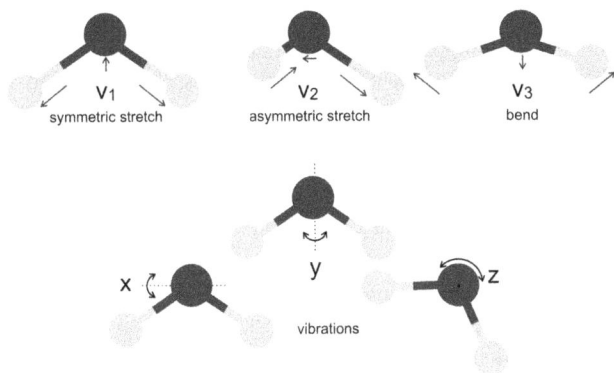

FIGURE 1.7 Typical modes of motion taking place within a molecule such as water. Adapted from http://www.acamedia.info/sciences/J_G/en vrad/microwaves/index.htm.

rotational degrees of freedom are much more restricted than in the case of liquids; 2) there are no interactions that correspond to intermolecular hydrogen bonds as would be found in water or ethanol; 3) metal structures are characterized by free electrons (typically one per atom) in the outermost orbital shells (called conduction bands) of their atoms. They are not bound to any atom strongly and move around freely amongst all the atoms of the lattice forming the 3D structure of a metallic object. These are primarily responsible for the excellent conduction of heat (and also electricity) in metals when compared to non-metallic solids, liquids, and gases. If one end of a metallic bar is heated, and the other end is cold, the electrons on the hot end have more thermal energy (associated with their increased kinetic energy) than the ones on the cold end. So as the electrons wander around, they carry energy from the hot end to the cold end, which is another way of saying they conduct heat. Since they move from a hotter region to a colder region, this process is in perfect agreement with the second law of thermodynamics. Of course, how fast these electrons conduct heat depends a lot on factors such as how many free electrons are around, how fast they move, and especially on how far they usually go before they collide with something within the metal causing them to change direction. This latter property defines the metal's resistance. Interestingly, those are the same factors that determine how well the metal conducts electricity. Thus, in general, *the thermal conductivity of a metal (at a given value of temperature) is proportional to the electrical conductivity.* In addition to the electron movement, the thermal energy associated with atomic vibrations at the hot end of the rod is transferred to adjacent atoms, which are in the cooler regions of the metal. However, unlike in the case of liquids, no part of that thermal energy is utilized to overcome additional bonding forces unless we raise the temperature of the metal close to the melting point, which represents a phase transition from solid to liquid. We discuss the topic of phase transitions later in this chapter.

In accordance with the definition of specific heat, 1 gram of water can be heated from 0° Celsius to 100° Celsius by a supply of 100 calories of heat. While the latter temperature is the boiling point of water, the water does not boil into steam at this temperature unless additional heat is supplied. Such heat is called the enthalpy of vaporization and is also known as latent heat of vaporization. Since enthalpy represents the amount of heat transferred at constant pressure and temperature, the enthalpy of vaporization relates to the amount of heat (usually expressed as calories/gram) needed to convert 1 gram of water to steam at 100° Celsius. For water, the conversion of 1 gram of boiling water to steam requires 548 calories (2,260 joules) of heat and reflects a phase transition from the fluid phase to the gaseous phase. Thus, the enthalpy of vaporization is significantly higher when compared to the amount of heat needed to raise the temperature to the boiling point. In the process of vaporization, all the intermolecular forces of attraction between the water molecules (hydrogen bonds) are overcome by the kinetic energy and the molecules are free of one another, resulting in expansion of volume. By the definition of boiling point, this expansion represents the work done to overcome the atmospheric pressure of the surroundings. A corollary of this observation is that when the surroundings have a higher pressure, boiling point temperature is elevated and lower pressure results in lowering of boiling point. For example, at a high-altitude place like Mount Everest or Mont Blanc, where the atmospheric pressure is much lower than at sea level, the reduced boiling point of water leads to undercooking of food, which consequently needs to be cooked for a longer time to achieve the same outcome. At lower atmospheric pressure, more of the molecules on the surface of water escape (evaporate) into the atmosphere against lower external resistance from fewer air molecules. Another interesting and instructive point is that distilled water may violently boil due to superheating. It has no impurities and consequently no nucleation sites for the formation of air bubbles, which become a trigger point for rapid evaporation and boiling. As a result, a glass of distilled water may be superheated above 100° Celsius without boiling. However, if impurities are then introduced into the superhot water such as a sugar cube, they provide the necessary factor for the water to boil rapidly, resulting in the so-called "exploding water". In the case of vaporization, as the phase changes from liquid water to steam, there is a large change in volume that accompanies the transition. For example, when 18 grams of liquid water is converted fully to steam, at standard temperature and pressure, the corresponding weight of steam would occupy 30.5 liters. Therefore, this also underscores why rapidly reaching the threshold of the heat of vaporization causes explosive behavior.

As discussed above, the intermolecular hydrogen bonds of water are much stronger than those of, for example, ethanol. This explains why water snuffs out the flames of a matchstick easily whereas ethanol and even more volatile gasoline, causes ignition to spread. Also, water's rapid expansion into the gas phase, when hitting hot surfaces or flames, causes its vapors to expand to nearly 4,000 times the volume of water that is sprayed. This leads to deprivation of oxygen that is needed for combustion as well as cooling due to the absorption of the heat from the fire (vaporization is an endothermic reaction, meaning it requires heat to proceed).

The molecular mechanism of heat transfer is understood through the theory of molecular collisions. *Thermal energy is created at the molecular level by collisions of particles, which provide the transfer of heat.* Heat is closely related to temperature; however, the latter is a measure of the mean kinetic energy of particles, regardless of collisions. Heat is not a property that is present in an object or system, but rather is transferred. It is a change in thermal energy or the change in the kinetic energy of particles that is responsible for the collisions rather than the mean kinetic energy of particles. This is because the spreading out of an unevenly distributed microscopic collection of particles creates the collisions, which largely underpin the heat transfer ultimately raising the temperature. However, once the particles are evenly distributed within the available space in the system it is their mean kinetic energy independent of their collisions that define temperature. *Heat transfer is a microscopic mechanism in contrast to temperature, which is a macroscopic property.* As mean kinetic energy rises due to heat transfer, for example when water is in liquid phase, the hydrogen bonds linking water molecules break, increasing the number of particles in given regions of fluid where bond breaking takes place. This leads to the collisions of neighboring particles representing the heat transfer that is responsible for increasing the temperature of the fluid defined by the new higher mean kinetic energy of the constituent water particles. Increased number of particles (particle concentration) and their pressure and velocity (and therefore temperature) affect the process of collisions during an expansion process (greatest for gases, less so for liquids, and least so for solids) as it evolves to distribute evenly the pressure and dissipate the heat. Water thermally insulates the body, i.e. it is a thermogenic insulator helping to optimize the efficiency of physiologic processes. The robust strength of its hydrogen bonds minimizes the rise in collisions and hence pressure and temperature.

The inventor of the internal combustion engine, Sadi Carnot, was especially interested in the interconversion between mechanical work and heat energy. According to the first law of thermodynamics, the internal heat content (ΔQ) of a closed system is defined to be equal to the energy input into or the internal energy of the system (ΔU) minus work done by the system (ΔW). This definition of ΔQ may be substituted into the equation $\Delta S = \Delta Q / T$ such that:

$$\Delta S = \left(\Delta U - \Delta W\right)/T \tag{1.2}$$

The insights gleaned from these equations provided the foundation for thermodynamics while at the same time the concept of entropy found application in physical chemistry. The interest in this concept to chemistry was to serve as a predictor for whether a reaction will occur spontaneously or not. Accordingly, the equation $\Delta S = \Delta Q / T$ or $(\Delta U - \Delta W) / T$ is translated to $T\Delta S = \Delta U - \Delta W$. The change in internal energy is the same as the change in enthalpy ΔH (related in turn to changes in heats of formation) for a system under constant pressure and volume. Under these conditions, the amount of work done by the system represents the change in Gibbs free energy or available energy to carry out work (such as expansion by a hot gas to compress the pistons of a car engine). In this light, the above equation can be re-written as $T\Delta S = \Delta H - \Delta G$. Moving the parameters around algebraically gives the equivalent equation $\Delta G = \Delta H - T\Delta S$. The significance of the above equation is that a negative ΔG (< 0) indicates a spontaneously occurring reaction, while a positive value of ΔG indicates

that the reaction will not proceed spontaneously unless external energy is supplied to the reaction. The Gibbs free energy change can be negative under two scenarios: a) ΔH is negative (as in an exothermic reaction in which the sum total of the heats of formation of the products is less than the sum total of the heats of formation of the reactants) and ΔS is positive. ΔG can also be negative in a reaction with negative values of ΔS, provided that the negative change of enthalpy (heat released) more than compensates for the negative change of entropy. b) ΔH is positive (as in an endothermic reaction in which the sum total of the heats of formation of the products is more than the sum total of the heats of formation of the reactants) and $T\Delta S$ is sufficiently negative to overcome the positive change in enthalpy. For example, a reaction that is non-spontaneous at lower temperatures can become spontaneous at higher temperatures.

1.12 Thermodynamics of Mechanical Engines

This section looks at various types of machine engines including internal and external combustion engines, which operate by heat transfer and illustrate the concepts of internal energy and entropy applied to ideal gasses (the term ideal refers to the hypothetical behavior that all collisions between particles of the gas are entirely elastic, i.e. no energy is lost in the friction of the collisions as heat in the form of entropy. Ideal gas laws serve as a good approximation of the real behavior of gasses). *Energy is most accurately defined by physics as the ability to do work. There are several forms of energy, three of which are very relevant to thermodynamics: 1) potential energy, such as the energy stored in a stretched object or gravitational energy stored in an object held at some height in either case there is "potential" for kinetic energy; 2) kinetic energy which is energy associated with movement and may be translational (linear), vibrational, rotational or oscillatory; and 3) heat or thermal energy.* Ideal gas laws are the simplest illustration of internal energy because there is no potential energy and therefore deal only with kinetic energy and heat. In liquids and gases, adding heat causes atoms to move faster. As the atoms move faster, they bump into one another with increased force. This pushes them further apart from one another, which increases their volume, that is, the system that they comprise becomes thermally expanded. If the volume is confined, the pressure increases according to the relationship expressed as $PV = nRT$ (where n is the number of atoms expressed as moles; R is the universal gas constant 8.3145 J/mol/K). This relationship invoked the innovation of combustion engines capable of transforming the kinetic and thermal energy of gases formed from the combustion of various hydrocarbon fuels into useful purposes of mechanical work. An alternative type of engine, however, that did not operate by heat transfer is the fuel cell. Instead, the fuel cell generates electrical energy that can be harnessed to do mechanical work. It is analogous to a battery. In contrast to mechanical heat engines, its energy sources are not fossil fuels, but only water (as a source of hydrogen) and even some of the same hydrocarbon nutrients that fuel biological organisms (*e.g.* glucose). The principle through which fuel cells transform their fuel sources into movement and the ability to do work is by the forces of electromagnetism, which is one of the four fundamental forces of nature. Fuel cells can eliminate pollution caused by burning fossil fuels; for hydrogen-fuelled fuel cells, the only by-product at the point of use is water. Fuel cells do not need conventional fuels such as oil or gas and can therefore reduce economic dependence on oil-producing countries, creating greater energy security for the user nation. The fuel cell is therefore a significant model for living organisms including humans. Like combustion engines, fuel cells constitute an equilibrium device. *Although biological organisms as metaphorical engines include fuel cells, they are regarded importantly as far-from-equilibrium and endowed with organized complexity that describes living beings.* Both fuel cells and far-from-equilibrium engines are potentially more efficient than combustion machines that rely on heat transfer. Far-from-equilibrium engines invoke 20th-century revolutions in theoretical physics including chaos and complexity theories and quantum mechanics. The latter in turn is invoked in the understanding of quantum metabolism and the evolving field of quantum biology.

The industrial steam engine of the late 1700s (Figure 1.8), which fueled the industrial revolution, is a classic example of an external combustion engine, such as a steam turbine or an old steam locomotive so described because combustion takes place outside (external) the cylinder in a separate "boiler" chamber (Figure 1.8). Pressurized steam is then created before being transferred to the piston cylinder. The pressurized steam acts on the piston cylinder, pushing it down that in turn is connected to the rotating crankshaft that translates the heat energy into mechanical work. Conversely, the combustion of gasoline or diesel in an internal combustion engine takes place inside (internal) the cylinder that compresses the air, burns the fuel, and expands the combustion gasses, which then connect to the rotating crankshaft that turns the wheels. Thus, the internal combustion engine, like the external combustion engine, also converts heat energy to useable mechanical energy. These engines that operate by heat transfer are equilibrium devices.

In an internal combustion engine, hydrocarbons (*e.g.* octane) mixed with oxygen (from air) are ignited by a spark and combusted to produce carbon dioxide (CO_2), water, and heat energy. The heat expands the gas that compresses the pistons (*i.e.* the system doing work on the surrounding) that provides mechanical energy transferred to the locomotion of the vehicle. The addition of heat to the system may cause an increase in the kinetic energy of the molecules of the system, *i.e.* causing the molecules to move faster. Alternatively, heat can be exchanged with the surroundings to allow for the expansion of the gas at a constant temperature so that its molecules maintain the same kinetic energy with the trade-off of increasing the entropy (*i.e.* the molecules have more places where they can be leading to an increased number of possible available states) as the gas expands. In addition to the effects of heat due to combustion of octane on the temperature, volume, and entropy of vaporization, it can also produce light in the form of either glowing or a flame that can be seen coming out of the exhaust pipes with a high-octane fuel as demonstrated for example in the old Batman episodes (Figure 1.9).

The combustion of hydrocarbons in gasoline fuel is remarkably similar to that of nutrient fuel such as carbohydrates and fat (*i.e.* metabolic oxidation). In each case, the fuel requires the presence of oxygen and activation energy or spark resulting in the products, which release heat to the environment along with carbon dioxide and water. Similarly, the burning of coal (mainly consisting of carbon atoms) in oxygen produces carbon dioxide, water, and heat that can be used to heat water to produce steam at high pressures. The expanded volume of the steam (gas) contributes to the work component of the internal energy that spins the turbines connected to the generator to produce electricity.

During the first half of the 20th century, simple batteries were applied to cars and approximately one-third of all automobiles on the road were battery-powered electrical vehicles. However, electric-powered vehicles subsequently became essentially antiquated because of their high cost, low top speed, and short range. However, the beginning of

FIGURE 1.8 The depiction of a steam locomotive. Source: Maksym Kozlenko (public domain).

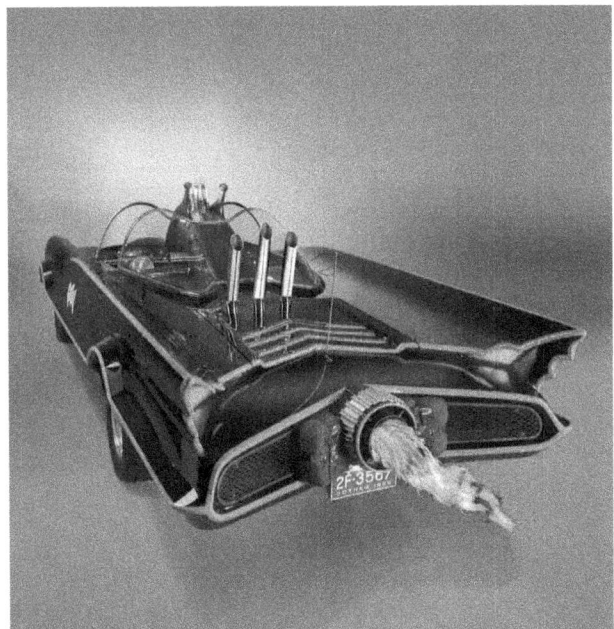

FIGURE 1.9 Flame exhaust from a Batmobile (seen in *Batman* episodes). Source: https://noveltystreet.com/authentic-1966-batmobile.

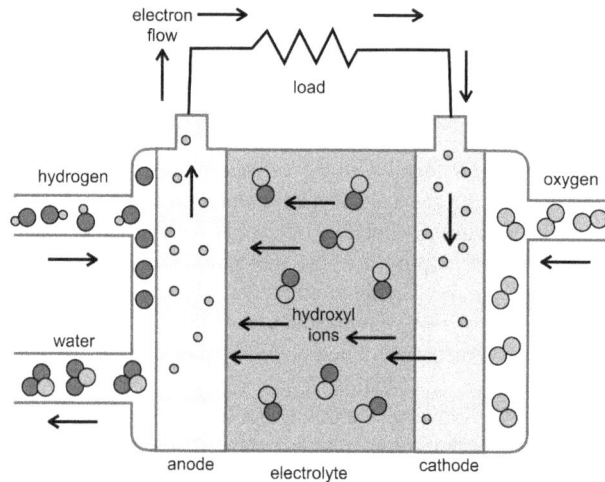

FIGURE 1.10 A schematic illustration of a fuel cell.Source: adapted from https://www.sciencedirect.com/topics/engineering/alkaline-fuel-ce ll. See Ref. [16].

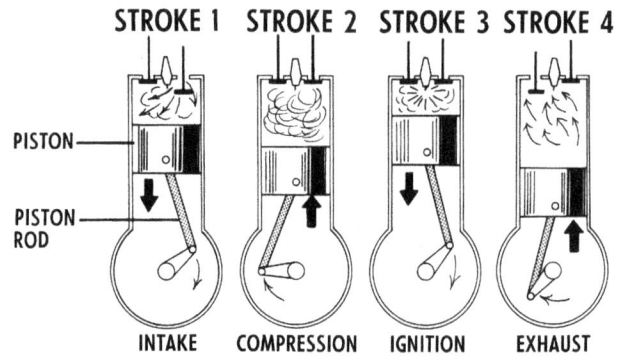

FIGURE 1.11 A schematic illustration of the four-stroke internal combustion engine. Source: Pearson Scott Forseman (public domain). Refer to http://www.blog-city.info/en/rudolfdiesel.php for additional information on the working of a diesel engine.

the 21st century has seen a revitalized interest in electrical and other alternative fuel vehicles because of the growing concern of overreliance on foreign oil and hydrocarbon internal combustion engine emissions creating environmental pollution with carbon dioxide and other exhaust. Gas batteries, or fuel cells (described earlier), have begun to provide an alternative to rechargeable batteries, even for motorized vehicles. This process involves a fuel processor that converts a host of possible substrates to hydrogen gas, which is supplied to the fuel cell. Oxygen from the air combines with hydrogen in the fuel cell to produce electricity (accompanying heat and the by-product of water vapor). A fuel cell is comprised of electrodes where reactions take place and electrolytes that carry charged particles from one electrode to the other (Figure 1.10). The fuel cell requires hydrogen and oxygen. The hydrogen atoms enter the fuel cell at the anode where they get stripped of their electrons forming protons. The electrons travel through wires providing electrical current that does work (such as providing the glow and heat of a light bulb). The oxygen, typically from air, enters at the cathode, picks up the electrons, and combines with hydrogen protons producing water as the exhaust product. An electrolyte ensures the appropriate ions passed between the electrodes across a semipermeable membrane.

A fuel cell produces energy electrochemically without combusting the fuel. The device converts chemical energy from hydrogen-rich fuels in an efficient, non-polluting process, into electrical power. The fuel cell is very analogous to the electron transport chain of cellular mitochondria. In both cases, energy-rich electrons are extracted from hydrogen-containing substrates in an electrochemical process that transforms the energy into useful purposes. There are many different types of fuel cells that can extract hydrogen from a wide variety of sources including fossil fuels (natural gas, coal, methane, *etc.*), water, and microbial nutrients such as glucose and acetate [15]. The electrical current produced by fossil fuels as well as other hydrogen sources has a capacity to shepherd mechanical locomotion as an alternative to combustion engines.

The Otto engine was a large stationary single-cylinder internal combustion four-stroke engine designed by Nikolaus Otto (Figure 1.11). In a typical four-stroke engine, the piston completes four separate strokes while turning a crankshaft and can be described as follows: 1) intake stroke when the engine starts the piston moves downwards in the cylinder because of which a region of low pressure is created in the cylinder above the piston. At this moment, the intake valve opens and the fuel mixture (petrol vapors and air) is sucked into the cylinder from the carburetor. 2) Compression stroke, when a sufficient amount of fuel is sucked into the cylinder, the intake valve closes. The piston is forced to move upward by the rotating crankshaft (rotation being promoted by the expansion of the air-petrol mixture) and as a result, the fuel mixture is compressed to 1/8 its original volume. This compression increased the temperature of the fuel mixture (the higher the compression ratio the more efficient engine). 3) Power stroke when the fuel mixture is completely compressed, the spark (electric spark) plug ignites the mixture. The petrol burns rapidly and produces a large volume of hot gases. The pressure exerted by these combustion products pushes the piston downward with a great force. The piston rod then pushes the crankshaft, which in turn rotates the wheels of the vehicle. (In diesel engines this power stroke is not required as the fuel ignition occurs during compression stroke as it is already highly compressed.) 4) Exhaust stroke when the piston has been pushed down to the bottom of the cylinder, the exhaust valve opens and due to the momentum gained by the wheels, the piston is pushed upwards. This upward movement of the piston expels the spent gasses into the atmosphere through the exhaust valve. It then gets closed; the intake valve opens and the four strokes are repeated.

The diesel engine (Figure 1.12) is another four-stroke internal combustion model. In this case, air is injected above the piston causing it to move downward (first stroke) resulting in the rotation of the crankshaft. In the second stroke, as the piston moves up (due to the crankshaft rotation), the injected air is compressed to a very high pressure, and diesel is injected into the cylinder with combustion occurring at constant pressure. This in turn causes an explosion resulting in expanding hot air-diesel mixture that pushes the piston down (third

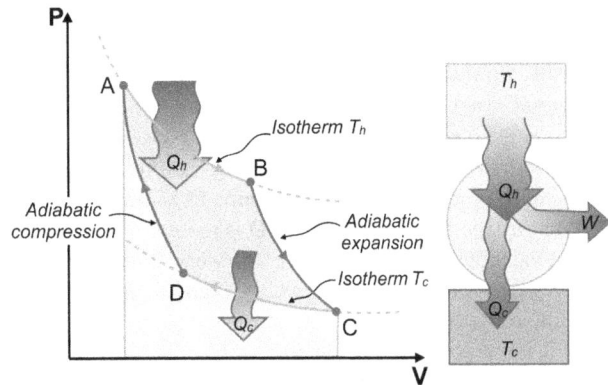

FIGURE 1.13 Illustration of the theoretical thermodynamic engine called the "Carnot cycle". Qh is the amount of heat entering the Carnot cycle from a hot reservoir during the isothermal expansion (A → B). Qc is the amount of heat exiting the Carnot cycle during the isothermal compression (C → D). Source: adapted from https://courses.lumenlearning.com/physics/chapter/15-4-carnots-perfect-heat-engine-the-second-law-of-thermodynamics-restated/.

FIGURE 1.12 A typical diesel engine. Source: TTTNIS (public domain).

stroke). Thus, in diesel engines, the release of energy is not mediated by a spark of energy, but by mixing the fuel with air at very high pressure. By contrast, in the gasoline engine combustion occurs at constant volume. *The thermodynamic laws concerning pressure, volume and temperature relationships are obeyed by heat engines as work is done by changes in the volume of the gas.* If a given mass of a gas is compressed or expanded at a constant temperature, the pressure varies inversely with the volume, or the product of pressure and volume is constant, i.e. $P_1V_1 = P_2V_2$. The resultant transfer of this heat carried by the vapors is transferred to mechanical work by the moving pistons. The heat energy not converted to work is dispersed in the form of entropy.

1.13 The Carnot Engine

To better appreciate the efficiency of the molecular machinery of life, we now review some basic principles of internal combustion engines, which were the main practical embodiment of 19th-century thermodynamics and still play a major role in how our technological-age society operates. The fundamental concepts of heat, work, efficiency, and entropy in the context of a heat engine are generally understood through the well-known ideal or theoretical engine known as the Carnot cycle or Carnot engine [1–4]. Unlike most practical engines (including biological engines), the Carnot engine, which deals with an ideal gas, is a completely reversible engine in the sense that each step of the changes that take place in the course of the cycle takes place in a quasi-static manner such that no heat is lost to friction or entropy. In a quasi-static change, the system is either at equilibrium or near-equilibrium state at all times of the cycle. Conceptually, this can be imagined to be feasible when the system is "moving" infinitesimally slowly leading to practically no loss of heat to friction in the ideal engine. Since friction is defined as the resistance encountered by one surface moving against another, it can be visualized

that when the two surfaces move infinitesimally slowly relative to one another, they are doing so without "encountering" one another and hence do not experience friction. The Carnot cycle (Figure 1.13) consists of four phases as follows: 1) isothermal expansion from state A to state B in which the ideal gas takes up heat (Q_h) from an infinite hot reservoir thus maintaining the temperature at a constant value (T_h); 2) adiabatic expansion from state B to state C in which no heat either enters or leaves the system (by definition of the term "adiabatic"), but the ideal gas expands further causing a drop in temperature to T_c; 3) isothermal compression from state C to state D which leads to the output of heat Q_c to the surrounding heat sink; and 4) adiabatic compression from state D to state A leading to an increase in temperature back to T_h. During the expansion phases (1 and 2), work is done by the engine on the surroundings by the expansion of ideal gas, while during the compression phases, work is done on the system. Since there is no loss of heat to friction, all the heat coming in and going out of the system is used to either do pressure-volume work on the surroundings or get work done by the surroundings. Thus, In the process of completing the cycle, the net amount of work done by the engine is represented by the area enclosed by the four states A, B, C, and D. Since this theoretical heat engine is completely reversible, it can also serve as Carnot refrigeration cycle going from A to D to C to B to A.

In light of the reversible nature of the Carnot engine, the changes in internal energy and entropy (both of which are state functions independent of the path) of the ideal gas system around a cycle are zero since at the end of the cycle the system returns to exactly the same state (state A in Figure 1.13). Thus, *the efficiency of the "ideal" Carnot engine would be the maximum possible physically achievable and any other engine, which would be non-ideal and suffer from at least some loss of heat to friction, would consequently be characterized by lower efficiency than the idealized Carnot engine.*

The efficiency of any thermodynamic engine would be defined as the ratio of the amount of work done by the engine to the amount of heat taken up by the system to carry out the work. As an example, if all the heat taken into the system is used to carry out work, the efficiency of such a system would be 100%. However, in practice, this cannot occur even for the idealized Carnot engine where some heat has to be thrown out to the "cold sink" during the compression phase of the cycle when work is done by the surroundings on the system, for the cycle to be completed (Q_c in Figure 1.13). Since the area enclosed by the cycle A – B – C – D is the net work done by the engine (obtained by subtracting the work done on the system in the two compression steps C → D and D → A from the work done by the system in the expansion steps A → B and B → C), it is represented by the net amount of heat taken up the system, i.e. $Q_h – Q_c$, since no heat enters or leaves the system in the adiabatic phases of the cycle. Thus, the efficiency of the Carnot cycle would be defined by the equation: η = net work done by the system/amount of input heat. Hence, $\eta = (Q_h – Q_c)/Q_h$, where $Q_h – Q_c$ is the net work done by the Carnot cycle and Q_h is heat taken up by the Carnot cycle. Simplifying the above equation yields the final formula for the *efficiency as:* $\eta = 1 – (Q_c/Q_h)$. In light of this equation, the efficiency of the Carnot engine can never be 100% since Q_c has to be positive for the cycle to complete. In the case of a non-ideal engine, the term Q_c would be higher than that in Carnot engine because of loss of heat due to friction which is not usable for doing any work by the engine. The efficiency of a Carnot cycle is also defined as $(I[T_c/T_h]) \times 100\%$ where T_c and T_h are the temperatures of the cold and hot isotherms as described in Figure 1.13. It must be noted that this definition of efficiency applies only to a Carnot engine, which consists of only reversible processes. In this light, it is important to recognize that a biological engine with its complex of several coupled biochemical reactions would by definition be operating at lower than maximal possible efficiency. The reader is referred to the chapter on the molecular machinery of life for the intricate details of the molecular motors that drive the mechanical demands of living cells. Note also that a combustion engine in an automobile would be operating at a low efficiency compared to a normal biological engine. Modern gasoline engines are believed to have a maximal thermal efficiency of about 25% to 30% which can be unfavorably compared to the *approximately 40% efficiency of energy utilization in human biology* (the far-from-equilibrium biological engine) [17]. It is interesting to note that in systems that deal with energy conversion from or to non-thermal forms (*e.g.* electrical energy into kinetic energy as in an electric motor, electrical energy into heat in an electric heater, electrical energy into radiation energy in incandescent bulbs, light-emitting diodes, metal-halide lamps, and low-pressure sodium lamps), the efficiencies of conversion are generally less than 100%. Thus, all physical systems are subjected to some wastage of input "fuel" as a part of deriving energy and work for their sustenance.

A natural consequence of the above discussions on the efficiency of the Carnot heat engine is manifested as an example in the accumulation of excess food in the form of fat leading to obesity. This is because if the system is losing energy as heat in mitochondria it is not making ATP for the physiologic needs of the body. *In the case of excessive heat production in the mitochondria, the associated excessive reactive oxygen species formation (which at moderate levels is adaptive to prevent unneeded ATP while fulfilling the requirements of thermogenesis) results in pathological oxidative and inflammatory stress that compromises the structural and functional integrity of mitochondria.* This process underpins insulin resistance and inefficiency in the ratio of the amount of work (the physiological needs of the body) done by the biological engine (the drivers of which are the motors that produce ATP and hydrolyze ATP liberating Gibbs free energy as well as other biochemical reactions that liberate Gibbs free energy and what seems apparent bio-photon energy and possibly the energy of electromagnetism) to the amount of heat taken up by the system to carry out work. That is, too much energy is lost as useless heat unavailable to do work that promotes entropy lost to the surroundings. Notably, a *sine qua* non-clinical feature and insulin resistance and type II diabetes is an appetite that is poorly sated. This relates to not only insulin resistance but leptin resistance (insulin and leptin being the two adiposity signals of the body) and reduced circulating GLP1 and other hormonal mediators. Detailed discussions on these topics are found in Volume 2 of this book.

It is interesting to note that if the Carnot cycle is reversed, the engine would now become a refrigerator instead of a heat engine. In such a cycle A → D → C → B → A, the first step would be adiabatic expansion, followed by isothermal expansion during which heat Q_c enters the system from the cold reservoir. The third step would be adiabatic contraction followed by isothermal contraction, during which heat Q_h is expelled into the heat reservoir. Thus, in this cycle, work is done "on the system" rather than "by the system" as would be the case of Carnot heat engine, described above (Figure 1.13). Analogous to the efficiency of a heat engine, a term called the coefficient of performance (COP) is used to characterize a refrigerator's performance and is defined as the ratio of the heat taken from the cold reservoir to the amount of work done "on the system" which would be Q_h-Q_c. Unlike the efficiency of a heat engine, the value of COP can be greater than 1, and hence, the term "efficiency" (which is normally deemed to have a maximum value of 1 or 100%) is inappropriate to describe a refrigerator. By this definition, a refrigerator that cools something down to very cold temperatures would have a lower coefficient of performance than the one that cools down to moderately cold temperatures. In a practical example, in the case of a refrigerator commonly found in every household kitchen, the cold reservoir from where heat Q_c is taken in is the inside of the refrigerator (where food is stored) and the hot reservoir is the kitchen into which heat is thrown out. The refrigerator is operated by a compressor, which is the engine that provides the work done to take the heat out of the system.

1.14 Enthalpy and Internal Energy—Compared and Contrasted

Enthalpy is a function defined by the thermodynamic state of the system; hence it is referred to as a state function. Entropy is also a state function. *Enthalpy is a measure of the heat content of a thermodynamic system. It includes the internal energy, which is the energy required to create a system, and the amount of energy required to make room for it by displacing its environment and establishing its volume and pressure*. The concept of enthalpy (*H*) evolved to explain chemical reactions and physical processes open to the atmosphere whereby heat can flow adiabatically allowing pressure-volume work to take place in the form of volume expansion, particularly applicable to gases. The definition of enthalpy is provided by the equation: $H = U + PV$, where *U*, *P*, and *V* are all state functions and correspond to internal energy, pressure, and volume. Effectively, the product *PV* in the above equation for enthalpy corresponds to the contribution by pressure-volume "work" to energy. By extension of this definition, the change in enthalpy of a system is the heat content of the system at constant pressure and constant volume. Since most chemical and biochemical reactions in nature occur at a roughly constant pressure of 1 atmosphere, it is logical to think of the change in enthalpy as simply the change in the heat content of a system as it undergoes a chemical reaction. The work done on or by the system most typically refers to pressure-volume work of gases. However, in chemical reactions, it often applies to the work associated with breaking or forming chemical bonds between atoms and molecules. *A negative value ΔH in a reaction corresponds to an exothermic reaction and a positive ΔH to endothermic reaction*. For example, the exothermic reaction of the formation of methane from its elements (carbon and hydrogen) has a *ΔH* of –74 kJ/mol indicating that heat energy of 74 kJ/mol is released upon the completion of the reaction. Stated differently, the heat of formation of methane is –74 kJ/mol. The positive *ΔH* of an endothermic reaction reflects the heat absorbed by the system because the products of the reaction have a greater enthalpy than the reactants. In the case of metabolic pathways of for example glucose and fat catabolism, that are exothermic reactions liberating heat to the surrounding used to produce ATP, the negative pressure-volume work by the liberation of carbon dioxide and water gaseous waste is vanishingly trivial, albeit does occur. In the case of the endothermic process of water vaporization, the volume of water vapor molecules liberated far exceeds that of the liquid phase.

1.15 Gibbs Free Energy and the Chemical Potential

The feasibility and spontaneity of a chemical reaction (and by extension biochemical reaction) are generally understood through the changes in the Gibbs free energy of such a reaction. Gibbs free energy is another state function and it is a measure of the amount of energy available to do work in an isothermal and isobaric (constant temperature and pressure) thermodynamic system. The term "free" refers to the amount of energy in a system that is available for usage. In other words, Gibbs free energy is a measure of the thermodynamic potential that can be employed to do maximal reversible work in a chemical reaction at constant temperature and pressure. This term (denoted as *G*) is defined as $G = H - TS$, where *H*, *T*, and *S* represent enthalpy, temperature, and entropy, respectively.

Gibbs free energy can be called a state function because all the components that enter into its formula are themselves state functions. As a consequence, *the value of G does not depend on the path by which a state is achieved*. As is often the case,

typically in considering the thermodynamics of a chemical reaction, we are interested in the change of the state function rather than its absolute value. Therefore, change in Gibbs free energy (denoted by *ΔG*) can be defined as $\Delta G = \Delta H - T\Delta S$, where *ΔH* is equal to the change in enthalpy (measured in units of joule/mole or calories/mole), T is the temperature of the reaction (measured in degrees Kelvin), and *ΔS* is the change in entropy (measured as calories or joules per Kelvin). *For a reaction that is in chemical equilibrium, ΔG is equal to zero, while it is less than for a spontaneous reaction (exothermic). By extension, ΔG > 0, for a non-spontaneous reaction, which requires an external energy input (endothermic)*. While *ΔG* provides an idea of the direction of the reaction, it makes no statement on the amount of time required for the reaction to proceed.

J. Willard Gibbs (after whom Gibbs free energy is named) was an American chemical physicist whose work laid the basis for the development of physical chemistry as a science and who was described by Einstein as "the greatest mind in American history." He introduced the concept of chemical potential, the "fuel" that makes chemical reactions work. The concept of free energy has been universally called Gibbs free energy. It relates to the tendency of a physical or chemical system to simultaneously lower its energy and increase its disorder, or entropy, in a spontaneous natural process. His approach allows the calculation of free energy or "available energy" from which work can be done in chemical and physical processes. In physical processes, such as the boiling of water, the changes in the internal energy of water do not involve any chemical rearrangement of the constituents involved. For example, water molecules remain as water molecules (that is they are not disintegrated into individual atoms of hydrogen and oxygen), albeit in a higher energy state. They are in a higher energy state because hydrogen bonds are broken and the water molecules are free to move about with higher kinetic energy (Figure 1.14).

On the other hand, entropy changes are a significant part of Gibbs free energy of vaporization since in the vapor state, the water molecules, which are no longer hydrogen-bonded and move about with many more degrees of freedom than in liquid state, have considerably larger numbers of energy states accessible to them. Naturally, a solid has less inherent entropy than a liquid, which has less inherent entropy than a gas or a vapor. The changes in entropy are manifested in terms of latent heat of vaporization or fusion at phase transition temperatures where two or more states of matter can co-exist at the

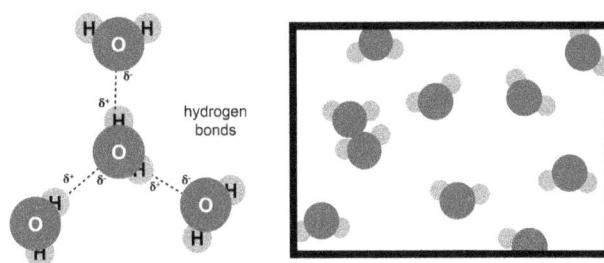

FIGURE 1.14 Hydrogen bonds in water (left) are broken in the water vapor phase (right). Source: adapted from https://ib.bioninja.com.au/standard-level/topic-2-molecular-biology/22-water/hydrogen-bonding.html.

same free energy value. The latent heat of a process or reaction allows absorption or release of energy without changes in temperature.

In a spontaneous process involving a thermodynamic system, free energy is released usually including heat exchange with the surroundings and the system resultantly moves to a lower, more thermodynamically stable, energy state. Gibbs free energy is the state function relevant for processes that occur under constant pressure and temperature conditions, prevalent in a large majority of real-life biochemical reactions. As mentioned above, *a negative Gibbs free energy change (ΔG < 0) defines the tendency of a chemical or physical process to occur spontaneously.* Although most chemical reactions and changes in physical systems that are spontaneously favored to occur have an entropic change that is positive and an enthalpy change that is negative (exothermic), a negative entropy change *(ΔS < 0)* may lead to a negative change in Gibbs free energy (ΔG < 0) and hence a spontaneously favored process when the reaction products release sufficient heat of formation (exothermic, *ΔH < 0*) and the temperature under which the process occurs is low enough. This describes the process of the evolution and maintenance of biological systems occurring in circumstances of negative entropy changes. The Gibbs free energy is essentially the available energy to do the work in a coupled reaction. It may be noted that a chemical reaction may proceed forward spontaneously after the activation barrier is overcome with the help of a heat source accessible to it. *The release of thermal energy into the surroundings from a system that reaches a lower energy state is an exothermic reaction and the heat released is called the enthalpy of the reaction.* A negative value for the change in enthalpy between the products and reactants implies that the net energy of the products is less than that of the reactants, the difference being liberated as heat to the surroundings. Some of this energy could be used for lowering the entropy of the products (for example by bringing order or organization to the structures of the products). A negative change in enthalpy, when reactions occur at temperatures that are low enough and a system that is having work done on it by the surroundings, describes the paradoxical tendency of the system to simultaneously lower its energy state while reducing its state of disorder (*i.e.* creating organized structures) allowed by the concurrent dissipation of heat as entropy. Conversely, when the enthalpy of the products is higher than that of the reactants (endothermic reaction), the change in enthalpy is positive, requiring an input of energy from the surroundings into the process allowing it to occur. This is an uphill chemical reaction or physical process, which is not spontaneous and requires input heat into the reaction.

The equation for the change in the Gibbs free energy involved in a thermodynamics process (ΔG = ΔH − TΔS) offers an understanding as to whether a process is capable of proceeding in a given direction without needing to be driven by an outside source of energy. It is used to refer to macroprocesses in which entropy increases, such as the diffusion or spreading out of molecules (as may occur as a smell diffusing in a room, ice melting in lukewarm water, salt dissolving in water, iron rusting, the oxidation of an apple core as it turns brown or the oxidation of tissues of an organism such as humans at end of life). In the case of decreased entropy, the

Gibbs free energy of that system, or energy available for useful purposes, *i.e.* work, is reduced. The time evolution of a system in which it releases free energy and moves to a lower more thermodynamically stable energy state, as the Gibbs free energy equation shows, is governed not only by the direction of entropy but temperature and change in enthalpy as well. A ball rolling from the top of a slide has greater enthalpy at the top of the slide (total potential, because of the ball's higher position and kinetic energy related to the ball's velocity before coming to a stop) than it does at the bottom of a slide. However, if friction is ignored, the energy of the ball at the top of the hill would be the same as at the bottom and the potential energy (*E = mgh*) would have been transformed into kinetic energy (*KE = (1/2)mv²*), so that *v = (2KE/m)^{1/2}*. However, if friction is taken into consideration, then heat is given off to the environment due to it and the final velocity v_f now is lower than v calculated above. The amount of heat given off is equal to *(1/2)m(v² − v_f²)*. That is, as the ball rolls down the slide it releases energy as a macroscopic example of an exothermic (or exergonic) process. The energy of the ball as it rolls down can potentially be useful to do work on another system (*e.g.* a stationary object that needs to be moved). A spontaneously favored reaction denoted by a negative Gibbs free energy change (ΔG < 0) occurs in the setting of an exothermic reaction, *i.e.* a negative change in enthalpy that releases the flow of heat from the system to the surrounding (this heat may potentially be used by surrounding systems in the conversion to work) and a positive change in entropy which comes in the form of dissipated energy that is not usable for any work. *Thus, all reactions with a positive entropy change and a negative enthalpy change are spontaneous (negative Gibbs free energy).* When change in enthalpy is positive (as would be the case in an endothermic reaction) or change in entropy is negative (as would be the case in a more complex and organized state of the products relative to the reactants), the arbitration of whether the reaction is spontaneous (change in Gibbs free energy negative) or not spontaneous (change in Gibbs free energy positive) is dictated by the temperature of the system assuming that *ΔH* does not depend on temperature. At higher temperatures, the change in entropy is the dominant factor since the entropic contribution to the Gibbs free energy change is proportional to *T*, i.e. TΔS. Conversely, at low temperatures, the change in enthalpy becomes the dominant factor. For example, the efficiency of photosynthesis increases with temperatures reaching a plateau at roughly 300 Kelvin. The idea here is that we are living at temperatures favored for the purposes of efficiency. In the conditions of negative change in entropy such as exhibited by organized complexity of life structures, the temperature must be low enough for the negative change in enthalpy to trump the value of the negative entropy change in order to promote a negative Gibbs free energy change allowing the reactions to be spontaneously favored. Conversely, the positive enthalpy that occurs with breaking bonds only occurs if the temperature is high enough so that the positive entropy value trumps the positive enthalpy value. *The overarching significance of inflammatory conditions and associated oxidative stresses that drive morbidity conditions and accelerate senescence is that it provides the heat that allows the increased entropy value of tissue breakdown to overwhelm the positive change in enthalpy value, hence promoting the randomization of our molecules, atoms and subatomic particles to the universe.* This is the explanation for how living organisms ultimately remain compliant with the second law of classical thermodynamics.

In thermodynamics, the term "chemical potential" is frequently used to describe the direction of chemical reactions and is defined as "partial molar Gibbs free energy". Since it is related to Gibbs free energy, its definition is qualified at a given condition of temperature and pressure. It is a form of potential energy that can be released or taken up during a chemical reaction or in phase transition. Just as heat flows from a hotter object to a colder one, in any chemical reaction, the direction of movement of molecules is from a state of higher chemical potential to a state of lower chemical potential. For example, at standard temperature and pressure (STP), the chemical potentials of ice, water, and steam are respectively –236.6, –237.1, and –228.6 kilo-Gibbs (1 Gibbs = 1 J/mol). In this light, under standard conditions ice would melt and water vapor would condense, since liquid water has the lowest of the three chemical potentials. On the other hand, when temperature is raised to 125° Celsius, the chemical potential of water vapor is –248 kilo-Gibbs, while the corresponding values for liquid water and ice are –244 and –241 kilo-Gibbs. Thus, at the higher temperature, water would be found in the form of steam unless pressure constraints are changed.

As outlined earlier, the Gibbs free energy is the free or available energy describing whether a process or reaction is a spontaneous one dependent on the variable of temperature and the initial and final values of enthalpy and entropy. *Entropy progresses to a state of greater disorder (positive ΔS), enthalpy progresses to a state that releases energy to its surroundings (negative ΔH), and the greater the temperature the more likely the available Gibbs free energy is negative and spontaneous.* The production of light energy from our Sun represents a fine example of a reaction that is spontaneous while being exothermic and with increasing entropy. If the Sun were fueled by the combustion of wood, oxygen, and gasoline (whose combined total mass would equal that of the Sun), it would burn out in a few millennia. How is it that the Sun is not burned out after 4.5 billion years? To answer this question, we must for a moment step outside the confines of thermodynamics and venture into both quantum physics and the special theory of relativity. The reason is that in its intensely high pressure and high-temperature core, it is fueled by nuclear reactions that fuse hydrogen atoms into helium releasing huge amounts of energy. Specifically, the nuclear reaction consists of four ionized hydrogen protons fusing through the mechanism of "quantum tunneling" (probabilistic phenomenon unique to quantum physics) to form helium, positron, and neutrino, while the mass lost in the process is released in the form of energy (as directed by the special theory of relativity equation $E = mc^2$). Since the protons carry a positive charge of +1, under normal circumstances they would undergo electrostatic repulsion with one another resulting in them being kept apart. *However, the "wave nature" of the particle enables two protons to come together under the constraints of intense gravitational pull and extremely high temperatures of the solar core using the tunneling effect.* As illustrated in Figure 1.15a particle in quantum physics is represented by a wave function,

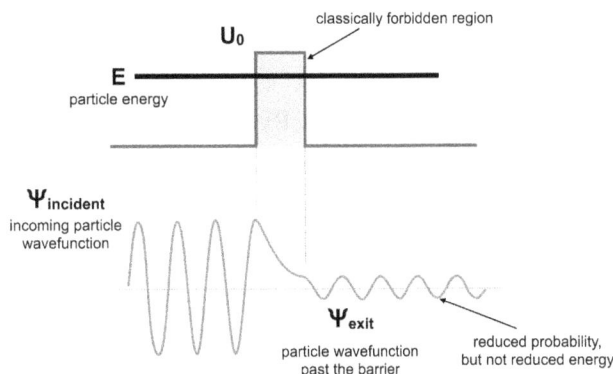

FIGURE 1.15 An illustration of the quantum tunneling effect. Source: adapted from Trixler, Frank. (2013). "Quantum Tunnelling to the Origin and Evolution of Life". *Current Organic Chemistry* 17: 1758–1770.

which is a measure of the probability of it being found at a particular point in time and space dimension. Consequently, as illustrated, the probability across the energy barrier and on the "other side" of the barrier is non-zero causing the so-called tunneling of the particle across the barrier. It should be noted that the probability in the barrier region suffers an exponential decay but does not actually become zero.

The quantum tunneling effect serves to bypass the normal activation barrier for the two protons to come together. Although the probability of such an event is small, it is non-zero and thus in the ambiance of the huge mass of hydrogen atoms in the extremely hot core of the Sun, even the small probability event becomes a possibility. Once the fusion of protons and neutrons is completed in the fusion reaction, the positively charged particles are held together by the so-called "strong nuclear forces" which operate at very short ranges as those found in the nucleus of a helium atom. The mass of a helium atom formed in the fusion reaction is lower than that of the four protons by 0.0297 amu (atomic mass unit), which is released in the form of energy (27.67 MeV), making it exoergic. This energy travels out of the Sun in the form of electromagnetic radiation (photons), which brings light and warmth to the planets of the solar system. Furthermore, the release of positron and neutrino contributes to the increase of entropy of the Sun as a system. In this way, the Sun slowly converts its mass to energy in the form of sunlight. While the above example is important to understand the source of energy for life on our planet, the concepts involving quantum tunneling and special theory of relativity, are also at play in various chemical reactions, with examples in enzymatic biochemical catalysis. Consequently, *both special theory of relativity and quantum mechanics, being the foundational theories of modern physics, cannot be dismissed or ignored not only in the context of biochemistry but, by extension, physiology.*

1.16 Thermodynamics of Biochemical Reactions

The many reactions of the Krebs cycle (also known as the "citric acid cycle") provide very illustrative examples of how uphill chemical reactions are driven by coupled downhill

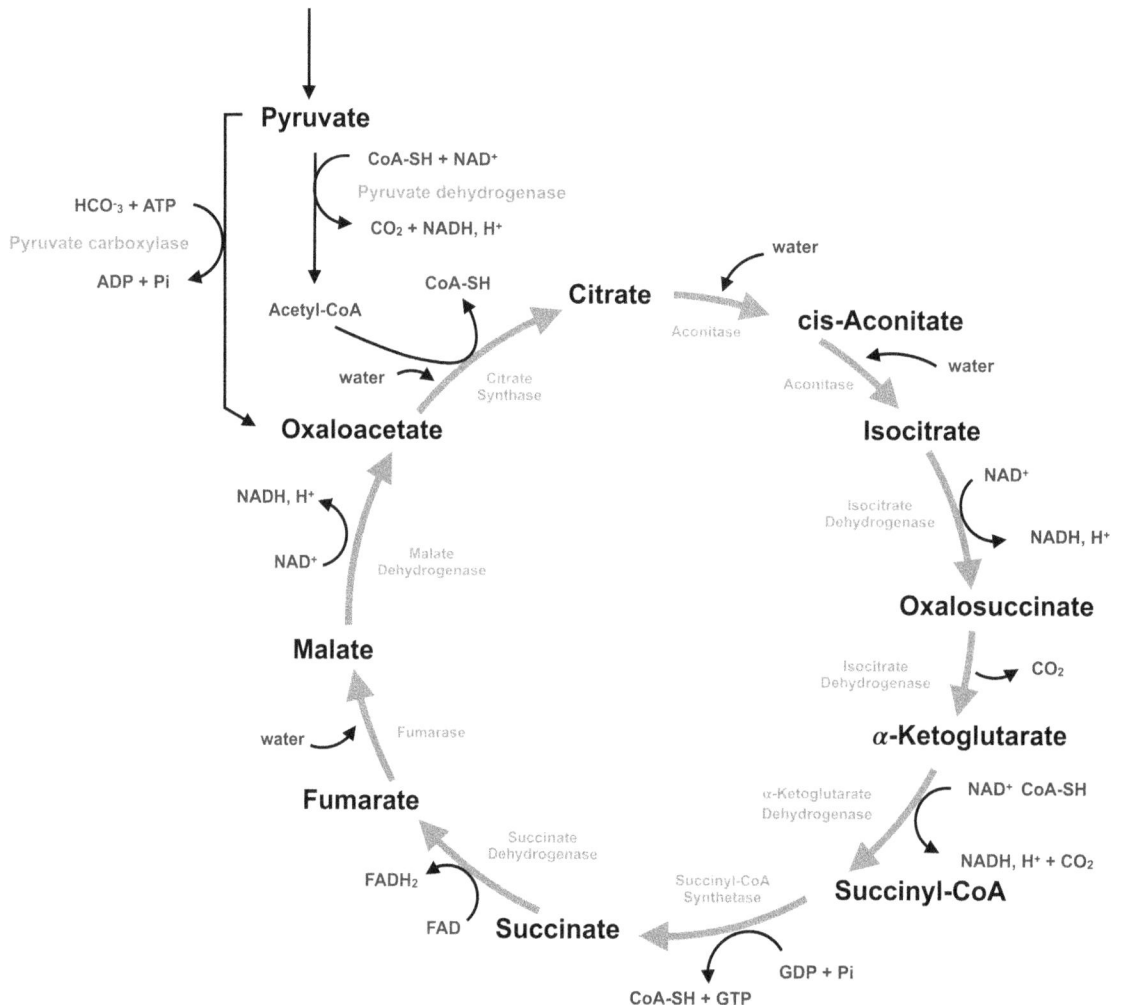

FIGURE 1.16   Reactions of the citric acid cycle.

spontaneous reactions that release large amounts of Gibbs free energy [18, 19]. Figure 1.16 shows the reactions of the citric acid cycle beginning with the entry of acetyl-CoA, which combines with oxaloacetate, in a reaction catalyzed by citrate synthase, to produce citric acid, which serves as a substrate for seven distinct enzyme-catalyzed reactions that occur in sequence and proceed with the formation of seven intermediate compounds, including succinic acid, fumaric acid, and malic acid. Malic acid is converted to oxaloacetic acid, which, in turn, reacts with yet another molecule of acetyl CoA, thus producing citric acid, and the cycle begins again. The Gibbs free energy changes associated with the various enzymes of the TCA cycle are shown in Table 1.2. As seen in this table, the reaction catalyzed by citrate synthase can drive the relatively uphill reactions catalyzed by aconitase. The subsequent five reactions are exothermic with negative values of DG and in combination they drive the endothermic reaction that converts malate to oxaloacetate, catalyzed by malate dehydrogenase.

As an extreme example of an exothermic reaction, nitroglycerin promotes an oxidation reaction that releases a huge amount of energy. It is very unstable and an explosion of nitroglycerin killed Alfred Nobel's brother. Nobel invented dynamite, which stabilizes nitroglycerin by wrapping it in a form of cellulose (nitrocellulose). It is used in mining and construction as well as in war, having caused some people to call Alfred Nobel "the merchant of destruction". For the past 130 years, nitroglycerin has also been used in medicine as a potent vasodilator important in the management of congestive heart failure and the prevention of heart attacks.

1.17 Information Energy

The basic components of energy, namely, heat (kinetic energy transferred to the atoms and molecules comprising a system) and work (the useful product of energy) in terms of classical thermodynamics, are coupled by free energy (Helmholtz free energy when pressure is not constant which is typical in experimental physical systems; Gibbs free energy, when pressure is constant, for example in biological systems which are at constant atmospheric pressure). The input of heat into a system that is not converted to work is transferred to thermal energy in the form of internal energy (molecular vibrations), which may raise the temperature (molecular collisions) of the system. Ultimately, heat that is not transferred to work is lost to entropy in accordance with the second law of

TABLE 1.2

Gibbs Free Energy Changes Associated with the Various Enzymes of the TCA Cycle

Step	Enzyme	$\Delta G°$ (kJ/mol)
1	Citrate synthase	−32.2
2	Aconitase	+6.3
3	Isocitrate dehydrogenase	−20.9
4	a-ketoglutarate dehydrogenase complex	−33.5
5	Succinyl-CoA synthetase	−2.9
6	Succinate dehydrogenase	0.0
7	Fumerase	−3.8
8	Malate dehydrogenase	+29.7

thermodynamics. *Entropy only stays the same in reversible processes while it increases under any other circumstances.* In a biological system, Gibbs free energy (or more precisely its difference between the initial and final thermodynamic states) determines whether a reaction will proceed spontaneously. A spontaneously occurring reaction involves a lowering of the Gibbs free energy which could take place with a lowering of entropy if the enthalpic contribution compensates for it so that the second law of thermodynamics is not violated. This allows the construction of organized complexity of biological structure and function while circumventing the term "negative entropy". This forms the basis for the physical chemistry and biochemistry of life forms although non-classical quantum phenomena and chaos theory/modeling are also contributory. Why is this important to a physician?

Informational energy is important to a physician because it represents a measure of a lack of capacity to be applied to something organized and useful and has many mundane applications (e.g. obesity). Accordingly, in this paragraph, an explanation is given how I see the information content of nutrition. In my opinion, information content is proportional to the energy that puts the chemical structures of biology into their functional form. In isolation, a snapshot of this information of chemical three-dimensional structures may have no inherent meaning. However, the interactions of these forms or information in a four-dimensional sense, that is as three spatial dimensions dynamically over time, ultimately demonstrates an adaptive value of the system as a whole. Hence, information per se is just "bits" in isolation, however, understanding the interactive nature provides a deeper meaning of the system as a whole. Approaches to thinking about and understanding biological systems most effectively engage new and often novel perspectives capable of finding the deeper meaning and value in terms of problem solving. Analogous to the deeper meaning of information by the interaction of the parts of a system into a whole is the connection of information as parts into knowledge. The latter reflects an understanding of the bigger picture, seeing the forest for the trees. *Nutrient-replete energy can be understood and distinguished from nutrient-depleted forms in that only the former can promote organized complexity in the consuming host.* If the calories we consume have low informational content, the energy contained within it has a low capacity to be converted to the biological currency of ATP and for that ATP to manifest the dynamic complexity of

biological structure and function. In other words, nutrient-poor dietary energy has a compromised capacity to put the biological chemistry of the host into an organized form of dynamic complexity. Sources of high energy (*e.g.* "empty calories" with low or no nutrition value) but low "informational content" are unable to provide the necessary tools for promoting biological work which requires a number of co-factors (*e.g.* minerals and vitamins) in addition to energy. Consequently, mental and physical processes are compromised in their ability to connect chemical biological structures to adaptive functions in complex, chaotic, or quantum coherent systems. Rather, energy is maladaptively deposited as adipose depots in the body. The notion of thermodynamic entropy here becomes applicable because the transfer of heat cannot be harnessed for useful purposes. This leads to the feed-forward maladaptive processes or inflammation and oxidative stress in a metabolically unhealthy state.

The bird's-eye perspective of why the concept of informational energy is in my opinion important to physicians for clinical problem solving is fundamentally rooted in the distinction between classical thermodynamics and non-classical physics. *From the perspective of thermodynamics, chemical systems evolve to an equilibrium state in which the entropy of the system and the environment in combination is maximized.* That is the change sum total of the changes in the entropy of the system and the surroundings (collectively termed the "universe") has a positive value. Consistent with the second law of thermodynamics, this change in entropy of the universe inherently reflects the dissipation of heat into configurational rearrangements and cannot be utilized as free energy that can be harnessed to do work. Implicitly, any biochemical system that reaches such a state is no longer viable or contributory to a viable or healthy state of the larger organismic system (the individual). In contrast, living systems are characterized as being in a far-from-equilibrium (metastable) state of organized complexity. Accordingly, they successfully harness energy, which is transferred into a hierarchical arrangement of systems and related networks of subsystems. Complexity theory involving non-linear chaos models is deemed to explain the top levels of such a biological organization, while the lower levels (encompassing detailed components such as enzymes, receptors, endogenous ligands, etc.) are best explained by models of quantum phenomena. A full appreciation of this requires another distinction, that between the bottom-up processes that lead to the emergence of unpredictable behavior and the top-down processes that are predictable. It can be argued that at some level of organization, living systems are machine-like but at a more subtle (and perhaps fundamental) level, they are less predictable since they are not completely algorithmic. An understanding of the former is a key to the capacity to manipulate a therapeutic response predictably and effectively or to prevent a pathogenic one. *As in physics, also in biology, systems under study tend to favor some form of stability. In physics, this means thermodynamic equilibrium determined by the condition of maximum entropy, or equivalently by the condition of minimum free energy. In biological systems, which are far from thermodynamic equilibrium, stability means steady-state dynamics, i.e. homeostasis. However, in both physical and biological systems, these stability conditions have limited ranges of existence.* This means that if a system is perturbed far enough from its attractor state, it may undergo a sudden and sometimes irreversible transition. This type of change is referred to as a phase transition and the area of physics dealing with these phenomena is called the physics of phase transitions and critical phenomena. In the language of mathematics, these transitions between stable states are called bifurcations. Below, we present a very brief outline of the concepts developed over the last century in this area of science. While there has been interdisciplinary cross-fertilization between the theory of phase transitions in physics and in such fields as sociology and even economics, much needs to be done to successfully implement this rich conceptual framework in biology and medicine. One of the key ideas that can help physicians understand their patients better comes from the theory of phase transitions and is called a fitness landscape.

1.18 Thermodynamic Stability: Phase Transitions, Order Parameters, and Susceptibility Functions

Earlier in this chapter, we discussed conditions for thermodynamic equilibrium, which are based on the second law of thermodynamics and a tendency of closed systems to reach a condition of maximum entropy. This, in turn, under various external conditions, such as fixed volume of pressure translates into a minimum of either the free energy or Gibbs free energy. These are necessary but not sufficient conditions from the point of view of thermodynamic stability, which is a crucially important concept. Additional conditions that need to be satisfied involve the second derivatives of these thermodynamic potentials with respect to all independent variables such as temperature, volume, pressure, magnetization, polarization, etc. The easiest way to explain this concept is to refer to an elastic system such as a rubber band. Its length depends on the force of tension applied to its ends and this is well described by one of the earliest laws of mechanics called Hooke's law. When the force is removed, the rubber band will spring back to its natural length. For moderate tension forces applied to it, the amount of extension is proportional to the magnitude of the force. Moreover, the energy stored in the expanded rubber band is proportional to the square of the extension. The elastic energy can be depicted as a parabola with respect to the extension whose shape depends on the elasticity coefficient. The higher the elasticity coefficient, the harder it is to expand the rubber band. For rubber bands, the elasticity coefficient is small, while for metal wires it is very high. This coefficient defines the proportionality between the force and its effect. In thermodynamics, there is an entirely analogous situation where the role of the energy function is played by each thermodynamic potential: the free energy, the Gibbs free energy, and the Helmholtz energy. The role of the extension is played by so-called intensive thermodynamic variables, *i.e.* the magnetization, mass density, polarization, etc. *The equivalent of the force in thermodynamics is an extensive variable that is a conjugate of the intensive variable. For example, a magnetic field is conjugate to magnetization, an electric field is conjugate to polarization, pressure is a conjugate of volume, etc. Therefore, these conjugate extensive variables can be seen as thermodynamic forces, which change the state of the system as described by the conjugate intensive variables. The coefficients of proportionality are called generalized susceptibilities and they must be positive by definition in order for the system to be stable. A larger value of susceptibility function corresponds to a more pliable state of the system. In other words, susceptibility is a ratio of the extensive to intensive variable or a second derivative of the thermodynamic potential with respect to the intensive variable.* Properties of systems close to instability points may be manifested by the creation of new ordered states that highlight broken symmetries in the system. As a result, long-range order may be established and the emergence of non-linear dynamical modes of behavior takes place. One can measure non-linear responses to external stimuli or perturbations such as hysteresis loops. These responses are an indication of the changes to the system's generalized rigidity, which means that average fluctuations in physical quantities may be more or less pronounced depending on the structure of the system. In the critical phase, an unbounded growth of fluctuations takes place. Near criticality, we select a finite number of most important (driving) degrees of freedom, which are referred to, following Landau, as order parameters [20]. The remaining "fast" degrees of freedom (so-called slaved modes) can either be incorporated through fluctuations, temperature-dependent coefficients, dissipative terms in the equation of motion, or ignored altogether. Landau proposed a theory of phase transitions and concentrated on a single thermodynamic quantity called the order parameter [20]. The effect of all the other degrees of freedom was incorporated in temperature-dependent coefficients of his famous free energy expansions.

One of the characteristics of systems undergoing phase transitions is the emergence of long-range order whereby the fluctuation of a physical quantity at an arbitrary point in the system is correlated with its fluctuation at a point located a long distance away from it. This property, only present near a critical point defining a phase transition signifies extreme sensitivity of the system to external perturbations. In thermodynamic systems kept at sufficiently high temperatures, thermal energy is bound to exceed that of the interactions between particles leading to the predominance of disorder. However, when thermal fluctuations are reduced with the lowering of the temperature at which the system is kept, order emerges and its onset is manifested by long-range correlations. *Long-range order can, in general, arise for two different reasons. The first is the existence of long-range forces that maintain coupling between particles and result in the "rigidity" of the system [21]. The second is due to the fact that when symmetry present in the disordered phase is broken, uniformity is adopted on a macroscopic scale.* This can be seen in many condensed matter phenomena, such as superconductivity, ferromagnetism, superfluidity, liquid crystals, *etc.* [22]. A local application of a short-range perturbation will be felt throughout the system as a result of interactions between its parts giving rise to the dramatic increase in the system's susceptibility when it develops an ordered phase.

For illustrative purposes, in the following section, we list and briefly describe a number of representative examples from condensed matter physics that are characterized by long-range order and related concepts. A typical example of long-range order is the perfect crystal where the mass density distribution has an infinite correlation length related to the onset of spatial periodicity. Crystal formation can be seen in the emergence of a regular diffraction pattern. Periodic distribution of atomic masses comprising the crystal can be seen as a defining feature of this phase transition from an amorphous solid to a crystal. Such quantities are referred to, following Landau, as order parameters [20]. Phase transitions involving crystal symmetries not only refer to crystallization from a liquid solution or through adsorption on a surface of gas molecules but also to transformations from one crystal type to another. Distortions resulting from temperature changes or an application of external stress may lead to so-called structural phase transitions. As a consequence of structural phase transitions between various possible symmetric phases of the crystal, new mechanical, electrical, and sometimes magnetic properties may be observed with their associated thermodynamic anomalies at the transition temperature.

Atomic spins in many compounds and alloys tend to spontaneously form domains within which their individual spins are either parallel or antiparallel. In addition to these two basic magnetic types of order: ferromagnetic and antiferromagnetic, respectively, more complicated regular spatial arrangements of localized magnetic moments are encountered in solids. Typically, lowering the temperature below the transition temperature T_c (called Curie temperature for ferromagnets) causes the magnetic interactions to overcome thermal fluctuations and an ordered state is adopted by the system. Thus, in these cases, the disordered phase is paramagnetic, which exists above the Curie temperature and in which spins are oriented randomly. Below the Curie temperature, all spins align with themselves, and the lower the temperature, the greater the level of alignment. Hence, for ferromagnets, the order parameter is the system's net magnetization.

A number of metals and alloys as well as ceramics (high-T_c superconductors [23]) exhibit an ordered state in terms of the conduction electron behavior when these systems are cooled down below their characteristic critical temperature T_c. This state has two very important features: ideal conductivity (*i.e.* zero resistance) and complete expulsion of magnetic flux lines. A trio of American Nobel Prize–winning physicists, Bardeen, Cooper, and Schrieffer demonstrated that the equilibrium stable state of a superconductor is formed by bound pairs of electrons called Cooper pairs which are mirror images of each other with respect to their spin and momentum orientations. This is a rare case of a physical system where the order parameter, is given by a quantum mechanical property, namely the Cooper pair's wave function forming a so-called condensate [21]. This is a macroscopic manifestation of a quantum phenomenon and it emerges as a result of a phase transition from a metal to a superconductor. Numerous other examples of ordering and phase transition phenomena exist, such as binary fluids, the metal-insulator transition, polymer transitions, spin, and charge-density waves. What is important to stress, however, is that all these systems show certain similarities in

their behavior, namely: a) the existence of order parameters; b) similar features of phase diagrams; and c) extreme sensitivity in their responses to external influences at and around their respective critical temperatures. Next, we elaborate further on these common features of critical systems.

Systems undergoing phase transitions lend themselves to phase diagrams, which delineate the regions of stability of equilibrium phases in the space of thermodynamic coordinates (pressure P, temperature T, density ρ, etc.). Boundaries of the regions of existence of a given phase are drawn as lines of two types: continuous and broken referring to continuous (second-order) or discontinuous (first-order) phase transitions, respectively. These are characterized as follows. First-order phase transitions are associated with abrupt changes of their order parameters at the transition temperature, for example, measured as a nonzero latent heat of transformation [24]. On the other hand, second-order phase transitions have a continuous function of the order parameter and hence no latent heat is associated with the transition temperature [20].

Some phase transitions (only second-order) can also be associated with a broken symmetry phenomenon. *The term "broken symmetry" refers to a situation in which the new ground state of the system does not possess the full symmetry group of the system's Hamiltonian (energy) function* [21]. A broken symmetry can occur either spontaneously (*e.g.* by lowering the temperature) or through the application of an external field or constraint. A classic example is the ferromagnetic-to-paramagnetic phase transition at T_c where the full rotational symmetry of the paramagnetic phase is broken by the axial nature of the new ground ferromagnetic state below the Curie temperature. When symmetry that is broken is continuous (*e.g.* translational invariance), then a new excitation may appear, which is called a massless Goldstone Boson whose frequency goes to zero at long wavelengths [22]. Examples of Goldstone Bosons include ferromagnetic domain walls and acoustic soft modes (phonons whose frequency vanishes when the temperature approaches the critical point) in structural phase transitions. There can be several different types of broken symmetries identified in critical systems, which we list here:

- Translational (e.g. crystal formation, structural transitions).
- Gauge symmetry (e.g. superfluidity, superconductivity).
- Time reversal (e.g. ferromagnets).
- Local rotational (e.g. liquid crystals).
- Rotational (e.g. some structural phase transitions).
- Space inversion (e.g. ferroelectricity).

Gauge symmetry is of special importance since it is a universal property of quantum systems with Hamiltonians (representing energy functions) for which the total number of particles or a generalized charge-like conserved quantity exists. In these cases, such as for superconductors, the order parameter ψ is a complex quantity (whose complex conjugate is ψ^*) and its local density can be defined as $\rho = \psi^*(r)\psi(r)$ (probability density) so that a phase shift of ψ according to $\psi \rightarrow \psi \, exp \, (i\phi)$ leaves the Hamiltonian invariant.

An inherent property of systems undergoing phase transitions is the decrease of their rigidity, or conversely, their increased softening. *Anderson [21] defined generalized rigidity as the propensity of the system to evolve towards a stable ground state in the low-symmetry phase. Consequently, work has to be expended in order to change this state to another equilibrium.* An associated property is the presence of small amplitude oscillations about the equilibrium state. The concept of an order parameter was first introduced by Landau [20] and until this day it does not have a precise definition. It is a thermodynamic macroscopic quantity, which is invariant with respect to the symmetry group of the low-temperature (and low-symmetry) phase and it is zero above the transition temperature while nonzero below. It is a quantitative measure of the amount and type of order that is built up in the neighborhood of the critical point. Below T_c, it signifies multistability of the system, that is the fact that there can be several possible manifestations of the bifurcation effect (e.g. all spins of a ferromagnet lined up or down). From its original application to second-order phase transitions the idea of an order parameter has been extended to first order. It has been generalized from a scalar to a time and space-dependent function. Examples of diverse applications of the order parameter concept to both equilibrium and nonequilibrium critical phenomena are listed in Table 1.3.

Since it is impossible to change the symmetry of a solid gradually, Landau [20] deduced that *second-order phase transitions are associated with symmetry breaking* and can be qualitatively described through the use of an order parameter ψ, which is zero in the symmetric (ordered) phase and nonzero in the dissymmetric (disordered) phase. The cornerstone of the Landau theory of phase transitions is the assumption that sufficiently close to critical temperature T_c, on both sides of it, the free energy of the system, F, can be expanded in a power series of ψ whose simplest form consistent with stability criteria is

$$F(T,V,\psi) \equiv F_0 + a\epsilon\psi^2 + A_4\psi^4 \qquad (1.3)$$

where $a > 0$ and $A_4 > 0$ for stability reasons. In addition, introducing a control parameter for spontaneous second-order phase transitions as the reduced temperature $\epsilon \equiv (T - T_c)/T_c$, Landau was able to show that a bifurcation takes place when changing the sign of ϵ, so that F transforms from a single ($T > T_c$) to a double-well shape ($T < T_c$). In other words, the system changes its stability conditions at $T = T_c$ and instead of having a single stable state above the critical temperature, it has two below it. This can be seen by solving the equilibrium conditions for F, which yields ψ given by

$$\psi = 0 \text{ for } \epsilon > 0 \qquad (1.4)$$

$$\psi = \pm\left(-\frac{a\epsilon}{2A_4}\right)^{1/2} \text{ for } \epsilon < 0 \qquad (1.5)$$

It is important to stress that the free energy profile (Figure 1.17) changes from a parabola above the critical temperature to a double-well below. This is associated with instability at the transition point. The profile of the free energy is an example of a fitness landscape where the valleys represent stable regions of the system's behavior while the peaks represent instability points.

One can create models of critical behavior with very complicated free energy (fitness) landscapes. For example, a sixth-power expansion has been widely used in the past to model first-order phase transitions [25]. In the language of catastrophe theory, this latter expansion describes a "butterfly catastrophe" [26] and a connection between catastrophe theory and the concept of structural stability can be found in the literature [27].

A more rigorous theory of systems at criticality followed Landau's ideas but included the concept of scaling, which is based on the fact that *at phase transitions, these systems manifest long-range order and are seen to be identical on all length scales.* This led to the introduction of scaling transformations to the mathematical descriptions of these systems, a construct which culminated in the development of the so-called

TABLE 1.3

Examples of Order Parameters in Physical Systems with the Associated Ordered and Disordered Phases Both in Equilibrium and Nonequilibrium Phenomena

Phenomenon	Disordered Phase	Ordered Phase	Order Parameter
Equilibrium			
Condensation	Gas	Liquid	Density difference $\rho_L - \rho_G$
Spontaneous magnetization	Paramagnet	Ferromagnet	Net magnetization $\mathbf{M}$
Antiferromagnetism	Paramagnet	Antiferromagnet	Staggered magnetization $\mathbf{M_1} - \mathbf{M_2}$
Superconductivity	Conductor	Superconductor	Cooper pair wavefunction ψ
Alloy ordering	Disordered mixture	Sublattice ordered alloy	Sublattice concentration
Ferroelectricity	Paraelectric	Ferroelectric	Polarization
Superfluidity	Fluid	Superfluid	Condensate waveform
Nonequilibrium			
Tunnel diode	Insulator	Conductor	Capacitance charge
Laser action	Lamp (incoherent)	Laser (coherent)	Electric field intensity
Super-radiant source	Noncoherent polarization	Coherent polarization	Atomic polarization
Fluid convection	Turbulent flow	Bénard cells	Amplitude of mode

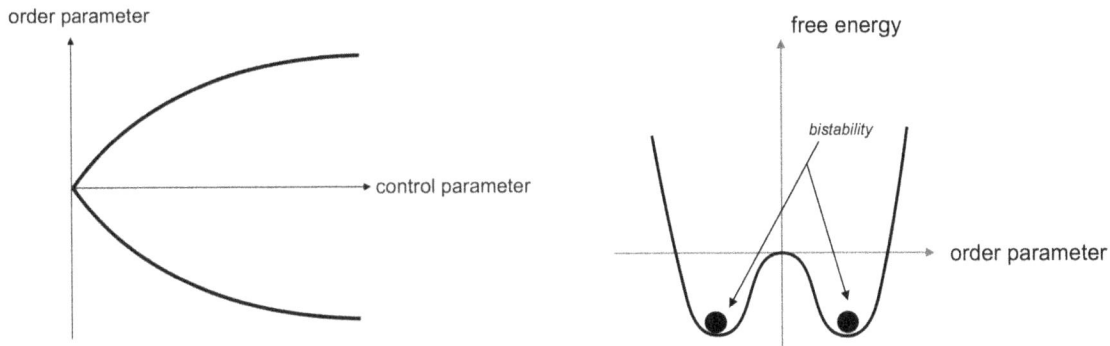

FIGURE 1.17 A canonical example of a bifurcation diagram and an associated change in the form of the free energy function from a parabola to a double well.

Renormalization Group (RG) Theory of phase transitions, for which Kenneth Wilson [28] received a Nobel Prize in Physics. Renormalization Group theory seeks a fixed point of the model Hamiltonian, H^*, which is invariant with respect to scaling transformations. Each step in the RG transformation chain reduces the number of degrees of freedom by a factor related to the size of the scale change [29]. Effectively, it replaces individual atoms by their clusters called blocks. What is extremely interesting and perhaps relevant to biology as well is that all types of systems undergoing phase transitions end up falling into classes with the values of parameters called critical exponents that are identical within each class. *As a consequence, numerous physical systems can be described by the same approximations. This statement is known as the universality hypothesis and it goes further by conjecturing that for any two physical systems with the same physical dimensionality, d, and the same number of order parameter components, n, all such systems belong to the same universality class so that each fixed point corresponds to one universality class.* If this were to apply to biological systems, it might imply that from a plethora of phenotypes, one may be able to construct universality classes with common characteristics, perhaps akin to the concept of kingdoms. This could provide an elegant mathematical description of the various observations in evolutionary biology and even embryology.

1.19 Expanded Concepts of Entropy and Information

Information theory founded in 1948 with a landmark paper by Claude Shannon [30], a mathematician and electronic engineer, is a branch of applied mathematics, electrical engineering, and computer science involving the quantification of information. It was developed to find limits on signal processing operations such as compressing data and reliably storing and communicating data. It has found application to a host of areas including neurobiology, evolution, and quantum computing. Information theory refers to entropy, i.e. information entropy that germinated from the origin of thermodynamic entropy. Shannon was the first to expand the concept of entropy into a novel domain. The value of entropy, as borrowed by its use in information theory, is determined by the number of variables of a system and the number of potential ways energy can be arranged in a system. Information entropy, like information energy, expresses a potentiality for "cracking the codes" that drive biological complexity and stability. It is the basis for organismic heterogeneity or variability through the process of entropy reduction. In a general sense, entropy is taken to refer to disorder or randomness but more accurately in a thermodynamic sense, it is the dissipation of heat in an isolated or closed system. Specifically, entropy represents the number of states a system can be in, given the probability distribution of that system in phase space. This refers to every way a system can be varied in terms of geometric and energy configurations. Rolling dice will have a greater spread in the probability distribution versus flipping a coin in terms of getting a certain result. Specifically, if tossing a coin yields one bit of information, then rolling a die (which has the predictability of rolling on one of six sides) yields approximately 2.58 bits of information. Thus, the information of rolling a die is greater than that of tossing a coin. In a limited sense, rolling dice may be described as informational entropy. However, it is important to emphasize that this is a one-parameter example that does not reflect the relevance of energy, and hence, refers only to information entropy rather than a thermodynamic entity. Information entropy is a conceptual progeny of thermodynamic entropy. It should be underscored that information entropy can be a highly confusing and seemingly paradoxical term. *This is because information in the sense of information theory when applied to biology is entropy reduction medicated by informational energy that serves as the glue to the chemistry of biological systems. The entropy component of the term information entropy refers essentially to the opposite implication of entropy in the sense of moving to a state of maximal stability of component particles. Entropy in the case of informational entropy refers to a potentiality for the arrangement of atoms and molecules into a stable macroscopic whole system.* Classical entropy has also been applied to the realm of quantum mechanics. The latter similarly deals with probabilities although there are differences in how they are calculated. There is also the concept of evolutionary entropy used in the field of evolutionary biology, which invokes a central notion of heterogeneity. Evolutionary entropy will be discussed in some detail in a later chapter.

1.20 How Information Is Connected to Energy

In 1948, John von Neumann, the legendary mathematician and the founding father of computer science, was approached by Claude Shannon, then a young IBM engineer, attempting to quantify information contained in a message. Von Neumann's answer to Shannon's question was as follows:

> Why don't you call it entropy? In the first place, a mathematical development very much like yours already exists in Boltzmann's statistical mechanics, and in the second place, no one understands entropy very well, so in any discussion you will be in a position of advantage [31].

Claude Shannon had already published a seminal article in 1948 entitled "Mathematical Theory of Communication" [30], in which he stated that the content of a message can be expressed through a set of binary decisions, or yes or no answers, in analogy to a game in which it is possible to identify a specific person in a group by a series of questions. Each of these answers has been termed a bit, which stands for binary digit. Assuming for simplicity that all the choices have the same probability of ½, and, if I choices are to be made on a set of Ω objects, it follows that $\Omega = 2^I$, and hence: $I = ln\ \Omega$ where I is the quantity of information required to identify the system. Obviously, the situation becomes more complicated if the events have different probabilities, in which case

$$I = -\Sigma_i p_i \ln\left(p_i\right) \qquad (1.6)$$

with arbitrary probabilities p_i. The latter equation has the same mathematical form as entropy, except for the sign and the lack of the Boltzmann constant k_B in front of the summation sign, hence it sometimes is called negative entropy. *Therefore, information is directly proportional to the uncertainty in our knowledge about the occupation of the Ω microstates of a thermodynamic system.*

Rolf Landauer was interested in the question of the energy cost of acquiring information. Indeed, information cannot be created or lost in abstraction from a physical situation, such as the generation of holes in a perforated paper tape, the creation of a spin-up or spin-down of a particle, or the flow of an electric current in a circuit. Furthermore, *every type of information is encoded, processed, and transmitted by physical means such as electrical circuits, optical fibers, or magnetic types.* To make a point, Landauer entitled his paper on the topic: "Information is inevitably physics" [32]. In it, he evaluated a minimum cost of information given by this formula:

$$I = \frac{S}{k_B \ln 2} = \frac{\text{energy}}{k_B T \ln 2} \qquad (1.7)$$

where S is the entropy of the system.

Hence, the minimum amount of energy associated with a single unit of information, that is one bit ($I = 1$) is equal to $k_B T\ ln2$. In agreement with the second law of thermodynamics

implying the presence of irreversibility, Landauer formulated the hypothesis that computation is necessarily connected with the dissipation of energy whenever a bit is changed. Then the only irreversible processes involved in computation are those in which information is erased, so that cancellation achieves a leading role and he showed that *the erasure of information causes the release of $k_B T ln2$ of heat per bit into the environment.* Charles Bennet, one of Landauer's students at IBM, analyzed every kind of computer, real or abstract, and confirmed that a great deal of computation can be done with no energy cost but energy dissipation always occurs when information is erased. Edward Fredkin, a collaborator of America's arguably most brilliant physicist, Richard Feynman, went even further and stated that information is more fundamental than matter and energy because atoms, electrons, and quarks consist ultimately of bits of information. Their behavior and thus the behavior of the entire Universe must, therefore, be governed by an algorithm. This idea, taken to its limit can be understood that we all live in a computer simulation similar to that depicted in the movie *The Matrix*.

Information in itself does not necessarily have any inherent meaning. As stated above, the unit of information is one bit, which is related to a binary element whose value can be either 0 or 1. Information can be degraded over time because it must be stored in some physical form, which is always subject to some external influences that can have an impact on the physical integrity of the recording device or method. When information takes on a deeper meaning, however, it becomes more enduring and harder to erase. This may be thought of as being translated into knowledge. Information becomes more enduring in terms of an informational bit in nature when it interacts with other bits to create a system whole that promotes the adaptive survival of the systems in the context of its environment. In this sense, this book endeavors to discuss these important issues from an organizational structure that frames not only how nature finds solutions for adapting to its surroundings, *i.e.* finding the "code", but how such a "code" creates a deeper meaning of information in the form of knowledge. *There is a deep parallel between understanding bits of information as knowledge and translating those bits into actual solutions (which is the goal of nature) to generate an enduring survival value. Thus, thinking about how we think helps in finding that deeper meaning or the "code". It also helps us to solve the problem of biological challenges in medicine by approaching them from the same perspective as that used by nature itself.* Accordingly, this writing hopes to be an exercise in metacognition, or thinking about how we think. The principles of how we think may be in the fabric of recognizing distinctions (what something is and what it is not), systems of parts and wholes, how those parts interact and relate to one another as well as to the larger system whole; and finally the perspective through which we understand such distinctions, relationships, and systems of parts and wholes.

1.21 Steady States and Homeostasis

All physical systems in nature are composed of atoms and molecules, and as shown by statistical thermodynamics arguments, all systems seek a state of equilibrium, which means a state whose free energy is a minimum under the given set of conditions that includes the presence of a heat reservoir that maintains a constant temperature. In the particular case of a system that is isolated from its environment both in terms of energy and matter transfer, the thermodynamic equilibrium condition implies achieving a state of maximum entropy, meaning maximum disorder under prevailing physical constraints (such as finite volume). *On the other hand, living cells, biological systems, and organisms, which are in an open thermodynamic state that is characterized by a constant exchange of material and energy with their environment, seek a steady state of homeostasis.* However, this is not a thermodynamic equilibrium stationary state (Figures 1.18a and 1.18b), but a dynamically stable steady state. Figures 1.18c and 1.18d illustrate the pancreas feedback homeostasis, which in a biological system is strictly in the so-called "far-from-equilibrium" state since its stability is connected to the stability of other biochemical activities proceeding throughout the organism. Thus, this example (and others similar to it) does not represent a state of thermodynamic equilibrium but a state of limited stability under external perturbations. *To reiterate, living systems are far-from-equilibrium, except when they approach death, which represents a transition from a biological state (metabolically active) to a physical state (metabolically inactive).*

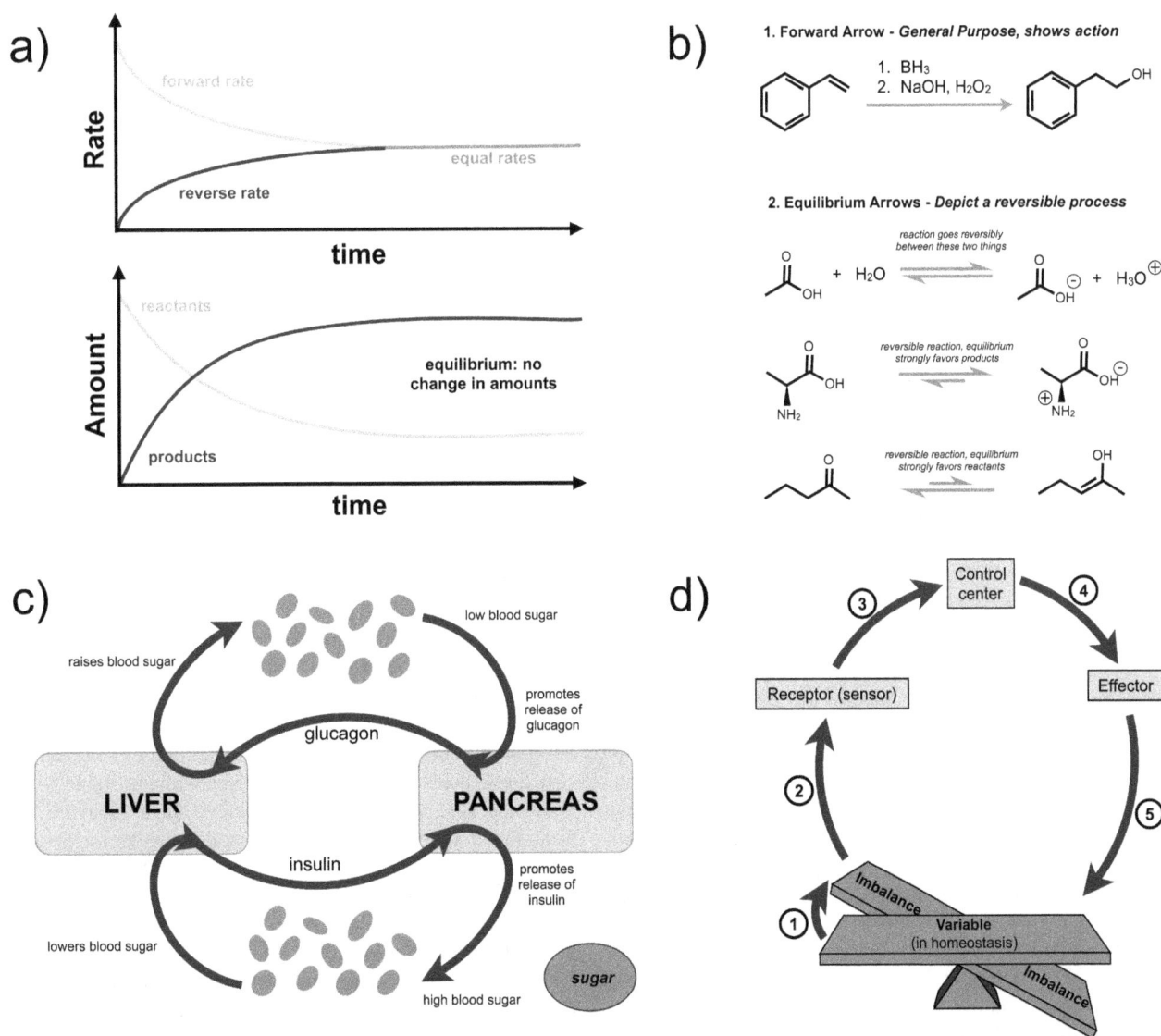

FIGURE 1.18 Examples to illustrate chemical equilibrium between molecules a) and b) and pancreas feedback loop homeostasis c) and d). The feedback loop in d) shows the following workflow: 1) stimulus produces change in variable; 2) change detected by receptor; 3) input: information sent along afferent pathway to control center; 4) output: information sent along efferent pathway to activate; 5) response of effector feeds back to influence magnitude of stimulus and returns variable to homeostasis. Source: adapted from the following: a) http://faculty.chem.queensu.ca/people/faculty/mombourquette/FirstYrChem/equilibrium/index.htm. b) https://www.masterorganicchemistry.com/2011/02/09/the-8-types-of-arrows-in-organic-chemistry-explained/. c) https://www.atrainceu.com/content/4-regulation-blood-glucose. d) https://www.apsychology.family/news-1/2018/3/20/positive-feedback-systems.

A healthy human body is a perfect example of an exquisitely constructed and highly efficient biological system with built-in strategies for adaptability and survival under various forms of stress and environmental challenges. Its organizational perfection can be seen through spatiotemporal symmetries. Symmetries refer here to a dynamic balance of a living system, which is adaptive and maintains the health of the organism.

Two main strategies in response to stress are: a) increased resilience, which maintains the state of symmetry, or b) symmetry breaking and establishing a new stable state with new symmetries. This adaptability is clearly seen in the mechanisms of homeostasis and allostasis, the former maintains symmetry while the latter may lead to new symmetries. Healthy allostatic changes maintain physiological homeostasis at the organismic level.

A canonical example of maintaining symmetry when exposed to stress and making the system even more resilient than before is called hormesis. Somewhat paradoxically, some responses leading to pathologies, such as insulin resistance, can be viewed as manifesting adaptation by breaking old symmetries and finding new ones, which are better suited to the prevailing conditions. Moreover, type 2 diabetes can be seen as an adaptation, not just a pathology. The human body automatically optimizes its response by finding the highest fitness state achievable. *Losing the ability to form new adaptive symmetries under stress represents a loss of robustness, flexibility of allostatic compensations, and hence resilience to external perturbations challenging the biological system. Aging and senescence gradually lead to a complete loss of adaptability when a biological system turns into its physical shadow incapable of entropy reduction and symmetry restoration enabled by metabolism, the essence of life.*

In physical systems, there are conventionally accepted four fundamental forces (gravitational, electromagnetic, strong nuclear, and weak nuclear) and 12 fundamental particles that are responsible for all interactions between the building blocks of both inanimate and animate matter and indeed for the behavior of the universe including its evolution. Let us imagine that these fundamental components can only make one inanimate thing, namely cement, resulting in a bucket of cement. Let us further imagine all the things that can be built with this one type of material, for example, bookshelves, driveways, homes, etc. By contrast, these same few fundamental forces and fundamental particles create a large number of genes (DNA—DeoxyriboNucleic Acid) in the human body (approximately 25,000). Now envision in another bucket the amino acids coded for by these genes and then factor in the extraordinarily complex and sophisticated genetic expression capable of explosive and intricate building potential to control the structure of the human body. *The number of possible DNA structures composed of the four nucleic acids is on the order of 4^N where N is the length of a DNA strand measured in base pairs. Since N can run into billions, the number of DNA sequences is not only astronomical but exceeds the immense*

number of 10^{110}, which provides a combinatorial barrier to even an inspection of such a set by any conceivable computer. Similarly, the number of protein structures is on the order of 20^n where n is the length of the peptide sequence representing the protein in question. Typically, n is on the order of hundreds and sometimes thousands meaning that proteins, too, represent an immense set.

1.22 Structures and Their Functions

The extraordinary diversity of structure, form, and function of animate systems is in contrast to inanimate matter and poses conceptual challenges to the scientists working in the life sciences. Another important aspect is that in addition to energy, information plays a crucial role in biology while in physics it is only now that the role of information is being appreciated. In fact, one can introduce the hybrid concept of information energy. The bits of such information (expounded in terms of the number of genes and the amino acid sequences coding the genes) constitute the basis of understanding the nature of various life forms. In this context, the term "information energy" which refers to the energy that is required for promoting the work of putting something into form, is often employed to emphasize the value provided by such bits of information. It is the purposeful application of energy into something useful such as the construction of biological organization using informational bits or building blocks of chemistry into three-dimensional structures that evolve dynamically over time. *Informational bits in isolation may have no inherent meaning. However, the energy that glues information bits into a system whole provides deeper meaning in the sense of value or durability. The amount of information stored in each gene is a magnificent and awe-inspiring masterpiece; for example, the nucleus of each human cell contains as much information as would be found in one million pages of an encyclopedia. In comparison, the significantly smaller number of nonliving structures (hence the information contained therein) created by the same four forces and 12 fundamental particles is remarkably striking.* The reason for this "combinatorial dissonance" between physics and biology is that out of the numerous possible combinations of fundamental particles into larger structures their vast majority are energetically unstable while all nucleic acid combinations are possible due to the relatively weak forces gluing them together and allowing an astronomical number of DNA sequences to be created. It is interesting to note that of all the fundamental forces, the electromagnetic force (for example that carried by solar photons) is believed to be most responsible for all the biochemical reactions, which allow the evolution of organized structures of living organisms from inorganic materials (made from the fundamental particles and forces). *Thus, the four fundamental forces and 12 fundamental particles are responsible to different extents for the full complexity of organic life.*

A central theme of this writing is to underscore the indistinct borders between animate and inanimate matter in terms of their building blocks as well as the laws and basis for how energy flows through these forms of matter. While there may be much to be learned from nonliving physical systems in nature noting parallels to certain biological systems, the hierarchical complexity and emergent behavior (that does not depend on individual components, but on their relationships to one another) of the latter makes it uniquely distinct. Conversely, *the major branches of classical physics such as Newtonian mechanics, thermodynamics, and electromagnetism, are characteristically deterministic and predictable in contrast to the typical nonlinearity of networks and systems biology.* The 20th-century scientific revolutions triggered by the theory of general relativity, quantum mechanics, and chaos theory (and the offspring field of complexity medicine), all hold intriguing and potentially important practical applications to biology at the macroscopic and human scale. These newer revelations in the physical sciences serve as an effective bridge across various disciplines of scientific understanding (*e.g.* scales of space and time). They frame new perspectives of complex interactions of systems that are responsible for adaptive self-organizing emergent phenomena, and the whole that these systems of parts create, that define our biology.

1.23 Negative Entropy and Self-Organization

In contrast to nonliving matter that obeys the laws of thermodynamics; living beings are self-organizing and seemingly violate the second law since they build complex well-ordered and information-rich systems during the course of their development. Thus, the local entropy of these systems actually decreases rather than increase. Rather than seeking out the lowest energy state, biological systems maintain far-from-equilibrium positive energy states of organized complexity. These states are locally stable, *i.e.* they are not affected by limited perturbations. They are not globally stable, which means that the lowest energy state is a state of death. *Conceivably, a state of disease may represent another metastable state that has an energy level lower than that of full health yet higher than the state of death.* We elaborate on this novel idea proposed by the author in sections dealing with the proposed Physiological Fitness Landscape. If left to its own devices a state of ill-health would most likely over time take the system to the state of death. To "lift it" to the state of health, an energy-boosting intervention must be made either by the system itself (change of lifestyle, rest, nutrition, etc.) or by an external effort (a health practitioner, pharmacological agents, *etc.*). However, such a development (*i.e.* maintenance of a higher-energy locally stable state with reduced entropy) takes place in the context of an "open" system which together with the surroundings defines the total closed system whose entropy increases. The thermodynamics of biochemical reactions utilize enzymes and ATP to mediate energy-requiring, or endothermic, reactions as well as harness the energy released by exothermic reactions to drive endothermic processes. Such non-equilibrium thermodynamics have been applied for explaining how the order and complexity of biological organisms can develop from disorder. Ilya Prigogine, a Belgian physical chemist and Nobel Laureate, called these systems "dissipative systems" because they are formed and maintained by the dissipative process that exchanges energy and matter between the system and its environment. Importantly, the

dynamic stability of these processes and pattern-forming phenomena sometimes associated with them are due to nonlinear interactions between several key components taking place in the chemical reactions. Linear systems do not have such properties and in view of dissipation would always succumb to the destructive role of entropy increase via the second law of thermodynamics. *While entropy is lower for an organized living system, it increases upon suffering from insults to normal physiological processes and reaches a maximum when senescence or demise of the organism occurs.* As discussed later, in addition to promoting the rise of the total entropy of the universe as the organism "breaks down", the actual localized Gibbs free energy, or negentropy, allows dissipation of heat from solar rays "into the cool".

The concept and phrase "negative entropy" was introduced by Erwin Schrödinger in his 1944 popular science book *What Is Life?* [33] in the context of discussing the evolution of life which is characterized by increasing order and complexity and perceived as moving from a less ordered state to a more ordered one. Since the building of a more ordered state of a system involves energy input (positive change of enthalpy ΔH) and negative change in entropy (ΔS), it can be argued that the evolution of organisms (maintenance of homeostasis, growth, and reproduction), as well as the evolution of species, would be characterized by non-spontaneous processes associated with positive changes in Gibbs free energy (ΔG). *Thus, the increasing organization and complexity seemingly run in the opposite direction to the second law of thermodynamics which simply states that the entropy change of the universe is positive. An obvious reason for the perceived discrepancy lies in the fact is that the evolution of both plants and organisms does not take place in a closed system but rather in an open one. The Sun provides the necessary input energy through its own spontaneous processes, as described above.* Energy associated with electromagnetic radiation provides the needed activation energy for the photosynthetic processes of glucose production characterized by a positive change in Gibbs free energy and a positive change in enthalpy. By implication, such processes would not occur if plants lived in a closed system cut off from the Sun. Plant life is at the base of the biological ecosystem and provides life for not only herbivores but also carnivores that get their energy indirectly from the prey animals who feed on the plant life for survival. As has been discussed earlier in this chapter, carbohydrates and fats are metabolized by living organisms dependent on plant life through the process of respiration deriving energy in the form of the universal currency of ATP whose bonds hold the potential to drive biochemical reactions that lead to the breakdown of glucose into carbon dioxide and water. *The molecules of ATP are utilized to capture energy released by exothermic reactions to drive downstream endothermic ones, which would otherwise be non-spontaneous. The work done by the utilization of ATP includes maintaining homeostasis of the body, growth and reproduction, muscle activity, and active transport systems such as driving the sodium potassium pump for neuronal transmission.* Regeneration of ATP is an endothermic (positive ΔH) reaction, the activation energy for which is provided by the negative ΔG and negative ΔH of glucose combustion. As additional examples, the activation energy provided by the breakdown of ATP into ADP promotes the conversion of glucose to glucose-6-phosphate by hexokinase, and conversion of fructose-6-phosphate to fructose-1,6-bisphosphate by hosphorfructokinase, early steps in the glycolysis pathway. As briefly described above, the localized negative entropy of biological structures in fact promotes the second law of thermodynamics by functioning as energy dissipators of the high pent-up energy of solar rays moving heat temporarily "into the cool". This is what Ilya Prigogine first described [34] and Jeremy England [35] more recently corroborated by mathematical equations. Ultimately, biological structures of localized organized complexity promote the entropy of the universe by returning the smallest scales of our somatic existence along with the heat locked inside our biological chemistry. In this capacity, organismic life becomes dissipated. Hence, biological structures both dissipate heat from the Sun as well as serving as so-called "dissipative structures" by returning that heat back to the environment. In the context of dissipative structures of Prigogine one needs to mention auto-catalytic feedback loops without which localization of structures and pattern formation is not possible. This is an example of nonlinearity that classical thermodynamics typically avoided. While clearly within the principles of thermodynamics, reaction-diffusion systems are very special examples of seemingly paradoxical entropy reduction in a subsystem.

In addition to thermodynamic entropy, the concept of evolutionary entropy is invoked in understanding the response of a resident phenotype to an invading genotype. *Evolutionary entropy is a measure of the variability in the age at which individuals in a resident population reproduce and die.* A more elaborate exposition of these ideas requires advanced mathematical concepts and is thus outside the scope of this book. The interested reader is referred to the work of Lloyd Demetrius [36]. While thermodynamic entropy, a central concept in physics, is a quantitative measure of the disorder in a system that is related to the geometry of the system, evolutionary entropy is a measure of the robustness (or alternatively demographic stability) of a population, a dynamical property which describes the rate of return of populations to its original size after a perturbation because of greater demographic stability as well as iteroparity. Robustness can be argued to require increased biological complexity of a system (which leads to lowering of the thermodynamic entropy of the system) that will be endowed with advantages in the course of natural selection. Thus, the concept of negative entropy (corresponding to increased order in biological systems) is deemed to parallel positive evolutionary entropy [36]. The latter is also deemed to represent a measure that describes the uncertainty in the age of the mother of a randomly chosen offspring. Thus, *this measure of entropy increases in systems with equilibrium species and decreases in systems with opportunistic species. It has been demonstrated that the concept of evolutionary entropy is critical for understanding the changes in the composition of populations as a function of mutation and natural selection.* It is empirically observed that every step in the evolutionary process provides selective advantage to the species. The evolution of the eye serves as an excellent example. An organism that just senses light, for example, the flatworm, only goes where there is darkness so it can avoid being dried out. If the area of photoreception is folded onto itself, it allows for the sensitivity of direction or motion. Each further development in the evolution of the eye including the retina, a transparent humor, a distinct lens and iris, and a separate cornea each give a natural selective advantage. There exist organisms in the evolutionary spectrum that have each of these types of eyes. This complexity that builds on itself is driven by the processes of mutation and natural selection is conceptually distinct. The idea that this represents a greater order of building on itself appears to trend opposite to the randomness and disorder that defines entropy from the perspective of physics. The question of whether the concept of evolutionary entropy will come to play as fundamental a role in biology as thermodynamic entropy does in physics, warrants further investigations.

1.24 Biological Engines as Metaphors of the Carnot Engine

A concept analogous to the internal energy of an internal combustion engine relates to the internal energy of biological engines, which may be deemed to be the agents that sustain all life-promoting activities in a living system. Such engines, like thermodynamic engines (similar to a heat engine or an internal combustion engine) are powered or driven by molecular motors that consume energy and turn it into work. Many protein-based molecular motors harness the chemical-free energy released by the hydrolysis of ATP in order to perform mechanical work. Some illustrative examples of molecular motors include cytoskeletal motors (*e.g.* myosin, kinesin, dynein), rotary motors (ATP synthase family), nucleic acid motors (RNA polymerase, DNA polymerase, helicase, topoisomerase, etc.), polymerization motors (e.g. actin, microtubules, dynamin). The efficiency of these motors translates directly, on a macroscopic scale, to the health and well-being of living organisms. In any event, the internal energy of a combustion engine and that of a biological engine reflects the total microscopic kinetic energy and potential energy of their component parts. Internal energy in biological systems is manifested in several forms: heats of formation of all cell or tissue components including the energy needed to form covalent bonds, noncovalent inter-atomic and intermolecular interactions; active transport energy-requiring process (including maintaining electrical gradient potentials across cell membranes of all cells of the body as a protective mechanism, other active transport pump mechanisms for charged particles and other molecules mediated by proteins embedded across cell membranes, or for neuronal transmission of the nervous system or cardiac conduction system); and energy for movement and mechanical work (for example for cell mitosis, cell motility or muscle contraction). *A motor may be defined as a device that consumes energy in one form and converts it into work (for example, mechanical motion) and in this sense, a biological engine may be deemed to be driven by molecular motors, biological engines, which utilize ATP as the fuel to drive various cellular processes by creating force and motion. Motor proteins are the driving force behind most active transport of proteins.* This includes the myosin motor proteins responsible for muscle contraction via the cross bridging with the thinner actin microfilaments. Kinesins are a class of intracellular motor proteins that are involved in the binding and transporting of organelles, e.g. mitochondria, vesicles, and protein complexes, along microtubule components of the cytoskeleton. Kinesins primarily move their cargo from the center of the cell in a linear direction toward the cytoplasmic membrane [18, 19]. Conversely, dyneins, another class of intracellular motor proteins, mainly transport cargo in the opposite direction. These motor proteins also enable cell motility and division, endocytosis, and exocytosis [18, 19]. ATP synthase is a rotary molecular motor protein located within the matrix side of the inner mitochondrial membrane of all living organisms. It is integrated into the inner mitochondrial membrane, which is essentially a power plant, which uses the electrochemical gradient of protons across the inner mitochondrial membrane to synthesize ATP [18, 19]. Furthermore, the activity of this molecular motor is reversible, that is, the rotary motor works in the opposite direction to hydrolyze ATP and utilize the energy released to generate the proton electrochemical gradient across the inner mitochondrial membrane. The direction of this rotary motor driving ATP synthesis versus hydrolysis depends on the concentrations of the three reactants ATP, ADP and phosphate molecules. Gibbs free energy value for ATP hydrolysis is proportional to the proton-motive force. Gibbs free energy for ATP hydrolysis to ADP is -48 kJ per mole (~ -11.5 kcal.mol^{-1}) [37]. The value of Gibbs free energy change for ATP synthase is consequently positive and hence the reaction is non-spontaneous (in the context of capturing the energy of the electrochemical gradient). Thus, ATP hydrolysis drives the electrochemical gradient across the inner mitochondrial membrane indirectly and may be the source for the specialized rotary molecular motors responsible for ATP synthesis. In other cases, energy is provided from the transfer of high-energy electrons of NADH and FADH2 molecules derived from the TCA cycle. The redox chain consisting of complexes I, II, III, and IV mediate the transfer of electrons between redox centers within the inner mitochondrial biomembrane and ultimately to the ATPase motor, which is involved in the phosphorylation of ADP to ATP. Each successive complex along this chain has a lower redox potential, the difference of which is liberated as Gibbs free energy (ΔG) to maintain the electrochemical proton gradient as electrons are passed from one complex to the next along this electron transport redox chain. This ultimately allows ATP synthase (ATPase) to use the flow of H$^+$ through a channel within the enzyme back into the matrix that combines with oxygen as the final electron acceptor to produce water; the energy released generates ATP from adenosine diphosphate (ADP) and inorganic phosphate. *ATP synthase is an energy-generating rotary motor engine that releases ATP. It works like a rotary engine. Driven by the energy of the proton electrochemical gradient the barrel-shaped rotator spins as component subunits bind to protons en route back across the inner mitochondrial membrane to the side where the proton concentration is lower transmitting the chemical energy into the connecting drive shaft at the base of this molecular machine which helps make ATP.* This drive shaft contains a specially placed bump which in the course of its spinning around causes subunit protein molecules at the bottom of the machine allowing ADP molecules to enter. The mechanical spinning motion of the drive

shaft and attached ADP molecules causes the ADP complex to bind to an additional phosphate group producing ATP. The ATP is released to the cell, available to power biomechanical reactions including other molecular proteins. Significantly, the ATP synthase rotator molecular protein motor simulates a microcosm to the technological mechanical motor designs of watches to automobile engines. In each case, they all include a rotor, a stator, a drive shaft, and other basic components of a rotary engine. ATP synthase is an eloquent example of intracellular biological nanoscale molecular motors brilliant engineered, that make life possible.

Once ATP is formed in a biological engine, the Gibbs free energy associated with its breakdown to ADP (adenosine diphosphate) is harnessed to drive forward a number of biochemical reactions (metabolism) that result in work being done by the biological engine. For example, two of the reactions in gluconeogenesis (reverse of glycolysis; Figure 1.19) are feasible through the breakdown of two ATP molecules to two ADP molecules per reaction. They are 1) conversion of pyruvate to oxaloacetate in the presence of pyruvate carboxylase

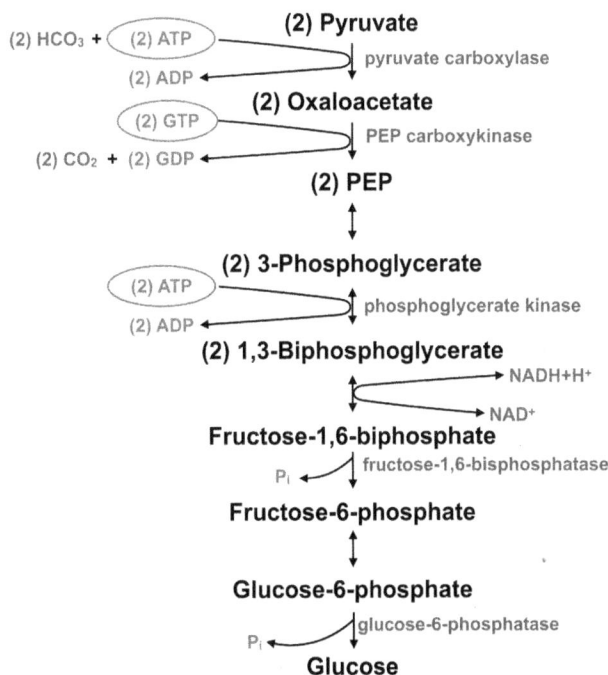

FIGURE 1.19 The pathway of glycolysis produces two ATP molecules from two ADP molecules per glucose molecule from the reaction step of phosphoglycerate kinase (moving 3 phosphoglycerate to 1,3-biphosphoglycerate) and two ATP molecules from two ADP molecules from the reaction catalyzed by enzyme pyruvate kinase (moving phosphoenolpyruvate to pyruvate). Together, these reactions produce a total of four ATP molecules per molecule of glucose. Conversely, early steps in the glycolysis cycle consume a total of two ATP molecules per glucose molecule: hexokinase (in skeletal muscle) or glucokinase (in liver) which phosphorylates glucose to glucose-6-phosphate (consuming one molecule of ATP) and phosphofructokinase which phosphorylates fructose-6-phosphate to produce fructose-1,6-bisphosphate (consumes one molecule of ATP per molecule of glucose). Together, the two reactions hydrolyze a total of two ATP molecules per molecule of glucose. The net ATP production by the glycolysis pathway is positive two ATP molecules per glucose molecule. Source: adapted from https://www.onlinebiologynotes.com/gluconeogenesis/.

and 2) the conversion of 1,3-bisphosphoglycerate to 3-phosphoglycerate mediated by phosphoglycerate kinase. The net reaction of glycolysis which converts glucose to pyruvate as per the reaction shown below is exothermic and has a Gibbs free energy change of –85 kJ/mole (–20.3 kcal/mole).

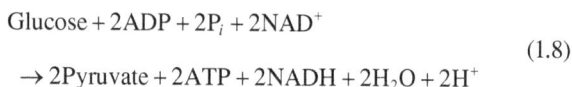

$$\text{Glucose} + 2\text{ADP} + 2P_i + 2\text{NAD}^+$$
$$\rightarrow 2\text{Pyruvate} + 2\text{ATP} + 2\text{NADH} + 2H_2O + 2H^+ \qquad (1.8)$$

However, gluconeogenesis is not merely the reversal of glycolysis as such a reaction would be quite endoergic (characterized by Gibbs free energy change of +85 kJ/mole). Also, three of the reactions (catalyzed by hexokinase, phosphofructokinase, and pyruvate kinase) in glycolysis are irreversible and need to be bypassed during gluconeogenesis, which is made spontaneous by the consumption of four ATP and two GTP molecules. The ΔG value for such a reaction is –38 kJ/mole, but in the process two GTP molecules and four ATP molecules have been lost.

Another example of utilization of Gibbs free energy from ATP breakdown to ADP is presented by the synchronized movement of actin and myosin protein filaments results in muscle contraction and mechanical work (Figure 1.20). In this case, ATP prepares myosin for binding to actin by moving it to a higher energy state and a "hooked" position. Myosin exemplifies a motor protein that requires the energy liberated from the hydrolysis of ATP for its function. Once the cross-bridge is formed, ATP is metabolized to ADP by the enzyme ATPase releasing the Gibbs free energy that is used by myosin to undergo a power stroke, reaching a lower energy state. ATP must bind again to the complex to break the cross-bridge between actin and myosin by releasing the "hooked" state of the latter and the cycle is repeated.

1.25 Metabolism: Life's Necessity

The term metabolism encompasses all chemical changes that take place within living cells needed to sustain life by allowing cells to grow, differentiate, repair damage, respond to environmental changes and divide. The term metabolism is derived from the Greek word "*metabolismos*" which means "change", or "overthrow". *Metabolism includes both the breakdown of some molecules and the synthesis of new ones.* All of this involves a complex network of interactions that transform the cells into a region of overcrowded molecular track. Two particular aspects of metabolism deserve special attention, *i.e.* metabolic energy and the nature of biochemical transformations. Metabolism can be defined as the sum of all physical and chemical processes in an organism by which its material substance is produced, maintained, and destroyed and by which energy is made available to the organism to carry out work (*e.g.* breathing, muscle contraction, *etc.*). Each of these processes is ultimately mediated by solar energy which is necessary for the generation of nutrient fuel in plants whose consumption and subsequent breakdown leads to the production of ATP.

The final products of glucose metabolic reactions are carbon dioxide and water. The release of biochemical free energy occurs through the hydrolysis reaction of ATP into ADP while the regeneration of ATP from ADP plus phosphoric acid restores the high-energy phosphate bond in the reverse reaction. On the other hand, if glucose is put in contact with oxygen, at a relatively high temperature, its oxidation results in the dispersion of the generated energy into the environment in the form of heat leading to an increase in the ambient temperature as is typical of combustion processes. This demonstrates that living cells cannot use thermal energy for fuelling their metabolism, since elevated temperature increases the rate of chemical reactions, which in turn leads to thermodynamic equilibrium. In fact, metabolic reactions provide the necessary conditions for living cells to maintain an out-of-equilibrium state. Therefore, the biochemical machinery of the cell must be constantly supplied with the free energy from various sources depending on the cell type, namely: 1) light, whose energy is harnessed through photosynthesis and trapped by plants, algae, and some bacteria and then used to generate carbohydrates plus oxygen in the following process:

$$6CO_2 + 6H_2O + \text{solar free energy} \rightarrow C_6H_{12}O_6 + 6O_2 \qquad (1.9)$$

and 2) chemical energy of some molecular bonds, particularly in the molecules of ATP, which are used in numerous subcellular processes such as the synthesis of nucleic acids, proteins, and sugars.

At a molecular level, the utilization of free energy increases the number of available states by facilitating the synthesis of new types of molecules and hence it contributes to the development and specialization of the system. While thermal energy mainly contributes to the kinetic, vibrational, and rotational motions of the molecules, the free energy instead affects the function of the system by breaking and making intramolecular bonds as well as increasing the number of available electronic states. Therefore, it can lead to an increased complexity of the system. This is all possible because of the presence of catalytic sites, which possess astonishing enzymatic properties so they can enhance the rate of some biochemical reactions by orders of magnitude. Metabolism would be impossible without enzymes and, as a consequence, life would not be able to be sustained. Enzymes behave as virtual ON/OFF switches, with efficient conversion to the production of specific compounds by promoting the activation of characteristic reaction networks. *Metabolism can be understood as a dynamic chemical engine that converts available raw materials into energy, as well as into the molecular building blocks needed to produce biological structures and sustain cellular functions. The utilization of the free energy produced by cellular metabolism is achieved by means of oxidation–reduction processes (redox), most of which occur in the mitochondria and can be aptly referred to as the power plants of the cell.* There, electrons are transferred from one molecule to another, which is associated with the changes in the oxidation states of the two interacting molecules. The electron donor is called a reductant and the electron acceptor an oxidant. These species, reductants and oxidants, work in pairs, and a sequence of oxidation and reduction reactions corresponds to the transfer of electrons along a chain of carriers similar to the flow of an

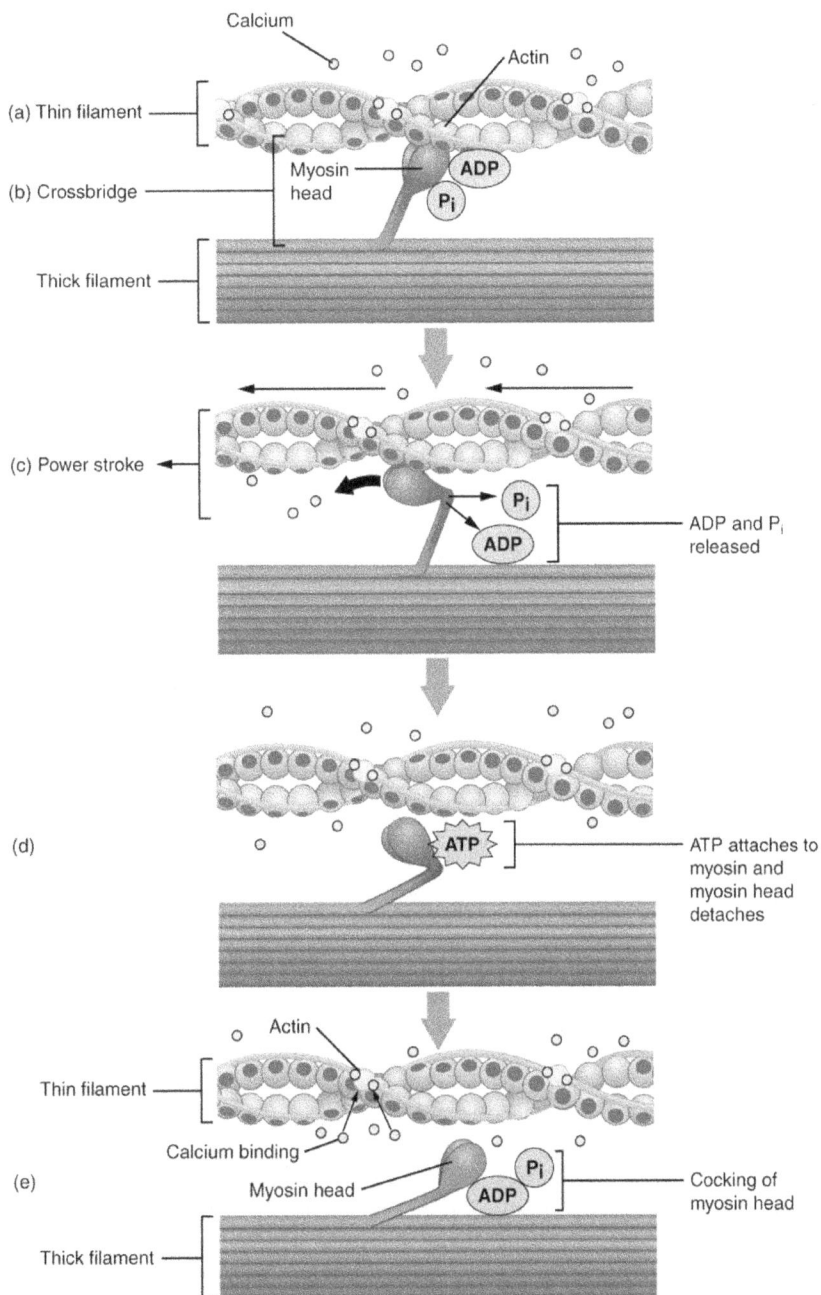

FIGURE 1.20 A schematic illustration of the actomyosin fibers providing contractile forces to living cells. Source: https://cnx.org/contents/FPtK1 zmh@8.25:fEI3C8Ot@10/Preface under Creative Commons license CC BY 4.0.

electrical current in a wire connected to a battery which provides a potential difference that drives the current.

Thus, metabolism is the sum of all biochemical reactions in a cell, carried out using ATP (or its analogue GTP) as the energy currency, coupling its hydrolysis to the synthesis or translocation of molecules in an endothermic process requiring energy. The sparkplugs in the combustion chamber of a gasoline engine provide the heat or activation energy of combustion much like the compressed air in diesel vapors of a diesel engine provides the activation energy from the heat generated by the collisions. Similarly, ATP provides the activation energy for many of the molecular motors of biological

engines. ATP is hydrolyzed to ADP and inorganic phosphate typically in the presence of a phosphatase. Alternatively, the rate-limiting enzyme of glycolysis, phosphofructokinase transfers a phosphate group from ATP to fructose-6-phosphate to yield fructose-1,6-biphosphate and ADP. Suitable molecular collisions are needed to occur between enzymes and their substrates, ATP, or other molecules to obtain the right orientation of the substrates such as ATP to release the energy harnessed in the substrates (*e.g.* facilitating the elimination of phosphate). Enzyme substrates may include ATP (*e.g.* ATPase) as the reactant or ATP may be supplemental to the reactant both of which require precise molecular collisions. This reduces activation barriers for the other components of the far-from-equilibrium biological engine. Notable is the importance of homeostatic body temperature and thermogenic mechanisms that maintain it (*i.e.* especially uncoupling ATP formation in the mitochondria instead of releasing heat) that provides optimal kinetic energy and collisions between enzymes and their substrates. It should not escape the attention of the reader to the fact that even ATP hydrolysis itself has an activation barrier that is reduced by its binding to kinase and phosphatase enzymes.

For any biochemical reaction to occur the reactants must collide and the colliding molecules must possess the correct amount of kinetic energy to overcome any repulsive interactions and to be oriented appropriately for the reaction to proceed. If the molecules are moving too fast, they may not have sufficient time to align themselves so that the reaction centers on each reactant are suitably aligned for the formation of the required chemical intermediates. On the other hand, if the molecules are very slow-moving, they may not find each other in the vicinity often enough for the reaction to proceed. Any forward movement in the reaction would be potentially accidental and thus unpredictable. Enzymes are relatively large proteins that bind reactant molecules and that stretch their bonds, reducing this activation energy hence making it more likely for the reaction to occur. Notably, however, as temperature increases, not only does the motion of substrate molecules increase and hence increasing collisions, but so does the molecular motion of enzymes that may reduce optimum enzyme performance. Hence, there is a temperature optimum for both enzymes and reactions to occur. In an aqueous solution, a dissolved molecule collides with surrounding water molecules billions of times per second. In this context, when a small molecule interacts with a protein (which is also solvated by a large number of water molecules), both these solutes need to be desolvated (at least temporarily) so that they can align suitably with respect to each other to form strong interactions such as hydrogen bonding, electrostatic attractions between charged centers, aromatic stacking, σ-hole effects, and non-specific hydrophobic packing. Such binding interactions would take place with the aid of energy released by ATP hydrolysis in its combustion chamber—the ATPase pocket binding ATP molecules. The bound enzyme-ligand complexes would potentially be resolved with rearrangement of water molecules that may participate in the biochemical reactions that result in downstream products formation. It should be noted that the energy input into such specific interactions could potentially lead to products that have much lower enthalpy and consequently release additional energy to drive further downstream reactions (as in the case of the TCA cycle reaction acetyl CoA and oxaloacetate → to citrate that concurrently provides the energy of the reaction catalyzed by malate dehydrogenase or be part of a feedback mechanism that signals the completion of a reaction cycle.

As discussed earlier in this section, whenever any system receives energy in the form of heat, part of such energy is converted into work done by the system on the surroundings, while another part may be utilized to alter the internal energy of the system and remainder may be dissipated to the surroundings in the form of entropy. The internal energy includes not only bond energies (expressed in the form of well-characterized enthalpies or heats of formation), but also the interactions of non-bonded energy between atoms that are connected to one another through three or more bonds. *Such atoms interact by exerting the so-called non-bonded forces, which include electrostatic components (e.g. ionic interactions between cationic and anionic centers [also termed "ion pairs"], electrostatic attraction between positively and negatively charged atoms, and electrostatic repulsion between like charges) and the weaker dispersion interactions (e.g. van der Waals forces, hydrophobic-aromatic interactions, π-stacking, σ-hole interactions between halogens and aromatic rings). The stronger ion-pair interactions are distinguished from the relatively weaker electrostatic interactions in terms of their different geometries of interactions as well as the strengths of charges carried by interacting atoms.* A typical ion pair commonly found in protein structures is between the guanidine group of an arginine residue and the carboxylate group of a glutamic acid residue (an amino acid residue is the basic unit of a protein structure). On the other hand, the hydrogen bonds that stabilize α-helix and β-sheets are good examples of electrostatic interactions, which positively reinforce protein structures. Non-bonded interactions have been modeled extensively to understand such phenomena as protein folding, binding of ligand to an enzyme, or binding of a control protein to a nucleic acid chain. Noncovalent interactions may also include an entropy contribution not useful as work. That is, electronic configuration changes of noncovalent interactions do not liberate enough free energy to drive reactions. *The adaptive value of noncovalent interactions may be rooted in for example quantum phenomena as may occur from induced dipoles in neuronal microtubules for example. More commonplace or less exciting examples of noncovalent interactions in biological systems which nonetheless have adaptive value include hydrogen bonding interactions, hydrophobic packing interactions. A noncovalent interaction may not liberate enough free energy to drive reactions, since the free energies associated with such interactions are typically of the order of 3 to 5 kcal·mol^{-1} and may be below the threshold of activation barriers needed to be crossed in bond formation reactions.* For example, nonpolar (more grease-like) molecules in aqueous solutions cannot participate in hydrogen bonding or ionic interactions with water molecules, and hence their interactions with water are not as favorable as are interactions between the water molecules themselves. Therefore, water molecules in contact with these nonpolar surfaces form "cages" around the nonpolar molecule, becoming more well-ordered (and, hence, lower in entropy) than water molecules free in solution. As two such nonpolar molecules come together, some of the water molecules are released, and so they can interact freely with bulk water and raising the entropy of the system.

The concept of quantum metabolism coined by Demetrius et al. [38] captures the fact that energy is utilized or released in terms of quantum packets of energy represented by *one molecule of ATP (there is no reaction that utilizes for example half an ATP molecule). Thus energy released in one event of metabolism of ATP is defined as the biological energy "quantum" and has a value of ~10^{-20} J.* On the other hand, the thermodynamic behavior of macroscopic systems is usually described in terms of state functions like temperature, volume, entropy, and pressure without the need to invoke quantum mechanical arguments. However, as is discussed elsewhere in this book, there are macroscopic quantum phenomena in physics, with the specific heat of solids giving a clue in the form of its dependence on temperature that cannot be explained in classical thermodynamics terms, but only through quantum statistical thermodynamics.

While the combustion of glucose into carbon dioxide and water is similar in the end products to the combustion of gasoline, there are considerable differences in the details of the pathways of the two classes of reactions. For example, the combustion of glucose is much more graded and complex. Just as combustion of gasoline is only useful to sustain pressure-volume work, combustion of glucose into carbon dioxide and water is meaningful only in the context of doing work needed to sustain biological activities in any organism. This would never happen by itself in any biological system due to the presence of activation barriers in many of the steps of the reaction. The metabolic combustion of glucose releases energy along the various reactions of the glycolysis and TCA pathways capturing the energy coming out of it in various molecules including NADH and $FADH_2$ that carry high energy electrons associated with hydrogen atoms that are passed off to the electron transport chain in a series of reactions that produce ATP in the process of oxidative phosphorylation (OxPhos). However, it should be noted that not all the energy generated in glucose combustion is transferred as useful energy to carry out "work" in the form of cellular metabolism. Some of the energy is released as heat to the surroundings as entropy, not usable to do any work, causing a lowering of efficiency of the biological engine.

FIGURE 1.21 My Porsche set ablaze by a fire caused by electrical problems. As the car accelerated toward its ultimate demise (death), the blaze released large amounts of unusable energy into the surroundings contributing to the overall increase in its entropy.

1.26 How Metabolism Is Linked to Aging

Since biological organisms (all living beings) survive in a thermodynamically open system, and because the second law of thermodynamics is defined by a closed or isolated system, localized entropy reduction within the boundaries of the organism is not only possible but represents the organized complexity that allows the capacity for life itself. The average healthy body temperature of a human being is 98.6° Fahrenheit (36.6° C) versus the typical ambient environmental temperature being approximately 70° Fahrenheit (21.1° C). Thermogenesis and isothermal body temperature, which are characteristic of all living systems, are driven by metabolic processes. Thermogenesis is the generation of heat within the body and the dissipation of heat within the body is responsible for the maintenance of isothermal body temperature. This heat generation and dissipation equate to entropy production. *Heat is defined as energy in transition, that when unable to be utilized for useful biological purposes and accordingly dissipates, this process equates to thermodynamic entropy.* Thermodynamic entropy in the sense of thermogenesis and isothermal physiologic body temperature is a complementary process to maintain organized complexity (for example it is the optimal temperature for enzymatic catalytic reactions) and life while at the same time maintaining fidelity to the second law via dissipation of heat from the body into the cool of the

surrounding environment. Acceleration of metabolic inefficiency with aging and the presence of disease hasten the entropy production rate back to the open system as an isolated system of the universe. Upon death, none of the heat present in the body is capable of being harnessed for biological work and this is entirely liberated to the surroundings to maintain the second law. Metaphorically, the high performance of organized biological complexity can be likened to an incredible high-powered piece of machinery, with deterioration the consequence of heat being lost as entropy to the environment.

Whatever mass of the body that is preserved stores energy, which potentially could be exploited as fossil fuel. Indeed, plants, animals, and human beings all share the element of hydrocarbon compounds that form the basis of life on our planet. An interesting example of resourceful adaptive utilization of metabolic heat to maintain life is the case of penguin colonies in cold Antarctica. The majestic birds huddle together into a tightly packed colony whenever a cold storm is prevalent. Interestingly, despite ambient temperatures being as low as –70° Fahrenheit (–56.7° C), the temperatures in the center of the huddle climb up to 100° Fahrenheit (37.8° C), no doubt caused by the release of heat from the metabolic needs to maintain their body temperatures. Consequently, the penguins are forced to exchange positions in a slow organized manner so that the birds in the perimeter can move in and those in the center can move to cooler areas.

Heat going into or coming out of a system is defined in terms of energy transfer that affects the molecular motion/kinetic energy in a substance. *The term thermal energy describes the total amount of heat stored in a substance by virtue of such molecular motion. It depends on the speeds of the constituent particles as well as the mass and number of particles. Conversely, temperature is a measure of the average motion of the particles of a substance.* Thus, the greater the number of particles in the system, the lower the temperature for a given amount of heat transferred. Thermal energy carried by a larger number of particles imparts a greater rise in temperature of the substance or system to which it is transferred (in comparison to heat carried by a fewer number of particles). The amount of heat transferred between two systems is directly proportional to the differences in their thermal energies and the flow of energy continues until a thermal equilibrium is reached. For example, if a block of ice were to be placed in a water bath, heat would flow from the water molecules to the molecules in ice whose melting would continue till thermal equilibrium is reached at which point the overall temperature of the water-ice mixture would be lowered, but all the ice would have converted to water. *The increase in internal energy of a system to which heat is transferred is at least partly put into increased linear translational, vibrational, and rotational motions, which result in an increase in temperature.*

In the classical regime of bioenergetic metabolism, an intrinsic property is energy conversion to heat that is lost from the system, unable to be used for useful purposes, i.e. ATP production or utilization to perform physiology, and often in excess compared to the requirements of physiological thermogenesis maintenance.

In the classical regime of metabolism, heat produced that is beyond physiological needs promotes inflammatory and redox stress.

Inflammatory and redox stress generated by metabolic reactions can be equated to entropy production rate (redox stress) with associated heat loss (inflammatory stress) of a physical system including the universe as a whole, thus tending a far-from-equilibrium biological system, a human organism, towards thermodynamic equilibrium.

Redox stress (and associated inflammation) is the most fundamental driver of an accelerated pace of aging and the chronic diseases of aging.

In a biological system, such as humans exercising, the production of reactive oxygen species at the mitochondrial level up-regulate uncoupling proteins that uncouple the electron transport chain proton pump from the production of ATP. Thus, instead of producing the higher energy triphosphate molecules for metabolic purposes that convert the transfer

of energy as work in muscles, the brain, and other systems, energy transfer occurs as heat. Heat transfer is utilized for physiologic thermogenesis and body temperature regulation.

However, when temperature rises to the limits of physiology with a corresponding increase in internal energy (and thermal energy) the body's protective mechanisms include the production and evaporation of sweat and to a lesser extent, by convection of heat out of the body via transfer of water vapor and other molecules to the exhaling lungs.

In cases where the body is unable to prevent excessive body temperature elevation, such as in elderly individuals with impaired thermal regulatory mechanisms, dehydration, and lack of air-conditioning during the hot summer, heatstroke may occur. It may also occur in young individuals who engage in strenuous physical activity for a prolonged period of time in a hot environment. The free radicals generated in the course of such activities could lead to thermal conditions where physiologically important protein molecules could denature and lose their natural functions causing downstream problems.

Fuel cells, particularly utilizing hydrogen as the fuel source may pose the most competition in the future to combustible engines for the automotive market. Essentially, it involves the burning of hydrogen (oxidation of hydrogen with oxygen) to produce water and electrical work. Engineering thermodynamics exploits the concept of energy conversion from one form to another. The goals of devices such as heat engines and fuel cells are to transform the energy contained in the chemistry of fossil fuels in the case of heat engines, or usually hydrogen in the case of fuel cells. In the case of thermal engines, the energy from the fossil fuel is converted into heat and ultimately mechanical work by a device such as a piston-cylinder or a turbine. In the case of fuel cells, the energy from the fuel source is converted into electricity and ultimately into work by an electrical circuit or an electromagnetic device such as a motor. Although the heat engine operates by the transfer of heat and the fuel cell by electricity, they both represent equilibrium devices (Figure 1.22). Whether the process directs towards thermal and chemical equilibrium (in the case of heat engines) or towards chemical equilibrium (in the case of fuel cells), once the state of equilibrium is reached no mechanical work can occur until the fuel source infuses more heat useful

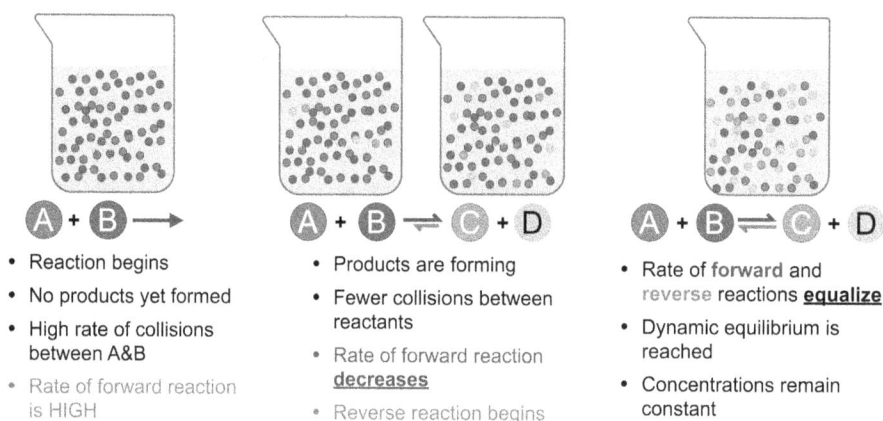

- Reaction begins
- No products yet formed
- High rate of collisions between A&B
- Rate of forward reaction is HIGH

- Products are forming
- Fewer collisions between reactants
- Rate of forward reaction **decreases**
- Reverse reaction begins

- Rate of **forward** and reverse reactions **equalize**
- Dynamic equilibrium is reached
- Concentrations remain constant

FIGURE 1.22 Molecular-level process leading to chemical equilibrium. Source: adapted from https://coolgyan.org/chemistry/energy-change-due-to-equilibrium/.

for producing the work. In the case of fuel cells, the energy source needs to be replaced.

1.27 The Ultimate Source of Life's Energy: Photosynthesis

The annual energy captured by photosynthesis in green plants accounts for about 4% of the total sunlight that reaches the Earth. Jeremy England, a physicist at MIT, recently provided a mathematical corroboration of the origins of life as dissipation-driven adaptations. This further validates the non-equilibrium thermodynamics characterizing biological organisms promoted by Prigogine. England describes "Darwinian evolution as a special case of a more general phenomenon" [39]. That is, the ability to harness the second law of thermodynamics is the most basic evolutionary adaptation promoting life forms.

Solar energy in the form of photons striking the surface of the Earth, when absorbed by chlorophyll molecules in plants, creates the potential energy of food mediated by the process of photosynthesis in plants (and in certain prokaryotic organisms such as photosensitive bacteria). The chemistry of fats, carbohydrates, and proteins represents the reconversion of this electromagnetic energy of light back into mass. However, we live in a world where we burn food to give rise to energy. The energy of food is harnessed as potential energy locked in the hydrocarbon bonds. Nature invoked the magic of quantum biology in the process of photosynthesis to generate the top of the food chain.

Furthermore, it enlisted quantum physics again to translate this mass, and the energy contained within it, to a form of energy the body can use to support its physiology, or in other words to maintain the living state. This extraordinary conversion of mass to energy is in the useful form of ATP molecules, the universal energy currency of biological systems, each of which contains a so-called quantum of biological energy, approximately equal to 10^{-20} J. *This multi-step energy conversion occurs in mitochondria ultimately through the process of oxidative phosphorylation. Hence, mitochondria may be considered a quantum mechanical energy transducer in eukaryotic cells.* Because these conversion processes occur cyclically with a characteristic turnover time constant, it can also be viewed as a biological quantum harmonic oscillator of sorts. Through quantum mechanical wave dynamics, mitochondria produce numerous quanta of ATP required for the living processes to occur. A huge number of these molecules or quanta of ATP on a macroscopic scale carry out the physiological requirements of maintaining life. In fact, the amount of ATP produced by a human organism every day is equal to the bodyweight although only 50 g of ATP is present in the human body at a given time. This phenomenon is termed quantum metabolism, a mathematical model developed and coined by Lloyd Demetrius [38] and promoted on a more conceptual clinical scale by Lloyd Demetrius and Jack Tuszynski [40, 41]. The mitochondria produce ATP along electron transport chains not by simply acting like a wire plugged into a battery or generator (a wall outlet) moving electrons along a chain, but rather by quantum mechanical wave dynamics. The priorities

for available useful energy (commonly known as Gibbs free energy) of the body primarily via the hydrolysis of ATP are to maintain homeostasis of redox and acid-base balance, which refers to the mechanisms the body uses to keep its fluids close to neutral pH in the range of 7.35 to 7.45 (that is, neither basic nor acidic) so that the body can function normally. These fundamental upstream metabolic requirements are necessary for maintaining the richness of complexity and homeostasis of the body's systems biology. The density of cellular mitochondria in our organs and tissues correlates to metabolic requirements. The heaviest lifting in terms of the metabolic demand that drives our physiology is carried out by the neurons in the brain. Accordingly, the brain represents only about 4% of body weight but is responsible for roughly 25% of the body's total energy production, i.e. 25 W. Given the amazing cognitive power of the brain, its power consumption is amazingly low, less than the power needed to run an electrical bulb on a night table. Neurons in general, both in the central nervous system and peripheral nervous systems, are comprised of cells with very high mitochondrial density. This is also the case with myocytes (skeletal muscle, smooth muscle, and cardiac muscle) and enterocytes. It may be noted that the numbers of mitochondria in various cells of the human body vary considerably from a few hundred to a few thousand. For example, liver cells can contain up to 2,000 mitochondria in a single cell. Mitochondria in the cells of the nervous system typically occupy 70% to 75% of the cell volume.

1.28 The Difference Between Quantum and Classical Metabolism May Be the Difference Between Health and Disease

Quantum effects are believed to be at the root of several underlying processes in biology that drive the development of living organisms. For example, the electron transport chain involves very rapid exchanges of electrons between various components of the mitochondrial membrane protein complexes and the only rational way to understand such mechanisms would be through the application of fundamental quantum phenomenon such as tunneling to rationalize the movement of electrons over distances of 20 to 30 Angstroms in very short periods of times. On the other hand, chaos theory is a field of study in mathematics that deals with the behavior of dynamical systems, which are very sensitive to initial conditions but are characterized by a lack of predictability. Chaos differs from randomness in that chaotic systems are characteristically deterministic; that is, they are entirely determined by a set of mathematical formulas and initial conditions, with no random elements involved. Fundamental to chaotic systems is their sensitivity to initial conditions; simulations of a chaotic system initiated at only slightly different states will quickly diverge so that predicting the state of one iteration of the chaotic system from a second is not possible. Chaos theory has been applied in cardiotocography (recording fetal heartbeat and uterine contractions). Also, models of lethal arrhythmias, epileptic seizures, warning signs of fetal hypoxia, have been obtained through chaotic modeling. The field is actively studying models to use for predicting lethal cardiac arrhythmias. Particularly intriguing are the same phase transition signatures shared by certain biological and physical systems, for example, the dripping faucet and cardiac conduction. This is useful since it is easier to manipulate a physical system to study the predictive signs of the dripping faucet transitioning into a chaotic pattern than it is to manipulate and study a heart's conduction. Importantly, there are many biological systems that display the fractal features of chaos theory, both spatial and temporal. Thus, chaos theory and quantum phenomenon form the root in informational energy or the energy of information required for understanding the three-dimensional organized structures in chemical biology. Such structures can change with time and affect biological function through chemistry. *Fundamental to both the adaptive strategies of chaotic and quantum phenomena are solutions attained unpredictably by the emergency of the systems as a whole by the interaction of their component parts.*

In light of the above discussion, *quantum randomness, which arises as a direct consequence of the fact that quantum theory computes probabilities for a particle to exist in a certain state, is recognized as a key driving force in the unpredictable nature of evolution of natural systems, which are deemed as a system whole.* There is a serious concern raised these days that technological singularity may supplant the need for much of the human workforce which has many uninvited consequences and potential unrecognized ones as well. Biomimetics, for example, an artificial pancreas, tries to mimic biological nature. On the other hand, the retrograde approach tries to understand biology through the proximity of technology, for example using quantum computing or information technology. Quantum computing may in fact be a particular aspect of quantum biology whereby out of the astronomical number of possible permutations and combinations of protein or DNA conformational states only the stable ones are rapidly selected by the use of parallel search algorithms afforded by quantum wavefunction spreading simultaneously into all available states. A classical sequential random search process would simply never work. The issue of quantum biology is an increasingly popular concept and its application to the understanding of mitochondrial metabolism and electron transport chain will be discussed later in this chapter. This approach rationalizes the optimum efficiency of metabolism through the ideas of wave function superposition, synchronous coherence, and optimum solutions by wave function collapse. It may be that the quantum phenomenon is the spark of life, the quantum ignition switch that needs to be implemented to achieve singularity. At the non-quantum, or classical, level it is unlikely that we will see anything as complex, difficult, and perhaps entirely unsolvable as living cells or human beings. Therefore, quantum biology may be considered part and parcel of singularity. In that sense, we may find that technology and biology may in fact up-regulate one another. Hence, irrespective of whether a quantum phenomenon represents the limits or boundaries of potential in terms of man and machine, the nontrivial nature and complexity of organismic biology underscores the inherent limitation of reductionist thinking and paradigms of clinical medicine for managing chronic disease. *Ultimately, understanding biological systems and bioenergetics from fundamental perspectives of classical and non-classical disciplines of physics, in my opinion, promises to sharpen the flexibility of clinical problem solving in healthcare.*

1.29 Thermodynamic Processes in Metabolism

What is a process? It can be envisaged to be a reaction in which energy is transformed between reactants on one hand and products on the other. Such an energy transformation may involve the transfer of heat out of the reaction or into the reaction. *Whenever energy transformations from one type to another allow dissipation of heat, the dissipated heat undergoes two possible outcomes: 1) it does not get taken into a coupled process of work (for example, in the electron transport chain production of ATP due to leakage of electrons causing superoxide formation and subsequent increase in the activities of uncoupling proteins that uncouple the energy from the proton pump engaged in ATP production); 2) it gets utilized as work to drive a coupled reaction (as would typically happen in reactions where the heats of formation of reaction products are lower than the heats of formation of reactants.* In general, the second law of thermodynamics provides guidelines for the direction of a biochemical process. In fact, most chemical reactions and processes transform the input flow of heat into internal energy changes (manifested as changes in enthalpy of the products of the reaction relative to the reactants) and work done on the surroundings (for example, to drive a coupled reaction). For example, the exothermic combustion of glucose to ATP, CO_2, and H_2O (products) results in a greater loss of heat (*i.e.* entropy) that is unusable to do work. The amount of heat used in the generation of high potential energy in the bonds of ATP is utilized to do the work done by the body (*e.g.* growth and organization, muscle movement during physical exercise, reproduction, immune system function). In other words, there is much more potential energy harnessed in the bonds of glucose ($C_6H_{12}O_6$) than there is in the bonds of carbon dioxide (CO_2) and water (H_2O) (both of which have lower heats of formation and thus contain lower potential energies), the difference in potential energy is either lost as heat or trapped in the bonds of ATP.

1.30 Two Paths to Metabolic Energy Production

Glucose molecules exiting glycolysis are converted from a six-carbon structure to pyruvate molecules (Figure 1.23) consisting of three carbon atoms that move from the cytoplasm into the mitochondria where pyruvate is converted to acetyl CoA (comprised of two carbons after liberating the third carbon as CO_2) before feeding into and powering the TCA cycle (Figure 1.24).

While still in the cytoplasm, the breakdown process of glycolysis of each glucose molecule into two pyruvate molecules creates a net of two ATP and two NADH molecules (Figure 1.23). The conversion of two pyruvate molecules into

GLYCOLYSIS

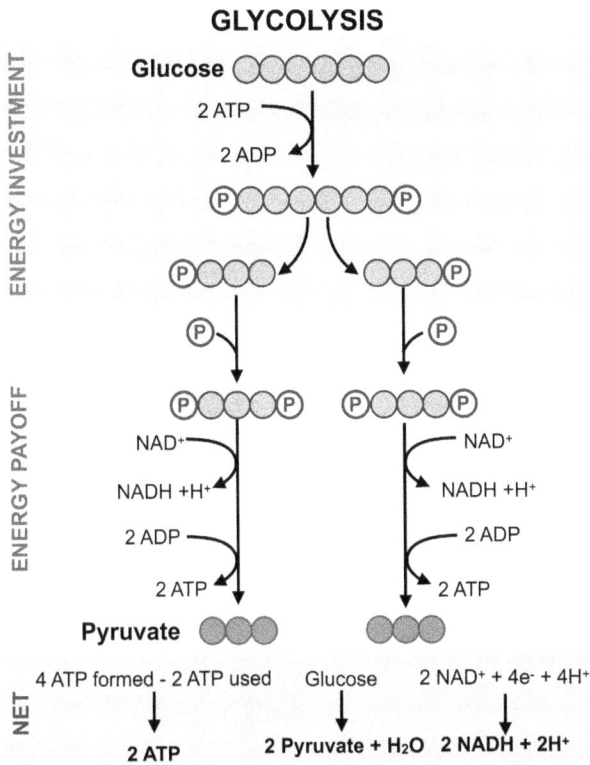

FIGURE 1.23 The glycolysis pathway. Source: adapted from references [18, 19].

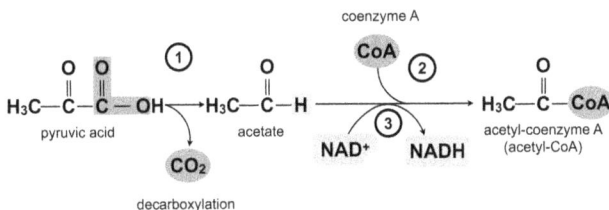

FIGURE 1.24 The generation of CoA prior to the entrance into the TCA cycle.

two molecules each of acetyl CoA, CO_2, and NADH prior to the entrance into the TCA cycle is catalyzed by the pyruvate dehydrogenase (PDH) enzyme complex (Figure 1.24). This critical step that transitions the metabolic process into the mitochondria, when impaired, such as in the process of insulin resistance, appears to be a crucial biochemical link to many chronic disease states including cancer. The gestalt of the oxidation of two acetyl CoA glucose-derived molecules as the starting point or initial substrates in the TCA cycle yields six molecules of NADH and two molecules each of FADH2, ATP, and CO_2 (Figure 1.25).

The FADH2 and NADH molecules store high-energy electrons that feed into the electron transport chain that produces the majority of the energy of glycemic fuel. This process is called oxidative phosphorylation because it requires oxygen as the final acceptor of electrons and protons (Figure 1.26). The exhaust products of the entire metabolic combustion of glucose are carbon dioxide and water, the same as in the oxidative combustion of gasoline and other fossil fuels (Figure 1.27).

Of the total 36 molecules of ATP produced in the combustion of glucose, the overwhelming majority are produced in the electron transport chain. In summary, each six-carbon glucose molecule is metabolized to produce the high chemical potential bonds of ATP that can be used to carry out the physiological work demands of the body, while carbon dioxide and water are exhaled and discharged as exhaust from the body in a manner analogous to the combustion of fossil fuels by steam engines (Figure 1.28).

The energy released from the combustion of nutrient fuel (glucose) is converted to bond enthalpy or bond energy that is potentially convertible to work but accounts for only about one-third of the combustion under normal circumstances, i.e. about two-thirds of the energy generated is liberated as heat. The vast majority of heat loss occurs along the electron transport chain. The energy not trapped as potential energy within the phosphate bonds of ATP is utilized for physiologic thermogenesis of the body. This is what keeps the body at its homeostatic temperature of 98.6° F (or 36.6° C). Under pathological conditions of mitochondrial dysfunction or excessive consumption of food (dietary energy) relative to mitochondrial capacity and energy utilization, heat is produced in excess of its physiological requirements for thermogenesis. This drives oxidative stress and inflammation. Similarly, only about 20% of the fossil fuel that is burned is harnessed in the pressure-volume work ($p\Delta V$) of the carbon dioxide and water–gas conversion from gasoline, while approximately 80% of the energy generated in the exothermic reaction is liberated as heat. Although the vapors of carbon dioxide and water are generated along the glycolysis, TCA, and electron transport chain pathways, the quantity of work required to "create room" for the expansion of the system is negligible, as the volume of the biological system where these reactions take place remains practically constant. Accordingly, there is no pressure-volume work of any significance in the process of biological combustion of energy. The two final products (CO_2 and H_2O) end up as waste products in the form of exhalation, sweat, urine, etc. When little or no pressure-volume work gets done, the change in enthalpy corresponds to either the heat added to the system or leaving it. That is, at constant atmospheric pressure, the typical conditions for biological reactions $p\Delta V = 0$ and so $\Delta H = \Delta U$. Therefore, the change in heat contained within a system is the change in the system's internal energy, which is the same quantity as the change in enthalpy (*i.e.* $\Delta H = \Delta U$). This quantity is the totality of potential energy and kinetic energy internal to the system. This occurs when reactions do not involve gases, the number of particles does not change or the change in volume (of particles and gas) is negligible. In contrast, the volume created by the combustion of gasoline in a combustion chamber that moves a cylinder piston is an example of pressure-volume work due to the production of heat that promotes the expansion of the volume of product particles of gas (discussed ahead). In either case, the transfer of energy as heat is in the form of variations in bond enthalpy, breaking old bonds and creating new ones. Nonetheless, there is a crucial distinction in the entropy of the universe, which must always increase with each process (which may be a reaction or a cycle of reactions as would happen in a cyclic engine), versus the entropy of a system, which may not always increase. A thermodynamic

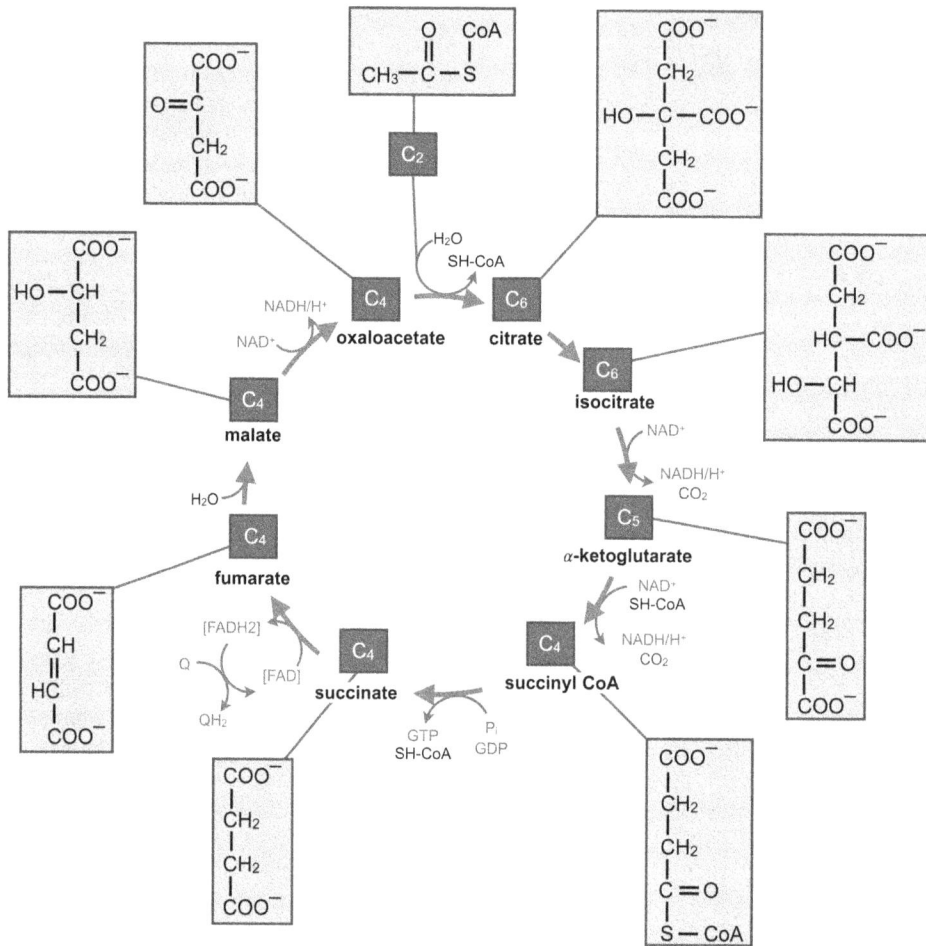

FIGURE 1.25 The TCA cycle: here the acetyl group from acetyl CoA is attached to a four-carbon oxaloacetate molecule to form a six-carbon citrate molecule. Through a series of steps, citrate is oxidized, releasing two carbon dioxide molecules for each acetyl group fed into the cycle. In the process, three NAD⁺ molecules are reduced to NADH, one FAD molecule is reduced to FADH2, and one ATP is produced (by substrate-level phosphorylation). Because the final product of the citric acid cycle is also the first reactant, the cycle runs continuously in the presence of sufficient reactants. Source: adapted from https://courses.lumenlearning.com/boundless-microbiology/chapter/the-citric-acid-krebs-cycle/.

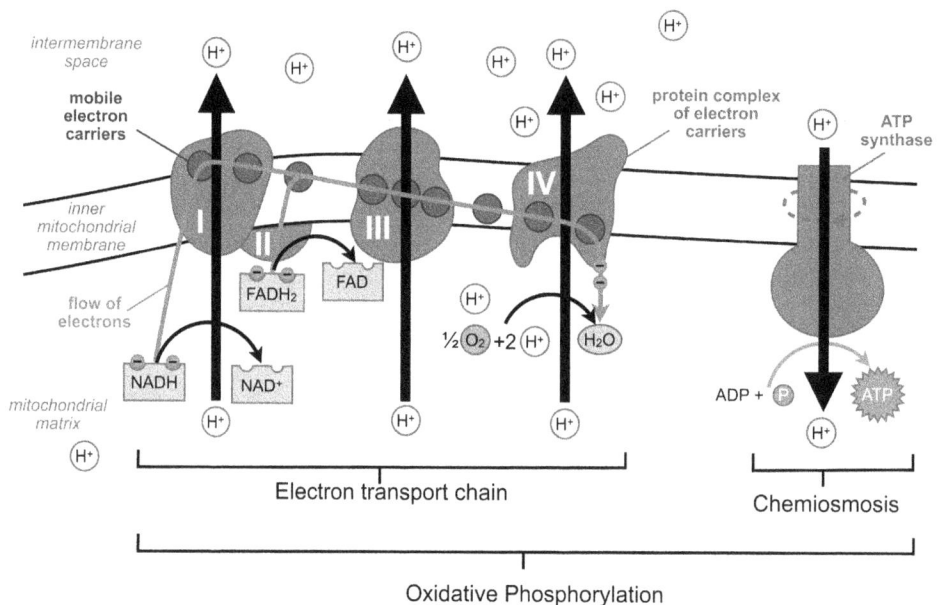

FIGURE 1.26 Schematic representation of oxidative phosphorylation. Source: adapted from https://alevelbiology.co.uk/notes/oxidative-phosphorylation/. See references 18, 19, 42, 43 for details.

$$2\ C_8H_{18}\ +\ 25\ O_2\ \rightarrow\ 18\ H_2O\ +\ 16\ CO_2\ +\ energy$$

$$octane\ +\ oxygen\ \rightarrow\ water\ +\ carbon\ dioxide\ +\ energy$$

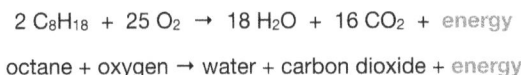

FIGURE 1.27 Summary reaction for the oxidative combustion of octane.

system is a quantity of matter of fixed identity around which a fixed or a movable boundary can be drawn. Everything outside the system boundary is termed as the "surroundings". The system and its surroundings together make up the universe. By corollary, when the universe is the system, it has no surroundings and is considered an isolated system. Work or heat can be transferred across the system boundary. In that sense, the system definition is very arbitrary and is guided by our interests and considerations. For example, a beaker of water with the boundary being the glass walls separating it from the rest of the universe could be defined as a system. Similarly, it may be a few molecules or atoms or may be defined as an individual human or other animal or organism. (Stated another way, a thermodynamic system may be anything that we want it to be for the purpose of study or intervention. It can be limited to an isolated reaction, to a biochemical cycle, or defined macroscopically such as an organ system, organism itself, or even broader such as an ecosystem or the entire universe.) *The energetic costs of producing organized structures such as well-developed organisms (e.g. human beings) within the surrounding universe are offset by an even greater amount of dissipation of heat into the surroundings in order to maintain fidelity to the second law of thermodynamics.*

1.31 Inflammation, Pathogenesis, and Obesity

The cardinal features of inflammation, calor, dolor, rubor, and tumor are manifestations of an acute localized infectious process designed to wall off, degrade and kill microbial invaders [44]. It is inextricably linked to the process of oxidative stress that affects the biological structures of proteins, lipids, and nucleic acids that contribute to the degradation of the microbes. *The chronic process of inflammation is much less fulminant than the classical acute process, but in contrast to being adaptive and physiological, it is maladaptive and pathophysiological. Moreover, it is fundamental to virtually all chronic diseases and accelerated senescence. Importantly, inflammation promotes*

oxidative stress, which also activates and promotes the inflammatory process in positive feedback or feed-forward fashion. The roots of systemic inflammation are typically the response to microbial pathogens in the gut. Allergen and toxicant exposures in the gut often facilitate disruption of the tight junctions between the lining epithelial cells that form the barrier between the GI tract lumen and the wall of the gut. A breach of this interface between the body and the environment is most relevant to the state of human diseases. Dysbiosis, or a pathogenic gut microbiome composition, appears to be the most predisposing factor to an overactive innate and adaptive immune system in the gut wall. It is also becoming evident to be an essential factor in the tracking of allergens in the GI tract permitting exposure of allergens (for example, gluten) to the lining epithelial cells and resultant cytotoxicity (discussed in further detail in a later section). Direct elaboration of toxins by microbes may also be responsible for cytotoxicity and immune system activation and inflammation. Secondary to this systemic auto-inflammatory process is the propagation of oxidative stress exerting its degradation effects on the three-dimensional organizational quality of biological structures and hence dynamic energy flow and associated biological function. The other major origin of chronic system smoldering inflammation in the body is rooted in oxidative stress as the primary process that occurs in mitochondria. Inflammation is secondary in this case and occurs via activation of the central transcription factor NF-kB. *Reactive oxygen species (free radicals) and oxidative stress in the region of the electron transport chain of mitochondria uncouples the production of ATP from the reversal of the electrochemical gradient across the inner mitochondrial membrane. This impairs the capacity to build energy-requiring bonds and nonbonded interactions of biological structures. Its constituent molecules are destabilized by free radicals resulting in a chain reaction in which electrons are extracted from neighboring molecules leading to the degradation of the membrane structure. This transition to a state of disorder that evolves from an ordered state involves the transfer of heat in an exothermic fashion and such energy cannot be used towards energy-requiring bond formation or other useful constructive purposes of work.* The process of inflammation results in heat in the form of elevated temperature as a subclinical more generalized form to that of the acute physiological process manifested as calor, rubor, dolor, and tumor; in either case, heat is ultimately dissipated to increase the entropy of the surroundings.

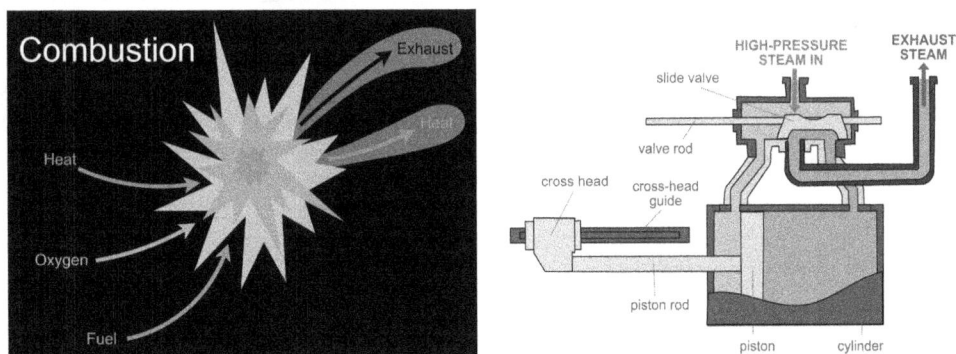

FIGURE 1.28 An illustration of the process of combustion and a combustion engine. Source: adapted from https://media1.shmoop.com/images/chemistry/chembook_stoich_graphik_22.png and https://cdn.hswstatic.com/gif/steam-labels-a.gif.

The immune response is also energetically highly expensive. Macrophages and other immune cells are richly comprised of mitochondria organelles. For example, macrophage response to lipopolysaccharide (LPS) with the production of protein cytokines engaging MAPK pathways is robustly energy consuming. A chronic stress response is often accompanied by gut epithelial intercellular tight junction disassembly allowing translocation of LPS into the portal circulation causing chronic or subacute endotoxemia underpinning a cascade of protein proinflammatory and psychogenic stress responses, which are of feed-forward and interlocking type. The inextricable processes of inflammation and oxidative stress ultimately promote the degradation of mitochondria and the transduction of the quantum mechanical oscillators. This leads to an impaired capacity to maintain cognitive function, cardiovascular health, reproductive function, muscle strength and musculoskeletal activity, immune function, and other aspects of good health. Consequently, *the incidence of metabolic disease rises including not only insulin resistance, metabolic syndrome, and type 2 diabetes but also autoimmune disorders, cardiovascular disease, neurodegenerative diseases, and cancers.*

The pathogenesis of disease is largely mediated promiscuously by the processes of oxidative stress and inflammation. These processes are inextricably linked and initiated by xenobiotics, impaired bio-detoxification of xenobiotics, or endogenously produced molecules such as hormones (notably estrogens), microbes, toxins, and allergens. It may be emphasized that the impairment of the organized complexity of the far-from-equilibrium state may (as described above) alternatively or concomitantly derive from the lack of sufficient building blocks, such as the appropriate balance of phytonutrients, minerals, and vitamins accompanying the consumption of dietary energy. This is particularly relevant in the context of diet and calorie-dominated modern lifestyles, which tend to emphasize dietary supplements to make up for the deficiency of vitamins, minerals, and microelements rather than obtaining them through normal food sources.

Prevalence of obesity is regarded to be one of the key factors in the so-called metabolic syndrome, which is a constellation of metabolic derangements including dyslipidemia and hypertension, involved in type 2 diabetes. Obesity is now considered a metabolic disease that strongly correlates to life-threatening conditions such as cancers and cardiovascular disease. The accumulation of adiposity, notably omental, mesenteric, and other visceral depots that derive from inherently inflammatory myeloid bone marrow progenitor cells, is deleterious to the growth of an organism. This should be distinguished as distinct from subcutaneous adipose tissue, which derives from noninflammatory mesenchymal cell precursor cells. Visceral adipose tissue has similar bio-toxification machinery (such as the cytochrome P450 oxidoreductase enzymes) and aryl hydrocarbon receptors as the liver. Interestingly, visceral adiposity is a modern evolutionary development coincident with the rise in dietary xenobiotics (John Kral, personal communication). This allows these adipose depots to protect the liver from xenobiotic toxicants originating from the gut. Furthermore, the expansion of these adipose depots (Figure 1.29) provides an adaptive capacity to store these lipophilic toxicants away from the rest of the body. Thus, there are multiple evolutionary-based mechanisms for the promotion of obese states

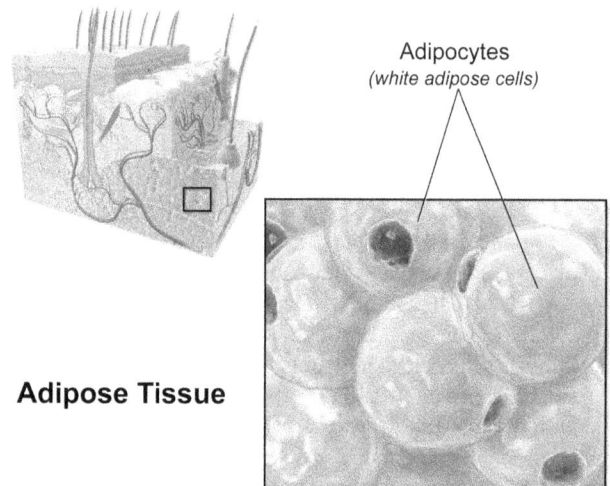

FIGURE 1.29 Schematic illustration of adipose tissue. Source: Blausen.com staff (2014). "Medical Gallery of Blausen Medical 2014". *WikiJournal of Medicine* 1(2). DOI:10.15347/wjm/2014.010. ISSN 2 002-4436.

including a) the need for greater soluble storage capacity of lipophilic toxicants; and b) for bio-detoxification of these toxicants. Another driver as described above is a poor-quality diet depleted of nutrients. *Progressive accumulation of adiposity resulting in adipocyte hypertrophy promotes hypoxia within the fat cells and a hypoxia-linked inflammatory cascade. Once adipocyte storage capacity is reached, in addition to the spillover of inflammatory adipokines (fat cell-derived hormones) and cytokines (molecular protein mediators) into the circulation, a surplus of circulating free fatty acids deposit ectopically in tissues throughout the body with characteristic pathogenicity. Such ectopic fat deposition is well characterized in the liver, skeletal muscle, pancreatic islet cells, brain, myocardium, and vasculature (see dedicated subchapter or elaborated annotation).* All this contributes to pernicious feed-forward decay in systemic health mediated by the interference of the protective adaptations explained by non-classical models of physics. Understanding systems from the perspective of parts and wholes recognizes the body's notion of a search for solutions to persevere the challenges of the environmental surroundings. It seeks to maintain a complex organized far-from-equilibrium state and to prevent the opposite state of thermodynamic equilibrium. Furthermore, understanding the complexity of interactions between the components of the systems seen as wholes appreciates the innate problem-solving abilities of living systems. This insight allows the extrapolation of the problem-solving clinical skills aimed at achieving effective preventative and interventional therapeutic designs.

1.32 Ecological Symbiosis of Plants and Animals

The energy emitted in the form of light (or electromagnetic radiation in general) provides the activation energy for photosynthesis occurring in plants with the conversion of carbon dioxide and water to glucose. This reaction is not spontaneous ($\Delta G > 0$) and requires energy input that is provided by the energy associated with light radiated from the Sun. The formation of glucose in this reaction is not spontaneous because the enthalpy of its formation is higher than the sum of enthalpy of formation of the reactants (CO_2 and water), *i.e.* it is an endothermic reaction that takes energy from the surroundings to create bonds in glucose. Although the glucose molecule is chemically stable at ambient conditions of temperature and pressure, its chemical potential is higher than that of its constituents. Furthermore, the process of photosynthesis occurs in a system, which achieves a higher order in its organizational structure when compared to the starting materials which start off in a relatively higher state of disorder—gases, liquids, and vapors. The plant takes in carbon dioxide from the air, water from the Earth as well as a small amount from water vapor in the air. From this disordered beginning, it produces the highly ordered and highly constrained sugar molecules, like glucose. The radiant energy from the Sun gets transferred to the bond energies of the carbons and the other atoms in the glucose molecule. In addition to making the sugars, the plants also release oxygen, which is essential for animal life. Thus, photosynthesis results in the lowering of entropy of the system. However, it should be noted both the endothermic nature of the reaction and lowering of entropy in the end products are achieved at the expense of energy from sunlight. In this context, while photosynthesis is not naturally spontaneous, it is driven to spontaneity by the infusion of external energy from sunlight in the open system.

The output of photosynthesis, viz. glucose and oxygen, becomes the input for cellular respiration in which glucose is combusted in the biological system to form carbon dioxide and water—photosynthesis in reverse. This combustion also generates the ultimate energy currency used for all life-sustaining work of the biological system including tissue growth, reproduction, maintaining homeostasis and an internal stable environment, and organizing, from the level of macromolecules and organelles to cells, organs, and organ systems as well as muscle activity and active transport systems. *The production of ATP by cellular respiration is an exothermic energy-releasing process. Cellular respiration involves three processes: glycolysis, the tricarboxylic acid cycle, and the electron transport chain.* Glycolysis makes two molecules of ATP as does the TCA cycle whereas the electron transport chain makes approximately 32 to 34 molecules of ATP. The electron transport chain (ETC) refers to a series of compounds that transfers electrons from electron donors to electron acceptors via redox reactions, and couples this electron transfer with the transfer of protons (H^+ ions) across the inner mitochondrial membrane [42, 43]. These transfers create an electrochemical proton gradient that drives ATP synthesis (Figure 1.30). The citric acid cycle in the mitochondrial matrix generates NADH (which feeds into complex 1 of the electron transport chain) and succinate molecules (with subsequent conversion to fumarate yields NADPH which feeds into complex 2 of the electron transport chain) are necessary for the production of the proton gradient.

The electrons in mitochondria are transferred to other protons and molecular oxygen that we breathe in to make the by-product of water. The protons actually come from the intermembrane space (each complex of the electron transport chain pumps protons outside to the intermembrane space creating a positive charge gradient), flowing through ATP synthase to combine with electrons and molecular oxygen, which

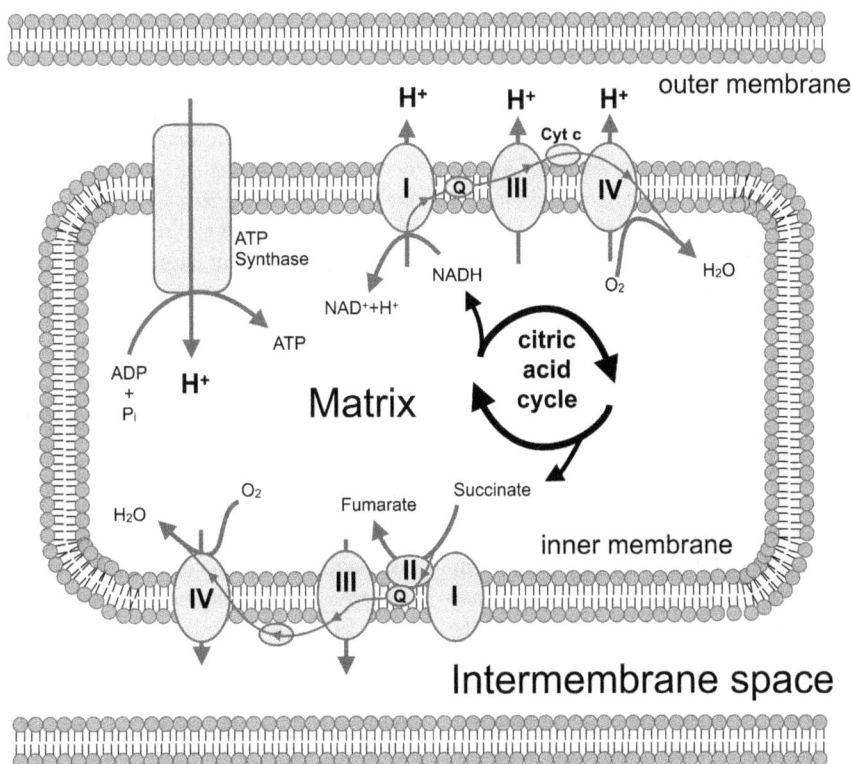

FIGURE 1.30 A schematic illustration of ATP synthesis in mitochondria. Source: adapted from https://en.wikipedia.org/wiki/Electron_transport_chain.

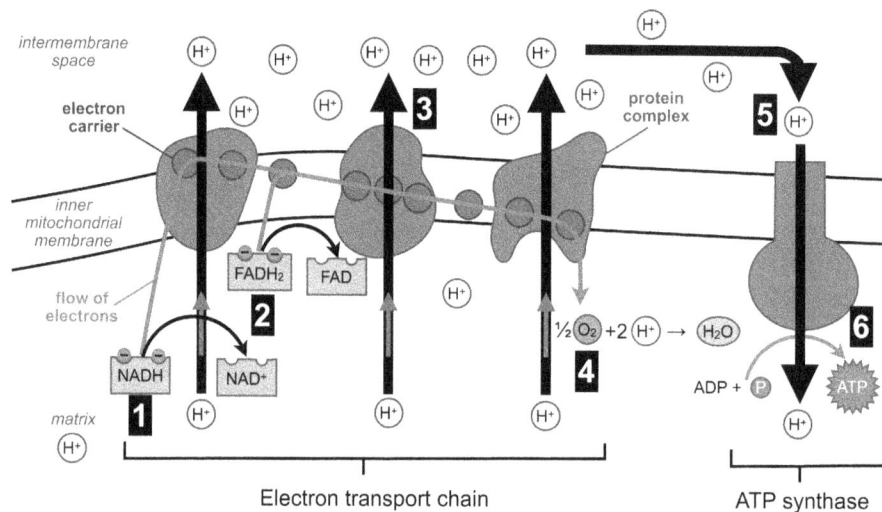

FIGURE 1.31 An illustration of the ATP synthase acting like a rotor. Source: adapted from https://microbenotes.com/electron-transport-chain-etc-components-and-steps/#electron-transport-chain-etc-components-and-steps.

is the final electron acceptor. As demonstrated in Figure 1.31 the protons from the intermembrane space can only move through ATP synthase, the site of ATP synthesis. It works similarly to an electrical engine's rotor. The electrochemical proton gradient allows ATP synthase (ATPase) to use the flow of H^+ through the enzyme back into the matrix to generate ATP from adenosine diphosphate (ADP) and inorganic phosphate. Every time a proton goes through, it switches on the enzyme, which attaches a phosphate group onto ADP to make ATP, in an endothermic process that stores energy in the added bond. This energy is released in a controlled fashion rather than as a ball of fire as would take place in a combustion chamber of a mechanical engine (which unlike biological engines carries out pressure-volume work). ATP in turn is converted to ADP in the process of performing the work duties of biology (organized, grow, reproduce, maintain homeostasis, and the work of muscle activity or active transport). *The conversion of ATP to ADP is also a spontaneous reaction with a negative Gibbs free energy value, hence it is an exothermic reaction that allows the necessary physiological functioning.*

The cascade of energy transfer from the Sun to plants (autotrophs) and subsequently to the heterotrophs, which feed on the plants is a functional design necessary for life. Any disruption in this energy balance will have an impact on the plants and the subsequent impact on the food web that relies on plant life. Easter Island represents an illustrative example of such a disruption caused by its inhabitants. This is an island off the coast of Chile which used to look like Hawaii with trees spread out prodigiously over the entire island. Easter Island Palm, a tree suitable for building homes and canoes, making ropes, and serving as fuel-wood, was an important mainstay for the inhabitants of the island, who were isolated from the rest of the world. However, excessive deforestation resulted in the extinction of the Easter Island Palm, which set off a chain reaction of events that led to the decimation of the population. Modern humans (Europeans) discovered in the early 18th century on the basis of pollen analyses, that once thriving people on Easter Island were virtually wiped out in large numbers because the paucity of palm trees led to fewer trips out to the sea in search of their main food source—porpoises. This implied that the land-based birds and animals were increasingly consumed by the inhabitants for their survival leading in turn to the increased desertification of the island creating an undesirable ecological imbalance.

1.33 Metabolic Dysfunction and Disease States

Generalizing our insights into the metabolic process beyond normal physiology, we can state that metabolic diseases disrupt the normal process of converting food to energy on a cellular level and affect critical biochemical reactions that process proteins, carbohydrates, and lipids. Thus, both healthy metabolism and metabolic disease are related to energy balance; accordingly, the system's biology of an organism calibrates energy-expensive processes to energy flow availability. For example, activation and recruitment of the immune system in the setting of inflammatory disease and oxidative stress, or the chronic stress response, is highly energy-consuming. Classic and common clinical presentations of these energy-consuming states include fatigue, poor social vigor, poor libido, exercise intolerance, mood swings, anxiety, depression, weight gain, and infertility. This gestalt of symptoms and findings relate to the central suppression of the gonadal axis, which is not surprising because reproduction is the most energy-requiring physiological process of the body. Evolutionary wisdom is designed to delay reproduction during periods of acute stress, however, the chronic stress response, both physical and psychogenic, are conditions for which we do not evolve. Acute stress is a pro-oxidative, proinflammatory, hypercoagulable, and insulin-resistant response. This so-called "fight or flight" physiologic response has a priority for immune function, wound healing, and delivery of nutrients and blood flow to the heart, brain, and muscles of the extremities. Ironically, the adaptive features of the acute stress response are the signature maladaptive features of type 2 diabetes, the central chronic metabolic disease state of modern times. It is characterized by the chronic manifestations of the clinical sequelae of the acute stress response including the chronic symptoms and findings of the hypogonadal state. *A fundamental pathogenesis of the chronic insulin-resistant state is the loss of mitochondrial structure and function. This promotes a feed-forward generation of free radicals and reactive oxygen species, inflammation, worsening mitochondrial impairment, and advancing insulin resistance.* There are many pernicious facets to this interplay of factors, for example, the inflammatory cytokines that promote exaggerated cognitive and emotional responses to perceived stress, heightened useless utilization of energy, and oxidative destruction of cell component structures such as lipid cell membranes, proteins, and DNA. In terms of tying in the theme of thermodynamics in this pathogenic process it is especially critical, in the opinion of this writer, to consume dietary surges of energy of high informational value (informational energy) to optimize the capacity to transfer that energy for useful biological purposes into the internal energy of our body's chemistry. Internal energy again may be for practical purposes used interchangeably with enthalpy and is widely considered a more intuitive term representing the entirety of the useful potential energy and kinetic energy of a system, which may be the human body as a whole. The goal physiologically is to maximize the change in Gibbs free energy. *The body is isothermal (maintaining a constant temperature) but the term TΔS is high due to the excessive wasteful energetic loss as heat with associated high change in entropy (ΔS). Inflammation, the fundamental underpinning to virtually all diseases if not all diseases, is by definition associated with excess heat.* It has protective value when localized to a site of injury—calor, dolor, rubor, and tumor are the cardinal features of this process. However, when the process is "subclinical", non-localized and chronic, it underlies the manifestations of any chronic disease state. Health or conversely frailty, senescence, and illness are mediated by the strength or paucity, respectively, of metabolic efficiency, plasticity, flexibility, and robustness of the pathways that make up the networks and systems of the body. On an evolutionary scale, these same strategies direct the gene pool of organisms and organism species.

1.34 Inflammation, Toxicity, and Reactive Oxygen Species

Vitamin and mineral cofactors for enzymatic activity lower the activation energy of reactions that promote organized biological anabolism through bond formation and the release of heat transferrable to free energy and subsequently to the work required for a healthy physiology. Such work includes active ion transport, motility of molecular structures, and mechanical muscle contraction. The inflammatory nature of phytonutrients (which are understood to be natural chemicals in plant food that help protect the plants from harmful agents such as bugs, germs, fungi) underscores their binding to antioxidant response elements of genes that promote the inextricable antioxidant and anti-inflammatory processes or responses of the body to these phytonutrients. The dynamic flow of energy along the pathways and networks of systems

biology with maximum complexity requires nutrient energy replete with vitamins, minerals, and phytonutrients. It is my conjecture that maximum information of chemical and biological structures corresponds to the highest level of complexity of interactions consequent to the effective and efficient transfer of free energy to perform biological work with limited heat loss. Reduced information then results in increased heat loss (entropy generation or entropy increase) from nutrient fuel to the work of forming molecular structures. *Oxidative stress reflects an imbalance between the systemic manifestation of reactive oxygen species and a biological system's ability to readily detoxify the reactive intermediates or to repair the resulting damage. Perturbation in the normal redox state of cells can lead to toxic effects through peroxides and free radical production that damages all components of the cell, including proteins, lipids, and DNA through denaturation and conformational changes. Oxidative stress from oxidative metabolism causes base damage, as well as strand breaks in DNA.* Base damage is mostly indirect and caused by reactive oxygen species (ROS) generated, *e.g.* O_2^- (superoxide radical), OH (hydroxyl radical), and H_2O_2 (hydrogen peroxide). Their oxidative damage is implicated in a number of diseases such as Asperger syndrome, ADHD, cancers, Parkinson's disease, Alzheimer's disease, atherosclerosis, heart failure, myocardial infarction, fragile X syndrome, Sickle Cell Disease, autism, infection, and chronic fatigue syndromes, autoimmune disorders, diabetes, and neurodegenerative diseases. Efficient functioning of antioxidant vitamins and phytonutrients helps prevent the formation of the reactive oxygen species which can potentially cause inflammation that is a high-energy requiring process disrupting normal biological structure and function over time. Hyperglycemia, even high levels of glucose within the normal range (hemoglobin A1c 5.4 to 5.8) leads to glycation of proteins and enhanced glycated end-products (AGEs) that bind to receptors of advanced glycated end products (RAGEs) that activate nuclear factor kappa beta (NF-kB) the hub of the inflammatory cascade.

1.35 What Can Einstein's Theories of Relativity Tell Us about Aging?

In the closing sections of this chapter, I wish to venture into somewhat more speculative ideas inspired by advanced physics, which may illuminate our deeper understanding of human physiology. Some of these profound concepts in physics are encapsulated by Einstein's theory of special and general relativity. An extraordinary aspect of the theory of special relativity that is worth exploring in the context of human physiology is the notion of time dilation/length contraction. As first explained by Einstein, as a physical object approaches the speed of light, the length of the object will appear foreshortened or contracted in the direction of motion, an effect referred to as Lorenz contraction. This contracted appearance is relative to stationary spatial dimensions. The length of the object is at a maximum in the reference frame in which the object is at rest. As the object approaches the speed of light the time measured in the moving frame will be running slow, or "dilated", relative to the time measured in the frame in which the object is at rest. Conversely, a moving object registers a different duration between any two time instances, which is called time dilation.

A highly intriguing although speculative application of this concept to human physiology is the relationship of biological age relative to chronologic age. Consequently, the more dilated or slower the time aging process, the more spatially unconstrained the system is. This also involves the notion of energy within the organism. As the energy flow within the organism/individual approaches its inherent limit analogous to the speed of light, time slows down and biological aging is reduced relative to chronological age. This idea can be viewed as a powerful metaphor since there are no important parts of the human body moving at relativistic speeds other than the core electrons of its atoms. However, there may be a deeper connection to quantum biological phenomena that make possible coherent, simultaneous, and virtually instantaneous orchestration of physiological behavior in a spatially unconstrained fashion. Such quantum biological behavior is a hallmark of the limits of optimal mental, emotional, and physical health and of mind-body potential. Conversely, as the aging process accelerates, the body's systems biology becomes spatially constrained and compartmentalized as a result of interruptions in the connectedness of energy flows. In this setting of accelerated aging, there exist roadblocks in the flow of energy and accordingly reduced complexity between pathways that comprise the body's networks of its systems biology. The outcome of these processes would be a reduction in the speed of energy propagation and an associated decrease in time dilation. Conversely, removal of such roadblocks would slow down the aging resulting in the biological age being lower than chronological age. Such a removal may be achieved for example through regular low to moderate intensity exercises such as yoga, slow jogging, brisk walking, or meditation, all of which can potentially contribute to stress reduction.

By contrast with special relativity, the theory of general relativity (proposed by Einstein a decade after the former) deals with space-time curvature whereby heavy objects such as the Sun cause space to bend creating a dent along which lighter objects such as the Earth, will orbit, altering the linear trajectory of the Earth. That is, in the absence of the Sun, the Earth would indeed have a linear trajectory. This gravitational deformation of the space-time continuum also causes light rays to bend compared to mass-free regions of space. Furthermore, not only is space curved, but space and time warp each other. Space-time as a composite four-dimensional quantity gives rise to what we perceive as gravitational force acting between two astronomical objects such as the Sun and the Earth. The sum of the component gravitational systems does not determine the whole because gravity begets further gravity, in a form of positive feedback. Hence, *both space-time as the field and the gravitational forces acting between two bodies (caused by space-time warping) are nonlinearly integrated.* While special relativity, which may be explained more simply by linear equations, has speculative musings which are potentially more relevant to biological systems, such as its effect on biological versus chronological aging disparity and hence promoting associated chronic diseases of aging, elucidation of general relativity theory turns out to be the first of the

nonlinear models of physics that evolved later in the century. Interestingly special theory of relativity has been successful combined with quantum mechanics giving rise to relativistic quantum mechanics and later relativistic quantum field theory where the Schrödinger equation is replaced by the Dirac equation. On the other hand, so far there has been no successful attempt at combining general theory of relativity with quantum mechanics or quantum field theory.

Einstein's theory of special relativity is important for readers to understand because quantum phenomena are by necessity associated with lower thermodynamic entropy or ideally 0 entropy (since heat collapses a superposition wave function). *In accordance with the second law of thermodynamics, entropy is the arrow of time. Thus, when entropy is reduced, such as when quantum biological processes supervene, time dilation occurs and the process of senescence slows. The biological processes of inflammation and oxidative stress are the sources of heat and hence, the drivers of entropy.* This conceptual notion is consistent with recent published studies showing a significant mismatch between chronological and biological age [45]. For example, at a chronological age of 70, an individual may have a biological age of 80 (accelerated aging) or 60 (slower-than-average aging). The theory of special relativity furthermore intersects with the notion of quantum metabolism as a biological quantum phenomenon and may also intersect with the notion of quantum electrodynamics if there is associated electromagnetism that explains the phenomenon. While Demetrius never explicitly demonstrated this in terms of data *per se*, this may be intuitively implied in his works.

1.36 Limitations of Scientific Reductionism and a Way out

The highly reductionist approach predominant in the traditional establishment medicine (which is increasingly compartmentalized into sub-specializations of clinical practice and foci of research) draws attention to its crucial inherent limitations. Although complex biological systems may at some levels be understood in terms of the laws of physics and chemistry, it is not the same as saying "we are nothing more than a blueprint that can be reduced to a top-down sum of its parts". This reasonable assumption is not equivalent to saying that biology is but a simple application of physics and chemistry. Accordingly, we are also not just a simple bottom-up blueprint representing the sum of our parts. Undoubtedly, we are greater than the sum of our parts. Unlike machines, such as a clock or a car engine, we cannot be easily reduced to our constituent parts such that a broken part can be replaced and we can be put back together. Unpredictability is an innate emergent quality of complex biological systems. It stems from the quantitative nature of the various progeny concepts of entropy such as Claude Shannon's information entropy and Lloyd Demetrius' evolutionary entropy [36], both central to the development and maintenance of organism species. Both forms of entropy imply an intricate, interconnected complexity whereby the system as a whole cannot be predicted from the individual parts. These biological systems evolve from their parts in a nonlinear fashion. Larger informational entropy of an ecosystem or a species implicates their greater biodiversity, robustness, and survival. *Evolutionary entropy, a measure of biological complexity, dynamically over time linearly correlates with adaptability and survival under certain ecological and demographic conditions (for example when resources are limited but constant). Similarly, entropy in the context of chaos theory ("bounded" or "constrained" chaos or entropy) is inherent to the notion of the search for solutions to biological (and physical) systems to survive in the context of perturbations from the environment.* Importantly, chaotic systems do not evolve in a predictable fashion due to the informational energy, which may have the quality of apparent randomness, hence the term information entropy. In the sense of informational energy, energy flows through the system providing the glue for the information bits, which constitute the work necessary to provide value, durability, or survival of the organism or system whole. In the remaining chapters of this book, we will delve deeply into the topics outlined here focusing on both the molecular machinery of life and its understanding based on modern physical concepts.

REFERENCES

1. Atkins, P. (2010). *The Laws of Thermodynamics. A Very Short Introduction.* Oxford University Press. Oxford.
2. Atkins, P., de Paula, J. and Keeler, J. (2018). *Atkins' Physical Chemistry*, 11th edition. Oxford University Press. Oxford.
3. Borgnakke, C. and Sonntag, R.E. (2013). *Fundamentals of Thermodynamics*, 8th edition. John Wiley and Sons, Inc. Hoboken, NJ.
4. Fischer, C.J. (2019). *The Energy of Physics, Part I Classical Mechanics and Thermodynamics*, 2nd edition. Cognella Academic Publishing. San Diego, CA.
5. Einstein, A. (1905). "Ist die Trägheit eines Körpers von seinem Energieinhalt abhängig?". *Annalen der Physik* 18:639–641.
6. Mamedov, B. A. and Esmer, M. Y. (2014). "On the Philosophical Nature of Einstein's Mass-Energy Equivalence Formula E=mc2". *Found Sci* 19:319–329. doi:10.1 007/s10699-013-9339-6.
7. For the Love of Physics. (2019, May 24). Can you Prove E=MC²?" [Video]. YouTube. https://www.youtube.com/w atch?v=VZEhmWFlrnM
8. Onicescu, O. and Stefanescu, V. (1979). "Elemente de statistica informationala cu aplicatii/ Elements of informational statistics with applications". *Editura Tehnica*, 11–31.
9. Rizescu, D. and Avram, V. (2014). "Using Onicescu's Informational Energy to Approximate Social Entropy". *Procedia Social and Behavioral Sciences* 114:377–381.
10. Boltzmann, L. (1974). The second law of thermodynamics. In *Theoretical physics and philosophical problems* (pp. 13–32). Springer, Dordrecht.
11. Harris, S. (2004). *An Introduction to the Theory of the Boltzmann Equation.* Dover Publications, Inc., Mineola, New York.
12. Jaynes, E.T. (1965). "Gibbs vs. Boltzmann Entropies". *Amer. J. Phys.* 33:391–398.
13. Gibbs, J.W. (1873). "A Method of Geometrical Representation of the Thermodynamic Properties of Substances by Means of Surfaces". *Trans. Connecticut Acad. Arts and Sci.* 2:382–404.
14. Feynman, R. P., Leighton, R. B., and Sands, M. (2011). *The Feynman lectures on physics, Vol. I: The new millennium edition: mainly mechanics, radiation, and heat* (Vol. 1). Basic books.
15. Logan, B. E. and Regan, J. M. (2006). "Electricity-producing bacterial communities in microbial fuel cells". *TRENDS in Microbiology*, *14*(12), 512–518.
16. Gharehpetian, G.B. and Mousavi Agah, S.M. (2017). "Distributed Generation Systems". (Chapter 5) *Fuel Cells*: 221–300.
17. Engine Efficiency. (2018, January 13). In *Wikipedia*. https://en.wikipedia.org/wiki/Engine_efficiency
18. Nelson, D.L. and Cox, M.M. (2008). *Lehninger Principles of Biochemistry*, 5th edition. W. H. Freeman and Company, New York.
19. Berg, J.M., Tymoczko, J.L., Gatto, Jr., G.J. and Stryer, L. (2015). *Biochemistry*, 8th edition. W. H. Freeman and Company, New York.
20. Landau, L.D. and Lifshitz, E.M. (1959). *Statistical Physics*. Pergamon Press, London.
21. Anderson, P.W. (1984). *Basic Notions of Condensed Matter Physics*. Benjamin/Cummings, Menlo Park, CA.
22. White, R.H. and Geballe, T.L. (1979). *Range Order in Solids*. Academic Press, New York.
23. Phillips, J.C. (1989). *Physics of High-Temperature Superconductors*. Academic Press, New York.
24. Stanley, H.E. (1972). *Introduction to Phase Transitions and Critical Phenomena*. Oxford University Press, Oxford.
25. Binder, K. (1987). "Theory of First-Order Phase Transitions". *Rep. Prog. Phys.* 50, 783.
26. Thompson, J.M.T. and Stewart, H.B. (1986). *Nonlinear Dynamics and Chaos*. John Wiley and Sons, New York.
27. de Alfaro, V. and Rasetti, M. (1978). "Structural Stability Theory and Phase Transition Models". *Fort. Phys.* 26, 143.
28. Wilson, K.G. *Rev. Mod. Phys.* 55, 583. 1983; *Sci. Am.* 241. 1979; *Phys. Rev. Lett.* 28, 248. 1972.
29. Ma, S.-K. (1976). *Modern Theory of Critical Phenomena*. Benjamin, New York.
30. Shannon, C.E. (1948). "A Mathematical Theory of Communication". *Bell System Technical Journal* 27(3): 379–423. doi:10.1002/j.1538-7305.1948.tb01338.x.
31. Avery, J. S. (2012). *Information theory and evolution*. World Scientific.
32. Landauer, R. (1961). "Irreversibility and heat generation in the computing process". *IBM J. Res. Develop.* 5:183–191.
33. Schrodinger, E. (1944). *"What is Life?"* Cambridge University Press, Cambridge United Kingdom.
34. Prigogine, I. (1978). "Time, structure and fluctuations". *Science* 201(4358):777–785.
35. England, J.L. (2015). "Dissipative adaptation in driven self-assembly". *Nature Nanotechnology* 10(11): 919–923. doi:10.1038/NNANO.2015.250. PMID 26530021
36. Demetrius, L., Legendre, S. and Harremöes, P. (2009). "Evolutionary Entropy: A Predictor of Body Size, Metabolic Rate and Maximal Life Span". *Bull Math Biol*. 71: 800–818.
37. Rosing, J. and Slater E.C. (1972). The Value of G Degrees for the Hydrolysis of ATP. *Biochim Biophys Acta*. 267(2):275–290. doi:10.1016/0005-2728(72)90116-8. PMID:4402900. (b) http://sandwalk.blogspot.in/2011/11/better-biochemistry-free-energy-of-atp.html

38. Demetrius, L.A., Coy, J.F. and Tuszynski, J.A.. (2010). "Cancer Proliferation and Therapy: The Warburg Effect and Quantum Metabolism". *Theor Biol Med Model.* 7:2–19.

39. Wolchover, N. (2014). A new physics theory of life. *Scientific American.*

40. Davies, P., Demetrius, L.A. and Tuszynski, J.A. (2012). "Implications of Quantum Metabolism and Natural Selection for the Origin of Cancer Cells and Tumor Progression". *AIP Adv.* 2(1):11101. doi:10.1063/1.3697850. Epub 2012 Mar 19.

41. Demetrius, L.A. and Tuszynski, J.A. (2010). "Quantum Metabolism Explains the Allometric Scaling of Metabolic Rates". *J. Roy. Soc. Interface* 7(44):507–514. doi:10.1098/rsif.2009.0310. Epub 2009 Sep 4.

42. Kracke, F., Vassilev, I. and Krömer, J.O. (2015). "Microbial Electron Transport and Energy Conservation—The Foundation for Optimizing Bioelectrochemical Systems". *Front Microbiol.* 6:575. doi:10.3389/fmicb.2015.00575. PMC 4463002. PMID 26124754.

43. Zorova, L.D., Popkov, V.A., Plotnikov, E.Y., Silachev, D.N., Pevzner, I.B., Jankauskas, S.S., et al. (2018). "Mitochondrial Membrane Potential". *Anal Biochem.* 552:50–59. doi:10.1016/j.ab.2017.07.009. PMC 5792320. PMID 28711444.

44. Freire, M.O. and Van Dyke, T.E. (2013). "Natural Resolution of Inflammation". *Periodontol.* 2000. 63(1):149–164. doi:10.1111/prd.12034.

45. Fedintsev, A., Kashtanova, D., Tkacheva, O., Strazhesko, I., Kudryavtseva, A., Baranova, A. and Moskalev, A. (2017). "Markers of Arterial Health Could Serve as Accurate Non-Invasive Predictors of Human Biological and Chronological Age". *Aging* 9(4):1280–1292. doi:10.18632/aging.101227.

2

Biological Engines and the Molecular Machinery of Life

Chapter Overview

Is the human body a machine? Life is all about movement at all levels of anatomical and cellular organization. How is mechanical work performed to move the various parts of living systems according to physiological demands? Where is the energy needed to carry out this work coming from?

Living systems use energy and generate mechanical force. Nutrition provides the fuel for all life processes. Cells can be viewed as information processing units but also as mechanical machines with many interlocking parts. Hence, physical forces and energy-transducing elements are important aspects of cell behavior. In this chapter, we reviewed the key physical forces and their origin. Importantly, life is sustained by molecules present in food and the energy stored in their chemical bonds. Once again, metabolism is the central aspect of the living systems seen as complex machinery driven by interlocking cyclical biochemical transformations. Structural stability of a cell is provided by tensegrity, which relies on maintaining tension, which in turn requires energy input from the cell's "power plants", the mitochondria. Equally importantly, both active and passive transport in and out of the cell maintains a balance of material fluxes and electric potential gradients across the many membranes in its compartments and around the cell itself. While mitochondria produce the quanta of biochemical energy in the form of ATP molecules, the cytoskeleton uses this energy as input to perform the work of cellular "bones and muscles", i.e. microtubules, actin filaments, and their complexes with other proteins such as kinesin and myosin, respectively. This molecular machinery of the cell includes both translational and rotational motors, some of which uncannily resemble man-made machines and even electrical engines. It is astounding that the efficiency of these biological machines surpasses that of man-made equivalents. Nature, through the eons of evolutionary refinement, achieved the level of a nanotechnology master. Although we now know a lot about biological motors, many mysteries remain unsolved. One of them is how cells, tissues, and organisms maintain functional coherence across all scales, facing constant thermal noise and various perturbations. Is it achieved by electromagnetic communication, quantum coherence, or something completely different that we still do not know? This spectacularly well-organized coherent whole is perhaps best illustrated in the case of the human brain, one of the most complex systems found in nature. However, when this exquisite functional integration is lost, pathological states ensue. In other words, we can summarize it by saying that coherence is a hallmark of health and decoherence correlates with disease states.

2.1 Living Systems Viewed as Machines

In this chapter, we address the issue of force production by living organisms, which requires energy generation and energy transduction into all the cells of the body transported by a complex network of blood vessels and coordinated by the central nervous system. These are very fundamental questions for understanding life processes at all scales. *Without metabolism and without the ability to move there is no sustainable life. This ability to generate force given by nutritional energy combined with intentional directionality of movement distinguishes living from non-living systems.* The latter, of course, can move, but need to be either directed or, when left to their own devices, simply follow external forces or move randomly. Living organisms have devised mechanical machinery at all scales from nanometers to meters that propels their motion in the direction they want to pursue either in search of food, to increase their mating chances, or to find more favorable living conditions. Here, we will explore the ingenious ways of producing and transducing energy by the machinery of life from an organismic level of muscles and bones, all the way down to the sub-cellular level of nano-scale motor proteins [1].

2.2 Physical Forces in a Biological Context

The four fundamental physical forces: gravitational, the weak and strong, and electromagnetic forces underlie the structure of and interactions within biological systems as they are all-present in physical matter that is the building block of nature [2]. Obviously, the weak and strong subatomic forces are not directly involved in biological processes as they pertain to the fine structure of matter in the universe, which applies to a much smaller spatial scale and a much shorter temporal scale of events. Gravitational forces are always present on Earth and we take them for granted because their strength is almost always the same. They provide our orientational anchoring with the surface of the Earth and the Earth's gravitational pull sometimes exerts its crushing effect towards the center of our planet. This central gravitational pull of the Earth oriented the humans strongly towards the planet they have inhabited for millions of years as homo sapiens. The electromagnetic forces, however, have a particularly significant but hardly recognized role in biology. We anticipate this role to be invoked by innovations in medicine and physiology in the coming years and decades. Particular attention should highlight three basic forms of electromagnetism in biological systems at the molecular level. These include electrostatic and electrodynamic interactions, and the Lorentz force coupling magnetic fields and moving charges. These forces are responsible for the active movements of ions and electrically charged molecules. It is the charged nature of the particles and molecules that constitute living matter, the magnitude of that charge, and the mass of the particles that determine the strength of the resultant electromagnetic forces.

Electrostatic interactions, generated by the Coulomb force, drive the motion of charged particles towards the opposite charge. Magnetic fields originate from either the result of the motion of charged particles (subatomic particles or ions, or combinations thereof, for example, pairs of these objects forming dynamic electric dipoles), according to Ampere's law, or from the unpaired electrons in atoms giving rise to net spin, which is a purely quantum phenomenon at the root of ferromagnetism. *According to the Lorentz force formula, magnetic fields exert a force and hence affect the movement of charged particles on which they act.* This may lead to the formation of ionic waves acting in part as counter-ions attracted to microtubules or actin microfilaments [3] as will be discussed in more detail later in this chapter. These ionic waves in turn may promote the movement of calcium within the cell. Calcium movement may further promote charged amino acids contained in protein neurotransmitters to be released into neuronal synaptic clefts, for example. Movement in biological systems at the biological level may also be generated by the Lorentz force due to the interaction between the magnetic fields of moving charged particles (representing analogs of electrical currents). An example of such an effect in biology may be the rotation of molecules resulting in the opening of ion channels into cells such as neurons which subsequently may drive cell depolarization and neuronal firing due to calcium influx. Alternatively, calcium influx is important for a host of effects such as activation of many enzymes. *In molecular level biology, motion may also be propelled directly by passive electrochemical gradients and indirectly by transfer of electrochemical bond potential energy from high-energy bonds,* such as between the second and third phosphate bonds in an ATP (Adenosine TriPhosphate) molecule as heat, or "energy in transit", in a process with a positive Gibbs free energy change. The best example of such transfer of "energy in transit" is from ATP hydrolysis to motor proteins or to the active transport of ion pumps. Again, more will be said about these effects later in the book.

Electrostatic or Coulomb interactions occur between so-called stationary charged particles, although the particles are not in actuality stationary, they are hypothetically "frozen in time". Their interaction forces are related but categorically different from the forces of magnetic fields acting on electric charges. The Lorentz force, in particular, relates to the interaction of magnetic fields on moving charges and produces a torque on a charge engaged in circular motion. In a nutshell, *electromagnetic forces are at play whenever electrostatic charges interact, which includes the formation of chemical bonds* because of the involvement of either valance electrons of the atoms forming a molecular bond, or protons as in hydrogen bonds, or permanent dipoles, or finally induced dipoles.

All these cases play prominent roles in biological structure formation. The geometrical distribution of these chemical bonds and their strength expressed as chemical affinity or the binding free energy determines key biological properties.

2.3 Force and Energy Generation at the Organismic Level

Various biochemical reactions within the body are responsible for storing, releasing, absorbing, and transferring the energy humans (and animals) need to move, breathe, pump blood, process nutrients, excrete waste products, etc. Those reactions, which require energy in the form of heat are called endothermic reactions while those that release heat are termed exothermic reactions. The energy source for endothermic reactions is contained in the food we eat. The energy required for muscle contraction is brought about through a network of chemical reactions within which there are two types (input and output ingredients). *Input ingredients come from the air we breathe and the food we eat. Food needs to be broken down into manageable components in order to be used and transformed into biological energy units.*

The two main molecules released as outputs in the breakdown of nutrients are carbon dioxide we exhale and water contained in urine and sweat we produce as by-products. Specifically, the lungs remove oxygen from inhaled air and transport it to muscle cells *via* hemoglobin in the bloodstream. The body digests food in the mouth, stomach, and gastrointestinal tract, processing some of it into glucose ($C_6H_{12}O_6$), part of which is transported to muscle cells where it is utilized to provide energy in the form of work performed by muscle contraction. In this process, which is an exothermic reaction, oxygen and glucose combine to form water, carbon dioxide, and output energy in the form of work performed denoted here as E_{out}. Thus, in terms of a chemical reaction, we can summarize it as

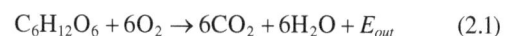

$$C_6H_{12}O_6 + 6O_2 \rightarrow 6CO_2 + 6H_2O + E_{out} \qquad (2.1)$$

Importantly for the timing of the metabolic processes, the body has the ability to store energy, mainly through the production of ATP molecules from glucose metabolism. Almost all metabolic energy of animals and humans comes from the conversion of ATP into ADP (adenosine diphosphate) and AMP (adenosine monophosphate). *The energy to convert ADP and AMP back to ATP is supplied by the oxidation of carbohydrates, fats, and proteins.* Therefore, these reactions can be seen as occurring cyclically, timed to correlate with the consumption of nutrients. In addition to the supply of nutrient molecules at regular intervals, metabolism requires a constant supply of oxygen, which can be used as a measure of the metabolic rate. Oxygen is supplied from the lungs to all cells of the body employing the cardiovascular system, which transports blood pumped by the heart at a variable rate depending on the type of activity pursued. For example, a person completely at rest consumes approximately 15L of oxygen an hour. The resultant energy production is correlated with the oxygen supply *via* the relationship: E = 2 × 10⁴ J/(L of oxygen supplied). It is straightforward to convert this relationship to one describing the metabolic power generated, which is approximately P = (15 L/h) × (2 × 10⁴ J/L) × (1h/3600s) = 83W. For the purpose of mental simplification, we can assume that the average resting power demand of the human body is comparable to that of an incandescent bulb with a 100 W power rating. *About 3/4 of the metabolic energy generated is converted to heat and the remainder can be used for physical activities such as walking and running and a fraction of this amount is utilized for mental activities.* Various organs consume different amounts of energy, but the human brain is characterized by the highest power consumption rate of all organs taking in almost 25% of the total power supply in spite of weighing only about 1.5 kg on average [4].

Food energy is the energy stored by chemical bonds and has traditionally been measured in food calories but in physics, different energy units, namely joules, are used as a measure of energy. *One food calorie is equivalent to 1,000 physics calories or 4,186 joules. To measure the energy content of food, the food is burnt in oxygen and the resultant heat produced is determined as the energy content. Assuming 100 W as an average metabolic rate for an adult translates into a dietary requirement of 2,600 cal a day.* During vigorous physical exercise such as cycling or running a sprint, the metabolic rate may increase five-fold to over 400 W or even ten-fold under extreme physical demand conditions [4].

Animal and human posture and motion are controlled by mechanical tension and compression forces generated by muscles. A muscle is composed of a large number of fibers, which can be contracted under direct stimulation by nerves. A muscle is connected to two bones across a joint and is attached by tendons at each end. The contraction of a muscle produces coupled action-reaction forces between each bone and the muscle. The molecular structure of muscles involves fiber-like bundles called fascicles, which are made up of very long cells, 0.4 mm in diameter and about 40 mm in length. The forces which the fibers can exert add up so that the total force generated is directly proportional to the number of fibers and hence the cross-sectional area of the muscle corresponds to the maximal force generated by it [4].

Later in this chapter, we discuss the molecular-level organization of force-generating molecules but here we simply state that muscle fibers come in two types of filaments (thin and thick). Thin filaments are called actin protein fibers and thick filaments are known as myosin protein fibers. Fibers are arranged in an alternating pattern of thick and thin filaments, which slide over each other. *When we contract our muscles, motor proteins such as myosin form molecular cross-bridges, which pull the thin filaments over the thick ones.* When the muscles are extended filaments do not overlap. Motor proteins are fixed to myosin fibers and move over the actin fibers, similarly to a collapsing telescope. The muscle force generated is larger the more activated motor proteins there are.

In order for the muscles to work, energy-giving molecules of ATP must be provided in sufficient quantities at a sufficient rate and oxygen must be supplied through the cardiovascular system powered by the pumping action of the heart that generates blood pressure. Blood pressure is controlled by osmoregulation, which is extremely important because eukaryotic cells cannot withstand large osmotic pressure differences. *Blood contains water, glucose, electrolytes, gases, proteins, red and white blood cells, as well as waste materials such as urea.* The blood has a high concentration of proteins compared with the surrounding tissues. Furthermore, blood capillaries are permeable to small molecules such as water, oxygen, carbon dioxide, and salt but not to proteins. The capillaries are pressurized by the blood pressure with respect to the tissue and the average blood pressure is close to 23 mm Hg. It turns out that the osmotic pressure difference is neatly offset by the hydrostatic pressure difference, but this delicate balance may be easily upset by such situations as high blood pressure or starvation. *High blood pressure can lead to the accumulation of fluids in tissues in a condition known as edema. When we either drink too much water or consume too much salt we may be in danger of upsetting our delicate osmotic balance, which can be prevented through a regulatory mechanism provided by ADH (an antidiuretic hormone).* This hormone rapidly alters the water permeability of tubes connected to the urinary duct in response to changes in salt concentration. In humans and other animals, the interstitial fluid is regulated by the exchange of substances with blood in capillaries. Many substances are moved across capillary walls, but water is more concentrated in the interstitial fluid than in the blood. Reverse osmosis occurs near the entrance of the capillary. Blood pressure near the capillary's exit is less than the relative osmotic pressure, and osmosis carries water from the interstitial region into the capillary. The overall result is an exchange of water while the total amount of water in the interstitial region remains constant.

As mentioned above, most of the physiological processes, such as muscle contraction or blood pressure maintenance, are controlled by the brain even without us thinking about them. *The human brain is arguably the most complex structure in the known universe, and it underwent a transformation over millions of years of evolution of life on Earth with some of the structures that have appeared relatively recently.* For example, only 100,000 years ago, the ancestors of modern man had a brain weighing roughly a third of the weight of present-day humans. Most of this increased weight is associated with the cortex, which consists of the two roughly symmetrical, corrugated, and folded hemispheres, which sit astride the central core. Almost all the complex tasks that can be performed by humans are associated with signal and information inputs followed by processing in the brain's cortex region. In fact, the human brain consists of three main parts. The first part called the brain stem consists of structures such as the medulla (controlling breathing, heart rate, and digestion) and the cerebellum (coordinating senses and muscle movement). The second segment is called the midbrain and it contains structures that link the lower brain stem to the thalamus (relaying information) and to the hypothalamus (regulating drives and actions). *Hypothalamus is part of the limbic system, which lies above the brain stem and under the cortex and consists of a number of interconnected structures, which are associated with the regulation of hormones, drives, temperature control, emotion, and one specific part, the hippocampus to memory formation* [5]. The nerve cells or neurons affecting heart rate and respiration are concentrated in the hypothalamus and direct most of the physiological changes that accompany emotional responses. *Aggressive behavior is linked to the action of the amygdala, which lies next to the hippocampus.* The latter plays a crucial role in processing information eventually being consolidated into long-term memory. Damage to the hippocampus can result in the inability to generate new stores of information (global retrograde amnesia). *Most parts of the lower and mid-brain are relatively simple systems designed to register experiences and regulate behavior that does not involve conscious awareness such as breathing.* Finally, the forebrain has evolved in humans into the walnut-like configuration of left and right hemispheres whose highly convoluted and complicated surface of the hemispheres, the cortex, is about 2 mm thick and has a total surface area of about 1.5 m². *Most of the complex functions performed by the human mind are performed in the cortex.* Some of its regions are highly specialized such as the occipital lobes located near the rear of the brain, which are associated with the visual system while the motor cortex helps coordinate all voluntary muscle movements [6].

Life is the ultimate example of a complex dynamical system or, in fact, a hierarchically organized set of dynamical systems. Above, we have briefly discussed the muscles that produce mechanical forces, the cardiovascular system that distributes oxygen and nutrients, and the brain that controls physiological functions and processes information inputs into commands giving rise to actions. *A living organism operates through a sequence of interlocking transformations involving an immense number of components, which are themselves made up of molecular subsystems. Yet when they are combined into a larger functioning unit (e.g. a cell), then so-called emergent properties arise.* For the past several decades, biologists have greatly advanced the understanding of how living systems work by focusing on the structure and function of constituent molecules such as DNA (deoxyribonucleic acid). Understanding what the parts of a complex machine are made of, however, does not explain how the whole system works. Nonetheless, all the components of the living system are maintained at a far-from-thermodynamic equilibrium state and this state requires a sustained input of energy in the form of nutrients whose biochemical bonds are transformed through metabolic reactions into simpler structures, which are easy to process.

2.4 Cell Energetics: The Cell as a Machine

In order to function, every machine requires specific parts such as screws, springs, cams, gears, and pulleys, which need to be interconnected in an intelligent fashion in order to perform the desired function. In addition, a steady supply of energy must be provided to convert it, with some level of efficiency, into useful work. Likewise, all biological cells, like machines, must have many well-engineered parts to work. Examples include units called organs such as the liver, kidney, and heart. These complex life units are made from still smaller parts called cells (Figure 2.1), which in turn are constructed from yet smaller machines known as organelles. Cell organelles include mitochondria (Figure 2.2a), Golgi complexes (Figure 2.2b), endoplasmic reticulum (Figure 2.2c), and the protein filaments of the cytoskeleton, in particular, microtubules (Figure 2.2d) and actin filaments. Even below this level of organization, there are machine-like parts of the cell, such as motor proteins and enzymes, that perform specific functions involving energy input and power output (*e.g.* transport).

A critically important molecule is ATP that is a complex nanomachine, which serves as the primary energy currency of the cell. ATP is used to build complex molecules, provide energy for nearly all living processes, such that it powers virtually every activity of the cell. Nutrients contain numerous low-energy covalent bonds but unfortunately, these are

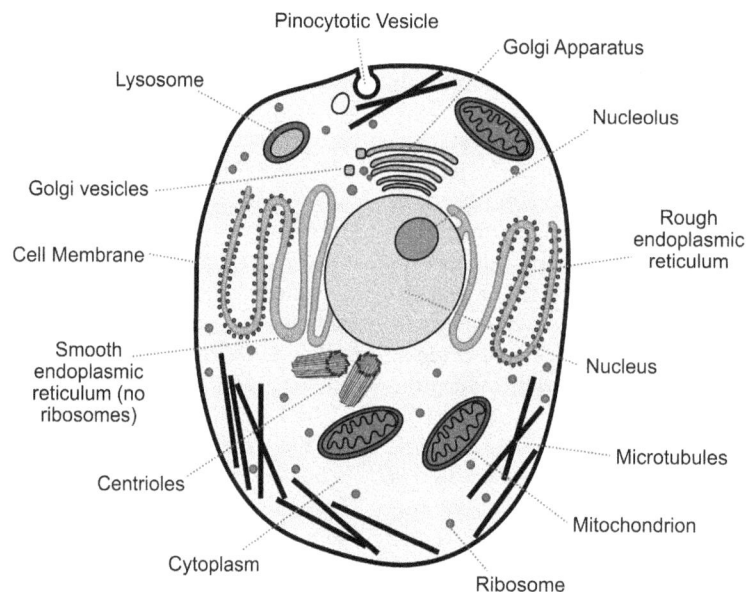

FIGURE 2.1 Representative illustration of a eukaryotic cell. Source: adapted from https://www.biocompare.com/Bench-Tips/361744-New-Method-Transforms-Cell-Fractionation/.

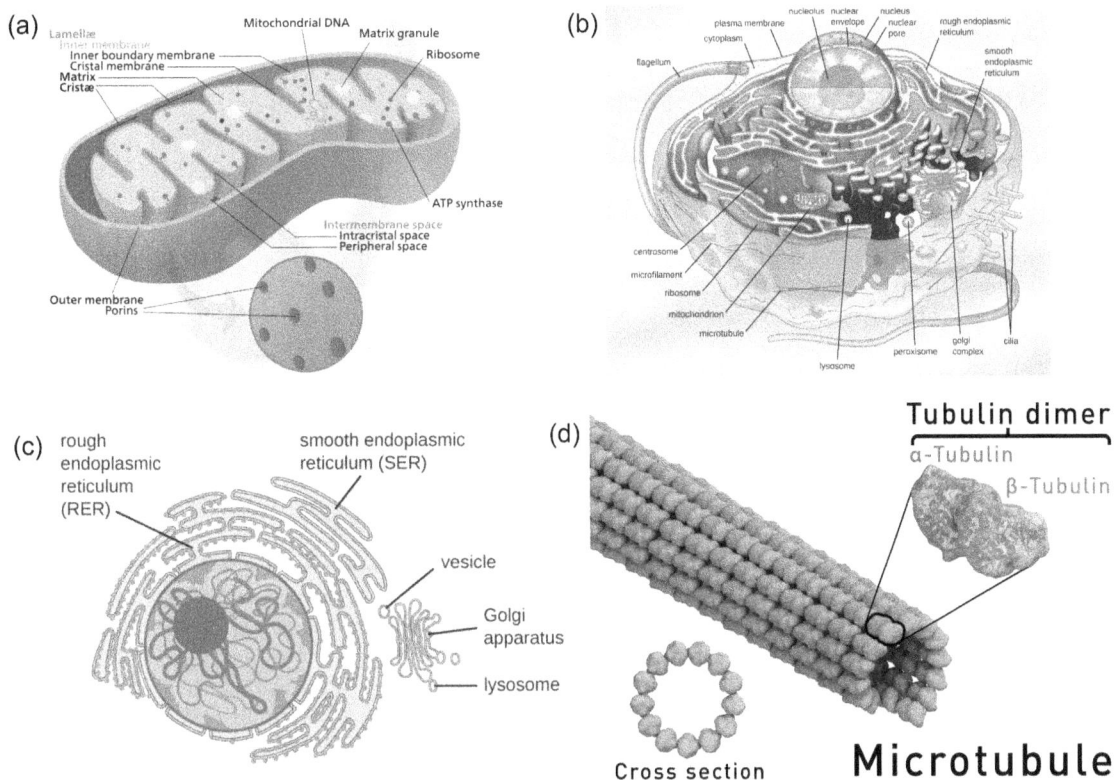

FIGURE 2.2 Schematic illustrations of various cellular organelles: a) mitochondrion, b) Golgi complex, c) endoplasmic reticulum, and d) microtubule. Sources: a) by Kelvinsong, modified by Sowlos, used under Creative Commons license, https://creativecommons.org/licenses/by-sa/3.0/legalcode; b) and c) https://cnx.org/contents/5CvTdmJL@4.4 under Creative Commons license and https://creativecommons.org/licenses/by/4.0/legalcode; d) by Thomas Splettstoesser (www.scistyle.com), used under Creative Commons license, https://creativecommons.org/licenses/by-sa/4.0/legalcode.

not very useful to do most types of work in the cell. Thus, low energy bonds must be translated into high-energy bonds using ATP energy by removing one of the phosphate-oxygen groups and turning ATP into ADP [7]. Subsequently, ADP is usually immediately recycled in the mitochondria where it is recharged and re-emerges again as ATP. At any instant, each cell contains about one billion ATP molecules. Because the amount of energy released in ATP hydrolysis is very close to that needed by most biological reactions, little energy is wasted in the process. A human being processes an enormous amount of ATP molecules, equivalent in weight to the total weight of the human body.

The geometry of mitochondrial energy "generator" is quite complex as one needs proton pumps with a pH gradient across the mitochondrial wall, an electron transport chain taking place along the length of the wall, and ATPases which pump out ATP molecules into the cytoplasm, so there are both gradients of concentrations into and out of the mitochondria as well as electron motion taking place (at least partly by tunneling) along the wall. This can be viewed as an electric circuit and there should be some magnetic fields generated in the process but probably very small. On the other hand, in glycolysis, the associated chain of enzymatic reactions is distributed throughout the cytoplasm and there is no defined electric circuitry associated with it [8].

The mitochondrial way of producing energy is not only very efficient but well engineered and requires a well-constructed structure, unlike glycolysis where basically all the reagents are "thrown together" into a bag (cytoplasm) and reacting due to their affinities in a diffusion-limited way without any synchronization. Diffusion by definition is subject to environmental noise (Brownian motion) and hence does not require fine-tuning in contrast to oxidative phosphorylation (OxPhos) in mitochondria. It is "noisy" and inefficient. On the other hand, OxPhos is efficient and smooth but could be disrupted by chemical, mechanical, or radiation damage. This is like comparing transport by an army tank to an electric car, say "volt". The tank will get you there no matter what but it will guzzle a lot of fuel. The "volt" will get you faster, cheaper, and in greater comfort but may get stuck in the mud. Currently, there are some experimental cancer therapies under development based on the concept of upregulating OxPhos and down-regulating glycolysis by using various pharmacological agents, e.g. DCA (Dichloroacetate) and 3-BP (3-Bromopyruvate), respectively. Some researchers believe that the *root cause of at least some cancers is metabolic dysregulation away from OxPhos and toward glycolysis, which then leads to genomic instability.* This has been referred to in the oncology literature as the Warburg effect [9].

2.5 Cells' Tensional Integrity: Tensegrity

As mentioned above, each cell's spatial integrity is delineated by its membrane, which allows for the selective passage of material through ion channels and ion pumps. The latter allow for active transport and utilize metabolic energy products to accomplish concentration gradients, which would otherwise disappear over a short period of time due to the inevitable entropy increase governed by the second law of thermodynamics. To understand fully the way living systems form and function beyond their "territorial" integrity (including the mechanism of cell division), we need to discuss the basic principles that guide biological organization and stability. In this connection, it is rather remarkable that many natural systems are constructed using a common form of architecture known as tensegrity [10], a term that refers to a system that stabilizes itself mechanically because of the way in which tensional and compressive forces are distributed and balanced within itself. A mechanical model of cells in the living creatures can be represented by tensegrity structures (Figure 2.3). The tensegrity architecture depends on tensional prestress that is a critical governor of cell mechanics and function, and consequently, the use of tensegrity by cells contributes to their mechanotransduction. *Tensegrity structures are mechanically stable as a result of the way the entire structure distributes and balances mechanical forces and torques and not because of the strength of individual members. Since tension is continuously transmitted across all structural members, a global increase in tension is balanced by an increase in compression within members of this mechanical structure distributed throughout the cell.*

Although the various filaments of the cytoskeleton are surrounded by membranes and penetrated by viscous fluid of the cytoplasm, it is this mechanically stable but flexible network of molecular struts and cables that stabilize cell shape in a way similar to pitching a tent. *The microtubules represent compressed, rigid elements within the network while the actin filaments are tensile elements. Intermediate filaments add to the complexity of this elaborate meshwork of dynamic structures that can be reorganized thereby allowing the cell to move or even divide while maintaining inherent stability.* Moreover, pulling on receptors localized at the cell membrane's surface produces immediate structural changes deep inside the cell including the DNA contained in the nucleus. Thus, cells and nuclei are far from behaving simply like containers filled with a viscous cytoplasmic soup. The existence of a tensegrity force balance provides a means to integrating mechanics and biochemistry at the molecular level. The concept of tensegrity explains why the structure of the cell's cytoskeleton can be changed by altering the balance of physical forces transmitted across the cell surface without compromising the integrity of the entire cell. This is important in view of the fact that many of the enzymes that control protein synthesis, energy conversion and growth in the cell are often bound by the cytoskeleton. Therefore, there is a dynamical interplay between the cytoskeletal geometry and the kinetics of biochemical reactions including such crucial processes as gene activation. In fact, it is entirely possible that by simply modifying their morphologies, cells may be able to switch between different genetic programs executing different functional roles. For example, cells that spread flat on a surface are known to be more likely to divide, whereas round cells that do not easily adhere to a surface have a tendency to activate a death program called apoptosis. *When cells are neither too spatially extended nor too retracted, they neither tend to divide or die but, instead, differentiate themselves in a tissue-specific manner. This specific mechanism of differentiation is, however, likely to be a function of external mechanical forces as well as other factors such as electrical field gradients, ionic concentrations, and pH, among other aspects.* This is still an active area of research and undoubtedly much more will be found about these processes in the coming decades. However, it is already safe to say that mechanical restructuring of the cell and its cytoskeleton informs the cell whether to grow, divide, perform other functions, or die by transmitting appropriate signals. Hence, mechanical forces are transmitted over specific molecular pathways in living cells. Because a local force can change the shape of an entire tensegrity structure, the binding of a molecule to a protein can cause the different, stiffened helical regions to rearrange their relative positions throughout the length of the protein. It is not a stretch of the imagination to conclude that *cancer cells are likely to have different mechanical properties, as do the surrounding tissues. Whether this can be exploited in therapeutic applications is too early to tell but in view of limited progress in improving standard therapeutic modalities, this is worth pursuing.*

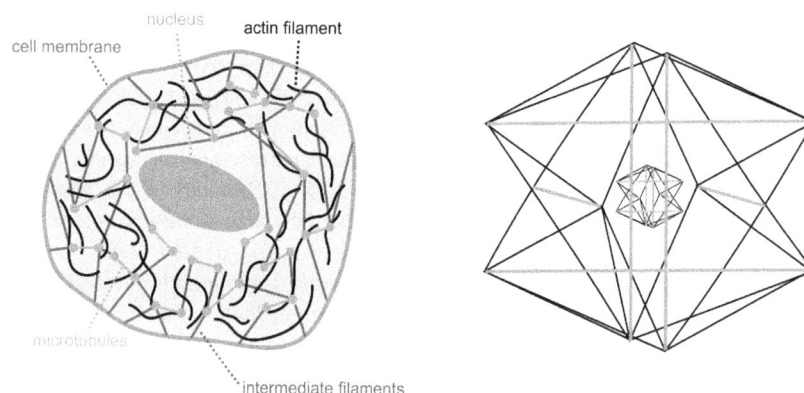

FIGURE 2.3 Illustration of a prototypic mechanical model of cells in living creatures (left) which can be represented by a tensegrity structure (right). Source: adapted from Gan, B.S. (2020). "Tensegrity in Biological Application: Cellular Tensegrity". In: *Computational Modeling of Tensegrity Structures*. Springer, Cham. https://doi.org/10.1007/978-3-030-17836-9_8.

2.6 The Mechanics of Cell Motion: Cell Motility

Cells not only need to maintain their morphology and structural stability but also, to varying degrees, need to be able to move from place to place (Figure 2.4). *Cellular movement is accomplished by complex molecular machinery [11], principally via cilia and flagella.* Cilia and flagella are hair-like extensions protruding from the cell membrane (Figure 2.5). Their structures are similar except that cilia tend to be smaller and numerous and flagella tend to be larger and fewer. They both execute beating motion back and forth rhythmically. In unicellular organisms, their job is to provide locomotion. In large multicellular organisms, their role is to move fluid past the cell.

Cilia are hair-like structures that can beat in synchrony causing the movement of unicellular organisms such as paramecium. Cilia are also found in specialized linings of eukaryotic cells. For example, cilia sweep fluids past stationary cells in the lining of the trachea and tubes of the female oviduct. Flagella are whip-like appendages that undulate to move cells and are longer than cilia, but have similar internal structures, which are invariably made of microtubules. Prokaryotic and eukaryotic flagella differ substantially from each other. Both flagella and cilia exhibit the same structure, a so-called 9 + 2 arrangement of microtubules. This arrangement refers to the nine fused pairs of microtubules on the outside of a cylinder, and the two unfused microtubules in the center. Dynein "arms" attached to the microtubules serve as the molecular motors that supply mechanical forces required for the movement of the entire structure. Naturally, the fuel for these motors is in the form of ATP molecules. The dynein motors undergo a cycle of activity, during which they form a transient attachment to the doublet and push it towards the tip of the cilium or flagellum. The dynein arms, in the presence of ATP, can move from one tubulin to another. They enable the microtubules to slide along one another so the cilium can bend. The dynein bridges are regulated so that sliding leads to synchronized bending. Because of the proteinaceous linkage called nexin and radial spokes, the doublets are held in place, so sliding is limited lengthwise. If nexin and the radial spokes are subjected to enzyme digestion and exposed to ATP, the doublets continue to slide and telescope up to nine times their length [8].

2.7 Energy Production and Energy Transduction

All processes involved in the creation and maintenance of life on our planet need a driving energy, which originally comes from the quanta of electromagnetic energy of visible light (photons), emitted by the Sun and absorbed by pigments of photosynthetic units. After this process of molecular excitation, the absorbed energy is accumulated and transmitted to other parts of the cell, to other parts of the plant, and finally to other organisms that ingest the plant, which are not able to obtain energy by photosynthesis directly. This process of biochemical energy passage in the food chain continues to all forms of life including humans but it must be kept in mind that the primary source of energy is the Sun. Below, we examine how energy is produced and transduced through work between parts of a living system.

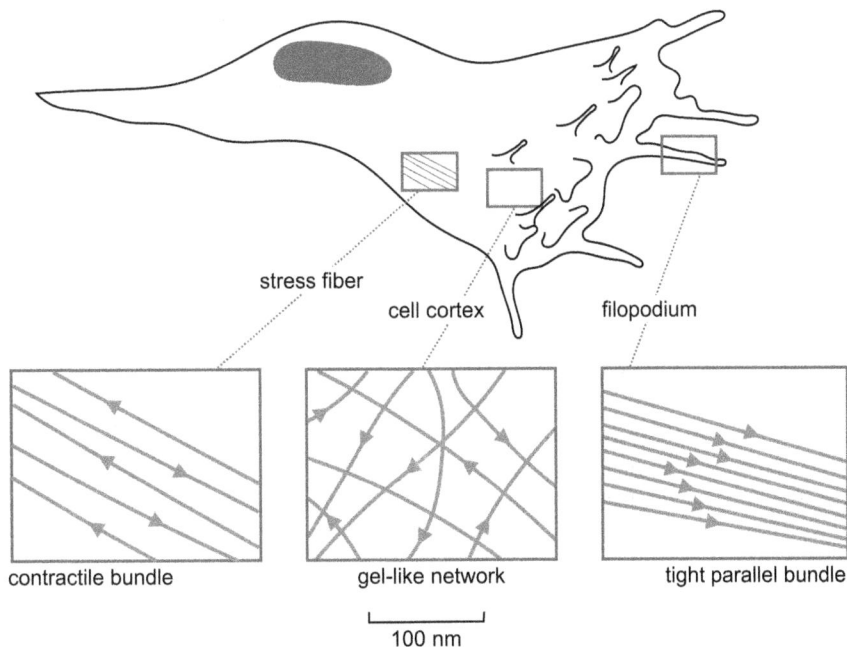

FIGURE 2.4 Schematic illustration of examples of actin filament arrays involved in processes of migrating cells. The contractile bundles, gel-like networks, and tight parallel bundles which form constituents of stress fibers, cell cortex, and filopodia enable cells to maintain their morphology and structure. Source: adapted from Alberts et al. *(2015) Essential Cell Biology.*

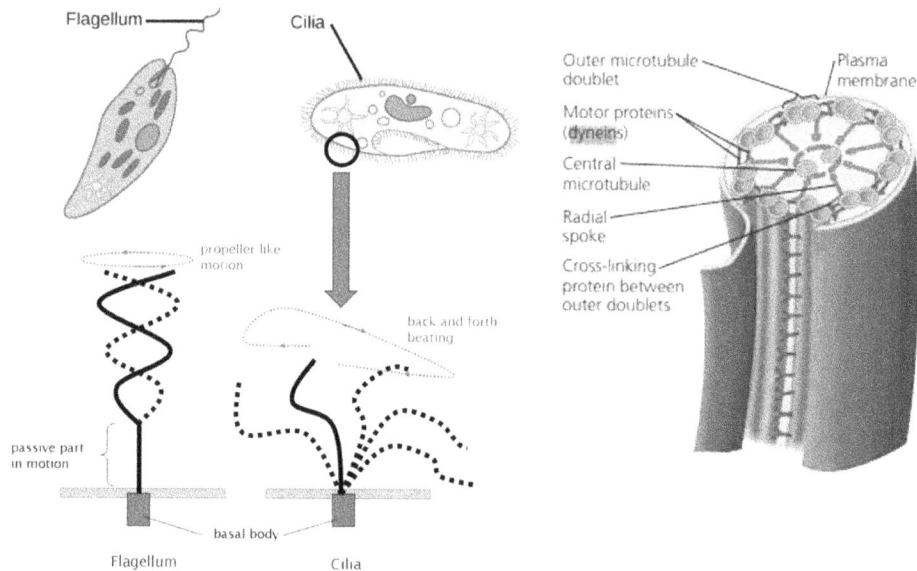

FIGURE 2.5 Graphical illustration of flagella and cilia and their role in cellular motility. Graphical illustration of flagella and cilia and their role in cellular motility. license https://creativecommons.org/licenses/by-sa/4.0/legalcode. Sources: top left modified from http://cnx.org/contents/GFy_h8cu @10.53:rZudN6XP@2/Introduction, used under Creative Commons license https://creativecommons.org/licenses/by/4.0/legalcode; bottom left from Kohidai, L. (2008) used under Creative Commons license https://creativecommons.org/licenses/by/3.0/legalcode; right from Biological Developer, modified, used under Creative Commons

2.8 Mitochondria

The mitochondrion is an organelle that is the site of aerobic respiration (Figure 2.2a). Most of the key processes of aerobic respiration occur across its inner membrane. One theory holds that they evolved from endosymbiotic bacteria. The mitochondrion itself functions to produce an electrical chemical gradient, somewhat like a battery, by accumulating hydrogen ions in the space between the inner and outer membrane. This energy comes from the estimated 10,000 enzyme chains in the membranous sacks on the mitochondrial walls. Most of the food energy for most organisms is produced by the electron transport chain in the mitochondrial wall (Figure 2.6).

Cellular oxidation in the Krebs cycle [12] causes an electron build-up that is used to push H+ ions outward across the inner mitochondrial membrane. As the charge builds up, it provides an electrical potential that releases its energy by causing a flow of hydrogen ions across the inner membrane into the inner chamber. The energy causes an enzyme to be attached to ADP, which catalyzes the addition of a third phosphorus to form ATP. Plants can also produce ATP in this manner in their mitochondria but plants can also produce ATP by using the energy of sunlight in chloroplasts as discussed later. In the case of eukaryotic animals, the energy comes from food, which is converted to pyruvate and then to acetyl coenzyme A (acetyl CoA). Acetyl CoA then enters the Krebs cycle, which then releases energy that results in the conversion of ADP back into ATP. The more protons there are in an area, the more they repel each other. When the repulsion reaches a certain level, the hydrogen ions are forced out of revolving-door-like structures mounted on the inner mitochondria membrane called ATP synthase complexes, which we discuss in more detail below. This enzyme functions to reattach the phosphates to the ADP molecules, again forming ATP. ATP is used in conjunction with enzymes to cause certain molecules to bond together. The correct molecule first docks in the active site of the enzyme along with an ATP molecule. The enzyme then catalyzes the transfer of one of the ATP phosphates to the molecule, thereby transferring to that molecule the energy stored in the ATP molecule. Then, a second molecule docks nearby at a second active site on the enzyme. The phosphate is then transferred to it, providing the energy needed to bond the two molecules now attached to the enzyme. Once they are bonded, the new molecule is released. This operation is similar to using a mechanical jig to properly position two pieces of metal, which are then welded together. Once welded, they are released as a unit and the process can begin again. Scores of other enzymes exist in order for ATP to transfer its energy to the various places where it is needed. *Each enzyme must be specifically designed to carry out its unique function, and most of these enzymes are critical for health and life.* The body does contain some flexibility, and sometimes life is possible when one of these enzymes is defective, but the affected person is often handicapped. Also, backup mechanisms sometimes exist so that the body can achieve the same goals through an alternative biochemical route. These few simple examples eloquently illustrate the concept of over-design built into the body. They also prove the enormous complexity of the body and its biochemistry.

2.9 Chloroplasts

Chloroplasts are the sites of photosynthesis in plants and some bacteria [13, 14]. It is here that plant cells trap the energy of light and use it to manufacture food proteins for the cell.

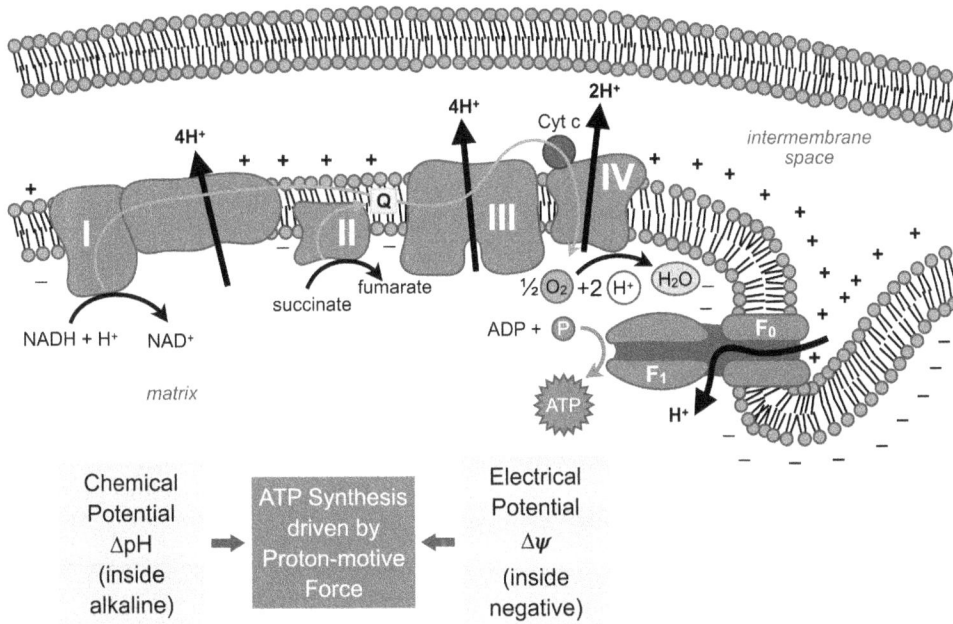

FIGURE 2.6   Schematic illustration of electron transport chain (ETC). Source: adapted from https://www.biosciencenotes.com/electron-transport-chain-etc/.

Chloroplasts are double-membrane ATP-producing organelles found only in plants. Inside their outer membrane is a set of thin membranes organized into flattened sacs stacked up like coins called thylakoids. The disks contain chlorophyll pigments that absorb solar energy, which is the main source of energy for all the plant's needs including manufacturing carbohydrates from carbon dioxide and water. *The chloroplasts first convert the solar energy into ATP stored energy, which is then used to manufacture storage carbohydrates, which can be converted back into ATP when energy is needed. The chloroplasts also possess an electron transport system for producing ATP.* The electrons that enter the system are taken from water hence the significance of water supply to plant life. During photosynthesis, carbon dioxide is reduced to carbohydrate by energy obtained from ATP. Photosynthesizing bacteria (cyanobacteria) use yet another system. Cyanobacteria do not manufacture chloroplasts but use chlorophyll bound to cytoplasmic thylakoids. The two most common evolutionary theories of the origin of the mitochondria-chloroplast ATP production system are: a) endosymbiosis of mitochondria and chloroplasts from the bacterial membrane system and b) the gradual evolution of the prokaryote cell membrane system of ATP production into the mitochondria and chloroplast systems. Both the gradual conversion and the endosymbiosis theory require many transitional forms, each new one must provide the organism with a competitive advantage compared with the unaltered organisms.

For metabolic purposes, the energy of molecular excitation needs to be transformed into the chemical energy of so-called high-energy compounds. As stated before, the most common accumulator of chemical energy in the cell is ATP, formed in the process of photosynthesis and used in nearly all processes of energy conversion in other cells. The hydrolysis of ATP, producing ADP and catalyzed by special enzymes, the so-called ATPases, allows the use of this stored energy for ionic pumps, for processes of molecular synthesis, for production of mechanical energy, cell motility, and many others. The amount of energy, stored by the ADP → ATP transformation in the cell is limited just because of osmotic stability. Therefore, other molecules, such as sugars and fats, are used for long-term energy storage. The free energy of ATP is used to synthesize these molecules. Subsequently, in the respiratory chain, these molecules are decomposed, whereas ATP is produced again.

2.10 Osmotic Work

If two solutions have the same osmotic pressure, we call them iso-osmotic but when the pressures are different, the one at higher pressure is called hypertonic and the one at lower pressure is called hypotonic. When cells are placed in a solution and neither swell nor shrink, we call the solution isotonic. *Every living system maintains a desirable osmotic pressure difference across its membrane through the activity of ion pumps involving the process known as osmoregulation.* Ion pumps require energy input for their functioning in the form of ATP molecules. The cell composition begins to shift away from its optimal mixture if the ion pumps are chemically or mechanically damaged. Across the cell wall, the osmotic pressure difference then rises, causing the cell to swell, which can become turgid and eventually explode. Note that the cells of bacteria and plants are not osmotically regulated since their cell walls are able to withstand pressures in the range of 1 to 10 atm. The minimum work performed when n moles of solute are transferred from one solution with a concentration c_1 to a solution with concentration c_2 is given by the formula [4]

$$W = nRT \ln \frac{c_1}{c_2} \qquad (2.2)$$

where c_1 is the intracellular salt concentration and c_2 is the extracellular salt concentration. If $c_2 > c_1$ the osmotic work done by the environment is negative, which gives the physical reason for the cell to expend energy to move salt molecules from a solution of low concentration inside to one of high concentration outside.

2.11 Energy and Material Transport in and out of a Cell

2.11.1 Passive Transport

Many molecular phenomena and biological processes such as diffusion, osmosis, reverse osmosis, dialysis, and reverse dialysis (filtration) involve the transport of substances on the cellular level in a passive manner. The driving energy for passive transport comes from molecular kinetic energy or pressure and results from the application of the second law of thermodynamics (Figure 2.7). There is another class of transport phenomenon, called active transport, in which the cell's membrane itself utilizes energy to generate the transport of substances across it (Figure 2.7).

2.11.2 Active Transport

Biological organisms sometimes need to transport substances from regions of low concentration to high concentration, which is the direction opposite to that in osmosis or dialysis, hence in seeming contradiction to the second law of thermodynamics. While sufficiently large back pressure causes reverse osmosis or reverse dialysis, there are known situations in which substances move in the direction precluded by the second law of thermodynamics where the existing pressures are insufficient to cause reverse osmosis or reverse dialysis. In these instances, active transport assists in overcoming an "uphill direction" of mass movement. In these cases, it is the biological membranes, which expend their energy to transport substances. Active transport can also assist and accelerate osmosis or dialysis and explains why some transport proceeds faster than expected from osmosis or dialysis alone.

Active transport is particularly vital for excitable cells such as nerve cells. For example, changes in the concentration of electrolytes across the membranes of the nerve cell are instrumental in creating nerve impulses in the form of action potential [15]. Following repeated nerve impulses, major changes in ionic concentration occur and active transport in terms of "ion pumps" returns the electrolytes back to their original locations.

2.11.3 Ion Channels and Ion Pumps

The cell membrane possesses pores or channels that allow a selective passage of metabolites and ions in and out of the cell [16] (Figure 2.8). They can even drag molecules from an area of low concentration to areas of high concentration by working directly against diffusion and the second law of thermodynamics. An example of this is the sodium/potassium pump. Most of the work done on the transport across membranes is done *via* ion pumps such as sodium-potassium pumps. The energy required for the functioning of the pump comes from the hydrolysis of ATP in which a phosphorylated protein is identified as an intermediate in the process. *The hydrolysis of the phosphorylated protein usually causes a conformational change that opens a pore that drives the sodium and potassium transport.* Some membrane proteins actively use energy derived from ATP hydrolysis in the cell. Here, the energy of a phosphate is used to exchange sodium atoms for potassium atoms. The free energy change in the hydrolysis of a phosphorylated protein with a value of 9.3 kJ/mol is sufficient to drive a concentration gradient as high as 50:1 uphill. The sodium/potassium ATPase pump pushes

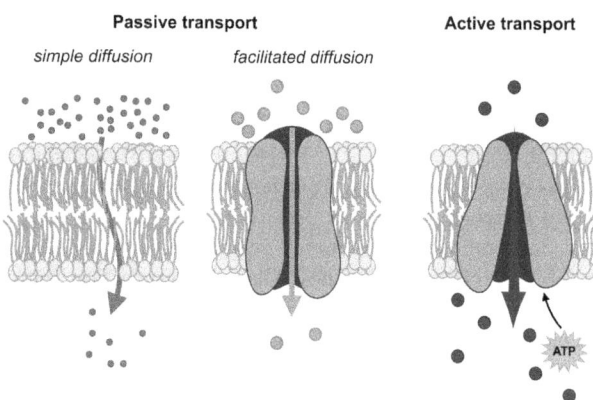

FIGURE 2.7 Schematic illustration of the passive and active transport of ligands across cell membranes. Simple diffusion-based and receptor-mediated (facilitated) diffusion transport of ligands is demonstrated in the left and central panels. Active transport, driven by energy input, is shown in the right-side panel. Source: adapted from Regents Biology: http://www.explorebiology.com/regentsbiology/.

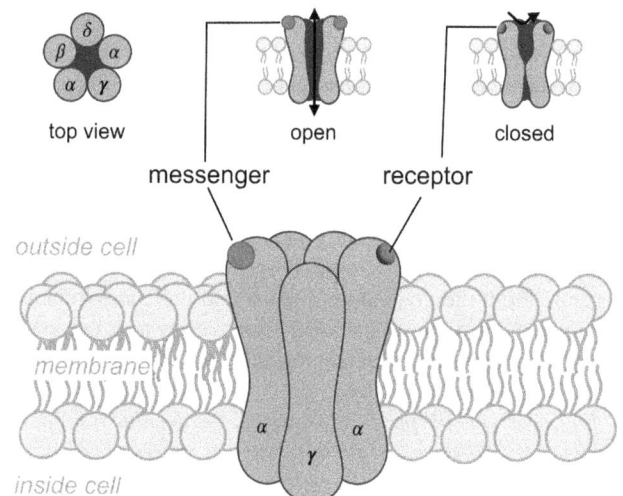

FIGURE 2.8 Illustration of the putative structures of ion channels embedded in cell membranes. Source: adapted from https://www.news-medical.net/health/Types-of-Ion-Channels-in-the-Body.aspx.

two potassium ions (K⁺) into the cell for every three sodium ions (Na⁺) it pumps out of the cell, so its activity results in a net loss of positive charges within the cell. *The sodium/ potassium ATPase produces a concentration of Na⁺ outside the cell that is some ten times greater than that inside the cell and a concentration of K⁺ inside the cell some 20 times greater than that outside the cell.* The concentrations of chloride ions (Cl) and calcium ions (Ca²⁺) are also maintained at greater levels outside the cell except that some intracellular membrane-bound compartments may also have high concentrations of Ca²⁺.

Ion channels come in three general classes: a) voltage-gated, b) ligand-gated, and c) so-called gap junctions. They differ not only in their design geometry but also in the use of physical and chemical mechanisms for the selection of ions for passage [17].

It is imperative and completely in accordance with the second law of thermodynamics that maintaining a proper composition inside the cell requires work to be performed by ion pumps located in the cell membrane using biological energy units in the form of ATP molecules produced mainly in the mitochondria. *The existence of a physical barrier, i.e. the cell membrane (as well as internal barriers delineating the various organelles such as mitochondria and the nucleus) built by every cell to protect is required for the cell's integrity.* Inside the cell membrane, there is complex and multifunctional machinery making sure the cell performs the tasks required for its survival: motility, division, metabolism, damage repair and waste removal, etc. The interior of the cell is contained in the cytoplasm. The cytoplasm, whose composition includes water, ions, salts, and proteins at regulated concentrations, includes critically important structural and functional components called the cytoskeleton and certain smaller compartments known as organelles, which are specialized to perform their respective functions as is briefly discussed below.

2.12 The Cytoskeleton

Interiors of living cells are structurally organized by the cytoskeleton networks of filamentous protein polymers with motor proteins interacting with these protein filaments and providing force and directionality needed for transport processes [18]. Unlike the rigid skeleton that structurally supports mammals such as humans, the cytoskeleton is a dynamic structure that undergoes continuous reorganization in response to the changing demands of the living cell it inhabits. *The cytoskeletal network of filaments is responsible for defining the cell shape, protecting the cell from changes in osmotic pressure, organizing its contents, and providing cellular motility. In the case of dividing cells, which comprise most types of cells other than neurons and muscles, the cytoskeleton is also the supramolecular structure, which is responsible for generating force to separate chromosomes during mitosis.*

The cytoskeleton is composed of three different types of molecular protein polymers known as actin filaments (AFs) or microfilaments, intermediate filaments (IFs), and microtubules (MTs). They are polymerized from different types of proteins at different rates and they form different geometries with different mechanical properties. Cells acquire and maintain their shapes as a result of their tensegrity, which is due not only to the cytoskeleton's protein filaments but also to the interplay with the extracellular matrix, the anchoring scaffolding to which cells are secured in the body. Throughout the cell, a network of contractile microfilaments exerts tension and pulls the cell's membrane and all its internal constituents toward the nucleus at the core. Opposing this inward pull are two main types of compressive elements, one of which is outside the cell and the other inside. The component outside the cell is the extracellular matrix; the compressive "girders" inside the cell can be either MTs or large bundles of cross-linked microfilaments within the cytoskeleton. The third component of the cytoskeleton, the intermediate filaments interconnect MTs and contractile AFs to the surface membrane and the cell's nucleus. Again, any and all of these processes are energy-driven and fuelled by either ATP or GTP molecules, hence the cell shape and structure stability are a result of metabolic energy supply. The cytoskeleton is unique to eukaryotic cells. It is a dynamic three-dimensional structure that fills the cytoplasm. This structure acts in a way that mimics both muscle and skeleton, for movement and stability. The long fibers of the cytoskeleton are polymers composed of protein subunits. *In addition to the filaments of the cytoskeleton mentioned above, there are also associated proteins that form complexes with the filaments as well as motor proteins that endow these structures with force and torque generation abilities. This is a set of molecular machines that make living processes possible.*

2.13 Work During Cell Division: Chromosome Separation

The separation of chromosomes occurring in mitosis is accomplished by the cytoskeleton's largest constituents, the microtubules (Figure 2.9). *During mitosis, MTs connect to each of the chromosomes and align them along the cell's equatorial plate. A mysterious balance of forces prevents the MTs from separating the chromosomes until all the chromosomes have become aligned and division may proceed in unison.* The discovery of microtubule-associated motor proteins with a clear involvement in mitosis and meiosis has attracted great interest since motor proteins could account for many of the movements of the spindle and chromosomes in dividing cells. Microtubule motors bind to and move unidirectionally on MTs and have been proposed to generate the force required for spindle assembly and maintenance, attachment of the chromosomes to the spindle, and movement of chromosomes toward opposite poles. Kinesin motors have been shown necessary for establishing spindle bipolarity, positioning chromosomes on the metaphase plate, and maintaining forces in the spindle. Evidence also exists that kinesin motors can facilitate MT depolymerization, raising the possibility that the motors modulate MT dynamics during mitosis. The force generated by MT polymerization and/or depolymerization is thought to contribute to, or underlie, spindle dynamics and movements of the chromosomes.

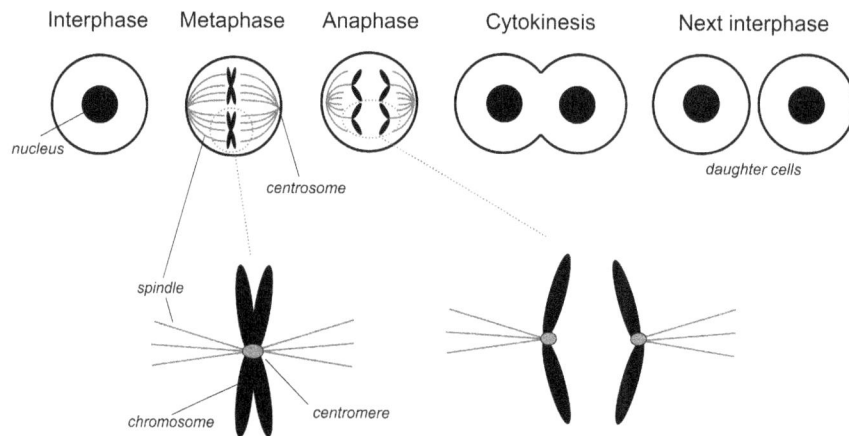

FIGURE 2.9 Illustration of chromosome separation during mitosis when spindle microtubules capture the kinetochore formed in the centromere. After mitosis, chromosomes are divided into daughter cells. Source: adapted from Hara, M., Ariyoshi, M., Okumura, E.I., Hori, T., and Fukagawa, T. (2018). "Multiple Phosphorylations Control Recruitment of the KMN Network onto Kinetochores". *Nat Cell Biol.* Dec 20(12): 1378–1388. doi: 10.1038/s41556-018-0230-0. Epub Nov 12, 2018. Erratum in: *Nat Cell Biol.* 2018 Nov 19; PMID: 30420662.

2.14 Microtubules

The thickest and perhaps the most multifunctional of all cytoskeletal filaments are microtubules (MT), which are comprised of the protein called tubulin. They are found in nearly all eukaryotic cells [19]. The elementary building block of MTs is an α – β tubulin heterodimer whose dimensions are 4 by 5 by 8 nm whose thousands of replicas assemble into a structure that typically has 13 protofilaments that join to form a hollow cylinder. The outer diameter of a microtubule is 25 nm and the inner diameter is 15 nm. *MTs are the largest and most rigid members of the cytoskeleton, more rigid than the actin filaments or intermediate filaments, and thus they serve as the major architectural strut within the cytoplasm. They are compression resistant but are fairly flexible. MTs act as a scaffold to determine cell shape and provide "railroad tracks" for cell organelles (including mitochondria) and vesicles to move on with the help of motor proteins that propel intra-cellular transport.* MTs are involved in a number of specific cellular functions: a) organelle and particle transport inside cells, b) signal transduction, c) when arranged in geometric patterns inside flagella and cilia, they are used for locomotion or cell motility, d) organization of cell compartments (*e.g.* positioning of ER, Golgi, and mitochondria), and perhaps most importantly, e) cell division. During cell division, they form mitotic spindles, which are indispensable for chromosome segregation, and in the process of daughter chromatid separation they exert pulling forces by depolymerization at the distal ends. MTs perform all of these tasks by delicate and precise control over their assembly and disassembly (MT dynamics is referred to as dynamic instability) and by interactions with microtubule-associated proteins (MAPs). The force generated during mitosis is partly due to motor proteins decorating the surface of MTs and partly due to MTs themselves. It is worth noting that in biology redundancy is almost always necessary to ensure failsafe accomplishment of all significant processes.

Normal cell organization precludes the existence of free MTs since MTs become attached to centrosomes or organized within minutes. Microtubule organizing centers (MTOCs) such as centrosomes contain MAPs that bind minus-ends of MTs and nucleate MT assembly. Within the cell body, the majority of the MTs emanate from a centrosome where the negative ends of MTs are anchored. The MTs *in situ* are interconnected and intra-connected by MAPs, which have a stabilizing effect on the dynamics of MTs. At the base of cilia and flagella, MTOCs are called basal bodies or centrioles. In interphase cells, the MTOC is found in the centrosome, which contains two centrioles plus pericentriolar material.

In eukaryotic cells, there are regulatory processes that appear to control MT assembly and disassembly. MT growth can be promoted, when needed, in a dividing or moving cell by providing a so-called GTP cap on the growing (plus) end of an MT, which is a region of the tip that contains tubulin dimers with two GTP molecules attached to them. Once the GTP attached to tubulin is hydrolyzed to GDP, the GTP cap begins to shrink, which eventually leads to MT disassembly. Note that disassembling MTs, which are connected to chromosomes by molecules of the kinetochore, exert pulling forces on chromosomes during mitosis. Conversely, when MTs assemble and grow, they can exert significant pushing forces on objects encountered on their path. A rigid barrier such as a cellular membrane can cause a subsequent buckling of the growing MTs that collide with it. It is worth noting that MTs and their individual tubulin dimers possess permanent dipole moments. It has been recently emphasized that *MT networks play an essential role in cellular self-organization phenomena, which include reaction-diffusion instabilities in the mechanisms of cytoskeletal self-organization.* We believe and discuss later in this chapter that dipole-dipole interactions are crucial in the mechanism of self-organization of the cytoskeleton, which is so prominent in mitosis. The various aspects of electrostatic, electromagnetic, magnetic, and conductive properties of cells and sub-cellular structures will also be discussed later in this chapter.

2.15 Actin Filaments (Microfilaments)

Actin is one of the most abundant proteins in eukaryotic cells. Globular (G-) and filamentous (F-) actin are reversible structural states of this protein and its assemblies. F-actin is a linear double-stranded polymer formed as a result of the assembly of G-actin. These two strands of F-actin are capable of forming both stable and labile components within cells. F-actin is the main component of thin filaments in muscle sarcomere. Actin can also be found in non-muscle cells. Actin is often associated with other proteins in the cytoskeleton referred to as actin-related proteins (ARPs). Actin provides a dynamic, tension-resistant component of the cytoskeleton. The polymerization of F-actin from G-actin is a largely monotonic process that is crucially dependent on the supply of ATP. F-actin assembles according to a standard nucleation-elongation mechanism. Once assembled, typical microfilaments have a diameter of about 8 nm. *F-actin can be found exhibiting characteristic forms as follows: a) thin filaments in striated muscle, b) individual filaments in cortical cytoplasm, c) bundled filaments (stress fibrils) in inter-phase fibroblasts, d) short actin bundles in microvilli, e) filament bundles ending in focal contacts, and f) filaments running parallel with the membrane at the membrane-cytoplasm interface.* Microfilaments are often found with the lattice configuration near the leading edge of growing or motile cells where they provide greater stability to the newly formed region. New AFs are nucleated at the leading edge of the cell's growth and trailing microfilaments are disassembled.

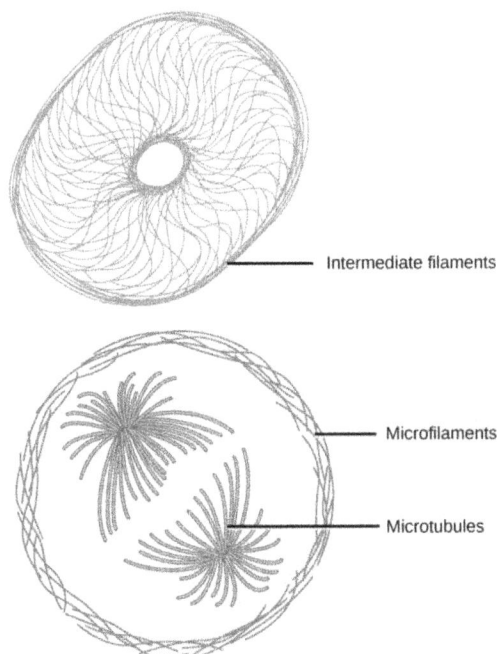

2.16 Intermediate Filaments

The pool of unpolymerized IF subunits is insignificant since newly synthesized proteins polymerize very quickly. Filaments appear to require no cofactors or energy input for their assembly. IFs appear to provide resilience and plasticity to the cell and may provide the main cytoskeleton form with complex arrays in cells. IFs provide tensile strength for the cell. They are formed by the polymerization of elongated fibrous protein molecules such as vimentin, keratins, desmins, and nuclear lamins. These molecules are quite unlike the globular molecules, tubulin and G-actin, that form MTs and AFs, respectively. *Vimentin, keratin, desmin, and lamin molecules form dimers through an anti-parallel association of two molecules in a coiled-coil configuration. Consequently, IFs are non-polar structures. Two dimers associate to form a symmetrical tetramer. The overlap between dimers allows for the filaments to stretch and, as a result, IFs are able to withstand large stresses without breaking.* An extensive network of IFs surrounds the nucleus forming what is known as the nuclear envelope (Figure 2.10). The formation and disassembly of the nuclear lamina are regulated by phosphorylation. This allows for the elimination of the nuclear envelope prior to mitosis. IFs also extend out into the cell periphery where they act to maintain cell integrity and may connect with the cell membrane. Specific cells, such as neurons, have their own distinctive IFs, known as neurofilaments.

However, an important force-generating family of proteins that has been mentioned but not discussed in detail yet deserves special attention due to their dynamic behavior and pervading

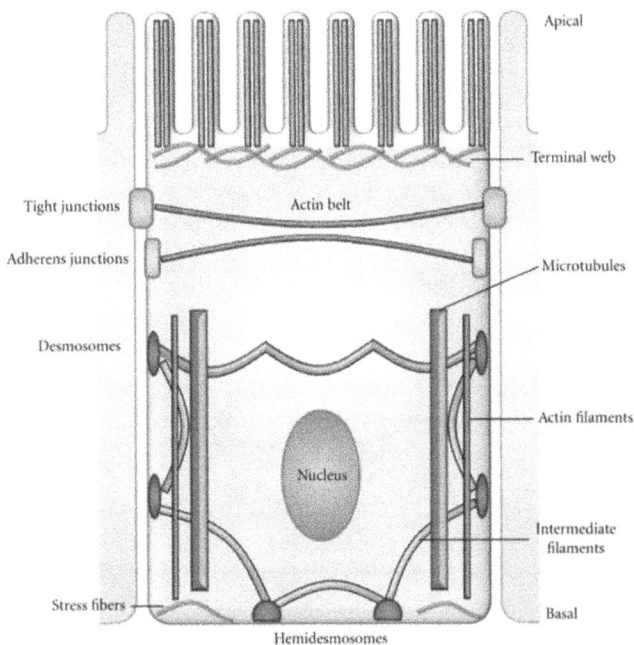

FIGURE 2.10 Illustration of the structure of a eukaryotic cell demonstrating the roles played by various structural proteins including intermediate filaments which are found throughout the cell and hold organelles in place. Sources: Left from http://cnx.org/contents/GFy_h8cu@10.53:rZud-N6XP@2/Introduction, used under Creative Commons license https://creativecommons.org/licenses/by/4.0/legalcode. Right from Silvia Chifflet et al. "The Plasma Membrane Potential and the Organization of the Actin Cytoskeleton of Epithelial Cells". *International Journal of Cell Biology* doi:10.1155/2012/121424, used under Creative Commons license https://creativecommons.org/licenses/by/4.0/legalcode.

presence in all living cells. This class is called motor proteins. *For motor proteins to function, a source of energy is required that acts as fuel for mechanical engines.* We discuss this fuel next before we characterize the types of motor proteins that use it for movement and force generation.

2.17 The Quantum of Biological Energy: ATP

Without ATP, life as we understand it could not exist. ATP is a perfectly designed intricate molecule that plays a critical role in providing the proper size energy packet for thousands of biochemical reactions that occur in all life processes. The ATP energy system is very fast, highly efficient, produces a rapid turnover of ATP, and is capable of rapid response to energy demands within a cell. ATP is a complex molecule that consists of an adenosine nucleoside and a tail consisting of three phosphates (Figure 2.11) and weighs ~507 atomic mass units (also called daltons). Metaphorically speaking, the ATP molecule is a machine with a level of organization on the order of an optical microscope or a radio set. It is a complex nano-machine that serves as the primary energy currency of the cell.

ATP is a molecule that is arguably second in importance only to DNA since it is the most widely distributed high-energy compound within the human body. This ubiquitous molecule is used to build complex molecules, contract muscles, generate electricity in nerves, and pump toxic molecules out of the cell. All living cells produce ATP from fuel provided by Nature and ATP in turn powers virtually every activity of the cell and by extension an entire organism. To the best of our knowledge, all organisms from the simplest bacteria to humans use ATP as their primary energy currency. The energy level it carries is just the right amount for most biological reactions. Nutrients contain energy in low-energy covalent bonds, which are not very useful to do most kinds of work in the cells. These low energy bonds must be translated to high-energy bonds, and this is a role played perfectly well by ATP. *A steady supply of ATP is so critical that a poison that attacks any of the proteins used in ATP production kills the organism in minutes.* ATP is used for many cell functions including transport work moving substances across cell membranes. It is also used for mechanical work, supplying the energy needed for muscle contraction. It supplies energy not only to the heart muscle

(for blood circulation) and skeletal muscle (such as for gross body movement), but also to the chromosomes and flagella to enable them to carry out their many functions. *A major role of ATP is in chemical work, supplying the needed energy to synthesize the many thousands of types of macromolecules that the cell needs to exist and function. ATP is also used as an on-off switch both to control chemical reactions and to send messages.* The shape of the peptide chains that produce the building blocks (proteins) and other structures used in life is mostly determined by weak chemical bonds that are easily broken and remade. These chains can shorten, lengthen, and change shape in response to the input or withdrawal of energy. The changes in the chains alter the shape of the protein and can also alter its function or cause it to become either active or inactive. The ATP molecule can bond to one part of a protein molecule, causing another part of the same molecule to slide or move slightly which causes it to change its conformation, inactivating the molecule. Subsequent removal of ATP causes the protein to return to its original shape, and thus it is again functional. The cycle can be repeated until the molecule is recycled, effectively serving as an ON–OFF switch. *Both adding a phosphorus-containing group (phosphorylation) and removing it from a protein (dephosphorylation) can serve as either an ON-OFF switch. Importantly, ATP synthesis occurs mainly via the electron transport system in the mitochondria.*

Generally, ATP is coupled to another reaction such that the two reactions occur nearby utilizing the same enzyme complex. The release of phosphate from ATP is exothermic while the coupled reaction is endothermic. The terminal phosphate group is then transferred by hydrolysis to another compound, via a process called phosphorylation, producing ADP, phosphate (Pi), and energy. Phosphorylation often takes place in cascades becoming an important signaling mechanism within the cell. Importantly, ATP is not excessively unstable, but it is designed so that its hydrolysis is slow in the absence of a catalyst. This ensures that its stored energy is released only in the presence of an appropriate enzyme. *The mitochondrion, where ATP is produced, itself functions to produce an electrochemical gradient, in a way similar to a battery, by accumulating hydrogen ions between the inner and outer membrane.* This electro-chemical energy comes from the estimated 10,000 enzyme chains in the membranous sacks on the mitochondrial walls. As the charge builds up, it provides an electrical potential that releases its energy by causing a flow of hydrogen ions across the inner membrane into the inner chamber. The energy causes an enzyme to be attached to ADP, which catalyzes the addition of a third phosphorus to form ATP.

Adenosine Triphosphate (ATP)

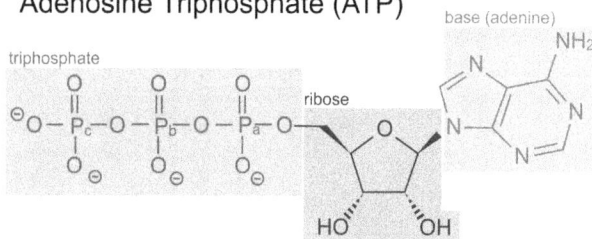

FIGURE 2.11 Graphical illustration of the chemical structure of ATP (adenosine triphosphate). The terminal phosphate group (left-most PO3 group with two anionic oxygen atoms) is cleaved during the conversion of ATP to ADP (adenosine diphosphate) to release energy that drives several biochemical reactions. Source: adapted from https://byjus.com/biology/energy-currency-of-the-cell/.

2.18 Molecular and Biological Machines: Motor Proteins

Biological motors are central force generation elements to the machinery of living systems just as mechanical motors are to a car engine. In both cases, the motor is responsible for converting one form of energy into a force or a torque that generates motion involving translational or rotational kinetic energy, respectively. In a limited but most common notion of a motor, it implies an electrical device as an arrangement of a stator (stationary device) containing a permanent magnet with a rotor (rotating device) containing conducting wire coil windings as an electromagnet connected to a power source (a battery or fuel cell). The rotor rotates hence turning the shaft as a result of the force generated by the interaction between the electromagnet of the rotor with the permanent magnet of the stator. Similarly, an engine in a limited sense refers to a piston-cylinder device. However, in a broader sense, a motor is a subtype of engine that produces motion. By distinction, an engine takes in energy per unit time and possibly raw materials and converts those into useful output. It is a useful transformation of one form of energy into another (*e.g.* chemical into mechanical). A biological system such as an organism may be considered not only a biological engine but also a complex structure comprised of various types of biological engines as well as motors intricately coordinated into a single larger system that involves its subsystems working in concert.

In general, *biological motors at a subcellular level can be divided into: a) linear motor proteins that move in a processive fashion along protein filaments, e.g. kinesin or dynein along MTs or myosin along AFs (Figure 2.12); b) rotational motors such as F-ATPases; c) beating complexes of motor proteins and filaments such as the microtubule-dynein arrangements in cilia and flagella; and d) contractile arrangements of interconnected long and short filaments involving actin-myosin complexes.*

The motor of a biological machine may be considered at many levels. At the largest scale, the body parts that cause locomotion are the legs. Hence, the legs may be recognized as mechanical motors. The arms may be considered motors as well, as they can cause the motion of objects in the external world. At an intermediate level, various organs perform work such as kidneys and liver selecting and removing harmful substances from the bloodstream. The human brain performs signal processing and information gathering as well as coordination of the body's functions using electrical signal generation. At the microscopic level are molecular motors,

which are responsible for the motion that drives our physiology of the well-engineered tissue and organ parts of the overall biological machine that constitutes our physical being. The energy that ultimately provides this potential for biological motors and engines is ATP (or GTP), electrical potential, and chemical concentration gradients (some of which are achieved and maintained by the use of ATP) and electromagnetism. Electromagnetism also helps to drive the production of ATP in mitochondria as well as mediate in-phase long-range coherence of ATP production that orchestrates the coordinated functions of the biological engines [20].

The fuel cell provides hydrogen atoms that are stripped of their electrons that in turn are the source of electricity that powers the engine. The electron transport chain of mitochondria in biological cells is similarly provided hydrogen atoms from bio-fuels. The stripped electrons generate an electrical current in a sense as well as provide the source of energy that is ultimately incorporated into the phosphate bonds of ATP, which power the biological work of the body. Fuel cells are an alternative to battery-operated electrical automobiles. Fuel cells create electricity from an electromagnetic coil of copper wire. Consider the coil of conducting wire that is wrapped around the rotor (the rotating part of the motor), which is positioned between the north and south poles of permanent magnets. The interaction of the magnetic field of the permanent magnet with the moving electrical current results in a Lorentz magnetic force that causes the rotating motion of the rotor. This is attached to the crankshaft, which rotates the wheels. Analogously, biology appears to co-opt the electromagnetic force generated in the direction perpendicular to the electrical current of the electron transport chain to propel the motion of hydrogen protons across the inner mitochondrial membrane (to the inter-mitochondrial space). This electromagnetic force that propels the motion of hydrogen protons is referred to as the proton-motive force. Additionally, it is intriguing to hypothesize that associated electromagnetic fields generated in the direction perpendicular to the electron transport chain may play a role in mediating synchronized coherent conversion of nutrient biofuel to ATP energy. Hence, a fuel cell and a quantum mechanical engine may both be considered biological engines. However, *quantization of ATP energy and quantum tunneling of electrons in the mitochondrial complexes provide an argument for a nano-scale quantum mechanical-biological engine.* We discuss the quantum mechanical aspects of metabolism in a separate chapter on quantum biology.

Motor proteins provide one widely available method of generating force and movement within the cell. The assembly and disassembly of the cytoskeleton in conjunction with force generation by motor proteins is thought to be the main mechanism for mitosis, as well as for organelle transport within the cell. *There are three main families of naturally occurring motor molecules: the myosins, the kinesins, and the dyneins. All function by undergoing shape changes, utilizing energy from the biological fuel adenosine triphosphate (ATP).* Each family has members that transport vesicles through the cell's cytoplasm along linear assemblies of molecules-actin in the case of the myosins and tubulin for both of the other families. The kinesins and dyneins move or "walk" along MT cylinders constructed from tubulin carrying their cargo. The molecular

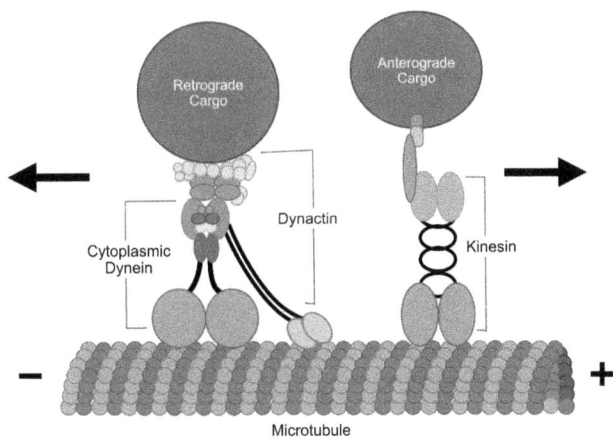

FIGURE 2.12 Schematic illustration of the roles played by dynein and kinesin. These two proteins transport cargo in the retrograde and anterograde directions toward the minus and plus ends of the microtubule. Adapted from Duncan, J. & Goldstein, L. (2006). "The Genetics of Axonal Transport and Axonal Transport Disorders". *PLoS genetics* 2: e124. 10.1371/journal.pgen.0020124.

motors are ATPase enzymes that move objects on the filament surface. The axonemal dyneins drive MT sliding in cilia and flagella; cytoplasmic dynein and kinesins move many objects in opposite directions in the cytoplasm. In general, kinesins move toward the plus ends of MTs (outward) and dyneins move toward the minus ends (inward). We describe these and the remaining major motors proteins next.

2.19 ATP Synthase

The ATP synthase (Figure 2.13) functions in a similar manner to a revolving door and resembles a molecular water wheel that harnesses the flow of hydrogen ions in order to build ATP molecules. Each revolution of the wheel requires the energy of about nine hydrogen ions returning into the mitochondrial inner chamber. Located on the ATP synthase are three active sites, each of which converts ADP to ATP with every turn of the wheel. Under maximum conditions, the ATP synthase wheel turns at a rate of up to 200 revolutions per second, producing 600 ATPs during that second.

In the opinion of this writer, *the molecular motor ATP synthase is particularly structurally and functionally analogous to the electrical motor. It contains both a stationary stator to which the rotor, the rotating component is associated. The stationary stator anchors the complex within the bi-lipid inner mitochondrial membrane including the knob, the site for catalyzing ADP to ATP, and the hydrogen ion channel for protons to be transported through the channel to participate in the conversion of ADP to ATP.* Notably, complex 5 of the electron transport chain, or ATP synthase, is known to be involved in iron metabolism. It is intriguing to consider if iron is mechanistically a controlling factor in the rotational component of ATP synthase. Iron, having a net magnetic moment, is polarized in a magnetic field, which is produced by the electric current

generated in the electron transport chain (from the moving electrons as well as the molecules containing iron) whereby the electrons ultimately combine with oxygen to produce water at complex 4. It appears conceivable to this writer that the interaction of the moving electrons, the electromagnet, and iron, the permanent magnet expected to be present in complex 5, ATP synthase, may be responsible for the force of motion of the rotor component of complex 5. Independent, however, of whether this rotational mechanism is due to the interaction of iron (a permanent magnet) and electron transport (a conductor analogous to a copper wire producing a magnetic field around it), a spectacular similarity of how nature has devised a molecular biological motor to that of a machine electrical motor's is stunning. It highlights a valuable lesson that studying nature provides the best blueprint for designing machines.

2.20 The Myosin Family of Motors

Myosins are a family of motor proteins that move along actin filaments while hydrolyzing ATP. Myosin is an ATPase enzyme that converts the chemical energy stored in ATP molecules into mechanical work. The motive power for muscle activity is provided by myosin motors, organized as thick filaments which interact with an array of thin actin filaments to cause the shortening of elements within each myofibril. This shortening is achieved by relative sliding of the myosin and actin filaments.

Myosin I, which is associated with the plasma membrane, may play a role in pulling the plasma membrane along the actin filaments, as microvilli grow by the addition of actin monomers at the tip. Myosins I and V, which bind to membranes, are postulated to have roles in the movement of organelles along AFs as well as movements of plasma membrane relative to AFs (Figure 2.14). They have been found to associate with Golgi

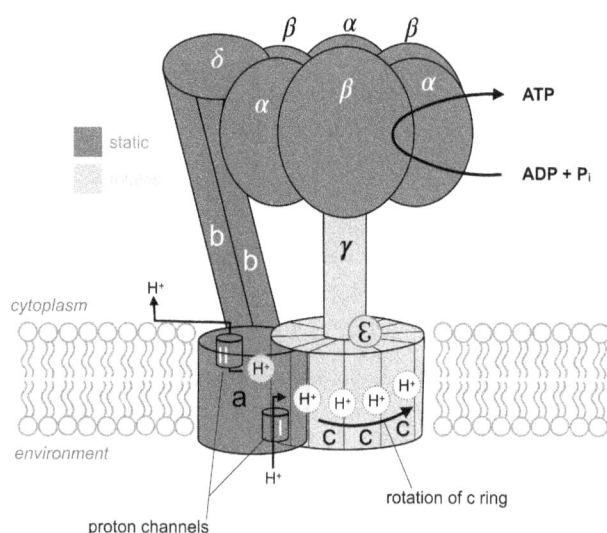

FIGURE 2.13 Schematic illustration of the ATP synthase structure in terms of its static and rotating components. Source: adapted from Jena, B.P. (2020). "ATP Synthase: Energy Generating Machinery in Cells". In: *Cellular Nanomachines.* Springer, Cham. https://doi.org/10.1007/978-3-030-44496-9_4.

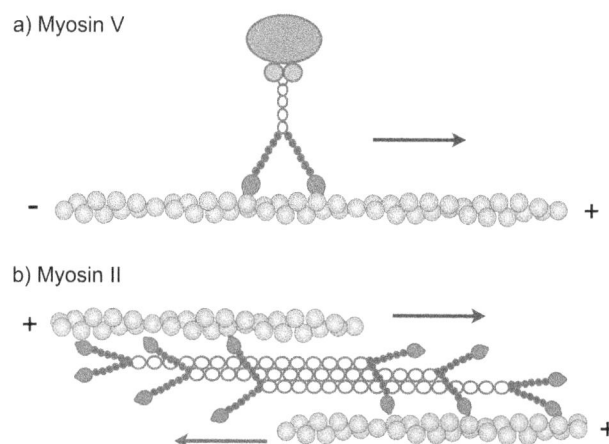

FIGURE 2.14 Schematic views of dimeric myosin V motor proteins step unidirectionally along actin cytoskeleton filaments (A) and a group of monomeric myosin-II motor proteins combined in the filament moving together along several actin filaments. Source: adapted from Kolomeisky, A.B. (2013). "Motor Proteins and Molecular Motors: How to Operate Machines at the Nanoscale". *J Phys Condens Matter* 25(46): 463101. doi: 10.1088/0953-8984/25/46/463101. Epub 2013 Oct 7. PMID: 24100357; PMCID: PMC3858839.

membranes, which give rise to secretory vesicles, including synaptic vesicles. Myosin II, the form found in skeletal muscle, is sometimes referred to as conventional myosin (Figure 2.14). Myosin II head domains interact with AFs in a reaction cycle that may be summarized as follows. ATP binding causes a conformational change that causes myosin to detach from actin. The active site closes, and ATP is hydrolyzed, as a conformational change (cocking of the head) results in myosin weakly binding actin, at a different place on the filament. Pi release results in a conformational change that leads to stronger myosin binding, and the power stroke. ADP release leaves the myosin head tightly bound to actin. In the absence of ATP, this state results in muscle rigidity called rigor. In non-muscle cells, myosin II is often found associated with actin filament bundles and is located mainly in stress fibers, bundles of actin filaments that extend into the cell from the plasma membrane. *The movement of myosin V along actin is processive, consistent with the role of myosin V in transporting organelles along AFs.* Each of the two myosin V head domains dissociates from actin only when the other head domain binds to the next AF with the correct orientation, about 13 subunits further along the AF. The actin-myosin interaction is controlled in various ways by Ca^{++} specifically to the tissue and organism type. Some myosins are regulated by Ca^{++} binding to calmodulin-like light chains, in the neck region. Some myosins are regulated by phosphorylation of myosin light chains, catalyzed by a Ca^{++}-dependent kinase or by a kinase that is activated by a small GTP-binding protein. A complex of tropomyosin and troponin, which includes a calmodulin-like protein, regulates actin-myosin interaction in skeletal muscle. Caldesmon, a protein regulated by phosphorylation and by Ca^{++}, controls actin-myosin interaction in smooth muscle.

2.21 The Kinesin Family of Motors

Kinesin is abundant in virtually all cell types, at all stages of development, and in all multicellular organisms. While a majority of kinesin appears to be free in the cytoplasm, some of these proteins are associated with various membrane-bound organelles, including small vesicles, the endoplasmic reticulum, and membranes between the ER and the Golgi apparatus. Kinesin's main role is to transport membrane-bounded organelles and macromolecular protein assemblies toward MT plus ends, which are also called distal ends. Kinesin is a motor for fast axonal transport in neurons, but it does not participate in mitosis or meiosis in dividing cells. *Kinesin is a force-generating motor protein, which converts the free energy of ATP into mechanical work, which is used to power the transport of intracellular organelles along MTs toward the plus, or fast-growing, end of the MT, hence kinesin is called an anterograde motor.* Kinesin has only one binding site per tubulin dimer and it takes 8-nm steps from one tubulin dimer to the adjacent one in a direction parallel to the MT protofilaments. Each step made by a motor domain of kinesin corresponds to one cycle of the ATPase reaction. An explanation of how kinesins generate force is that their motor domains contain an elastic element, akin to a spring, that becomes strained as a result of one of the transitions between chemical states involving ATP

hydrolysis. This strain exerts the force that the motor puts out, and the relief of this internal strain is the driving force for the forward movement.

2.22 Dynein

Dyneins move towards the minus-end of the MT, which tends to be anchored in the centrosome of the cell. Dynein motors also cause sliding between MTs that form the skeleton of cilia and flagella. Other ciliary structures resist this sliding with the result that bends form along the length of the cilium or flagellum and propagate from base to tip or from tip to base. The propagation of these bends requires coordinated action of the several types of dynein motor present in the cilium or flagellum and the structures providing the resistance to sliding. *Dynein drives fast retrograde vesicle transport in axons and other cells toward the centrosome.*

As described above, the two principal motor proteins that attach to MTs are kinesin and dynein. Each of these proteins consists of a globular head region and an extended coiled-coil tail section. These motor proteins and also myosin have a similar structure and their long tail is able to increase the force generated by the molecular motor. However, the essential components for force generation are located within the globular head region. The efficient propagation of these proteins, often involving pairs of molecules such as kinesin and dynein moving in opposite directions simultaneously and seemingly avoiding collisions led to the logical assumption that they may be directed by electrostatic interactions with the MT, which is supported by the fact that the binding of kinesin to MTs has been shown to be primarily electrostatic. The processive linear motion of the motor protein is fuelled by the hydrolysis of ATP. The motor protein has two or more distinct states where at least one conformational change occurs and is driven by ATP hydrolysis. Phosphorylation of the motor protein such as in the case of myosin may lead to subsequent conformational changes. Thus, *the protein may be viewed as walking along the MT powered by ATP.* It is interesting to note in a metaphorical manner that the use of GTP to control MT dynamics and ATP to control motor protein motion along MTs, allows a cell to have control over both the cars (motor proteins) and the track (MTs) individually.

2.23 Energy Combustion Similarities between Cells and Automobiles

An engine becomes flooded when too much liquid fuel enters the combustion chamber just like when too much fuel enters the mitochondria. This may occur with carbureted engines when cool and cranked full throttle. Consequently, the engine does not start. This can also occur if the fuel mixture is an incorrect grade for the engine. In either case, the engine may be damaged. When the engine of a car is damaged, the car does not drive right. The fuel for an automobile is as vital a form of energy as food is for humans and animals. When too much molecular fuel enters the mitochondria of cells, they too become flooded, and in addition to poor quality food (impurities in food), which can directly or indirectly damage mitochondria, physiological processes do not function right. The associated pathological process of insulin resistance is both a driver and a result of impaired mitochondrial structure and function. *The fundamental metabolic hallmark of insulin resistance is metabolic inflexibility, impaired switching from fatty acid to carbohydrate substrates during transition from fasting to fed states. ATP, the universal carrier of chemical energy is an end-product of catabolic pathways.* It is a substrate for biosynthetic anabolic pathways in addition to representing the potential energy to perform the work of life processes and involuntary movement.

Interestingly, gasoline and fat, the main energy storage systems in automobiles and animals/humans, respectively, are roughly equivalent producers of energy upon combustion. Gasoline releases less than 12 kilocalories per gram of energy in comparison to fat that releases 9.5 kilocalories per gram upon combustion. In both examples, hydrocarbons combine with oxygen to liberate CO_2 and water as by-products of the energy it creates. The advantage of storing fat over gasoline is that fat is solid. One distinct disadvantage for the theoretical use of gas as storage fuel would be an outdoor exposure to a thunder and lightning storm! All cases of energy derive from the conversion of mass. The automobile combusts gasoline to generate energy-promoting functions such as maintaining a charged battery, the movement of the fan belt, and locomotion of the vehicle. Fat is utilized in the body for metabolic energy (healthy metabolism = normal energy balance). Stored energy is potential energy, like a rock sitting at the top of a hill. Once resistance is removed, the rolling rock has kinetic energy (*i.e.* the energy of movement). Combustion results from fuel being oxidized, but it must first be activated. Activation energy is the energy required to ignite an engine, remove the resistance to the rock that allows the kinetic energy of rolling downhill, or the minimum amount of energy necessary to generate chemical reactions in cell metabolism. In biology, the energy contained within covalent bonds (*i.e.* molecules and atoms that share electrons) is released when the bonds are broken (i.e. catabolism), which may be harnessed to produce the work of biosynthetic processes (i.e. anabolism), for example, growth, or forming new bonds, and reproduction, or cell division. Alternatively, it may be harnessed to perform the work of exercise or movement, or other physiological processes. *Biological catalysts, i.e. enzymes, increase the rate of energy-requiring reactions by lowering the activation energy.*

The mitochondria may be considered the engines of cells that transform the reducing potential energy derived from breaking the covalent bonds contained within carbohydrate and fatty acid molecules into the higher energy bonds of ATP, the metabolic currency of the body. It does this by the transfer of electrons across an extraordinary quantum-mechanical transducer device (*i.e.* the electron transport chain). The final acceptor of each electron pair is oxygen resulting in the by-product of water (H_2O) to the production of ATP. This process is known as oxidative phosphorylation and represents the oxidative combustion of food allowing the capacity of the body to perform the work of exercise or movement and to generate heat, light, and electrical activity (see Figure 1.25 in Chapter 1 of this volume). Biological catalysts, *i.e.*, enzymes, increase

the rate of energy-requiring actions by lowering the activation energy. Conversely, oxidative stress refers to the dysfunctional oxidation of protein, lipid, carbohydrate, and nucleic acid (i.e. genetic) molecules within or outside of cells, that is, uncoupled to the production of reduced high energy compounds, especially ATP. As any given atom is oxidized the energy of opening the mass within the molecule containing the atom may be transferred to forming bonds of potential energy, by reducing atoms within molecules, that may subsequently be used to perform biologically constructive work. Otherwise, this energy is released as pollution-grade heat. Dysfunctional oxidized molecules promote the growth of inflammation and hence deplete the biological performance of cells and systems of the body.

Earlier in this book, we elaborated on the importance of the two laws of thermodynamics. The first law of thermodynamics states that energy cannot be created or destroyed, only converted from one form to another. This speaks to a quantitative uniformity of energy. We have abundantly illustrated energy transduction on various examples within the context of cell biology. *The second law of thermodynamics states that in an isolated (thermodynamically closed) system, its entropy can never be reduced spontaneously. In other words, entropy tends to increase by producing homogeneously dissipated energy states.* Another consequence is that in the universe, entropy is perpetually increasing because as more energy is utilized, more energy is liberated as heat. Heat spontaneously spreads from hot systems to cold systems. This speaks to the qualitative deterioration of energy following usage whether it be for biological processes or to propel an automobile. However, living systems are characterized by entropy reduction, which is not in contradiction with the second law of thermodynamics because they are open, and they dissipate heat into the environment by simultaneously performing metabolic work. By the same token, refrigerators reduce entropy and temperature locally by performing work (using electric power) and expelling heat into the environment.

It is the opinion of this writer that *senescence and disease states represent a progressive and, in fact, accelerating loss of energy-driven constructed organization activity with both normal aging and disease.* This occurs as energy is lost as heat with the spiral of inflammation related to escalating tissue loss especially the loss of muscle and central nervous system tissues (tissues with the highest density of mitochondria and metabolic/energy demands) or to disease states *per se*. In the latter case, pathological immune system activation promotes to conversions of energy-consuming inflammation to distinct regions of the body, hence redirecting organizational activity that can do work in the form of health, to a destructive, chaotic pathological process whereby energy falls to a pollution grade before it leaves the body.

Energy can also travel in the form of electromagnetic waves such as visible light, ultraviolet radiation, infrared radiation, radio waves, and microwaves. *Disturbed energy, either too much or too little energy stores, may result in a spiral of oxidative stress, whereby excessive reactive oxygen species accumulate in biological cells and tissues. The electronegative potential of these molecules promotes "stealing" electrons from neighboring molecules often overwhelming the body's antioxidative protective capacity to detoxify these reactive* intermediates and repair resulting damage. That is, a state of pathological redox leads to oxidative damage to proteins, DNA, and lipids. Consequently, chronic disease states ensue that characteristically affect the vascular tree and tissues that are most concentrated with mitochondria concordant with their higher energy requirements, especially potential nervous systems, heart, and skeletal muscle.

2.24 Molecular Motors and the Laws of Thermodynamics

The free energy contained in biological molecules results from the chemical energy that holds the atoms of the molecules together. The stronger and more stable the bonds the more energy is required for their formation. The formation of weaker bonds requires less energy. Conversely, breaking stronger and more stable bonds, which have lower free energy, releases more energy than breaking weaker, less-stable bonds. The formation of stronger more stable bonds results in the protons and electrons going from structures of higher to lower free energy of the reaction while the surrounding molecular kinetic energy increases as the heat released in these chemical reactions. Analogously, the macroscopic gravitational potential energy of a ball, which has already fallen to the bottom of a hill, has very little potential energy to fall further and hence little potential to generate kinetic energy. The Gibbs free energy that drives anabolic processes coupling exothermic to endothermic reactions also includes the hydrolysis of ATP to drive uphill reactions. Accordingly, the phosphate atoms of the ATP molecules have weak bonds with relatively little stability but high potential energy stored in these bonds. The bonds break quite easily requiring minimal energy input followed by the formation of new bonds in the products of ADP and inorganic phosphate, which have much lower potential energy having more stable bonds than the original ATP molecule. The formation of the new bonded structures of ADP, an inorganic phosphate, may release energy in the form of heat to the surrounding. This reaction can then be coupled with thermodynamically unfavorable reactions to give an overall negative spontaneous Gibbs free energy for the reaction sequence (the G for ATP hydrolysis as an exothermic process releasing 30 to 50 kJ/mol of energy, the more negative value in the presence of sufficient magnesium availability). The heat released from the hydrolysis of ATP in the context of the energy function of *Gibbs free energy results in the production of useful work such as muscle contraction, the active transport and establishment of electrochemical gradients across membranes and biosynthetic processes necessary to maintain life.*

In Chapter 1 of this volume we explained that the first law of thermodynamics says that the change in a system's internal energy is the sum of the heat added to or taken away from it, and the work it either does or is done upon it. For example, the system that is represented by the hydrolysis of ATP results in lower potential energy of the products of ATP, an inorganic phosphate, relative to the reactant ATP liberating heat to the surroundings. The internal energy of this system is lower; hence the internal energy of the system is reduced because the bonds between the atoms of the products have lower potential

energy being more stable, and the kinetic energy of the vibrations and rotations of those bonds is also reduced transferring that kinetic energy to the surrounding system with the transfer of heat. Consequently, heat added to the system to which it is coupled allows the potential for further transfer of that heat into useful work such as muscle contraction and other energy-requiring processes mentioned above. In human physiology, roughly 60% of the energy released from the hydrolysis of ATP produces metabolic heat not utilized for useful purposes of work. For example, the myosin biological motors cannot produce muscle contraction in the absence of the ATP energy that drives it. The survival value of contracting muscle represents a metaphorical comparison to the information of muscle that converts heat as an energy source to internal energy, which can, in turn, create something more enduring. According to the second law of thermodynamics [21], some of the thermal energy in transit is lost to randomized motions outside the system as entropy, serving to increase the kinetic energy of the universe, as energy that cannot be harnessed for useful purposes.

Power is the rate at which work is performed. Work is the product of force x distance and power is the product of force x distance over a period of time or equivalently, the product of force x velocity. The totality of work may be considered in terms of thermodynamics. In the absence of work done, objects seek states of lowest energy or highest entropy (which is a variation on the second law of thermodynamics). In biology, this force of nature that provides the existence of life of all forms may be termed negative entropy, borrowing thermodynamic terminology. However, negative entropy in a closed system is not allowed thermodynamically. Therefore, its reference must be in the context of an open an adiabatic system, whereby there is a transfer of matter and solar energy from the surroundings. The end result is structure and function formation, both of which lower the entropy and, as we discuss below, create information.

The magnetic fuel cells create electricity and accordingly an electromagnet that is positioned between permanent magnets provides the energy that promotes vehicle locomotion via rotation of the wheels. Analogously, the electron transport chain is also an electromagnet utilizing hydrogen atoms as its immediate fuel source (similar to fuel cells), stripped of their electrons to generate electrical current. In an intriguing analogy to fuel cells, biological engines not only generate electromagnetism, but they appear to transition into a quantum mechanical engine (or their components become quantum mechanical engines) wherein the electron transport chain behaves like a quantum mechanical oscillator. The fuel cell provides hydrogen atoms that are stripped of their electrons that in turn are the source of electricity that powers the engine. The electron transport chain of mitochondria in biological cells is similarly provided hydrogen atoms from bio-fuels. The electrons that are stripped are the source of energy ultimately funneled into the phosphate bonds of ATP, which power the biological work of the body. Fuel cells of automobiles that create electricity produce accordingly an electromagnet from a coil of copper wire. This coil of wire is positioned between the north and south poles of permanent magnets that result in rotation of the wheels from the Lorentz electromagnetic force (in simple terms). Analogously, biology appears to co-opt the electromagnetic force generated perpendicular to the electrical current of the electron transport chain to propel the hydrogen protons across the inner mitochondrial membrane (to the inter-mitochondrial space) (the proton-motive force). Additionally, it is intriguing to consider that associated electromagnetic fields generated perpendicular to the electron transport chain may play a role in mediating synchronized coherent conversion of nutrient biofuel to energy and ATP. Thus, it is tempting to think of the biological engine as both a fuel cell and a quantum mechanical engine. This quantum mechanical engine, which is the mitochondrion, can be viewed as a quantum mechanical oscillator, that when functioning as such, represents nature's expression of perfection that provides a selective advantage to the organisms in which it is endowed. This idea requires careful experimental validation.

The geometry of mitochondrial energy "generator" is quite complex as one needs proton pumps with a pH gradient across the mitochondrial wall, an electron transport chain taking place along the length of the wall, and ATPases which pump out ATP molecules into the cytoplasm, so there are both gradients of concentrations into and out of the mitochondria as well as electron motion (at least partly tunneling) along the wall. This can be viewed as an electric circuit and there should be some magnetic fields generated in the process but probably very small. On the other hand, in glycolysis, the associated chain of enzymatic reactions is distributed throughout the cytoplasm and there is no defined circuitry. More discussion on the quantum mechanical aspect of mitochondrial function can be found in the chapter on quantum biology.

2.25 Analogy between Mechanical and Biological Engines

An engine is any entity that does work. A motor is a particular type of engine that generates movement by a cycling mechanism. Electric motors invoke principles of electricity and magnetism based on providing an electric current through a metal wire coil and forming an electromagnet by putting a metallic solid at the core of the coil of wire. This electromagnet is placed between the north and south poles of a permanent magnet of ferric metal (soft iron). This arrangement generates motion mediated by a magnetic force. The essence of a rotating motor makes use of the three-dimensional relationships of magnetic fields between the north and south poles of permanent magnets, and the electrical current moving through an electromagnetic wire running at right angles to the magnetic field between the opposite poles of the permanent magnet. The direction of the magnetic field is determined by the right-hand rule whereby the current runs in the direction of the thumb with the magnetic field generated at right angles from the current in the direction of the fingers pointing on the right hand. This forms two of the three dimensions of the relationships between the magnetic field, the electrical current described above. The third dimension of this relationship represents the direction of the magnetic force that is responsible for the movement of the rotating motor, which is at right angles to both the magnetic field and the electrical current in accordance with

Fleming's left-hand rule. Fleming's left-hand rule uses the thumb and first two fingers of the left hand to illustrate the directions of these three dimensions mentioned above. The thumb points in the direction of the motion or magnetic force. The first finger represents the direction of the magnetic field while the second finger indicates the direction of the electrical current. Importantly, the thumb, first and second fingers of the left hand are held in such a way that all three digits are at right angles to one another, hence forming the three-dimensional relationship.

The coil of wire of the electromagnet is connected to a battery via a rotor or commutator. The rotor, or commutator, ensures that the current and associated magnetic field continues to move in the same direction when the U-shaped wire coil is flipped over from one side to the other. As the wire coil flips over the rotor (or commutator) causes the current generated from the battery to alternate to allow the magnetic force to promote full cycles of rotation rather than being canceled out at each half turn. This describes the functional design of an electric motor. On the opposite end from the battery/rotors (commutators) connected to the other end of the electromagnet coil of wire, can be a fan or any other moving cycling device (such as the rotating axel of a car) generated by the magnetic forces of the electric motor. The speed of rotation of the motor is proportional to the current generated by the battery, the magnetic field between the north and south poles of the permanent magnet, and the number of turns of the wire (each turn being a U-shape).

Richard Feynman stated that this classical description of electromagnets is flawed in the sense that electromagnetism is a quantum mechanical phenomenon fundamentally that includes quantum tunneling properties of spin and sometimes super-fluids and superconductivity (formation of Cooper electron pairs below critical temperature thresholds for a given metal or composite system, responsible for the so-called Meissner Effect, *i.e.* perfect diamagnetism). A similar design invoking electromagnetism as well as associated quantum mechanical phenomena appears to be responsible for mitochondrial metabolism in the process of oxidative phosphorylation in biological organisms. This is consistent with the notion of quantum metabolism described by Lloyd Demetrius and Jack Tuszynski [22, 23]. It is not exactly understood how this apparent quantum mechanical oscillator mediates the energy currency of ATP. However, the comparison to the design of the electric motor described above may be envisaged as follows. Electron reducing power is provided to the complexes arranged in the three-dimensional inner mitochondrial membrane. There are five complexes in total, which are cytochrome oxidase enzymes that contain iron. These iron-containing complexes may parallel or correspond to the permanent magnets of the electric motor. The electrons delivered from NADH to complex 1 (or $FADH_2$ to complex 2) represent the fuel source for the electrical current that is passed along the inner mitochondrial membrane at right angles to the iron-containing enzymes. *The magnetic force (Lorentz force) is perpendicular to both the direction of electric current (electron transport) and magnetic field and is the driver of the proton motive force that moves protons across the inner mitochondrial membrane into the inter-mitochondrial space creating an energy-dependent electrical potential.* The energy of this active transport and electrical gradient is trapped further down in the electron transport chain by the depolarization process whereby the protons travel back across the inner mitochondrial membrane into the mitochondrial matrix via the fifth complex of the electron transport chain. ATP synthetase is the only one of the five complexes, which does not contain iron. The protons along with the electrons combine with oxygen to form water in addition to the formation of ATP from ADP, an inorganic phosphate. It appears that the phenomenon of electron tunneling occurs at least at the level of complex 1 highlighting a quantum mechanical component to this process of ATP production. Additionally, as Demetrius demonstrated by mathematical formalism that invoked the same equations Peter Debye used to describe the specific heat of solids, a quantum phenomenon, the process of ATP production by oxidative phosphorylation is likewise quantized. We return to this aspect in the chapter on quantum biology.

2.26 Biological Thermodynamics

Closely related to the topic of molecular motors is the issue of biological thermodynamics. Biological thermodynamics concerns energy transformation in living matter, which in fact is the title of the German-British medical doctor and biochemist Hans Krebs' 1957 book [24]. Famously, the mitochondrial TCA cycle was named after and described by Hans Krebs. The reader is referred to Figure 1.24 in Chapter 1 of this volume for a schematic illustration of the TCA cycle. The focus of biological thermodynamics is on principles of chemical thermodynamics in biology and biochemistry. It is worth emphasizing that both classical and quantum systems of many particles are amenable to analysis using statistical thermodynamics. Classical systems follow the formulas of the Boltzmann probability distribution while quantum systems are either described by the Fermi-Dirac statistics or the Bose-Einstein statistics [25]. Chapter 1 of this book is dedicated to the topic of thermodynamics. Nonetheless, some important concepts in thermodynamics and applied to molecular motors are worth repeating. These concepts include the first two laws of thermodynamics, Gibbs free energy and the chemical potential. *Accordingly, thermodynamic biological engines may be considered to include far-from-equilibrium thermodynamic states in which the concept of the Gibbs free energy is invoked to create entropy reduction or, equivalently, an increase in information.* Such information gain in biology is an essential characteristic of living systems where it is generated across many hierarchical organizational scales ranging all the way from elementary particle physics (to accommodate physical principles) to the scales of molecules and macromolecules subsequently leading to the transition to life's building blocks such as cells, tissue, and organs.

Finally, on the scale of biological organization, information is further produced by entropy reduction [26] due to structure formation enabling information processing as all prokaryotic and eukaryotic organisms, either directly or indirectly, are dependent on the energetic input from the Sun as mentioned earlier. We discuss information theory later in this chapter.

One of the key results of the theory of special relativity is the famous equation $E = mc^2$, which is a linear relationship between mass and energy. Energy, which was initially produced from mass by the process of nuclear fusion in the Sun is initially reconverted from the form of electromagnetic rays back to mass in prokaryotes and plants by the quantum mechanical process of photosynthesis [27]. Subsequently, the spectacular phenomenon of quantum mechanical transduction again appears to manifest in specialized cells or organelles, mitochondria, the powerhouse of cells. *The mitochondria are the cell's most efficient energy-producing factory whereby nutrient hydrocarbons are stripped of electrons with the ultimate transformation to the universal biological currency of energy, ATP.* In this sense, the dependence of biological systems on mitochondria and quantum mechanics provides a fundamental basis for considering organismic life as a quantum mechanical-biological engine as is discussed at length in the chapter on quantum biology. Furthermore, there are additional contexts in which higher-level organisms possess quantum mechanical thermodynamic engines. A discussion of thermodynamic engines requires distinctions between machines, which may be understood by reductionistic reasoning, breaking systems down to their smallest component parts and rebuilding them such as may be done with a heat engine, or an electric motor. The latter may be a battery- or fuel cell-powered engine. However, biological engines cannot be understood entirely by reductionistic reasoning due to high degrees of nonlinear complexity.

Fuel cells and batteries are similar because they use a chemical reaction to provide electricity (which is nothing more than a continuous flow of electrons). *A battery stores the chemical reactant, typically metal compounds such as lithium, zinc, or manganese, which once used up must be recharged. A fuel cell creates electricity through hydrogen and oxygen stored externally, which when used up needs to be replaced rather than recharged.* In a fuel cell, a hydrogen atom is stripped of electrons (a negative stream of electric charge) converting the remaining hydrogen atom into a proton, or a positive electrically charged hydrogen atom. The electrons create the electricity that generates the energy of the car's motor. The proton bonds with oxygen atoms in the vicinity to form water molecules. Water molecule formation is an exothermic reaction, releasing heat, which becomes the exhaust (as steam) from the fuel cell engine. The energy chemistry in fuel cells is analogous to that in mitochondria. In both fuel cells and mitochondria, electricity is converted to water from hydrogen and oxygen atoms in the presence of catalysts.

Much of the fuel we burn is used to generate electricity, typically at approximately 30% efficiency. In addition to the combustion release of carbon dioxide (and putatively associated global warming) incomplete combustion products also include carcinogenic and polluting products such as particulates, nitric oxides, and polycyclic aromatic hydrocarbons. This process of burning fossil fuels is often required to create the energy that the fuel cell and battery provide. This is notable since it is often considered that fuel cell and battery-generated electric motors are environmentally friendly. *Mitochondria-generated energy is approximately 90% efficient.*

The intersection of the physical and biological realms of science may enhance the efficient application of biological energy to fuel cells that may not require fossil fuels. This may be especially useful for mobile applications of fuel cells. Conversely, integration of the *application of electromagnetic energy of engines offers conceptual insight metaphorically to electromagnetic phenomena in biology.*

Man-made electric battery or fuel cell motors invoke electromagnets. This may be most simply described by the following experiment. Attach a permanent magnet to the positive terminal of a battery. Attach a bent wire from the positive terminal of the battery to the negative terminal with the wire configured so that it runs up along both sides of the battery. The current will flow out of the battery from the positive terminal and into the magnet along both ends of the wire back into the negative terminal of the battery. *A magnetic field is created that is perpendicular to the current flow in the wire and a force (Lorentz force) is exerted by the magnetic field.* The Lorentz or magnetic force is oriented perpendicularly to both the magnetic field and the electrical current flow in the wire. Notably, once you have a current flow in the wire, you have magnetized the wire and that has made it an electromagnetic device. This electromagnetic force, Lorentz force, causes the wire to rotate in the magnetic field applied to it. This is a simple experiment, however, understanding these simple fundamentals allows us to understand how an electric motor works which is the secret to how all electric machines from wristwatches to large heavy equipment work.

Electrical current and electromagnetic induction have analogous applicability at the biological level similar to the automobile engine. The armature or rotor is the rotating part of the motor. This is the electromagnet analogous to the wire connected from the positive to the negative terminals of the battery. The stator in an automobile engine represents the permanent magnet in this case typically steel, which lies on either side of the armature or rotor. Rather than using a wire as in the prior simple illustration, the rotor consists of rings that in the case of the conduction metal (brushes) typically consist of aluminum or copper. Permanent magnets such as in the case of an automobile engine the stator, typically consist of iron (or soft iron) or steel (or hard iron). Just like the Lorentz force of the magnetized wire in the prior illustration causes the wire to rotate (according to the right-hand rule) the position of the rotating part of the motor, the armature or rotor, between the north and south poles of permanent magnets (the stator) promotes the rotation of the rotor and ultimately the locomotion of the automobile due to the simple relationship between a magnetized conduction metal, that is an electromagnet positioned between permanent magnets promoting the Lorentz force or magnetic force that rotates the motor. Stated another way, electric motors involve rotating coils of wire, which are driven by the magnetic force exerted by a magnetic field on an electric current. The electromagnetic force bears analogy at some level to the proton motive force that pushes protons across the inner mitochondrial membrane along the electron transport chain. The electrical current is provided by the reducing molecules of NADH and $FADH_2$ generated by the TCA cycle in the mitochondria. These are known to contain so-called high-energy electrons. *The electron transport chain from one complex to another may be considered metaphorically a conduction wire*

with the iron-containing cytochrome complexes (the enzyme oxidases/reductases representing the permanent magnets. This is not an entirely clear analogy since the process of oxidative phosphorylation along the electron transport chain is far more complex and involves a quantum phenomenon (electron tunneling). The metaphor between a man-made machine, a car engine, and a biological engine has many potential parallels. One is the caloric consumption of petroleum or carbohydrate by the process of combustion in a heat engine or in the mitochondria of our body's cells both requiring oxygen as a substrate and releasing carbon dioxide in water vapors as exhaust. In each case, a gram of petroleum fuel or carbohydrate nutrient fuel burns in the vicinity of 4 calories.

Even more intriguing is the analogy of how the separation of electrons from hydrogen atoms promotes the stream of protons and the continuous flow of electrons that occur in cell mitochondria compares to what occurs in fuel cells accompanying oxygen in the vicinity in each case that in turn provides the necessary energy and water vapor exhaust. Mitochondria are the organelles (components or "organs" of cells) that play the role of the power plants of the body. Rather than man-made, these highly complex organelles are self-organizing structures. They are actually rooted in archaic unicellular organisms (bacteria) that evolutionarily became subsumed into the cells of eukaryotes (multicellular organisms such as human beings). *Biological existence is predicated on the self-organizing evolution of complex adaptive systems. Accordingly, organisms at all hierarchical scales along the food and predator chain are characterized by a dynamic equilibrium whereby the entire internal environment has characteristic steady states that maintain homeostasis essential for an organism's survival. The establishment of steady states requires a continuous input of free energy into the system.* Over time, conditions become stable and the system is maintained at a higher level of order than its surroundings leading to an entropy reduction paradox that is discussed at length elsewhere in the book. Accordingly, the energy provided by mitochondria as the universal biological currency of ATP, represents the metaphorical biological engine. The engine that drives an organism's function may be fundamentally considered to be the collective mitochondria, which metaphorically represent the combustion chamber where nutrient fuel is oxidized to the products carbon dioxide and water vapors and the energy currency in the form of synthesized molecules of ATP.

Mechanistically, mitochondria represent at least to some degree molecular quantum mechanical oscillators whose functioning involves the process of quantum tunneling. It makes sense that the physiological source of the calorie is rooted in Einstein's equation $E = mc^2$ predicated by the process of plant photosynthesis. Just like chloroplasts in plants that are capable of transforming energy from one form to another in accordance with the first law of thermodynamics, mitochondria in cells carry out an analogous function. In each case, the phenomenon of tunneling is thought to be the mechanistic quantum "magic" responsible for carrying out this remarkable transformation. Additionally, *the production of ATP is quantized according to the same statistical quantum mechanics formalism described by Einstein for gases and expounded upon by Debye for solids in the 1920s.* Essentially, Debye demonstrated that crystals use vibrational quantized energy, which is manifested in the dependence of their specific heat as a function of temperature. In an analogous fashion, biological metabolism depends on mass, which is a hallmark of quantum statistics. That is, Demetrius used the same equations of solid-state physics enlisted by Debye in his theory of specific heat of solids (the amount of heat required to raise the temperature of the solid by 1° Celsius) to show that our cellular metabolism is the manifestation of the quantum nature of energy production, hence the term quantum metabolism. In qualitative terms, heat introduced into a solid crystalline lattice increases molecular quantized vibrations that correspond almost exactly to the enzymatic oscillations embedded in the mitochondrial bio-membranes (from nutrient heat). The amount of internal energy in the crystal lattice is quantized to the rise in temperature and corresponds in mathematical terms almost exactly to the thermal energy that drives the body's metabolic rate as a function of weight. Since organized complexity corresponds to anabolism, requiring the biological currency of ATP, the quantum nature of biological energy production was used by Demetrius to demonstrate that basal metabolic rate correlates to body size according to the allometric scaling laws of physiology [22, 23] investigated almost a century ago by Kleiber.

This ATP-dependent organized complexity of biological systems represents a far-from-equilibrium thermodynamic state. This is created by the coupling of the molecular quantum mechanical engines of mitochondria to the mechanical machinery of the cell in the form of motor proteins and the proteins of the cytoskeleton (actin filaments, intermediate filaments, and microtubules). Batteries, fuel cells, and quantum mechanical processes are represented by models rooted in physics, and their analogous biological engines may be considered to synergistically satisfy the ATP energy needs of the body. Furthermore, additional models in physics rooted in Chaos theory provide the basis for how mitochondrial energy is harnessed to create and maintain dynamic far-from-equilibrium steady states and homeostasis. This is characterized by nonlinear complexity often including "chaotic" systems capable of carrying out the flexible adaptive emergent phenomena necessary for organismic survival.

MRC = the rate of energy production (substrate to product) per unit time in a classical regime versus MRQ = the rate of energy production per unit time as a product of mass in a quantum regime.

Accordingly, in classical regime (and classical teaching of metabolism) a fast metabolic rate in terms of efficiency of energy production engages OxPhos.

Glycolytic production of ATP is much faster than mitochondrial OxPhos per cycle of substrate conversion to product. However, it is much less efficient in terms of the amount of ATP produced per molecule of glucose consumed.

The efficiency of nutrient fuel mitochondrial oxidation is gauged by the P:O ratio, *i.e.* the amount of ATP produced per oxygen consumed. This ratio is higher for glucose oxidation (coupled to glycolytic metabolism) than for fatty acid oxidation.

Principles of classical thermodynamics remain very important in explaining many aspects and components of biological function in concert with the principles of physics. They remain particularly useful at the level of designing machines. However, investigating the fundamental nature of matter one must go beyond the "billiard ball collisions" level of classical representation at the microscopic level and instead use the concepts of quantum mechanics, which is especially true of molecular biology. Understanding how mitochondria work, how various enzymes operate, or how recognition operates at the subcellular level invokes the insights of quantum mechanics, which are explored in detail in a separate chapter on quantum biology.

Finally, on the scale of biological organization, *information is further produced by entropy reduction due to structure formation enabling information processing as all prokaryotic and eukaryotic organisms, either directly or indirectly,* are dependent on the energetic input from the Sun. We discuss information theory later in this chapter but first delve into a related topic, namely biological signaling.

2.27 The Many Types of Biological Signals

The coordination of cellular activities in the human body involves two systems engaged in biochemical and electrical signaling [28]. These are: *a) the endocrine system which employs hormones, i.e. chemicals secreted into the blood by endocrine glands and carried by the blood to the responding cell, and b) the nervous system which employs electrical impulses passing from the central nervous system to muscles and glands. We elaborate on these signaling and coordinating activities below.*

Signaling by varied means involves energy transduction and is required to regulate the complex behavior of living systems including humans. For any particular cell of the organism, in addition to intracellular signaling, it must both transmit and receive extracellular signals. In order to interpret signals from other cells, a given cell requires specialized membrane receptors, which can detect the presence of signaling molecules in the extracellular fluid. In addition to biochemical signaling, there could be other forms of physical interactions that cells can use for signal propagation, processing, and integration. Albrecht-Bueller [29] proposed an "intelligent cell" model that claims that living cells are able to sense light through the use of centrioles (Figure 2.15). Since the centrosomes are always found with a perpendicular orientation of their two constituent centrioles, this would allow the cell to discern directional information about a signal through latitude and longitude measurements. Furthermore, mitochondria may generate light signals at infrared frequencies.

A living cell has several other methods of communication at its disposal through the use of its varied signaling molecules. The molecules are first packaged and then expelled from the cell. In the simplest type of messaging, the chemical signals are dumped outside the cell and carried diffusively. This method of communication is effective for only the signaling of nearby cells and is known as paracrine signaling. Synaptic signaling is a refined version of the paracrine model where the signal

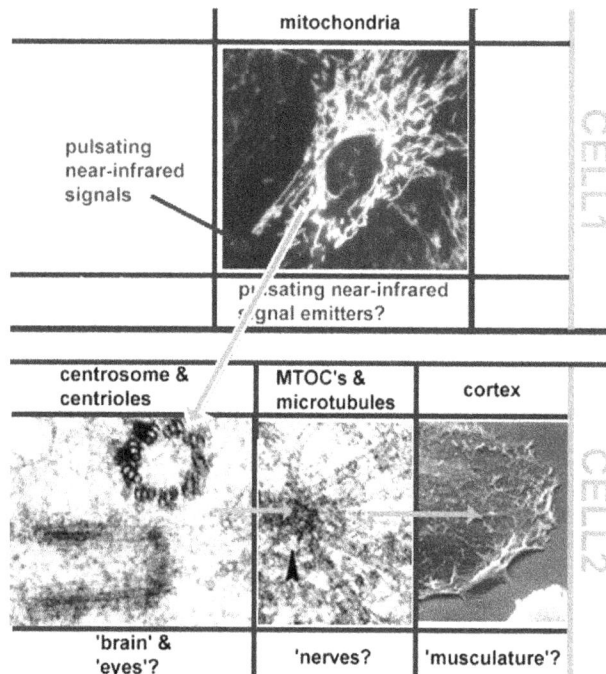

FIGURE 2.15 A pictorial illustration of cellular movement using centrioles as light-sensing devices.Source: www.basic.northwestern.edu/g-buehler/htmltxt.htm.

molecule, a neurotransmitter [30], is released at a specifically designed interface providing intimate contact between the source and target cells. This allows quick and direct signaling but still relies on fairly slow and imprecise molecular diffusion to carry the signal molecules across the narrow junction. The last main type of signaling, known as endocrine signaling, is used when the target cells of the signal are either more distant or more widespread. The molecular signaling molecules used in this case are referred to as hormones and are secreted by the cell into the circulatory system. Although the process of molecular diffusion is used yet again in this case, propagation in the bloodstream enables these molecules to carry the signal over a long distance. Due to the dilution effect of the circulatory fluid, hormones must be effective even at concentrations as low as 10^{-8} M (moles per liter).

Intracellular signaling includes mechanisms such as the action potential, which is electrical in nature and driven by chemical gradients. Sensitivity of individual cells to concentration and potential gradients is necessary if the cell is to respond to gravitational or electric fields. Intra-cellular signaling coordinates the orchestra of cellular processes to ensure that the entire cell works in harmony. In mitosis, chromosome segregation to each pole of the mother cell is mediated by MTs. However, the simultaneity of the separation must be explained and requires some kind of signal to be mediated by the MTs, which may well be electrical or electromagnetic in nature.

2.28 Neuronal Signal Propagation

Nerve cells can be viewed as conducting cylinders surrounded by insulating sheets of bio-membrane. From the viewpoint of

electrostatics, such a sheet behaves like a capacitor, but it is not much of an insulator since it leaks electrical currents *via* ion channels. While the extracellular fluid is rich in sodium ions (Na^+), and Cl^- ions, the intracellular fluid, *i.e.* cytoplasm, is rich in potassium (K^+) ions. The cytoplasm also contains mostly negatively-charged proteins. Since a living cell has a selectively permeable membrane, there is a small buildup of negative charges just inside the membrane and an equal number of positive charges on the outside. This gives rise to a resting potential, which causes an electric charge separation across the plasma membrane. The size of the resting potential varies, but in excitable cells such as neurons and muscles, it is typically held at about –70 mV. The influx of sodium ions and the efflux of potassium ions occurs due to biased diffusion. However, certain external stimuli reduce the charge across the plasma membrane, namely: a) mechanical stimuli (*e.g.* stretching, sound waves) activate mechanically gated sodium channels, and b) certain neurotransmitters (*e.g.* acetylcholine) open ligand-gated sodium channels. As a result, the facilitated diffusion of sodium into the cell reduces the resting potential at that place on the cell creating an excitatory post-synaptic potential (EPSP). If the potential is reduced to the threshold voltage (about –50 mV in neurons), an action potential is generated in the cell, which ranges from –40 to –90 mV, but is typically –70 mV. When a neuron is in its "resting" state it is not conducting an electrical signal. It is the change in the resting potential, which is so important in the initiation and conduction of a signal. "Gates" in the membrane open when a sufficiently strong stimulus is applied to a given point on the neuron and sodium ions rush into the cell [6].

The Na^+ ions are driven into the cell by attraction to the negative ions on the inside of the cell as well as by the relatively high concentration of positive sodium ions outside the cell. This large influx of Na^+ ions first neutralizes the negative ions on the interior of the membrane and then causes it to become positively charged. If depolarization at a location on the cell membrane reaches the threshold voltage, this results in the opening of as many as hundreds of voltage-gated sodium channels in that portion of the plasma membrane. The channels remain open for about 1 ms allowing some 7,000 Na^+ ions to rapidly enter into the cell. The sudden complete depolarization of the membrane opens up more of the voltage-gated sodium channels in adjacent portions of the membrane creating a wave of depolarization that sweeps along the cell membrane giving rise to the action potential. This electrical signal propagates down the axon at speeds between 0.5 and 130 m/s, to the next neuron or to a muscle cell. *As an action potential propagates along a neuron it undergoes the following phases: a) resting state, b) depolarization state, c) repolarization state, and d) undershoot which occurs because the potassium gate stays open too long.*

Each neuron is capable of receiving up to 10^5 inputs from its neighbors. A long extension of a neuron is called an axon and its length ranges from several mm up to a meter for motor neurons. At the far end each axon branches out into axon terminals that release neurotransmitter molecules into the so-called synaptic clefts or into gap junctions. The material released at the synapse has to be recycled back into the neurotransmitter vesicles for release. This is achieved by transport through axonal microtubules at speeds reaching 400 mm/day.

Transport along the return direction, called retrograde transport, occurs at roughly half the speed of anterograde transport. Simultaneously each nerve can accommodate many signals in an analogous way that a number of telephone calls are transmitted along a single telephone cable.

In addition to electrical signaling, coordination by the nervous system is also chemical. *Most neurons release chemicals called neurotransmitters, which are sent to a receiving cell that may be one of the following types: a) a "post-synaptic" neuron, b) a muscle cell, and c) a gland cell.* Note that neurotransmitters are chemicals that act in a paracrine fashion. Hence, both nervous and endocrine coordination involves chemical signaling but nervous coordination is faster and more location specific.

The junction between the axon terminals of a neuron and the receiving cell is called a synapse (Figure 2.16). A synapse can be either the gap between one neuron and another or between a nerve terminal and a muscle fiber across which electrical signals pass in only one direction. Conduction across a synapse is not the same as that along the neuron, in which the spike potential travels in both directions from the point of stimulation. In most cases of *synaptic conduction, the signal transmitted is a chemical involving neurotransmitter molecules. The neurotransmitters are a diverse group of chemical compounds ranging from simple amines such as dopamine and amino acids such as g-aminobutyrate (GABA) to polypeptides.* Assuming that a given neurotransmitter has a net charge of several elementary charges, the current flow across the synapse is of the order of 10^{-8} A (or 10 nA).

Each neuron's dendritic tree is connected to thousands of neighboring neurons. When one of those neurons fire, a positive or negative charge is received by one of the dendrites. The strengths of all the received charges are added together through the processes of spatial and temporal summation. Spatial summation occurs when several weak signals are converted into a single large one, while temporal summation converts a rapid series of weak pulses from one source into one large signal. The aggregate input is then passed to the soma. However, the soma and the enclosed nucleus do not play a significant role in the processing of incoming and outgoing data. The strength of the output is constant, regardless of whether the input was just above the threshold, or many times greater. This uniformity of signal strength is critical for the proper functioning of the brain where small errors can propagate and become magnified, and where error correction is more difficult than in a digital system, for example, a computer chip.

A single neuron may have thousands of other neurons with which it exchanges signals. Some of these connected neurons release activating (depolarizing) neurotransmitters; others release inhibitory (hyperpolarizing) neurotransmitters. The receiving cell is able to integrate these signals. There are in general two types of synapses: a) excitatory (the neurotransmitter at excitatory synapses depolarizes the postsynaptic membrane) and b) inhibitory (the neurotransmitter at inhibitory synapses hyperpolarizes the postsynaptic membrane). The mechanisms by which neurotransmitters elicit responses in both pre-synaptic and post-synaptic neurons are diverse. The physical and neurochemical characteristics of each synapse determine the strength and polarity of the new input signal. Changing the constitution of various neurotransmitter chemicals can increase or decrease the amount of stimulation that the firing axon imparts on the neighboring dendrite. Many

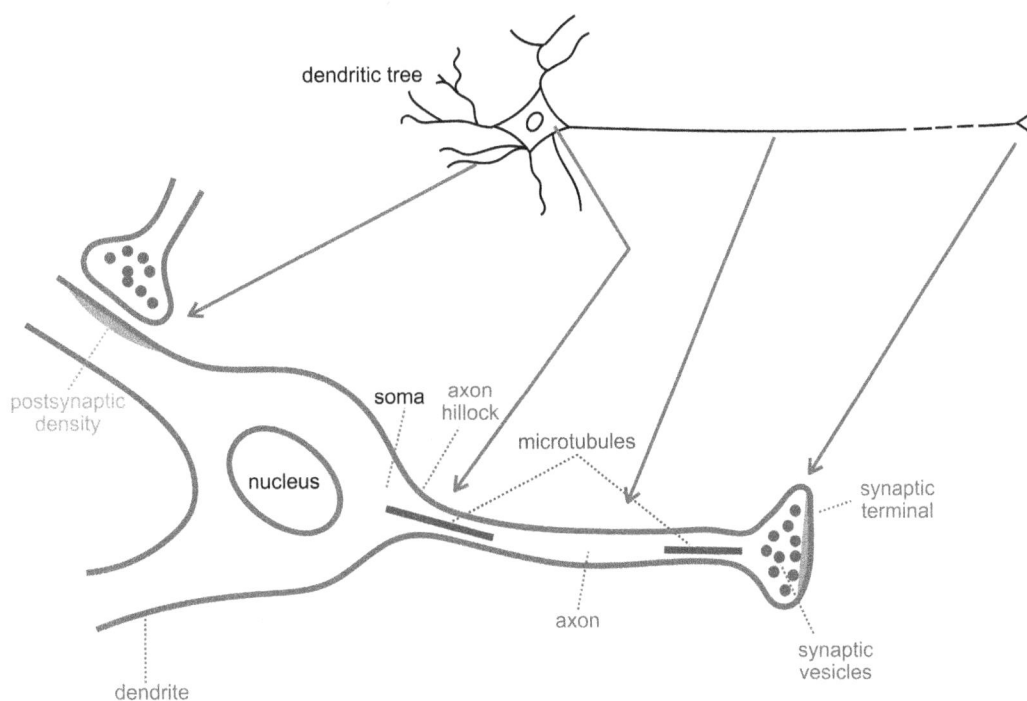

FIGURE 2.16 Illustration of a synaptic cleft (synapse) between a sending neuron and a receiving neuron that communicate with each other via electrical currents called "action potentials" and chemical neurotransmitters, in the context of an ultrastructure of the neuron. Source: adapted from Levitan, I.B. and Kaczmarek, L.K. (2003). *The Neuron Cell and Molecular Biology.* Oxford University Press, New York NY.

drugs such as alcohol and psychoactive compounds have dramatic effects on the production or destruction of these critical chemicals.

2.29 Electromagnetic Energy across Scales of Biology

An integration of electromagnetic energies shows its potential as the link across several scales of biology. Even at the single level of the cell, for example, the interaction of electromagnetic energies can manifest at much higher levels of organization. For example, magnetic fields influence microtubules' self-assembly which in turn influence biological processes like cell growth and differentiation. This may hold true for a single cell or a cluster of cells and hence increased levels of organization.

Electromagnetic phenomena external to a biological system appear to generate oscillatory patterns characteristic of the physiological activity of the cell as well as higher levels of organization such as the brain. Accordingly, it follows that living cells are, in a sense, antennae that resonate with their environments. These receptive structures include organelles (the cell nucleus and ribosomes with oscillatory patterns of transcription), microtubules and their constitutive elements, tubulin dimers, and even water itself (in the form of the so-called exclusion zone water around hydrophilic surfaces). *The electromagnetic environment resonates or interferes with the healthy oscillatory patterns of the cells and the tissues it comprises, the latter causing pathophysiology.*

Therapeutic approaches in medicine, which aim to normalize pathological states attempt to do so by surgical or chemical means. However, iatrogenic complications of some type are unavoidable. *Changing electromagnetic fields therapeutically may provide the external signal to a damaged signal receiver, the cell that can be processed clearly.* That is, the abnormal cells are incapable of processing the electromagnetic signals otherwise. Thus, modified signals promote normal functional states in the absence of healthy organic tissue. That is, the modified signals create virtual structures in the absence of the structures capable of reading and modifying the usual signals.

Physical, chemical, biological, and even social systems are connected within the electromagnetic environment. Here we are particularly interested in the function of electromagnetic interactions at the level of individual cells. The following mechanisms of intermolecular energy transfer should be considered in the context of cellular activities:

- Energy transfer by radiation.
- Energy transfer by charged waves.
- Energy transfer by hopping of individual charges.
- Energy transfer by elementary excitations such as phonons or excitons.

In the subsections below we discuss in more detail these specific forms of electromagnetic interactions.

2.29.1 Bioenergetics: The Davydov Soliton

As explained in detail earlier in this chapter, the currency of biochemical energy is the molecule ATP or its analogs such as GTP. ATP binds to a specific site on a protein such as kinesin, reacts with a water molecule, and releases a quantum of biological energy in a process called hydrolysis.

While single energetic events are reasonably well understood, energy transport on the length scale of protein filaments, say microtubules, or along DNA strands is still a challenge. In the 1980s Aleksander Davydov [31] developed a non-linear theory of biological energy transfer which focuses on the so-called amide I bond of C = O in a peptide chain containing the H-N-C = O group and its vibrational and dipolar coupling. Now suppose that the amide I vibration is excited on one of the three spines of a helix by the energy of the ATP hydrolysis. The oscillating C = O dipole with dipole moment interacts with the dipoles of the neighboring peptide groups of the spine and is the interaction energy between two parallel dipoles positioned at a distance R from each other. Thus, the energy of the initial excitation does not remain localized, it propagates through the system. However, the excitation of the amide I vibration results in the deformation of the H – O bond since it is much weaker than the covalent bonds of the helix. The deformation of the hydrogen bond can be described as the deformation of a spring of length K. The initial deformation of a spring will cause the deformation of neighboring springs of the same strength and the disturbance will propagate along the chain of coupled springs (H – O bonds) as a longitudinal sound wave. The longitudinal sound wave can be dynamically coupled in a nonlinear way to the amide I vibration providing a potential well trapping the energy of the amide I vibration. The energy remains localized and travels along the helix chain as a soliton [32]. Davydov treated the amide I vibrations quantum mechanically while the longitudinal sound wave classically. He then used a continuum approximation resulting in the non-linear Schrödinger equation describing this effect. The resultant solution is *a localized pulse-type wave called a soliton, which travels without energy loss maintaining its shape.* This is a very attractive hypothesis, which would explain an amazingly high efficiency with which energy is distributed in living systems. However, so far experimental evidence still remains elusive.

2.29.2 Biological Coherence: The Fröhlich Model

While Davydov tried to find spatial localization of vibrational energy in biological systems such as DNA and peptides as we have outlined in the previous section, another famous physicist, Herbert Fröhlich sought evidence for frequency selection in biological systems. Starting in the late 1960s and continuing until his death in 1991, Fröhlich developed a theory of biological coherence that was based on quantum interactions between dipolar constituents of biomolecules [33], in particular the polar head groups of cell membranes. Fröhlich advocated momentum-space correlations within a living system such as an enzyme, a membrane, a cell, or an organism. This dynamic order emerging in such a system would be a characteristic feature that distinguishes living systems from inanimate matter. Fröhlich assumed the following:

a) A continuous supply of metabolic energy (also referred to as energy pumping) above a minimum threshold level required to achieve synchronization of dynamical membrane dipole oscillations.

b) The presence of thermal noise due to physiological temperature.

c) Internal structural organization of the bio-system that promotes functional features.

d) The existence of a large trans-membrane potential difference.

e) A nonlinear interaction between two or more types of degrees of freedom.

As a result of these nonlinear interactions, in addition to the global minimum that characterizes a biological system in its non-living state, a metastable energy minimum was predicted to emerge in the living state.

The resultant effect of biological coherence is due to condensation of quanta of collective polar vibrations. It is a non-equilibrium property due to the interactions of the system with both the surrounding heat bath and a metabolic energy supply. This externally supplied energy is channeled into a single collective mode that becomes strongly excited. Most importantly, it relies on the non-linearity of internal vibrational dipolar interactions. Associated with this dynamically ordered, macroscopic quantum state is the emergence of electric polarization due to the ordering of dipoles in biomolecules. *Fröhlich predicted the existence of coherent modes of excitation such as dipole oscillations operating in the microwave frequency range of $10^{11} - 10^{12}\,Hz$* [33]. Nonlinear interactions between dynamic degrees of freedom were predicted to result in the local stability of the polarized state and in the long-range frequency-selective interactions between two identical systems such as two cells or two enzymes [34]. In this resonant frequency case, the effective interaction energy between two oscillating dipoles exhibits long-range dependence on distance dropping off as $r^{1/3}$ (cube root of r). Consequently, due to the resonant dipole-dipole coupling in a narrow frequency range, *the entire biological system can be seen as a giant oscillating dipole* [35, 36].

The degrees of freedom leading to the coherence of the phospholipid head groups forming giant dipole oscillations described by Herbert Fröhlich represent optical phonons that occur within membranes of cells due to their strong electric potential gradients on the order of 100 mV across the thickness of 5–10 nm giving an electric field intensity of $1 - 20 \times 10^6$ V/m. The resultant dipole-dipole interactions were predicted to propagate with a velocity of about 10^3 m/s. However, it is still unclear which molecular groups of polar heads of membrane phospholipids may lead to coherent dipoles oscillations in the microwave range of frequencies as predicted by Fröhlich [33]. The polar heads refer to the phosphatidyl groups (phosphatidyl choline, phosphatidyl serine, and phosphatidyl inositol. However, what specific molecules are theorized to oscillate is unclear, presumably these are the double-bonded phosphorous-oxygen or oxygen-carbon structures whereby the atoms of these structures coherently oscillate giving rise to optical phonons. The vibrational motions of the atoms represent the displacement of both the nuclear and electronic components, which alternate in a periodic fashion moving further apart and closer together in each half-cycle. These are coherently oscillating adjacent like molecules across the cell's membrane and possibly a similar mechanism can even involve like cells in the vicinity. This coherent activity can result in an increase in the active transport of calcium ions and associated oxygen molecular complexes across the cell membrane. The proposed role for this is to mechanistically explain the emergence of coherence in cell biology.

The cell membrane as described above in terms of Fröhlich long-range coherence with a given oscillation frequency, dependent on its structural integrity or rigidity can modulate the oscillation frequency of protein receptors embedded within the cell membrane. *Cell membrane properties are affected by the state of health of the cell or even by cell death.* This would affect the rigidity of the cell membrane and hence the receptor embedded in the membrane, which becomes united. The receptor itself would have its own characteristic frequency not initially in resonance with its ligand, for example, a hormone. The receptor's characteristic frequency is distinct from that of the cell membrane in which it is embedded. The receptor may perhaps modulate the frequency of an unattached hormone or other ligand analogously to how a cell oscillation correlates and modulates the frequency of the embedded receptor's oscillation. Once a receptor and ligand become bound, the receptor and attached ligand share a heavier mass and oscillate in unison, each with a shifted dampened frequency. The heavier mass of the bound state of a ligand receptor can be reflected in a frequency shift downward of the oscillating atomic molecular component of the receptor and ligand now oscillating in unison. For example, a phosphorous-oxygen double bonded structure consists of two atoms that are oscillating in unison with other like atomic molecular structures that generate an optical phonon. Each oscillating element per se does not change mass just the totality of all the oscillating components taken together does so. The shift in frequency can be useful by preventing the inefficiency of having another hormone or, like ligands, attracted to a receptor already bound to another ligand. Prior to binding, a long-range interaction may occur for example between a hormone ligand and receptor separated in distance by a few nanometers based on the size of the two interacting component objects. This interaction is a form of electrostatic attraction that promotes the translational movement of the ligand towards its receptor with ultimate docking. Subsequently, hydrogen bonding may occur to stabilize the complex.

In addition to membrane dipoles, several other candidates for Fröhlich coherence were hypothesized, namely: double ionic layers, dipoles of DNA and RNA (ribonucleic acid) molecules, plasmon oscillations of free ions in the cytoplasm, etc. Applications of the Fröhlich theory were subsequently made to cancer proliferation where a shift in the resonant frequency was seen to affect the cell-cell signaling, the brain waves, and enzymatic chemical reactions to name but a few examples.

The standard electrodynamic dipole-dipole interactions are short-range molecular interactions and hence of limited utility in biology. However, the dynamic interactions of atomic and molecular vibrations as described by Fröhlich are long-range because they involve resonance in the oscillating dipolar degrees of freedom. The macroscopic character of these oscillating giant dipoles relates to their biological effects on structure and function mediated by inducing the movement of charged particles in electric fields [37]. Fröhlich described the process of collective oscillation of giant dipoles, which represent a quantum process in the sense of a Bose-Einstein condensate to require the continued input of energy, which he compared to the "pumping action in a laser". It may be presumed that the source of energy input serving to induce dipole oscillations (either Fröhlich type or induced dipole-induced dipole type) are either the electric fields that exist across cell membranes, a high potential energy gradient, *i.e.* the Nernst potential, or biophotons. In either case, the electric field or biophotons would seem to provide the initial non-thermal activation. Recently, different types of Fröhlich condensation have been classified as weak, strong, and coherent. In particular, the coherent laser-like oscillation described in cell membranes would require a higher energy input than would the type of condensation, for example in the dipole states of microtubules which is of the weak type. In this case, the optimal frequency is found to be only around 8 MHz, which might be preferable for mediating quantum computation. It may be worth mentioning that Hameroff and Penrose invoked dipolar quantum oscillations of neuronal microtubules as key processes underlying consciousness in the so-called Orch OR theory [38]. The oscillating motions of molecular dipoles of tubulin involve atomic-level electron cloud movements requiring two or more connected or closely associated benzene ring molecules. This is a quantum phenomenon derived from the basic force of electromagnetism between the positive charge of protons and the negative charge of electrons. *However, microtubule oscillations appear to involve multiple physical mechanisms, which could convey different types of information or signaling depending on the biological context. These include the Fröhlich dipolar oscillations, which involve atomic or molecular optical phonon vibrations, and the electronic oscillating states of the hydrophobic aromatic rings with underlying London forces of induced dipoles.* The latter is a more rapid frequency oscillation due to the substantially smaller mass electrons versus the nuclear component of protons and neutrons engaged in oscillations. Note that the oscillations that involve electron cloud atomic-level movements are quantum phenomena derived from the basic force of electromagnetism between the positive charge of protons and the negative charge of electrons. As electrons move in one direction asymmetrically, the positively charged protons move in the opposite direction. This positive charge exerts a force of attraction on oppositely charged (negatively charged) particles causing them to move toward the positive charge.

While Fröhlich was not a biologist, his theory has important biological implications in terms of structure and function due to this quantum coherent laser-like state promoting synchrony and entanglement of cells. These *specific resonant electromagnetic interactions can drive motion and organize coherent biological functions within and between cells. Fröhlich interactions are resonance interactions requiring matching frequencies of oscillation at the micrometer range, about the size of a cell or larger.* Since long-range coherence is understood to be coherence at distances longer than the size of the object itself, the long-range Fröhlich interactions refer to cell-cell interactions. Moreover, Fröhlich's long-range coherence may be applicable to interactions between a circulating hormone, growth factor, cytokine, or other endocrine messenger ligand that seeks binding to a cell membrane-based protein receptor. Notably, a capillary comes within the distance of four cells of any cell in the body. The circulatory hydrostatic pressure pumps ligand-containing blood through capillaries into the extracellular fluid before the oncotic pressure of the circulation draws it back into the cell. This allows a period whereby the circulatory fluid containing hormones comes within a few cells distance of its receptors, hence essentially representing the conversion of an endocrine system into a paracrine one. An example of an endocrine messenger ligand having a paracrine interaction may be dipole oscillations within tyrosine and other aromatic amino acids of insulin with phase-coherent oscillations of one another within the insulin molecule that matches the frequency of its interacting protein receptor, for example of the aromatic ring containing amino acids of the insulin receptor. These matching frequencies bring the ligand closer to its receptor as a result of a form of electrostatic attraction. *The frequency of the hormonal and receptor oscillations is in the range of hundreds of kilohertz (kHz), which is consistent with the mass of the amino acids whereby the electron subatomic particles are oscillating along the aromatic ring structures. It should be pointed out that this is not a long-range interaction of the giant dipole described by Fröhlich, although there may be endocrine messengers and protein receptor interactions that do involve Fröhlich-type oscillations.*

The dynamic dipole oscillations that occur in microtubules as well as in cell membranes and by extension oscillations in circulating hormones or other endocrine messengers and their protein receptors embedded in cell membranes, as moving charged particles, may be considered analogs of an electricity-conducting wire. A given combination of oscillating charged particles may interact as a result of their emitted electromagnetic field in a number of ways. They promote the movement of charged particles. This movement may be translational movement of freestanding particles such as a hormone ligand being drawn toward its receptor, or movement of ions across the cytoplasm of a cell. Conversely, magnetic field interactions may accelerate the motion of subatomic charged particles such as electrons of atoms and molecules for example increasing an electron dipole oscillation frequency. This latter effect may also be the mechanism for modulating and coordinating a matching frequency between ligand and receptor which may in turn cause a magnetic field translational effect on freestanding particles such as on a free ligand pulling it towards its receptor. That is, a magnetic field induced by the moving subatomic particles (electron dipoles) of the ligand or of the receptor shifts the frequency of the moving electron dipoles of the other so that the frequencies match. Moving charged particles in the cell also exhibit an interaction of their magnetic fields that create a Lorentz force which may act on for example surface cell membrane receptors making it geometrically accommodating for its binding ligand, or alternatively these electromagnet-like biological structures may interact by the Lorentz force to open ion channels. The opening of ion channels is fundamental to very many cell processes. For example, calcium influx promotes the movement of neurotransmitters to the synaptic cleft by creating an electrochemical gradient potential from presynaptic to postsynaptic neurons. In dendritic processes of postsynaptic cell bodies, calcium promotes phosphorylating activation of calcium calmodulin kinase II (CaMKII) enzyme, which is in turn responsible for depositing synaptic information into microtubules on whose surfaces they move [39] (Figure 2.17). *On the other hand, Lorentz force effects on ligand molecular docking and complimentary geometrical optimization as in the case of insulin receptor activation may have profound bioenergetic and physiological effects ranging from glucose uptake into cells to lipid and cholesterol utilization, blood pressure regulation, satiety, and body weight maintenance and even normal cognition.*

Olfaction is the best example thus far of vibrational resonance of aromatic (in two senses of the word, aroma [as in olfactory] as well as benzene or indole-type) rings. It is perhaps serendipitous that both meanings apply to the molecules recognized by their olfactory receptors. There is a mutual attraction and shift in frequencies of these ligands and their receptors [40]. This is quite consistent with the Fröhlich theory although with a different application to biology since Fröhlich was primarily concerned with cell-to-cell interactions. The extension of this concept to the endocrine system is hence plausible with a high level of expectation. Other molecules, such as serotonin and dopamine, containing the benzene rings of tryptophan that bind to neuronal cell receptors and strengthens signals, which are part of cognition, serve as another excellent example [41]. This particular example is especially intriguing because it concerns the notion of quantum coherence. The quantum wave function of a molecule is very small in this case. However, in special cases such as *macroscopic collective quantum states, referred to as condensed states, this type of interaction becomes very long-range*, in fact macroscopic. Quantum coherence is also believed to play an important role in the functioning of the photosynthetic reaction center II [42]. Although it is completely unexplored territory it is conceivable that one can amplify the effect of, for example, protein receptor interactions, such as insulin-insulin receptor by involving water vibrations surrounding these molecules. *Conglomerates of water molecules surrounding any protein-ligand hormone, enzyme, etc. could be vibrating at the same frequency and then the dipole moment could become proportionately larger and the strength of interaction could increase dramatically. This describes another possibility for why water is so important in biology.*

It must also be said that various experiments appeared to demonstrate the sensitivity of metabolic processes to certain frequencies of electromagnetic radiation above the expected Boltzmann probability level. While some of these experiments illustrate non-thermal effects in living matter that would require non-linear and non-equilibrium interactions for an explanation, no unambiguous experimental proof has been furnished to date to support Fröhlich's hypothesis.

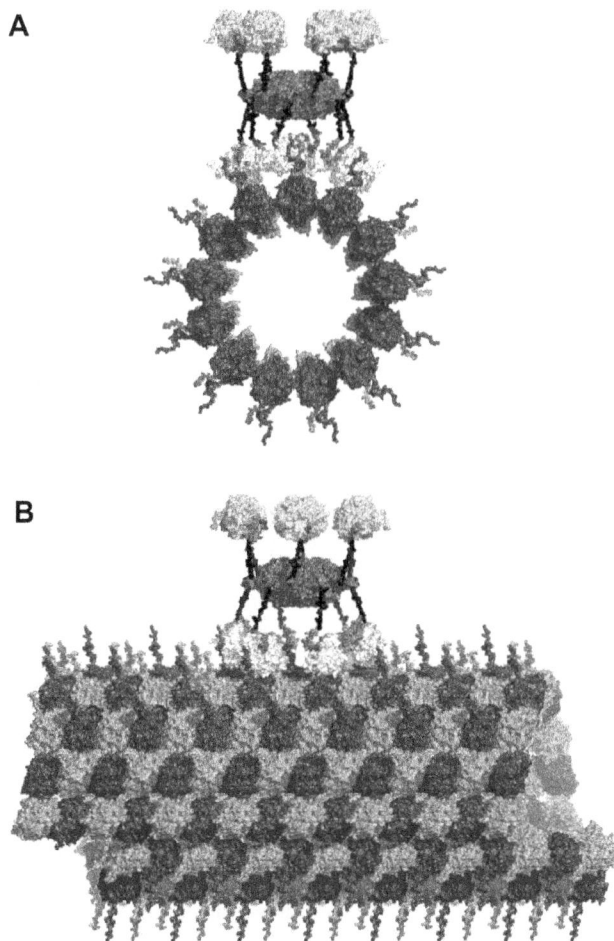

FIGURE 2.17 Graphic illustration of synaptic memory encoded in microtubule lattices (gray and black regions) by CaMKII phosphorylation demonstrating the role of this enzyme in cytoskeletal signaling. a) Face view; b) side view of molecular surface. Source: from [39], used with permission.

2.30 Electrodynamic Interactions in Biology

Electrostatic interactions occur between so-called stationary charged particles (although the particles are not in actuality stationary, they are hypothetically "frozen in time"). The Lorentz force is different in that it relates to the interaction of magnetic fields that exert a torque causing circular motion on molecular structures possessing a charge in moving in the presence of a magnetic field. In the context of biochemical reactions, an additional Lorentz force may lead to a rotational influence that allows an optimal conformational orientation between the complementary surfaces of the ligand and its receptor. An example of such molecular docking likely allows the exposure of tyrosine residues of an insulin receptor upon insulin binding to be phosphorylated coincident with the activation of tyrosine kinase enzyme of the insulin receptor. Hence, insulin binding through the various forms of chemical interactions results in autophosphorylation of insulin receptors upon binding of the insulin ligand.

The larger the atom, the stronger the dispersion force (and hence polarizability), thus inducing a dipole because its electrons are further from the nucleus as well as because a large number of electrons present has a greater likelihood of an asymmetric distribution. Charge separation represents one form of asymmetric distribution and leads to the formation of electrostatic moments such as a dipole and quadrupole moment. *Electrostatic interactions are screened due to the presence of counter-ions and water but nonetheless, they can contribute to structural complementarity, which is the basis of molecular recognition in biological systems.* The greater this complementarity, the stronger the dispersion forces. That is, increased dispersion forces generate increased charge separation of the complementary molecules allowing a stronger attraction between the opposing charges of the receptor and ligand.

Movement in biological systems at the molecular level may be generated by the Lorentz force due to the interaction between the magnetic fields of moving charged particles (representing analogs of electric conductivity in metals); electrostatic interactions between so-called stationary charge particles, and the long-range Fröhlich oscillations of dipoles. Phase coherence as a result of electrodynamic forces allows electrostatic interactions that otherwise would not occur. They are a special form of electrostatic interactions, however, with the distinction that they are represented as oscillating vectors rather than scalar point charges. *While Fröhlich oscillations are atomic or molecular vibrations, the electronic oscillating states of the hydrophobic benzene rings involve London forces of the induced dipoles.* The latter is a more rapid frequency oscillation due to the significantly smaller mass of electrons versus the nuclear component of protons and neutrons. These mechanisms may occur separately or in coexisting fashion.

2.31 Charge Transport

Charge transport in biomolecular systems is quite different from the free flow of electrons or holes in metals and semiconductors. A single charge in a biomolecule strongly polarizes its environment acting like a polaron in a solid. In fact, *environmental distortions caused by charge transfer in proteins can, at low temperatures, occur by quantum tunneling effects. In biomolecules, electron transfer can occur over long distances reaching tens of angstroms. This usually takes place as hole tunneling along the covalent backbone of these molecules.* The term, organic semiconductor refers to organic compounds that exhibit properties inconsistent with electrical insulators. These organic materials may be grouped into three categories: molecular crystals (characterized by van der Waals bonds), charge transfer complexes (with covalent and coordinate bonding present), and polymers. *Charge transfer complexes play an important role in biology as they are responsible for the generation of useable energy through the processes of respiration and photosynthesis* [27, 43]. In some cases, biological electron donors and acceptors are isolated from each other and hence the carriers reside in the vicinity of the same molecule for a long time before hopping over to a neighboring molecule in a process, which is governed by activation energy.

2.32 Electric Field Effects Present in Cells and Acting on Cells

The study of the effects of electrical fields on living cells dates back to 1892 when Wilhelm Roux subjected animal eggs to electric fields and observed a pronounced stratification of the cytoplasm [44]. Since then, a number of observed effects implicate electric fields and/or currents in the cytoskeletal or cytoplasmic self-organization processes. Growing tissues and organs exhibit sensitivity to magnetic fields, and to electric currents. Coherent polarization waves have been seen in mitosis to play a key role in chromosome alignment and their separation. *Regulation of cellular growth and differentiation, including the growth of tumors, may be directly modulated by electromagnetic fields*, especially low-frequency fields in the range of 50–75 Hz but also at around 100 kHz [45] and possibly also very high-frequency THz (10^{12} Hz) fields [46].

The electret state of hydrated biopolymers has been found to be a general property of these systems. Moreover, bound water was found to strongly contribute to the dielectric polarization of biomolecules in solution. A nonlinear bioelectret may stimulate ferroelectric hysteresis curves which introduce both memory and irreversibility to the behavior of these systems. A particular example worth mentioning is the case of microtubules. An overall electric polarization of an MT may strongly couple with the net dipole moment of a nearby tubulin which could aid in the assembly and polymerization of MTs [47]. Since this is a distance-dependent effect, high-concentration assemblies of MTs should exhibit different growth dynamics than individual MTs. This could provide a better understanding of the assembly/disassembly behavior, especially during cell division, and the effects of externally applied electric fields on dividing cells. However, when an oscillating electric field is produced and maintained within a microtubule, an electromagnetic force is generated which acts on oscillating dipoles located in its vicinity. The force can be either repulsive or attractive depending on whether the filament's frequency is lower or higher, respectively, than the molecule's frequency. Thus, this mechanism could be instrumental in selective biological processes such as those associated with mitosis. The strength of the force is regulated by the gradient in the electric field between the filaments and the surrounding cytoplasm.

2.33 Ionic Current Flows through Intra-Cellular Electrolytes

Electrical current in cells and organisms is not typically carried by electrons. Instead, the electrical current is transported by the mobile ions of electrolytic solutions. The proportional relationship between the electromotive force, E, and the electrolytic current, I, follows Ohm's Law, which remains valid for electrolytic conduction. The electrolyte's resistance depends on the dimensions of the cell. The typical order of magnitude of the resistivity for body fluids is about 1 Ωm. which is almost 100 million times larger than the resistivity of copper, so unsurprisingly electrical conduction by ions is much lower than electrical conduction by electrons. The conductivity of an electrolytic solution is proportional to the ionic concentration for low salt concentrations. According to the so-called Kohlrausch Law, the molar conductance of salt is the sum of the conductivities of the ions comprising the salt. Overall, electric current effects in cells have been related to various biological phenomena, most specifically to growth and differentiation processes as well as cell division. Most notably, *current fluxes were measured in cell growth regions and they were detected to flow across the cytoplasmic bridge in the process of cell division.*

2.34 Proton Transport

Protonic conduction differs from electronic conduction in several respects. First, *protonic conduction involves a positive charge, which is three orders of magnitude more massive than an electron.* Secondly, it is abundant in water complexes that surround all bio-polymers in cells. Protons also constitute mobile units in the commonly encountered hydrogen-bonded structures in peptides, proteins, carbohydrates, and DNA. Finally, protons are freed in the hydrolysis reaction of the energy-giving molecules of ATP and GTP. It is, therefore, expected to find protonic conduction in biochemical processes at a sub-cellular level. Protonic conduction usually manifests itself through mechanisms of fault and defect migration in linear and closed (ring-like) organic polymers. An example of proton conduction that drives membrane-bound ATP synthase to make ATP from ADP during photosynthesis is demonstrated in Figure 2.18.

2.35 Electron Conduction and Tunneling

Energy transfer by charge carriers is the most common reaction in metabolic processes. It is a process, which can take very different courses depending on the situation. The redox process consists basically of the transfer of one or two electrons from the donor to the acceptor system. Consequently, the donor becomes oxidized and the acceptor reduced. However, this seemingly simple process can involve a number of complicated sub-routines, which have not yet been completely resolved. *For electron transfer to take place, donor and acceptor molecules must be in precisely defined position with respect to each other and at a minimum distance in order for the overlap of electron orbitals to occur.* This process of electron-transfer complex formation sometimes requires steric transformations of both molecules and takes place at lower rates than energy transfer by induction. Hence, the charge-transfer complex is an activated transition state that enables redox processes to take place between specific reaction partners in the enzyme systems of the cellular metabolism. Due to the oscillating nature of electron transfer, this coupling of two molecules is strengthened by additional electrostatic forces referred to as charge-transfer forces. Also, differences between the energetic potentials of donors and acceptors play a very important role. An uphill transfer of electrons is only possible with an input of external electromagnetic radiation energy at a level of approximately 1.5 eV. For comparison purposes, a potential of 0.82 V exists between a hydrogen electrode and an oxygen electrode, when generating water. Many bio-molecular complexes, such as peptide chains, alpha-helices, and protein filaments represent complex polymer structures with periodically located structural units, which are expected to exhibit semiconducting properties. These bio-molecular complexes are commonly characterized by a periodically repeated hydrogen-bonded unit. Hydrogen-bonded chains allow for another type of charge transfer mechanism, namely an electron-proton couple in a process that takes place in the following steps:

a) Transfer of proton impurity H^+.
b) Acceptance of electron ejected by donor molecule.
c) Formation of charged radical H^+ and transport of the radical H^+.
d) Ejection of electron from the protein molecule with the charged radical H^+.

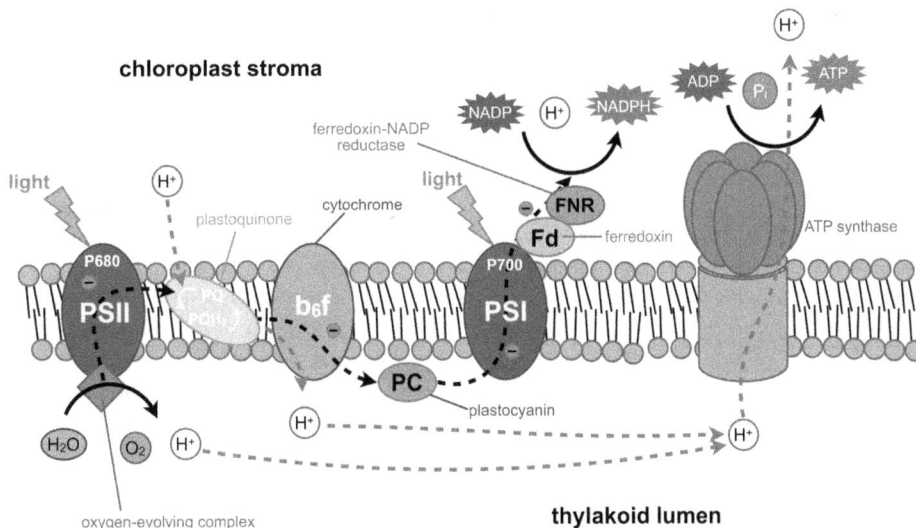

FIGURE 2.18 Graphical illustration of proton uptake during photosynthesis to ultimately drive ATP synthase in the preparation of ATP from ADP. Source: adapted from http://en.wikipedia.org/wiki/File:Thylakoid_membrane.png.

The electron transfer mechanism has found a very important application in the functioning of enzymes such as cytochrome oxidase. The process of electron transfer between cyt c1, cyt c, and cytochrome oxidase consists of the following stages:

a) Orientation.
b) Docking.
c) Relaxation to the functioning configuration.
d) Release and rotation.

These mechanisms are certainly at the heart of biological sensing in naturally occurring sub-cellular structures such as enzymes.

2.36 Interactions of Biological Systems with Electromagnetic Radiation

In 1922 A. Gurwitsch provided evidence of a weak photon emission [48] of a few counts/(s.cm²) in the optical range from biological systems, pointing out that it stimulates cell division. Later, the use of photomultiplier techniques revealed "ultra-weak light emission" from various living tissues. It is believed that spontaneous biophoton emission originates from radical reactions within the cells. F. Popp claimed that biophoton emission is due to a coherent photon field within the living system, responsible for intra- and intercellular communication and regulation of biological functions such as biochemical activities, cell growth, and differentiation. Indeed, biophoton emission has been traced back to DNA as the most likely source, and that delayed luminescence, which is the long-term afterglow of living systems after exposure to external light, corresponds to excited states of the biophoton field. All the correlations between delayed luminescence or biophoton emission [49] and biological functions such as cell growth, cell differentiation, biological rhythms, and cancer development, appear to be consistent with the coherence hypothesis [50]. Furthermore, numerous biochemical reactions are known to be regulated by light energy. Every biochemical reaction of this type takes place if and only if at least one photon excites the initial state of molecules reacting to a transition state that decays finally into the stable final product by releasing the overshoot energy of at least one photon. The availability of suitable photons determines the reaction rate. *Of particular importance are photo-activated enzymatic reactions in which enzymatic components exist that link parts of the enzyme via covalent bonds. They can be activated by photon absorption and de-activated by photon emission changing the functional characteristics of the enzyme in a dramatic way. Energy transfer by radiation can be envisioned in the following way: the excited molecule emits fluorescent radiation which matches precisely the absorption spectrum of the neighboring molecule and, and consequently, excites it. Such mechanisms are capable of transferring energy over large distances compared with the other processes described in this biological context.* However, the efficiency of this process is quite low and it decreases sharply with distance. Transfer of energy by an inductive process such as resonance transfer is especially important in

photosynthesis [27, 43]. This form of molecular energy transfer is a non-radiant energy transfer since fluorescent light does not occur in this process. This mechanism can be described as coupling between oscillating dipoles. The excited electron of the donor molecule undergoes oscillations and returns to its ground state thus inducing excitation of an electron in the acceptor molecule. This process requires an overlap between the fluorescent bands of the donor with the absorption band of the acceptor, i.e. it relies on the resonance of both oscillators. The smaller the difference between the characteristic frequencies of donor and acceptor, the faster the transfer will be. It is worth mentioning that strong dipole-dipole couplings are possible at distances of up to 5 nm. Finally, we should refer the reader to additional discussion on photosynthesis in the chapter on quantum biology since photosynthesis is a prime example of a quantum phenomenon in biology.

Endogenous biophotons are generated from various cell regions, most notably the mitochondria, which are the powerhouses of the cell producing ATP, the biological energy currency. However, it is positionally perhaps the most important endogenous source of biophotons. ATP that is produced by mitochondria is hydrolyzed to release heat or electromagnetic waves in the red or near-infrared wavelengths represents energy that can do work. Such waves may be considered to be a form of short-range biophotons. Another source of energy that potentially can come from mitochondria in the form of biophotons includes heat that is released from the mitochondria but which is classically considered to be a function for providing the thermogenesis of the body, hence for maintaining homeostatic body temperature. However, this heat formed as a result of the uncoupling of the proton electrical gradient across the inner mitochondrial membrane to ATP synthesis, may in part as infrared waves promote biological work, in which case these waves would be considered biophotons. A third type of electromagnetic energy produced from mitochondria may be the high-energy biophotons [51] as a result of the recombination of reactive oxygen species, which occur as a result of electron leakage along the electron transport chain. Such *biophotons may be in the infrared range as the lower range of frequency and as high as the ultraviolet range of electromagnetic rays.* Finally, two biophotons at the lower frequency of the infrared range may act in a cooperative fashion to provide the necessary energy exciting the various types of dipole oscillations. Such infrared-range low-frequency biophotons may occur as a result of recombination of reactive oxygen species as well as a result of the heat released due to the uncoupling of proteins which uncouples the energy of the proton gradient to ATP synthesis.

In summary, it may be considered that at least the above-mentioned *four types of biophotons generated from the mitochondria may provide the thermal activation of the initial energy for dynamic dipole oscillations within cell membranes, microtubules, and other cytoskeletal structures. These biophotons, via the initial energy given off for microtubule oscillations, may be responsible for the storage of cognitions, feelings, emotions, awareness of experience and learning received from synapses. This may include the initial energy for the Fröhlich coherent laser-like oscillation. These Fröhlich oscillations occur at the level of cell membranes* [52] that may promote phase coherence between cells responsible for biological coordinated activity. There are apparently three types of Fröhlich condensations as mentioned above, weak, strong and coherent. The weaker the condensation the greater the number of possibilities of dipole oscillations. Although by definition these condensations are incoherent, they contain far fewer possibilities than the ground state which represents the lowest or basal energy state with maximum possibilities, in other words, maximum entropy or maximum randomness. The type of condensation in MTs that is optimal appears to be around 8 MHz. This frequency was indicated as preferable for mediating quantum computation [53], which requires only synchrony and entanglement but not a single synchronous entanglement with a unitary possibility of information processing mediated by maximum coherence. That is, a stronger or coherent condensation would not allow such quantum computation. The quantum coherent laser-like state is a single state of all the phospholipid head groups forming a giant dipole and dipole wave may oscillate along the connecting cell membranes of the tissue or organ. Unlike the lower frequency condensations that may occur in atomic molecular groups of microtubules, the goal of having a metastable coherent state is to synchronize and entangle the orchestrated biological function of many cells within a tissue or organ type. Conversely, a less coherent Fröhlich condensation in microtubules has a different purpose of providing many possible states that allow cognitions and the conscious experience. Notably again, Fröhlich type condensations do not involve the electronic oscillating dipoles of aromatic rings which could lead to a different type of information processing or signaling. *The Fröhlich oscillations, whether in cell membranes or within microtubules, represent atomic nuclear vibrations in contrast to the electron dipole states of hydrophobic aromatic rings with underlying Van der Waals forces.* Hence, microtubule oscillations may involve multiple mechanisms, which may occur independently or in a coexisting fashion. Both mechanisms however do share the property of operating under a quantum regime since they cannot be described by the classical laws of physics.

The biophoton thermal activation of Fröhlich coherent oscillations in cell membranes may interact with circulating hydrophilic hormones or other endocrine messengers with the same oscillation frequency. Although Fröhlich did not specifically apply his concept of long-range coherent oscillations to that of interacting receptors with their ligands, this notion is consistent with an application to such a biological realm. We stress again, that the above descriptions are still in the realm of hypotheses since no direct experimental evidence has been provided yet.

2.37 Bioelectricity and Biomagnetism

Magnetic fields are created by moving charged particles (sub-atomic particles for example dynamic electric dipoles or ions) and represent electromagnetic effects that can promote or accelerate the movement of other charged particles. This may lead to the formation of ionic waves acting, for example, in part as counter-ions surrounding microtubules or actin microfilaments (which can be additionally affected by dynamic induced dipole-induced dipole interactions). The ionic waves in turn may promote the movement of calcium within the cell. Calcium movement may further affect charged amino acids containing protein neurotransmitters to be released into neuronal synaptic clefts for example.

The movement of charged molecules in biological systems at the sub-cellular level may also be generated by the Lorentz force involving the interaction between the magnetic fields due to ionic currents and moving charged particles (representing analogs of electricity). An example of such an effect in biology may be the rotation molecules resulting in the opening of ion channels in cells such as neurons which subsequently may drive cell depolarization and neuronal firing due to calcium influx. Alternatively, *calcium influx is important for a host of effects such as activation of many enzymes. In molecular-level biology, motion may also be propelled directly by passive electrochemical gradients and indirectly by transfer of electrochemical bond potential energy from high-energy bonds,* such as from between the second and third phosphate bonds in an ATP molecule as heat, or "energy in transit", to a process with a positive Gibbs free energy. The best example of such transfer of "energy in transit" is from ATP hydrolysis to motor proteins or to the active transport of ion pumps.

Dry proteins appear to be useless from a functional point of view. On the other hand, *hydrated proteins appear to be critical for the functionality of proteins.* In fact, experiments dating back to the 1990s show that when the level of hydration of protein enzymes and hormones is gradually increased, at and above a critical threshold the proteins became functional. There may be circumstances where the dipole moments of water molecules actually compete with proteins and more work needs to be done to elucidate these specific mechanisms. There is work demonstrating that water may reduce electromagnetic interactions because the dielectric constant of water is very high. However, this effect may be completely obliterated by the fact that water molecules have their own dipole moment that can be oriented and coordinated in a special way giving rise to strong long-range correlations, both static and dynamic. Water molecules' dipolar dynamics could be correlated to the proteins or it could coordinate the effects of, for example, hormones and receptors. This can be exemplified by a recently demonstrated coordination by a single ion of three million water molecules' dipolar dynamics. Therefore, the high dielectric constant of water likely has a rather minor effect on reducing magnetic interactions of proteins compared to the effect of correlated dipole moments of water molecules that can oscillate in synchrony with each other and with the proteins. The water molecules are far smaller than the proteins, however, coordinating thousands or even millions of them could lead to a gestalt effect. In fact, the number of atoms comprising all the water molecules surrounding a single protein in a cell is much greater than the totality of atoms of the protein molecule itself. Although water molecules have a permanent dipole moment, that is a non-oscillating dipole, an entire ensemble of water molecules due to the rotational degrees of freedom of water, this can result in dynamic oscillating dipole moments correlating with each other and with the dipole moments of the protein.

Although not known with certainty, Gerald Pollack contends that the range of interaction between endocrine ligands and their receptors may easily be up to tens of nanometers, even hundreds of nanometers for a single protein. *The notion of the Exclusion Zone (EZ) for biological water in biological systems as posited by Pollack et al. [54], contends that there is essentially no free water inside cells and significant exclusion zones of water in the extracellular space. Accordingly, ions and proteins correlate water molecules such that a single ion correlates a couple of hundred water molecules in each direction which in turn correlate with another ion. Thus, the network of water molecules extends interactions with other ions and molecules roughly 30 to 50 nanometers away.* However, a protein has many charges and thus its correlation with water molecules should easily be capable of interacting with other proteins over a hundred nanometers away, which could indeed be mediated by the permanent dipoles of water. These estimates, however, need careful spectroscopic measurements to be verified.

Dynamic oscillating water dipoles decorating and correlating with a collective quantum state of an oscillating hormone such as insulin with roughly a half dozen aromatic amino acids may also interact with the coherent oscillations of the atomic molecular vibrations in the case of both hormone and water molecules as coherent condensations. As an example of a macroscopic correlation in biology, microtubules may be interacting with neuronal firing across potentially billions of neurons across large regions of the brain. This may integrate exclusion zone water with microtubules and quantum metabolism, which provides the supply of ATP for the ATP-requiring processes in addition to the associated mitochondrial released biophotons, which induce dipoles as the initial thermal activation energy for inducing dipoles in microtubules as well as in cell membranes. *The modulated oscillation frequency of receptors embedded in cell membranes by the cell membrane interactions reflects the health of the cell and hence the ability of receptor activation to coordinate with a commensurately capacity of the cell to respond to the signaling effects of receptor activation to hormones and other ligands.* While the calculations of a single protein correlating with water molecules may extend more than one hundred nanometers, there are thousands of proteins and millions of charged molecules including copies of proteins and DNA in a given cell. Consequently, the water surroundings of all of these structures, under optimal healthy conditions of having no free water inside the cell, i.e. one hundred percent of water molecules correlated to ions, protein, or DNA, the entirety of functional interactions in the cell is potentially long-range correlated and mediated by water molecules. For example, in the case of tubulin correlating water molecules, there does not appear to be any intracellular water that is not correlated with some tubulin. The analogous long-range interaction between endocrine ligands and receptor apply to the extracellular space, and the distance of long-range interactions depend on a number of factors. These factors include *a) the presence of collective oscillations (possibly coherent quantum condensation), b) the size of the interacting charged particle-containing molecules, and c) the health of the alignment of water molecules around these charged particles.* The last of these factors could be intuitively

dependent on the health status of the cell in terms of biophoton generation that provides initial thermal activation energy for inducing dipoles in cell membranes and receptors embedded in cell membranes. This underscores the importance of the likely correlation between quantum metabolism and biophoton generation inside the microtubule organelles. Intuitively, we recognize the significance of the extent of water molecules with proteins inside and outside of cells, which highlights the importance of having an adequate volume status for the individual in order for these functional proteins to behave in a normal and healthy fashion.

2.38 Biological Engines and the Quantum Biological Processes Explaining Cognition

Structured water was first described by Mae Wan Ho [55] as having quantum mechanical properties in biological systems [56]. Hence, *the so-called fourth phase of water defined by Gerald Pollack may transfer information between disparate regions of the body, such as from the microtubules in the neurons of the brain to the metabolic enzymes* in the cells of visceral organs or of skeletal muscles of the extremities or perhaps even to the motor proteins myosin present in skeletal or cardiac muscle [57] (Figure 2.19).

Mae Wan Ho proposed that water aligns in a highly constrained fashion within the nanotubes of the triple helices of collagen that span throughout the entirety of the extracellular space of the body [56]. It is hypothesized to transfer the information of specific resonance frequencies from where it originated, for example, the microtubules in the neurons in the central nervous system, to intracellular domains elsewhere in the body. Upon transfer of information from the extracellular to the intracellular space the water molecules could align as the propagation of positive electricity or rather the conduction of electricity with the propagation of hydronium ions

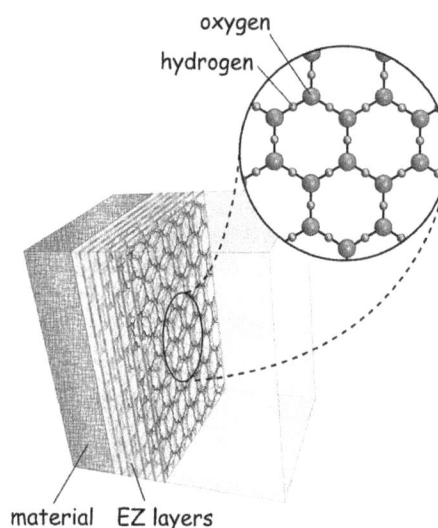

FIGURE 2.19 Illustration of "structured water" constituting the "fourth phase" of water defined by G. Pollack [54]. Source: reprinted with permission from Gerald Pollack, https://www.pollacklab.org/.

constrained within the nanotubes of double-stranded cytoskeletal filaments of actin (in the intracellular space) until they reach and transfer their information to specific targets. The targeted structures may be those with the same frequency. The amplification of that frequency may be what is required to impact the function of those structures. Importantly, Mae Wan Ho described the transmission of biological water along the nanotubes of collagen to occur in so-called positive electricity with the sequential formation of hydronium ions propagated along the linearly lined water molecules. *In this fashion, the so-called positive electricity represents a flow of information approaching the speed of light so that it underpins a potential synchronized and coherent organismic level integration of function from mind to body. The concept of biological water can hence be recognized as a quantum mechanical-biological engine.* Perhaps it is a well-orchestrated symphony integrated with the metabolic activity of mitochondria and the process of oxidative phosphorylation and quantum metabolism on a macroscopic scale.

Critically, structurally organized and stabilized microtubules in neurons of the brain appear to be key players in the formation of consciousness, definable as an awareness of being aware, or as higher-level cognitions. Microtubules, as the mediating structures of sensory input, are information processors, translating information into an output response. It has long been pondered but remains an unanswered question, where does consciousness come from and where does it go when we die? Well, microtubules in the brain's higher center neurons may lie at the focal point of this quandary. Information comes in from our sensory neurons of the optic, auditory, olfactory and somatosensory nerves, which ultimately synapse with neurons in higher-level centers of the brain within which microtubules integrate this sensory input processing and coupling it to mental cognition. These cognitive processes are the manifestations of awareness. The depth of that awareness may be considered a non-intellectual or emotional IQ, a perceptiveness. Simple awareness of any sentient organism is what is derived from the senses independent of cognition. Awareness of being aware is a proposed minimum criterion defining consciousness. An awareness of being aware may be exemplified by the pride that one may feel or an intellectual achievement or the anxiety and depression one perceives how others may think negatively of him or her. Higher-level degrees of consciousness may be the awareness that one is aware of being aware and so on, as may be exemplified by transcendental meditation.

Although intellectual cognition may also be mediated at the level of neuronal microtubules in the brain as information processors as well as storage and retrieval of memory such as from the hippocampal neurons, this is not a defining feature of consciousness *per se* [5]. The notion of consciousness equates to many on an intuitive level to the "soul" or identity of an individual. It is this soul or identity of consciousness, which is imponderable in terms of where it comes from that defines us and where it goes when we die. One interpretative description of consciousness includes both an awareness of being aware particularly but not exclusively limited to human beings, and fundamentally rooted in quantum mechanics [58]. The origin of this quantum phenomenon has been posited by Hameroff and Penrose [38, 59], as well others who followed in their footsteps, to be the microtubules of neurons in the brain. *Although information processing occurs in neurons [60], the quantum nature of microtubules is inherently stochastic. Indeed, there is a daunting impossibility of estimating the origin of a quantum process of cognition beyond the microtubule as well as of predicting the destiny of our identity beyond our mortal lifespan. Quantum neurobiological processes are characteristically coherent, i.e. information is shared between sites effectively. Additionally, the sites have waveforms that are synchronous, that is, are in the phase whereby the peaks and valleys happen at the same time in both waveforms. There is an in-phase timing relationship. The flow of energy and information is unconstrained spatially and temporally such that manifestations of quantum phenomena may occur instantaneously and simultaneously between spatially distant areas.* Mae-Wan Ho [55] argues this in the case of information carried by electrically-activated water through the nanotubes of collagen throughout the extracellular space of the body. In this sense time dilation may be coincident with the unconstrained spatial domain to the flow of information [61, 62] and/or energy. Nonetheless, the above concepts are still nebulous in view of the lack of convincing experimental data but are worth exploring.

Another, more mechanistic, question arises naturally in regard to how may the integration of quantum metabolism, microtubules, and consciousness or quantum cognitions be assimilated? Although the electric fields generated by the tubulin arrangements comprising microtubules cannot be easily explained by classical science, the neural sensory input into the brain from the surrounding environment in the presence of neuronal influences such as emotional memories all contribute to the formation of cognition. Classical components of neuronal activity include in part, the release of neurotransmitters from the terminals of neurons into synaptic clefts resulting in the opening of cation channels, for example, sodium channels. This promotes a change in the membrane potential with depolarization and firing of action potentials. The influx of cations generates ionic waves which propagate along both actin filaments and microtubules although the mechanism of wave propagation is different, and they play different roles in directing neuronal behavior. *Microtubules or indeed other biomolecules such as the DNA or actin filaments, as well as cell membranes, may have a large number of degrees of freedom, some of which can operate in the classical realm, e.g. vibrational dynamics, while others, e.g. tryptophan excitations may operate in the quantum domain.* The two classes of degrees of freedom may even interact with each other such that classical dynamics may affect quantum behavior by forming classical fields that bias or influence quantum dynamics. Cations may, as an example, represent counter-ions attracted by the negatively charged amino acids of tubulin or to the surface of actin double-helical filaments. Ionic waves of these cations form and propagate as soliton-like wave ionic clouds along actin filaments virtually without loss of wave amplitude and without slowing down. When the wave reaches the end of the actin filament, it can trigger a polarization event, for example causing neurotransmitter release from a synaptic bouton. Similarly, cations representing counter-ion waves are attracted by the negatively charged amino acids of tubulins present on the surface of microtubules. *However, the conduction of ionic waves along microtubules may lead to helical currents around their surfaces, which may then induce magnetic fields in a solenoidal fashion creating conditions for microtubules to behave like magnetic needles that are very sensitive to magnetic fields.* It has already been demonstrated experimentally that counter-ion waves along microtubules elicit transistor-like behavior in the sense that it significantly amplifies the wave signal generated at one of the microtubules [63]. Undoubtedly many new phenomena will be uncovered that shed light on the electromagnetic properties of sub-cellular structures such as MTs and AFs.

Moreover, microtubules can propagate ionic waves into contact with actin filaments, which can then relay these waves all the way to ion channels. It is also known that microtubules interact with some ion channels directly. Interestingly, microtubules closely associate with mitochondria, which move within the cell along microtubule tracks using the energy they produce in the form of ATP molecules as fuel for motor proteins such as kinesin and dynein that propel them along these microtubule tracks. There may also be additional forms of interaction between microtubules and mitochondria since mitochondria produce electric fields and emit weak electromagnetic radiation during their activities. *Electric fields are produced by mitochondria established by the proton gradient across the inner mitochondrial membrane.* Additionally, weak electromagnetic radiation is emitted in the form of biophotons produced by the recombination of reactive oxygen species (see discussion of biophotons above) [64]. Microtubules, in the immediate proximity of mitochondria, may absorb these signals, transducing them into bits of information stored in the subtle structural changes such as C-termini states or electronic transitions [65]. This information can possibly affect the way microtubules conduct ionic waves and hence may influence the whole cell's response to electric and electromagnetic stimuli. In terms of ionic waves, albeit they exhibit classical behavior per se, although they may directly or indirectly play a role in positively or negatively influencing quantum effects occurring at a lower spatial scale. This would be of special significance in neurons where a change in the cell's threshold for firing an action potential could have an effect on the synaptic connection between neurons. Ionic waves may for example exert a number of differential effects that may potentially enhance or inhibit quantum regime, which consonantly improves or impairs respectively the quality of mental and emotional cognition. For example, ionic waves may affect motor protein binding to microtubules. Furthermore, because motor proteins are responsible for moving mitochondria along MTs, the notion of quantum metabolism may be relatively integrated with neuronal cell quantum processes. Alternatively, ionic waves may promote microtubule disassembly and accordingly discourage microtubule-mediated neuronal quantum effects. *Microtubules whether as part of a quantum-mediated function or a classical function, the latter possibly related to ionic waves independent of microtubule function per se, may influence response to neurotransmitters and hence depolarization and neuronal firing.* These mechanisms can connect groups of neurons and make them more or less active, eventually effecting cognitive processes in either a positive or negative way depending on the particular mechanism and the specifics of these interactions.

2.39 Connections between Electricity, Magnetism, and Energy Generation

Magnetic induction as a property refers to magnetic flux or flux density. As a process, it is one by which an object or material is magnetized by an external field. Such an external field may be a permanent magnet (a ferromagnetic substance) or conversely a temporary magnet, i.e. an electromagnet. Note that a paramagnet is also a temporary magnet, although it is temporary in the sense of the alignment of electron spins, which are determined by the type of material and respond to externally applied magnetic fields. For example, nickel, iron, and cobalt when subjected to a magnetic field of a permanent magnet also become permanent magnets, whereas certain other metals and oxides or iron are paramagnets. In both cases, whether ferromagnet or paramagnet, it has to do with the alignment of electrons in the outer shell of the atoms. *Permanent magnets are ferromagnetic and are stronger magnets than are paramagnets. Furthermore, both paramagnets and ferromagnets are stationary in contrast to moving charged particles, which are electromagnets.*

An external magnetic field, whether it be due to a stationary magnet (ferromagnet or paramagnet) or a moving charged particle (an electromagnet, a conducting wire, or ionic flow) or oscillating or propagating charged particle-wave (for example Fröhlich dipole oscillations and amino acid aromatic dipoles) in either of these three cases is generated in the surrounding space. These magnetic fields are all capable of one of the three following effects: a) acceleration of the angular velocity of charged particles or accelerating or preventing linear motion of another charged particle from coming to a halt; b) interacting with another magnetic field and exert a Lorentz force; or c) to modulate the frequency of another oscillating charged particle. In this latter case, perhaps it is the oscillating or propagating charged particle wave that has the ability to modulate another oscillating or propagating charged particle wave in phase coherence or synchronously with one another.

Phase coherent oscillating electromagnetic biomolecular structures such as ligands and receptors may exhibit interactions, which are a function of attractive electromagnetic fields, and the fact that the ligand is translationally free to move (not structurally tethered) may result in drawing the ligand closer to its receptor. This is a different phenomenon than the attraction of two stationary magnets (ferromagnet or paramagnet) or stationary electric charge, the latter are normally screened over short distances due to counterions.

Earlier in this chapter, we have discussed rotary motors such as ATP synthase molecules, which operate in a fashion highly reminiscent of electrical DC motors. Another application of an electric motor of an automobile is found in a microphone. Sound is picked up by a diaphragm. The vibrating diaphragm attached to a coil (wrapped around a permanent magnet) moves the coil, which in turn results in the induction of current in the wire. Thus, sound waves transmit electronic signals which act as amplifiers through a speaker. While we have not identified directly a biological equivalent of this type of electromagnetic-to-acoustic energy transduction, it is possible that centrioles (as an example) can use this principle in the process of capturing electromagnetic waves coming from the environment and transducing this energy into the mechanical motion of their microtubule triplets as explained in the work of Gunter Albrecht-Buehler who likened the centriole to a cell's eye [29].

In the case of electrical engines, we have described the electromagnetic interactions of machines and their metaphorical extrapolation to biological systems. These electromagnetic interactions describe the motor *per se* because it is what is causing movement. In a biological engine, the motors are molecular motors, moving charged particles, for example, ions, electron dipoles, Fröhlich oscillations, and hydronium ions. Metabolism that supplies energy for these motors (for example ATP hydrolysis and biophotons [actually ATP hydrolysis releasing infrared waves may also be considered a biophoton]) is part of the overall engine. The quantum molecular and electrical engines together represent the far-from-equilibrium engine. When functioning properly and in a healthy fashion, this biological engine maintains homeostasis. The most fundamental homeostatic parameters are redox, Gibbs free energy, and an acid base. As will be elaborated on in Volume 2 of this book, key organizing principles of the body ensure the maintenance of homeostasis. *This is achieved through physiological processes of allostasis (maintaining stability through change) whereby perturbations of endocrine, including hormonal, exocrine, and paracrine functions are observed. These latter functions may be in a state of homeostasis when not perturbed. However, when the body's homeostasis is challenged many paracrine, endocrine, and even exocrine functions exemplify allostatic changes in order to maintain homeostasis of the most fundamental and precious functions of Gibbs free energy, redox, and an acid base upon which metabolic viability and life depend* [66].

2.40 Connections between Microtubules, Molecular Motors, and Mitochondria: Toward a Molecular Explanation of Free Will

The concept of consciousness is notoriously ambiguous. It is often referred to in medicine in terms of a patient, often in the ICU, being awake versus asleep or in a coma. Consciousness is perhaps the most central issue in the current philosophy of mind. Although a hard definition of consciousness is fiercely debated, it is generally agreed to include sentience but goes beyond that and includes the perception of these experiences in a context of a sense of self. All we derive from our senses including perceptions and feelings are a measure of our sentience. The integration of sentience and sense of self is our consciousness. *Roger Penrose describes consciousness as an awareness of awareness. The strength or depth of this consciousness defines our emotional intelligence, our identity.* The intriguing issue that questions the permanence of consciousness, where it comes from, and where it goes when we

die, was briefly discussed in Chapter 1. It should be recognized that no one knows what consciousness is. The concept of energy is analogous and perhaps rooted in the same fabric of consciousness. Thus, it is not surprising that no one knows what energy is. We can define what both consciousness and energy do but we do not know what either of them is at a fundamental level. The elucidation of these properties may escape the primacy of science and further may be inextricably part and parcel of the same entity representing the agents of "soul" and immortality within the universe.

It is theorized by Penrose and Hameroff that consciousness appears to be rooted in the neuronal microtubules of the higher centers of the brain, probably the limbic system, the emotional center of the brain. It is thought to relate to the quantum nature of the microtubules at the level of certain amino acids comprising the tubulin dimer, which is the building block of microtubules. More specifically, it is likely a function of double-well energy landscape within these amino acids, most likely corresponding to positively charged amino acids allowing electrons to oscillate between the two energy minima. Quantum properties of MTs may germinate from a variety of allowed quantum superpositions of dipole states leading to either a standing wave oscillatory pattern or a wave-like propagation of patterns across the amino acids of tubulin dimers and indeed an entire microtubule which is composed of them. One possible quantum superposition of dipoles is the double-well structure resulting in an electron-hole exciton. This coherent quantum wave within a given tubulin dimer molecular structure may have a variety of effects such as inducing dipoles of neighboring amino acids (within the same or adjacent tubulin molecules) to promote an oscillating or propagating cascade of induced dipole waves. The electric field generated by these sequential induced dipoles may promote electronic reconfigurations within the microtubule structures responsible for mechanical conformational changes. These conformational changes in turn may promote attachments of microtubule-associated proteins (MAPs) such as MAP2 or MAP tau which play key roles in the stability of the neuronal cytoskeleton. Incidentally, almost all neurodegenerative diseases such as Alzheimer's and Parkinson's are associated with the impairment of the neuronal cytoskeleton. This is also directly linked to motor protein tracking, such as the carrying of cargo by kinesin or dynein motor protein, for example transporting RNA or even mitochondria. Since mitochondria are directly responsible for quantum metabolism, this could be a link between two different quantum biological processes: energy production and dipole oscillations. The electric field generated by these propagating or oscillating induced dipoles is also responsible for opening ion channels in the postsynaptic cell bodies contributing to neuronal firing. *The junction of a quantum exciton wave within microtubules transitioning to classical induced dipoles may represent a feasible interface between quantum and classical mechanisms related to higher-order cognitions.*

Moreover, there are additional important communication channels between classical and quantum mechanisms of conscious cognitions of the central nervous system. Using the double-well structure as the example among the possible quantum superpositions of dipoles, it is plausible to speculate that mental discipline and focus that remove extraneous neuronal input, or noise, and allowing strong assimilation of our feelings and perceptions with our identity of self, has the capacity to open ion channels of the particular central nervous system neuronal pathways that mediate free will, a function of our individuality. The influx of ions could then result in ionic flow along the exterior of microtubules that sets up electrical potential and voltage that may lower one potential well relative to the other providing a bias whereby electrons will be much more likely to be in the lower energy state than the higher one simply because, like all systems, it favors a lower energy state. In contrast, if the two wells are symmetrical with the same height then tunneling oscillations may occur back and forth between the two states assuming there is no bias in one direction or the other. This latter case highlights the truly random nature of quantum transitions though the former case underscores how this can be biased and thus the quantum phenomenon can be co-opted to the advantage of free will. *Notably, the notion of free will equates to controlling outcomes. Accordingly, this is a critical and fundamental element to success in human society.* Success may be defined in many forms including professional success, athletic or academic achievements, as well as the capacity to meditate. The capacity to meditate is not only essential to the parameters of success just mentioned here but relates to emotional disturbances of anxiety, depression, adapting to change, and coping with circumstances of tragedy.

Another way to interpret the potential to bias quantum transitions in the human brain is to recognize the dual nature, classical and quantum, of the tubulin dimer. The ion channels of neurons and movement of ions into the neuronal cytoplasm promoting ionic waves and external electric fields along the microtubules represent classical interactions that bias quantum transitions. Perhaps superimposed on this proposed mechanism for classical biasing of quantum transitioning, environmental, sensory, and perceptive input and integration of neuronal pathways promote more robust classical ionic flow and electrostatic interactions with tubulin dimers that trigger the 'pre-existing' biased quantum transition and manifest free will. A plausible speculation may be the mental discipline and focus remove extraneous neuronal input or noise that allow strong assimilation of our sentience with our identity of self that opens ion channels enough to depolarize the neurons to allow lowering of one well of a double well more than the other. This may lead to primary microtubule-mediated opening of more ion channels and ionic waves along the surface of microtubules and actin filaments in a particular direction of dipole-dipole interactions. This may also lead to a particular pattern of neuronal synapses extending from the neuron terminals to interconnecting pathways of neuronal networks mediating spatio-motor and cognitive functioning. *This hypothetical framework suggests that quantum mechanics underscores consciousness, which may be rooted in the microtubules within the neurons of emotional centers in the brain. Consciousness subsequently could play a crucial role in determining information processing, spatio-motor, and cognitive response to environmental challenges.*

Our total life experiences, our sensory input, perceptions, knowledge, feelings, emotions, cognitions, and our free will can be plausibly stored in the microtubules (mediated by CaMKII) [39]. Our free will is what gives us control rather than simply reduced to deterministic living beings. Free will is more than just imagining we have some control in who we become as people, for example in terms of our character, our successes, or failures. The totality of this information stored in microtubules may be considered a representation of consciousness.

Mitochondria are moved as cargo along microtubules playing a crucial role in the spatial organization within the cell. *The microtubule's role in mediating the free will component of consciousness along with their role in the spatial arrangement of mitochondria couples with the phenomenon of quantum metabolism that may result in the orchestration of mind-somatic function as an integrated and correlated quantum phenomenon.* Somatic function includes the brain and the peripheral components of the body outside the central nervous system. Microtubules may provide the biological interface with the information from outside the body as photons to the material components of the body. Quantum biological processes make possible the notion of the free will of humans, the capacity to consciously make decisions not deterministically controlled by the laws of physics in chemistry in our brains. Free will is the capacity to control the outcome based upon a unique and fundamental meaning and purpose of life and core values. It is scientifically credible to be rooted outside the body in the brain in a deeply entangled intangible fabric of energy connected at the level of a collective unconsciousness. According to Penrose and Hameroff, the biological physicality of quantum processes involved in consciousness and free will is rooted in the microtubules of the brain. Furthermore, the translation of free will into control of cognitions, emotions, and physical actions is most effectively and efficiently mediated by the quantum bioenergetic phenomenon of quantum metabolism. The spatially organized mitochondria generate energy as ATP correlated macroscopically and transforms into the biological work of hard-wired neuronal circuits of the brain including the limbic system (emotional center), the prefrontal and other cognitive regions of the cortex, and the pyramidal system. The coherent integration of those regions in neuronal tracts of the brain translates into a powerful capacity to control outcomes. For example, the pyramidal system provides the highest order of motor function in humans capable of the artistic movements that may be lifesaving against predatory attack in special circumstances and may be responsible for incredible displays of athleticism. Hence, in the latter case, free will may make the difference between winning and losing a competitive sports confrontation. Furthermore, the notion of free will based on quantum processes in neuronal microtubules and anchored to our cognitions, emotions, autonomic output, and motor functions may give us the potential and the opportunity to control our outcome in the world such that we are not simply predetermined by the laws of physics in chemistry that constitute the biology of our body and including the brain. *Motivation in free will tethered to the bioenergetic potential quantum metabolism in the brain and body makes possible our ability to control our own successes, whatever they may be, or however we define them.* They may be athletically directed or directed towards success in school or professional success. They may be the capacity to cope with tragedy or even simply to enjoy the passage of time.

2.41 Collective Unconscious and Society

The notion of "simple rules" based on local interactions as a bottom-up process achieves extraordinary feats in the absence of a central planner. However, this process conceptually may be applied to political biases and voting patterns. In this sense the collective intelligence to which so-called "simple rules" of local interactions apply may represent the underpinning to which many of our political biases are evolutionary rudiments. However, in contrast to the wondrous adaptive value it has provided, many biological species considering an analogous collective unconsciousness in human populations and society in terms of political views may have an enormous perverse potential for our current political system.

Inherently, agents in a system of collective or swarm intelligence that pose a barrier to the greater good as a collective whole would not be accepted by the rest of the community. There must be some mechanism of keeping all the agents "in line". Since this is collective intelligence, no single individual agent knows the outcome or understands the big picture. If there were such an agent, rather than the process being bottom-up, it would be top-down such that one or a few agents dictate the pattern of behavior of the larger group. If a single or few individuals of a group control the masses, it would suggest a consciousness or awareness of the consequences or outcome of a pattern or different patterns of behavior that in general is superior to lower scale agents being directed. In contrast, the bottom-up process of swarm intelligence is a collective unconsciousness tending toward a minimum energy state that will allow for the survival of the colony and moreover species. It is compelling to find similitude of this to the modern-day political process. However, there are two major parties in the US to which the collective intelligence tends to whereby each major party represents an attractor state. Likening this to swarm intelligence roots it fundamentally to being about survival at an unconscious level. Transitions may occur but they are relatively difficult because it requires overcoming an energy barrier. This can actually bring about a phase transition to the opposite attractor state. It is emotional because it is fundamentally equated to survival at an unconscious level. This asserts the emotionally charged argumentative nature of political discussions. There is decided danger when politics becomes an exploitation of this ancient evolutionary force intended to favor survival. The dual processes of mutation and natural selection guide adaptation in a very different context than that of choosing a leader of the country. The evolutionary purpose of swarm intelligence, collective intelligence, or collective unconsciousness is rooted in directing bottom-up emergent self-organizing systems. By contrast, the paradox here in the application to a political context is that nature's will is violated by transferring the bottom-up control to a top-down control electing leaders who are capable of exploiting a collective unconsciousness [67]. Most of the individuals in a given population being exploited remain unaware of that exploitation. *Critically, it is inherent that individuals within the reaches of a collective intelligence are incapable of understanding the big picture beyond the scope of local interactions. Indeterminacy refers to parameters of a system, which are not known.* Of course, we are often unaware of what we do not know. This gets to the core of unconsciousness, and unawareness of being unaware, the converse of the definition of consciousness as described by Roger Penrose, an awareness of being aware. It is the local interactions between similar agents of a system, or conforming individuals, that account for near-monolithic voting blocks. The elected officials then have the potential to do nefarious things within a population that has turned over control to a commander. The paragon example of this was Nazi Germany when Hitler was able to win over the support of the German voters including enough Jews. He did not do this by telling the truth but rather by propaganda and charisma.

REFERENCES

1. Sweeney, H.L. and Holzbaur, E.L.F. (2018). "Motor Proteins". *Cold Spring Harbor Perspect. Biol.* 10(5):a021931. doi:10.1101/cshperspect.a021931. PMID:29716949; PMC ID:PMC5932582.
2. Davies, P. (1986). *The Forces of Nature*, 2nd edition. Cambridge University Press. Cambridge, England.
3. Craddock, T.J., Friesen, D., Mane, J., Hameroff, S., Tuszynski, J.A. (2014). "The Feasibility of Coherent Energy Transfer in Microtubules". *J. Roy. Soc. Interface.* 11(100):20140677. doi:10.1098/rsif.2014.0677. PMID:2523 2047; PMCID:PMC4191094.
4. Tuszynski, J.A. and Dixon, J.M. (2002). *Biomedical Applications of Introductory Physics.* J. Wiley & Sons, New York.
5. Priel, A., Tuszynski, J.A. and Woolf, N.J. (2010). "Neural Cytoskeleton Capabilities for Learning and Memory". *J Biol Phys.* 36(1):3–21.
6. Squire, L., Berg, D., Bloom, F.E., Du Lac, S., Ghosh, A. and Spitzer, N.C. eds. (2012). *Fundamental Neuroscience.* Academic Press. Cambridge, Massachusetts.
7. Berg, J.M., Tymoczko, J.L., Gatto, Jr., G.J. and Stryer, L. (2015). *Biochemistry*, 8th edition. W. H. Freeman and Company, New York.
8. Roberts, K., Alberts, B., Johnson, A., Walter, P., & Hunt, T. (2002). Molecular biology of the cell. *Garland Science.* New York.
9. Liberti, M.V. and Locasale, J.W. (2016). "The Warburg Effect: How Does it Benefit Cancer Cells?" *Trends Biochem Sci.* 41(3):211–218. Erratum in: *Trends Biochem Sci.* Mar;41(3):287. (b) Schwartz, L., Supuran, C.T., Alfarouk, K.O. (2017). "The Warburg Effect and the Hallmarks of Cancer". *Anticancer Agents Med Chem.* 17(2):164–170.
10. Ingber, D.E. (2003). "Tensegrity I. Cell Structure and Hierarchical Systems Biology". *J Cell Sci.* 116(7):1157–1173.
11. Boal, D. and Boal, D.H. (2012). *Mechanics of the Cell.* Cambridge University Press. Cambridge, England.
12. Kornberg, H.L. and Krebs, H.A. (1957). "Synthesis of Cell Constituents from C2-units by a Modified Tricarboxylic Acid Cycle". *Nature* 179(4568):988–991.
13. Scholes, G.D., et al. (2011). "Lessons from Nature about Solar Light Harvesting". *Nature Chemistry* 3(10):763–774.
14. Dörnemann, D. and Horst, S. (1986). "The Structure of Chlorophyll RC I, a Chromophore of the Reaction Center of Photosystem I". *Photochem. Photobiol.* 43(5):573–581.

15. Hodgkin, A.L. and Huxley, A.F. (1952). "A Quantitative Description of Membrane Current and Its Application to Conduction and Excitation in Nerve". *J Physiol.* 117(4):500.

16. Cotterill, R. (2003). *Biophysics: An Introduction.* John Wiley & Sons Ltd., West Sussex, England.

17. Hille, B. (2001). *Ion Channels of Excitable Membranes.* Sinauer, Sunderland, MA.

18. Howard, J. (2001). *Mechanics of Motor Proteins and the Cytoskeleton.* Sinauer Associates, Sunderland, MA.

19. Dustin, P. (2012) *Microtubules.* Springer Science & Business Media. Springer-Verlag, Berlin, Heidelberg.

20. Preto J. (2016) "Classical Investigation of Long-Range Coherence in Biological Systems". *Chaos* 26(12):123116. doi:10.1063/1.4971963. PMID:28039969.

21. Ben-Naim, A.Y. (2013). *Statistical Thermodynamics for Chemists and Biochemists.* Springer Science & Business Media. Berlin/Heidelberg, Germany.

22. Demetrius, L. (2006). "The Origin of Allometric Scaling Laws in Biology". *J Theor Biol.* 243(4):455–467.

23. Demetrius, L. and Tuszynski, J.A. (2010). "Quantum Metabolism Explains the Allometric Scaling of Metabolic Rates". *Journal of the Royal Society Interface*, 7(44):507–514.

24. Krebs, H.A., Kornberg, H.L. and Burton, K. (1957). "A Survey of the Energy Transformations in Living Matter". *Ergeb. Physiol.* 49:212–298.

25. Grin, A., Snoke, D.W. and Stringari, S. (1996). *Bose-Einstein Condensation.* Cambridge University Press, Cambridge UK.

26. Schrödinger, E. (1992). *What is Life?: With Mind and Matter and Autobiographical Sketches.* Cambridge University Press. Cambridge, England.

27. Sension, R.J. (2007). "Biophysics: Quantum Path to Photosynthesis". *Nature* 446(7137):740–741.

28. Friesen, D.E., Craddock, T.J., Kalra, A.P. and Tuszynski, J.A. (2015). "Biological Wires, Communication Systems, and Implications for Disease". *Biosystems* 127:14–27.

29. Albrecht-Buehler, G. (1992). "Rudimentary Form of Cellular Vision". *Proceedings of the National Academy of Sciences* 89(17):8288–8292.

30. Beck, F. and Eccles, J.C. (1992). "Quantum Aspect of the Brain Activity and the Role of Consciousness". *Proceedings of the National Academy of Sciences* 89:11357361

31. Davydov, A.S. (1979). "Solitons, Bioenergetics, and the Mechanism of Muscle Contraction". *Int. J. Quantum Chem.* 16(1):5–17.

32. Scott, A. (1992). "Davydov's Soliton". *Physics Reports* 217(1):1–67.

33. (a) Fröhlich, H. (1986). "Coherent Excitation in Active Biological Systems". In: *Modern Bioelectrochemistry* (pp. 241–261) Springer, US. (b) Fröhlich, H. (1968). "Long-Range Coherence and Energy Storage in Biological Systems". *Int. J. Quantum Chem.* 2(5):641–649.

34. Paul, R., Chatterjee, R., Tuszyński, J.A. and Fritz, O.G. (1983). "Theory of Long-Range Coherence in Biological Systems. I. The Anomalous Behaviour of Human Erythrocytes". *J. Theor. Biol.* 104(2):169–185.

35. Nardecchia, I., Spinelli, L., Preto, J., Gori, M., Floriani, E., Jaeger, S., Ferrier, P. and Pettini, M. (2014). "Experimental Detection of Long-Distance Interactions between Biomolecules through Their Diffusion Behavior: Numerical study". *Physical Review E* 90(2):022703.

36. Lundholm, I.V., Rodilla, H., Wahlgren, W.Y., Duelli, A., Bourenkov, G., Vukusic, J., Friedman, R., Stake, J., Schneider, T. and Katona, G. (2015). "Terahertz Radiation Induces Non-Thermal Structural Changes Associated with Fröhlich Condensation in a Protein Crystal". *Structural Dynamics* 2(5):054702.

37. Preto, J., Pettini, M. and Tuszynski, J.A. (2015). "Possible Role of Electrodynamic Interactions in Long-Distance Biomolecular Recognition". *Phys. Rev. E. Stat. Nonlin. Soft Matter Phys.* 91(5):052710. doi:10.1103/PhysRevE.91.052710.

38. Hameroff, S. and Penrose, R. (2014). "Consciousness in the Universe: A Review of the 'Orch OR' Theory". *Physics of Life Reviews* 11(1):39–78.

39. Craddock, T.J.A., Tuszynski, J.A. and Hameroff, S. (2012). "Cytoskeletal Signaling: Is Memory Encoded in Microtubule Lattices by CaMKII Phosphorylation?" *PLoS Comput Biol.* 8(3):e1002421.

40. Franco, M.I., Turin, L., Mershin, A. and Skoulakis, E.M. (2011). "Molecular Vibration-Sensing Component in Drosophila Melanogaster Olfaction". *Proc. Natl. Acad. Sci. USA* 108(9):3797–3802.

41. Tonello, L., Cocchi, M., Gabrielli, F. and Tuszynski, J.A. (2015). "On the Possible Quantum Role of Serotonin in Consciousness". *J. Integr. Neuroscience* 14(03):295–308.

42. Romero, E., Augulis, R., Novoderezhkin, V. I., Ferretti, M., Thieme, J., Zigmantas, D. and van Grondelle, R. (2014). Quantum Coherence in Photosynthesis for Efficient Solar Energy Conversion. *Nat phys.*, 10(9):676–682. doi:10.1038/nphys3017

43. Engel, G.S., Calhoun, T.R., Read, E.L., Ahn, T.K., Mancal, T., Cheng, Y.C., Blankenship, R.E., Fleming, G.R. (2007). "Evidence for Wavelike Energy Transfer Through Quantum Coherence in Photosynthetic Systems". *Nature* 446(7137):782–786.

44. Hamburger, V. (1997). Wilhelm Roux: Visionary with a Blind Spot. *J Hist Biol.*, 30:229–238.

45. Kirson, E.D., Dbalý, V., Tovarys, F., Vymazal, J., Soustiel, J.F., Itzhaki, A., Mordechovich, D., Steinberg-Shapira, S., Gurvich, Z., Schneiderman, R., Wasserman, Y., Salzberg, M., Ryffel, B., Goldsher, D., Dekel, E., Palti, Y. (2007). "Alternating Electric Fields Arrest Cell Proliferation in Animal Tumor Models and Human Brain Tumors". *Proc. Natl. Acad. Sci. USA* 104(24): 10152–10157.

46. Titova, L.V., Ayesheshim, A.K., Golubov, A, Rodriguez-Juarez, R., Woycicki, R., Hegmann, F.A., Kovalchuk, O. (2013). "Intense THz Pulses Down-Regulate Genes Associated with Skin Cancer and Psoriasis: A New Therapeutic Avenue?". *Scientific Reports* 3:2363.

47. Brown, J.A. and Tuszynski, J.A. (1999). "A Review of the Ferroelectric Model of Microtubules". *Ferroelectrics* 220(1):141–155.

48. Beloussov, L.V., Opitz, J.M. and Gilbert, S.F. (2004). Life of Alexander G. Gurwitsch and his Relevant Contribution to the Theory of Morphogenetic Fields. *Int. J. Dev. Biol.* 41(6):771–777.

49. Popp, F.-A. and Yan, Y. (2002). "Delayed Luminescence of Biological Systems in terms of Coherent States". *Physics Letters A* 293(1):93–97.

50. Szent-Györgyi, A. (1977). "The living State and Cancer". *Pro Natl Acad Sci.* 74(7):2844–2847.

51. Cifra, M., Brouder, C., Nerudová, M. and Kučera, O. (2015). Biophotons, Coherence and Photocount Statistics: A Critical Review. *J. Lumin.* 164:38–51.

52. Jelínek, F., Cifra, M., Pokorný, J., Vanis, J., Simsa, J., Hasek, J., Frýdlová, I. (2009). "Measurement of electrical oscillations and mechanical vibrations of yeast cells membrane around 1 kHz". *Electromagn. Biol. Med.* 28(2):223–232. doi:10.1080/15368370802710807. PMID:19811404.

53. Vedral, V. (2006). *Introduction to Quantum Information Science.* Oxford University Press on Demand. Oxford, England.

54. Pollack, G.H. (2013). *The Fourth Phase of Water: Beyond Solid, Liquid, and Vapor.* Ebner and Sons Publishing. Seattle, Washington.

55. Ho, Mae-Wan. (2008). *The Rainbow and the Worm: The Physics of Organisms.* World Scientific.

56. (a) Ho, Mae-Wan, Yu-Ming, Z., Haffegee, J., Watton, A., Musumeci, F., Privitera, G., Scordino, A. and Triglia, A. (2006). "The Liquid Crystalline Organism and Biological Water". In: Pollack G.H., Cameron I.L., Wheatley D.N. (eds) *Water and the Cell* (pp. 219–234). Springer, Dordrecht, Netherlands. (b) Ho, Mae-Wan. (2015). "Illuminating Water and Life: Emilio Del Giudice". *Electromag. Biol. Med.* 34(2):113–122.

57. Mayburov, S.N. (2011). "Photonic Communication and Information Encoding in Biological Systems". *Quant. Com* 11:73.

58. Poznanski, R.R., Tuszynski, J.A. and Feinberg, T.E., eds. (2016). *Biophysics of Consciousness: A Foundational Approach.* World Scientific.

59. Hameroff, S. and Penrose, R. (1996). "Orchestrated Reduction of Quantum Coherence in Brain Microtubules: A Model for Consciousness". *Mathematics and Computers in Simulation* 40(3–4):453–480.

60. Kumar, S., Boone, K., Tuszyński, J., Barclay, P. and Simon, C. (2016). "Possible Existence of Optical Communication Channels in the Brain". *Sci Rep.* 6:36508. doi:10.1038/srep36508. PMID:27819310; PMCID: PMC5098150).

61. Shannon, C.E. (1949). "Communication theory of secrecy systems". *Bell Labs Technical Journal* 28(4):656–715.

62. Casagrande, D.G. (1999). "Information as Verb: Re-conceptualizing Information for Cognitive and Ecological Models". *J. Ecol. Anthropol.* 3:4.

63. Priel, A., Ramos, A.J., Tuszynski, J.A. and Cantiello, H.F. (2006). "A Biopolymer Transistor: Electrical Amplification by Microtubules". *Biophys J.* 90(12):4639–4643.

64. Yu, W., Naim, J.O., McGowan, M., Ippolito, K., Lanzafame, R.J. (1997). "Photomodulation of Oxidative Metabolism and Electron Chain Enzymes in Rat Liver Mitochondria". *Photochem Photobiol.* 66(6):866–871.

65. Priel, A., Tuszynski, J.A. and Woolf, N.J. (2005). "Transitions in Microtubule C-termini Conformations as a Possible Dendritic Signaling Phenomenon". *Eur. Biophys. J.* 35(1):40–52.

66. Oschman, J.L. (2015). *Energy Medicine: The Scientific Basis.* Elsevier Health Sciences. Amsterdam, the Netherlands.

67. Jung, C.G. (1936). "The Concept of the Collective Unconscious". *Collected Works*, 9(1):42.

3

From Quantum Biology to Quantum Medicine

Chapter Overview

What is different between quantum physics and classical physics? If the most fundamental understanding of matter requires quantum physics, does it not apply to biology? Is metabolism a quantum biology effect and if so, wouldn't this be of major significance? If metabolism is a quantum phenomenon and metabolic impairment is a root cause of many diseases, does it mean that quantum medicine may soon emerge out of quantum biology?

There is a paradigm shift occurring across the scientific landscape. Linear reductionist thinking going back over three centuries to René Descartes and Isaac Newton seems to be running out of steam and is unable to provide a proper account of the functioning of complex systems that abound in biology. Physics has undergone several scientific revolutions and it seems that with the dawn of Big Data swamping medical researchers with enormous amounts of information about living systems, the time has come to rethink the foundations of the research methodology appropriate for this field today. One of the key discoveries that changed the face of physics in the 20th century was quantum mechanics, which shattered the myth of Cartesian certainty and replaced it with the world described by probabilities and strangely entangled particles, which are also behaving like waves. It took chemistry to embrace the quantum reality a few decades and biology now appears to be reluctantly accepting that at least some phenomena in the warm, wet, complex, and heterogeneous hierarchical living systems need to be explained using quantum mechanical principles. The list of such effects is short but non-trivial and growing. Photosynthesis, vision, olfaction, bird navigation, and perhaps even molecular recognition are all examples of quantum physics at work in the service of biology. It is not surprising that Mother Nature found uses for sophisticated quantum algorithms, perhaps even in driving gene mutations to favorable configurations in a self-propelled evolutionary race to never-ending optimization. After all, biology has had two billion years and countless replicates of its experiments to find the best solutions for its problems. Importantly, strong evidence exists that metabolism, which is, arguably, the most crucial attribute of biological systems, is itself a quantum phenomenon. It was mathematically demonstrated that the empirically discovered allometric scaling laws of physiology can be very well understood using the methods of the quantum theory of solids within the theory called quantum metabolism. This chapter presents the reader with a kaleidoscope of insights and elaborations on the theme of quantum physics informing biology and also on the potential of these advances making their way to the field of medicine. Once again, quantum coherent states achievable in such physical systems as lasers, if applicable to the living state, could shed light on the coherent unitary state of self in biology. This would be especially significant in the context of brain dynamics. Conversely, loss of coherence could be directly linked to pathologies and disease states, for example, metabolic diseases when the human body's energetics is disturbed, or to mental diseases such as bipolar disorder when the brain's coherent state is perturbed beyond the range of stability.

3.1 On the Cusp of a Quantum Biology Revolution

Traditional establishment medicine in the West has achieved amazing successes diagnostically and therapeutically by applying strategies according to linear models, which reflect direct relationships between cause and effect, especially in acute care settings. Algorithms for treating seizures, cardiac arrest, and respiratory failure as well as the application of antibiotics and analgesics are examples of linear models that have been extremely successful in medical practice. These successes, however, have now plateaued and new ones are harder to come by. Particularly in the area of chronic diseases (and especially cancer) further strives will require engaging the perspectives predicated on nonlinear models and behaviors, which are non-trivial but can be predictive in nature.

Living systems have evolved by breaking and forming new symmetries resulting in complex and sometimes chaotic organized structures. The notion of symmetry is a foundational framework perspective from which to understand not only biology, both mathematically and qualitatively (i.e. macroscopically), but the universe as a whole. Physical or biological systems may be defined according to conventional wisdom or arbitrarily. However, if a system is defined in nature bounded by a natural hierarchical scale or disciplinary field of science, the relationships between the component parts in terms of their interactions, then the system as a whole may be described from the perspective of symmetry.

Symmetry defines operations performed on the system or its parts that leave the system invariant. The enduring nature of human biology (as well as of all organismic biology) relates to the human flexibility and robustness that underscores adaptability, as complex adaptive systems, which provides a different perspective than that of symmetry and symmetry breaking. However, both perspectives provide useful and complementary insights.

DOI: 10.1201/9781003149873-3

The purpose of life may be found at the interface of philosophy and material science, such that biology exists not in the chemical and physical interactions per se, but in the goal-oriented positive and negative feedback loops, which are the hallmarks of complex adaptive systems providing them with directionality.

This may be a concept that can be related to the physical equivalent of purpose. It is somewhere between the boundaries of random disorder and predictable regularity that the elegance of biology and ecosystems emerges. Laws and rules represent the regularities and patterns of fundamental molecular anatomy and interactions that ultimately promote a higher-level purpose for biology. Strategies that embrace such nonlinear biological designs, including a perspective on the motivation of biological systems, will allow potentially explosive and game-changing technological, diagnostic, and interventional innovations that will catapult clinical health care to new tiers. Accordingly, it is the aim of this writing to highlight perspectives on the flow of energy that underpins the efficiency of metabolic designs and their converse that is often associated with diseases. Most importantly, this approach can offer a challenge to those charged with resolving problems in health care and medicine to be metacognitive and purposeful about the strategies we select in order to advance our society and the human condition. Recently, much progress has been made at a fundamental level of our understanding of biological systems in terms of physical mechanisms including atomistic descriptions of proteins, DNA, and other biomolecules.

Systems biology has introduced networks of genes, proteins, and metabolic enzymes as dynamical, self-regulating, and self-organizing structures functioning at a cellular level and above.

Finally, after a century of extraordinary successes in explaining the nature of physical reality, quantum concepts are slowly making their way toward addressing the most intriguing issues in biology. These are all game-changing developments in the life sciences that will undoubtedly affect the future of medicine.

The challenges of difficult problem-solving in health care and medicine are best engaged by first distinguishing the system being studied as either linear (i.e. complicated) or nonlinear (i.e. complex). In cases of both the complicated and the complex systems, computational modeling of their behavior would either be helpful or necessary to enable predictions regarding their properties.

It should be noted that linearity generally means in physics a proportional relationship between a (deemed to be an independent variable) cause and effect (deemed to be a dependent variable), in which case, one applies linear mathematical equations and formulas to describe such systems and uses the results of these mathematical operations to make predictions regarding their behavior as a function of new input value(s) of the independent variable(s). Linear systems are as a rule much more predictable than non-linear systems. For nonlinear systems (and almost all systems are nonlinear at least over some range of their behavior), one can use as a starting point

qualitative algorithms or protocols stratified by statistical outcome possibilities of interventions predicted on the relationships of cause and effect.

Complicated problems that have linear cause–effect relationships typically invoke and are well served by reductionist top-down strategies. On the other hand, in the description of complex systems one seeks patterns or relationships between the interacting parts of a system that may or may not be premised by rules, referred to as simple rules intended to "simplify complexity".

These rules are then inputted into computational models describing such systems. The relationships between the interactive components in this case are nonlinear in terms of cause-effect dependences and consequently describe the system as a whole in terms of some new and often unexpected qualitative behavior. Hence, the characterization of a system as either linear or nonlinear defines the relationships between the parts and the system as a whole. Analyzing the system's response to external stimuli provides a crucial perspective by which to understand and critically problem solve the vexing quandaries and toll of suffering in health care and medicine. Although a great deal of reduction helps define the simple rules, that is the interactions between elements of a system at a basic level, before it evolves into an escalating level of complexity due to inherent nonlinearity.

The relationship of the ultimate emergent behavior cannot be easily reduced to the individual components.

Hence, the bottom-up approaches enlist reductionist models only partially. An example of a bottom-up approach may involve the description of molecular causes of a disease, as is the case with the result of mutations in the huntingtin protein leading to its misfolding and subsequent aberrant aggregation in neurons leading to the onset of Huntington's disease. In this case, the understanding of the disease is offered by a bottom-up approach linking gene mutations to protein expression and cellular damage. However, the development of a therapy requires a top-down approach since a dose of an inhibitor of the aggregation process must be found together with proper scheduling and an investigation of pharmacokinetic and pharmacodynamic properties of the drug molecules in a concentration-dependent manner, all of which require an organism-wide understanding.

The translation of bottom-up understanding to top-down therapy at the individual (or whole organism) level for complex systems must rely on empiricism rather than strict reductionism.

In practice, an interplay between the two approaches and a hybrid combination of empirical and reductionist model development is the most successful strategy, which allows for a boot-strapping of model parameters in a multilevel model development. While it is true that empirical management may be regarded as the "art of science", computational models hold the potential to be surprisingly predictable, providing the optimism for future progress. This is especially true of the recently implemented artificial intelligence models that use big data analytics to derive algorithmic predictions based on training sets with no underlying hypothesis, only empirical basis provided.

For over two billion years of evolution involving life on planet Earth, nature has been "performing" countless experiments whose objects were all organisms that ever lived. This amounts to an astronomical number of possibilities that have either led to successful improvements or failures. The latter has been mostly discarded such as dinosaurs or woolly mammoths. This massive process of natural selection obviously overshadows the number of experiments that scientists have conducted in their labs over the past century or two.

> *Quantum mechanics, special theory of relativity, or quantum search algorithms may have been "discovered" and found of use by Mother Nature long before we humans have.*

Hence, nature seems to have acquired an almost infinite wisdom that basic and applied scientists can learn from by examining as many perspectives as possible. This logic has fundamentally inspired this writing.

We can find inspiration in all aspects of human creativity such as art, architecture, science fiction, and, of course, music. For me, some of the most energizing pieces of popular music have been created by the British rock group Led Zeppelin. The universally most enjoyed piece of music ever produced by the band Led Zeppelin was "Stairway to Heaven". In my opinion, this song is the most timeless piece of their 11 years' body of work. It took me almost as long to put together the contents of this chapter—that is, the book, which covers a huge swath of science and practical applications of metabolism. Within this comparison between the musical output of a rock band and a scientific body of knowledge on metabolism, the analog of the song "Stairway to Heaven" would be the chapter on quantum metabolism and quantum processes. This chapter provides the conceptual underpinning for many, if not all, of the properties characterizing optimal clinical health. These include the arrow of biological time relative to chronological time (i.e. slowing, or dilating, the aging process) and a satisfying, vitalizing psychological state. These physical and cognitive components of health are inextricably connected and inversely correlated to the level of systemic inflammation as the biological equivalent of thermodynamic entropy. *Biological information is greatest when the interface of energetic glue in the form of metabolic efficiency with chemistry, in terms of molecular building block formation, is maximal. Degradation of biological complexity, i.e. information, occurs in the setting of inflammation.* In other words, when negative entropy, which is information, is broken down it transitions to positive entropy in a thermodynamic sense. It has been demonstrated that under favorable conditions our metabolism reflects the quantum nature of energy production. In this case, inflammation and the arrow of time are minimal and negative entropy is maximal. *Quantum metabolism in particular and quantum biology in general is the metaphorical stairway to heaven right here on earth and may be the key to unlocking the door to biological information, enabling maximal enjoyment of the* *passage of time, prolonging that time and empowering all facets of well-being.* Just like "Stairway to Heaven" was part of a musical explosion orchestrated by Led Zeppelin, quantum metabolism and quantum biology are part of a stunning impact physics is likely to have on the future of medicine. But first of all, we should ask what is quantum physics and why should we even bother with it?

3.2 A Historical Perspective on Physics

Physics has evolved over the past 400 years from an empirical science to a fundamental basis of human knowledge codified by sophisticated mathematical formulas. This took place as a result of several scientific revolutions in our understanding of nature that were ushered by the discovery of new organizing principles: Newton's laws of mechanics, Maxwell's theory of electromagnetism, quantum mechanics as embodied by the Schrödinger, Heisenberg, and Dirac equations for the time-dependent evolution of states and their representations and, finally, Einstein's theories of special and general relativity all of which profoundly changed the way we understand the world around us.

> *Classical physics is concerned with the behavior of material objects having defined positions and velocities, with masses (and sometimes electrical charges) subjected to forces and torques acting on these objects.*

Physics, whether classical or quantum, involves the indestructible quantity called energy, a conserved quantity that cannot be created nor destroyed but can be transformed from one form to another, or transferred from one object or place to another. Various forms of energy will be discussed here in connection with biology, in particular, *biological energy transformations through metabolism.* However, there are other conserved quantities in physics, such as electric charge, momentum, angular momentum, etc. and they all play crucial roles in the foundations of physics leading to the definitions of symmetries under which physical laws are invariant. These symmetries are of cardinal importance to how we view matter, space, and time in the universe. The concepts of energy, position, mass, charge, and velocity (momentum) needed to be redefined in quantum mechanics in a very abstract, counterintuitive, and mathematical way in order to form a complete and self-contained theory of physical reality, which applies to the tiniest particles referred to as elementary particles (electrons, neutrons, protons) [1]. In a radical departure from the classical (also called Newtonian or Hamiltonian) mechanics paradigm, quantum mechanics does not consider position, velocity/momentum, angular momentum, and even energy in terms of precisely measurable quantities characterizing a physical system.

> *In quantum mechanics, physical properties are replaced by abstract mathematical representations of physical states of particles described by probabilities (rather than deterministic properties).*

A solid or fluid in classical physics may take a less well-defined localized representation in quantum mechanics as a probability "cloud" lacking defined confines and spreading over time.

> *Quantum mechanics (and its modern extension, quantum field theory) is the most fundamental theory of matter.*

The intellectual *tour de force* afforded by quantum physics is ultimately responsible for much of the technological advancements of the last and present centuries. Inventions such as MRI machines, lasers, cell phones, plasma screens, and laptop computers represent but a few examples of the applications of quantum physics to modern technology. In quantum mechanical formalism, many of the variables regarded as continuous by classical physics take on discrete values. Initially, the term quantum was used to denote a discrete packet of electromagnetic radiation. In a more generalized definition, a quantum (Latin word for "amount") refers to the smallest possible discrete unit of a physical property especially the energy or matter but also physical properties such as momentum, angular momentum, or spin.

The birth of modern quantum mechanics emerged almost accidentally at the end of the 19th century as physicists struggled to find a solution to the problem faced by classical physics when heat emission spectra were experimentally determined for physical objects called "black bodies" at various temperatures. The black body effect may be explained as follows. As the temperature of any physical object increases, the intensity of electromagnetic radiation emanating from it increases but so does the peak of the absorbed and emitted radiation, which shifts to shorter wavelengths (or higher frequencies). An illustration of this behavior is provided in Figure 3.1.

For example, at a temperature of 3,000 K the color of the black body is red; at 4,000 K is yellow and at 5,000 K it is blue. This property was empirically described by the so-called Stefan-Boltzmann law, which was well established in the second half of the 19th century. Different temperatures result in different peak frequencies, given by the so-called Wien's displacement law. White light has a very limited spectrum of frequencies ranging from blue (the shortest wavelength and highest frequency) to red (the longest wavelength and lowest frequency). Max Planck at first tried and failed to reconstruct the empirical formula for the frequency distribution using classical concepts based on a continuum of energy and a continuum of frequencies. At higher temperatures, the base of wavelengths involved in the emission spectrum becomes broader, extending beyond the infrared range to the visible spectrum. At higher frequencies, the intensity gradually tends to 0. That is, as the temperature of the black body is increased, the intensity of infrared frequency is also increased but the base of the electromagnetic radiation absorption is simultaneously broadened, extending beyond the infrared frequency. Therefore, as the temperature of the black body is increased, observing the black body, it initially appears black but then starts to absorb frequencies within the visible range ultimately appearing white. This was termed the black body effect whereby absorption spectra and emission spectra involving the quanta of light (called photons) are absorbed by atoms and molecules of the physical object resulting in raising the energy levels (which is referred to appropriately as an excitation) of its electrons in the valence band. These excited electrons then return to their low-energy state (called the ground state) and then emit photons with their frequency *corresponding to the energy difference between the excited state and the ground state of the electron. This black body effect was eventually correctly described in quantitative terms by Max Planck only when he assumed that the energy values can come in integer multiples of hf where "f" is the frequency and "h" a constant, later called the Planck constant.* This amazing and unexpected result is the first manifestation of the quantum nature of matter in the universe and also a simple linear relationship between energy and frequency. It is noteworthy that the mechanism of electron excitation and then de-excitation when it is falling back to the ground state was

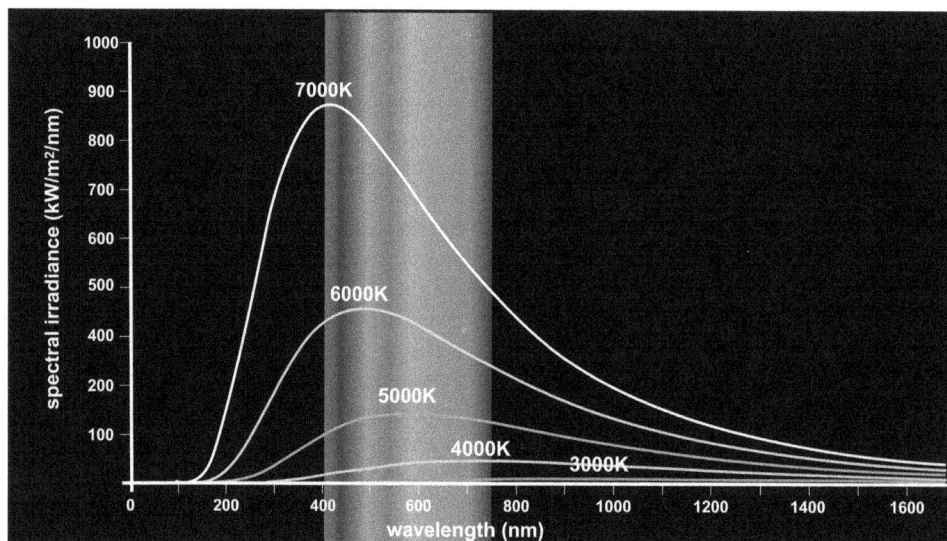

FIGURE 3.1 Intensity of the black-body electromagnetic energy spectrum as a function of wavelength shown for five different temperatures: 3,000° K, 4,000° K, 5,000° K, 6,000° K, and 7,000° K. Source: adapted from http://howthingswork.org/physics-qm-black-body-radiation/.

not yet understood at the time that Planck described the black body effect. Planck assumed that the radiation in the black body cavity was emitted (and absorbed) by some sort of "oscillators" contained within the walls of the cavity. He, in fact, made another modification to the classical theory accepted at the time, namely that the oscillators of electromagnetic origin can only have certain discrete energies. This importantly became the basis for the introduction of a quantum harmonic oscillator, which is a fundamental concept both in classical and quantum physics and can be best understood as a pendulum that oscillates back and forth around its vertical position. In classical physics, all values of the pendulum's amplitudes are allowed while in quantum physics only specific ones can exist as per Planck's quantization equation. Planck's introduction of an *ad hoc* quantization formula for these energy levels was expressed as:

$$E_n = (n + 1/2)hf \qquad (3.1)$$

where n is an integer number characterizing the energy levels and h = 6.6 × 10^{-34}J.s is the value of the Planck constant obtained to fit the experimental data. The Planck constant turned out to be a fundamental constant of nature, one of such numbers that are characteristic properties of the universe we live in, and it cannot be reduced or explained by anything else in physics. The product of the Planck constant with frequency represents that smallest quantity of energy, an energy quantum, which can only be multiplied by n (= 1, 2, 3, ...), i.e. an integer number to result in a physically acceptable number of energy units. No physical system can have a fraction of its energy quantum just like a person cannot step on a half-rung of a ladder. This seemingly accidental discovery led to *a revolutionary transformation of physical principles departing from the mechanistic laws of Newtonian physics to wave mechanical descriptions of quantum physics* developed in the first decades of the 20th century following Planck's epoch-making explanation of the black body radiation. Eventually, Newton's equations of motion that govern the mechanics of classical objects (relating force and acceleration) were replaced by Schrödinger (in the non-relativistic case) and Dirac (in the relativistic case) equations. These quantum physics equations govern the time evolution of the motion of microscopic objects such as elementary particles by providing the time-dependent probabilities for the evolution of their states.

The term quantum mechanics signifies quantization (or granularity) of the energy levels of microscopic objects.

This can be metaphorically understood in terms of finite-size pixels forming an image on the TV screen. Each pixel has a different color and is finite in size. However, from afar the human eye can only see a continuous palette of colors forming an image. Similarly, up close quantum systems are described in terms of granular or quantized energy levels they occupy, but when we add more and more particles to a quantum system or add more and more quantum particles, eventually a smooth transition to the familiar classical world of the continuum energy values is achieved.

Unfortunately, there is no precise answer to the key question: where do the quantum world end and the classical world begin?

In this chapter, we argue that in spite of the fact that biological systems are by and large macroscopic, some of their aspects involve quantum physical principles.

According to quantum mechanics, a physical particle such as an electron or a proton may manifest both particle-like behavior and wave-like behavior depending on the experimental probe or method used to interrogate its properties. Hence, an electron can undergo diffraction like a beam of light when it encounters a double slit. Conversely, a photon, which is a quantum of electromagnetic radiation, may behave as a particle or a wave depending on the experiment used to examine it. Further, it can collide with a particle and transfer part or all of its energy to it as if both of these objects were billiard balls. Inherently, in contrast to classical physics, quantum mechanics is characterized by a lack of certainty. According to the determinism of classical mechanics, as first formulated by Simon de Laplace over 200 years ago, given the knowledge regarding the precise positions and velocities of every atom in the universe, their past and future values for any given moment in time could be precisely calculated using Newtonian mechanics. *Quantum mechanics, on the other hand, uses the language of probabilities, which changed from the language of position-momentum predictability.* With the advent of quantum mechanics, our perception of physical reality has forever changed by abandoning determinism and adopting the imperfect understanding of the universe which can at best provide us with the evolution of probabilities. The certainty of classical mechanics was critically important in the 19th century for engineering purposes and allowed the launch of the Industrial Revolution. The new industrial revolution of the modern world is based on information technology where solutions are obtained for problems involving optimization based on probabilities rather than certainties.

A quantum state replaced the position and momentum used in classical mechanics and represents a superposition of a number of possible outcomes of measurements used to determine the physical properties of a quantum system.

It is important to note in this connection that any measurement or observation of a system unavoidably affects the outcome of the system's behavior hence there is never a possibility of knowing everything about any single system due to the experimenter's interference with the system probed. Further, the behavior of a quantum system is fully described by an abstract mathematical object called a wave function, denoted Ψ, which depends on both spatial coordinates (position) and time and which has an amplitude (or magnitude) defined by the product Ψ*Ψ, where the star quantity indicates the complex conjugate of the wave function Ψ*. The amplitude determines the probability of finding the system in a particular state at a given moment of time. *The time evolution of this system is prescribed by the Schrödinger equation for the wave function, which replaced Newton's equations of motion of classical* *mechanics*. In simple terms, the wave function can be thought of as a probability cloud describing the spatio-temporal evolution of the quantum object that has a positional extent that spreads over time according to the forces acting on it.

> *Quantum objects are no longer mass points but probability clouds that can never be precisely "captured" by any measurement, hence they are elusive.*

The so-called Copenhagen interpretation of quantum mechanics suggests the following:

> *All the information about a quantum system of particles is contained in its wave function.*

As stated above, according to the Born rule formulated by Max Born, the amplitude of the particle's wave function can be interpreted as the probability of finding a particle in space at a point in time, thus information about the particle is described probabilistically rather than deterministically. In particular, wave functions behave like waves and hence can diffract, and interfere (which is similar to light waves in optical experiments involving diffraction and interference) with each other forming superpositions meaning that they can be added up. They can interfere constructively leading to increased amplitudes and destructively leading to decreased or even canceled amplitudes, which is similar to the formation of bright and dark fringes of light seen in a double-slit light diffraction experiment.

> *Quantum particles exist in multiple spatial locations or states simultaneously.*

This is counter-intuitive to say the least, especially considering that these tiny particles have masses and hence would be easier to imagine as tiny ping-pong balls than dynamical clouds. When a measurement is made, one of the multiple states is selected (or revealed) and the quantum superposition of states ends up leaving a classical state in a process known as the collapse of the wave function. *After the measurement of the properties of quantum mechanical particles, their wave functions collapse to a localized outcome given by a particular quantum state and the system is no longer in a state of uncertainty.* Prior to it, every allowed outcome is possible, each with a given statistical probability. We say that the observation or measurement of this quantum object collapses its wave function, i.e. the possibilities of different observational measurements or outcomes are altered by the very act of the observational measurement, an outcome, or reality itself.

The collapse of the wave function is an instantaneous "reality check" with inputs (measurements and observations), which may not be spatially limited to regions of organized structures or to contiguous matter but even involving far separated systems if they are quantum entangled. *Quantum entanglement refers to a situation where two or more wave functions of independent particles are not independent of each other but form a composite state whereby acting on one of these particles affects the other as well.* Using an everyday analogy, a married couple is legally, financially, and emotionally entangled because affecting one of the partners has an effect on the other. The famous Schrödinger's cat thought experiment was meant to illustrate the strange nature of quantum superpositions. A quantum system such as an atom or a photon exists as a combination of multiple states corresponding to different possible outcomes. This can be extrapolated to the scenario of a real-life cat concocted by Erwin Schrödinger as a metaphysical example of quantum entanglement connecting an atomic level of a physical object with a macroscopic level of a living creature. The idea involves a cat in a box rigged in such a way that if the cat were exposed to cyanide planted in the box it would be dead, if not it would be alive. The release of poison was linked to a trigger involving a quantum stochastic event such as radioactive decay. Hence, both macroscopic states defining the cat's fate, that is being alive or dead, are possibilities requiring further observations to determine the actual reality. The question posed was when exactly quantum superposition ends and reality collapses (collapse of the wave function) into one possibility or the other. It is not until the box is opened that the observer determines the fate of the cat by collapsing its wave function. However, paradoxically before this happens the cat is in a quantum (but macroscopic) superposition state of being both dead and alive. Or is it? This "Schrödinger's cat" paradox highlights that both possibilities exist in the pre-observation period of uncertainty, however, the intervention by the observer collapses the wave function into a single reality. This interpretation, which is consistent with the Copenhagen Interpretation, argues that there is only one reality in contrast to the Many Worlds Theory, which theorizes that many simultaneous alternative realities may coexist. This latter theory proposes a situation with parallel universes whereby each branch point of alternative outcomes is real from the perspective of different dimensional life experiences. However, these alternative universes cannot communicate with each other due to destructive interference, i.e. they possess wave functions, which are out of phase with each other. In order for the different realities to progress simultaneously, time *per se* as an absolute quantity, consistent with Einstein's theory of special relativity, is premised not to exist. As such, the past, the future, and the present are all in a sense coexisting, whereby "sequential" lives are experienced in parallel. It is considered by many quantum physicists that each major branch point of possibilities that define reality over the course of an individual's lifetime exists in parallel universes. Existence within each universe in different dimensions simultaneously allows the superposition of all possibilities and collapsing the wave function, which results in a different isolated outcome experienced in any given reality. Accordingly, the alternate parallel universes may be connected by the concept of entanglement. However, the quantum world and the associated phenomenon of superposition and quantum entanglement are traditionally described at the atomic and subatomic levels. The paradox of Schrödinger's cat and macroscopic extension of the Many Worlds Theory are most plausibly just metaphorical extrapolations with no empirically demonstrated validity other than theoretical musings. However, in the strange world of quantum physics, almost everything is possible.

When two consecutive measurements are made on certain pairs of quantum variables (representing physical properties such as angular momentum and spin) called *complementary variables*, there is a fundamental limitation on the precision of the two measurements. Specifically, there is no state in which both complementary variables can be determined simultaneously with arbitrary accuracy. This phenomenon is known as the Heisenberg uncertainty principle and it has fundamental scientific and philosophical consequences regarding the nature of physical reality and our ability to describe it.

One cannot unambiguously specify the values of a microscopic particle's position and its momentum at the same time.

That is, one cannot precisely specify where an object is and simultaneously how fast it is moving. Position and momentum are examples of two mutually exclusive, complementary quantum variables. If one of these properties is known precisely, we know nothing about the other one. However, both can be known imprecisely to within the Heisenberg limit. This Heisenberg limit is given by the Planck constant that first appeared in connection with the black body radiation explanation described above. There are a number of other pairs of complementary variables that follow the same rule of the Heisenberg uncertainty principle.

One can never know precisely both the energy of a system and time (where the system was at that energy level).

That is because to determine the energy, a measurement must be performed, which displaces the system out of that energy level by perturbing it in some way. Analogously, one cannot know the precise values of how rapidly an object is spinning (angular momentum) and what angle the object is at (angular position) at a given point in time. Any two components of the spin variable are also complementary variables, so in essence, one cannot precisely determine the spatial orientation of the spin vector in quantum mechanical systems.

The next important contribution to the development of quantum physics was Einstein's explanation of the photoelectric effect. In this case, photons in the visible range of the electromagnetic spectrum strike electrons in the valence band of atoms on the surface of a metal electrode, and the momentum transferred in the collision between the photon and the electron exerts a force analogous to the collision between two billiard balls. Note in this connection that the classical definition of momentum (p = mass x velocity) is not applicable for particles traveling at relativistic speeds. The energy of the photon is then absorbed by the metal's conduction electrons provided it exceeds a threshold value determined by the difference between the ionization energy and the valence energy level of the electrons denoted by U. Given Planck's formula for the energy of photons: $E = hf$, one can conclude that if E exceeds this threshold value U, an electron will be ejected from the metal surface and will travel to the positively charged electrode (anode) triggering a current flow that can be measured by an ammeter. The amount of current generated in this process is proportional to the intensity of light. However, when the frequency of light is below this threshold, $f < U/h$, independent of the intensity of light, no current will ever flow since electrons won't be ejected from the metal surface due to insufficient energy absorbed by them to free them from the attractive potential of the metal's nuclei. This corroborated the idea that *light energy is quantized and under some conditions, photons behave like particles by colliding with electrons*. The latter aspect can also be observed in the Compton effect, which describes how photons transfer their momenta to massive particles such as electrons in a photon-electron collision, resulting in the motion of the electrons. This causes an increase in the wavelength of the light wave by reducing its energy in proportion to the amount of energy lost by scattering the electron. Next, the so-called double-slit experiment demonstrated that electrons, protons, and neutrons also behave like waves, namely given the choice of passing through one or the other of the two slits placed in their path, they paradoxically pass through both and their interference pattern is generated on a screen positioned behind the slits showing the same image as that produced by the diffraction of electromagnetic or acoustic waves which was known to science since the 19th century. The fact that seemingly solid objects such as protons exhibit this typical wave-like behavior was shocking to the physics community at the time and it still baffles the minds of physicists and non-physicists alike since the tenets of quantum mechanics are in stark contrast with our intuition based on everyday life experiences.

The existence of magnetic spin was demonstrated in the Stern-Gerlach experiment where particles such as electrons and protons subjected to magnetic field gradients change their trajectory traveling up or down, depending on the magnitude and orientation of their spin.

Spin is an entirely quantum property.

In classical mechanics, there is no spin variable and there can be no ferromagnetism without spin. Hence, spin is a necessary condition for the emergence of permanently magnetized systems such as iron oxides. We have been using magnets in compasses for more than 3,000 years not realizing this is a quantum phenomenon, which occurs even at room temperature and on a macroscopic scale. In quantum biology, we talk about the applicability of quantum processes to large-scale objects. One argument against this applicability is that quantum processes cannot occur at room temperature. However, this is not true in view of this latter example we discussed.

Quantum mechanics can apply to bulk properties of matter even at room temperature and is not limited to individual microscopic particles at temperatures close to absolute zero.

There are other such examples and they will be mentioned below.

The wave-particle duality property of subatomic particles such as electrons and electromagnetic radiation (photons) highlights a radical departure from classical physics concepts where particles are massive objects with well-defined locations in space and shape while waves have no mass but are spatially extended and propagate through space. In quantum mechanics, there is no such distinction.

> *An electron is both a particle and a wave simultaneously. A photon is a wave and a particle at the same time. Wave-like and particle-like properties can be manifested in both cases under specific circumstances.*

For example, a particle-like property is manifested in collisions where momentum and energy can be transferred from one quantum object to another. A wave-like property is the ability of quantum objects to undergo diffraction and interference and in fact, the superposition principle is a typical characteristic of waves, such as light waves. The superposition principle became one of the postulates of quantum mechanics.

The quantization of energy is another important property of microscopic particles, which was already mentioned before. There is a distinct ground state, or the lowest allowed energy level for a quantum system. There may be a number (sometimes infinitely many) of higher, excited states of a quantum particle, which may eventually lead to a continuum of energies at which point this particle is free from a confining potential.

> *Quantum physics transitions smoothly at the macro scale to classical physics when energy quanta become insignificantly small.*

As stated above, the problem is that it is not well defined where a strict boundary between quantum and classical worlds exists, either in terms of energy, mass, length, or time. Roughly speaking quantum mechanics applies to small molecules, atoms, and elementary particles while classical mechanics to large molecules and everything above on the scale of sizes. At the macroscopic scale, quantum processes have been thought to become irrelevant. However, this position is harder to defend these days in view of the growing number of macroscopic quantum phenomena such as superconductivity, superfluidity, and laser action, not to mention permanent magnetism, which is an inherently quantum property that cannot exist without spin, a quantum variable introduced into physics by Wolfgang Pauli. *Magnetism has two bases for exemplifying quantum behavior. It is required to include both relativistic properties and the quantum nature of the electron spin in order to explain the magnetic behavior of a ferromagnet.*

Tunneling in quantum mechanics refers to the ability of microscopic particles that may not have enough energy to cross over a potential barrier and so may instead penetrate through the barrier like a ghost. Strictly speaking, it is their wave function whose extent exceeds the confines of the potential well where the particle is trapped. It is one of the quirky phenomena of quantum mechanics analogous to, albeit at the subatomic microscopic scale, throwing a tennis ball over a wall, not hard enough and so rather than bounce back from the wall, the ball would reach the other side by partially penetrating through the wall. Most of the tennis ball's mass would still be trapped behind the wall but a part of it would magically appear on the other side of the wall. Recall that quantum particles do not have a defined shape or size. Tunneling implies

penetration through an energy barrier for a particle that does not have sufficient total energy to simply cross over the barrier (strictly speaking its energy, denoted E, is less than the potential barrier height, denoted V). Imagine a particle trapped in a potential well of depth V whose total energy $E = T + U$ (kinetic plus potential energies, respectively) is less than V ($E < V$). Classically, this particle would forever bounce around between the walls of this well (ignoring friction). Quantum mechanically, it has a probability that it will penetrate through the wall even if $E < V$. This tunneling probability decreasing with the thickness of the wall, denoted x, and is proportional to the exponential function:

$$\exp\left[-x\left[\left(2m(V-E)m\right)/h\right]\right]^{1/2} \qquad (3.2)$$

where m denotes the mass of the particle. It can be concluded that light particles penetrate through the barrier more than heavy ones. Coherent tunneling would require a coupling mechanism between two or more particles involved by mutual attraction or simultaneous stimulation by electromagnetic radiation (photon). An intriguing application of this property in cell biology would be the problem of potassium or sodium tunneling across ion channels in the cell membrane. Moreover, since there are thousands of ion channels in a single cell separated by reasonably small distances, coherent ion channel tunneling could explain the observed phenomenon of synchronization of ion waves in cells and even between cells.

Coordinated tunneling may also happen when one particle leaving its location and vacating a locality, would create what is called in solid state physics a "hole". A hole will attract a neighboring particle due to the opposite charge and cause it to tunnel there. In the case of a system composed of many particles and a regular array of holes, the effect would be similar to a set of falling dominos, the first falling domino will hit and topple the next one, and so on, resulting in a coordinated process of particles hopping from hole to hole. A hole represents an electron vacancy on some atom in the molecules that make up the crystal of the solid. It represents a region that another nearby electron can fall into due to electrostatic attraction, which then effectively moves the "hole" to where that other electron was. An apt metaphor could liken it to an auditorium whose seating area is full, except for one seat. If a person sitting next to an empty seat moves in, the seat where the person was becomes vacant. So, in essence, the empty seat has moved but in the opposite direction. Using the concept of a hole to describe the situation is akin to describing the motions of all the people sitting in an auditorium and shifting their seating arrangement to accommodate a latecomer whereby instead of looking at the people one is looking at the motion of an empty seat, even though the hole is not really an object that moves, but it acts like one when it moves. This is why a hole is called a quasi-particle; not a real particle, but it acts like one. These metaphors are not only interesting but also important because the concept of hopping electrons and moving holes is behind our understanding of semiconductors, another area where quantum physics at a macroscopic scale and at room temperature has led to a technological revolution ushering us into the era of modern electronics. This has been exploited in semiconductor-based electronic devices such as diodes and transistors. *Hopping of electrons and electron vacancies (called holes) is more general than tunneling because it does not necessarily involve insufficient energy to overcome the potential barrier and allows for a repetition of the tunneling process, hence the propagation of a microscopic object over significant distances.* Hopping may involve tunneling or may be simply traveling without tunneling from one localization center to the next. However, for it to be localized, there must be some local attractive potential, so it has to jump out of it, which could be thermally stimulated, spontaneous or due to electromagnetic radiation absorption, all of these effects assist the microscopic particle in making an individual hop or a series of hops.

> *Almost all concepts in quantum mechanics, virtually by definition, are counterintuitive relative to our thoughts, perceptions, and experiences.*

Quantum ideas were developed through mathematical abstractions and generalizations based on empirical observations and not through our intuitive understanding of the universe around us. The microscopic world of quantum physics is not accessible to human senses and hence has not been described the same way as in other branches of science, which developed by our direct contact with physical, biological and societal reality. Quantum physics is replete with strange properties and behaviors that appear bizarre to a person grounded in classical physics or relying on the intuitive understanding of the world around. We have already described a few examples of such bizarre behavior. Schrödinger had stated that it is not electron hopping, superposition, wave-particle duality, or other property that is the fundamental property of quantum mechanics but rather that of quantum entanglement.

> *Quantum entanglement is a state where two quantum particles are linked despite a lack of physical contact or indeed an interaction force.*

It concerns the behavior of tiny particles, such as electrons that have been specifically prepared to form a pair with opposite spin orientations in order to produce a net spin value of zero at some point in time. Having then been moved apart, even thousands of miles or much farther, they are still bound together by this spin compensation requirement so that the reality of one particle's state is instantaneously and simultaneously coordinated with that of the other particle (the particle's partner). That is, if one such particle is perturbed by the action of measuring its position, momentum, or spin, its partner, wherever it is, "responds" in a coordinated fashion. John Stuart Bell some 50 years ago provided a mathematical proof and formalism for this strange connectedness of quantum entanglement. Using pairs of photons, he and others further showed with great success that nature behaves according to his equations. It was experimentally demonstrated by Alain Aspect and his colleagues [2] that the correlations between the measurements of the property of polarization of photons were accurately predicted by Bell's formulas. A host of such experimental tests of entanglement have provided a vanishingly low probability

that such findings may be explained by some alternative theory distinct from quantum theory. *Quantum particles, like photons, are linked due to the principle of superposition, whereby they exist in every theoretically possible state at the same time. Hence, the different polarizations of the photons, their spin, may occur horizontally and vertically at the same time.* However, when one member of a quantum particle entanglement is measured, its wave function collapses causing it to fix on a single state forcing its partner(s) to do the same. The notion of entanglement requires that its "estranged" partner instantaneously and simultaneously assumes the opposite state. That is, if one particle spins horizontally, the other does so vertically, representing its "polar" opposite. Spin-spin (magnetic) or Coulomb (electrostatic) interactions provide a physical entanglement between the two systems. Note that two electrons, one with its spin up and the other with its spin down as illustrated in Figure 3.2, as in Bell's famous thought experiments, are kept in a down and up position together due to a spin–spin interaction when they are placed next to each other. When released to travel in opposite directions, the interaction between these electrons disappears with the distance between them but they remain entangled. In other words, an important aspect of this experiment is "preparing the two particles to be in an entangled state". This does not happen for all particles at all times, and hence the validity of philosophical arguments extrapolating this situation to all objects in the universe is very limited.

Elementary particles are subatomic objects, *i.e.* electrons, protons, neutrons, etc.

At the next level of organization of physical matter, a crystal lattice is formed from atoms or molecules (including biomolecules such as proteins) as elementary units of the lattice (nodes). *Phonons are the collective vibrations of crystal lattices or atomic displacement waves propagating on such lattices.* Phonons are not quantum particles in a strict sense but are solid-state equivalents of elementary particles called quasi-particles with their attendant quantum properties. It is important to stress that this is a direct application of quantum mechanics to systems composed of arbitrarily many particles, not to just several as in the case of subatomic physics. Collective quantum phenomena, where numerous elementary particles are involved, such as is the case with phonons, may therefore be seen at a macroscopic scale in solid-state physics. Collective modes of many particles give rise to other

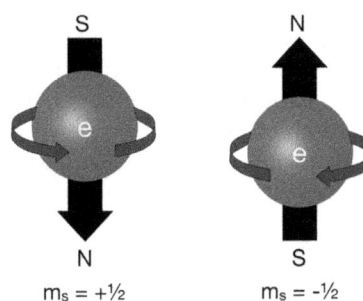

FIGURE 3.2 A schematic illustration of a spin-down and spin-up state of an electron. Source: adapted from http://web.fscj.edu/Milczanowski/psc/lect/Ch8/slide9.htm.

quasi-particles with exotically sounding names, such as excitons (collective modes of electronic excitations); magnons (collective modes of magnetic spins in a crystal lattice); and polarons (collective models of lattice dipolar polarization). These collective modes exist on a macroscopic scale, and their introduction into the physical theory was a major departure from the conventional quantum mechanics perspective, which previously considered subatomic phenomena only at very small, microscopic scales and for few objects.

> *There is in principle no size limitation to quantum effects and hence biological systems being relatively large remain acceptable candidates for the application of quantum physics concepts.*

Phonons belong to a class of quantum objects called bosons, which are allowed to occupy any and all quantum states in arbitrary numbers without any restrictions, unlike in the case of another class of quantum objects called fermions. Other examples of bosons include gluons, the famous Higgs boson, muons, quanta of light, i.e. photons and many other collective quantum excitations, such as magnons with an integer value of spin. In other words, thinking in terms of quantum states as a stepladder with the lowest rung representing the ground state (lowest energy state) and quasi-particles occupying these states, each step on this energy ladder can be occupied by an arbitrary number of quasi-particles. States other than the ground state are called excited states. By a physical law called the spin-statistics theorem, bosons must have an integer value of spin (0, 1, 2, …). This is in contrast to fermions, such as electrons, neutrons, or protons, whose states can be occupied only by one (or none) of these particles at a time. This restriction is called *the Pauli exclusion principle. According to this principle, one cannot have two or more electrons (or other fermions) occupying the same quantum state.* This, however, still allows for some room due to the fact that fermions have a half-integer spin (1/2, 3/2, 5/2, …) which permits two fermions to occupy the same energy level as long they have different spin states, one with spin up and the other with spin down. In the presence of orbital motion such as in the case of electrons bound to the nucleus of an atom, their orbital angular momentum allows for an even greater number of fermions occupying the same energy state due to the fact that for a particle with orbital angular momentum denoted L, there are *(2L + 1)* orbital angular momentum projections on a quantization axis. This is called "degeneracy". Hence, every electronic state in an atomic "orbit" can be occupied by up to *2(2L + 1)* electrons without violating the Pauli exclusion principle (the factor 2 is due to the spin states: up and down). These rules were instrumental in our understanding of the periodic table of the elements and gave rise to quantum chemistry by explaining atomic valence states and the formation of chemical bonds between atoms.

We should stress here the importance of the statistical nature of quantum phenomena. *Although all aspects of quantum theory are probabilistic even when applied to a single particle, there is also a distinct statistical aspect for systems of many quantum particles.* There is a major difference between probabilistic approaches to the quantum mechanics of systems of few particles (e.g. as in the Copenhagen interpretation applied, say to a single electron in a harmonic potential) and statistical quantum mechanics of many particles (as for example is the case for lattice vibrations in a crystal or helium particles in a superfluid). This is a distinction that has not been sufficiently understood and amplified in applications to quantum biology. In the former case, we cannot know for sure what the exact state of a particle is; we only know within some probability values its state and its time evolution. In the latter case, we deal with macroscopic averages over an ensemble of many particles. Metaphorically, this is akin to the difference between individual variability for the state of health of a person and statistical variability for a population of patients studied by epidemiology.

> *These averages provide us with hallmarks of the quantum states of each of these particles.*

For example, the specific heats of solids or their magnetic susceptibilities reveal some exotic properties due to their quantum nature.

> *At the scale of macroscopic averages, where there is the greatest potential for finding quantum effects in biology, not at the level of a single particle or a single atom, which by and large is not relevant for living systems.*

Another intriguing property of quantum states involving many particles is the so-called Bose-Einstein condensation [3], which may occur for bosons (or integer-spin particles) viewed as either individual particles or collective quasi-particles, i.e. quantum excitations, which must be confined in space to manifest this strange behavior. This effect was predicted to occur at a very low temperature, called the condensation transition temperature. What this effect means is that a macroscopic number of these bosons can occupy the same ground state, making the system globally coherent meaning that all its members behave as one. *When a condensate is formed, it represents a huge (possibly macroscopic) quantum state, which could explain the synchronization of diverse parts of a biological organism.* Bose-Einstein condensation has been discovered for superfluids and more recently in more exotic atomic gases in confined spaces. In principle, other solid-state quasi-particles could also undergo Bose-Einstein condensation, e.g. phonons, magnons, excitons, polarons, *etc.* but the conditions for creating and observing such exotic macroscopic quantum states may be technically too difficult at present to achieve [3]. There have been speculations that biological systems may provide examples of Bose-Einstein condensates by Herbert Froehlich, as this would explain some of their most enigmatic features such as organism-wide synchronization of functions and a unitary sense of self and we will discuss it in more detail later in this chapter.

There are three levels of quantum formalism, which are determined by the extent to which the continuous variables of classical physics are converted to discrete variables, otherwise known as quantization. Quantum mechanics is a first-quantized or semi-classical theory of physics in which particle properties given by the quantum numbers are discrete but field properties (*e.g.* electromagnetic) and interactions (e.g. electrostatic) are not. Quantum field theory is a so-called second-quantized theory in which all particle properties, field properties, and interactions are discrete except for those due to gravity. Quantum gravity is an incomplete third-quantized theory in which gravity is expected to be also made discrete. The long-awaited integration of gravity with quantum field theory would lead to the construction of a so-called Theory of Everything (TOE).

> *The notion that macroscopic-scale quantum mechanics can be applied to biological systems is the basis for the slowly but steadily growing field of quantum biology.*

Its implications are potentially ground-breaking. Biology studies living systems, largely comprised of water molecules and biomolecules, existing at the macroscopic scale and at relatively high temperatures. However, a number of examples of biological, biochemical, and physiological processes appear to be manifestations of quantum mechanics at the level of living systems. As hinted at above, there is nothing in principle that prevents the use of quantum mechanics in the realm of biology.

> *Neither their sizes, nor their complexity, nor the degree of fluidity, nor even the pervasive presence of thermal fluctuations prevents the applicability of quantum mechanics within some scope of biological organization.*

A so-far-limited but growing set of examples underpins a new and evolving scientific discipline known as quantum biology. The biological processes that we know to be quantum are very subtle and depend on a single energy quantum. However, we experience these processes daily. We will discuss some of these examples in more detail below in this chapter.

3.3 The Dawn of Quantum Biology

Historically, life has been described in terms of either functionalism or vitalism. Functionalism views life forms through their behaviors, such as self-organization, homeostasis, metabolism, growth and reproduction, adaption and robustness, evolution, and extinction. Conversely, a group of 19th-century biologists called vitalists searched for a unifying energy field, or life force (elan vital) animating living systems. With the major development of reductionist biology, vitalism fell from favor and has since been viewed as "unscientific". However, early vitalism was based either on electromagnetism or on some unknown forces outside science, so its demise is rather understandable. When vitalism first emerged, quantum mechanics had yet to be discovered. On the other hand, generally accepted notions in biology are based on the motion of molecules in a liquid medium being stochastic (random Brownian movements), which ignores long-range ordering effects in molecular living systems that are absolutely critical

to the functioning of organisms as one. *Biology today is at a point in its development, which appears to be equivalent to physics a century ago as a scientific discipline.* It has accumulated a massive amount of information and data but it is struggling with providing a conceptual framework and an organizing principle for both the descriptive and quantitative information. However, biological systems are much more complex than physical systems and it will likely take longer to understand living systems than nonliving systems.

> *A major paradigm shift is occurring in biology that will transform it from a mainly descriptive area of knowledge to a quantitative branch of science with its own methodology and conceptual framework.*

Today molecular biology and sister fields such as genetics, cell biology, and others have collected reams of real-life data that only computational methods are able to sift through, visualize, and organize. No human being, no matter how clever, is able to understand let alone analyze huge data sets without resorting to computer-assisted methods.

> *What is still sorely missing is the lack of an organizing principle such as the law of conservation of energy or maximum entropy principle.*

This leaves researchers at the mercy of computational tools. Erwin Schrödinger wrote a book entitled *What Is Life* [4], which exposed some of the main challenges found in biology from a physics perspective and paved the way for the birth of molecular biology in the 1950s and is still influential today. He also remarked that on its own, classical physics seems to be insufficient to explain the robust organization and functional efficiency observed in biology suggesting that life's unitary oneness may be derived from quantum coherence in biomolecular structures.

> *Life represents "order arising from order" rather than "order emerging from chaos".*

One of the main puzzles he posed was the pervasive reduction of entropy in living systems, which seemingly contradicted the second law of thermodynamics. An answer to this puzzle involves not only quantum physics but also information science since entropy is negatively correlated with information as defined by the great computer scientist Claude Shannon. This will be discussed later in this chapter. However, first, let us discuss the possibilities of quantum effects being present in biology.

Sixty years ago or so it was expected that quantum mechanics would easily solve the mystery of life. Having successfully explained some of the most difficult properties of non-living matter, the founders of quantum mechanics hoped their theory was both weird enough and powerful enough to explain the peculiar living state of matter, too. A number of great physicists with Nobel Prizes to their credit, *e.g.* Erwin Schrödinger, Niels Bohr, Werner Heisenberg, and Eugene Wigner all offered speculations in this regard. However, more than half a century later, the dream that quantum mechanics would straightforwardly explain life as it had explained the properties of matter so perfectly, has not yet been fulfilled.

> *Certainly, quantum mechanics is needed to explain the sizes and shapes of chemical molecules and the details of their chemical bonding as well as the chemical reactions they are involved in.*

This takes place mainly in a pairwise fashion describing individual interactions between molecules. Sadly, no clear-cut "life principle" has yet emerged from the quantum realm that would single out the living state in any special way. Furthermore, classical ball-and-stick models of proteins, DNA and other biomolecules augmented by Newtonian mechanics descriptions of their constituents so far seem adequate for most explanations in molecular biology. In spite of this, there have been persistent claims that quantum mechanics can play a fundamental role in biology through the use of properties such as coherent superpositions, quantum tunneling, and wave function entanglement, referred to earlier in this chapter.

> *Superpositions of quantum states have been speculated to exist at the level of protein folding phenomena and even gene mutations.*

Entangled states have been most frequently invoked in terms of brain dynamics and human cognition. Tunneling may be at play whenever electrons traverse some distances in biological matter as in photosynthetic complexes or in mitochondrial walls. As these examples illustrate, quantum processes may not only pertain to different scales of biology (from genes to proteins to cells to organs), but possibly across scales which, if true, would explain in a fascinating way how biology manages to synchronize and coordinate its many diverse activities across multiple sizes and time scales. The biological applications mentioned here include plausible ideas like quantum-assisted protein conformational changes which are important in its biological function (e.g. activation/deactivation of MAP Kinases). In this phenomenon, while a protein structure may potentially adopt a multitude of possible conformations only one of which is stable and corresponds to the lowest free energy, it refolds rapidly to such an optimal conformation from its unfolded state by exploring a multitude of other conformations in parallel by a quantum search process. This can also lead to more speculative suggestions, such as the genetic code having its origin in the so-called Grover algorithm of quantum computation, or quantum-mediated gene expression. Unfortunately, biological systems are so complex that it is hard to separate "pure" quantum effects from the various examples of essentially classical processes with some quantum or quantum-like aspects that are also present in biology. For example, there could be a quantum (based on a vibrational resonance) component to a ligand-receptor interaction, which becomes significant only over a very short range of the distance separating the two structures while most of the longer-range interactions are entirely classical in nature (*e.g.* van der Waals interactions or screened Coulomb forces).

> *There is plenty of scope for disagreement about the extent to which life utilizes non-trivial quantum*

> *processes other than those involved in biochemical reactions.*

Undoubtedly the next few decades will bring this debate to the fore of scientific discourse.

Why should quantum mechanics be relevant to life, beyond explaining the basic structure and interaction of biomolecules?

> *Quantum effects can serve to facilitate or accelerate processes that are either too slow or impossible to attain according to classical physics.*

As already mentioned, *protein folding occurs very rapidly in nature but a classical explanation using the combinatorial space of possible conformations is woefully inadequate as opposed to a quantum wave function exploration approach.* Quantum physics introduced such concepts as discreteness, quantum tunneling, superposition, and entanglement, which produce novel and unexpected phenomena. For example, electron transport in terms of hopping from one potential well to the next by tunneling through a barrier is completely impossible in classical physics. The formation of conduction bands in semiconductors and metals with forbidden energy zones has led to the development of modern electronics. The ejection of electrons into their conduction states by sufficiently high-energy photons in the photoelectric effect has resulted in the development of photosensors. Given that the basic processes of biology take place at a molecular level, harnessing quantum effects does not seem a priori implausible.

Rather straightforward extensions of quantum mechanics to chemical compounds and chemical reactions can be used to properly describe the behavior of chemical bonds between molecules using such semi-classical formalisms as the density functional method. *All chemistry including biochemistry (studying molecules found in living systems) is based on the creation and destruction of bonds between atoms in molecules and hence it can be argued to rely on quantum interactions.* However, the unitary oneness and ineffability of living systems have suggested that higher-level quantum properties such as Bose-Einstein condensation, quantum coherent superposition, coherent tunneling, resonant energy transfer, and entanglement may also operate in biology under certain conditions. Hence, this suggestion links "quantum weirdness" inextricably within biological systems and organismic biology due to the need for a mechanism that can explain biological synchronization and coherence. However, the main challenge in this regard is to explain how the enemy of coherence, namely decoherence, can be kept at bay for long enough in a living system. We discuss this topic next.

3.4 Decoherence

The advantage of quantum coherence is enhancing coordination in both space and time. Biological systems need it even more than physical systems due to their heterogeneity and complexity, as well as due to the relatively high value of physiological temperature, at least in the case of warm-blooded animals. How coherent behavior emerges at the level of a cell, tissue and organism is not well understood at present and poses a major intellectual challenge.

Quantum coherence involving its parts, most likely in a hierarchical way, is a viable explanation of a biological "unitary sense of self" and purposeful activity of an organism.

An even more perplexing issue is how these quantum coherent modes become amplified from a microscopic to a macroscopic level. How a single photon or two, triggering an electronic excitation in the rods and cones of a retina, ends up causing a reaction in the brain that is perceived by the "owner" of this brain as a flash of light is still a key mystery. Propagation of coherence over the biological space can involve quantum mechanisms at several levels. Multiple entities can be coordinated within a unit and then they become synchronized amongst themselves. In other words, this mechanism can and should be hierarchical. This conclusion can be reached because of the vast span of time and frequency ranges over which biological systems operate in concert. From ps (10^{-12} s) oscillations of H-bonds to μs (10^{-6} s) frequencies of conformational states of proteins to ms (10^{-3} s) operation of action potentials, to second-scale for the human senses to an hour-scale of cell division (10^4 s) to circadian rhythms, *etc.*, up to the life-span scale (10^9 s), *biology encompasses at least 21 orders of magnitude on the time scale, whose synchronization is mind-boggling.* Therefore, the time scale alone has to involve nested cycles with feedback loops in a boot-strapping fashion so there is nonlinearity embedded in responses while within each autonomous network one can develop quasi-linear quantum descriptions. Similarly, on the scale of spatial dimensions from atoms on the angstrom scale (10^{-10} m), to proteins on the nm scale (10^{-9} m) to cells on the μm-scale (10^{-6} m) and finally to the organism on the meter-scale, *physical dimensions of living systems span at least ten orders of magnitude that need to be synchronized and coordinated for the system to work as a whole.*

While organism-wide coherence is observed in biological systems, quantum coherent effects have been expected to be washed out on scales only slightly larger than individual atoms or complexes thereof, at warm physiological temperatures, and in aqueous media, in which biological systems are embedded coherence should disappear fast. Thus, the likelihood of quantum states playing functional roles at mesoscopic, let alone macroscopic, scales in "warm, wet and noisy" biological systems has been viewed as highly problematic due to environmental decoherence in the sense of noisy thermal environment, which has been invoked on numerous occasions as a serious impediment to quantum biology.

Quantum decoherence represents the loss of information both in terms of frequency and phase.

Frequency is important because almost all biological processes are cyclical. Phase is important when coupling cyclical processes together since being out of phase causes destructive interference. This underlines the fact that quantum coherence includes two aspects: frequency coherence and phase

coherence. Frequency coherence means that various subsystems are synchronized to oscillate at the same frequency, akin to the laser light being monochromatic. Phase coherence means that there is an additional level of synchrony since the systems are now oscillating with the same phase, like musicians in an orchestra not only playing to maintain the same tempo but also to the same cue. The opposite is called dephasing. If one loses frequency coherence, inherently phase coherence is not possible anymore. According to the Heisenberg uncertainty principle, mentioned above, in quantum systems there is a limit on the combined phase (ϕ) frequency (ω) precision due to the uncertainty criterion given by: $\Delta\phi\Delta\omega > h/2$. This means that if we manage to pinpoint the frequency precisely, phase precision is completely lost and vice versa. Hence, there is never perfect coherence possible in both frequency and phase, but thermal (due to heat) decoherence just makes it worse and worse. In fact, laser light achieves the minimum value of uncertainty in the product, namely h/2 where h is the Planck constant. White light is the opposite since it has neither frequency nor phase coherence. Laser light is an example of a so-called coherent superposition state. *Decoherence means a loss of this combination of states and gradual dephasing* into a Gaussian probability distribution (meaning a bell-shaped curve) with a zero average, i.e. no particular frequency or phase has been selected. We can think in this context of a red laser light pointer that gradually loses its coherence and starts showing different colors eventually turning into an incoherent source of white light, *e.g.* an ordinary flashlight.

As stated above, the warm and wet interior of a living cell hardly seems to be a promising playground for quantum coherence, and "quick and dirty calculations" suggest unimaginably short decoherence times of less than 10^{-13} s for most biochemical processes at body temperature. However, there are reasons why real biological systems might be less susceptible to decoherence than their simplistic models predict.

Biological organisms are highly non-linear, open, driven systems that operate away from thermodynamic equilibrium, which means that coherence is maintained by a constant energy supply.

This is in the form of metabolic energy due to nutrients, also called energy pumping. The physics of such systems is not well understood and could conceal novel quantum properties that life has discovered before humans have. More sophisticated calculations indicate that simple models generally greatly overestimate decoherence rates and underestimate coherence times. There are two other ways in which decoherence could be diminished to enable biologically important processes to occur. The first is thermal noise shielding by diminishing random noise. If the system of interest can be quasi-isolated from the decohering environment, then decoherence rates can be sharply reduced. One such possibility could involve structured water, which is much less dynamic and resembles two-dimensional ice more than fluid. Organisms may exploit thermodynamic gradients by acting as heat engines converting heat into useful work and thereby drastically reducing the effective temperature due to the fact that partially converting heat energy into useful work will not contribute to a rise in the temperature of the system. For example, ATP (adenosine triphosphate) molecules in some complexes behave as if they experience a temperature well below the physiological temperature, which is due to environmental shielding effects causing the relevant degrees of freedom of the complex to be unaffected by thermal fluctuations surrounding them, effectively freezing them. Additional mitigating effects have been found in maintaining long-term long-range coherence such as the ordering of water dipoles, the far-from-equilibrium states of living systems due to constant energy supply either from sunlight (plants) or nutrient metabolism (animals), as well as functional and structural coordination of macromolecules. The final argument is that natural evolution through billions of years of experimentation and countless attempts of trial and error may have solved the problem of decoherence beyond our current level of understanding so that mesoscopic/macroscopic quantum states are essential features of biological systems. If organized quantum states exist in living cells, they are presumably integrated among their components, organelles, or even organs (*e.g.* the brain).

The Heisenberg uncertainty principle, due to the fundamental limits on measurement precision, sets a lower bound on the fidelity of individual molecular processes and hence, by extension, on the accuracy of collective molecular processes. This has implications not only for subatomic physics but also cell biology. For the cell to perform its functions properly, it is crucial that the right parts are in the right place at the right time.

> *We might expect some of life's processes to evolve to the "quantum edge" sometimes referred to as quantum criticality or poised realm where a compromise is struck between speed and predictability since quantum processes are fast but probabilistic (hence noisy) as opposed to classical determinism.*

However, intuition gained from macroscopic classical physics mechanisms and lessons learned from the nano-scale science where quantum effects dominate can be misleading on a mesoscale where most of the cell biology operates and where we still need to discover numerous intricate properties of matter in between strictly quantum and classical physics regimes.

3.5 Quantum Weirdness and Biology

Quantum phenomena are often called spooky and weird. It is hard to understand them intuitively as they do not appeal to our everyday experience. Although it is not the way our brain teaches us, it probably is how the brain operates. Since both physics and chemistry crucially depend on the power of quantum mechanics to provide fundamental insights into the world around us, it is natural to inquire whether biology offers examples of phenomena where quantum mechanics is the most appropriate physical representation. In fact, a more appropriate approach would be to ask whether the same principles and phenomena that apply to quantum physics, such as complementarity, entanglement, non-locality also apply at macroscopic scales. Given the small value of Planck's constant, one would not necessarily expect to get identical situations as the scales

applicable to biological systems would be much larger than the microscale of cell biology or the nanoscale (a nanometer is expressed as 10^{-9} m = nm and a nanosecond is 10^{-9} s = ns) used in molecular biophysics, let alone picoscale of atomic physics, (picometers 10^{-12} m and picoseconds 10^{-12} s) for atomic-level processes. On the other hand, as already pointed out before, in biology we deal with sizes ranging from meters (organism) to micrometers (cells), and time periods of years (life cycles) down to microseconds (protein conformational changes). In many instances perhaps a better term to use would be quantum-like rather than quantum. While this is not always appreciated, it is important to draw the distinction. For example, the Einstein-Podolsky-Rosen (or EPR) paradox involves non-locality, which applies to faster than light correlations between two entangled quanta and it would not be expected that it plays any role in biological systems but one could still have correlations that are revealed to be traveling much faster than those afforded by chemical signaling due to their diffusion-limited propagation speeds. It is of course the task of experimental scientists to show that such situations exist, in which case quantum-like generalizations of quantum phenomena would apply to biology.

In its most profound sense quantum biology refers to the application of quantum mechanics to biological objects and processes. The term particularly applies to nontrivial applications such as 1) superposition; 2) tunneling; 3) entanglement; and 4) non-locality. An example of quantum tunneling with biological relevance concerns the chemistry of proteins. In particular, the action of ion channels has been described using quantum tunneling. In 1992 physicist Friedrich Beck and physiologist Sir John Eccles, (a 1963 Nobel Laureate in Physiology or Medicine) jointly developed a quantum mechanical model [5] of exocytosis and neurotransmitter release at neuronal synapses in the human cerebral cortex. The model supports interactionist dualism by postulating that human consciousness affects the functioning of synapses in the brain through quantum tunneling of electrons between the lipid bilayers of the synaptic vesicle and the presynaptic membrane. The tunneling of electrons was envisaged in this model as triggering the process of exocytosis and initiating the transmission of information from the presynaptic to the postsynaptic neuron.

Some proteins contain active sites contain residues that are involved in hydrogen bonding interactions, and to reach the sites, the hydrogen atom has to negotiate an elaborate and shifting potential energy landscape. Quantum tunneling can speed up this process. Studying just how important tunneling might be is highly challenging because many complex interactions occur due to protein dynamics in response to thermal agitation resulting in its shape changes. Recently, Vattay et al. [6] suggested the following:

> *Life has evolved via protein behavior existing at the edge of quantum criticality, i.e. at the very thin edge of the edge separating quantum and classical worlds.*

Recent results using deuterium seem to confirm that quantum tunneling is indeed significant and they raise the fascinating question of whether some proteins have actually evolved to take advantage of this, making them in effect "tunneling enhancers".

In evolution, even a small advantage in speed or accuracy can bootstrap into overwhelming success, because natural selection favors the relative proportion of the winners over many generations in an exponential manner that eventually leads to the extinction of the "losing" population.

Mutations are the driver of evolution, so in this limited sense, quantum mechanics can be a contributory factor to evolutionary change, e.g. in environmentally stressed bacteria, which could boost their survivability.

It is indeed becoming increasingly clear, although examples of quantum effects in biology can so far be considered only a minor part of life processes as we know them, that macroscopic manifestations of quantum phenomena exist. Nonetheless, it is worth noting that the situation with physical systems a hundred years ago was not very different where only a handful of exceptions to the rules of classical physics were initially found. In an effort to illustrate the present-day situation in biology, we provide a brief overview of attempts to apply quantum (or quantum-like) principles to biology. First of all, however, we will discuss the pros and cons of using quantum concepts in biology in the first place.

3.6 Can Objections to Quantum Biology Be Overcome?

Quantum biology is doubted because of the following premises:

1) *The building blocks of life are large*, occurring at the cell level versus microscopic elementary particles usually associated with quantum mechanics. However, a number of examples highlight that quantum mechanics do exist at the macroscopic level. First, magnetism is a quantum phenomenon, which occurs at a macroscale. Second, quantized crystal lattice vibrations occur in arbitrarily large solids. Specific heats of crystals were measured and demonstrated to follow quantum laws below the so-called Debye temperature. Third, laser action is a quantum phenomenon. For example, laser pointers and scanners function according to quantum principles at room temperature. Again, energy is forced into the system and emission spontaneously occurs in the form of coherent light with a single frequency and color. As described above, this coherence manifested by preserving a single frequency and in-phase behavior is an example of the superposition of place and states at the same time.

2) *Life survives in hot conditions as opposed to quantum physical systems*, which typically require cold conditions. Most observed quantum phenomena occur only at a few degrees Kelvin. At warmer temperatures "thermal decoherence" predominates, which is a technical term describing the loss of coherence

due to thermal fluctuations. However, superconductivity exists at a much higher temperature (at up to roughly 200° Kelvin) than was originally predicted. Superconductivity is a macroscopic quantum phenomenon exhibiting zero electrical resistance also involving complete expulsion of magnetic fields (the so-called ideal diamagnetism).

3) Quantum biology is also doubted based on the premise that *life is wet in contrast to quantum mechanics, which is not found typically in wet environments* except in superfluids at extremely low temperatures. Superfluids are materials such as liquid helium ^{3}He that behave like fluids with zero viscosity; they also exhibit the ability to self-propel in a way that defies forces of gravity and surface tension and a quantum condensation with in-phase behavior. A potential answer to how quantum behavior may occur despite the premise that quantum mechanics is rarely found in wet environments is that biological water is not tap water. Bulk water is different from the water that is present in living cells. The water present in cells is structured water that is not free to rotate with three-dimensional degrees of freedom. More than 50% of water molecules in cells form quasi-two-dimensional water structures with dipoles oriented hexagonally resulting in a long-range order. Consequently, *biological water is structurally and dynamically ordered* and resembles a liquid crystal or ice. Gerald Pollack [7] went as far as to describe biological water as a fourth phase of water (in addition to the standard three phases, namely: ice, liquid, and vapor). Hence, structured water, as described by Mae Wan Ho [8] in her "proton jump conduction" theory, may be amenable to quantum mechanical properties in biological systems as it may transfer information between disparate regions of the body, such as from the microtubules (MTs) in the neurons of the brain to the metabolic enzymes in the cells of visceral organs or of skeletal muscles of the extremities or perhaps even to the motor proteins myosin present in skeletal or cardiac muscle. Wan Ho proposed that water aligns in a highly constrained fashion within the nanotubes of the triple helices of collagen that span throughout the entirety of the extracellular space of the body. She also suggested the following:

Organized water may transfer biological information using specific resonance frequencies from where it originated to intracellular domains elsewhere in the body.

Upon transfer of information from the extracellular to the intracellular space the water molecules may align and facilitate the conduction of electricity via propagation of hydronium ions constrained within the nanotubes of double-stranded cytoskeletal filaments of actin (in the intracellular space) until it reaches and transfers its information to specific biological targets. The targeted structures may be those with identical characteristic frequencies. The amplification of that frequency may be what is required to impact the function of those structures. In this fashion, the so-called positive electricity (because it is carried by positively charged proton species) represents a rapid flow of information that could be approaching the speed of light so that it underpins a potential synchronized and coherent organismic level integration of function from mind to body. The concept of biological water can hence be viewed as a type of quantum mechanical engine. Perhaps it also signifies a well-orchestrated symphony integrated with the metabolic activity of mitochondria and the process of oxidative phosphorylation and quantum metabolism on a macroscopic scale. We will discuss it in more detail later. Moreover, to support this synchronization, it has been very recently demonstrated that a single ion can coordinate dynamically up to three million water molecules.

4) Quantum biology is also doubted by many because of the premise that *life proceeds on a slow time scale* (milliseconds to hours or perhaps down to a microsecond time scale for conformational changes of molecules) compared to quantum mechanics, which is measured in time units on the order of femtoseconds (10^{-15} seconds) or lower. Therefore, it is assumed that at slow time scales quantum states collapse to their classical manifestations. Furthermore, life is complex in comparison to quantum mechanics as it requires billions of particular relationships between tens of thousands of different proteins in comparison to fewer than 100 particles (usually identical or perhaps coming in only several types) described by quantum mechanics.

5) *Life is not fuzzy, it either exists or does not* in contrast to quantum mechanics, which explains things based on probabilistic occurrences of multiple states. However, life may initially be fuzzy until it is forced to take a definitive position such as left or right, divide or not, survive or not, take flight or fight.

6) Finally, quantum biology is doubted based on the premise that *life is real, local, and stable in contrast to Heisenberg quantum mechanics, which is uncertain* and occurs at nonlocal realms. Anybody who has had a brush with death understands both the robustness and fragility of life.

3.7 The Appeal of Quantum Mechanisms to Biology

However, as stated earlier, *Mother Nature has had a long time to experiment and is adept at finding solutions*. As we learn daily from structural biology research, nature is the master of nanotechnology, which has created proton pumps, nanoscale rotary F-ATPase enzymes, motor proteins, and DNA polymerases, among other nanomachines. Undoubtedly in the years and decades to come, we will see more and more marvels of nano-biotechnology at work in living cells. Hence, having raised some serious objections to the use of quantum physics in the context of biology, we should be open-minded enough to realize that biology is on the cusp of a quantum revolution. Today, nano-biotechnology is allowing us a level of insight and understanding into the inner world of living cells that was unprecedented before. The examples of nanoscale biological systems and processes that have been closely examined include photosynthesis with its complex light capture structures in the chlorophyll, motor proteins forming the machinery of a living cell, as well as effects of metabolic enzymes, ion channels, *etc.* It is expected that the number of examples will accelerate in the near future.

There are many problems with the current explanation of what is going on inside the cell. One of the primary problems is the efficiency with which molecular partners for biological reactions are found in a crowded and dynamically changing environment of the living cell. The efficiency in celerity by which molecules find their way and position themselves to interact with one another is not well explained by the current classical model of molecular and cell biology. It, no doubt, would be the greatest elucidation of biology if the current and increasing understanding of quantum mechanics manifesting at a macro scale were to directly demonstrate that systems at the level of proteins or enzymes use quantum recognition algorithms. This would be an affirmation that living systems are actually quantum coherent. Accordingly, disturbing this delicate balance has consequences in terms of disease states including cancer for which there is currently available support in the literature in terms of disruption of quantum coherence at the cell level.

> *Biological behavior of organisms of all forms and scales, both plant and animal, prokaryotes and eukaryotes, including humans, requires the understanding of both their components and their interactions that may follow both classical thermodynamic and quantum regimes depending on the circumstances.*

Particularly, quantum biology recognizes the entangled parts that form the whole and cannot be isolated. The phenomenon of entanglement, among other quantum effects, is considered by quantum biophysicists to be the unifying principle that coherently supplies energy to living systems and integrates that energy throughout the entire sum of the cells into a functioning organism. The notion of quantum phenomena in biology is both intriguing and provocative.

The concept of entanglement in quantum mechanics describes shared functions and information describing two objects that are far away but are quantum entangled, feel each other's presence even though they are not contiguous. Entanglement is meant to represent two or more particles, which interacted at some point in time in such a way that they create a conserved quantity (e.g. net spin or linear momentum) but they do not need to have identical energy levels. Distant particles that are quantum entangled affect one another without energy transfer.

Quantum consciousness and psychological phenomena are concepts that are likely to incorporate quantum entanglement.

This especially resonates with the idea of telepathy, precognition, and related mental states. *Interference with quantum behavior, or quantum decoherence, likely underpins compartmentalization of physiology that correlates with the processes of disease states and senescence.* Interference with quantum behavior, or quantum decoherence, occurs in the setting of mitochondrial disease and the associated loss of quantum metabolism at a macroscopic scale in classical metabolic and other tissues is promoted by the dual and reciprocal processes of inflammatory cascades and oxidative stress, the biological equivalents of thermodynamic entropy. Heat is known to cause decoherence in quantum computers. An analogous effect is likely to be the case for decoherence of quantum metabolism within and across cells whereby heat directly, due to increased reactive oxygen species that increase uncoupling proteins (annotation) or to angiogenesis due to the effect of inflammatory cytokines (particularly notable in cancer states) or directly related to inflammation as a result of the parallel upregulation of oxidative stress and oxidatively modified constituents of mitochondria that impair mitochondrial function. Furthermore, insulin resistance results from pro-inflammatory cytokines that further compromise mitochondrial function promoting further heat and oxidative stress. Chronic and subclinical pervasive systemic-wide inflammation within an individual is a display of thermodynamic entropy in a biological system. It represents insidious degradation of biological complexity that correlates with the processes of disease states and senescence. In the case of insulin resistance, a loss of physiological information and increasing compartmentalization accompanying thermodynamic entropy and likely decoherent quantum metabolism and other quantum phenomena are mediated by the following factors: a) increased neurohumeral tone, i.e. increased sympathetic to parasympathetic autonomic nervous system balance, b) disturbed gut microbiome, c) disruption of insulin signaling in metabolic tissues, d) high circulating insulin levels, and e) a host of adipocyte-derived inflammatory mediators and hormones along with several gut-derived hormones. To reiterate, *when two systems are entangled we have complete information about their joint state but no information about their individual states.*

Non-locality explains the wave-like characteristics of quantum algorithms that are capable of finding solutions virtually instantaneously representing enormous shortcuts to an otherwise very large combinatorial space.

Non-locality is the apparent ability for information to travel instantaneously across vast distances, not limited to subatomic realms challenging our concepts about the everyday world.

There are intriguing claims that genetic evolution has been driven by quantum paradigms accelerating evolutionary advances by quantum search algorithms. The so-called Levinthal's paradox demonstrates the impossibility of real-time protein folding by a random exploration of conformational space of a peptide that is necessary to make a protein. It would take the age of the universe to fold one protein, but it typically happens in a fraction of a second. Hence, there must exist shortcuts, which may be explored by quantum algorithms, which employ wave functions that survey the entire space capable of finding a solution virtually instantaneously. The same is true concerning genetic mutations (DNA sequence variations) and epigenetic variations (e.g. post-translational modifications). It is not likely that adverse mutations are caused by quantum coherence resulting in the transition to disease states. Conversely, it is more likely that the opposite is true, i.e. *disease states occur due to quantum decoherence or loss of quantum correlations.*

In quantum mechanics, it is the superposition wave function that is able to simultaneously examine all the possibilities for solutions to how the parts of a system can fit together to achieve an energy-optimized system as a whole by unlocking the code and finding the answer to complex and complicated problems.

This implies an inextricable inter-connectedness of biological structures to a larger system of which they are part. Notably, the general notion of superposition states of a wave function and its application to quantum search algorithms serving as shortcuts to finding biological solutions allows us to address the challenges of parallel processing at amazing speeds. This underscores the essential nature of understanding systems as parts and wholes and supports the quest for developing quantum biology foundations. That is, without the parts being tethered to the whole, there would be no reference point or context to how the folding of an individual protein provided an adaptive value to the organism as a whole.

The implications of the quantum biology paradigm for potential practical clinical applications are as immense as they are unexplored. As a prominent example, it might be feasible to harness the ability to manipulate wave function superpositions across the energetic circuitry of biological systems to yield the most efficient and optimal solutions, which may even apply to conscious thought and cognition. In another possible application, one can consider the onset of metabolic diseases. This process is analogous to reactive oxygen species that could be causing the collapse of a wave function in biological systems such as in the human brain.

The central axes of hermetic neuroendocrine and immune responses that are maximally adaptive, modulate oxidative and inflammatory responses so that chronic disease states of all forms are most favorably modified at the epigenetic level.

As an example, the epigenetic environment favors the expression of genes for weight set-points, which likely involve the triumvirate of bile acid metabolism, endogenous gut hormone GLP-1 (glucagon-like peptide-1), and a predisposing gut microbiome. All of the components of this triumvirate are regulated in a complex nonlinear fashion, the gestalt of which in turn correlates to obesity and metabolic disease states such as diabetes. This process is analogous to reactive oxygen species that could be causing the collapse of a wave function in biological systems such as in the human brain.

> *In addition to quantum biology that could modulate chronic diseases mediated by pathogenic oxidative stress and inflammation, quantum algorithms may play a more direct role in preventing the development of cancers in the future.*

As a metaphoric example, the observation of an electron using a photon (analogous to observing an object in a dark room by shining light on it) disturbs the quantum state of the former; hence heat arising as a result of their interaction causes a wave function collapse of the said electron, which might be utilized for read-out purposes in future technological applications such as quantum computing. This process is analogous to reactive oxygen species that could be causing the collapse of a wave function in biological systems such as in the human brain.

> *Reactive oxygen species and the inextricably linked process of inflammation are associated with the release of heat that cannot be harnessed for useful purposes of biological work but can trigger pathological transformations at a cell and organ level by promoting quantum decoherence.*

Mitochondria of eukaryotic cells exhibit very complex levels of organization. There are five complexes embedded within the inner mitochondrial membrane that involve electron hopping (meaning electron transport between two neighboring atomic sites with a vacancy or hole in one site). These complexes are close enough for quantum tunneling to take place, that is, less than a distance of about 2 nanometers. Distances further apart are not conducive to the phenomenon of quantum tunneling due to exponential damping of the wave function. Quantum tunneling has been shown in mitochondrial energy transduction in complex 1, which leads one to conclude that energy production in most cells involves at least partially some quantum processes.

There is only indirect evidence of mergers of quantum processing in biology, painting a metaphor between physics and biology. As discussed above, the paradigm of quantum mechanics was established in the early part of the 20th century for elementary particles. The two important examples were the black body effect (explained by Max Planck) and the photoelectric effect (explained by Albert Einstein). The photoelectric effect shows that photon energy correlates with the frequency of light and not with the intensity of light. The biological parallel is photosynthesis, which involves the capture of light and the transformation of light into biochemical energy through electron transfer processes. It is a quantum biological application where a single photon under certain conditions acts like a particle. *A photon of light exerts momentum (defined as h/λ where λ is the wavelength) that is transferred into chromophore molecules resulting in knocking out electrons away from the atoms* to which they are normally bound in the benzene rings of the light-harvesting complex (LHC) structure such as the famous FMO (Fenna-Matthews-Olson)

complex in *C. Tepidum* bacteria [9]. These electrons are then transferred and harnessed in the reaction centers for further biological uses.

Quantum physics transitions smoothly at the macro-scale to classical physics. At the macroscopic scale, quantum processes have been thought to become irrelevant. Biology is an example of the macroscopic scale. However, it has become increasingly apparent that this limitation is only the case for some but not all biological processes. This underpins a new and rapidly evolving scientific discipline known as quantum biology. The biological processes that we know to be quantum are very subtle and depend on a single quantum of biological energy. Just as physics extended the sphere of quantum applications beyond individual particles to their macroscopic ensembles in solids (*e.g.* propagation of lattice vibrations called phonons in crystals), it is expected that biology will soon demonstrate the same transformative change. In this context, we first examine the role of photons in living systems.

3.8 Biophotons: Light in Cells

Obviously, photons are emitted by all living and non-living systems in the form of black body radiation due to the heat absorbed and emitted by the harmonic oscillators comprising these systems so long as these systems are at a finite temperature in equilibrium with the environment. However, this is not what was meant by biophotons when in the early 20th century, Alexander Gurvitsch introduced this concept in an effort to elucidate embryology through the action of so-called morphogenic fields, an as-yet-unproven hypothesis.

> *Biophotons are defined as electromagnetic waves emitted by living systems due to their metabolic activity and possibly due to other functions (e.g. cell division) related to the living state that are absent in non-living systems.*

Hence, this is the amount of radiation that is over and above the expected black-body intensity at a given temperature. As an example, infrared cameras can detect the presence of humans even in a dark environment while a dead person will not be seen by an infrared camera. Following in his footsteps, Fritz Popp [10] demonstrated that photons or electromagnetic energy quanta can be both absorbed and emitted by DNA molecules and this involves low-intensity ultraviolet ranges of the spectrum. Guenter Albrecht-Buehler [11] showed that living cells detect and respond to infrared electromagnetic waves with a peak of their sensitivity close to the wavelength of 1,000 nm, which is in the near-infrared part of the electromagnetic spectrum. He hypothesized that mitochondria, which employ a proton transfer mechanism, are involved in energy production leading to the release of photons as by-products of metabolism. This seems to be particularly a function of the recombination rate of reactive oxygen species that fall to a lower energy emitting the energy difference in the form of biophotons. Centrioles (Figure 3.3), dubbed by Albrecht-Buehler "the eyes of the cell", are intricately structured from nine triplets of microtubules (Figure 3.3) which may enable them

FIGURE 3.3 A schematic illustration of a centriole that consists of nine microtubule triplets arranged in the shape of a cylinder with two centrioles making up one centrosome. Source: adapted from http://cnx.org/contents/GFy_h8cu@10.53:rZudN6XP@2/Introduction.

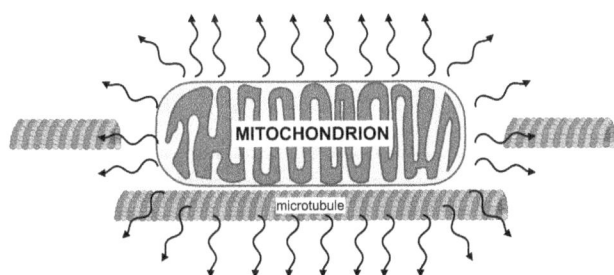

FIGURE 3.4 Schematic illustration of mitochondrial production of biophotons in the cell and their further transmission via microtubules. Source: adapted from Meijer, D.K.F. and Raggett, S. (2014) "Quantum Physics in Consciousness Studies. The Quantum Mind Extended". In: *Quantum Mind UK*, ed. Raggett, S., p. 113.

to absorb these photons and trigger an electronic signaling cascade (as will be discussed later in this chapter) propagating through the cell, communicating to the cell that another living cell is in its immediate vicinity. Biophotons could be seen as a quantum manifestation of bio-electromagnetism. *Electromagnetism is inherently quantum mechanical and it seems quite plausible that electromagnetic fields could be responsible for a long-range coherent effect of mitochondrial electron transport chain ATP production.*

The sources of biophotons could be diverse and they may involve free radical recombination as mentioned above. *At the cellular level, living systems emit ultra-weak spontaneous biophotons without external excitation.* These photons are produced via various biochemical reactions, but principally from bioluminescent (in the visible range of the spectrum from 400 to 700 nm wavelengths, see Figure 3.1) radical reactions of reactive oxygen and nitrogen species and the relaxation of excited electronic states. The oxidative metabolism of mitochondria and lipid peroxidation (which is pathological) appear to be primary sources for this activity. Neurons also continuously produce photons during their ordinary metabolism, and it has been shown in vivo that the intensity of photon emission from rat brain correlates with cerebral energy metabolism, electrical activity, blood flow, and oxidative stress. *Furthermore, ultra-weak bioluminescent photons can propagate along neural fibers [12] and can be considered a means of neural communication.* Specific mechanisms that can arise as a consequence are speculative at best and are under research investigation at present (Figure 3.4). Radical

recombination within mitochondria can lead to the emission of photons in the ultraviolet (UV) range required to excite the chromophoric network within microtubules. Presumably, biophotons promote adaptive protective effects in living cells. However, the Fenton reaction, which is inhibited by superoxide dismutase is described by the biologically deleterious reaction between hydrogen peroxide and ferrous iron to produce hydroxyl radicals and the oxidized form of iron (ferric iron) [13]. Furthermore, weak magnetic fields (that are associated with the movement of electrons, protons in the electron transport chain) can affect the rate of recombination of radical pairs induced by iron ions or other pro-oxidative metal ions resulting in oxidative processes such as DNA damage (both nuclear and mitochondrial) and lipid peroxidation. *Biophotons may not be just a by-product, but rather, can be linked to precise signaling pathways of reactive oxygen species and reactive nitrogen species, which may produce biophotons within cells including neurons.* This may be responsible for free radical recombination with iron within electron transport chain complexes and correlate with biochemical processes (e.g. reduction of oxygen to water and subsequent ATP production). Mitochondrial electron transport chain (see Figures 1.26, 1.30, and 1.31 in Chapter 1) contains several chromophores, especially the porphyrin rings of cytochrome oxidase (complex 4). The absorption of biophotons by these photosensitive molecules can produce an electronically excited state, which has very different chemical and physical properties compared to their electronic ground state. In this case, the absorption of biophotons within mitochondria occurs in very close proximity to the origin of photon emission. Presumably, biophotons in this scenario assist in driving the metabolic production of ATP for the energy needs of the body. However, the short range of action between its site of production and that of absorption questions the role of biophotons in this very limited set of considerations in promoting synchronized coherent states for biological functions.

The classical counterpart to the quantum photon is the electromagnetic wave. The movement of electrons along the electron transport chain is an electrical current that creates an electromagnetic field. Metaphorical to the fuel cell analogy, the entrainment of electron transfer along electron transport chains across mitochondria of the same cell, across cells, and even across tissue types underscores a synchronization of bioenergetics and potentially as an organismic whole analogous to the electromagnetic phenomenon responsible for the coordinated motion of an electrical car. Thus, the quantum photons in biology, or biophotons that seem to play a role in mediating the correlated state of the phenomenon of quantum metabolism, are interwoven with the classical properties of electromagnetism, another overlooked but fundamental basis of biological evolution in human biology and behavior.

Additionally, biophotons emitted in the infrared or visible electromagnetic spectrum may excite the UV aromatic chromophores (e.g. in microtubules) in a multiphoton excitation manner (vide infra). This may cause induced dipole-induced dipole interactions propagating along a subcellular structure. Such multi-photon excitation exploits nonlinear quantum effects to allow a molecule to be photo-excited with much lower energy and a longer wavelength compared to what would be predicted from its conventional single-photon excitation spectrum. Thus, a chromophore molecule can absorb multiple low-energy photons simultaneously, whose total energy equals or exceeds its one-photon excitation energy, causing it to excite. More is stated about microtubules in connection with various aspects of cognition below (*vide infra*). It suffices to say here that the potential for biophotons to underpin synchronized coherent processes may relate to the documented interactions between mitochondria and microtubules. In this case, mitochondrial-derived biophotons interacting with microtubules can correlate with the strength of electrical α wave activity in the brain, for instance. In addition to biophoton emission from reactive oxygen and nitrogen species such as in mitochondria (due to the free radical recombination with iron), visual perception and imagery within visual areas of the brain in response to visible light spectra induce membrane depolarization, which results in biophoton emission from neural tissues. Visual photons through the retina cause neuronal depolarizations ultimately reaching the visual cortex. This induces free radical recombination in mitochondria of neurons. This leads to the interaction with microtubules resulting in microtubule excitation/oscillation or propagation. This emits magnetic fields, which open cation channels and subsequent neurotransmitter release or neuronal depolarization depending on whether it is the presynaptic nerve terminal or the postsynaptic dendrite/cell body [14]. *The membrane depolarization here is an energy-requiring process whereby the biophoton emission may arise from within the mitochondria through free radical recombination.*

The synergistic coherence of intracellular mitochondria in the process of quantum metabolism may be mediated by electromagnetic waves, which may be in the form of biophotons or alternatively in the form of electromagnetic fields emitted from an electrical current generated by the electron transfer along the electron transport chain; a process which is coupled to a proton movement. Thus, electromagnetic waves may promote the entrainment of electron transport chains across mitochondria. An alternative hypothesis is that the electromagnetic waves serve as superposition wave functions whereby they are present at many places at the same time with the capacity of using quantum algorithms to find optimized solutions. For example, an optimal solution could be in-phase coherence of NADH electron transfer to complex 1 or FADH2 electron transfer to complex 2 initiating synchronous propagation of electrons and ATP production to do the physiological work of the body with optimal efficiency. The collapse of such a wave function underlies the notion of committing to a single place in time that represents an actualized solution for a system as a whole. The idea of entrainment seems more plausible in the context of mitochondria-synchronized coherence. As mentioned above, decoherence may be caused by excessive heat generation. In the case of biological systems, heat is linked to reactive oxygen species generation that is fundamentally linked to inflammation. In the case of mitochondria, reactive oxygen species cause up-regulation of uncoupling proteins in adipose tissue and may do the same in all cell types. The uncoupling proteins uncouple the energy gradient along the electron transport chain from being trapped in the bonds of ATP; instead, the energy is released as heat. The proposed biological significance of biophotons is still unclear other than to enhance inter-individual perceptions (that is, magnetic resonance frequencies emitted by some organisms and perceived by others) for bonding, herding, or social interactions as well as in animals for predatory-prey advantage.

It is becoming more and more plausible that the molecular machinery of living cells emits and absorbs photons.

Most of these photons are emitted and reabsorbed within the biological system with only a small proportion being radiated out into the surrounding environment, which makes it difficult to detect. Recent evidence has been published that this light is a new form of cellular communication [15]. Biophotons are usually emitted at or below a very low rate of 10/sec/cm² of cell culture, which explains why biophoton activity as a form of cellular communication has been doubted by mainstream science as too controversial. However, Russian biologist Mayburov [15], having spent countless hours recording the patterns of biophotons fish egg cells emit, observed biophoton streams consisting of short quasi-periodic bursts, which are remarkably similar to those used to send binary data over a noisy channel. That might help explain how cells can detect such low levels of radiation in a noisy environment. In several experiments, biophotons from a growing plant were seen to increase the rate of cell division in other plants by 30%, which is significantly higher than is possible with ordinary light that is several orders of magnitude more intense. He also reported that biophotons from growing eggs enhance the growth of other eggs of a similar age. However, the biophotons from mature eggs can hinder and disrupt the growth of younger eggs at a different stage of development. This is suggestive of resonant phenomena that were discussed above in connection with biological coherence.

The above discussion about biophotons opens up a new range of possibilities for medicine, which could be termed photo-medicine.

There is a small but rapidly growing body of evidence in the scientific literature that photomodulation with specific frequencies of the electromagnetic spectrum can lead to clinical benefits, especially in the area of neurodegenerative diseases and cancer.

Photobiomodulation therapy with infrared light has been recently reported [16] to have major clinical benefits in clinical trials for patients with dementia. A number of research papers have also reported clinical benefits of light-level photobiomodulation therapy for cancer, which uses red light with a wavelength of 670 nm and intensity in the 5 J/cm² range [17]. One of the possible mechanisms implicated in this process is enhanced oxidative phosphorylation in mitochondria [18], which would nicely link with the theory of quantum metabolism that has been prominently highlighted in this chapter.

3.9 Quantum Nature of Vision, Olfaction, and Bird Navigation

In quantum biology, a crucial issue is the applicability of quantum processes to large scales and physiological temperatures as well as to systems operating under far-from-thermodynamic-equilibrium conditions. One argument against this applicability is that quantum processes cannot occur at and above room temperature. However, this is not true in general as explained above citing the examples of magnetism and laser action. Increasingly, evidence has been accumulating that quantum phenomena may and do occur at the macro-biological scale. This has been shown for the control of biological processes including photosynthesis, the migratory flocking behavior of birds, and the human sense of smell. These examples of functional quantum mechanisms in living systems all occur at ambient, or warm temperatures as elaborated on below.

At the level of organs and tissues, it has been demonstrated that the human eye is capable of detecting light at an extremely low threshold, as few as two to three photons at a time, perhaps even a single photon [19]. These photons strike receptors of the retina, which eventually results in action potentials that travel along an axon of the optic nerve to the visual perception area of the brain. The question, however, remains regarding a specific mechanism of amplification of individual quantum events involving photon-receptor interaction, which leads to classical macroscopic effects at the level of the human brain. How does the brain perceive a flash of light that was caused only by a few photons at the threshold of "visual perception"? Similarly, the work of Luca Turin and his colleagues [20] has provided strong arguments to support the claim that the sense of smell (olfaction) is based on a quantum resonant energy transfer mechanism involving vibrational degrees of freedom of aromatic molecules and receptors in the membranes of olfactory nerves. This also leads to still poorly understood signal amplification effects from a receptor level to the human brain. However, this mechanism is very different from the traditionally accepted lock-and-key mechanism of receptor–ligand interactions. As a consequence, at low temperatures, the acuteness of the sense of smell would be reduced due to lower vibrational intensity.

The second recent development in quantum biology concerns bird navigation. It is well known that some birds perform amazing feats of navigation using a variety of cues including the local direction of the Earth's magnetic field. The nature of the magnetic sensor has remained a mystery until recently. It should be added as a clarification to the non-physicists reading this book that an external magnetic field tends to align magnetic moments, such as electron spins, with its orientation. Hence, when one of the two spins coupled to form a singlet state is misaligned with the field of the Earth, it will undergo a transition and flip to the properly aligned orientation. This electron transition involves lowering the energy of the system and is called the Zeeman effect. Also, a gradient (or difference over a small distance in space) of a magnetic field exerts a force on the magnetic moment such as that of an electron spin. These two simple observations applied to the electron-pair part of the magneto-receptor system of the migratory birds provide the basis of the physical mechanism involved. However, a quantum mechanical model providing a quantitative explanation of this mechanism is not only very complex but also still debatable.

This model has been recently explained in terms of chemical magnetic reception *via* the generation of a pair of free radicals that suggests that these migratory birds (such as the European robin) are highly sensitive to weak magnetic fields of the Earth and are capable of using sophisticated quantum mechanical processes as a microscopic compass. Accordingly, birds' navigation mechanism has been proposed as based on quantum entanglement that persists for at least 20 μs which is longer than the currently maintained laboratory experiments performed on physical systems. *The specific mechanism appeals to photo-activation above the thermal noise of a two-dimensional array of aligned proteins,* producing radical ion pairs involving singlet (spins pointing in opposite directions adding up to a net-zero value) and triplet (spins aligned and adding up to a value of one) two-electron states. While electrons are in an entangled state and experience identical magnetic environments, their quantum coherence is preserved and their spins precess (or rotate around the magnetic field axis) in synchrony. However, in addition to the Earth's ambient field, there are nuclear magnetic fields present producing so-called hyperfine splitting of nuclear energy levels. If one of the electrons in a singlet pair shifts from its position slightly, the combined nuclear and terrestrial magnetic field it experiences will change, causing the singlet state to oscillate with a triplet state with a periodicity depending in part on the strength and relative orientation of the Earth's field. The system may then de-excite in stages and initiate a reaction that in effect acts as a chemical compass because the relative proportion of the reaction products can depend on the singlet-triplet oscillation frequency. It has been argued that in fact a highly anisotropic but weak nuclear hyperfine interaction involving the nuclear spins of the radical pair species is preferable and produces the overall optimum value for radical-pair lifetimes.

> *Mother Nature can teach us a thing or two about quantum mechanics.*

How is the angle of the field relative to the bird translated into neural information? This works through ocular photoreceptors or cryptochromes that are the site of photochemical reactions creating these radical pairs of electrons transiently as fleeting intermediates [21]. In migratory birds, photoreceptor molecules in the bird's retina such as cryptochrome contain the above-mentioned free radical electron pairs in the photoreceptor molecule (see Figure 3.5).

The excited states of the free radical electron pairs within the photoreceptor molecules then promote signaling to the brain leading to the bird changing its direction of flight until there is a de-excitation of the electron pair back to the ground state with the electron spins oriented in opposite directions. We should caution the reader that a generally accepted microscopic mechanism that fully explains the migratory bird magnetic field sensing is still under development. Moreover, very little is known about the bio-electronic circuitry connecting the radical-pair yield to a neurological signal except for the link to the role cryptochromes may play. However, a deeper understanding of the chain of events connecting the singlet/triplet yield to a neurological signal is very much needed so we can understand how the magnetic sense can be so sensitive to small changes in the field intensity and direction.

(a)

(b)

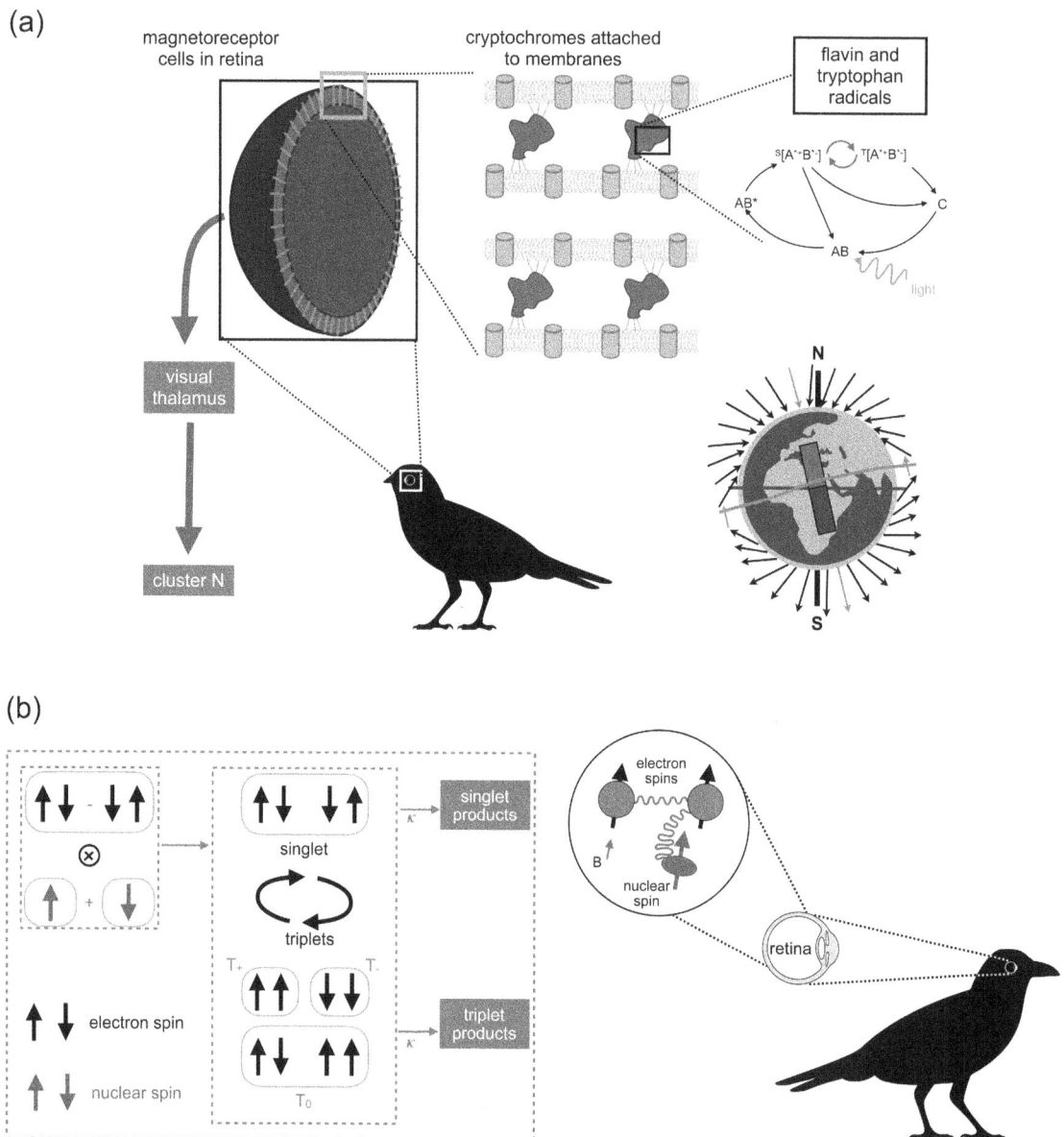

FIGURE 3.5 Schematic illustration of the molecular mechanism of singlet-to-triplet oscillation of the electron pairs in the free radicals associated with cryptochromes of the photoreceptor molecules in birds using magnetic fields for their navigation. Source: adapted from a) Mouritsen, Henrik. (2015) "Magnetoreception in Birds and Its Use for Long-Distance Migration". In: *Sturkie's Avian Physiology*, pp. 113–133. Academic Press, London. b) Yin, Zhang-qi and Tongcang Li. (2017) "Bringing Quantum Mechanics to Life: From Schrödinger's Cat to Schrödinger's Microbe". *Contemporary Physics* 58(2): 119–139.

3.10 Photosynthesis: Quantum Metabolism of Plants

The biological parallel of the photoelectric effect is photosynthesis, which involves the capture of light and the transformation of light energy into biochemical energy. Photosynthesis is a highly complicated and sophisticated mechanism that harvests light energy to split water by using individual photons to create a cascade of reactions (Figure 3.6). The process is extraordinarily efficient and represents a classic example of how evolution has fine-tuned the design of a physical system to attain near-optimal performance. *An important feature of photosynthesis is that the molecular architecture involved is structured in a highly specific and compact manner, suggesting it has been "customized" to exploit long-range quantum effects* (see long-range coherence which is here applied to quantum states). *This particular configuration could be uniquely efficient at preserving coherence for relatively long durations of time (on the order of several nanoseconds), enabling the system to speed up the delivery of energy to the reaction center. Photosynthesis involves reaction centers that capture individual photons and transfer exciton energy by tunneling to avoid decoherence even at room temperatures. The theory describing coherent energy transfer in plants involves the capture and transfer of photon energy via the interaction*

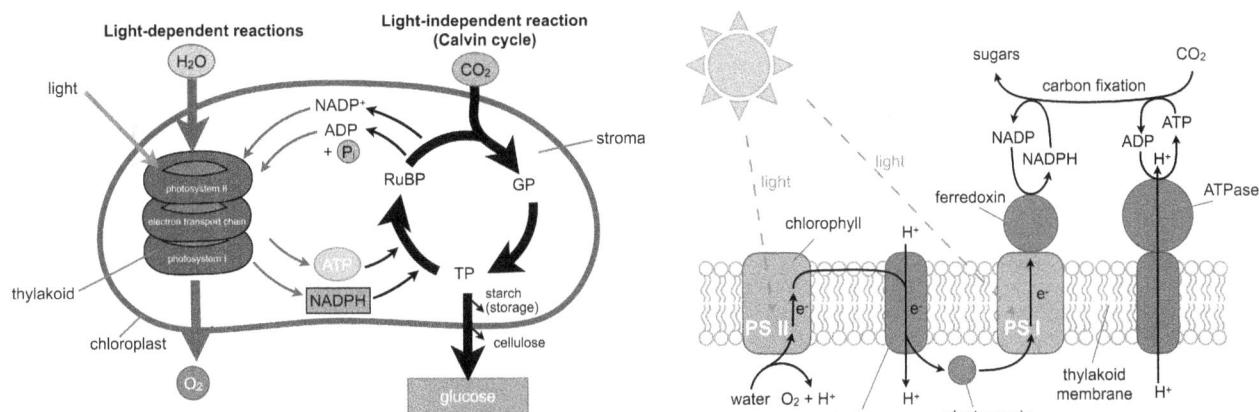

FIGURE 3.6 Schematic illustrations of an overview of photosynthesis and its reactions. Source: adapted from a) https://www.commonlounge.com/discussion/38eb93bbb6b8479f9abe8441a43c6734 and b) https://newunderthesunblog.files.wordpress.com/2013/07/light-reactions.jpg.

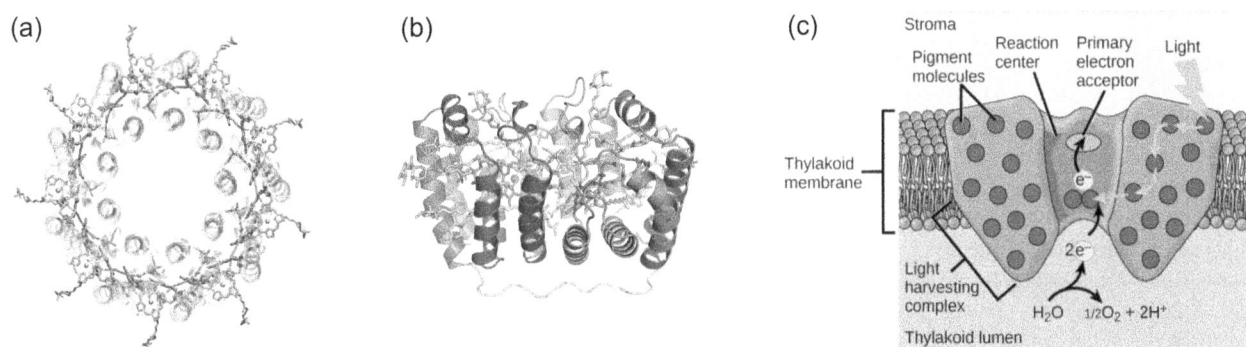

FIGURE 3.7 a) Light-harvesting complex (LH2) of purple bacteria. Top view. By aegon. b) Crystal structure of the soluble peridinin-chlorophyll-protein complex from the photosynthetic dinoflagellate *Amphidinium carterae*. Rendered using PyMol from PDB ID 1PPR. Sources: a) used under Creative Commons license https://creativecommons.org/licenses/by-sa/3.0/legalcode. b) From Hofmann E., Wrench P.M., Sharples F.P., Hiller R.G., Welte W., and Diederichs K. (1996) "Structural Basis of Light Harvesting by carotenoids: Peridinin-Chlorophyll-Protein from *Amphidinium carterae*". *Science* 272(5269): 1788–1791. DOI 10.1126/science.272.5269.1788. Uploaded by Opabinia regalis and used under Creative Commons license https://creativecommons.org/licenses/by-sa/4.0/legalcode. c) Photosystem II (P680). Modified from http://cnx.org/contents/GFy_h8cu@ 10.53:rZudN6XP@2/ Introduction and used under Creative Commons license https://creativecommons.org/licenses/by/4.0/legalcode.

of transition dipoles in chlorophyll molecules. This involves a sequence of excitation, de-excitation events with the associated photon capture by the neighboring molecules [22]. Light-harvesting complexes (Figure 3.7) in plants have a unique and exquisite geometrical arrangement of chromophores in chlorophyll pigments [23]. This arrangement allows a photon to be efficiently captured and funneled from the external environment to a reaction center where the energy is processed and stored in chemical components. This biological light-harvesting complex operates at nearly 100% efficiency, which is amazing in its own right. *Counter-intuitively, the efficiency of photosynthesis increases with temperature reaching a plateau at roughly 300° Kelvin (or 27° C). Hence plants may be living at temperatures favored for purposes of efficiency.* In terms of technological application, this may inspire a whole new wave of biomimetic research and development. This area of quantum biology, however, is somewhat retrograde because we are trying to understand nature through proximity to current technology. We may indeed find that nature inspires and influences technology. For example, Velcro was inspired by the tiny hooks on the surface of burs.

The remarkable performance in photosynthesis is achieved through quantum mechanisms, which allow the photon to be shared across the chlorophyll network (through the superposition of wave function).

The initially excited donor chromophore transfers its electronic energy to acceptor chromophores via electrodynamic coupling of their transition electric dipole moments. This energy is shared due to the close relation, or "resonance", between their energy levels. The key to this mechanism is the optimal packing of light-sensitive chromophores that possess a significant transition dipole upon excitation. The direct observation of quantum beats (analogous to the beats found in sound waves that are formed by two waves with slight differences in their frequencies) in the photosynthetic light-harvesting complex is a clear indication of electronic quantum coherence in biological systems. Similar coherent quantum phenomena have been seen in the light-harvesting complexes of plants, bacteria, and phycobiliproteins. Chlorophyll is one of a number of biologically available chromophores. Other bio-chromophores

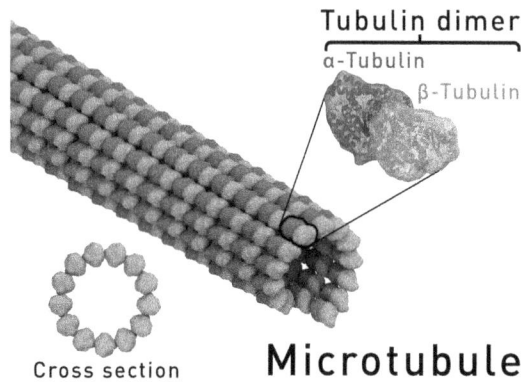

FIGURE 3.8 Schematic illustration of microtubule and its components. Source: by Thomas Splettstoesser (www.scistyle.com), used under Creative Commons license https://creativecommons.org/licenses/by-sa/4.0/legalcode.

include pyrroles, porphyrins, cytochromes, carotenoids, ferredoxins, flavins, heme, and melanin. Additionally, among naturally occurring amino acids three possess fluorescent chromophoric properties: tryptophan, tyrosine, and phenylalanine. These amino acids also contain resonance ring structures in which electrons are delocalized and can be used as sites for electronic energy transfer. Such intra-molecular energy transfer processes involving tryptophan, tyrosine, and phenylalanine residues are known to occur in protein structures leading to efficiently transfer energy between chromophoric residues [24]. Similar processes may take place in natural proteins, including the polymers found in the cytoskeleton, specifically microtubules (MTs) (Figure 3.8) and actin filaments (AFs) [25]. *It is hard to believe that the use of quantum processes in plants and bacteria would not be retained or indeed perfected through evolutionary achievements into higher-developed life forms such as mammals. Hence, it would not be surprising to see further adaptations of this process in higher organisms, especially humans.*

> *Energy production mechanisms in plants involve light harvesting through chlorophyll while animals predominantly utilize glucose metabolism in mitochondria.*

Molecules that are similar to the chromophores, which trap the energy derived from sunlight within the chlorophyll in the leaves of plants are also present in mitochondria but instead of excitons, electron transport chain mechanisms of energy production leads to the generation of ATP from ADP. It is, therefore, expected that at least some of the many steps comprising oxidative phosphorylation in mitochondria may involve quantum energy transfer, which would attest to the evolutionary retention of achievements by Mother Nature. Quantum energy transfer certainly qualifies as an achievement in the field of nanobiotechnology. Mitochondria of eukaryotic cells exhibit very complex levels of organization. There are five complexes embedded within the inner mitochondrial membrane that involve electron hopping. These complexes are close enough for quantum tunneling to take place, that is, less than a distance of about 2 nanometers. Distances further apart are not thought

to allow the phenomenon of quantum tunneling because particles would not have enough energy for the process of tunneling. Quantum tunneling has been shown in mitochondrial energy transduction in complex 1. The specific relevance in this case in the animal kingdom including humans is not yet entirely clear. However, it is a highly intriguing area of active research. Indeed, demonstrating proteins to have quantum recognition algorithms will be most exciting and probably one of the greatest clarifications in biology. In particular, tryptophan is an amino acid that is very susceptible to the absorption of light. Tryptophan and similar structures (*e.g.* serotonin, melatonin, *etc.*) may play an important role in light absorption in the human brain mediating quantum effects of conscious activities [26]. Hence, the use of quantum mechanics in the brain may help us understand cognition or even consciousness. Although the foregoing examples have been in the literature for a number of years, they have not yet led to a widespread acceptance that quantum physics is important for biology.

> *One of the most convincing demonstrations of the importance of quantum physics in biology is quantum metabolism.*

3.11 Quantum Metabolism

The main difference between living and nonliving systems is energy transduction. Living systems live at far-from-thermodynamic-equilibrium conditions whereas nonliving systems exist at thermodynamic equilibrium.

> *Living systems require the injection of energy in the form of light (plants) and/or nutrients (animals) and, as chemical engines, they transduce energy available from the external environment into chemical, structural, osmotic, electrical and mechanical work.*

In plants, the source of energy is, of course, light, which is converted by photosynthesis into chemical energy as described above. The sources of energy in animal cells are plant products, which are oxidized to extract the chemical energy and convert it into work. However, body size and structural organization impose certain constraints on the rate and intensity of these energy transformations. Elucidating the process of energy transformation results in the empirical rules, which relate to metabolic rate and body size, which are called allometric laws of physiology. They describe how metabolism is quantified within and across the different species of organisms. This issue was first addressed empirically by Rubner using domesticated animals, which showed that metabolic rate and body size satisfied a mathematical relation with a scaling exponent value close to 2/3. A simplistic explanation in terms of surface-volume relations of a spherical model organism was therefore proposed. Later, Kleiber studied the relation between metabolic rate and body size using both domesticated and undomesticated animals showing that the scaling exponent is closer to 3/4. However, recent investigations based on unicellular organizations, plants, and animals revealed that the scaling exponent, dependent on taxa and body size, ranges between 2/3 and 1. The proportionality constant in the allometric laws can assume different values depending on the particular organism studied. A quantum mechanical explanation of these phenomena was developed by Lloyd Demetrius [27], which is described in this section in some detail.

Energy flow in living organisms is mediated by concentration gradients of various molecular species, pH gradients, and electric potential gradients but not by thermal gradients. This leads to the emergence of cyclical biochemical reactions including metabolic reaction chains. Empirical studies indicate that the characteristic cycle time of these metabolic processes is related to the metabolic rate, i.e. the rate at which the organism transforms the free energy of nutrients into metabolic work. The production of ATP, which is the common energy unit of living organisms (sometimes referred to as the metabolic energy quantum), involves the coupling of two molecular reaction chains as proposed originally by Mitchell in 1966 (who was awarded a Nobel Prize in Medicine and Physiology): a) the redox chain, which describes the transfer of electrons between redox centers within the electron-transport chain, and b) the ATPase motor, which phosphorylates ADP to ATP. The above process based on electron transport is called oxidative phosphorylation. In addition to it, there is a purely chemical process of ATP generation called substrate phosphorylation. Both of these are cyclic processes whose transit time, denoted τ, determines the total metabolic flux, that is, the number of proton charges released by the redox reactions. *Oxidative phosphorylation takes place in the mitochondria where the electron transport between redox centers is coupled to the pumping of protons across the mitochondrial membrane generating an electrochemical gradient. Substrate phosphorylation, on the other hand, is confined to the cytoplasm and is driven by enzymes that couple ADP phosphorylation to the electron transport chain. Hence, the energy generated by the redox reactions can be stored in the form of coherent oscillations of metabolic enzymes.*

The theory of quantum metabolism proposed by Demetrius is based on the Einstein-Debye methodology (which treats a solid as a system of coupled harmonic oscillators) of the quantum theory of specific heats of solids and offers the only available molecular-level explanation of allometric scaling laws. Quantum metabolism makes a conceptual and mathematical parallel with Planck's quantization principle and states the following.

The metabolic energy stored by an enzymatic oscillator with frequency ω is quantized according to the formula $E_n = n\kappa\omega$, where κ is an analog of Planck's constant, called here the biological Planck constant.

Quantization of metabolic energy is due to the fact that only integer multiples of ATP molecules are produced in the cell's mitochondria and they have a relatively low energy content (in terms of the biological energy quantum, i.e. 10^{-20} J) compared to most biological and physiological processes that operate on much larger (mesoscopic to macroscopic) scales. The

equations of solid-state physics (Einstein-Debye formalism) have been accurately applied to energy production in organismic physiology. The main difference between these two methodologies is that while the Einstein-Debye model of solids describes thermodynamic equilibrium, *quantum metabolism refers to steady-state non-equilibrium thermodynamics of biochemical metabolic energy conversion.*

The notions of physiological time and metabolic rate are two fundamental properties, which have been used in the analysis of energy transduction in living systems.

Physiological time represents the duration of events such as heartbeat, duration of breath, or the time required for a nervous impulse to reach the brain.

The lengths of these times depend on the size of the organism and this typically involves the 1/4 power law that correlates physiological time with the weight of the system. Metabolic rate is the rate of heat production in a living system and it is a property that changes depending on the conditions, which gives rise to three general types of conditions, namely: standard metabolic rate, field metabolic rate, and maximal metabolic rate.

The standard metabolic rate is the steady-state rate of heat production by a whole organism under standard conditions.

The field metabolic rate is the average rate over an extended period of time when the animal is kept in its natural habitat. The maximal metabolic rate is the highest steady-state metabolic rate measured during hard exercise over a period of time. The maximum metabolic rate defines at a molecular level the characteristic cycle time, denoted t*, which is the shortest period of time over which metabolic enzymes can be restored to their active conditions (which is why this is also called a relaxation time). This corresponds to the maximum turnover rate for these enzymes, which is a rate-limiting activity for cell metabolism. Standard metabolic rate is most relevant and it satisfies allometric scaling laws, namely a large animal follows a 3/4 power law while most plants scale linearly with size. There have been two limiting behaviors found for metabolic scaling laws. First, in the quantum limit the cycle time t is much longer (its inverse, i.e. the rate, f, is low) than the characteristic cycle time t*. In the quantum limit the standard metabolic rate scales with size according to the d/(d+1) power law where d is the dimensionality of space occupied by the biological system of metabolic oscillators (d = 1 means a linear system while d = 2 is a planar system). Mathematically, this is expressed as: $P = \alpha W^\beta$ where β is the exponent that represents the relationship of metabolic rate P to weight W, and α is the proportionality constant [28]. In the classical limit the actual cycle time t is much shorter (the rate f is much higher) than the characteristic cycle time t* that corresponds to the maximum rate of enzymatic turnover. Here, *in contrast to the quantum limit, the metabolic network is over-supplied with nutrients, and the system is in a saturation limit corresponding to a pathological situation when electron leakage takes place leading to increased reactive oxygen species and oxidative* *stress, inflammation, acidity, and impaired energy production as ATP, the signature pathogenesis of disease.*

> *In the quantum limit, on the other hand, the system is seriously under-supplied.*

Most of the time the metabolic enzymes are "running on empty", i.e. they cycle faster than the rate of nutrient delivery 1/t. While these are the two extreme cases, most of the time the biological system operates in between these two limits ($\tau > t > t^*$), perhaps at the thin edge of the wedge between the two. The quantum limit signifies greater efficiency since the metabolic rate increases at a rate lower than the increase in the size of the system. For example, a person who weighs 150 lbs. and consumes 3,000 calories a day might increase his/her weight to 230 lbs. and his caloric requirement would be $(230/150)^{3/4}3000 = 4167$ calories, not $3,000 \times 230/150 = 4,600$. The reason for the lower caloric demand is due to the collective efficiency of organs and tissues considered to be societies of well-functioning cells. When this coordination of cellular activities within a tissue is lost, then this requires a high nutritional demand as is the case with malignant tumors. This can be compared to well-organized societies such as Switzerland where energy efficiency results in high productivity. On the other hand, chaotic societies of failed states are characterized by low productivity and waste due to corruption and inefficiencies. Similarly, a quantum biological system is much more efficient than a metabolic system operating in the classical regime. Another metaphor may be considered the vigorous cardiac output from myocardial contraction in normal sinus rhythm, which is a coherent and correlated contraction of muscle fibers with an intact conduction system, which may be considered the analog of the quantum regime. Conversely, when the conduction system is arrhythmogenic, the individual muscle fibers no longer contract in lockstep and the fibrillating desynchronized contraction may be considered the analog of the classical regime, a much less efficient function.

> *Moving from the quantum to the classical limit corresponds to a shift from fractional to linear dependence on size (unless the dimensionality of the system approaches infinity, which is hard to fathom). The classical limit applies to a case when a huge number of nutrient molecules renders quantization of energy irrelevant.*

This essentially mirrors an analogous situation in solids when thermal energy at high temperatures overwhelms the quantum energy spacing in the physical system. Since the scaling exponent for plants is close to 1, and since plants' chlorophyll complexes are two-dimensional structures, we conclude that they must operate largely in the classical regime (statistically speaking since each photosynthesis event is a quantum process) and this is due to the oversupply of nutrient in the form of photon quanta. We have previously used photosynthesis as an example of a biological quantum process, which appears paradoxical compared to the conclusion above. However, the experiments performed on the quantum nature of photosynthesis involved individual photons and not beams of light flooding the plant with light energy whose tiny fraction is effectively utilized by them.

> *The so-called "take-over" threshold refers to the point at which the quantum regime is followed below this threshold. Conversely, a system is characterized by the classical regime above the "take-over" threshold.*

This refers to metabolic systems whereby the cycle time (the time it takes to convert dietary substrate to product) equals the characteristic or relaxation cycle time (time to return to baseline metabolic rate from maximal metabolic rate). A quantitative description of this process is presented later in this section.

> *When cycle time is much longer than characteristic cycle time, metabolism follows the classical regime. Conversely, when cycle time is much shorter than characteristic cycle time then the metabolic rate follows the quantum regime.*

Notably, the faster the maximum metabolic rate the longer the characteristic cycle time and the longer it takes to return to baseline metabolic rate. "Take-over" threshold seems to correspond to the Debye temperature as the maximum temperature for which a specific heat applies to a physical solid (in the quantum regime).

The most important variables in quantum metabolism are the metabolic rate, the entropy production rate, and the mean cycle time, τ. The inverse of τ is the metabolic rate f. This latter quantity describes the mean turnover rate of the redox reactions within the cell. The fundamental unit of energy is here given by: $E(\tau) = g/\tau = \kappa f$, where the value assumed by g depends on the specific mechanism (electrical or chemical) involved in the electron transport chain coupling to ADP phosphorylation and g represents the magnitude of power generated in the process. Since biological systems operate far from thermodynamic equilibrium (albeit close to steady states), their bio-energetic quantities involve fluxes, i.e. rates of change of energetic values.

> *The molecule of ATP is a virtually universal energy currency in biological systems, producing work through the process of hydrolysis using a quantum mechanism.*

A mole of ATP synthesis whether generated through the electron transport chain of mitochondria or absorption of photons and chloroplasts requires roughly 60 kJ of nutrient or electromagnetic energy. The cycle time for energy production in the biological membranes ranges between 10^{-6} to 10^{-3} seconds with the energy of ATP generated per cycle being roughly 10^{-20} J, a value, which has been coined the quantum of biological energy. A mole of ATP is generated in 3 to 7 seconds. The hydrolysis of a mole of ATP releases roughly 30 kJ (thus about 50% efficiency of converting the energy delivered to energy available to do work, i.e. free energy, the remaining roughly 50% being liberated as heat). The human body requires the production of its weight in ATP per day in order to function, which translates into 10^{21} molecules of ATP per second. Interpolating this number of ATP molecules by the approximately 1000 mitochondria per cell and roughly 3.5×10^{13} cells in the human body, each mitochondrion produces roughly 3×10^4 (30,000) molecules of ATP per second. This is generated by the conversion of one molecule of glucose into 38 molecules of ATP through oxidative phosphorylation. The characteristic cycle time of this network of pathways is approximately 10^{-3} seconds (a millisecond), which corresponds to a frequency of about 1 kHz. Quantum metabolism focuses not on a single particle but rather particles in lockstep, huge, coordinated units of quanta of ATP on a macroscopic scale. They are collective modes of many particles working in unison for greater efficiency of purpose, which is typical of emergent phenomena. This may be considered to contain high informational energy and hence negative entropy.

This entropy reduction effect is due to the coherent behavior of the system's components. Systems defined by quantum metabolism have a correlated energy production in terms of coherence and synchronicity on a macroscopic scale of ATP energy to biological work and entropy reduction.
This amounts to optimal efficiency of energetics and transformation of Gibbs free energy to do useful work.

Since each mitochondrion produces 3×10^4 ATP/s and each ATP synthase operates at a rate of 600 ATP molecules/s, we estimate that each mitochondrion has on average 50 ATP synthase enzymes. Since the frequency of the oxidative phosphorylation reaction is approximately 1,000 cycles per second for each enzymatic complex, using the energy quantization identity $E_0 = \kappa f$ we conclude that the biological equivalent of Planck's constant is $\kappa = 10^{-24}$ J s and it is approximately 2×10^{10} larger than the physical Planck constant. This may have a simple explanation [29]. While the physical Planck's constant corresponds to a single atom, the biological constant corresponds to a mitochondrion. Various cells have different morphologies and volumes ranging from a micron cubed for bacteria to several thousand microns cubed for lymphatic cells. There are approximately 1.9×10^{13} atoms per cell and approximately 1000 mitochondria per cell, which gives 1.9×10^{10} atoms per mitochondrial "sphere of influence" within the cell. This is reasonably close to the ratio κ/h, which reflects the correspondence between biological and physical energy scales.

When metabolism relies more heavily on the glycolytic mode than on oxidative phosphorylation, the oscillation frequency of metabolic enzymes would be slower than if metabolism were predominantly be driven by mitochondrial activity. The oscillation frequency of metabolic enzymes that determine overall metabolic rate would be an average of a cycle time for substrate-level phosphorylation (ATP production through glycolysis and oxidative phosphorylation, i.e. mitochondrial ATP production through the electron transport chain). Accordingly, using a metaphor with black body radiation in physics, low amounts of nutrient bio-fuel consumption are analogous to exposing the physical system to electromagnetic radiation at low temperatures. This figurative black body absorbs the heat

and emits it in the form of ATP production. As calorie consumption increases up to the point of the "take-over" threshold it would be analogous to a black body exposed to electromagnetic radiation at higher temperatures, above which the analogy would collapse because the metabolic rate and oscillation frequency of metabolic enzymes (cycle time) would dramatically decrease.

Highly developed biological systems such as mammals are largely isothermal and hence temperature does not change to any significant degree to sustain the living metabolic state.

Biological systems exist and age according to the dimension of time, which is the manifestation of the presence of activity cycles. Our body temperature is normally 98.6° Fahrenheit (or 36.6° C). If that deviates over a sustained period of time to say 88.6° Fahrenheit or to 108.6° Fahrenheit, the time-dependent function of metabolism will soon come to a halt as the manifestations of cycles will no longer exist.

At the point of death, the far-from-equilibrium biological structure and function transitions to a state of thermodynamic equilibrium, and metabolic cycles grind to a halt. The matter of a living system at that point becomes inanimate matter as temperature replaces time as an organizing principle of its fundamental nature.

Consequently, temperature as a physical variable is not as relevant in biology as in physics since it is by and large kept constant, at least in mammals. While time is a characteristic of living systems, the greater the quantum nature of that living system defined by quantum metabolism, the more dilated time becomes, i.e. the more slowly it moves. We will discuss the consequences of this statement on health, disease, and aging later in this chapter.

3.12 Consequences of Quantum Metabolism

A nutrient contains some amount of glucose (or a substrate for glucose production). Glucose produces ATP in a series of coupled reactions (glycolysis or oxidative phosphorylation—OxPhos) resulting in free energy stored in the form of ATP molecules (the last phosphate group to be specific). Most of it can be transformed into useful work (e.g. one ATP hydrolysis per one step of the kinesin motor protein movement). Some amount of this energy is dissipated into the environment to maintain physiological temperature, i.e. it produces heat. *OxPhos produces a much higher ratio of useful energy compared to dissipated heat than glycolysis.*

To understand it better, it is important to state some basic aspects of metabolic efficiency. Metabolic rate is defined as the rate of energy production (substrate-to-product conversion) measured per unit time, which is commonly used for analysis in the classical regime. Alternatively, metabolic rate can be measured in terms of the rate of energy production per unit time per unit mass of the organism. Accordingly, glycolytic mode of ATP production is much faster than mitochondrial OxPhos production per cycle of substrate conversion to product. However, glycolysis is much less efficient in terms of the amount of ATP produced per molecule of glucose consumed. The efficiency of nutrient fueling mitochondrial oxidation is gauged by the P:O ratio, i.e. the amount of ATP produced per oxygen consumed. This ratio is higher for glucose oxidation (coupled to glycolytic metabolism) than for fatty acid oxidation.

Furthermore, *in comparison to the classical regime of bioenergetics, in the quantum regime, the metabolic rate utilizing OxPhos (coupled to glycolysis when glucose is the substrate) is much slower.* This is because the energy production is coherently synchronized in a correlated fashion across a larger mass and because it includes additional steps in the electron transfer chain in the mitochondria. The greater efficiency of energy production in the quantum regime of metabolism is present on two levels: it is a function of a higher P:O ratio in addition to there being more of the infrared energy released from the hydrolysis of ATP captured in biological work. Thus, less energy is lost as unusable heat in the process of producing ATP and its utilization.

Regarding metabolic enzymes, their operating frequency is related to the turnover rate of substrates and products. It is something very tangible and determined by the machinery of a living cell. The metabolic rates are functions of specific modes of energy production (with different rates for OxPhos and glycolysis). *The metabolic enzymes can operate in either the quantum or classical regime, which boils down to high or low turnover rates leading to quantum or classical metabolic rate limits.* This is the most important finding in the theory of quantum metabolism (in addition to the role of spatial dimensionality) since there is a potential for transitioning between classical and quantum regimes of operation in biological systems. While in the theory of solids a high specific heat requires a high amount of energy input into the physical system to raise the temperature by 1° Celsius (or approximately 2° Fahrenheit), a high metabolic rate requires less energy (because it is more efficient) to produce a given number of quanta of ATP over a unit of time. Moving from a quantum regime to a classical regime, relatively speaking, the quantum levels become very close to each other because they are being populated with a huge amount of energy and the system explores a very large space of available states. A similar phenomenon occurs in biology when the rate of energy production is very high and one molecule of ATP makes a small difference. In other words, a single energy quantum becomes insignificant.

The cycle time replaces temperature as an organizing parameter because living systems support dynamic steady states and are isothermal.

Cycle time is described to be the time interval for both acquiring resources in addition to the time to metabolically convert substrate nutrients to the product of ATP. However, the component of cycle time, which is the metabolic conversion of substrate to product (of ATP), seems to be a very small amount of time in comparison to the time interval for nutrient acquisition. Notably, time scales change on going from one level of organization to the next. At a mitochondrial level, a turnover rate may be on a millisecond rate, at a cell level, it may be on a second level, at a tissue level on the order of minutes, and finally at an organism level on the order of an hour. These time scales refer to characteristic periods governing each structure. *In terms of the allometric scaling law relationship between metabolic rate and weight of the organism, the larger the organism of a species the higher the metabolic rate but this relationship is nonlinear and scales with the β exponent.* So, if plants have a β exponent close to 1, it means the metabolic rate correlates to the weight of the plant linearly. As we discussed briefly above, since the dimension of the metabolic bio-membrane in plants (thylakoid) is two-dimensional (which would indicate a 2/3 exponent in the quantum regime), the reason for the classical regime of plant metabolism is not the geometry but the rate of energy supply that under most circumstances is enormous. Hence, what matters most here is the rate-limiting process, i.e. the longest time scale involved in the entire process. The oscillatory frequency of metabolic enzymes is hastened (shortened) as nutrient heat activates the metabolic enzymes.

This has consequences from a clinical disease perspective in terms of overloading mitochondrial capacity and a feed-forward process of mitochondrial dysfunction and reliance on glycolysis for the needs of ATP. However, there continue to be mitochondrial means of ATP production although not in the optimally efficient synchronous and coherent quantum regime. As a result, overload of nutrient consumption cannot feed into the mitochondria, and energy production is reduced to the energy by the glycolysis pathway. So, above the "take-over" threshold, ATP production in mitochondria follows a classical metabolic regime, which is less efficient because it is less coordinated.

Cytosolic substrate-level phosphorylation is much less efficient than oxidative phosphorylation.

Certain opportunistic organisms rely more on glycolysis for energetic needs because they are less complex, smaller and glycolysis (although has a lower metabolic rate and metabolic efficiency in terms of ATP production per unit nutrient per unit time) provides the necessary smaller amount of energy output in a more immediate fashion (often required for initial spurts of energy to escape predators or track down prey).

So, whereas substrate-level phosphorylation is adaptive in some organisms, in others it is not because it does not support the organism's mass and complexity.

In terms of the allometric relationship between metabolic rate and weight in humans, as weight increases excessively, the β scaling exponent may be reduced, which could signal a slower metabolism and thus contribute fundamentally to the mitochondrial dysfunction associated with metabolically unhealthy obesity. The latter is characterized by insulin resistance metabolic syndrome and often type 2 diabetes whereby reduced metabolic rate parallels less reliance on mitochondria (and ultimately a well-defined relationship between insulin resistance and diabetes with cancer and Alzheimer's disease (AD) may also be found). This could be a working hypothesis for future clinical investigations.

Both cancer and Alzheimer's disease (AD) have a major metabolic aspect while being metabolic "opposites" of each other due to the different directions of the metabolic shifts involved (the Warburg effect for cancer and what Lloyd Demetrius calls "anti-Warburg" for AD [30]).

The Warburg effect is defined as a shift in energy production from a predominant mitochondrially-based oxidative phosphorylation mode to a glycolytic mode that utilizes cytoplasmic enzymes. The former utilizes the electron transfer chain while the latter is entirely dependent on glycolytic enzymes. The anti-Warburg effect is a shift in the opposite direction, i.e. from glycolysis to oxidative phosphorylation.

Cancer as a metabolic disease invoking concepts such as "take-over" threshold and associated Warburg effect underscores the importance of dietary restriction.

There are data supporting this statement and an epidemiological link can be found to longer lifespan. Mechanistically, dietary restriction preserves mitochondrial function and favors the more efficient energy metabolism and energy output. The role of mitochondria is important not only in cancer, but also in Alzheimer's disease (AD). Metabolic disturbances such as insulin resistance and type 2 diabetes are associated with disturbed mitochondrial structure and function. Hence, it makes sense that cancer and AD are recognized to be associated with insulin resistance in type 2 diabetes (DMII). The maximum efficiency of mitochondria may well be rooted in quantum mechanisms.

All living systems struggle to overcome the second law of thermodynamics and must expend energy to lower their entropy since the natural tendency in all material systems is to increase entropy over time due to damage and degradation. A departure from a healthy to a pathological state (especially cancerous) means entropy increase.

This can be demonstrated mathematically for cancer cells at all levels of biological organization, namely from the proliferation of DNA mutations to the complexity of protein signaling networks, to the disordered cell morphology and even tissue (dis-) organization. This also translates into an inefficient way of metabolizing nutrients into ATP. The by-product of metabolism is heat dissipation, i.e. entropy production, so metabolism has a prominent influence in this framework of a progressive entropy increase culminating in the state of death from which point entropy increases monotonically with time. Organismic biology may even be considered a heat engine in the sense of being an energy dissipator so that rather than violating the second law of thermodynamics, it preserves the law. This also gets to the intriguing nature of the indistinct borders between living and nonliving systems and the corresponding relationships between thermodynamic variables and biological parameters.

On a clinical level, consider the example of a critically ill person admitted to the intensive care unit or a person with cancer. In both cases, the traditional reflex in the hospital setting has been to "force-feed". However, in nature, the last thing these individuals want to do is eat. Sure enough, nature seems to have greater wisdom than hospital policies. Studies have shown that such "force-feeding" in the example of the critically ill patient indeed increases mortality. Only within the past few years has this practice changed.

Finally, it should be mentioned that Demetrius' quantum metabolism theory focuses exclusively on metabolic energy production by combining Mitchell's chemiosmotic model of mitochondrial metabolism with Debye's collective mode excitations and applying it to metabolic enzymes. However, *the other side of the metaphoric metabolic coin is energy consumption and both production and consumption constitute a totality of metabolic processes. In this context, we propose the existence of both classical and quantum modes of energy consumption which differ in the absence of collective synchronization in the process of transducing ATP molecules' energy into biological activities, respectively.* An example of synchronization of energy consumption is muscle contraction utilizing the actin-myosin complex of thin and thick protein fibers. Without such synchronization, no effective force generation can be produced. On the other hand, utilization of glycolysis production of ATP in cancer cells for the purpose of their uncontrolled and unsynchronized cell division is an example of a classical type of energy consumption.

3.13 Synchronization of Cellular Activities

The quantum superposition of elementary particles' wave functions and the quantum entanglement effect of these wave-particles may offer a conceptual metaphor for what happens across cells, tissues, and organ systems throughout the body. These phenomena underpin instantaneous and simultaneous transfer of energy into the work of biological functions occurring in synchrony as an organismic whole. The transfer of electrons extracted from the breakdown components of fatty acids and glucose entering the mitochondrial TCA cycle to the electron transport chain is responsible for this phenomenon. Traditional molecular biochemistry describes how the transfer of electrons from these metabolized nutrients to the electron transport chain along the inner mitochondrial membrane leads to the subsequent transfer of electrons along an electron gradient produced by the coupling of redox cycling enzymes. The final acceptor of electrons is oxygen, which combines with hydrogen protons to form water molecules. The energy made available in this process is captured in the high-energy phosphate bonds with the conversion of ADP into ATP. However, independent of this molecular biological currency of energy in the form of ATP, there is a quantum mechanical phenomenon that occurs along the electron transport chain. As mentioned above, the transfer of electrons in mitochondrial protein complexes exhibits a quantum tunneling phenomenon.

The relevance of this macroscopic scale of quantum phenomena manifested by cellular metabolism and beyond underpins a greater efficiency of purpose as an emergent property of biological systems. This may be a result of an evolutionary advantage to biology that has been obtained by harnessing quantum mechanics on a macroscopic scale. Its specific manifestation is the quantum coherence of mitochondrial ATP production as opposed to glycolytic pathway utilization in metabolism, which is strictly classical and based on chemical kinetics combined with Brownian motion. However, there is possibly a much broader significance of quantum superposition that may find other uses in biology and that is mediated by mitochondria. This involves the superposition of various wave functions that creates a huge combinatorial space for the purpose of finding solutions in a parallel very efficient quantum search algorithm. When the optimum solution is found, the wave function can collapse into a single outcome. One of the best illustrations of this process would be perhaps the folding of a protein that finds the optimal ways to do this in a relatively rapid fashion in the context of the near infinite possibilities that exist in the protein-folding pattern. If these possible pathways to a folded state were to be explored sequentially one by one, the age of the Universe would not suffice to arrive at the folded state solution. This is called Levinthal's paradox in protein biophysics. However, a quantum search algorithm can achieve it in a very short time. This concept differs from the macroscopic scale of quantum biology using the analogy of 20,000 soldiers marching in lockstep. In the former case we have a wave function that spreads and through tunneling is able to explore numerous possible states at the same time in a quantum search algorithm. *Quantum coherence is rather opposite where many possible quantum representations converge on one, namely a coherent state, a very special superposition state, which minimizes fluctuations and hence introduces dynamical order.*

A hallmark of quantum strategies in biological systems is metabolic efficiency in terms of driving energy-requiring processes in the least amount of time while demanding the least amount of energy resources and producing the least amount of heat loss in the process. There are two separate strategies for achieving this hallmark as follows: 1) synchronized coordination of quantum particles coherently in lockstep together representing a macroscopic-scale effect and 2) divergent superposition wave function covering a broad range of possible outcomes. Both strategies involve quantum wave function. In the former strategy, it is possible that biophotons, originating from reactive oxygen species in the mitochondria, mediate synchronization across the component electron transport chains within the same mitochondria as well as between mitochondria of the same cell and of different cells. In the latter strategy, quantum coherence is not a desired feature since it would limit the scope of possibilities. Instead, a superposition of numerous wave functions forming a composite quantum state allows the system to rapidly explore the range of possibilities and then undergo a collapse on the most desirable state. This may be applied not only to the protein folding or DNA packaging problem but also to the genetic evolution and the search and retention of favorable mutations that endow the system with superior properties.

There may also be synchronization involving the lattice vibrations of mitochondrial walls that most definitely could affect the electronic transitions within the protein complexes of the electron chain. This may occur by the coordinate regulation by enzymatic activity of the electron transport chain,

and of TCA cycle enzymes that feed electrons into the electron transport chain, within the linear phase of enzyme activation related to nutrient intake (i.e. below the "take-over" threshold). Complimentary to the case of a quantum coherence superposition state of mitochondrial ATP production, a divergent superposition quantum wave function manifests a search mechanism or algorithm for the optimal solution(s), which could be likened to "cracking the code" and may be occurring at multiple levels of biological organization. This could take the form of electron tunneling within electron transport chain complexes and achieving great efficiencies in the process. This could further be manifested in the form of electromagnetic waves such as biophotons that mediate electronic transitions within the biological organizational structure and culminating with the collapse of the wave function. All of the above are hypothetical scenarios requiring careful experimental validation.

3.14 The Orchestra of Life: Biological Coherence

An early idea about quantum effects in biology was proposed by Fröhlich [31], who suggested that the *modes of vibration of phospholipid groups in cellular membranes might exhibit the phenomenon of a Bose-Einstein condensation*, in which many quanta settle into a single quantum state with long-range coherence. Bose-Einstein condensates are normally associated with very low-temperatures, but Fröhlich proposed that nonlinear coupling between a collection of dipole oscillators in the cellular membrane driven by a thermal environment and metabolic energy pumps could channel the system's dynamics into a single coherent dipolar oscillator even at biological temperatures. The advantage an organism would gain from this mode of energy storage could be to control chemical reaction and synchronize cellular activities. Recently, there has been renewed interest and some experimental support, at least for weak *Fröhlich condensates, which could play a dramatic role in chemical kinetics of far-from-equilibrium biological nano-systems*, and perhaps quantum coherence is an inherent property of living cells. Unfortunately, so far only scant experimental evidence exists to support these claims but new results from various labs around the world emerge that can drastically change this somewhat inconclusive assessment.

Low-frequency electromagnetic radiation may affect biological structure and function by increasing the oscillation frequency of the atomic molecular structures on one hand or, on the other hand, may change the direction of spin of an unpaired electron in a free radical. Free radicals are important for cell signaling in normal physiology. However, they also have the potential for promoting oxidative stress (by stealing electrons) and the associated process of inflammation, which again is of a feed-forward type. While Fröhlich did not consider the mitochondrial membranes in his theory, only cell membranes as the source of dipole oscillations, any suitably constructed phospholipid bilayer including the mitochondrial wall or nuclear wall can lead to Fröhlich condensation phenomena. From the cell biology point of view, there are of course major differences between these membranes but from the physics point of view the main issues are only: their thickness, electric potential gradient across their thickness, rigidity and permeability. In each case there are differences in this regard, especially between health and disease situations. Electric charge separation occurs within the lipid bilayer between two distinct sets of phospholipid head groups. Usually charged groups are hydrophilic, hence they are exposed and face the aqueous environment, which would be the cytoplasm and the extracellular matrix, respectively, for the cell membrane. The tails of phospholipids are hydrophobic and hence they do not carry electrostatic charges. The potential difference is due to charge separation and typically amounts to between 70 and 120 mV over a distance of 4 to 5 nm, which gives rise to very strong electric fields on the order of 10 million V/m. Dipolar oscillations refer to two charges equal in magnitude and opposite in polarity, say $+q$ and $-q$, held at a distance from each other, d, give rise to a dipole moment, p, with a magnitude $p = q \times d$. These charges will oscillate with respect to each other so that the distance varies between $d + x$ and $d - x$ where x is small, say a few percentage points of d. These oscillations occur with a frequency f (estimated to be in the $10^{11} - 10^{12}$ Hz range) leading to sinusoidal waves, which can propagate along the membrane structure as postulated by Fröhlich. This oscillatory behavior of dipole moments in membrane could be sufficient to emit electromagnetic radiation much like an antenna transmitting a radio signal. The condensation aspect is a separate issue. Since membranes may have a huge number of different types of dipolar oscillations, Fröhlich hypothesized that under special conditions these oscillations might condense (or reduce) to just one dominant wave, the other ones would be "extinguished" or made insignificant. This is needed to create a strong signal that might lead to cell-cell communication via a narrow frequency band, much like a selection of a channel on a radio dial.

Specific electromagnetic frequencies can be used by biomolecules for molecular recognition to promote aggregation. The same can be done for cell-cell communication albeit at different (probably lower) frequencies.

One particular example that was worked out within the Fröhlich theory illustrates the formation of rouleaux structures composed of the red blood cells in the human body [32].

The electromagnetic activity of biological molecules is an emerging field of research that is still in its infancy in spite of a long history of conjectures and speculations. It is very likely that this field, when properly researched will lead to breakthroughs in our understanding of biology and physiology that will translate into new medical technology. For example, it is believed that enzymes generate specific frequencies corresponding to their function and hence can be regulated by specific-frequency electromagnetic waves. There are indeed a growing number of indications for the existence of resonant frequencies in different ranges of the spectrum that correspond to different biological mechanisms. Effects of THz electromagnetic radiation are currently analyzed in such proteins as BSA (bovine serum albumin) affecting its diffusion and aggregation [33], lysozyme (causing some conformational changes especially in the α-helix region) [34], and transcription factors

(affecting their activation) [35]. This is just starting to be systematically evaluated by sensitive experimental techniques in Canada, the USA, Sweden, France, and Italy. THz frequency means 10^{12} Hz. Then, in a much lower range of 100s of kHz, and more specifically between 100 and 200 kHz, microtubules were shown to absorb electric field energy, which, as a result of this become distorted in mitotic spindles of dividing cells [36]. This new technology is already being used to successfully treat brain cancer by an Israeli company called Novocure. Other reports from Japan point to resonances in the GHz and MHz regions of the electromagnetic radiation spectrum [37]. Mapping these various resonant frequencies to structure and from structure to function will probably lead to a new field of science, which could be tentatively called bio-electro-medicine and there are great hopes for potential revolutionary changes to the way we treat human diseases. It is very likely that this can augment standard pharmacological approaches. There could also be direct interventions into metabolic diseases where we start to see now how oscillator frequencies can be tuned to change the body's response. By changing the characteristic frequency of a biomolecule, one may either enhance or hinder biological activity.

It is likely that biology evolved using quantum coherence in ways that we still do not understand but which can be found responsible for the evolution of the species. There are two aspects that might be useful for evolution.

First, in terms of survival mechanisms, efficient means of electromagnetic communication could give an edge to a species, even as primitive as bacteria, in their response to danger. Second, in the evolution towards more complex multicellular organisms, lower levels of organization could be associated with higher frequencies of communication/organization channels while higher levels would lead to slower, lower frequency means of communication.

These could represent interlocking hierarchies where the fast signals at the lower level are more or less seen as noise by the higher level except when there is a new average or baseline emerging due to a bifurcation, which could signal a system's instability representing a pathological change. A relevant analogy would be a functional organization of our society. At a personal level, we have some chatter that appears to be noisy and irrelevant to national politicians. However, when at a local state a new party appears with a challenging platform this is no longer noisy chatter but a message that is sent and received at a higher level. We have this kind of information filtering all the time when news is produced at a local level and then only some of it is filtered to the national level and even less so into the international arena. Local news only reaches an international level when something extraordinary happens. The same may be true of biological hierarchies of organization and communication.

Biophotons may be correlated with the activity of the oscillating enzymes. If so, the intensity of increasing nutrient consumption parallels the temperature and hence the higher frequency of the oscillations (at lower ranges of nutrient intake). According to Quantum Metabolism theory, mitochondrial enzymes are indeed correlated, that is, they are coherent in terms of the frequency of their behavior and coupling and also metabolic reactions are correlated in the whole tissue. This effect would be a logical evolutionary design for the purpose of optimizing efficiency.

> *In tumor cells, both the level of activity and the correlation between mitochondria is blocked in parallel to the degree of malignancy.*

This is called the Warburg effect where a metabolic shift away from oxidative phosphorylation and toward glycolysis signals a malignant transformation. Eventually, the mitochondria become de-correlated and become very inefficient from an energy production point of view. This may be viewed to imply that cancer collapses a superposition of the wave function that is responsible for coherently synchronized mitochondria and hence the efficiency of metabolic activity working in unison is gradually degraded and then destroyed.

Free radicals constitute one of the important sources of biophotons in the human body. They are byproducts of metabolism and their recombination leads to the emission of biophotons. Hence, it would make sense if free radicals, produced predominantly in mitochondria, at a low or optimal level, would enhance quantum metabolism via the production of biophotons through recombination, whereas when overproduced due to over-nutrition, can be associated with inflammation. Thus inflammation, being a source of excess heat, may result in the collapse of a superposition wave function that degenerates synchronous coherent oxidative phosphorylation metabolism.

The phenomenon of biophoton generation and utilization in biology promises many potential applications whose understanding may ultimately advance practices in health care and medicine.

> *Molecules and cells represent systems that emit and absorb external electromagnetic waves within the body and from outside the body, and in that sense serve as antennae that resonate with their environments.*

This could represent a fundamental characteristic of biological systems in mediating biological coherence. Such coherence includes the oscillatory patterns of transcriptional activities of cells, secretory patterns, and cell replication. The in-phase resonance between ligands and their receptors appears to represent another example of physiologic activity whereby such coherent resonance facilitates ligands aligning with their receptors. While the predominant concept in ligand-receptor interactions is based on the lock-and-key principle, i.e. shape matching between the two structures, examples emerge where this idea is clearly not sufficient and maybe even inapplicable. This has been shown to occur in the phenomenon of olfaction where resonant recognition of the vibrational frequencies of the interacting molecular species is a more plausible concept. Similarly, anesthetic molecules do not conform to the lock-and-key approach and many more examples may be found in the future.

It is also likely that *the phenomenon of quantum metabolism involves metabolic coherence between mitochondria within a cell as well as between cells.* As stated above, perhaps the most significant source of biophoton emission is through the recombination of reactive oxygen species, which occur along the electron transport chain in the process of ATP production. This could be both beneficial and, if not produced in a sufficient quantity or over-produced, could be detrimental to the overall synchronization and coordination of biological functions. It, therefore, makes sense to propose that ATP generates fundamental biological quanta that are produced coherently in a correlated fashion on a macroscopic scale, which may be mediated by phase-coherent oscillations of enzymatic activity with the same frequency of the biophotons.

An example of direct research translation to medicine follows.

> *In diabetes, reduction of biophoton emission could relate to the loss of mitochondrial structure and function, which is known to occur in diabetes mellitus.*

Such a loss occurs in both forms of diabetes related to high levels of glucose feeding into the mitochondrial electron transport chain. Furthermore, in type 2 diabetes insulin resistance serves as an independent cause of mitochondrial disease

3.15 Biological Motors

Biological motors are central force generation elements to the machinery of living systems just as mechanical motors are to a car engine. In both cases, the motor is responsible for converting one form of energy into a force or torque that generates motion involving translational and/or rotational kinetic energy. As stated before, the most common notion of a motor implies an electrical device involving an arrangement of a stator containing a permanent magnet with a rotor containing wire coil windings as an electromagnet connected to a power source. However, in a broader sense, a motor is a subtype of engine that produces motion. By distinction, an engine takes in energy per unit time and possibly raw materials and converts those into useful output. It is a useful transformation of one form of energy into another (e.g. chemical into mechanical).

> *A biological system as an organism may be considered not only a biological engine, but a complex structure comprised of several types of biological engines as well as motors intricately coordinated into a single larger system.*

The fuel cell provides hydrogen atoms that are stripped of their electrons that in turn are the source of electricity that powers the engine. The electron transport chain of mitochondria in biological cells is similarly provided hydrogen atoms from bio-fuels. The stripped electrons provide an electrical current and, in a sense, as well as provide the source of energy that is ultimately incorporated into the phosphate bonds of ATP, which power the biological work of the body. Fuel cells are an alternative to battery-operated electrical automobiles. Fuel cells create electricity from an electromagnetic coil of copper wire.

Analogously, biology coopts the electromagnetic force generated perpendicularly to the electrical current of the electron transport chain to propel the motion of hydrogen protons across the inner mitochondrial membrane (to the inter-mitochondrial space). This electromagnetic force that propels the motion of hydrogen protons is referred to at the proton-motive force. Additionally, it is intriguing to consider that associated electromagnetic fields generated perpendicular to the electron transport chain may play a role in mediating synchronized coherent conversion of nutrient bio-fuel to ATP energy. Hence, a fuel cell and a quantum mechanical engine may both be considered biological engines. However, quantization of ATP energy and quantum tunneling of electrons in the mitochondrial complexes provide an argument for a nano-scale quantum mechanical-biological engine.

In general, biological motors at a subcellular level can be divided into: a) linear motor proteins that move in a processive fashion along protein filaments, e.g. kinesin or dynein along MTs or myosin along AFs, b) rotational motors such as F-ATPases, c) beating complexes of motor proteins and filaments such as the microtubule-dynein arrangements in cilia and flagella, and d) contractile arrangements of interconnected long and short filaments involving actin-myosin complexes. All biological motors contribute to the various mode of cell motility such as crawling, sliding, blebbing, beating, and retracting. Directed intentional motion is what distinguishes living systems from inanimate matter.

One molecular motor, ATP synthase, is particularly structurally and functionally analogous to the electrical motor. It contains a stationary stator to which the rotor, the rotating component is associated. The stationary stator anchors the complex within the bi-lipid inner mitochondrial membrane including the knob, the site for catalyzing ADP to ATP, and the hydrogen ion channel for protons to be rotated through the channel to participate in the conversion of ADP to ATP. Notably, complex 5 of the electron transport chain, or ATP synthase, is known to be involved in iron metabolism. It is intriguing to consider if iron is mechanistically a controlling factor in the rotational component of ATP synthase. Iron, having a net magnetic moment, is polarized in a magnetic field, which is produced by the electric current generated in the electron transport chain (from the moving electrons as well as the molecules containing iron) whereby the electrons ultimately combine with oxygen to produce water at complex 4. It is conceivable that the interaction of the moving electrons, the electromagnet, and iron, the permanent magnet expected to be present in complex 5, ATP synthase, may be responsible for the force of motion of the rotor component of complex 5. Independently, however, of whether this rotational mechanism is due to the interaction of iron (a permanent magnet) and electron transport (a conductor analogous to a copper wire producing a magnetic field around it), a spectacular similarity of how nature has devised a molecular biological motor to that of a machine electrical motor is stunning. What is also amazing is the incredibly small scale of biological motors that operate at a nanoscale, a billion times smaller than man-made motors. It

highlights a valuable lesson that *studying nature provides the best blueprint for designing machines not only since Nature came first but also because it had countless real-life experiments to perform over billions of years to reach perfection.*

Closely related to the topic of molecular motors is the issue of biological thermodynamics. Biological thermodynamics concerns energy transformation in living matter, which in fact is the title of the German-British medical doctor and biochemist Hans Krebs' 1957 book [38]. Famously, the mitochondrial TCA cycle was named after and described by Hans Krebs. The focus of biological thermodynamics is on principles of chemical thermodynamics in biology and biochemistry. It is worth emphasizing that both classical and quantum systems of many particles are amenable to analysis using statistical thermodynamics. As stated earlier, classical systems follow the formulas of the Boltzmann probability distribution while quantum systems are either described by the Fermi-Dirac statistics or the Bose-Einstein statistics. *Thermodynamic biological engines may be considered to include far-from-equilibrium thermodynamic states in which the concept of the Gibbs free energy is invoked to achieve entropy reduction or, equivalently, an increase in information. Such information gain in biology is an essential characteristic of living systems where it is generated across many hierarchical organizational scales ranging all the way from elementary particle physics (to accommodate physical principles) to the scales of molecules and macromolecules subsequently leading to the transition to life's building blocks such as cells, tissue, and organs. Finally, on the scale of biological organization, information is further produced by entropy reduction due to structure formation enabling information processing as all prokaryotic and eukaryotic organisms, either directly or indirectly, are dependent on the energetic input from the sun.* We discuss information theory later in this chapter.

The burning off of carbon atoms from glucose molecules inside the metabolic pathways that occur in cell mitochondria couple metabolically un-favored reactions (with a positive change in Gibbs free energy, i.e. lowering the Gibbs energy of the resultant system) with favored reactions (with a negative change in Gibbs free energy). That is, the heat liberated from metabolically favored reactions is harnessed to drive the unfavorable ones, thus allowing the cycle to continue. These reactions do not discretely occur sequentially as represented in drawings of the cycle but rather in reality simultaneously couple to one another. For example, the negative change in Gibbs free energy of the fumarate dehydrogenase catalyzed reaction allows the positive change in Gibbs free energy (that is otherwise spontaneously not favored) of the citrate synthase catalyzed reaction. More commonly, the biological energy currency ATP is hydrolyzed which releases the free energy stored in the high energy bonds of ATP (relative to ADP) that enables otherwise non-spontaneous reactions or mechanical processes to occur by increasing the internal energy of a receiving system of bonds and molecules (as increased potential rotational and vibrational energy, kinetic energy or both). The result of this reaction has a high potential for unstable bonds to break and hence release heat with the coincident formation of bonds with relatively high stability of the product molecules formed. Hence, in addition to the change in the potential energy component of internal energy, heat is transferred as thermal energy, which at the molecular level represents the translational and rotational kinetic energy of the molecule. For example, *the release of carbon dioxide vapor accelerates the motion of molecules resulting in an increased temperature of the system due to their molecular collisions. Alternatively, the transfer of energy for useful purposes in the form of work in physiological processes is largely driven by the hydrolysis of ATP into mechanical processes of active transport, muscle contraction and biosynthesis.*

> *Principles of classical thermodynamics remain very important in explaining many aspects and components of biological function in concert with the principles of physics.*

They remain particularly useful at the level of designing machines. However, *when investigating the fundamental nature of matter one must go beyond the "billiard ball collisions" classical representation at the microscopic level and instead use the concepts of quantum mechanics, which is especially true of biology.* Understanding how mitochondria work, how various enzymes operate, or how recognition operates at the subcellular level invokes the insights of quantum mechanics.

One of the key results of the theory of special relativity is the famous equation $E = mc^2$, which states a linear relationship between mass and energy. In the present context of metabolism, the primary source of biological energy is light (or photons) generated by the process of nuclear fusion in the sun. This is then converted from electromagnetic rays to mass in prokaryotes and plants by the quantum mechanical process of photosynthesis. Subsequently, the spectacular phenomenon of quantum mechanical transduction again appears to manifest itself in specialized cells or organelles, mitochondria, the powerhouse of cells.

> *The mitochondria are the cell's most efficient energy-producing factory whereby nutrient hydrocarbons are stripped of electrons with the ultimate transformation to the universal biological currency of energy, ATP. In this sense, the dependency of biological systems on mitochondria and quantum mechanics provides a fundamental basis for considering organismic life as a quantum mechanical-biological engine.*

3.16 Classical and Quantum Molecular Motors and the Laws of Thermodynamics

The free energy of biological molecules results from the chemical energy that holds the atoms of the molecules together. The stronger and more stable the bonds, the more energy is required to break them while the breakage of weaker bonds requires less energy. The structural flexibility of most biological molecules utilizes weak bonds, such as hydrogen bonds while stability requires strong bonds such as covalent bonds.

The formation of stronger more stable bonds results in the protons and electrons moving from structures of higher to lower free energy of the reaction while the surrounding molecular kinetic energy increases as the heat is released in these chemical reactions. Analogously, a macroscopic ball, which has already fallen to the bottom of a hill, has very little gravitational potential energy to fall further and hence, little potential to generate kinetic energy. The Gibbs free energy that drives anabolic processes coupling exothermic to endothermic reactions also includes the hydrolysis of ATP to drive uphill reactions. Accordingly, the phosphate atoms of the ATP molecules have weak bonds with relatively little stability but high potential energy stored in these bonds. The bonds break quite easily requiring minimal energy input followed by the formation of new bonds in the products of ADP and inorganic phosphate, which have much lower potential energy having more stable bonds than the original ATP molecule. *The formation of the new bonded structures of ADP, an inorganic phosphate, may release energy in the form of heat to the surrounding. This reaction can then be coupled with thermodynamically unfavorable reactions to give an overall negative spontaneous Gibbs free energy for the reaction sequence* (the ΔG for ATP hydrolysis is an exothermic process releasing 30 to 50 kJ/mol of energy, the more negative value in the presence of sufficient magnesium availability). *The heat released from the hydrolysis of ATP in the context of Gibbs free energy results in the production of useful work such as muscle contraction, active transport, and establishment of electrochemical gradients across membranes and biosynthetic processes necessary to maintain life.*

> *In human physiology, roughly 60% of the energy released from the hydrolysis of ATP produces metabolic heat, which is not utilized for useful purposes of work.*

For example, the myosin motor cannot produce muscle contraction in the absence of the ATP energy that drives it. The survival value of a contracting muscle represents a metaphorical comparison to the information of muscle that converts heat as an energy source to internal energy, which can, in turn, create something more enduring. According to the second law of thermodynamics, some of the thermal energy in transit is lost to randomized motions outside the system generating entropy as energy that cannot be harnessed for useful purposes.

> *Power is the rate of work performed over time. Work is the product of force and distance and hence power is the product of force and velocity.*

The totality of work may be considered in terms of thermodynamics. *In the absence of work done on them, objects seek states of lowest energy or highest entropy (which is a variation on the second law of thermodynamics). In biology, this force of nature that provides the conditions suitable for life of all forms may be termed negative entropy, borrowing thermodynamic terminology.* However, negative entropy in a closed system is not allowed thermodynamically. Therefore, its reference must be in the context of an open adiabatic system, whereby there is the transfer of matter and energy from the surroundings. *The end result is structure and function formation, both of which lower the entropy and, as we discuss below, create information.*

The geometry of mitochondrial energy "generator" is quite complex as one needs proton pumps with a pH gradient across the mitochondrial wall, an electron transport chain taking place along the length of the wall, and ATPases which pump out ATP molecules into the cytoplasm, so there are both gradients of concentrations into and out of the mitochondria as well as electron motion (at least partly tunneling) along the wall (see Figure 3.9). This can be viewed as an electric circuit and hence there should be some magnetic fields generated in the process but probably very small. On the other hand, *in glycolysis, the associated chain of enzymatic reactions is distributed throughout the cytoplasm and there is no defined circuitry that would produce magnetic fields.*

> *The mitochondrial way of producing energy is not only very efficient but well-engineered and requires a structure, unlike glycolysis where basically all the reagents are "thrown together" into a bag of fluid (cytoplasm) and reacting due to their affinities in a diffusion-limited way without any synchronization.*

Diffusion by definition is subject to environmental noise (Brownian motion) and hence does not require fine-tuning in contrast to OxPhos in mitochondria. Diffusion is "noisy" and inefficient but OxPhos is efficient and smooth. However, it could be disrupted by chemical or radiation damage. This is like comparing transport by an army tank to an electric car, say "volt". The tank will get you there no matter what but it will guzzle a lot of fuel. The volt will get you faster, cheaper, and in greater comfort but may get stuck in mud or may stop working altogether if one cuts the wire to the battery. Some researchers believe that the root cause of at least some cancers is metabolic dysregulation involving mitochondria (the Warburg effect), which as a consequence leads to genomic instability. Consequently, there are some experimental cancer therapies under development based on the concept of upregulating OxPhos and down-regulating glycolysis by using various pharmacological agents, e.g. DCA and 3-BP, respectively. In other words, the idea is that cancer cells have traded the energy efficiency of a volt for the robustness of a tank.

The quantum mechanical engine, or motors forming its parts, exists outside of the mitochondria *per se* in other organelles such as the endoplasmic reticulum or Golgi bodies but not in the cytoplasm. There may be quantum coherence domains that could particularly exist in the nucleus. Biological activities of molecules such as enzymes have specific frequencies characterizing the rate at which they function. By changing their frequency values by binding to ligands or forming complexes, or perhaps also by externally applied electromagnetic waves, one may either enhance or hinder biological activity. This is consistent with the Fröhlich theory of biological coherence, which envisaged specific oscillator frequencies to be used for molecular recognition, attraction, communication, and adhesion between similar cells. So these enzymatic effects may be considered motors of the quantum mechanical

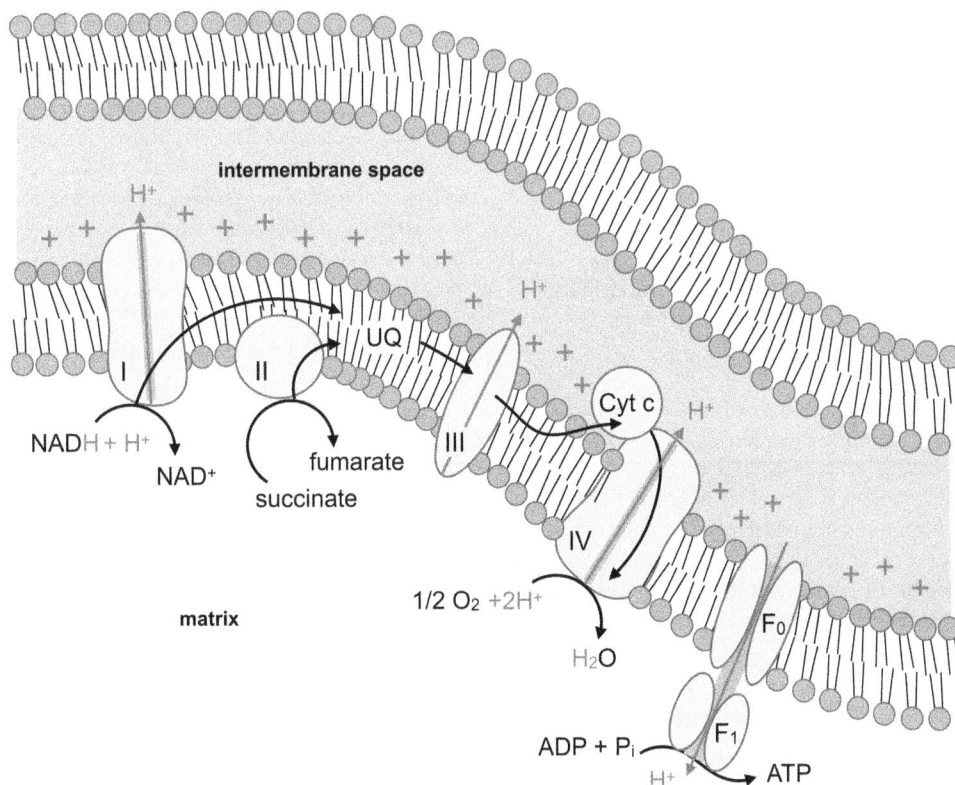

FIGURE 3.9 An illustration of the electron transfer chain complexes embedded in the inner mitochondrial wall.

engine analogous to the molecular motors in the mitochondria. Then the host of molecular motors found both in the mitochondria and throughout the cell can be viewed as representing the cell's quantum machinery when the substrates and products are included. Furthermore, it is logical to extend the metaphor to not just quantum mechanical motors and biological engines but, in concert with fuel cell motors in an engine, as the larger far-from-equilibrium biological engine consisting of multiple engines within the biological system. *Loss of coherence and synchronization is usually associated with pathological transformations.*

> *Disease states can be seen as quantum decoherence phenomena.*

For example, loss of quantum coherence in cancer has been recently proposed by Stuart Hameroff [39]. It is worth mentioning that mitochondria move throughout the cell by being literally transported by microtubules with the use of motor proteins, so there is an intriguing association between these structures and some researchers claim that the interaction is promoted or enhanced by oscillating electric fields produced by both microtubules and mitochondria.

3.17 Energy and Information: A Marriage of Physics and Information Science in Biology

Information theory was developed to find limits on signal processing operations such as compressing data and reliably storing and communicating data. In 1949 Claude Shannon [40] described information as any type of pattern that influences the formation or transformation of other patterns. Fifty years later David Casagrande [41] described information in similar terms. Gregory Bateson [42], an English anthropologist, described information as "A difference that makes a difference". Information comes from a Latin root meaning, "to put into form". It is conceived as a process that is fragmented and in transit, flowing (metaphorically) across time and space. *Conversely, knowledge is a state, constructed from the building block of information.*

> *Knowledge is structured, coherent, enduring, and is temporally and spatially expansive.*

Stoner said, "Information exists, it does not need to be understood to exist". "It requires no intelligence to interpret it". "It does not have to have meaning to exist" [43]. Analogous to the flow of information into knowledge is the flow of energy, in the form of heat, which is also transitory. *Ultimately, the energy transferred for work transforms information into more enduring knowledge.*

Information theory refers to entropy as an underlying concept.

> *The greater the entropy, the lower the information and vice versa.*

Since entropy is deemed to be a measure of randomness in a system, its larger value implies greater uncertainty in the systemic order, which can be deemed to be a measure of "information". For example, a coin flip has less entropy, providing less information than will the roll of a die. The greater the number of potential outcomes or results, the greater the entropy or the lower the information about the system. Conversely, selecting one of the many outcomes offers a high amount of information. A winning super-lotto ticket garners a multi-million-dollar prize while a winning bingo card carries only a small monetary value. Information is not only connected to probability and hence to entropy but also to energy.

> *Erasing information requires energy to perform this operation.*

In its simplest form, information is defined through the use of probability p that describes the extent of our incomplete knowledge. If $p = 1$, our knowledge about an event or process is complete, and if $p = 0$, the corresponding knowledge is totally incomplete. Hence, when p goes up, entropy S goes down and information I goes up, and vice versa. The actual formula for I was given by Shannon in terms of negative entropy as $I = -S$. Shannon was able to describe the notion of information without invoking quantum mechanics. In classical mechanics, in principle, there can be a state of zero entropy (at absolute 0 K temperature where everything freezes its motion, which is also called the third law of thermodynamics). In quantum mechanics that is not possible because even at 0 K there is so-called zero-point motion, such that particles at their ground states are still in motion with the ground state frequency f_0, and the ground state energy $E_0 = (\frac{1}{2})hf_0$. Hence, the uncertainty or randomness of information is inherently rooted in quantum mechanics as the element of surprise and the ultimate original creator of information. As mentioned earlier in this chapter, quantum mechanics does not allow for complete information due to the Heisenberg uncertainty principle. *In biology, information can take a very concrete form because there is a coding part (e.g. DNA) and a decoding part (e.g. transcription and translation molecular machinery using RNA and protein synthesis machinery).* Hence, in order to understand biological information, one has to "possess access to its code", e.g. the genetic code translating the triplets of DNA base pairs into amino acids which are the building blocks of proteins. Furthermore, biological information bits interact or relate to one another to form a larger more meaningful system whole. *In cell biology, this results in gene-gene interactions and protein-protein interactions eventually leading to a systems biology network of networks of the cell.*

> *Nature seeks solutions and has a certain infinite wisdom. There are, therefore, many layers of biological information coding and decoding, not only at the genetic code level.*

There also post-translational modifications that complicate matters beyond the genetic code and include a layer of epigenetic information. In addition, *there is cell-cell communication, receptor-antibody recognition, immune system surveillance, self-nonself recognition, etc., up to the pattern recognition and the fight-or-flight mechanism, which is really about receiving information and reading or decoding it at all levels of biological organization.*

Information that is not harnessed to do intellectual work of constructing a deeper meaning as a more complex expansive and coherent integration of knowledge is remarkably difficult to recall at a later time. That fragmented, transitory information, rather than becoming enduring and constructive, is lost in a process that resembles heat being lost to the randomized motions of entropy. Furthermore, the transfer of energy as heat from one system to another across the complex pathways of networks and systems is analogous to the flow of information. *Again, because information and the energy that runs through it is part and parcel of the same system, it makes it meaningful and valuable.* Energy represents the potential to do work. *The greater the flow of information capable of work requiring functions that drive the robustness and flexibility of an individual or organism, the greater the biological complexity. The greater the capacity of the system's biology to extract the energy flow of information into work, the greater the efficiency of this biological engine and hence the less that is lost to entropy.* Entropy production is typically associated with work done by what are called non-conservative forces (such as friction), whose work is not stored as potential energy. Energy that is transferred in a system as heat is lost to disorganized random motions unavailable to do useful work.

Furthermore, the purpose of this writing is to provide both the content knowledge of energy itself from appropriate perspectives of chemistry, physics, biology, and clinical medicine to facilitate a usable understanding of energy. Information metaphorically exists in many forms and recognizing it as such catalyzes the intellectual work of requiring a deeper and clearer understanding of the inextricable nature of the flow of information and energy. In the absence of energy, machines cannot function, and similarly life cannot exist.

The thinking that translates information into knowledge is work, which is paralleled by the energy flowing through and bringing out the potential information.

> *Knowledge may be considered the product of work on the substrate which is information. Furthermore, knowledge alone is not sufficient as it requires additional input of thinking and analysis to reconfigure the initial data into creative applications that later on promote the construction of adaptive resources.*

Just like biological adaptations impart survival advantage as coded information in DNA translated into phenotypic traits as solutions of ecological niches to challenging environmental conditions, knowledge coupled to creative imagination makes possible improved quality and longevity of human life. It takes work to understand the raw information of electrical principles analogous to the work in biology to construct the protein molecules of actin and myosin from the information encoded in DNA. The notion of an electrical motor, such as an electrical hydraulic lift was at one time a creative and innovative idea, the thought process itself required work. Taking the analogy a step further, machine motors and biological/molecular motors

are both the products of work and the conductors of work ultimately responsible for survival adaptations and in the case of machine motors, sometimes quality of life adaptations. *The work of complex reasoning and extrapolation of knowledge to innovation ("original" thought) imparts greater value than the prevailing existing knowledge.* For example, the fundamental information of negative electron and positive proton electrical charges accompanying the properties of current, voltage and resistance, provides the necessary objective substrates from which to derive an understanding and knowledge of electricity. *Further reconfiguring of that subsequent knowledge provides novel application of the principles of electricity beyond the value of already existing electrical appliances resulting in creative innovations* (Table 3.1).

> *Heat is the amount of kinetic energy of molecules undergoing collisions resulting in the transfer and equilibration of their energy.*

This energy flow is also partially responsible for the flow of information into knowledge. The transfer of heat causes internal energy changes, which then can be put into the work of reconfiguring the information into knowledge. A metaphorical extension of information is that any raw materials or substrates when heated and energized are directed to produce something of greater value. Accordingly, our DNA, our muscles, the food we eat, and the petroleum fuel we put into our automobiles are all examples of information and information transfer, sometimes resulting in knowledge creation.

> *Matter can be viewed as trapping in the form of potential energy of the chemical bonds linking the atoms in the formed structure.*

This is true for both inanimate and biological forms of matter. The original source of all energy on Earth is in the solar process of nuclear fusion and nuclear fission as shown via Einstein's famous relationship $E = mc^2$. This is also at the root of the physiological energy unit, i.e. the calorie. Solar energy promotes the configuration of nutrients such as glucose in the process of photosynthesis. The ability to aerobically metabolize glucose due to the evolution of cellular mitochondria

TABLE 3.1

Energy and Information, the Co-Building Blocks of the Structure and Function of Machines and of Biological Complexity, Mechanistic Comparisons

Information	Useful Purpose
Some examples:	Some examples:
1. Sensory or content information	1. Knowledge/innovation
2. Petroleum fuel (machine engine fuel)	2. Mechanical motion of an automobile, engineered mechanical motor
3. Food (biological engine fuel)	3. Muscle contraction (mechanical energy/myosin molecular motor)/ active transport including electrical potential and nerve conduction/ biosynthetic processes

germinated as a result of the photosynthetic process that liberates oxygen into the atmosphere. This coincided with the evolution of life on Earth. *The flow of energy through systems biology pathways and networks epitomizes the link between the derivation of matter from energy and vice versa.*

> *One can see both individual and collective life processes as the flow of energy that assembles material substrates, but it also creates enduring information. The information stored is the building block for the growth, complexity, and reproduction of humans and all organismic life.*

In the case of gasoline, it serves as a mechanical fuel, the potential energy locked up in the bonds of the petrochemicals. Just like food is information created by the biomolecules containing energy, so are fossil fuels. Moreover, not only does energy serve as the glue that holds the information together, but it gives the potential for the acquisition of knowledge, knowledge translation, and greater meaning, value, and usefulness. The energy, however, must be released from the information that harbors it in order to create new information. Furthermore, it requires activation energy (spark) in the product of the new information (e.g. the enthalpy of bond formation, or energy released from the formation of carbon dioxide and water vapor molecules) to unlock the energy from the petroleum. Biological systems require digestive and metabolic enzymes, available ATP, and a host of conditions that represent a composite system of information capable of unlocking the energy held within the bonds of ingested food. *Hence, energy and information run inextricably in a dynamical tandem.*

Sensory (visual or verbal) and content information is acquired in the cells of the mid-brain, specifically the thalamus. Glucose oxidative metabolism in these cells takes the information from these nutrient molecules and harnesses from it energy transferred as heat and ultimately the high internal energy bonds of ATP as potential energy. This in turn allows the necessary work of active transport involved in neuronal transmission that acquires the sensory and content information across the specialized sensory organs capturing it within the cells of the thalamus. Subsequent neuronal transmission communicates electrically with higher cortical regions of the brain. It is here that conscious processes organize and interpret the information by the four components of thought allowing consolidation of the information as knowledge and long-term memory stored in the hippocampus. The flow of information is sparked by the activation energy provided by the hydrolysis of ATP that phosphorylates glucose upon entry into the cell, trapping it inside the cell. This is followed soon thereafter by the activation energy provided by ATP catalyzed phosphofructokinase (PFK) enzymatic rate-limiting step of glycolysis. The remaining reactions comprising the glycolysis and TCA cycle catabolism of glucose into ATP synthesis occurring at the matrix side of the inner mitochondrial membrane occur without requiring coupling energy released by any further hydrolysis of ATP. Rather, the reactions occur spontaneously with or without the help of coupling to coinciding reactions within the pathway that have a negative enthalpy of formation (or reduction in Gibbs free energy). Hence, the spark of

an internal combustion engine that initiates the combustion of petroleum fuel is analogous to the activation energy provided by ATP hydrolysis. The final product of ATP production along the inner mitochondrial membrane is analogous to the heat released in the combustion chamber in each case as a result of the breaking and reconfiguration of the bonds of molecules with the transfer of heat. The transfer of work from the potential energy of ATP to muscle contraction (mechanical energy) is analogous to the transfer of work due to compressing of the cylinder piston by the heat-induced expansion of carbon dioxide and water vapor molecule products of combustion. *Thus, the metaphorical flow of heat through systems allows information of diverse forms to energetically transfer and empower the content of that information to work. Accordingly, the transfer of energy as heat and information are inextricably linked, as is the transfer of energy as work.*

> *Energy, whether transferred as heat or as work, allows the transformation of the information to something of greater usefulness.*

Information, as described in the above paragraph, may be exemplified to include: sensory or content information; food (for biological systems); fuel (for mechanical systems); biochemical pathways of physiology to metabolize nutrient fuels such as glucose; and the complex arrangement of components of any system at the macroscopic and microscopic levels (e.g. the musculoskeletal system and the myosins). Coordinate and requisite to information translated to knowledge is the energy that powers it. The information reorganized as knowledge becomes more enduring and increasingly more useful depending on the energy applied to it and in what regions of the brain it occurs. A further reorganization and consolidation of information along the neuronal circuitry of the brain harness the available raw molecular and biological machinery, which itself becomes information represented in the form of specific synaptic connections. This information has the available tools for the flow of energy to bring out its potential just as the extraction of energy requires the precise set of tools and conditions. ATP hydrolysis provides the energy to power the orchestration of the body's structure and function. This speaks to the critical role of extracting energy from the information contained in food to supply the information of the biological engine to keep it running. Fundamental to muscle contraction is establishing electrochemical potential across the myofibrils of actin and myosin by ATP hydrolysis, which also is needed for the mechanical sliding motion whereby myosin fibrils pull AFs inward that produces the contraction. ATP is required for the active transport process that generates the hyperpolarized electrochemical potential ($Na^+ - K^+$ ATPase) of all cells in the body. In neurons, electrical transmission is the mode of cell communication with other neurons or other types of cells (e.g. cardiac or skeletal muscle). In endocrine cells such as the pancreas, it is important for insulin release. The electrical potential of cells is also a protective mechanism that keeps harmful molecules outside of the cell. ATP hydrolysis also provides the energy necessary for many biosynthetic processes (e.g. DNA and RNA synthesis and the synthesis of the actin cytoskeleton of cells). ATP hydrolysis provides the activation energy for the phosphorylation of glucose that traps the glucose molecules inside the cell. It also provides the activation energy for the rate-limiting step of phosphofructokinase. Thus, ATP hydrolysis provides the spark for the combustion process of the analogous biological pathways not histologically demarcated as it occurs partly in the cytoplasm (glycolysis pathway and partly in the mitochondria TCA cycle). The coupling of negative change in Gibbs free energy (downhill reaction with increased internal energy) to positive change in Gibbs free energy (uphill with decreased change in internal energy) reactions underpins non-oxidative coupling of ATP production in the cytoplasm that produces a positive net of 2 molecules of ATP. The coupling of negative to positive change in Gibbs free energy reactions to drive otherwise unfavorable reactions of glucose metabolism also involves the formation of ATP by oxidative coupling in the electron transport chain.

3.18 Classical and Quantum Information in Biology

Information is fundamentally rooted in probability theory and so is quantum mechanics at least as far as the Copenhagen interpretation is concerned where predictions are made with probabilities of experimental outcomes represented by wave function amplitudes. Quantum mechanics provides not only an increased combinatorial space in which a quantum system is defined, but it also offers a relative reduction of a time scale such that energetic expenditure involved in the inspection of the information space of probabilities occurs relatively fast compared to classical information searches. Stated another way, *time is maximally dilated and space is minimally bounded when quantum information processing is compared to its classical counterpart. Hence, if biological phenomena were to fully exploit the possibilities contained in quantum physics, they could occur virtually instantaneously with extraordinary synchronously coordinated structure-function relationships.*

Quantum systems, especially those capable of quantum coherence, relative to thermodynamic systems lose very little energy to heat and hence entropy production making them very appealing as blueprints for biological function.

Although rooted in classical physics thinking, an apt metaphor for biological information is that of a book, which consists of many strings of letters. Each letter in a sentence has a sound assigned to it. A group of letters put together forms a word, which has a meaning in terms of symbolic representation of an object, an action, etc. A string of words put together to form a sentence has a meaning in terms of some activity in the past, present, or future. A paragraph formed of several sentences describes a situation or a short story. A chapter containing many paragraphs tells a longer story and it is necessary to read the entire book to understand not only the entire story but also its moral, subtext, and context. *The book may have deeper layers of meaning, symbolic, allegoric, mystical, and historical. A similar situation occurs in biology where DNA encodes for proteins, proteins assemble in a cell to make a*

functioning unit of a tissue or an organ. A collection of properly organized and orchestrated organs and tissues make a living organism. Each part of the organism contains structural and functional information representing a piece of the entire "story" of the organism to which it belongs. Obviously, there is both specialization and redundancy making sure nothing is left to chance when it comes to the survival of a living system. With a hierarchical structure of biological information, one can see a certain level of self-similarity or fractality in the assembled structures as each part reflects the properties of the whole. This is best seen in branching structures such as trees, ferns, or the vasculature of the lungs, the liver, or kidneys. In each case, a small part of a biological structure structurally and functionally resembles the whole, obviously with an account taken for the quantitative differences due to the distinct spatio-temporal scales. *Therefore, there is a clear analogy between the information content of a book and that of a living system. However, the difference between the two is in the dynamic restructuring of the information content of a living system as opposed to the static information content of a book. Therefore, a biological system is like a living book.*

There is a deeper meaning or significance of knowledge in contrast to raw information.

Knowledge is more enduring than information because it carries meaning and can be passed on while information may be simply erased, even if it is copied. Knowledge is the ability to translate symbolic information into actionable items or to use it in order to extract meaningful information from seemingly unrelated patterns.

For example, *medical knowledge is required in order to diagnose a disease from vague symptoms or recognize warning signals of an impending stroke, e.g. an aura or an atypical headache or some other symptom that a person unskilled in the art of medicine would not be able to recognize as a medical emergency. This recognition based on knowledge can then lead to action, i.e. a procedure aimed at stroke prevention.* The acquisition of such knowledge is vitalizing, empowering, and sometimes comforting. However, on an important level, it exists only as a potentiality. The critical step, therefore, is to apply additional work to this potential by creative thinking and innovative or at least original thought. *Sufficient knowledge allows its reconfiguration into ideas that shape the constructive evolution of the interface between human beings and their environment.* Furthermore, it makes possible the quality and longevity of human life.

To understand any kind of information (the genetic code, the content of a book, road signs, even traffic lights) *one needs to have "translation machinery" and a "dictionary" or "a codebook". A dictionary or a codebook provides an information transfer algorithm,* which by itself is useless unless utilized by somebody who can provide an input and generate an output. The DNA's genetic code was there for billions of years but only in the last 50 years or so have we been able to translate the code of nucleic acid sequences into amino acids. To actually understand it is one thing (an algorithm) but to extract this information one needs a DNA sequencer, which is quite another (technical) problem and it took quite a bit of time to solve it. Once all of this is in place, the actual decoding and understanding of the messages require an energy input. The same can be said about the human brain and language translation. *In childhood, we develop an understanding of the world in terms of our native language. Understanding a foreign language takes many hours of learning and the development of an internal dictionary or a decoding algorithm. The rules or codes (the internal dictionary or decoding algorithm) have to be learned first, requiring time and energy, in order to be applied later.* Prior to putting those coding instructions to full use requires practice or learning, which takes time and energy that in many cases decreases as the user acquires the knowledge and is able to utilize it almost automatically with little effort. To carry a conversation in a foreign language having mastered it requires the use of this "internal dictionary" that was built over the years of learning and practice, and activation of inputs and outputs, both of which are fueled by metabolic energy. The translational machinery of the brain contains the neuronal structures and their communication networks within the cortical area of the brain that enables putting the rules of cognition into work and producing the useful value of communication. A similar process applies at the molecular level where the "translational machinery" of the cell performs the coding and decoding of the genetic information in the sequences of the nuclear and mitochondrial DNA. The molecular transcription and translation machinery involves a multitude of transcription factors, polymerases, and various types of RNA, as well as an energy input in the form of ATP necessary to carry out the translation program faithfully at every step of the process. *In each case, at every level of these processes, an algorithm ("dictionary" or "code book") is put in place whether we know it or not and it performs what can be termed "cracking the code".* If such an algorithm or dictionary is not provided to the reader, understanding the code and acting on the instructions contained in it is essential and can be acquired by learning and practice. In either case, some form of work must be expended to either find or simply learn the code or its rules in order to subsequently use it to do something useful with it. Nevertheless, some form of work must be expended to either find or simply learn the code or its rules in order to subsequently use it to do something useful with it.

What types of information or messages are encoded and decoded in the human body or other biological systems?

Conventional message transmission and signaling processes in biological organisms include both chemical and electrical signals propagating between synapses and gap junctions, as well as autocrine, paracrine, and endocrine signaling. In addition, there are various means of communication in the human body, which include electrical signaling by the central nervous system, chemical signaling contained in blood plasma and propagating through the vasculature with hormones, and cytokine chemical signal transmission.

Over shorter distances, one can identify specific inter- and intra-cellular communication that can take the form of a

structurally and mechanically connected matrix linking the cytoskeleton, cell nuclei, and the extracellular matrix. Part of this subcellular scale communication employs mechanotransduction, which still offers much faster message transmission speeds than traditional chemical signaling. The subcellular structures involved in mechanotransduction include microtubules, actin, and collagen, and they form an extensive mechanical tensegrity (tensional integrity) matrix connecting all parts of the body. *Moreover, it has been hypothesized that these protein filaments can also be involved in an even faster electronic, protonic, and ionic signaling.* Bioelectrical signaling has been well recognized for a long time to be instrumental in limb regeneration, healing processes from injuries, neurite growth, and ordering structures within the organism. The main theories for an organism-level information network based upon bio-nanowires originate from the tensegrity work of Donald Ingber [44], James Oschman [45], and Mae-Wan Ho [46], who proposed the existence and functioning of sub-molecular (mechanical, electronic, protonic, and ionic) communication systems in the body. The semiconduction aspect of proteins and protein complexes was pioneered by the work of Albert Szent-Györgyi [47] who envisaged that while proteins are not good conductors of electricity, they can be made weakly conducting by promoting some of their valance electrons into the conduction band by photoexcitation or other external means. It is quite feasible that *an entire organism can be both mechanically and electrically linked in an interconnected electromechanical matrix that allows for long-range mechanical and electrical signal transduction. In this connection, it is important to also appreciate the role of the extracellular matrix, which defines the space between cells and forms connections to the cell's interior through integrins. Integrins are membrane proteins connecting to the intracellular cytoskeleton via receptor-based signaling processes. The cytoskeleton, in turn, connects between the cytoplasm and the cell's nucleus forming a continuous network of mechanical cables that may also serve as electrical wires.* Mechanical stress-wave propagation and electrical ionic wave flows, through these subcellular elements, can create signals that travel many times faster than chemical signals that typically are limited in their speed by diffusion or translocation. These effects represent very slow means of signal transmission. This connective matrix suggests the possibility of extremely high-speed communication systems connecting the various distant parts of the entire organism [48]. The range of possibilities does not end here.

Specifically, non-chemical information transfer systems may involve much more intricate and exotic physical mechanisms including those firmly based on quantum mechanics.

Signaling and messaging modes can be supported by electrons, ions, protons, excitons, mechanical vibrations, solitons, phonons, and photons.

These modes of signal transmission while rather exotic in the field of biology, are not unusual at all in the field of condensed matter physics. Discoveries of such mechanisms in biological systems have largely been slow due to the technological challenges of probing nanoscale events in biological systems. However, this has been changing in recent years due to the use of modern experimental tools such as fluorescence imaging, confocal microscopy, scanning tunneling microscopy, transmission electron microscopy, and atomic force microscopy. *While organisms conduct electricity, as evidenced by medical equipment such as electrocardiograms, electroencephalograms, and transcutaneous electrical nerve stimulation, questions still remain as to the nature of the charge carriers and the functions of these electrical signals in biology at various scales.* Action potentials that allow for neural firing and the control of the nervous system are well known and characterized. *A closer examination of the body reveals the exquisite role of membrane potentials, ionic currents, and endogenous electromagnetic fields in cellular proliferation, morphogenesis, and regeneration. Also emerging in the investigation is the potential crucial role of water in the proliferation of biological signals due to the structure imparted by hydrophilic surfaces.*

The analogy between information (which has a transient nature) and heat (which is energy in transit), is intriguing by virtue of their transitory character although on a different time scale. Important biological information such as the genetic information encoded in our DNA is largely protected and has long-term endurance. Other forms of biological information such as phosphorylation events at a the cell level can be very transient as they regulate events within a cell cycle. Furthermore, both concepts can be described using the language of probability and uncertainty. Information is rooted in uncertainty almost by its definition because it implies knowledge of something that has an element of probability and not certainty.

> *The lower the probability of an event or outcome that we have information about, the greater the information content.*

This is well known to those who operate hedge funds and invest billions of dollars in the New York Stock Exchange. Having detailed knowledge about a company whose stock is being traded could make a difference between rags and riches, especially if this information is not accessible to all. *Information that is public is rapidly discounted and has no value.* However, the uncertainty in regard to heat is more indirect. *Heat arises from the conversion of free energy (from a downhill exothermic reaction, including hydrolysis of ATP) and its integration with biological chemistry via coupled biochemical reactions that result in structure formation. This represents informational energy. Entropy is useless since it is produced by all physical systems, all the time without direction, energy expenditure, or purpose. To reduce entropy one needs to perform work against the natural tendency of the world toward disorder, to overcome the homogenization of all matter.* Another way of entropy reduction is to create its opposite, namely information. However, there is also *an energetic cost to information because it has to be stored and/or encoded somehow to be preserved, be it in a DNA sequence or an amino acid sequence, or a particular set of neurotransmitter electric signals in the brain.*

In information science, there is a formal cost analysis in terms of energy needed to erase a bit of information.

Information storage, other than the initial investment in encoding it (as printed matter or an electronic signal) may be stored at no additional cost. Of course, eventually, it will dissipate due to material deterioration of the information carrier, e.g. a magnetic tape that will disintegrate after a number of years. *A common experience of working with a computer is that the computer gets hotter the more it is used. This is consistent with the principle argued by physicist Ralph Landauer that in order to erase one bit of information we need to increase the entropy of the environment in a parallel fashion (i.e. dissipate heat into the environment).*

> *We can conclude from the second law of thermodynamics that erasing information must result in the generation of heat.*

This is described by the so-called Landauer formula [49]. One bit of information erasure must cost no less than 1 kT of energy input where k is the Boltzmann constant where T is the ambient temperature in Kelvin. To boot, Charles Bennett in the late 1970s showed that *if one could erase information (or create information by erasing a bit in some location) without generating heat, then one could construct a perpetual motion machine, which is against the law of conservation of energy, a fundamental physical principle* [50]. Consequently, to change a bit of information into another, there is always a minimum amount of energy input required. In other words, to flip a bit one needs to perform some work. Even a coin toss or a roll of dice requires mechanical energy. The above discussion regarding the energetic cost of computing has major repercussions on both the limitations of the human brain in terms of information processing power and predictions such as Moore's law in computer science.

> *The density of electronic elements in processors doubles every two years on average.*

This empirical rule has been followed for almost 50 years now. This means that there must be a limit to this exponential increase in computing power density since the heat generated by bit flips would eventually result in the melting of the electronic circuitry that is inevitably associated with information erasure. Correspondingly *there is an inherent limit on the processing power of the human brain since at some point processing huge amounts of information would literally lead to the brain being "on fire" due to the heat generated by its neurons.*

The notion of information seems to tie in with the concept of quantum superposition. Quantum information theory is very different from classical information, if only because of the qubit versus ordinary bit distinction. This distinction underscores a greater capacity for qubit versus bit informational signaling. That is, a qubit provides a greater capacity for "cracking the code" or finding a biologically most efficient solution as a result of the divergent wave function (spreading) with increased combinatorial space allowing the system to explore many different states simultaneously. Similarly, however, the other form of superposition, a convergent coherent state allows for simultaneous information propagation across the coherent domain. Hence, quantum search algorithms provide enormous speed-ups and time savings while quantum coherence provides the ability to coordinate and synchronize behavior over spatial domains, both of which are advantageous biologically. These concepts were briefly discussed earlier in this chapter in connection with quantum coherence, superposition of wave functions, and wave function collapse. It is important to emphasize that wave functions left to themselves spread out over time, in effect losing their information content because asymptotically there is nothing left to say about where a given particle is located if its wave function has the same probability everywhere. On the other hand, *there can be events akin to signal amplification, which reinforce the information content of a wave function by resulting in its convergence. The ultimate convergence of a wave function is its collapse in the process of taking a measurement and forcing the quantum system to "declare" its state.*

The structural biological information can be viewed as biochemistry that interacts with metabolic energy. The dynamical rearrangement of that structural information is functional information. For example, neural synapses from certain parts of the brain perform specific functions (e.g. visual information processing) and provide this information to other parts of the brain (e.g. topographical orientation area) and to the rest of the body acting as information inputs. The information content is simply a state of being "on" (firing) or "off" (not firing) by the neuron. The messenger is an ion wave causing an electrical potential change, the message is a conductive state switch for the neuron (excitatory or inhibitory) and the output is some type of perception but it requires a collective transformation at the level of neuronal networks. Hence *individual activity changes of single neurons can be viewed as being equivalent to letters in the alphabet. Sentences and chapters are created by firing patterns across millions of neurons forming various domains of the brain.* This description applies to the receiving neurons in the brain or peripheral innervated tissues such as skeletal muscle of the voluntary nervous system input or visceral tissues in the autonomic system input. It is easy to see the potential for an enormous amount of information being generated by noting that even if only a two-state approximation for the neuron is considered, since there are 100 billion neurons in the brain, the total number of possibilities is $2^{100,000,000,000}$ which is an astronomical number. As is elaborated on elsewhere in this chapter, the actual number of possible states of the brain is much, much greater if one considers the fact that each neuron is not simply an on/off switch but a complex computational device in its own right. This combinatorial complexity of brain states literally "boggles the mind"! Even more amazingly, *this complex network may also interact directly and indirectly not only with chemical inputs (neurotransmitters) but also with electromagnetic waves.*

DNA, hormones, and kinases/phosphatases are sources of structural information that act as functional messengers. Some of these messengers operate within a cell (e.g. DNA) and some others across the cell membrane by communicating with the surroundings (e.g. extracellular hormones). However, intracellular hormones may interact with transcription factors that bind to DNA response elements. In this case, the extracellular hormone (e.g. cortisol) is the message, the DNA the messenger, and effects such as vasoconstriction or bronchodilation can be viewed as the reader (instruction output). The sequence of the integration in the information processing from the message to instructional output may be recognized in the context of the stress response and the psycho-neuro-endocrine-immune systems (the "super systems"). Notably, the majority of the immune system is located in the wall of the gut (gastrointestinal-associated lymphoid tissue or GALT). Another type of messaging system involves external agents that may penetrate into the human body and affect its functioning negatively. As a case in point, the gastrointestinal tract is the primary portal of exposure to pathogenic microbes, toxins, and allergens that mediate disease but they may also be necessary for the healthy functioning of the digestive system. *The microbes, when pathogenic (i.e. pathogens) induce pathogenesis (disease), while a balanced microbiota that includes bacteria, archaea, protists, fungi, and viruses is increasingly seen as crucial for immunologic, hormonal, and metabolic homeostasis of their host, which highlights the significance of the microbiome. Interestingly, the microbiome contains tenfold more bacteria than we have somatic cells in our body. The surface area of the distal small bowel, a geometric fractal, is the size of a double tennis court. To digress, the human body only has about 25,000 genes, less than the number of genes a worm contains and about ten-fold fewer genes than a kernel of rice.*

> *What makes us humans so complex and sophisticated is that we are a composite of our own somatic genes and those of our microbiome in a complex interactive fashion. Therefore, information transfer between the microbiome and the host organism is a central axis of the host's health and disease dynamics.*

At the level of an individual cell, information processing is mediated by cell membranes, which allow a signal to be sent into the cell through membrane receptors acting as messengers providing instruction for a particular cellular transformation representing output manifestations. For example, insulin, the message, binds to cell surface insulin receptors promoting autophosphorylation of the insulin receptor. Hence, the insulin receptor exemplifies structural information that upon insulin binding results in dynamical rearrangement of that structural information, which represents functional information when acted upon by the message. This simple example shows how information processing translates extracellular messages into the cell through cell membranes. However, intracellular steroid hormones bind to their receptors present within the cells. The bound receptor may interact with transcription factors that in turn bind to DNA response elements. In this case, the hormone, let's say cortisol, is the message, or the cortisol receptor contains structural information. Once bound to cortisol the information content of the cortisol receptor changes resulting in dynamical rearrangement of this structural information, and as a consequence, functional information that is responsible for ultimately transfer from the cytoplasm into the nucleus and subsequent binding to the cortisol response element of the DNA. Consequently, transcription of anti-inflammatory cytokines within immune cells may be suppressed while pro-inflammatory cytokine transcription is activated. Such transcriptional output may be considered instructional output from the message, or cortisol mediated by the functional information.

So, in one sense, information may be rooted also in quantum mechanics by the statistical relationship to macroscopic collective modes of either elementary particles or collective quasi-particles. *However, there are two distinct manifestations of superposition in quantum mechanics: a) tunneling with multiple possible states at the same time and b) coherent collective modes with the macroscopic manifestations of their behavior. These are two layers of superposition. A single particle may be in a superposition state but also a collection of particles may be in a product superposition state, which is a superposition state of individual superpositions.* The latter is referred to in quantum physics as forming a so-called Fock space, which is an abstract space of multi-particle superposition states. A coherent state is a special quantum superposition state characterized by a minimum allowed uncertainty as is exemplified by monochromatic and phase-coherent laser light. White light is a random superposition state of frequencies and phases of the contributing photons or quanta of electromagnetic radiation in the visible range. Laser light carries more information than white light because it has been selected for frequency and phase coherence. This information content required an energy input in the form of pumping and optical material preparation.

We stated above that Landauer's formula puts an energy "price" on information in classical thermodynamic systems. In quantum physics, however, as stated by Vlatko Vedral [51], *we can erase information and cool the environment simultaneously, which means generating negative entropy. This is conceptually different from the negative entropy created by the dissipation of energy following the second law in classical thermodynamics and is not equivalent to the generation of perpetual motion machines.* Erasing the state does not generate entropy because the ground state of a quantum system has zero entropy introducing no disorder into the system and because *quantum physics allows erasing information while actually taking away heat from the environment rather than adding heat. This can all be done simultaneously by manipulating the processor and computer memory creating entanglement between the two that allows erasing computations making room for new ones.* Adding negative entropy (creating order within the system) is the same as taking entropy away because of the phenomenon of entanglement.

The connections between entanglement and negative entropy are rooted in Schrödinger's view of quantum mechanics. If we erase subsystems individually we contribute to entropy generation but if we erase the state as a whole we need not increase entropy. *The difference between global and local entropy is negative entropy. It is crucial to know that the information we erase arises from the entanglement with another system. By invoking the other system, we can actually erase heat added to the environment. The eraser therefore acquires a new dimension to account for the rise in entropy while remaining compliant with the second law of thermodynamics.* Nonetheless, the entropy of the whole universe cannot be decreased even when accounting for the "magical" phenomenon of quantum

entanglement. Hence, perpetual motion machines may be subject to identifying and manipulating a new dimension (or a parallel universe), which would be virtually impossible even for the most innovative minds. This may be necessary to account for the phenomenon of quantum entanglement allowing the prevention of a machine's motor from getting hot and hence releasing heat into the environment. However, erasing subsystems individually will contribute to entropy generation. The difference between global and local erasing is negative entropy. If systems are entangled, by invoking the other system in erasure, the environment can lose entropy. Implications of these speculative ideas for biological organisms treated as quantum entangled may be that we will be able to manipulate entanglement between the microprocessor and computer memory and erase computations to make room for new ones but keeping the environment cool.

> *Biological function should, therefore, be able to benefit from the exploitation of the concepts of quantum mechanics and quantum chemistry being in accord with the laws of thermodynamics and electromagnetism.*

We have seen how quantum entanglement may be used to overcome the energetic cost of information processing. When two well-defined quantum systems are entangled we have a complete understanding of the entirety of their composite state, which conceptually should be extended to a many-body system such as that defining the entirety of an organism. Conversely, understanding the entirety of an organism or state that involves the entanglement of multiple subsystems does not necessarily ensure that we know anything about the individual states or systems, i.e. parts. *Entanglement is an integrative principle that allows us to see the whole but not isolate one product from another so that when looking at the living system coherently involving the sum of its cells into a functioning organism, its constituents are not singled out.* Quantum entanglement very much appears to be the missing link between the system and its constituents. *This is especially relevant to the idea of quantum consciousness and psychological phenomena that resonate with the notion of telepathy and other related concepts such as the zeitgeist. For example, the Jungian idea of the collective unconscious gains traction when combined with the quantum entanglement so that each individual, without realizing it, is embedded into the overall societal set of norms and social constructs of the time that form the global wave function for the society as a whole. This also relates to the collective dynamics of quantum biology assimilating the traditional quantum physical microscopic scale to the macroscopic level and behavior such that a single cell becomes subsumed in tissue by the collective, synchronized behavior of the cells constituting that tissue.*

An example of the collective unconscious can be the failed predictions by pollsters in the run-up to the 2016 US presidential election, which only considered factors within established paradigms of empirical political research. However, a larger question is whether statistical models used in opinion polls employ the wrong lenses through which to view mass psychology. One answer, likely echoing sentiments shared by most established experts, is that the standard concepts of probability remain relevant but simply need tweaking. Such a perspective would seek to consider hitherto ignored variables by incorporating more accurate data, such as that mined from social media. However, sticking to this kind of program may lead to a repetition of blunders made by Hillary Clinton's consultants since there will always be unanticipated "hidden" variables missed by any future efforts. A second answer is that no empirical statistics can or will ever capture the shifting moods of crowds just like a measurement process of a quantum system always affects the state of the system. The raw feel of political passion did indeed unexpectedly turn out to resonate throughout the Trump campaign. Those who fear demagogues might learn a lesson from Trump's dismayingly triumphant intuitions. Hence, anti-Trump analysts seeking to deepen their understanding of Trump's worst supporters might do well to consult the qualitative insights of Sigmund Freud, Erich Fromm, and Wilhelm Reich regarding the collective psychodynamics of fascism. A third answer, perhaps more balanced than the first two, is that the wrong kinds of statistics, based on implicit analogies to 19th-century physics, have been misapplied to human cognition and behavior, but that newer statistical paradigms may offer improved predictive power. A disparately received theoretical paper [52] published recently by Emmanuel Pothos and Jerome Busemeyer, cogently argued that quantum-probabilistic formalisms, abstracted from their physical foundations, might generate more succinct, robust, and accurate psychological predictions than do standard statistical approaches undergirding traditional empirical assays. These assertions are part of a larger perspective, called "quantum interaction", whose advocates have developed quantum-formal models of not only individual but also group psychology. "Quantum interaction" embraces both empirical strengths sought by Democratic pollsters and, through its unorthodox logical algebra, a possible ability to grapple with the id-like "irrationality" of politically impassioned movements.

The idea of collective unconscious forming an invisible information field can be taken to the extreme by hypothesizing that the whole universe is at the most fundamental level informational in nature.

> *All things physical are information-theoretic in origin and this is a participatory universe.*

Observer participation gives rise to information. The great American physicist John Archibald Wheeler coined an apt phrase: "it from bit" to describe this concept.

3.19 Aging and Senescence

All isolated or closed systems evolve over time towards greater entropy. Although the human body is neither entirely closed nor isolated, its parts and compartments are partially separated from the external environment selectively allowing the flow of matter and energy into and out of their interiors.

> *The process of senescence progresses towards greater universal entropy and reduced local organization and complexity of structure and function, which translates into reduced negative entropy. Negative entropy represents information hence senescence results in biological information loss over time due to both natural and pathological degradation processes.*

However, it takes energy to break bonds. Heat provides additional (thermal) energy that facilitates bond breaking. The formation of new bonds is then a downhill process once the energy is available for breaking the old bonds of the reactants and allowing them to make new, often accidental associations. In the case of dietary fuel, oxygen, concentration gradients of products and reactants, vitamins, minerals and phytonutrients, enzymes at physiological body temperature, etc. are all required for biological purposes. The extraction of energy as heat and ultimate translation of information into work requires all the correctly balanced components of information to be present. The proper formation of structures that can perform desired biological functions contains an enormous amount of information. On the other hand, breaking bonds is an incineration process due to excess heat and it results in biological information degradation and eventual loss of structure and function. This is an outcome of chronic inflammation and oxidative stress, which disrupts the integrity of information systems leading to entropy increase. Accordingly, energy that flows as the transit of heat across these systems and is ultimately transferred as work, underpins the complexity of structure and function, which in parallel with health and survival becomes increasingly impaired when excessively accumulated and transformed into bond breaking.

> *The decoupling and separation of energy from information parallels and epitomizes the disease and senescence that characterize the progression from life to death.*

That is, the metaphoric marriage between energy and information continues until "death does them part". This is a complex relationship (like all marriages) where too much heat energy can lead to the degradation of structure and loss of function, i.e. information. However, energy is needed to sustain and repair structure, which would otherwise be degraded by the natural process involved in a constant increase of entropy. Hence, the proper flow of information and energy is the key. One might say that the secret to any properly functioning relationship is the exchange of energy and information. This applies to living cells as much as to human beings and even societies.

The work of transforming information into simple and complex conscious thoughts, the former shared by many animals including mammals and birds, the latter perhaps unique to humans and certain mammals, is no doubt mediated by proteins and serves as an equivalent function for survival advantage as the musculoskeletal work of the fight or flight (stress) response. In fact, the highest energy requirements are overwhelmingly represented in the brain and muscle that work in concert, orchestrated by central command hair-trigger cognitive (as well as autonomic nervous system and neuroendocrine) responses coupled to skeletal (cardiac and smooth) muscle contraction.

The comparison of information processed by our brain and body in the form of work for absolute survival and quality of life can be extended by the analogy of molecular to machine motors serving to produce motion or power for performing work.

Complex thoughts promote innovations such as the electrical hydraulic motor that facilitates the lifting of heavy materials against gravity in the construction of buildings. In the modern human era, this contributes to survival under harsh weather conditions as every year unfortunate homeless individuals die of heatstroke or frostbite. Certainly, the construction of homes and buildings for a host of reasons contributes to the comfort and quality of life. In contrast, earlier in our pre-civilization evolutionary development, we relied on manual resources for shelter, fending off predators, and capturing prey (e.g. Neanderthal or Cro-Magnon stages, as hunter-gathers).

Nonetheless, although our ancestors were not the strongest or fastest in the jungle, they were the smartest.

This allowed them to innovate tools and other resources for survival. This more solely relied on the biological, or molecular, motors coded for in the DNA. Myosin is the prototype molecular motor, a protein that converts chemical energy in the form of ATP into mechanical energy that generates the force and movement of muscle contraction. We refer the reader to a chapter on biological motors where the intricacies of these subcellular-scale protein-based machines are elaborated on.

Information may be that which is encoded in our DNA or may be any variety of building block materials needed for any physical construction project. It may be the fuel source itself that supplies energy, whether in the food we eat or the petroleum we put in our cars.

The Nobel Prize–winning theoretical physicist Richard Feynman wrote in 1963 that *"we have no knowledge of what energy is"* [53]. The idea that there are different forms of energy, such as thermal energy, mechanical energy, and chemical energy, is misleading, as it implies that the nature of the energy in each of these manifestations is distinct when in fact *they all are, ultimately, at the atomic scale, some mixture of kinetic energy, stored potential energy, and radiation (electromagnetic energy).*

What we do know about energy, however, is what it does, that is, the capacity to take information in an abstract sense, in all its forms, and transform it to the diverse physicality and quality manifestations of living beings, their conscious and reflex interpretation of the world necessary for survival and their utilization of physical resources to construct projects. Energy interfaces with information capable of constructing something of value and which is useful. Therefore, a usable understanding of energy is critical.

3.19.1 Machine versus Biological Engine Analogy

Relevant is the notion that sources of energy, or fuel, such as food that we ingest, or petroleum that fuels our cars, contain information with which initial energy is required to extract the potential energy (that is the "potential" for conversion to other forms of energy, especially kinetic energy that is usable). *It is important to recognize the common misconception that energy is simply contained or stored in fuel or in a given amount of food. If petroleum fuel for motor vehicles, or the macronutrient fuel in our diet, are combusted with very limited available oxygen, the combustion process is incomplete. Accordingly, there is less usable energy extracted from the respective fuels, and in parallel, less CO_2 and H_2O byproducts are produced.*

Gasoline and other fuels made from petroleum liquids are ignited and burned in the combustion chamber of an automobile in the presence of oxygenated air, producing the oxide products, CO_2 and H_2O. In this combustion reaction, the hydrocarbons of petroleum molecules are reconfigured to form these oxides. CO_2 is a very stable molecule; consequentially its formation requires no applied energy. Thus, the energy released from the breaking bonds of hydrocarbon molecules of petroleum fuel liberates heat in the process. This exemplifies an exothermic reaction, i.e. the energy in transit, or heat, released from the reactants (that require an energy source, such as ATP in a biological system, or a spark plug ignition in this case) is greater than the consumption of energy by the products formed in the reaction. *In a living system, the heat liberated by an exothermic reaction is often coupled simultaneously to another energy-requiring reaction. Biological evolution is typified by its pursuit to minimize waste.* Otherwise, heat is dissipated from the body, unable to contribute to useful structure and function. The concept of capturing the heat from an exothermic reaction for useful purposes is exploited in the chemical and mechanical engineering of internal combustion engines. The heat generated in the combustion chamber/cylinder causes the expansion of CO_2 and H_2O vapor, exerting pressure against the piston inside the cylinder, forcing its motion and consequently converting pressure into motion. This is the so-called power stroke. Its force is transmitted to the connecting rod, and consequently the crankshaft causing it to spin.

In general terms, any device that imparts motion via the transfer of energy from one form to another is a motor. The internal combustion engine exemplifies a machine motor, as do the electric motors discussed in Chapter 1. Biological motors are discussed in detail in Chapter 2. These molecular machines also exhibit force generation via power strokes.

Notably, petroleum is a fossil fuel derived from the decomposition of organic matter, that is, it is derived from dead plants, animals, and other organisms. An analogy can thus be made between the combustion of the hydrocarbon molecules of petroleum in the machine engine described above with that of the hydrocarbon molecules of macronutrients, such as sugars and fats in the diet. In both cases, the conversion to CO_2 and H_2O is accompanied by a transformation to usable energy.

In a living system, different metabolic pathways are recruited depending on the relative presence or absence of oxygen. In its presence, mitochondrial pathways coupled to the process of oxidative phosphorylation transform much more of the energy from the bonds of the hydrocarbon macronutrient molecules to the biological currency of ATP. Conversely, the anaerobic (non-oxidative) glycolytic pathway is far less capable of extracting energy from the diet.

Taken together, in both a machine engine and its metaphorical biological engine counterpart, *the presence of oxygen promotes a more complete breakdown of hydrocarbon molecules utilized as fuel, into molecules of CO_2 and H_2O, as well as a greater transformation of energy to serve a useful purpose.* The efficiency of fuel burned by an internal combustion engine correlates with the amount of oxygen-containing air available. Excess fuel relative to the usable ratio of petroleum fuel, such as gasoline, to oxygen is emitted through the exhaust pipes unburned. It follows that petroleum fuel combusted with very limited oxygen, liberates very little CO_2, H_2O and heat. Accordingly, only minimal pressure is mounted by the expanding gas, as is the conversion of pressure to the power stroke and its induced motion of a vehicle.

Analogously, the strength of performance of a biological engine, and thus health and disease, may be equated to metabolic efficiency, defined by the ratio of oxygen consumed to ATP produced.

> *Metabolic diseases of aging, such as cancer, heart and vascular disease, Alzheimer's disease, and other etiologies of accelerated cognitive decline, correlate with the severity of mitochondrial dysfunction.*

As oxidative bioenergetic pathways of mitochondria decline, oxygen delivered to cells is increasingly unable to be utilized for useful purposes of ATP production. Rather, molecular oxygen combines with electrons that are leaked from the electron transport system when the functional capacity of mitochondria cannot accommodate energy demands. *The resulting generation of reactive oxygen species and free radicals further degrade the molecular integrity and function of mitochondria in self-amplifying loops of pathogenicity.* Increasingly, glucose metabolism via the glycolysis pathway is uncoupled to mitochondrial oxidation. Hence, the breakdown of glucose to the byproducts of CO_2 and H_2O is incomplete, and the amount of ATP produced is only a fraction of that in the context of complete nutrient combustion. Similar observations are true in the case of fatty acid (and protein) oxidation when mitochondria are dysfunctional. *Indeed, mitochondrial dysfunction, and associated redox and inflammatory stress inflammatory stress, and insulin resistance are foundational to the chronic diseases of human aging.* These topics are discussed in greater detail in Volume 2.

3.19.2 Non-Redox Mediated Causes of Dysfunctional Oxidative Metabolism

In addition to pathogenic structural modifications induced by redox stress, the loss of mitochondrial function may be the result of other disturbed elements of systems biology, that occur along the journey of the aging process. These may include inadequate oxygen supply, mitochondrial enzymes, or cofactors. Altered composition of intestinal microbiota, for instance, is often a critical perturbation resulting in dysfunctional mitochondria.

Consider that *Firmicutes* and other environmental gut-inhabiting microbes including microbes such as *Proteobacteria, Bacteroidetes* are a source of *Pantothenic acid, vitamin B5, coded for in the microbe DNA but not in human DNA. Pantothenic acid is necessary for the synthesis of Coenzyme A (CoA), which is critical for the complete combustion of glucose and fatty acids in mitochondria.* Coenzyme A is an important carrier of acyl molecules (carboxylic acid-containing hydrocarbons, the smallest of which is acetate) within the cell that allows oxidative metabolism to occur. Accordingly, the formation of acetyl CoA groups within mitochondria makes possible the coupling of glycolytic glucose metabolism and fatty acid b oxidation to maximum combustion and ATP production.

The end product of glycolysis, pyruvate, is transported into the mitochondria where the decarboxylase action of pyruvate dehydrogenase complex enzyme (PDC) converts pyruvate to acetyl CoA. This step is crucial to its entrance and further oxidation via the tricarboxylic acid (TCA) cycle. Similarly, the process of β oxidation of fatty acids involves the initial activation of fatty acids, typically palmitate, to palmitate CoA in the mitochondrial matrix, where it is sequentially broken down into two carbon units of acetyl CoA prior to entering the TCA cycle where acetyl CoA combines with oxaloacetate (OAA) to form citrate, liberating CoA in the process. Within the TCA cycle CoA is also important for the transformation of α-ketoglutarate to succinyl CoA (see Figure 1.25 in chapter 1 of this volume for a schematic illustration of the TCA cycle).

The conversion and synthesis of fatty acids from carbohydrate also require CoA in the sense that the acetyl CoA is derived from the glycolysis of glucose. Acetyl CoA, derived from the glycolytic breakdown of glucose, similar to that produced from the b oxidation of fatty acids described above, condenses with OAA to form citrate within the TCA cycle. For fatty acid (FA) synthesis to occur, citrate is transported out of the mitochondria into the cytosol, where it is cleaved back to the products of OAA and acetyl CoA. While OAA returns to the mitochondria, acetyl CoA in the cytoplasm is carboxylated by acetyl CoA carboxylase (ACC) to form malonyl CoA, a rate-regulating step of fatty acid synthesis. FA synthesis follows the reverse pattern of FA oxidation, albeit in the cytosol and utilizing different enzymes. This *de novo* lipogenesis occurs in the liver and adipose tissue.

Taken together, vitamin B5 is a vital cofactor for many steps in metabolic processes of oxidative metabolism and fatty acid synthesis. The microbiota is a crucial source of this cofactor. As described in Volume 2 of this book series, *factors that impair the healthy and diverse compositional pattern of microbiota include chronic stress response, poor diet, and disturbances in circadian behaviors. Further, these control parameters of metabolic disease are all inextricably linked, with self-amplifying feed-forward loops. Thus, not only does each of these 4 broad parameters independently promote dysfunctional mitochondria and metabolic disease, but they potentiate one another's pathogenicity.* In the case of intestinal dysbiosis, inadequate availability of vitamins, such as pantothenic acid, may diminish healthy levels of adiposity and hepatic lipogenesis. Accordingly, the synthesis of triglyceride containing very-low-density lipoprotein (VLDL) particles, may not be sufficient to meet the metabolic demands for delivery of fatty acids to tissues, such as skeletal and cardiac muscle during periods of fasting. Further, the lack of coenzymes (e.g. CoA) and cofactors (e.g. vitamin B5 to assist in the biosynthesis of CoA), may compromise the necessary ATP production for preserving the structural and functional integrity of muscle, leading to sarcopenia and cardiomyopathy. Moreover, *muscle weakness and easy fatigability with exercise, and even with normal activities of daily living, may ensue. Additionally, these factors may underpin unexplained weight loss.*

Motor protein functions require ATP hydrolysis and accordingly impaired motor protein function such as myosin leads to reduced strength of muscle contraction. Thus, in addition to reduced muscle mass, muscle weakness is also present. *All motion within the cell is mediated by molecular motors including the spatial arrangement of mitochondria transported along the actin cytoskeleton, a separation of DNA contained in chromosomes attached to microtubules by kinetochores in the process of mitosis, to name a few. Disturbances to these motions may perturb quantum metabolism and predispose the cells to cancerous states.*

Depleted levels of vitamin B5, including due to an altered microbiota composition, may be responsible for other pathologies linked to inadequate ATP production or fatty acid metabolism, are also worthy of mention. *The energy stored in ATP is required for active ion transport, such as is necessary for maintaining the electrical potential of cells. Lower cell electrical potentials have likewise been linked to cancer risk. Reduced fatty acid synthesis diminishes this component of the phospholipid structure of cell membranes.* Phospholipids have both stabilizing and dynamical roles that are vital for cell protection as well as for the transport of many essential molecules. *Importantly, the brain is comprised of about 60% fat as the predominant constituent of cell membranes and synapses of neurons. It contains roughly 25% of the body's total content of fatty acid. Consequently, inadequate dietary fat or fatty acid biosynthesis may lead to cognitive impairment.* Another important role for vitamin B5 regulated CoA production is in the biosynthetic pathway of cholesterol, also crucial for the integrity of cell membranes. Cholesterol forms lipid rafts which are required for mechanical rigidity and signal transduction through cell membrane receptors. Additionally, *CoA is important for the synthesis of the neurotransmitter acetylcholine. When this synthesis is impaired, the ratio of adrenergic to cholinergic activity is elevated to a level that results in manifestations of the stress response as well as consequent metabolic disturbances of insulin resistance and mitochondrial dysfunction.*

3.19.3 Energy Transfer and Transformation of Information: Defense against Biological Aging

It appears that *prolonged vitamin D deficiency fundamentally predisposes to an altered microbiome due to an essential role that vitamin D serves in regulating the innate immune system. Conversely, insufficient nutrient composition in the diet, which although influences the gut microbiome, independently represents a punitive epigenetic modulator of the host gene and microbiome expression by lacking necessary enzyme cofactors.*

Such micronutrient cofactors are also necessary for bioenergetic pathways. Hence, the input of energy accompanying the specific assortment of reactants determines the capacity of energy extraction from the macronutrient dietary fuel source. The extraction of energy from reactants is a consequence of all of the nature of the reactants and products in the systems rather than simply one of the reactants. This reality reflects the complexities of biological systems that exceed applications of standard principles taught in physics or chemistry. The flow of energy, in any event, once extracted from the information of one form, is carried in the new form of information, that being the kinetic and potential energy of the bonds and forces between atoms and molecules of organic as well as inorganic subsystems. *The energy flows with the information contained in the machinery of the cell and determines the amount of complexity of information associated with the system. Therefore, understanding information as negative entropy we can consider molecules as information-directed energy fields, consistent with Linus Pauling's model of structure-function relationship* [54] *in the form of the lock-and-key principle that represents shape-contained information of receptors and ligands. In other words, the lock-and-key principle is not only a reflection of the energy minimization aspect of physical systems but also can be viewed as a form of information encoding through pattern recognition due to shape matching.*

However, the transformation of information contained in reactants to usable and greater information and complexity of products requires an associated and compatible expenditure of energy in the form of work. In this sense, converting the transfer of heat to internal energy (i.e. energy in transit from one system to another) can, in turn, create something more enduring, namely a properly functioning structure. This is a product of the transfer of energy into work that produces structures and functions of greater complexity (e.g. cilia and flagella that propel single-cell organisms in their search for food).

Endothermic reactions in biological systems represent anabolism whereby the balance of heat absorbed by breaking bonds and released by forming bonds, favors the former such that more heat is absorbed than released. When the molecular bonds of reactants require more energy to break than the energy released from the formation of the products the reactant bonds are more stable and stronger than the bonds of products. Accordingly, it would take more energy to break the bonds of the molecules formed (the products of the reaction) than those of the substrate molecules of the reaction.

Two strategies allow endothermic reactions to occur driving bioenergetic pathways and anabolism. The first is manifested by the Gibbs free energy that couples exothermic (the opposite of endothermic) reactions to endothermic reactions. The second and most fundamental mode of energy exchange in biological systems is the hydrolysis of ATP harnessed to drive processes such as muscle contraction. The coupling of endothermic reactions to exothermic reactions is based on the principle that a reduction in the Gibbs free energy is spontaneous whereas an increase of Gibbs free energy is non-spontaneous, the latter referring to uphill endothermic processes (e.g. ion channel-mediated active transport to create ion gradients across membranes). *Accordingly, non-spontaneous uphill and anabolic reactions (endothermic reactions with a positive Gibbs free energy change) are driven by utilizing the energy or heat in transit from exothermic reactions with a negative Gibbs free energy change occurring in the vicinity.*

Performing work is not only good for you, but necessary in order to overcome the natural process of entropy production through reorganization. The author of this book believes that there is a relationship between longevity and quantum metabolism. *Many people who live to be 100+ years old are both physically and mentally active. This can be viewed as representing adaptive hormesis levels in given individuals with an optimal degree of stress response activation. This is in support of the hypothesis that the extent of staying in quantum metabolic states is correlated with health since quantum metabolic regime corresponds to no excess entropy production. The person who utilizes quantum metabolism eliminates excess entropy by the work of transforming energy and information into useful cognitive and physical processes.* Since the free energy is defined as $F = U - T S$ (see Chapter 1), lowering the free energy (natural tendency) can be achieved either by lowering internal energy (expended work) or by an increase of entropy (heat production). *We must always look at both sides of the equation: input and output. Input is in the form of nutrition and information, while output occurs as heat and work.* One can lower entropy by increasing information. This implies that *to overcome the damage done to us by the passage of time, one should never stop learning* because by learning (which may actually mean new experiences such as travel to Argentina or learning a new skill, picking up a painting class, etc.) we increase information in our brain and other organs of the human body (playing the guitar is not only a matter of reading the notes but using nerves and muscles). Being physically active helps remove excess stores of nutritional energy that can otherwise be transformed into destructive heat dissipation or structural degradation, both of which create entropy.

3.20 Can Special Relativity Be of Relevance to Biology?

One of the greatest unexplained conceptual conundrums in human physiology is that the communication time from one end of the body to the other demonstrated by power reflexes is virtually instantaneous and far exceeds the maximal time explainable by neuronal transmission.

Typical speeds involved in neuronal transmission are on the order of m/s, so signals connecting distant parts of the body would take more than a second to travel assuming perfect efficiency and no delays on signal amplification and relay.

However, many examples, especially in terms of professional athleticism, defy this relatively slow

time scale for signal transmission, reception, analysis, and response that would at least double the previous estimate.

Top hockey players, tennis pros, and basketball players demonstrated reaction times as short as a millisecond when millimeter distances are involved in reacting to a fast-moving puck or returning a serve in excess of 160 miles an hour. At this speed, the ball traverses the tennis court in a fraction of a millisecond, which requires both the perception and reflexes involving cognitive processes.

This feat of athleticism, in order to be explainable, must invoke nonstandard mechanisms, possibly quantum mechanisms.

Numerous other examples of human performance defying explanation can be found in books such as the Guinness Book of World Records.

This includes also examples of fantastic skills at complex mathematical operations or premonitions as well as telepathy, which indicates that simplistic mechanical explanations using classical physics methods are woefully inadequate. A case in point is that of Srinivasa Ramanujan, an Indian self-taught mathematical genius, who independently compiled almost 4,000 mathematical results, many of which he obtained without proper derivations but simply by direct access to the solution with his mind. Nearly all of his mathematical claims have been proven correct.

He stated: "An equation for me has no meaning, unless it expresses a thought of God" [55]. Might it be that his brain operated at a quantum level much more so than that of an average person? A much more common, yet almost equally challenging phenomenon is that of an idiot savant, which is typically an autistic person with a special gift of being able to perform complex mathematical operations almost instantaneously and without error, defying the normal rules by which our brains are taught to solve problems algorithmically.

The intersection between physics and biology via mechanisms of quantum phenomena and synchronized physiology plausibly translates to the domains of athletic performance and success in general in life's pursuits (if you believe in yourself you can make it happen) in the context of the quantum nature of free will and consciousness.

On the other hand, it should be noted that the flow of energy and information is unconstrained spatially and temporally within the sphere of quantum phenomena and, as stated earlier, in cases of quantum entanglement may occur instantaneously and simultaneously between spatially-distant areas (via a so-called action at a distance). This, in a very intriguing way, also resembles the length contraction-time dilatation aspect of the theory of special relativity. If one stands still, this corresponds to passing through time at maximum speed.

Conversely, moving at the speed of light one is maximally dilating time. Stated another way, time is maximally slowed down for travel at the speed of light.

Although it is impossible for the human body to move at the speed of light, that may not be the case for the flow of energy through the body.

Indeed, energy or signals that propagate through the space of the human body in a manner that is unconstrained and temporally non-constricted (that is time-dilated) may be effectively moving at or close to the speed of light.

That is, time slows down relative to the rate of energy input. This parallels the idea that it takes energy as work to transfer that energy to information of biological systems that enhances connectedness, i.e. complexity, and hence it is information that ultimately has usefulness in terms of survival value.

This also highlights the biological connection to quantum phenomena and adds an extra dimension to the segregated compartmentalization of systems biology when applied to pathophysiology.

The loss of Gibbs free energy and the rise of entropy may further connect to the notion of decoherence of quantum metabolism and the reduced generation of ATP due to the inextricable processes of mitochondrial oxidative stress and inflammation. In fact, some of the top athletes have commented on their feats of unbelievable athleticism that when they are in the "zone", they experience a feeling as if time was standing still or slowed down. Therefore, it is not unreasonable to look for explanations of some of the exceptional human physiology and psychology phenomena in the realm of quantum mechanics and special relativity theory.

In the context of biological structures and function, spatial constraints that pose limitations on the speed of communication are relatively minimized and as such biological time can be assumed to move more slowly, perhaps in the subjective sense of the observer "in the moment", i.e. the person experiencing this particular effect.

In such cases, optimal biological complexity and maximal informational energy (the interface of energy with the body's chemistry) can be claimed to exist across the organization of systems biology at the organismic scale. More will be said about systems biology precepts in a later chapter.

When the complexity of interactions both within and between systems breaks down, the energy flow through these systems is reduced and slowed, underpinning the acceleration of both senescence and time. Quantum decoherence increasingly appears to be a plausible factor in the development of pathologies and aging.

In order for such relativistic quantum phenomena to occur, it requires the virtual absence of noise in the system.

In the case of a quantum computer, or hypothetical perpetual motion machine, the slightest amount of heat in the system results in the decoherence of the quantum wave function describing this system. In the case of biological systems, heat refers to free energy that cannot be captured as work or harnessed for useful purposes.

Such heat dissipates as entropy and is said to cause a collapse or decoherence of a quantum state returning the human physiology to its sub-optimal performance level.

These destructive processes are inextricably linked. Reactive oxygen and nitrogen species bind to DNA, protein, and lipid components of biological information (negative entropy) disrupting its structural and functional integrity, and in the case of DNA, potentially causing mutations. Oxidative stress elicits an inflammation response by up-regulating the expression of NF-κB, a central hub transcription factor whose role is to amplify open processes of inflammation and reactive oxygen species. Hence, the "regulation" is actually a feed-forward (that is, positive feedback) dysregulation.

Mitochondrial structure and function decline as a hallmark of the reduction of metabolic stability and loss of quantum metabolism as well as other likely integrated quantum biological processes.

Furthermore, these processes contribute noise that prevents the capacity for delocalized biological function at increasingly more constrained regions at the level of organismic systems biology. Commensurately, the spatial range for quantum correlations of biological processes could become increasingly reduced.

The integrated complexity of structure and function of an organism becomes increasingly compartmentalized and compromised. Accordingly, the progressively spatially constrained and in tandem time-constrained flow of energy within a given tissue or organ system results in accelerated biological aging.

3.21 Information and Nutrition

In earlier sections, we have stressed the importance of unobstructed energy and information flows within the human body. Additionally, we will relate the less classical manifestations of micronutrient deficiencies as obesity and other chronic disease states of cancer, cardiovascular disease, Alzheimer's disease, and accelerated cognitive decline as will be described herewith. The following are some classical disease states of micronutrient deficiencies: vitamin A (retinol) to xeroethalma (night blindness); vitamin C (ascorbic acid) to scurvy; vitamin D (calciferol) to rickets; vitamin B1 (thiamin) to beriberi; and vitamin B3 (niacin) possibly to pellagra. Vitamin B1 (thiamin) classically leads to beriberi. Some classical mineral deficiencies include those of iron and iodine and lead to anemia and thyroid goiter, respectively. In addition to the classical less common disease states, chronic disease states of aging, such as Alzheimer's disease, accelerated cognitive decline, cancers and cardiovascular disease are a function of mitochondrial dysfunction and often insulin resistance.

Twenty-one different minerals are required for mitochondrial function, which is, in turn, the powerhouse of cells and is what makes efficient energy production possible. In addition to minerals, such as magnesium, zinc, selenium, and many others, vitamins and other antioxidants, amino acids and fatty acids feed into the process of mitochondrial bioenergetics of the TCA cycle and electron transport chain of oxidative phosphorylation and hence ATP production.

For example, the amino acids cysteine, glutamine, and glycine together make glutathione, the major antioxidant of the body to manage oxidative stress, preventing excess oxidative damage. The electron transport chain is the most common source of oxidative stress often due to the overload of energy providing electrons into the electron transport chain. This is often found with dietary excess and obesity. Perhaps insulin resistance, which prevents the decarboxylation of pyruvate to acetyl CoA in the mitochondria for further oxidative metabolism represents a protective mechanism to prevent a pernicious feed-forward process. a-Lipoic acid is an important antioxidant necessary for optimal mitochondrial function. Coenzyme Q10 is necessary for electron transfer within the electron transport chain. Carnitine, a trimethylated amino acid that may be produced by de novo biosynthesis utilizing lysine and methionine, is required for b oxidation of fats allowing acetyl CoA to feed into the TCA cycle. Carnitine may also be derived from exogenous sources such as red meat, fish, and dairy products. The mechanism by which carnitine transports long-chain fatty acids across the inner mitochondrial membrane is discussed further in Volume 2 of this book. The AC:FC (acylated carnitine to free carnitine) ratio is a useful metric for assessing carnitine availability and mitochondrial function or energy production. An increased value of this ratio points to carnitine deficiency and reduced mitochondrial function. Conversely a low AC:FC ratio suggests healthy mitochondrial function and adequate carnitine availability. Evidence does show that reduced carnitine is associated with mitochondrial dysfunction [56]. This may play a role in the loss of nitric oxide signaling and the development of endothelial dysfunction associated with cardiovascular diseases [57].

The stripping of bio-nutrients as a result of overconsumption of white flour, white sugar, white rice, or white wine can be considered to represent the loss of information in the sense of informational energy reduction to decrease biological organized complexity.

In this sense excessive intake of a nutrient-poor diet has in general a lack of potential to promote organizational work mediated by biological free energy and instead this useless energy gets deposited as adipose stores. *Without nutrient informational energy the deficient minerals and vitamins are unable*

to serve as cofactors that catalyze enzymes of the bioenergetic pathways and metabolic reactions. Additionally, the lack of phytonutrients impairs the capacity to promote an antioxidant effect mediated by their binding to the antioxidant response elements of genes that attenuate the inflammatory and oxidant stress in cells.

A phytonutrient-rich diet and robust antioxidant response promote the flow of free energy that serves to reorganize biological chemistry as work while inversely reducing the manifestation of entropy, the "evil twin" of free energy.

Accordingly, informationally directed energy and construction of biological information is increased in association with greater value and usefulness in terms of health and survival. From the perspective of information being rooted in quantum mechanics, and the meaning that information has the potential to impart, it is inherently unpredictable with uncertainty in how that information will be expressed and utilized. Hence, *the information that free energy generates as it interfaces with biochemistry can only be predicted in terms of probability. The greater the biological complexity, the greater the quantum nature of the system, and accordingly, the greater the informationally instructive free energy within the system. Alternatively, the lack of phytonutrients results in inflammatory heat, and subsequent, associated entropy can be related to the destruction or degradation of biological information.* Informational entropy described in the context of quantum mechanics has functional value in the sense that complete order has no meaning as does complete disorder. *The optimum of information or meaning lies somewhere between maximum entropy, or chaos, and complete order (i.e. zero entropy), underpinning the notion of functioning at the "edge of chaos" as has been argued for various biological systems such as brain waves, cardiac rhythms and even protein dynamics.* A higher level of nutritional information is required by those tissues with the greatest bio-energetic and metabolic demands.

Notably, obesity itself promoted by an informationally depleted diet, such as a phytonutrient-deficient diet, is a process whereby lipid accumulation within adipocytes outgrows the blood supply resulting in an inflammatory process when the lipid outstrips its blood supply within the adipocyte.

It should also be noted that the anabolism of muscle requires a high dietary informational content. Furthermore, the contraction of muscle promotes the release of a thermogenic and fat-burning hormone called irisin [58], which causes the transdifferentiation of white adipose tissue to a fundamentally different type of adipose tissue, brown adipose tissue (more correctly termed "beige" adipose tissue). *Instead of storing fat as in the case of white adipose tissue, beige adipose tissue burns it.*

Biological information and complexity made possible by informationally directive nutrition are typically characterized by a high ratio of muscle to fat. Conversely, an informationally depleted diet with high energy content typically promotes sarcopenic obesity.

Additionally, the brain has enormous informationally directive energy dietary demands, and therefore it makes sense that obesity, sarcopenia, and accelerated cognitive decline correlate together. Calorie-dense but nutrient-depleted dietary consumption is insufficient to drive muscle anabolism and cognitive processes and therefore the excess energy gets deposited as fat in adipose stores. The same can be said of immune system regulation. *Pathophysiologically, nutrient depletion promotes mitochondrial dysfunction and relative lack of production of the energy currency as ATP in mitochondria. This results in insulin resistance, which impairs the satiety centers of the brain to a greater extent than it does the enzyme lipoprotein lipase responsible for the uptake of circulating triglycerides and storage in fat tissues.*

The so-called "nutritious" foods and "empty calories" may refer to the ease of incorporation of these molecular components into our cells as well as their functional role and the body's needs. To maintain body temperature, the human body needs simple foods that can be easily metabolized (glucose, fructose, etc.). However, for the *complex functionality of enzymes, we may also need micronutrients. They would carry a high information value and a low probability of general diffusiveness.* These molecules need to find their way to specific locations, hence high information content due to their specificity of function. *The specificity of metabolic processes is both spatial and temporal. For example, characteristic oscillation frequencies of enzymes are dictated by the informationally directive energy responsible for their structure, configuration, and function. That is, the optimal structure and function of enzymes allow them to have a cyclicity of action that is optimal for the physiologic needs of the organism. Particular mineral and vitamin cofactors are needed for the catalysis of enzymatic activity.*

> *Informationally-directed energy is essential for the synthesis of structurally and functionally normal enzymes. Additionally, the informational content of nutrients is required for mitochondrial structure and function and thus for oxidative phosphorylation to proceed unimpeded.*

A sufficient rate of efficient ATP production is needed for the synthesis of structurally and functionally healthy enzymes. Quantum metabolism could be the mechanism that ensures this in a spatially and temporally correlated way.

Micronutrients are substances required in very small amounts, in milligrams or micrograms that mostly function as cofactors of enzymes. Examples include vitamins and trace minerals iron, chromium, zinc, selenium, manganese, molybdenum, and cobalt as well as fluoride, iodide. *Conversely, macronutrients are chemical substances required in substantially larger amounts.* Examples here excluding energy content include the minerals sodium, potassium, chloride, magnesium, calcium, sulfur, and phosphorous. *In general, if some biomolecules found in food are very specific regarding their structure/function, their information content is very high (low entropy).*

Conversely, generic molecules (e.g. glucose, fructose, and amino acids) that are needed everywhere but nowhere specifically have very low information content (high entropy). The issue of their energy content is separate. This can be very easily calculated from the calories they carry. Consequently, paying attention to only the caloric value of food misses a major aspect of their information content. Only satisfying (or worse exceeding) the energetic demands of the body without adequate informational contribution results in inflammation and a subsequent associated entropy increase related to biological destruction in contrast to the constructive direction of biology promoted by informational energy.

A very useful physical concept introduced in the theory of phase transitions is that of an order parameter. *An order parameter is assumed to be zero in a normal state and increasingly non-zero in a pathological state, its value quantifying the severity of the latter. A possible starting point would be to accept the average (standard) physiological parameter values of a healthy person as those that define the order parameter of the non-obese state whereas the values exceeding the average of the healthy non-obese individual (designating a "critical state" a hypothetical departure from health to disease) would be used to describe the order parameter for pathological states such as obesity. Since healthy parameters have an acceptable range, say within a standard deviation of the average, exceeding this range would signal a transition to a diseased state.* These order parameters for health and disease states, respectively, may describe attractor states for health and disease viewed as dynamic phenomena, i.e. fluctuating constantly but within a limited range on either side of the average. Starting from a healthy set of parameters, as the deviation from normal healthy ranges gradually increases, a new attractor state may be approached, and eventually a transition may occur for the fluctuations to center around the disease state attractor. *Both healthy and disease state attractors have a local stability range meaning that the system has a tendency to regain stability by returning to its average order parameter values. The stability around the attractor states corresponds to the free energy landscape with a local minimum corresponding to these attractors separated by potential barriers that prevent an easy transition between attractor states. This is why it is hard to lose weight once a person has (usually gradually) transitioned to an obese state.* Conversely, a healthy individual may resist gaining weight as he/she is satiated having eaten a large meal and resists additional food even the next day after a big meal. These order parameters may be used to describe healthy homeostatic and allostatic functions versus unhealthy (prolonged and/or exaggerated) allostatic functions that compromise homeostasis. Hence, they are associated with new attractor states that follow a trajectory across criticality zones into chaotic states or pathological diseases where allostatic load and overload result in impaired capacity to maintain homeostasis. Let us assume for the sake of simplicity (perhaps oversimplifying the problem but making it easier to understand) that we consider BMI (Body Mass Index) to be the order parameter for a given age and sex. We can properly define it as $\varphi = (BMI - BMI^*)/(BMI^*)$ where BMI^* is (somewhat arbitrarily) chosen to be 30, which defines the boundary of a healthy state. In addition, we can say that if $\phi < 0$ we assign it a value of 0, so that the order parameter for non-obese people is 0 while for obese people it gradually increases from 0 to 1 as their BMI goes up from 30 to 60 where we have chosen 60 to be a maximum value. Here, we have standardized the order parameter for obesity to have values conveniently ranging from 0 to 1. Thus, these parameters are considered order parameters for health within range as having a zero value, and conversely, they are order parameters for disease states when outside that range, with a value between 0 and 1. The greater the deviation from normal values of this particular order parameter, the more dangerous the state of disease progression.

A critical state is a condition at the threshold between health and disease just like in the theory of phase transitions, a critical state (or criticality zone) is at the boundary between two phases of matter.

A free energy fitness landscape can be constructed using a simplified representation of the homeostatic/allostatic response of the human body whereby an order parameter of 0 corresponds to a healthy attractor state and 1 to the morbidly obese state. The specific shape of the "valleys" surrounding these attractor states and the barrier between them depends on each individual and can be mathematically determined based on physiological measurements. The shape of the valley in each state corresponds to the generalized susceptibility, which is the second derivative with respect to the order parameter, i.e. the curvature of the valley around the attractor state. In terms of human physiology, this corresponds to the resistance to change in the BMI of the individual. It can be asymmetric meaning that for some people it is easier to gain weight than to lose weight, which is probably true of most but not all people. In fact, the Minnesota Starvation Experiment is a very interesting demonstration of a highly nonlinear character of the human metabolism in relation to dietary intake. It was a clinical study performed at the University of Minnesota between 1944 and 1945 [59] designed to determine the physiological and psychological effects of severe and prolonged dietary restriction. The study used 36 male volunteers and it was divided into three phases: a) a 12-week control phase, where physiological and psychological observations were collected to establish a baseline for each subject; b) a 24-week starvation phase, during which the caloric intake of each subject was drastically reduced causing each participant to lose an average of 25% of their pre-starvation body weight; and finally c) a recovery phase, in which various rehabilitative diets were tried to re-nourish the volunteers. Among the conclusions from the study was the confirmation that *prolonged semi-starvation produces significant increases in depression, hysteria, and hypochondriasis. Participants exhibited a preoccupation with food, both during the starvation period and the rehabilitation phase. Sexual interest was drastically reduced, and the volunteers showed signs of social withdrawal and isolation.* There were marked declines in physiological processes indicative of decreases in each subject's basal metabolic rate, reflected in reduced body temperature, respiration, and heart rate.

As another example of the choices available for order parameters related to obesity, we may look at potential nutritional components that predispose to obesity. This could be the accepted average value for blood magnesium concentration whereby a person's deviation of magnesium concentrations beyond the accepted average range could be viewed as a measure of pathological changes in the body. Perhaps *one needs to consider a whole slew of different metabolic indicators and generate an order parameter vector (as opposed to a single number, a scalar) consisting of many components. In this vein, zinc, copper, and manganese, all cofactors for the antioxidant enzyme superoxide dismutase, as well as the person's age may be considered component order parameters for the obese state.* The actual definition could be based on the difference between the average accepted value within the standard deviation values for these minerals and the actual measured concentrations in the patient. Furthermore, *additional order parameters for obesity may be determined by the levels of folate, vitamin D, vitamin B3, carnitine, coenzyme Q10, α lipoic acid, vitamin C, and vitamin K (all necessary for mitochondrial function), which further illustrates a need for multicomponent order parameters and not simply a reliance on one number such as the BMI value. Taking all these indicators together, one could better define a predisposing state to obesity as the total combined magnitude of the above order parameter components.* Similarly, another system descriptor, or order parameter, may be systolic or diastolic blood pressure (differences between actual values and normal ranges). This would be an order parameter for cardiovascular health and disease in addition to other indicators such as cholesterol levels.

In addition to order parameters, which characterize the physiological system's state of health or disease by quantitative internal properties, one can introduce control parameters, which typically represent external means of affecting the state of health and disease of an individual. Some control parameters are relatively easy to define. These include the caloric intake of a person, the nutritional content of the food ingested, or the amount of exercise performed on a daily basis. Examples of corresponding order parameters include the difference between the BMI and the accepted range of normal values, the concentration of micronutrients in blood plasma, or the lean muscle ratio to the body weight. *However, a clear distinction between order parameters and control parameters becomes a bit blurred in some cases as both these characteristics depend on each other, especially order parameters on control parameters. As this is still a nebulous concept in physiology, it is important to establish how an order parameter should be properly defined.* In particular, order parameters based on the systolic or diastolic blood pressure (and other indicators of cardiovascular health) are highly dependent on a host of factors that are hard to quantify (e.g. stress level or anxiety). On the other hand, order parameters for obesity should be defined by values that are clearly outside a physiologically healthy range. One could argue that a non-obese state is an equilibrium physiological state under a low-calorie regime (let us say below 2,400 for a 70 kg male but one could again standardize it per kg of body mass). Conversely, an obese state is an equilibrium state for a high-calorie intake regimen.

One could in fact derive a relationship between the value of an order parameter and the caloric intake per kg. In physics, such relations are called scaling relations and in biology, they are often referred to as dose-response relationships. However, physicists typically consider power law representations for these relationships while biologists commonly adopt exponential function approaches, which allow for saturation effects to be seen. In the case of scaling laws one uses a log-log plot for the two properties that are linked to each other while for exponential relationships one uses a semi-log plot. *If we introduce this nutrient intake control parameter (akin to temperature in physics whereby temperature controls the phase positions between solid, liquid and gas), and call it c with a critical value of c, let us say c_0 = 3000 calories for a 70 kg male or 43 calories/kg), then we expect that:* $\phi = (c - c_0)^{\alpha}$, where α *is a characteristic scaling exponent for obesity as a function of overeating. The critical value c_0 corresponds to the transition point between normal physiology and morbidly obese state physiology.* One would expect a square root behavior with α close to 1/2. Of course, there are other factors to consider, such as lifestyle, and exercise, but everything else being equal, the deviation of the BMI from the critical value, should scale in some way with the caloric intake per kg of body mass. *Variations in the scaling relationship reflected in differences in the scaling exponent α may be classified and quantified according to genetic inheritance, microbiome gene products, or an interaction of the two, i.e. epigenetic factors. Accordingly, more or less energy may be extracted from the macronutrient energy sources.* For example, a gram of fat roughly carries a maximum of about 9 kilocalories of potential energy within the hydrocarbon bonds. However, the actual extracted amount of calories may be 6 kilocalories hypothetically, or the full 9 kilocalories, depending on the person's metabolism. Also, as mentioned earlier, sedentary versus active lifestyle results inchanges to the scaling exponent. In fact, these values can be calculated specifically to the particular activity or exercise regimen.

It is worth mentioning that BMI as an order parameter for obesity (without additional order parameter components) is itself antiquated in terms of it being a risk factor for pathologic disease states such as cardiovascular disease, cancers, Alzheimer's disease, or accelerated cognitive decline. Rather, it is the visceral adipose tissue, normal or healthy storage levels being roughly less than 2 liters, above which this becomes an order parameter for a susceptible state and a control parameter for cancers, cardiovascular disease, Alzheimer's disease, and accelerative decline. Visceral adipose tissue appears to derive from a different embryological cell type (i.e. bone marrow myeloid progenitor series, inherently pro-inflammatory cell types) versus subcutaneous adipose tissue, which largely derives from mesenchymal cells. *Accordingly, there will be specified scaling relationships between obesity as an order parameter, defined precisely in terms of visceral adipose tissue or caloric intake.* It should be mentioned that adipose cell number (directly) and cell size provide a stronger link to the chronic diseases of aging than does total adiposity mass (including visceral adiposity). An even more direct link is the presence and extent of ectopic lipid within the cells of affected tissues. However, the measurement of these parameters requires invasive testing and would not be easily performed on a widespread clinical scale.

Defining several macroscopic order parameters for cardiovascular disease and health or for healthy body weight versus obese states may indeed be needed to describe the poly-factorial risk factors. *This should account for the interdependence of the ranges of the various variable potential risk factors that may lead to disease when the standard deviation has a non-zero value.* The ranges of values of these interconnected systems (the blood concentrations of the minerals listed above, each representing a system) should fall within a certain age range to maintain homeostasis of cardiovascular health or body weight. Note that body weight may also be a risk factor or an order parameter component for cardiovascular disease. *The person's age and gender should probably be a coefficient that some of the relationships depend on as there is no such thing as a "standard human",* so we need to have the two genders and age as determinants of the order parameter versus the control parameter (scaling) relationship with specific coefficients of proportionality and possibly scaling exponents. In the case of cardiovascular health, the macroscopic order parameter could be diastolic or systolic blood pressure (within a given range) whereas in the case of healthy body weight the macroscopic order parameter could be the various mineral concentrations described for a given age range. *Conversely, a macroscopic order parameter, which has a non-zero value increases as a result of control parameter settings (e.g. consistent nutritional oversupply) and eventually reaches critical values leading to a disease state. At this point, homeostasis of the order parameters as interconnected systems may move from healthy values for body weight or cardiovascular health to that linked with obesity or cardiovascular disease, respectively.*

This suggests that there may be both coupled and uncoupled order parameters forming a composite variable that defines the state of health or disease.

For example, a diabetes order parameter could be defined as deviation of glucose levels from the norm but this disease may also result in downstream complications such as cardiovascular problems or poor eyesight, each of which would be characterized by their own order parameters. *An interesting and important question arises if and when are these transformations reversible, that is whether reversing the control parameter trends (by, for example, adopting a diet) will lead to a restoration of the state of health* as manifested by a zero value of the order parameter for obesity and cardiovascular disease. As the Minnesota starvation experiment demonstrated, a mechanistic type of reversibility may not be easy to achieve following a long period of disease development.

As the physiologic state of the human body characterized by the various macroscopic order parameters, such as body weight and cardiovascular health or disease, moves to a new attractor state, the associated fitness landscape may become flattened (see detailed discussion on the Physiological Fitness Landscape in Chapter 5 of this volume). This metaphorically corresponds to an increased susceptibility to disease (cardiovascular disease in the case of blood pressure and cancer/Alzheimer's disease/cardiovascular disease in the case of obesity) prior to reaching the critical states (criticality), which is the threshold, allostatic load, just preceding the disease state, i.e. allostatic overload (e.g. obesity or cardiovascular disease).

> *Once the order parameters go beyond the critical state into the disease state, this represents the failure to maintain homeostasis of body weight or of cardiovascular health.*

The further the difference of standard deviation from average normal values (increased order parameter), the greater the risk of progression of the disease state and a dimmer the hope for a reversible change in the body's physiology. Let us assume that a healthy human who is overweight at a susceptibility stage for obesity can eat an extra steak for a few days but it is very likely that he/she can go back to normal eating habits without ballooning in weight in the process. An obese (or non-obese person with increased susceptibility for obesity but still below the criticality threshold) person may take off on this trajectory and never return to a stable weight range (even if above norm). In this case, the system has transitioned to irreversible behavior. Obesity is a disease state that can return to a stable weight that is immediately responsive to a reduced calorie intake, and thus may be definable as a reversible phase of disease state before a new adiposity setpoint is established. *Moreover, reversibility may be further characterized by what types of control parameters are capable of inducing a reversal of the phase transition. For example, diet and exercise (but without extraordinary caloric restriction for the degree of exertion that compromises nutrient sufficiency because in the latter case mitochondrial disturbance and resultant insulin resistance prevent the establishment of a lower-body adiposity set point, even the prior baseline adiposity), pharmacological and bariatric surgery all seem to have the capacity for reversing the phase transition to obesity respectively at more advanced disease states along the spectrum of the free energy landscape.*

> *Far exceeding the range of order parameter values associated with the critical state, however, may result in irreversible changes. This is acutely and depressingly clear in the case of cancer where the threshold of reversibility is marked by the onset of metastases.*

It is the neuroendocrine-immune system axes, including sympathetic and parasympathetic activities, which regulate gut motility (which plays a role in microbiome composition) and immune system (which also plays a role in microbiome composition in part by mediating tolerance or reactivity). A sort of sliding scale relationship of flexibility and robustness between microbial and host cell symbionts is also analogous to the relationship of mutated cancer and ancestral cells. Accordingly, immune tolerance to commensal gut microbes may very well reflect an evolutionary stage currently in progress such that these microbes are becoming subsumed into our native eukaryotic cells, much the way bacteria evolved into eukaryotic mitochondria. In any event, chronic disease appears to be promiscuously mediated by the processes of inflammation and inextricably reciprocal oxidative stress.

> *It seems increasingly evident that the origin of chronic disease is rooted in exposure to pathogenic microbes in the gut.*

It appears that what determines why one individual gets a particular autoimmune disease while another individual gets cancer, is genetic predisposition. However, the process is of feed-forward type whereby the more disease states one has the more likely it is to have another. Immune system inflammatory cytokines act in the limbic system of the brain to reduce the threshold of perceiving stress (psychogenic stress). These inflammatory cytokines primarily act in the prefrontal cortex regions, which are important for rational thought, empathy, compassion, social awareness, and intuition, to put the brakes on the limbic system, the emotional centers of the brain. This in turn, via the overactive neuroendocrine-immune system axis promotes enhanced inflammatory cytokine production, which subsequently exacerbates the reception of psychogenic stress in a feed-forward fashion.

As an example of allostatic change that occurs as new attractor states evolve, a significant difference of individual order parameter values from accepted average values for obesity may include insulin resistance and hence elevated blood glucose and lipid levels. These allostatic levels themselves have attractor states within adaptive ranges for a given duration of time. However, when manifested as chronic, i.e. when the stress as a result of excess caloric intake does not remit, it becomes maladaptive. Accordingly, allostasis becomes an allostatic load (which correlates to the new attractor states just prior to critical states) and becomes allostatic overload once critical states are surpassed entering the zone of the chaotic state in clinical disease such as obesity per se. The description of the dynamical states at the level of an organism can be also extrapolated to the level of an individual cell and its metabolism. The protein enzyme complexes of the electron transport chain function as quantum harmonic oscillators. The oscillations of these enzyme complexes are correlated with the quantity of nutrients consumed. As nutrient bio-fuel intake increases (from low levels of intake), more and more enzymes of the entire metabolic machinery including the glycolytic enzymes, those of the TCA cycle as well as the electron transport chain become activated. It is conceivable that at a lower level of nutrient consumption, nutrient concentrations correlate linearly with enzyme activation, which is associated with a quantum harmonic oscillation, and hence most efficient manifestation, of the final pathway of metabolic oscillators, the electron transport chain. The correlated nature of energy production mediated by the electron transport chain in oxidative phosphorylation represents an example of quantum harmonic oscillators at a subcellular level. These complexes are pulling protons and electrons from glycolysis and the TCA cycle pathways. However, these biochemical pathways themselves may not be mutually correlated. Particularly, glycolysis is driven only by Brownian diffusion of the substrates and enzymes forming the products in the sequence that ultimately produces pyruvate that enters the mitochondria. From there the cyclical pathway of the TCA cycle produces the NADH and FADH2 that is brought to the electron transport chain complexes 1 and 2, respectively. In addition, fatty acids undergo fatty acid oxidation producing acetyl CoA that combines with oxaloacetate that produces citrate in the initial step of that cyclical pathway. Finally, amino acids feed in at various substrate levels of the TCA cycle. Hence, the only component of the metabolic production of ATP that represents a quantum harmonic oscillator is the electron transport chain. Nonetheless, this situation is perhaps common in biology where most of the processes or structures operate in the classical regime while only the essential degrees of freedom such as receptor sites in enzymes involve quantum processes.

Carbohydrates, particularly glucose, feed into glycolysis, which ultimately produces pyruvate, which is decarboxylated to acetate and forms acetyl CoA, which then moves into the mitochondria. Similarly, fatty acids as described above produce acetyl CoA, which converges into the TCA cycle. Amino acids as mentioned above also feed into the cycle at the various substrates depending on the particular amino acids. In any event, the greater the nutrient consumption, the greater the enzyme activation. In the case of carbohydrates, it is the enzymes of the glycolytic pathway and fats and proteins macronutrient consumption that correlates with enzyme activation. *When the amount of macronutrient consumption promotes less than 50% enzyme activation in these pathways and has not surpassed the so-called take-over threshold and the amount of subsequent NADH and FADH2, it then feeds into the electron transport chain associated with a maximally efficient process of oxidative phosphorylation. That is, the electron transport chains are correlated as quantum harmonic oscillators. There is no excessive degree of electron leakage along the inner mitochondrial biomembrane.* Conversely, when macronutrient consumption is excessive such that the amount of NADH and FADH2 feeding into the electron transport chain surpasses the take-over threshold, the quantum harmonic oscillators of the electron transport chain across mitochondria within the same cell and even across cells become de-correlated and accordingly metabolism functions in the classical regime, which is less efficient.

Accordingly, the quantum harmonic oscillator model for metabolic enzyme activities represents metaphorically the army of soldiers marching in lockstep across many mitochondria. Conversely, in cancer cells according to the Warburg effect, an increasing amount of energy production is carried out using the glycolytic pathway at the expense of OxPhos. This shift correlates with the stage of malignancy and progression of the disease. At the same time cancer cells become desynchronized from one another and from the host tissue or organ. The notion of "take-over threshold" in quantum metabolism represents the threshold where nutrient excess causes electron leaking, formation of reactive oxygen species, and inflammation with resultant disruption of mitochondrial structure and function. This in turn leads to greater reliance on the glycolysis pathway of energy production. Although ATP is a quantized metabolic energy unit, the process of metabolism involves numerous biochemical pathways in a multitude of cells. Therefore, the outcome of this process is obtained in the form of quantum statistical ensembles, which were discussed earlier in this chapter. As a consequence, *depending on the dimensionality and the characteristic turnover times, there will be different allometric scaling laws, including a potential transition to the classical limit if the nutrient supply exceeds the availability of the metabolic enzymes to perform their activity.*

Neuronal cells predominantly use the glycolysis mode of metabolism. First, the brain has the highest density of mitochondria of any organ or tissue, consuming about 25% of the body's energy needs despite only representing approximately 5% of total body weight. Secondly, under healthy conditions, it is highly synchronized to afferent and efferent activity with the rest of the body, not to mention the integration of consciousness awareness. At least 50% of the brain's mass is made up of glial cells, which prefer OxPhos. The so-called anti-Warburg effect in Alzheimer's disease (AD) is linked to the fact that AD neurons shift their metabolism from glycolysis to OxPhos as opposed to cancer cells. While network activation triggers a significant energy metabolism increase in both neurons and astrocytes, questions of the primary neuronal energy substrate (e.g. glucose versus lactate) as well as the relative contributions of glycolysis and oxidative phosphorylation and their cellular origins (neurons versus astrocytes) are still a matter of debates.

Informational energy provides useful work for a protein (or a hormone such as insulin) that ultimately helps maintain the homeostasis of glucose within the whole organism. Information flow occurs in both directions: bottom-up (from proteins to cells to tissues, e.g. in terms of the work of the cell such as muscle contraction fueled by ATP) and top-down (from organism to tissues to cells, e.g. in terms of nutrient and oxygen distribution).

> *The bottom-up arrangement has a "noise filter" built in to perform coarse-graining selectivity filtering so that minute fluctuations in the activity of individual cells are averaged out since there are so many of them. However, in some specific scenarios such as a slow initiation and progression of a disease such as cancer, this averaging out and filtering may not be perfect resulting in permanent changes.*

The degradation of H-bonds can happen as a result of thermal excitations. Of course, a single thermal energy unit, kT (= 0.6 kcal/mol), is not enough to break an H-bond but there is a small probability that occasionally two or three such units may arrive at a single place and break a bond, a bit like a rogue wave by the seashore. These are rare events but over time may accumulate the damage and "erase" the information stored in the structure. There may also be other rare events that cause degradation of structure such as radiation damage (from UV or cosmic rays, etc.). Over a long period of time, all of it contributes to the gradual destruction. It is now believed that *as much as 2/3 of all cancers occur as a result of random mutations in genes coding for key cellular proteins.* On the other hand, top-down changes result in major effects on the functioning of all the lower-level structural units. For example, a viral infection spreads to most of the cells in an affected organ weakening its resilience. This can be referred to as a change in the overall "fitness landscape". Some of these changes may be temporary and reversible (e.g. a curable infection), some may be gradual and irreversible (a progressive disease such as Parkinson's) and some may be fatal (e.g. an inoperable cancer).

The electromagnetic field emission is an additional aspect of cell metabolism. This involves the proton pump in the cytochrome enzyme whereby a passage of a proton from a higher energy state to a low energy state causes a photon emission in the near-infrared range, which is a mechanism of biophoton generation distinct from that involving the recombination of reactive oxygen species that Gunther Albrecht-Buehler presented a comprehensive exposition of these arguments on his website Cell Intelligence [60].

If it is plausible that very weak magnetic fields are generated by the electrical current of electron motion along electron transport chains, then this may at least be partly responsible for synchronous quantum metabolism. In the case of mitochondrial membranes, there are two perpendicular currents: electron hopping in the electron transfer chain along the mitochondrial wall, which produces a magnetic field around it in a radial orientation. Then there is a proton gradient across the mitochondrial membrane and this also produces an electric current as a result of the flow of positively charged protons. As such it produces a magnetic field around the proton pump, which is in the plane of the wall, again in radial orientation. Both of these effects are dynamically coupled and the magnetic fields produced by these currents will be small and localized spatially in the vicinity of the current flows but may be sufficient to trigger electronic/protonic transitions in the neighboring complexes. This mechanism may be the closest biological counterpart to the rotating rotor of an electric motor and it is based on the motion of protons across the inner mitochondrial membrane and electrons in the electron transfer chain. The ATPases, which pump out ATP molecules into the cytoplasm can be metaphorically equivalent to the rotor. While the ATPase has the right geometry (a rotating cylinder, roughly), a single proton pump involves a linear charge flow, so that is why the analogy is not perfect with an electromagnetic engine where closed circuits are at work. In quantifying these effects, we conveniently assumed that the rate of proton transfer corresponds to 0.1 ms over the thickness of the wall, say 5 nm. Both numbers are approximate and subject to revision. This leads to a current flow per proton pump on the order of $I = 10^{-15}$ A. From this one can calculate the magnetic field $B = \mu_0 I/2r$, which at a distance of $r = 1$ nm away from the pump would be on the order of 10^{-13} T. Then, assuming that the perpendicular electron hopping rate gives rise to a velocity of approximately 1 m/s, leads to a Lorentz force $F = qvB$, which is only 10^{-32} N. This is too small to make an impact because if we compare it to electrical forces on charges in membranes, then we find $F = qE$, with E around 100 mV so that $F = 10^{-20}$ N, 12 orders of magnitude larger than the Lorentz force above. Even if electrons were moving at the speed of light (clearly impossible), the Lorentz force would be weaker than the Coulomb force. However, there is also a circular current of protons in and out of the mitochondrial membrane and this gives rise to a magnetic field that may affect the electrons' motion along the inner mitochondrial membrane. Consequently, *biophotons and electromagnetic effects in cells may synchronize bioenergetics involving numerous mitochondria within a cell, cells within tissues, and eventually different organs and tissues within the body.*

3.22 Chemical Potential of Physical Biological Systems

The chemical potential is a generalization of the concept of potential energy to the case of systems with variable numbers of particles. Basically, it represents the cost of adding a particle of one kind to the system under consideration.

Therefore, this cost may differ depending on the chemical species and the state of the system considered. It may be very easy to add a charged ion to a neutral "bag" of molecules, but it may be very hard to do so for a system that has a net charge which is the same as that of the ion. Conversely, if the ion has the opposite charge to that of the system, the energy cost will be negative for adding it to the system. This is just an illustrative example as the chemical potential may have nothing to do with electric charge but the equilibrium values of molecular concentrations of the chemical species in a particular system at thermal equilibrium.

Particles or molecules tend to move from a region of higher chemical potential to that of a lower chemical potential. Hence, chemical potential is a generalization of "potentials" in physics such as gravitational potential. When a ball rolls down a hill, it is moving from a higher gravitational potential (higher elevation) to a lower gravitational potential (lower elevation). In the same way, as molecules move, react, dissolve, melt, etc., they always tend naturally to go from a higher chemical potential to a lower one, changing the particle number, which is the conjugate variable to the chemical potential. This means that the product of the chemical potential typically assigned the symbol m and a change in the number of particles dN in this system represents the change in the Gibbs free energy of the system, i.e. $dG = \mu dN$. Simply stated, the energy cost of adding a small number of particles (or molecules) to the system is proportional to the chemical potential.

A simple example is a system of dilute molecules diffusing in a homogeneous environment. In this system, the molecules tend to move from areas of high concentration to low concentration, until eventually the concentration is equilibrated to be the same everywhere. For a given temperature, a molecule has a higher chemical potential in a high-concentration area and a lower chemical potential in a low-concentration area. The movement of molecules from higher chemical potential to lower chemical potential is accompanied by a release of free energy. Therefore, it is a spontaneous process. The concept of chemical potential is valuable in the nutritional/metabolic discussions, and it should explain why this energetic cost varies for different food groups and minerals depending on a person's metabolism, genetic make-up, maybe even mental state. It might be helpful to elaborate on what we understand as energetic cost. On the energy production side of the equation, we can consider the amount of ATP produced by the human body per nutritional calorie of the food ingested. This also speaks to the efficiency of the system in terms of transforming nutritional energy into useful biological work. As discussed elsewhere, a so-called glycolytic shift present in cancer cells exemplifies a loss of biological energy due to loss of efficiency (mitochondrial dysfunction). *At an organism-wide level loss of energetic efficiency may manifest itself by a greater need for food consumption to achieve satiety. However, there may also be other causes of this effect (e.g. psychological).* Note that this is different from the definition of the metabolic rate which is the number of food calories required by a person over a period of time (typically a day) to maintain basic functions required to stay alive. An individual amount of food intake may differ depending on the level of physical activity and here individual efficiency emerge when we compare the nutritional needs and physical outputs in close analogy to mechanical machines that may be more or less efficient in terms of fuel consumption. These *inter- and intra-individual differences can be understood using the concept of the chemical potential as discussed below. A change in the chemical potential translates into the new equilibrium number of particles preferred by the system, very much like a new set point for an individual's weight that may be correlated with an altered level for satiety.*

The chemical potential of a chemical species is the slope (derivative) of the free energy with respect to the number of particles of that species. It reflects the change in free energy when the number of particles of one species changes. Each chemical species, be it an atom, ion, or molecule, has its own chemical potential in a given system. At thermodynamic equilibrium, free energy is at its minimum for the system, that is, $dG = 0$. It follows that the sum of chemical potentials is also zero. The use of this equality provides the means to establish the equilibrium constant for a chemical reaction.

Chemical potential can be thought of as a measure of the "escaping tendency" for a component in a solution, a measure of the reactivity of a component in a solution. Therefore, components with higher internal energies are destabilized relative to those with lower energies, and components with lower entropy values are destabilized relative to those with higher entropy values. Chemical potential determines the equilibrium number of particles per volume, which means equilibrium concentrations of the individual concentrations of the chemical species in the system, e.g. a gas mixture or a fluid with various solute molecules. The greater the chemical potential, the greater the energetic cost to the system per particle per unit volume. The variables that describe the thermodynamic state of a system that can exchange energy and molecules with its environment include pressure and volume, temperature and entropy, and chemical potential and number of particles. These variables listed above are pairs of conjugate thermodynamic variables such that knowing one, we can calculate the other based on the free energy formula and partial differentiation such that:

$$\mu_i = \left(\frac{\partial G}{\partial N_i}\right)_{T,P,N_{j\neq i}} \quad (3.3)$$

where i stands for a chemical species with the corresponding chemical potential μ_i.

We can take as an illustrative example a simple quadratic function (Fig. 3.10 where we assume a = 1): $G = a(N - N_0)^2$ which represents a tendency of the system to have a fixed number of particles, namely N_0. The chemical potential is obtained by definition as $\mu = dG/dN = 2a(N - N_0)$ and hence it depends

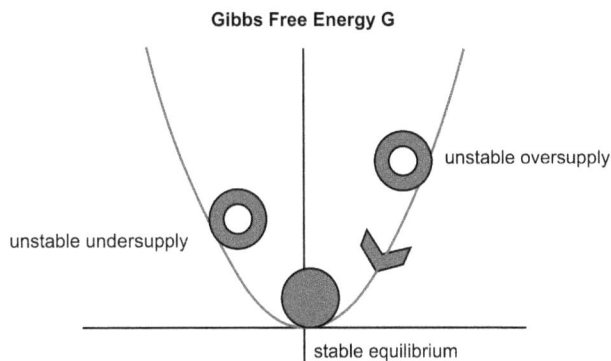

Gibbs Free Energy G

unstable undersupply

unstable oversupply

stable equilibrium

Deviation between the actual number of particles N and the optimum number of particles N₀

FIGURE 3.10 The Gibbs free energy G as a function of $N - N_0$ with $G = a(N - N_0)^2$ where we set a = 1 for illustration purposes.

on the number of particles in the system. For example, when the system is close to its equilibrium, $N = N_0$, and so adding another particle costs no extra energy when one is close to equilibrium (in this particular idealized example).

On the other hand, this cost grows rapidly when you have $N > N_0$ while it is negative when $N < N_0$, which means you will lower the free energy of the system by adding particles to it in the latter case. The system always tends to be close to its equilibrium. This is exactly what happens with food intake and nutritional ingredients as well. *Although the human body is out of thermodynamic equilibrium as being an open dynamic system, at every moment in time it is in a steady-state near its thermodynamic equilibrium.* The situation above illustrates a Physiological Fitness Landscape discussed at length elsewhere in this book and we see here a metaphorical fitness valley where the system achieves optimum fitness with the nutritional intake at a given rate. Both too little and too much nutritional intake puts the systems out of the valley and out of optimum fitness. Therefore, this is a relevant example of homeostasis.

One could go further and imagine a hypothetical Gibbs free energy landscape where $G = a(N - N_0)^2 - b(N - N_0)^3$ (as in Fig. 3.11). In this case, the chemical potential is easily found to be a quadratic function of the number of particles, as $\mu = 2a(N - N_0) - 3b(N - N_0)^2$. This is a little bit more complicated but at $N = N_0$, we still have no cost of adding extra particles. However, it is easy to see that for large values of N the Gibbs free energy G is decreasing and the chemical potential becomes more and more negative indicating that the system is unstable with respect to the addition of particles.

Metaphorically speaking, the system has exceeded its stability range which would represent allostatic overload. In terms of food intake, we can interpret it as a situation where it can never be satiated since beyond a certain number, specifically $N^* = N_0 + 2a/3b$, there is no lower limit on the Gibbs free energy. On the other hand, as long as N is close to N_0, the system tends to return to a stable attractor, which could represent homeostasis, but from N^* up, the system is never "satiated". This simple mathematical metaphor illustrates a transition from homeostasis to allostasis and beyond it, to allostatic overload. As always, staying close to a "sweet spot", referred to mathematically as

Gibbs Free Energy G

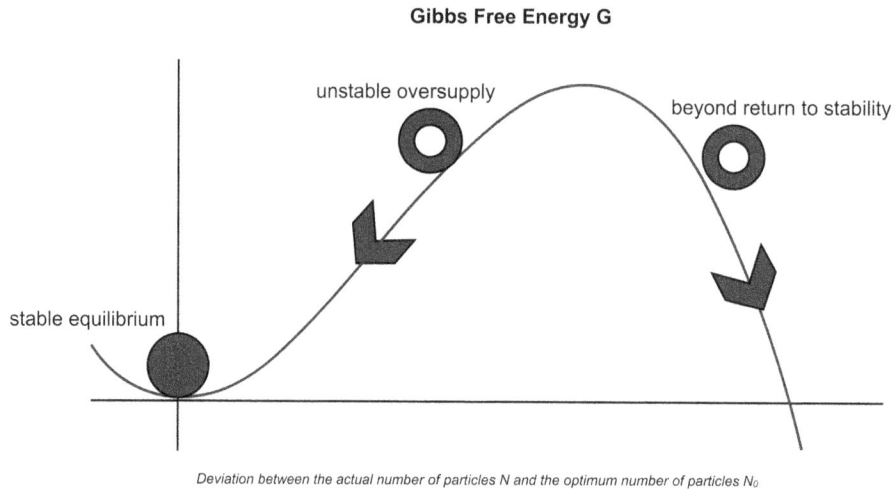

Deviation between the actual number of particles N and the optimum number of particles N₀

FIGURE 3.11   The Gibbs free energy G as a function of $N - N_0$. Here $G = a(N - N_0)^2 - b(N - N_0)^3$. We choose $a = b = 1$ for purpose of illustration.

a basin of attraction, ensures that the system maintains its stability. The same principle is used when setting a thermostat to a particular value of temperature in the room. As soon as the temperature exceeds a set point, air conditioning will kick in. On the other hand, when the temperature drops below a certain level from the setpoint, heating will automatically turn itself on. Our healthy bodies are programmed for self-regulation in terms of most physiological parameters, including temperature and also satiety. We can call it the "Goldilocks Principle". It is only when we destabilize the body beyond its range of internal controls that pathological transformations occur, many can be irreversible.

> *In an obese individual, the energetic cost is greater relative to the healthy lean state in terms of ATP and other energy requirements of bringing hunger to the sated state. That is, in an obese person, greater dietary consumption is required to promote sufficient satiety factor (e.g. insulin and GLP-1) signaling as well as sufficient dietary substrate to meet bioenergetic demands. In both regards, the state of insulin resistance, which is associated with metabolically unhealthy obesity, compromises these processes. In this state, in addition to dietary overconsumption due to an increased threshold for satiety, total body energy expenditure is impaired as a consequence of reduced mitochondrial activity that leads to increased glycolysis, reduced oxidative phosphorylation, and increased glycolysis. Taken together, a disturbed metabolic homeostasis is manifested as a positive total body energy balance with increased deposition of energy as fat.*

In the pathogenesis of obese individuals, glycolytic enzymes become saturated, and beyond that threshold dietary energy cannot be converted to ATP and consequently deposits as fat. Reduced mitochondrial activity leads to increased glycolysis. This results in glycolytic enzymes becoming saturated and beyond that threshold dietary energy cannot be converted to ATP and consequently deposits as fat.

The term energy cost is awkward but this is what the chemical potential means, the free energy cost of adding an extra particle to the system. In the case of metabolism, we should be moving away from the free energy and towards free energy consumption/acquisition rates per day, so it makes sense in terms of a balance of energy flow into/out of the living system.

The notion of chemical potential would be a correct framework since living systems are strictly speaking far from thermodynamic equilibrium. *Therefore, three questions can be posed in this light: First, what is the net gain/loss of free energy per day? Second, what is the rate of free energy consumption (or utilization) by the person and its breakdown into various uses* (e.g. maintaining constant body temperature, rebuilding proteins, performing work)? These rates could be catalyzed by microelements whose value is "informational" rather than strictly energetic due to their catalytic functions. This leads to the third question: *what is the rate of free energy input and its breakdown into nutritional components? This brings us back to the issue of chemical potential. In essence, what we need is a so-called detailed balance equation.*

> *Ideally, we achieve a dynamic equilibrium where the energy input equals the energy output. The former is in the form of nutritional calories while the latter in the form of heat generated and work performed by the body. In a diseased state, energy is not balanced.*

We may be depleting the body of valuable nutrients and causing weight gain due to poorly metabolized food, i.e. not converting food catabolism into the energy currency of ATP, or because of an unbalanced diet. One may also go in the opposite direction and cause a constant weight loss (cachexia or anorexia) resulting from other issues (tumor growth with its glycolytic metabolism or inadequate energy supply, respectively). This part should be almost an "exact science" due to an energy balance requirement. The only complication arises due to feedback loops and nonlinearities due to the catalytic role of various micronutrients. There is also the gradual change in the metabolic rates with aging as well as psychosomatic (*e.g.*

anorexia nervosa or stress-driven overeating). Additionally, changes in the microbiome can cause increased energy extraction (depletion of the microbiome) with associated weight gain or conversely (repletion of the microbiome), decreased energy extraction and weight loss [61].

To better understand the connection with the chemical potential in biological systems, let us use an example of steak consumption by a normal healthy individual versus an obese individual afflicted with a "disease" state (e.g. diabetes). By comparison to the chemical potential of physical systems, one could consider a person healthy if he/she would be satiated by eating one steak, whereas an obese person would require an additional steak to achieve the same effect. The additional steak metaphorically represents adding an extra particle to an open thermodynamic system. However, each "particle", such as the steak, provides less chemical potential energy usable in the form of work in the non-healthy state than it does in the healthy state. Part of this nutritional chemical potential energy can be regarded as the "energetic cost" to the system not only by building inefficiencies into the structure but also in terms of moving it away from its equilibrium state (in physical systems) or from its quasi-steady-state in biological systems (Figure 3.10). Eventually, this excess amount of non-usable energy moves the system away from the basin of attraction of the healthy physiological state toward a new set point which may correspond to the onset of metabolic disease (see Figure 3.11). Another illustrative metaphor that we can use to better understand optimum metabolic efficiency is that of the pressure in a car tire. At the recommended rating for pressure in the tire, the ride is smooth and the friction with the road at a minimum ensuring the best gas mileage. This corresponds to optimum food consumption in a healthy person. An underinflated tire compromises gas mileage by wasting a lot of energy to excess friction. An overinflated tire causes not only a bumpy ride but also poses an increased risk of puncture. By analogy with these physical examples both undernourishment (in the extreme case: anorexia or bulimia) or overeating (morbid obesity) result in serious system breakdown and changes in the naturally established set points for satiety.

Metabolic rate is measured in caloric intake over a period of time (a day), hence it is a measure of power consumption of a living system. The greater the energetic cost to the system per particle, the less ATP is produced per particle and the less work can be done per particle per unit time, i.e. the lower the power. In order to bring this reasoning closer to conventional physiology, one should think of the opposite side of this coin, namely the net energetic benefit of converting macronutrient resources or caloric intake into ATP or caloric output. The energy cost we discussed above has to be subtracted from the total energy content to give us the free energy output resulting from metabolizing the nutrients. Power is defined as work performed over time. Therefore, if we assume an identical period over which energy is consumed by a living system, then metabolic rates may be considered in terms of the chemical potential so the energetic cost of chemical potential equates to metabolic rate multiplied by the time period over which it is measured (one day). The metabolic cost in terms of the chemical potential energy for each particle, for example, a steak, in a healthy individual is greater than that

in an obese unhealthy individual since the metabolic rate in the former case is increased to a greater extent than it is in the latter case. However, *the metabolic rate by "caloric intake" would be lower in the healthy person because the chemical potential energy is greater so the number of particles overall would be lower because each particle would have the capacity of moving the system further away from its quasi-steady state.* Therefore, the cost of every "particle" of steak relates to the metabolic rate increase, which is required to move the system from its baseline quasi-steady-state of the fasting hunger state to the postprandial satiety state. *The energy cost of maintaining satiety by ATP hydrolysis is achieved by the metabolic flexibility to meet metabolic demands of maintaining homeostasis of the broader system (the organism).* Again, the baseline dynamic steady state is the hunger state. One steak in a healthy person moves further away from the state in the direction of satiety because the chemical potential is greater and may be measured in terms of ATP produced or "cost" to the system (in this case of maintaining satiety). ATP production per "particle" of nutrients is the net effect, which determines the corresponding work that can be performed using ATP in terms of establishing satiety. The greater the amount of work done by ATP produced per particle of steak, the greater the potential energy and hence the lower the cost to the system. Thus, the cost corresponds to the net amount of ATP produced per unit particle of food. The processes required to do this are organism-wide and include the secretion of adiposity signals (leptin and insulin primarily) and other satiety hormones (e.g. GLP-1), as well as the signaling activity of these hormones in the arcuate nucleus of the hypothalamus. ATP is also required to set up the polarized electrical potential for neuronal depolarization to occur. It is further required for the synthesis of POMC precursor molecule, subsequent cleavage into offspring neuronal transmitter hormones (e.g. ACTHα, α-malanocyte stimulating hormone (αMSH), βMSH and γ lipotropin hormone γLPH). These hormones are secreted into downstream hypothalamic regions where they bind to MC4 receptors [62]. This step may be mediated by energy provided by ATP or by magnetic fields including the Lorentz force. Additionally, the insulin signaling in peripheral metabolic tissues, particularly skeletal muscle that promotes glucose uptake into the cells and requires ATP, and contributes to the total number of particles required to achieve satiety. The insulin secretion and glucose uptake represent metabolic flexibility switching from the primary fatty acid metabolism of fasting to the primary carbohydrate metabolism in the postprandial state that maintains euglycemic homeostasis. It makes sense that sensitivity to insulin signaling in the satiety center of the hypothalamus would be correlated to peripheral insulin signaling because of the impaired ability to achieve the sated state that represents a lower chemical potential energy per particle of food or of metabolic rate. This equates to less ATP production to meet the metabolic demands of the body. In the obese individual (disease state), the cost in terms of metabolic rate is less because each "particle" of steak is less able to move the dynamic quasi-steady-state of hunger to a new attractor and basin of attraction representing the quasi-steady-state of satiety. That is, in the obese disease state, a single "particle" of steak is unable to provide sufficient chemical potential energy to move the patient all the way from the hunger state to a new attractor and basin of attraction quasi-steady-state of satiety. If it requires two steaks to do that, then a single steak has half the chemical potential energy in an obese disease state in comparison to a healthy lean state. The ability in a healthy non-obese state for a single "particle" of steak to move from the hunger to the satiety dynamic quasi steady-states corresponds to a higher maximum postprandial metabolic rate and hence ATP production in comparison to the obese disease state. This equates to a higher metabolic rate in terms of food intake required to produce the same amount of ATP as the lean healthy state. Though this would not apply directly to steak, which is composed of protein and fat since there is no carbohydrate or glucose content to go through glycolysis. Tied in with the energetic production required to achieve the postprandial sated state is the maintenance of normal blood levels of glucose and lipids. Normal postprandial fluctuations of blood glucose (less than 140) and triglycerides (less than 170) is considered healthy allostasis, that maintaining homeostasis through change.

> These order parameters of allostasis are kept within a relatively narrow range by the upstream control parameters of adiposity and satiety hormonal factors mentioned above, themselves being allostatic parameters as well.

Postprandial suppression of growth hormone and glucagon should also be considered part of the allostatic hormonal pattern that prevents exaggerated blood glucose and lipid excursions. The totality of interactions is extraordinarily complex far beyond the understanding of modern science and medicine. Even the postprandial suppression of gonadotropins, which to my knowledge has not been explained from a theological perspective, helps calibrate energetic demands away from energy-expensive hormonal behavior in favor of the system for achieving satiety. In any event, when these postprandial levels are elevated compared to normal, i.e. in the insulin-resistant range, but not in the diabetic range (blood glucose 140 to 200 and/or triglyceride greater than 170 [there is no lipocentric defining value for diabetes]) it represents allostatic load, typically corresponding with a state of being overweight, obese, or sarcopenic obese, and a hallmark of a susceptibility state of type 2 diabetes, metabolic syndrome, cancer, cardiovascular disease, accelerated cognitive decline, or Alzheimer's disease. These latter states are hallmarks of what has been described as allostatic overload. The states allostatic load and allostatic overload of obesity and type 2 diabetes, respectively parallel the extent of the chemical potential energy of metabolic rate in the sense described above. The greater degree of insulin resistance and associated relative insulinopenia (state of type 2 diabetes) promotes or corresponds to lower chemical potential energy per "particle" of food.

The further the distance one particle of nutrients will take one away from the quasi-steady-state of hunger to the new sated steady state, the greater the cost to the system. Since moving a biological system further away from equilibrium imposes an energetic cost, that chemical potential energy of a given particle (e.g. unit of macronutrient) may be measured in terms of ATP, the biological energy currency. The higher the metabolic rate of the system, the lower the cost to the system and the lower the amount of ATP produced by the system per particle of macronutrient, and hence lower the chemical potential energy per particle of macronutrient. When bioenergetic enzymes are 50% saturated, the rate of the reaction declines, leading to a reduction in mitochondrial function. As has been noted earlier, *reduced mitochondrial function leads to increased glycolysis. This results in glycolytic enzymes becoming saturated and beyond that threshold dietary energy cannot be converted to ATP and consequently deposits as fat.*

Concerning the metabolic demands of moving the quasi-steady state of hunger away from that state of "equilibrium" to a new attractor state of satiety, the following comments are worth noting. First, we need to include the ATP requirements of promoting or inhibiting neuronal transmission. This provides the necessary energy for the hyperpolarized state of a pre-firing electrical potential or of inhibitory electrical potential for neuronal firing. Energetic demands also include synthesis of parent neurotransmitter hormones, cleavage into the offspring neurotransmitter hormones, and secretion into synaptic clefts of hypothalamic neurons. There they either inhibit depolarization of orexigenic neurons or promote depolarization of satiety neurons depending on the particular neurotransmitter and the region of the hypothalamus. The secretion of hormonal neurotransmitters into the synaptic cleft may perhaps be mediated by magnetic fields involving the Lorentz force occurring at the neuron terminals.

3.23 Is Consciousness a Quantum Phenomenon?

Arguably, the concept of consciousness is more complex and intractable than any other topic discussed here in connection with quantum biology. There have been numerous attempts to describe and define consciousness applying spiritual, philosophical, and scientific terms of reference.

> Consciousness is still one of the most difficult and enigmatic problems to grapple with within the confines of empirical science.

In the early 20th century, the American philosopher and psychologist William James popularized the concept of consciousness. However, this was later overshadowed by the rise of behaviorism, which places emphasis on the measurable actions of humans and animals. Some 25 years ago consciousness once again became a topic of rigorous scientific discourse largely due to the efforts of Roger Penrose, Stuart Hameroff, and David Chalmers, among others. Currently, the generally accepted definition of consciousness is as the "condition of being aware of one's surroundings and one's own existence, i.e. self-awareness" or to put it in other words: awareness of awareness. Clearly, to the best of our ability to discern it, no artificial intelligence algorithm or device has achieved this quality yet.

> Neuroscience considers consciousness to be a function of the brain and hence can be attributed to

complex synaptic computation involving brain cells, neurons, which are largely treated as binary units. Present-day neuroscience also accepts that consciousness is a highly non-linear, emergent property arising from neuronal features of the brain that are fully compatible with the laws of physics and hence sees no need to seek its explanation outside the realm of physics.

It is further assumed that when a critical level of complexity is reached, interacting neurons operate in synchrony to generate some form of conscious experience. Indeed, classical physics treats large objects such as the brain, as an aggregate of smaller functional units, which mainly interact with their nearest neighbors. Synaptic connections facilitate the formation of larger functional systems. However, a serious question still remains unanswered, namely how spatially distributed brain activities bind together to produce the unitary sense of self in a conscious organism. While classical physics approaches appeal to the brain's non-linear and deterministic chaotic behavior, which leads to non-computable results characterized by the unpredictability of consciousness, this still falls short of a satisfactory theory of consciousness.

Two prominent neuroscientists have proposed differing views to explain how consciousness occurs. Baroness Susan Greenfield holds the view that there is no central brain region housing consciousness and there are varying degrees of consciousness, which correlate with transient neuronal assemblies. Christof Koch, on the other hand, suggests that there are specific sets of neurons in different brain regions that associate with each conscious event [63].

Information enters the brain via our sensory neurons of the optic, auditory, olfactory and somatosensory nerves, which ultimately synapse with neurons in higher-level centers of the brain. The processing of these inputs forms the basis of cognitive functions, which are the manifestations of awareness. Simple awareness of any sentient organism is what is derived from the senses independently of cognition.

Awareness of being aware has been proposed as a minimum criterion defining consciousness.

An awareness of being aware may be exemplified by the pride that one may feel or an intellectual achievement or the anxiety and depression one perceives how others may think negatively of him or her. Higher-level degrees of consciousness may be the awareness that one is aware of being aware and so on, as may be exemplified by transcendental meditation. The notion of consciousness equates for many on an intuitive level to the "soul" or identity of an individual. It is this soul or identity of consciousness, which is imponderable in terms of where it comes from that defines us and where it goes when we die. The interpretative description of consciousness as awareness of being aware, which is particularly but not exclusively limited to human beings, appears to be fundamentally rooted in quantum mechanics due to the non-computable aspects of numerous cognitive processes as elaborated on by Sir Roger Penrose in his masterpiece *The Shadows of the Mind*. The subcellular substrate of this quantum phenomenon has been

argued by many, including Penrose and Hameroff, to be based on the microtubules of neurons in the brain.

While much is known about the architecture of brain cells and their individual activities in terms of the action potential, ion channels, and ionic currents, much less is known about such issues as where memories are stored (other than recognizing that hippocampal neurons are involved which is largely due to correlational studies of memory assessment and hippocampal size), which molecular mechanisms are involved in information processing and cognitive functions. Speculations about molecular mechanisms of cognition abound and range from mundane to exotic. In particular, much has been speculated recently about the possibility of some cognitive functions requiring the operation at a level of quantum physics. It is tantalizing to consider the brain as a biological implementation of quantum computing. Matthew Fisher has recently argued that the only plausible repositories for quantum information in the brain are the Phosphorus-31 nuclear spins in phosphate ions. Because these nuclei have spins 1/2, their corresponding long coherence times are on the order of a second. Moreover, phosphate ions can be bound with calcium ions into clusters containing six P-31 nuclei whose coherence times can be greatly enhanced to as long as days or even weeks by an effect called motional narrowing. Since ATP sometimes releases diphosphate ions, which are later broken into two separate phosphate ions, each with a single P-31, they can be quantum entangled due to their spin states. This indicates the possibility that biological energy quanta carry qubits of quantum information. Thus, wherever metabolically produced ATP molecules are consumed by the molecular machinery of the cell, there is a possibility of creating entangled qubit pairs, which would link metabolism with consciousness providing an exciting relationship between the essential conditions for life's continuity with consciousness. However, if the phosphate molecules remain unbound, this entanglement will decay in about a second, but it is very different when the phosphate ions group together rapidly enough into so-called Posner clusters, with central Ca^{2+} ion surrounded by six phosphate PO_4^{3-} anions which in turn are surrounded by eight further calcium ions [64]. This allows the entanglement to survive for a much longer time. If the two members of an entangled qubit pair are snatched up by different Posner clusters, the clusters may then be transported into different cells, distributing the entanglement over relatively long distances.

Quantum entangled systems are not always physically coupled. For example, the active brain states of two individuals may be separated by long distances and may still be connected psychologically [65]. The two brains are deemed to be quantum entangled due to electromagnetic waves of the two brains being "in tune" with each other, a bit like a radio transmitter and a radio receiver. This, however, requires an exchange of electromagnetic signals.

From the perspective of classical physics, the intriguing question regarding where our consciousness goes when we die may be answered in terms of the second law of thermodynamics. That is, it may be simply dissipated as heat that increases the entropy of the universe. Moreover, as our consciousness, or individuality is lost, along with the other biological manifestations of quantum metabolism, the loss of metabolic stability and the passage of time becomes more accelerated. Analogous to a quantum computer, which has vast cryogenic requirements to present quantum decoherence, perhaps the heat associated with the natural inflammatory stated aging can be thought of in the same sense. However, it is understandable that this explanation will be unsatisfying and disquieting to those who believe the soul does not die. It is quite possible that it does not, and the alternative views do not necessarily need to be mutually exclusive. Indeed, there are plausible theories to suggest that the universe is deeply entangled at scales far more macroscopic than the boundaries of an individual. There are views such as the Many Worlds Theory, which posit essentially that an individual's "soul" or "consciousness" is not destroyed by the processes of inflammation and oxidative stress, but rather transferred to an entangled state in other dimensions to which our awareness in the current existence does not have access to, i.e. it is "blind" in this regard. Since the simplest definition of consciousness states that it represents an awareness of awareness, consciousness must be segregated to a given definable set of dimensions within a parameter of time. This is notwithstanding the notion that time, as an absolute quantity, does not exist. This sense provokes *the intriguing possibility that multiple lives and associated states of consciousness may exist consecutively while also being simultaneous*, albeit seemingly paradoxical.

3.24 Brain's Processing Power: How Many Flops and How Many Watts?

Most of the brain's functional roles involve motor control, cognition, and information storage and processing. Physics has until recently treated information with some neglect focusing instead on entropy and its physically measurable manifestation, heat. It is now known that entropy reduction in biological cells takes place at the cost of metabolic energy consumption and dissipation into the environment, which are concepts related to information storage and processing. The cost of such processes is worth exploring. Simple yet powerful analysis involving metabolic energy consumption, the maximum amount of information processing rates, as well as analysis of spatial and time scales, will allow us to conclude with a high level of confidence which subcellular structures, if any, can be possibly involved in information processing at the rates contemplated in various hypotheses. The protein tubulin and microtubules figure prominently in this picture since they have been for almost two decades touted as the key to understanding consciousness and cognition by the more exotic proposals [66, 67].

Below we discuss the bioenergetics of the brain's energy production in terms of both the amounts of energy and the numbers of ATP molecules [68]. The human body requires approximately 100 W (of course with individual variations) of metabolic power being consumed on average for its functional demands. Of all the organs in the human body, the brain has the highest rate of energy consumption, accounting for roughly 25% of the total energy demand, or 25 W. The vast majority of biochemical energy supply is provided by ATP molecules, each of which gives off on the order of 10 kT (where k is the Boltzmann constant) of free energy, or 4×10^{-20} J using SI units.

Undoubtedly, the main role of the human brain is to store and process information provided by sensory inputs. A vast percentage of the metabolic energy, approximately 70–80%, is used to maintain the constant body temperature in mammals (since constancy of body temperature is not a feature in non-mammalian systems). Most of the remaining free energy is used in the protein synthesis machinery, i.e. by the ribosomes. It can be therefore safely assumed that less than 5% of the metabolic energy of the brain is utilized by information storage (memory) and processing (cognition) demands. We can use a conservative estimate of 4 W being consumed for these purposes.

It is important to reiterate, following an earlier discussion in this chapter, that any form of information, physical or biological, always comes at an energetic cost. Landauer first stated that the minimum energetic cost of one bit of information is $\epsilon = kT \ln(2) = 4 \times 10^{-21}$ J, which basically comes from the thermodynamic formula defining the free energy as $F = U - TS$ and the fact that entropy S is equivalent to negative information, $-I$ and vice versa. Shannon's formula for information is $I = -kT \ln \Omega$ where Ω is the number of equal probability states in the ensemble representing the choices available for information storage at a given step (e.g. the number of cards in a deck from which one is drawn). The value ϵ is the physically determined minimum energetic cost of creating a bit of information, therefore it corresponds to $\Omega = 2$ and hence $\epsilon = kT \ln 2$. In the context of biological systems such as the brain's neurons, it is unlikely that simply the minimum energy cost is used in information encoding. It is more likely that the cost of information encoding in biology is higher than in physics and it is equivalent to the free energy value for ATP (or GTP), which is at least an order of magnitude greater in biological systems than the physical energy minimum given by Landauer's formula. This can be understood in view of the biological emphasis on the faithfulness of information encoding that trumps energy efficiency which is secondary. In order to relate the metabolic energy expenditure to the information processing rate and estimate the maximum value of the latter, the time scale of the predominant "bit switching" processes (i.e. flipping from 0 to 1 or vice versa), Δt, must be estimated. Therefore, $\Delta I / \Delta t = P/\epsilon = 10^{20}$ bits/s, which must be stressed is the absolute upper-limit estimate on the processing power of the brain. This can be very favorably compared to the processing rate of the most powerful modern-day supercomputer clusters such as the BlueGene, for which the processing rate is on the order of 10^{19} bits/s but the corresponding power consumption is enormous, on the order of 10 MW (megawatts) underlining a huge difference in the efficiency of information processing in biology compared to present-day technology.

The characteristic time scales of information processing units in the brain are crucial in determining the maximum information processing rates given the tiny power consumption of the brain. Seminal experiments conducted by Libet [69] determined the pre-processing time (which is roughly equivalent to the reaction time) of the human brain to be approximately 500 ms. Therefore, it can be inferred that for the whole brain, Δt is on the order of 1 s or less consistent with the frequency of brain waves, f, in the 10–100 Hz range, as recorded by EEG. Once again, for the whole brain, it appears that the maximum value of the information processing rate is 10^{20} bits/s. Within the human brain, the next level of the structural and functional hierarchy involves neurons. There are approximately 10^{11} neurons in the brain, hence assuming all of them being equal, a single neuron is expected to have an information processing rate on the order of 10^9 bits/s. The time scale of these processes is largely determined by their firing rates, hence with a time scale on the order of 1 ms, the corresponding frequency, f, is in the 1 kHz range. Assuming that the computational elements within neurons are microtubules (MTs), and estimating the number of MTs per neuron as 10^3, the average processing rate per MT would be 10^6 bits/s. The corresponding time scales are therefore on the order of 1 ms, which is a typical time for a conformational change to occur in a protein. The corresponding frequency, f, is expected to be 10 MHz. Finally, the lowest level of information processing discussed in the literature is that of the constituent protein, tubulin. For a typical 10μm-long MT, there are on the order of 10^4 tubulin dimers, hence if a tubulin dimer is the basic computational element (corresponding to a bit of biological information), then its maximum information processing rate is 100 bits/s. The latter estimate, however, leads to a contradiction with the various claims of tubulin being a biological qubit that processes information on much shorter time scales. The conclusions reached here indicate tens of millisecond transitions, which is consistent with microtubule coupling to action potential but most definitely not with the sub-nanosecond electronic transitions and GHz frequency rates. Consequently, if information processing taking part in the brain does so in a hierarchical manner with neurons forming the second layer and MTs the next one with final steps being made by tubulin dimers, the maximum theoretically possible information processing rates can be: a) 10^{20} bits/s per brain, b) 10^9 bits/s per neuron, and c) 10^6 bits/s per microtubule and 10^2 bits/s per tubulin. These numbers may be a reasonable quantification of biochemical processes occurring in the brain in the classical regime, consistent with various possible biochemical scenarios. For example, as discussed below phosphorylation events of tubulin by calmodulin kinase II (CaMKII) enzyme [70] would possibly occur on a time scale of tens of milliseconds, as could interactions of MTs with action potential fields. It challenges the present-day knowledge about biochemistry and cell biology as well as quantum physics to claim that such slow processes can be performed in a quantum mechanical regime.

3.25 The Human Brain: Its Structural Complexity and Amazing Efficiency

The brain is a fascinating and hugely complex organ containing approximately 1.5 kg of soft tissue that is composed of a collection of highly organized electrically excitable cells, mainly neurons and glia. The approximately 100 billion neurons, comprising approximately 50% of the brain, connect and communicate with one another via 1,000 trillion synaptic connections to control the central nervous system, regulating autonomic functions, neuroendocrine function, energy homeostasis, circadian rhythm, cognition, and behavior.

Conscious activity is known to correlate strongly with coherent neural oscillations using still largely unknown mechanisms of synchronization.

Can these mechanisms possibly be linked to quantum mechanics? If consciousness arises from the complexity of the collective action of approximately 100 billion neurons acting cooperatively/synchronously, then the collective action of 100's of billions of sub-neural components can only add to the rich complexity and exquisite sensitivity of these processes.

From quantum coherence in photosynthesis and magneto-reception in bird navigation to quantum olfaction and individual photon effects in vision, the field of quantum biology is rapidly entering into the mainstream of life sciences.

Therefore, it makes sense to ask if the human brain also finds uses for quantum effects, and if so, how?

Higher cognitive functions including perception, attention, decision-making, learning, analytical thinking, abstract reasoning, memory are associated with synchronized oscillations of large groups of neurons.

This effect is commonly referred to as neuronal synchrony. Importantly, interruptions and irregularities of these patterns of oscillations are found as hallmarks of various brain diseases such as epilepsy, Alzheimer's, and Parkinson's disease. In a healthy brain, this large-scale coherent pattern of synchronization induces a plethora of cognitive processes including the maintenance of memory, and perceptual stabilization. Periodic firing patterns serve as a temporal reference for neuronal signaling. Rapid signaling events conducted via gap junctions coupled with postsynaptic potential inhibition create temporal gaps, allowing time for the generation of action potentials in excitatory neurons (Figure 3.12).

However, it still remains largely unknown how groups of neurons produce synchronization domains, while rapidly switching between coherent and incoherent states that correlate with cognitive function. Integration of inputs implies merging and consolidation of numerous and diverse information sources. *At the level of individual neurons, integration is approximated as the linear summation of synaptic membrane potentials.* However, integration in branching dendrites and soma requires logic, amplification of distal inputs, branch point effects, and signaling in dendritic spines (Figure 3.13).

According to the generally accepted model proposed in the 1950s by Hodgkin and Huxley [71], all such factors are reflected in the dynamics of transmembrane potentials, specifically in the propagating action potentials, and thus the neuron should behave completely algorithmically and deterministically. *For a pre-determined set of inputs, synaptic strengths, and firing threshold, a fixed output in the form of axonal firings, or spikes, is expected to take place.* However, a theory like this leaves no place for such behavior as subjective and non-algorithmic action or free will representing

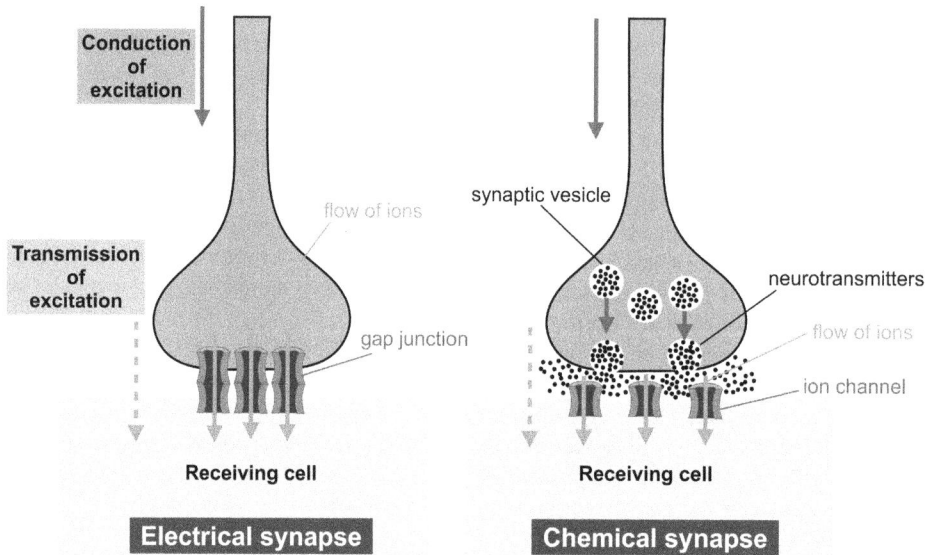

FIGURE 3.12 Illustration of synaptic roles in chemical and electrical signaling across synapses, ion channels, and gap junctions. Source: adapted from https://deskarati.com/2014/02/06/whats-the-difference-between-a-chemical-synapse-and-an-electrical-synapse/.

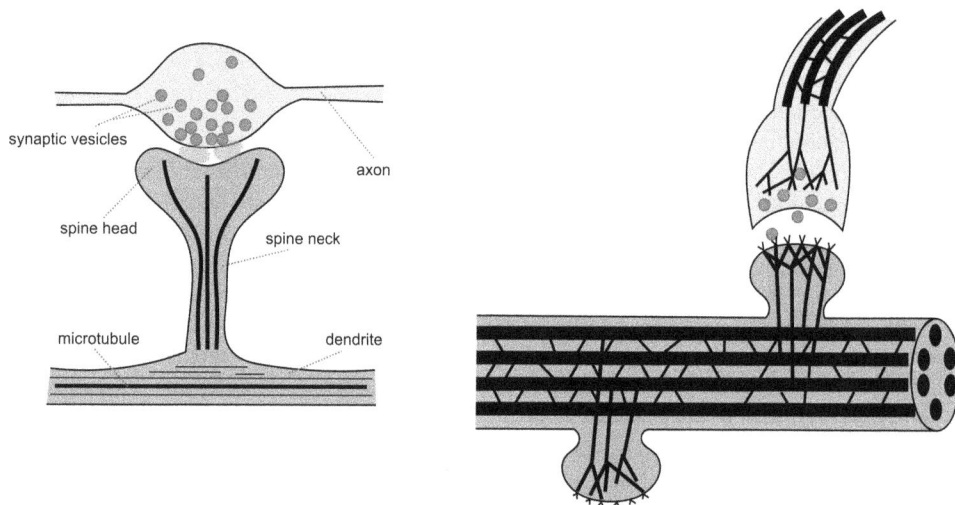

FIGURE 3.13 Schematic illustration of synapse with microtubules and related cytoskeletal structures in axon (top) and dendritic spines (bottom).

the human brain as a complicated computational device, in essence treating it as a machine. Obviously, real neurons differ from these hypothetical mathematical models of neurons. It is apparently clear that we as humans do have free will as we can do as we please and decide on the course of our actions. However, this observation has been put in question in light of a study that has revealed that unconscious activity in the supplementary motor area comes before conscious awareness of that decision. Thus, decisions about movement in this context were already decided upon prior to the subjects' awareness of the decision, demonstrating the possibility that not all decisions are consciously made [72]. While some animals appear to be largely driven by instinct and simply respond to stimuli as is the case for zebrafish, much remains to be determined for other animals such as corvids and cephalopods that exhibit

high-level intelligence, reasoning, and decision-making capabilities [73, 74] Therefore, *factors other than inputs, synaptic strengths and the collective dynamics of transmembrane potentials must contribute to and affect the firing decisions.* It should be noted that the complexity of an individual neuron parallels the overall neuronal complexity of the brain as a microcosm. Consequently, we believe that neuroscientists should investigate deeper layers of the structural and functional hierarchy of the brain delving toward ever-smaller size and time scales.

The neuronal cytoskeleton, especially microtubules and motor proteins should hold a lot of clues regarding the brain's inner workings and this should bring neuroscience closer to quantum physics.

3.26 The Neuron: Its Architecture and Central Role in the Brain's Activities

Every neuron is a highly structured matrix of nucleic acids, proteins (ion channels and membrane receptors included), ions, and water molecules, in which these structures are bathed, that work in concert to produce normal cellular function. Consequently, this complicated intra-cellular machinery should be investigated as a whole, especially regarding its modes of communication within and between cells of the brain. Conventionally, cell signaling is considered to include autocrine, paracrine, and endocrine chemical communication, and electrical signals (mediated by ions) propagating through synapses and gap junctions. These processes include binding and activating receptors. The traditional view of the cell interior can be oversimplified as a bag of randomly diffusing ions, proteins, and other biomolecules, which in the course of their diffusive transport, bind to and unbind from other molecules. In fact, the conventional, reductionist view adopted by cell biology is that of intra-cellular signaling and signal processing taking place in cytoplasmic liquid composed of randomly diffusing ions, proteins, and enzymes driven by gradient forces and thermal motion. However, this is a very inaccurate representation since the cytoplasm includes the cytoskeleton, which is to a large degree in a solid state since some of its components such as microtubules, microfilaments, internal membranes, and organelles are fairly rigid and the ordered water surrounding them is likely to be dynamically and structurally ordered. Hence, while the cell is an extremely crowded environment vastly different from the dilute, idealized conditions found in most biophysical experiments, it is not a random collection of biomolecules drifting aimlessly in this aqueous environment. The sophisticated arrangement of internal membranes and cytoskeletal networks has a strong effect on the diffusivity of the biomolecules and our view of the cell is that of a highly ordered collection of molecular machines operating mostly coherently, with some amount of thermal noise, giving rise to coordinated cell function.

The cytoskeleton spans the range from the centrosome near the nucleus to the cell membrane, defining cell architecture both structurally and dynamically. Microtubules and AFs as well as intermediate filaments self-assemble and disassemble when needed by the changing state of the cell, as in mitosis or cell motility. In neurons, which do not divide, dynamical processes are mainly related to neuronal development, cell motility, differentiation, synaptic generation, and synaptic plasticity. Active transport of intracellular material is carried out in all cells by motor proteins such as kinesins and dyneins, which bind to MTs, or myosins, which bind to AFs. In neurons, motor proteins transport synaptic components such as neurotransmitter molecules along MTs to their final destination guided in this task by microtubule-associated proteins ("MAPs"). Calcium influx into the presynaptic neuron terminal activates calcium calmodulin-dependent kinase II, CaMKII, which causes the vesicles to dissociate from the cytoskeleton and move towards the presynaptic cell membrane. The vesicles dock to the membrane through the process of exocytosis when released to neurotransmitters into the synaptic cleft. Structural

defects occurring in the cytoskeleton due to genetic mutations or traumatic insults often lead to neurodegeneration with serious consequences for health and disease. For example, in Alzheimer's disease, MAP tau proteins are detached from MTs, aggregating into neurofibrillary tangles causing MTs to destabilize and disassemble leading to neuronal malfunction and eventual death [75].

Various forms of MT disruption occur in nearly all neurodegenerative diseases confirming the key role of the neuronal cytoskeleton in the cognitive functioning of the brain. The cytoskeleton's architecture allows for long-range communication both within the neuron and between neurons or other cells, with the active participation of such structures as gap junctions. Ionic and neurotransmitter signals have the ability to alter cellular architecture affecting both its functioning and plasticity.

For example, changes in synaptic strength can regulate and modify neural firing patterns altering such cognitive functions as learning and memory. Importantly, we believe that neuronal functioning is unlikely to operate via a command-and-control mechanism since both ion channels and membrane receptors transduce signals bi-directionally between the cell and its exterior. Pharmacological agents and signaling molecules can lead to induced alterations of MTs by modulating calcium channels in neurons and influencing calcium signaling. The $GABA_A$ receptor, which is essential to anesthetic function, can be directly affected by MTs.

In summary, the cytoskeletal matrix (composed of MTs, AFs, intermediate filaments, and a host of proteins that interact with them) organizes structurally and functionally cell interiors. However, it is unlikely to be simply a mechanical structure needed for the tensegrity of the cell since it is electrically sensitive to ionic, protonic, and electronic conduction in addition to mechanical signals and chemical processes. Its electrical sensitivity offers the possibility of very fast electrical and even electromagnetic signal propagation across the cell and between cells potentially being synchronized across an organ or a tissue.

The cytoskeleton can be viewed as forming a dual electromechanical signaling network responsive to electrical, electromagnetic, chemical, and mechanical stimuli that can propagate and influence the cell's interior as well as other cells in the vicinity.

Furthermore, it is highly probable that some synchronization, which is needed for coordination within such a network may include quantum mechanical mechanisms.

3.27 The Special Role of Neuronal Microtubules and the Cytoskeleton

Among the various subcellular biomolecules within the human brain, one of the most important elements is tubulin, which is a building block of microtubules. Tubulin is a globular protein whose stable form is a heterodimer composed of β and α tubulin monomers. MTs are cylindrical polymers of the tubulin dimers with a 25-nm outer diameter and a 15-nm inner diameter forming a hollow cylinder. MTs are typically comprised of 13 longitudinal protofilament chains of tubulin dimers (Figure 3.8) [76]. Both α and β tubulin monomers are negatively charged but their charges are largely screened by the surrounding positively charged counterions. As a result, in polyelectrolyte solutions such as the cytoplasm, each tubulin dimer as well as entire MTs possess very substantial net electrical dipoles. Within each cell, tubulin isoforms are expressed by many different genes and may structurally and functionally differ as a result of genetic mutations, post-translational modifications, phosphorylation states, binding affinities for ligands, and MAPs. This can cause changes in both their net dipole value and its orientation. This leads to an enormous number of combinatorial possibilities of storing biochemical and biophysical information over both long- and short-term using the various modes of information encryption in MTs. This may be particularly important in the human brain where MTs are very abundant (with an average of a billion tubulins/neurons). Hence, MTs appear to be exquisitely suited for information processing, encoding, and memory as argued forcefully by Penrose and Hameroff [64, 65] over the past two decades of research into the physical basis of consciousness.

These highly symmetric and fairly stable neuronal microtubules could well be the key players in the formation of consciousness states. Microtubules could also work as the mediating structures for sensory inputs, as information processors, translating information into an output response. It has long been discussed and debated but remains an unanswered question, "where does consciousness come from and where does it go when we die". Well, microtubules in the brain's higher center neurons lie at the center of this quandary.

The highly negative surface charge of MTs attracts positively charged counter-ions in the cytosol, which condense around it and form highly moveable ionic clouds. An externally applied voltage gradient (for example by a charged macromolecule or an action potential) can set this cloud in motion along the MT, which could then be used as a signal or stimulus at a destination within the cell that co-localizes with an MT end. Similar properties were observed for AFs showing extremely coordinated dynamics of ion flow along their lengths propagating in a soliton-like fashion [77]. Hence, electric fields generated by firing neurons can strongly affect the neural cytoskeleton by influencing the movement of counterions along AFs and MTs, which could then lead to rapid inter- and intra-cellular signaling and communication. Furthermore, as discussed earlier moving charges create magnetic fields around them and this generates a Lorentz force that could promote the opening of ion channels, for example, calcium ion channels, in the presynaptic nerve terminals that promote the release of neurotransmitters into the synaptic cleft.

Since the neuronal cytoskeletal network effectively connects synapses throughout the cell, electric signals can propagate across the network in response to multiple inputs affecting the neuron's response and hence can play a major role in cognitive functions.

Conversely, the interactions between electromagnetic fields generated by firing neurons and the cell interiors can turn the cytoskeleton into a network of protein-based antennas able to amplify the neural response to exogenous or endogenous electromagnetic fields. An alternative (or parallel) channel of information storage and processing could involve electronic, protonic, or ionic conductivity. Electronic conductivity, while yet unproven, could occur through the protein itself via electron hopping based on semiconducting bands of tubulin that are formed when tubulin assembles into an MT lattice, which would validate the hypothesis put forth by Albert Szent-Györgyi more than 75 years ago!

Unlike other cells, neurons once formed, do not divide, and so neuronal microtubules in principle could remain assembled indefinitely, in the absence of subcellular damage caused by disease or trauma, providing a stable medium for memory encoding.

In non-neuronal cells, MTs are organized radially extending from the hub-like centriole outward toward the cell membrane, having the same polarity which means that they all point in the direction of their plus ends (Figure 3.3). However, MTs in neurons are arranged in parallel bundles, which are of uniform polarity in axons but in dendrites and cell bodies/soma, they are short and aligned in mixed polarity networks. In neurons, MTs are also interconnected by MAPs for enhanced stability. These dendritic-somatic MTs are very well suited for memory encoding and information processing due to their key roles in cell organization and functioning (e.g. by providing transportation networks for the movement of motor proteins and mitochondria) and due to their regular, geometric lattice structure. These processes might involve electromagnetic field interactions. It was suggested in this context that each tubulin in an MT lattice may be represented by a particular conformational state, e.g. binary "bits", which can be assigned values of either 1 or 0. This bit of biophysical information can represent either a dipolar state, an electronic state, a conformational state, or a phosphorylation state. Such states could also physically interact with neighboring tubulin states (e.g. by dipole-dipole coupling for dipolar states or by electrostatic interactions for electronic states) and this can be modeled using a molecular "automata" algorithm in order to generate interactive patterns that describe information processing that may occur. This can be further extended to include quantum states representing qubits. For instance, induced dipole states of the tryptophan residues of tubulin can be photo-excited leading to the generation of quantum exciton states that can represent specific quantum bits ("qubits") which could then be utilized in quantum computational MT lattice networks.

Another important question that arises is in regard to the integration of quantum metabolism, MTs, and consciousness or quantum cognition. Although the electric fields generated by the tubulin arrangements comprising microtubules can be explained by classical science, the neural sensory inputs into the brain from the surrounding environment in the presence of accompanying neuronal influences such as emotional memories all contribute to the formation of cognition. Classical components of neuronal activity include the release of neurotransmitters from the terminals of neurons into synaptic clefts resulting in the opening of cation (positively charged ion) channels, such as sodium channels in the postsynaptic neurons. This promotes a change in the transmembrane potential with an attendant depolarization and firing of action potentials. The influx of cations generates ionic waves, which propagate along both AFs and MTs involving slightly different mechanisms of ionic wave propagation and playing different roles in affecting neuronal behavior. MTs, or indeed other biomolecules such as the DNA, may have a large number of degrees of freedom, some of which can operate in the classical realm, e.g. vibrational dynamics, while others, such as the aforementioned tryptophan excitations may operate in the quantum domain. The two types of degrees of freedom are expected to interact with each other such that classical dynamics may affect quantum behavior and vice versa. Ionic waves of the cations surrounding biofilaments propagate in the form of clouds along AFs virtually without loss of wave amplitude and speed, which is characteristic of soliton-like waves whose properties include loss-less energy transport at a constant velocity and a localized amplitude. When this electric potential wave reaches the end of an AF, it can trigger a polarization event, for example causing neurotransmitter release from a synaptic bouton. This may occur as a result of magnetic fields generated by counterion waves that accelerate the movement of charged particles in the cell. Multiple ion waves associated with different AFs may cause the accelerated movement of multiple ions such that magnetic fields from the two charged particles may interact causing a Lorentz force that opens calcium channels that promote neurotransmitter release. Similarly, cations as counterions can form and propagate along and around the surface of MTs. However, the conduction of ionic waves along MTs may also take the form of spiral-shaped signals (due to the helical organization of the MT protofilaments formed on their cylindrical surfaces) that generate a solenoid-like effect with magnetic fields produced in the MT lumen. This effect could be viewed as shielding the magnetic fields from environmental noise, hence providing a "safe" communication channel. Moreover, *ion conduction along MTs was shown to exhibit an electron-hole-type phenomenon typical of a transistor in the process of amplifying the wave signal and could indeed represent a quantum effect.*

Furthermore, MTs can propagate ionic waves into a point of contact with AFs, which can then relay these waves all the way to ion channels. It is also known that MTs interact with some ion channels directly, which may help explain the role of MTs in consciousness using more conventional neuroscience approaches.

Moreover, an integrated model of the quantum nature of consciousness, i.e. of cognitions, with the notion of quantum metabolism mediated by coherent mitochondrial production of ATP on a macroscopic scale should be considered.

Interestingly, MTs associate with mitochondria which move within the cell along MT tracks using the energy they produce in the form of ATP molecules as fuel for motor proteins such as kinesin and dynein that propel them along these MT tracks. There may also be less well-studied forms of interaction between MTs and mitochondria since mitochondria produce electric fields and emit weak electromagnetic radiation during their activities.

Electric fields are produced by mitochondria due to the proton gradients across the inner mitochondrial membrane and these fields directly affect charged molecules, such as ions, in their immediate vicinity. Additionally, weak electromagnetic radiation is emitted in the form of biophotons produced by the recombination of reactive oxygen species. MTs, in the immediate proximity of mitochondria, may absorb these electromagnetic signals, transducing them into bits of information that can be stored in the subtle structural changes in tubulin such as C-termini conformational changes. This information can in turn affect the way MTs conduct ionic waves due to different boundary effects and hence may influence the whole cell's response to stimuli.

In terms of ionic waves, albeit they exhibit classical behavior per se, they may play a direct role in positively or negatively influencing quantum effects. This would be of special significance in neurons where a change in the cell's threshold for firing an action potential can have an effect on the strength of the synaptic connection between neurons. Ionic waves may also potentially either enhance or inhibit quantum regime, which consonantly improves or impairs, respectively, the quality of mental and emotional cognition. For example, ionic waves may enhance motor protein binding to MTs by exposing C-termini to which motor protein head domains bind. Because motor proteins are responsible for moving mitochondria along MTs, the notion of quantum metabolism may be consistently integrated with neuronal quantum processes. Alternatively, *ionic waves (especially calcium waves) may promote MT disassembly and accordingly discourage MT-mediated neuronal quantum effects.*

MTs, whether as part of a quantum-mediated function or a classical function (the latter possibly related to ionic waves independent of MT function per se), may influence response to neurotransmitters and hence depolarization and neuronal firing. These mechanisms can connect groups of neurons and make them more or less active eventually affecting cognitive processes in either a positive or a negative way depending on the particular mechanism and the specifics of these interactions. The quantum processes involving MTs may be able to spread throughout bundles of neurons to effect cognitions in a nonlocalized coordinate fashion that must accompany correlated bioenergetics for this energetically expensive process. Synaptic functions including synaptic transmission, ion pumping, maintaining resting electro-potential and biochemical pathways underlying synaptic neurotransmitter, and vesicle recycling are extraordinarily energy-requiring. In fact, these processes represent the most expensive energy-consuming processes of the human body under normal healthy circumstances.

> *The arrangement of mitochondria along microtubules provides a spatial correlation with the phenomenon of quantum metabolism allowing it to be coherently correlated with the quantum process of neuronal cognitions. Together, these effects could manifest the most favorable solutions to the issue of free will.*

Although intellectual cognition may also be mediated at the level of neuronal MTs in the brain as information processors as well as storage and retrieval of memory such as from the hippocampal neurons, this is not a defining feature of consciousness per se. While information processing occurs in neurons, the quantum nature of MTs is inherently stochastic and how precisely this translates into the activity of neurons and neuronal clusters is impossible to quantify at present. Neurobiological processes are characteristically coherent, i.e. information is shared between sites effectively. Indeed, there is a daunting impossibility of estimating the origin of a quantum process of cognition beyond the MT. At the same time, it may be construed that normal (non-pathological) quantum neurobiological processes are characteristically coherent, i.e. information is shared between different sites in the brain effectively.

> *Somewhere between quantum stochasticity and quantum coherence is the realm of sub-neuronal functioning, perhaps at the "thin edge of the wedge" between the quantum and classical worlds.*

Making a bold extrapolation, we could say that predicting the destiny of our identity beyond our mortal lifespan is an impossibly daunting task.

3.28 Where Is Memory Stored in the Brain?

Memory is generally attributed to synaptic plasticity. However, synaptic proteins are transient, lasting hours to days while memories often last lifetimes. How can these vastly divergent time scales be reconciled? The currently most commonly accepted experimental model for memory is due to long-term potentiation (LTP) in which synaptic activation results as a result of calcium influx, which triggers activation of the hexagonal holoenzyme CaMKII. Each activated CaMKII contains two sets of six kinase domains, each of which can encode up to six bits of synaptic information by phosphorylation in an appropriate hexagonal lattice for storage and functional processing. Craddock and his colleagues [68] showed how each CaMKII molecule can encode six bits of memory in hexagonal lattices of MTs. If this were the molecular basis for memory encoding in neurons, it would be a very efficient process that could easily mediate with other mental states and cognitive functions. As discussed above in connection with

the limitations of the power consumption of the brain, we can assume that tubulin states could act simply as binary bit-switching units but at very low frequencies of 100 Hz while coherent states of entire MTs could operate at frequencies in the MHz range. However, reduction of conscious processes to this lower level of the hierarchy appears to move further away from global unity and binding taking place in the human brain between the "grey matter" and the mind, which are essential features of life and consciousness. Direct experimental verification of these hypotheses is needed for the field to advance.

In the context of understanding the molecular basis for memory encoding in neurons, it is intriguing to investigate the role of electrostatic interactions at the quantum-chemical level in neuronal functions which extensively involve the movement of charged species. In fact, most drugs, neurotransmitters, and hormones are electrostatically charged, soluble in water (and blood), and bind due to Coulomb interactions in the polar, aqueous medium inside cells and on intra- and extra-cellular surfaces, e.g. to membrane receptors and to surfaces of the proteins of the cytoskeleton. However, many structures in cells also contain non-polar "hydrophobic" regions. These are typically found within membranes, proteins, and other biomolecules, shielded from the hydrophilic charged and polar regions. Amino acids, from which all proteins are constructed, can be divided into either polar and water-soluble ("hydrophilic") or non-polar and soluble in oil-like media ("hydrophobic"). Non-polar, hydrophobic amino acids cluster together bound due to van der Waals attraction, excluding water, forming non-polar "hydrophobic pockets". Such non-polar interiors are found in membranes, proteins, and nucleic acids. Nonpolar regions are comprised mainly of π electron resonance clouds, e.g. of the aromatic rings of amino acids phenylalanine, tyrosine, and tryptophan. Similar to chromophores mediating quantum coherence in photosynthesis proteins (as discussed earlier in this chapter), these π resonance clouds involve the benzene ring, C_6H_6, a hexagonal ring with 3 π electron resonance bonds. Electrically neutral and non-polar, π electron resonance structures are polarizable and interact with other such molecules via induced quantum dipoles resulting in the so-called "π stacking" effect. Moreover, π resonance structures are optically active, able to fluoresce by absorbing photons, which induce excited electron states and then get de-excited to lower states, emitting lower-energy photons. These energy quanta can be efficiently transferred non-radiatively via excitons over individual distances of less than 2 nanometers using the so-called Förster resonance energy transfer (FRET) mechanism [78]. This mechanism was mentioned above in connection with the quantum coherent energy transfer in photosynthesis, upon which life on Earth depends. Interestingly, a similar process is highly possible to take place in MTs as we discuss below.

3.29 Are There Quantum Excitations in Microtubules?

A more complete and in-depth exposition of the ideas presented below can be found in a recently published book [79]. It is important to note that *geometrically organized π resonance molecular structures in hydrophobic regions of biomolecules, when properly shielded from the polar environment, create suitable conditions for the emergence of quantum coherence in biological systems.* It is also worth mentioning that tubulin dimers have 8 tryptophan indole rings arranged in a geometry similar to that found in photosynthesis complexes discussed earlier in this chapter. The fluorescence quantum yield (i.e. the number of photons emitted per unit of time) for the tryptophan amino acid is comparable to that of bacteriochlorophyll providing additional support to the hypothesis that tubulin and MTs can behave similarly to photosynthetic systems in regard to quantum exciton generation and propagation of these states over distances as long as the length of MTs, i.e. over micrometer (μm) ranges that span the length of a cell. In fact, energy transfer between neighboring tryptophan residues has already been observed in tubulin. The tryptophan networks in MTs span the length of each protofilament and hence form 13 parallel conduction pathways for quantum processes (Figure 3.14).

This was demonstrated recently in computer simulations based on the model developed for photosynthetic light-harvesting complexes [80]. These studies found coherent quantum beats (which means that two frequencies of quantum oscillations with similar values are superimposed) lasting less than 1 nanosecond that can travel the length of the tubulin dimer, which indicates that it is possible for an entire MT to effectively transfer exciton energy along its length. In the geometrically structured neuron cells, various MAPs interconnect MTs into networks, forming scaffolds for both intra- and intercellular architecture. Hence, tubulin's tryptophan indole rings can provide a suitable substrate for coherent quantum oscillations within MTs leading to the emergence of quantum states in the neuronal cytoskeleton. Rapid coherent quantum communication through electronic conduction pathways through aromatic groups may coordinate the complex functioning of the neuronal cytoskeleton such as neuronal firing, motor protein tracking, and cell motility. While this quantum excitation

dynamics in MTs has only been modeled computationally but not confirmed experimentally yet, there is a high probability that MTs exhibit the predicted effects. Moreover, MTs have been shown to reorganize after exposure to UV light, which can possibly link it to mitochondria that generate such electromagnetic waves. Furthermore, aromatic amino acids such as tryptophan, tyrosine, and phenylalanine, which are abundant in tubulin, can be used as acceptors for hydrogen bonds that may be perturbed upon excitation. They could play a critical role in controlling or regulating both the conformations and motion of proteins in general. These mechanisms may explain the observations of an apparent UV mediated cell-to-cell influence on cell division.

These biophotonic and bioelectronic activities are not independent of biological events in the nervous system, and their synergistic action can play an important role in cytoskeletal ionic signaling, neural signal transduction, and even neuron-to-neuron co-ordination.

Alteration of these activities via physical disruption (i.e. mechanical vibration, ultrasound, electromagnetic fields), or the chemical effects of drugs (e.g. anesthetic molecules or psychoactive compounds) can alter this signaling cascade. Note that *the cytoskeleton is a target for many drugs, some of which alter mood and subjective experience.* For example, the antidepressant effects of fluoxetine (Prozac) take several weeks, apparently because of the need for cytoskeletal reconfiguration. A more comprehensive enunciation of the ideas proposed in this section is available in [66].

3.30 Is Anesthesia a Quantum Process?

One of the greatest medical discoveries of all time is still an enigma in the 21st century. The nature of the molecular mechanism by which anesthetic molecules cause a reversible loss of consciousness, muscle control (paralysis), perception of pain (analgesia), and memory (amnesia) still remains unknown today. The initial discovery of anesthetic molecules, which enable all of modern surgery, sparing the patient the agony of pain, occurred, as is often the case, due to serendipity when in the mid-19th century, gases like diethyl ether and nitrous oxide were accidentally found to have "anesthetic" properties. It was observed that inhaling these gases caused animals and humans to lose consciousness or exhibit some associated effects like catatonic states in response to stimulation. Exhaling an anesthetic gas, the subjects woke up and regained consciousness, with their mental and physiological functions essentially unchanged. Anesthetic potencies for each molecule were gradually quantified as correlated with the gas concentration at which half of the subjects lose consciousness and the other half stay awake. It was discovered to great surprise that each anesthetic acted at the same concentration in all animals or humans subjected to it. In the intervening decades, dozens of anesthetic gases were found and characterized. It is hard to understand and explain on the basis of a molecular

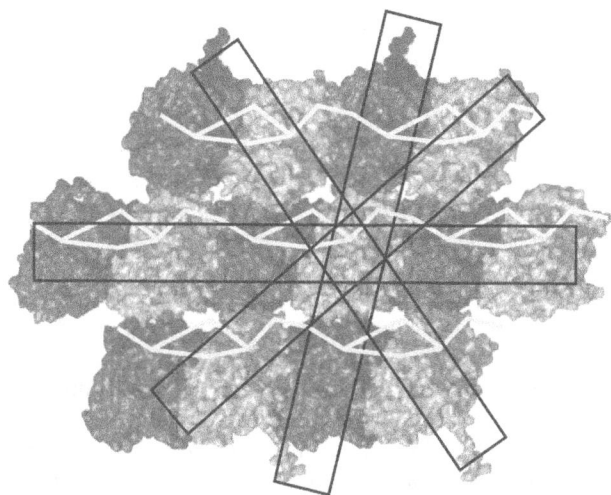

FIGURE 3.14 A lattice of seven tubulin dimers as found in the microtubule lattice. The lines connect tryptophan residues, and rectangles show four possible winding patterns for exciton transfer processes.

type "lock-and-key" principle that they represent a wide range of dissimilar chemical structures such as ethers, halogenated hydrocarbons, nitrous oxide, and even the inert gas xenon, which is the smallest-size anesthetic. Despite these significant chemical differences, each molecule has a particular potency for all animals and humans. For these reasons, *the mechanism of action attributed to membrane or protein receptors in living cells, i.e. the lock-and-key principle does not seem to apply to anesthetics.* However, a clue regarding a molecular mechanism of action involves solubility in a particular non-polar medium. This is based on an observation made by Hans Meyer and Charles Overton who, over a century ago, ranked anesthetic potency for many anesthetics and tested their solubility in various solvents.

> *The potency of all anesthetics correlates very closely with their solubility in a particular non-polar, lipid-like, hydrophobic environment.*

This finding led to the hypothesis that anesthetics act in lipid regions near and at neuronal membranes in the brain and hence disrupt membrane fluidity, which subsequently affects neuronal signaling. However, this idea still leads to many inconsistencies. For example, increasing body temperature disrupts membrane fluidity to an extent comparable to anesthetic action, but this physical effect does not result in anesthesia. Moreover, there are compounds with structures and lipid solubility, which should lead to a prediction by the Meyer-Overton correlation to act as general anesthetics but they only exert convulsive effects and no anesthetic action. Chemically homologous anesthetics, such n-alkanes, exhibit increasing potency with increasing molecular size, which is consistent with the Meyer-Overton hypothesis. However, above a critical molecular size the series loses anesthetic potency despite increasing lipid solubility Anesthetics have also been found to bind to various proteins, most importantly to tubulin and they are known to act on MTs. In view of the prominent role played by the cytoskeleton in all eukaryotic cells, especially neurons, this merits close examination.

> *Recently, it has been hypothesized that the mode of action of anesthetics is indeed crucially linked to the binding affinity for tubulin, not exclusively membrane proteins as has been previously assumed. How exactly anesthesia is induced by various anesthetic molecules binding to microtubules is a major research question that should be investigated by modern experimental and computational methods. An answer to this question may well implicate quantum processes in the mechanism of anesthetic action. If this is supported experimentally, a new mechanism of action involving microtubules will undoubtedly lead to the development of new and improved anesthetic molecules with fewer side effects.*

3.31 Relevance of Quantum Biology to Health and Disease

It may be that quantum biology is the spark of life, the quantum ignition switch that needs to be implemented to achieve singularity, not only technologically in man-made machines, maybe even leading to a perpetual motion machine, but more importantly in terms of biological beings. Quantum biology may be considered as part and parcel of singularity. Classical principles, which are used to create machines, can never be as complex and profound as living cells or human beings and especially the human brain. In terms of technological application and the comparison of the efficiency and complexity of human beings to machines, a new wave of bio-mimetic material is sweeping the technological landscape trying to mimic nature. By comparison, the area of quantum biology is somewhat retrograde because it aims to understand nature through the proximity of technology rather than the other way around. Nanotechnology is allowing us a comprehensive understanding that is unprecedented in the history of science by probing biological systems at space and time nanoscales. Such an understanding brings us close to the quantum realm description of biological systems. *The birth of modern nanotechnology was inspired by the development of a scanning tunneling microscope that enabled "visualization" of individual atoms. The field of quantum biology has been elucidated by recognizing the quantum nature of photosynthesis and mitochondria, the indirect effects of metabolic enzymes, and the allometric scaling laws of physiology.* The latter demonstrates mathematically that basal metabolic rate is a function of size, especially in optimally functioning healthy cells. Such correlations were drawn by applying equations of solid-state physics, specifically, those Debye utilized to show that crystals use vibrational energy to demonstrate the quantum nature of energy production. What happens when pathological states set in is a problem that can also be better understood within the context of decoherence and desynchronization, which are at the center of the quantum biology debates.

In particular, inflammation and oxidative stress are the driving forces of clinical diseases that start as a subclinical issue but progress to the potential for a feed-forward culmination of disease states. The process of disease progression is a promiscuous one independent of the type of tissue or organ system involved. Hence, the current health care model of compartmentalized and sub-specialized medicine is too narrowly focused and should probably be revisited.

> *The ideas of quantum metabolism and quantum biology provide a new perspective for understanding disease, which could lift the veil of opaqueness and provide a potential for at least preventing disease, if not treating it.*

The breakdown for quantum metabolism for example appears largely rooted in mitochondria and their energy-producing networks with super-physiologic levels of oxidative stress and heat generation associated with the process of inflammation. This is commonly initiated and sustained by dietary excess relative to physiological requirements whereby too many electrons overburden the electron transport chain. Many of these electrons leak and are scavenged by oxygen molecules to form superoxides and free radicals (see free discussions on radicals in Chapter 2). This signals the generation of the NFkB complex leading to an inflammatory response that underpins loss of mitochondrial structure and function. This insight in part helps to explain the Warburg Effect and the association between diabetes and cancer, the latter by impairing nutrient metabolism by oxidative metabolism and hence providing a selective advantage to competing cancer cells, which favor extra-mitochondrial glycolysis for the energy needs of cell replication. The reader is referred to Volume 2 of this book for more detailed discussions on the role of the Warburg effect in cancer and diabetes.

> *Mitochondrial dysfunction also lies at the center of all chronic diseases of aging. Fundamental to this issue is the reciprocal relationship between mitochondrial dysfunction and insulin resistance.*

Further, perhaps most striking in the context of the biological significance of quantum behavior and mitochondrial dysfunction is the connection with neurodegenerative processes of accelerated cognitive decline and Alzheimer's disease. Also, the connection between mitochondria and MTs, viewed as quantum oscillators, encoding and storing the information of cognitions or consciousness, resonates with the idea of mind-body connections and even the notion of free will.

Systems that are supported by glycolytic metabolism rely on a much lower efficiency of energy production and hence the lower energy requirements translate into less biological complexity in terms of interconnectedness. *For example, cancer cells are autonomous and require enough energy for replication but not for functional differentiation. Normal tissues that are overburdened with excessive dietary consumption are characterized by impaired mitochondrial structure and function via the inextricably linked processes of oxidative stress and inflammation.* These processes lead to the already mentioned Warburg Effect with the anaplerotic process of using mitochondrial substrates for the reproductive needs of cancer cells. Hence, this can be the root cause (or one of several causes) of cancer initiation and progression and as such deserves serious preventative and therapeutic consideration.

Insulin resistance that underpins type 2 diabetes is characterized by increased pyruvate kinase with reduced pyruvate dehydrogenase enzyme complex, which when functioning normally transitions the product of glycolysis (pyruvate) from the cytosol into the mitochondria. Insulin resistance and type 2 diabetes represent the classic metabolic disease states that may be characterized by a breakdown in quantum metabolism to a metabolically classical regimen. In this case, the metabolic rate is reduced in the sense that less ATP is produced per unit nutrient of glucose per unit time. The amount of energy production of ATP in mitochondria from one mole of glucose ranges between 32 to 34 moles in comparison to only 2 moles produced in the cytosol from glycolysis.

> *The transition from quantum to classical regime of energy production occurs when the mitochondrial capacity for energy production is exceeded by nutrient supply.*

In the case of a crystal lattice, the harmonic oscillation of the molecular structure of each node produces a quantized acoustic wave (phonon) whereas in the case of mitochondria the wave function may be hypothesized to be electromagnetic in the form of biophotons or electromagnetic fields. The latter are predicted to emerge on the basis of the electron current generated by the transfer of electrons across the electron transport chain complexes as well as the iron-sulfur core of the enzyme complexes 1 through 4 (Figure 3.9). In addition, three copper atoms in conjunction with two heme groups facilitate electron transfer in complex 4.

In a biological organism such as a human, having a central nervous system may be sufficient to maintain coherence throughout the body as downstream signals could then be transmitted through the neuroendocrine system, the autonomic and peripheral nervous systems. There is empirical evidence that whole-body coherence exists and it functions in the quantum regime. Such evidence includes the virtually instantaneous time from visual stimuli in the head to the acral extremities, which cannot be explained by nerve conduction times or classical models of physiology. Cognitions generated in the central nervous system may be propagating synchronously in phase through long-range coherence with distant body parts, both in an autonomic fashion mediated by the neuroendocrine and nervous systems. However, in order for such long-range coherence to occur, classical physiology should be augmented by quantum metabolism, long-range correlated bioenergetics, and other quantum biological phenomena. In particular, the mechanisms involved in the tubulin dimer's induced dipole-induced dipole interactions may function to promote coordinated neuronal depolarizations that synapse across wide bands of neurons involving gap junctions. These mechanisms may be spatially and temporally correlated via the necessary energy supply from ATP hydrolysis at a physiological scale to the energetic demands of neural, as well as neuroendocrine functioning. It is also intriguing to invoke organized structured water, both intracellular and extracellular, since it can facilitate electromagnetic signal propagation at the speed of light and hence much faster than the consecutive synapsing relay of neural transmission. This would allow virtually simultaneous transmission of signals between disparate parts of the body and their organism-wide synchronization. Principles of synchronization of interacting parts within a biological system are likely universal across scales. Phase locking of a team (e.g. the extraordinary feats of hockey and football teams) may be analogous to a school of fish or a flock of birds whereby the individual organisms are functionally a part of the greater whole. Analogously, the interactions within a small protein and between proteins are part of the entirety of the cell that contains copies of this particular protein. In all, these examples involve amazing interactions that while still inexplicable by the classical models of science, must invoke mechanisms of quantum entanglement and superposition states or at the very least their metaphorical representations.

> *All biological systems are inherently composed of electromagnetic entities, bound by the fabric of electronic configurations and fields generated by and interacting with charged particles or magnetic spin. The quantum nature of these interactions is by definition delocalized.*

Although this delocalized quantum nature of matter has traditionally pertained to the subatomic scale, macroscopic scales of non-classical electromagnetism in biology are increasingly recognized. The real elucidation of how parts of a greater whole may function together in unison at widely different scales, as mentioned in the examples above with such unimaginable and surreal beauty, appears to be fundamentally rooted in the science of quantum electromagnetism.

More specific examples can be found illustrating a general biological theme that in the human body electromagnetic energy produces electronically excited states, which correlate with qualitatively essential biochemical processes that promote metabolic health. For example, melanin in the skin is a chromophore, which absorbs sunlight to produce vitamin D. This is very analogous to the heterocyclic aromatic rings, chlorins. Magnesium-containing chlorins are called chlorophyll, the central photosensitive pigment in chloroplasts of plants. In both cases, a certain wavelength of physical light is absorbed by the chromophore by exciting an electron from its ground state into an excited state. (Conjugated chromophoric molecules have alternating single and double bonds creating the delocalized excited electron state between energy levels extended along orbitals.) Vitamin D is associated with significant salutary effects primarily on the innate immune system, which have been correlated with benefits in terms of breast cancer, colorectal cancer, and a host of other cancers. For example, leukemic cells bathed in vitamin D solutions turn normal [81]. Vitamin D is also associated with benefits in terms of cardiovascular disease, diabetes, obesity, and neurodegenerative processes. In terms of autoimmune disorders, using high doses of vitamin D at the initial onset of type 1 diabetes, schizophrenia, Graves' disease, or inflammatory bowel disease, there has been a significant number of reported cases resulting in complete therapeutic remission. Furthermore, lymphoma cells bathed in vitamin D solution have been shown to revert to their normal state [82]. Recently, there have been some interventional studies, which have demonstrated that vitamin D reduces the progression from pre-diabetes to diabetes. It appears that the benefits of vitamin D in terms of sunlight exposure occur with 10 to 20 minutes of direct exposure in the summer months between the hours of 10 a.m. and 3 p.m. in the northeast United States. It would be expected that the same is true throughout the year in geographical areas south of about 70°s latitude. In any event, it is intriguing to consider vitamin D in the context of photon absorption by melanin, which is analogous to aromatic porphyrin rings in cytochrome oxidase of mitochondria, richly endowed in the cells of the innate immune system due to their high energetic demands or in hemoglobin. Clearly, *the electromagnetic effects of sunlight have physiological significance.*

The most interesting source of biophoton production is the proton pump and it has a relationship with the mitochondrial electron transport chain enzyme complexes. Protons that have been pumped across the inner mitochondrial membrane into the inter-membrane space of the mitochondria seek to reenter back across the inner mitochondrial membrane through proton gates. As the protons impinge on the proton gates, their movement is decelerated and some of that reduced kinetic energy is transferred into the gate and utilized to excite the heme group (the porphyrin ring containing iron which facilitates electron tunneling) pulling an electron from the valence and subsequently into the conduction band of iron. Consequently, conformational changes occur in the structure of the gate promoting its opening. As the protons pass through the gates in complexes 1, 2, and 4, the excited electron of the corresponding heme group is then returned to its valence shell at which point a photon in the near-infrared range with the energy of approximately one electron volt is released as a biophoton. Molecules containing iron are contained in complexes 1, 2, and 3 while both iron and copper are contained in complex 4, which is formally considered cytochrome oxidase. It is intriguing that these roughly 1,000 nm wavelength biophotons have an energy level equivalent to about two ATP molecules. Consequently, they may be important in facilitating enzymatic catalytic activities in the cell. Biophotons may be responsible for the spatio-temporal correlation of quantum metabolism. That is, they may provide the energy for motor proteins to move mitochondria along microtubules in a spatially correlated manner. Biophotons may also catalyze the movement of other electrons from their NADH or FADH2 donors of other electron transport chains within the same cell across mitochondria and even between cells in a temporally correlated manner. This new insight may support the idea that electromagnetic fields projected by the electron transport chains may have a role in synchronizing metabolic activity between mitochondria and even between cells and tissues. While an electron transport chain emitted magnetic field is likely to be weaker than a biophoton, it is again unlikely that nature would not have intended to have a purpose for it. In fact, a multipurpose maximally-efficient organizational value of such mechanisms is likely to have been retained over two billion years of evolution. Due to very low intensities of the biophoton emission compared to the flood of photons bathing the environment, these effects are extremely hard to detect.

In experiments performed by Daniel Fels at the University of Basel in Switzerland, dividing cells were kept in separate quartz containers (in order to allow unimpeded transfer of electromagnetic energy) so that there was no physical or chemical contact between them [82]. However, they were shown to synchronize their cell cycles when the containers were close enough indicating that there must be some kind of electromagnetic means of communication between these cells. When the distance between them was increased, or an obstacle was put between them (e.g. a screen), these cells desynchronized. So, these experiments appear to highlight that photons may actually be helping to synchronize cellular activities in an efficient manner. This can be similar to the role of ATP molecules, whose free energy amounts to about 0.5 electron volts and is utilized to promote the activation energy of enzymatic catalytic reactions (acid-base, electrostatic, covalent, and proximity/orientation).

Biophotons, not unlike ATP, could provide an ample supply of energy that can be utilized by cellular processes. While biophotons can provide a similar amount of energy to that which ATP provides, they are less localized in space and hence play a greater role in orchestrating synchronized coherent activity within and across cells. For example, this could assist in transporting electrons from reactions in the TCA cycle to NADH or FADH2 to the electron transport chain in complex 1 (see Figure 3.9). However, in the context of metabolism and biology, such as the role of biophotons as promoting synchronous, coherent ATP production, nuclear transcription, and other metabolic activities, much still needs to be explained. Perhaps if many biophotons are released simultaneously, electron transport chain oscillation frequency may be maintained in synchronous coherent fashion such that the collapse of the wave function with each particular biophoton affects only mitochondrial enzyme oscillations that otherwise would not keep pace.

In summary, *there may be simultaneously up to four electromagnetic mechanisms operating inside the electron transport chain.* First, biophotons may be generated from the recombination of reactive oxygen species or alternatively from the heme excited state of electrons making their transition back to the valence orbital from the conduction orbital. Second and third are the magnetic fields (always associated with movement of charged particles), which may in this case be the vertical motion of protons or the horizontal motion of electrons (notably two electrons moving at a given time) in the inner mitochondrial wall. Finally, magnetic fields may be due to the spin magnetic moment of iron ions present in complexes 1, 2, 3, and 4.

3.32 The Feasibility of Encoding the Totality of the Human Experience and the Information Field of the Brain

It was recently shown that a 53,000-word book can be encoded in our DNA. In fact, 5.5 petabytes (a quadrillion bits) can be stored per cubic millimeter of DNA. Conversely, the numerous bits of genetic information encoded DNA can be stored in hard drives and magnetic tapes. Hence, the transfer of information between living and nonliving matter is not only possible but offers significant technological advantages. Imagine the possibilities of enriching living matter with the information of nonliving media. This has many nuances of implications for how information flows not only to and from the systems biology of individuals but how it is transferred across systems at every length and time scale of relevance to biology. As stressed earlier in this chapter when discussing biophotons, principles of electromagnetism are profound and fundamental to living systems. Without question in the coming decades, when these principles are fully explored in the context of biology, we will be poised to profoundly transform how we understand human life in addition to having explosive implications for future medical research and clinical medicine. *Electromagnetic phenomena of significance to a biological system appear to generate oscillatory patterns characteristic of the physiological activity in the cell as well as higher levels of organization such as the heart or the brain.* Accordingly, it follows that cells, or indeed some subcellular components (e.g. microtubules) are in a sense electromagnetic antennae that resonate with their environments.

The human brain can store at least 10^{20} bits of information, as estimated more than 50 years ago by the famous physicist John von Neumann, and possibly much more. What is truly remarkable is that encoding this massive amount of information represents a minuscule energy requirement for human physiology. The question that arises is whether the totality of an individual's life experience can be stored without loss and eventually transmitted to a memory storage device for an indeterminate amount of time after the individual's passing. This is one of the dreams of the movement called trans-humanism.

Indeed, this author is of the opinion that this huge amount of information does not have to be lost. Moreover, at least in principle, it can be transformed into an electromagnetic form of information. There is no deep mystery here. Even today our modern electronic communication devices are based on the principle of wireless information transfer via electromagnetic wave encoding, and its decoding at the receiver and storage in a hardware device such as a desktop computer, a laptop, or a cell phone. The biological analogy involves biological bits of information being encoded into the various classical and/or quantum states of tubulin dimers in microtubules, which are distributed among the 100 billion neurons of the brain. Moreover, each tubulin dimer (or even one degree of freedom of the many possible ones utilized for memory and information processing by each tubulin dimer) in the brain can emit one photon. Each photon can carry a bit or several bits of information if additional aspects are included such as amplitude or frequency modulation. Then, *the photon field released from the human brain in the form of an electromagnetic information field could easily contain all of the information stored in a human lifetime.* This would be analogous to sending wirelessly a movie file to be downloaded by an external device. The difference is only in the size of the file. While a 90-minute movie can be stored in a gigabyte file, a human life's experience would require many petabytes of information. However, the principle is virtually the same.

> *The implications for this still futuristic but entirely feasible technology are truly game-changing. We hypothesize here that the brain is capable of encoding all our experiences, perceptions, and sensory inputs and storing them in several possible states of tubulin in neurons.*

Some of it may be damaged or destroyed by various biological, chemical, and physical interactions. Our memory is not one hundred percent intact but, under normal physiological conditions, a lot is accumulated and can be preserved long-term. What is truly awe-inspiring is that at the time of death, this information encoded in the electronic structure of various tubulin states may be released by emitting photons. Each tubulin state can release a photon that carries that specific bit of information so that a photon field from the entire brain may be released as a biophotonic information field. Hence, we start at infancy by developing the brain's structure (hardware) and filling it with information and instructions on how to retrieve and manipulate the information (evolvable software due to synaptic plasticity). As we mature and age, we continue uploading data and improve the speed and efficiency of manipulating it as we learn and experience life in all its forms. The brain's hardware structure is constantly being prepared, regenerated with fresh protein replacing damaged ones, restored and preserved.

Moreover, all these processes including information encoding and processing require very small but non-zero energy inputs in the form of ATP molecules (biological energy quanta). Alas, at death these mechanisms grind to a halt. Since a high energy state of being alive and metabolically active transitions to the absolutely thermodynamically stable state of being dead, there is an energetically coupled possibility of emitting a biophoton "storm" that can carry the totality of the information stored in the brain. This is analogous to an emission of a photon when a valance electron of a hydrogen atom falls from an excited to a ground state. In the case of the human brain, this would require on the order of 10^{18}–10^{19} photons to carry all this information now encoded in electromagnetic fields so that basically the entire human experience of a lifetime could be released as a single photon field. The energy involved in generating such a biophoton field is relatively small, only between 0.5 and 5 joules, barely enough to be recorded by a photo-detection device. For comparison, *the emission of biophotons from a living human amounts to 1–1,000 photons per second per cm² of surface area of the body. Consequently, an entire body emits up to ten million photons a second.*

This biophoton information field plausibly may be used to preprogram another field, another brain. In other words, it may be absorbed and integrated into another substrate. Brains are the substrates for cognition. But what happens when we die? This information field can possibly be transformed into an electromagnetic field, which is not bound to a physical substrate anymore. It is released into space. Where it goes from here is unclear, unknown, entering the realm of spirituality. However, staying within the scientific realm of physics, the process described above is entirely plausible. We are not talking about delusion or fantasy. Electromagnetic waves are used for radio transmission or TV transmission. So, we can easily imagine that our brain is a large-scale movie production. We started the digital revolution with digital memory, using memory sticks, encoding a gigabyte. We now have digital terabyte memory sticks, which are electronically encoded. This is not decoded until inserted into a USB port. When we put it in a computer and press a start button, this will produce imagery on the computer monitor, which represents data transformed into electromagnetic fields. It is exactly the same principle that we employed for the human brain. John von Neumann estimated the number of bits of information that enter the human brain through sensory inputs over a lifetime. We showed that one can actually store the entire lifetime of data encoded in the brain in the states of its tubulin structures forming the cytoskeleton of the neurons in the brain. What we know fairly confidently are the upper limits of what can be encoded. We do not know exactly how much information each one of us stores in our brains. Numbers can vary depending on life histories. However, while we scratch the realm of possibilities here, this is not science fiction. It is entirely conceivable that a single neuron has the potential for storing and processing one gigabyte of data and we have one hundred billion neurons in the brain. Hence with very simple math, one can conclude that 1,020 bits of information stored in the human brain should not be a stretch. What this is also telling us is that *a neuron, contrary to the current paradigm of neuroscience, is not a single-bit device. It is not a simple "on"/"off" switch but a complex*

memory storage and information processing device performing complex and fast operations.

A neuron can be closer to a computer than a binary switch.

To support this statement, recent research demonstrated that a single cell can learn and respond according to the experience gained through learning. An on/off switch could not possibly do that!

There may also be some additional information compression at play. A two-hour movie could be stored in a gigabyte file, which corresponds roughly to one neuron's capacity for information storage. Therefore, if one neuron could store two hours of life experience, and we have a hundred billion neurons, then you can imagine that it is entirely trivial to encode our entire experience into the memory of a neuron since even a 95-year life span would only amount to 400,000 two-hour-long "movies". These "movies", if properly "packaged" into a compressed wave, could at the moment of passing be released in the form of an electromagnetic field. This is even irrespective of a possible modulation of these waves. For example, in AM radio we have amplitude modulation and in FM, radio frequency modulation. Hence, the brain has a lot of potential for powerful encryption of information into the waveform of these photons that could additionally possess an internal spatio-temporal structure.

Since the time scales of the dynamic processes taking place in the brain range from a femtosecond (10^{-15} s) for atomic motions in proteins to microseconds (10^{-6} s) for conformational changes in proteins, to milliseconds (10^{-3} s) for ion channel activity, to seconds for conscious perception, to days for circadian rhythms, to months for seasonal changes to years for aging resulting in a lifespan of approximately 3×10^9 s. This means that the time scale of relevance to the human brain spans 24 orders of magnitude, hence some form of fractality would be a useful way of encrypting deeper, faster time scales inside slower processes. Hence, there may well be a fractal structure of the biophoton field. In other words, using an earlier metaphor, it is almost as if one could unfold the entire book of life whereby the long-wavelength photons would have inside them higher frequency and lower wavelength photons within the envelope, on many levels where nanosecond events are hidden inside millisecond events, which are embedded into second events, etc., just like letters make up words and words make up sentences, which form paragraphs leading to chapters and eventually an entire book is written. Of course, it is not a trivial exercise to be able to encode information in a frequency domain this way but is completely doable using the principles of Fourier analysis and self-similar algorithms. In summary, to be able to release information storing a lifetime experience in the form of an electromagnetic field, let's call it, a quantum field of life, approximately 10^{18}–10^{19} photons need to be released in some time- and space-coherent electromagnetic field structure. The energy of such an electromagnetic wave packet would not be exorbitant, not exceeding a few joules. Since the term cloud computing is currently becoming fashionable and serves as a useful metaphor, we could call it a bio-information cloud. To be used by a distant receiver, however, it must be first decoded. Similarly, when you send an email it is broken up into fragments and these fragments are reassembled at their destinations. Hence, it is also conceivable that the biophotons can be released separately as information fragments rather than as a unitary coherent structure. In this case, there must be a decoder to reassemble the separate bits back into a coherent structure. We have arrived at a point in our discussion where a frontier of science has been reached. We hope that what is currently at the edge of science fiction, will soon become reality and a new era of communication will be initiated. This will not only affect the way we understand our brain but also how we view health and disease, an integrated aspect of the energy-information field. Below we discuss these consequences for medicine.

3.33 An Integrated Perspective of Energy and Information Flow in Health and Disease

It appears that the major pitfall of traditional establishment medicine is the increasing specialization and compartmentalization of its subfields. This is as if the heart, lungs, kidney, and thyroid, for example, all lived in their own separate parking lots. The reality is they share a common soil and a common purpose. Moreover, they are interconnected and interacting with one another. However, if quantum synchronized coherence were to occur in the brain initiated in conscious awareness, this is the essential organ from which all other organ systems receive information and instructions for coordinated activities.

The role that quantum biology plays is the integrative principle of entanglement. In the absence of quantum biology insights, we are often unable to see the integrated whole because one cannot isolate a product from another as they are often intangibly entangled.

There is an entanglement of energy in systems biology within the confines of the individual organism including long-range coherence (superposition states), which can be driven by solar energy supply in plants and food supply in animals, as well as coherence inter-individually due to electromagnetic interactions or quantum entanglement or both. The latter could explain such phenomena as telepathy, premonitions, and conscious thought, intuitively the synchronicity seen at various levels in competitive sports, flocking behavior of migratory birds, schools of fish such as herring. This may also apply to a fairly common experience of perceiving being watched by another individual crossing a parking lot and looking back in response to a sensory perception, which did not involve any actual visual input. Perhaps this latter phenomenon is a residual evolutionary adaptation that engages the protective social interactions, or that warns prey of stalking predators.

Another fascinating example where the quantum nature of our memory may be playing out is manifested by reports of the remembrance of past lives. Dr. J. Tucker has studied US children claiming to remember past lives [83]. He reports that in about 70% of the cases of children, the deceased died from an unnatural cause, suggesting that traumatic death may be linked to the hypothesized survival of self. He further indicates that the time between death and apparent rebirth is, on average, 16 months, and that unusual birthmarks might match fatal wounds suffered by the deceased. Although critics have argued there is no material explanation for the survival of self, Tucker hypothesized that quantum mechanics may indeed explain these strange phenomena by which memories and emotions could carry over from one life to another arguing that since the act of observation collapses quantum wave functions, the self may not be merely a by-product of the brain, but rather a separate entity that impinges on matter. Tucker argues that viewing the self as a fundamental, nonmaterial part of the universe makes it possible to conceive of it continuing to exist after the death of the brain.

Tucker provides the analogy of a TV set and TV transmission; the device is required to decode the signal, but it does not create the signal. In a similar way, the brain may be required for awareness to express itself but may not be the source of awareness.

Quantum entanglement is real but requires a very special set of conditions not to disrupt it, i.e. the isolation of a system from interactions with other systems which would also become entangled with these systems, and they are entangled with others, etc. Once we start perturbing these systems with other entanglements, the clean argument disappears and it is considered that the nature of "every day" reality is far from clean. However, this is not to discredit that there may be some situations where the type of behavior is manifested.

For example, it is quite possible that our dream world may open a window on this. Some premonitory dreams or very vivid dreams of actual events taking place far away in space and/or time and involve people with whom we have a meaningful connection (say our loved ones) are likely real manifestations of quantum entanglement [84].

Our brains, or other aspects of our consciousness, may be fundamentally connected with other people at special moments in time. Probably most people have had such personal experiences of this type of phenomenon. There are some examples in the biography of Carl Jung (1875–1961) who was a Swiss psychiatrist and psychotherapist and founded analytical psychology. He extensively corresponded with the famous physicist Wolfgang Pauli and both of them were purposefully interested in the potential of quantum explanations of the world of the human psyche.

Entanglement may also speak to the relationship of the four-dimensional coordinate system of space-time linking quantum mechanics to Einstein's theory of special relativity.

For an object traveling through space at or near the speed of light, time appears to be dilated (slowing down) and space appears to be contracted which means that for a signal that travels fast, stationary distances appear shortened. This analogy, applied to the human body would mean that in a well-integrated biological system, signals travel fast and space between their end points shortens in a relative sense.

On the other hand, in poorly synchronized organisms, separate compartments are not integrated and signals slow down when traveling between them. Consequently, coordination is impeded. This is a perspective similar to that of a very fast traveler compared to a stationary observer.

Taken to the extreme, the past, the present, and the future all occur simultaneously and time stands still, so to speak, which would be a metaphorical achievement of the fountain of eternal youth where the effects of the passage of time on the human body are eliminated.

As a result, the superposition of spatial and temporal features simultaneously could be seen as a quantum mechanical phenomenon.

This is important because the properties of light are relative to the observer and unobserved quantum events therefore exist only as superpositions of possibilities but are not real until measured.

The actual observation of events collapses the wave function; this decoherence reduces the range of possibilities of superposition to a single reality. In terms of translating this observation into clinical practice, it is an *intriguing possibility that such a mechanism may explain the placebo effect.* Patients on blinded clinical trials are not informed if they are given a real drug or are on a placebo pill, which may *affect their body's response since in their minds there is a superposition of both possibilities present* and can with some probability elicit an effect. Similarly, the power of suggestion, conviction, or alternatively religious or spiritual belief may lead to the equivalent of an observation resulting in real outcomes.

The power of a positive belief system is hence conceptually rooted in quantum theory.

This is a fascinating hypothesis that can explain a number of paradoxical observations.

Hence, the living organism as a metabolically functioning system is an integral part of a larger system bathed in the energy supply that created it, an expression of the first law of thermodynamics of energy transformation. That may be the missing link; as one could state that biology has been walking blindfolded before the advent of quantum biology, which offers a unifying principle.

The application of this idea to medicine and biology in the context of disease states is that *the complexity of interactions within systems biology may lead to a breakdown of coherence in a pathological state whereby systems become isolated and compartmentalized.* This latter notion parallels the idea of an accelerated arrow of time, that is, an increase of the thermodynamic entropy and the resultant senescence. Conversely,

dilation of biological time is part of a superposition of space and time whereby this quantum mechanical state involves energy flow at close to the speed of light and biological processes that are nonlocal and deeply interconnected and entangled. An additional nuance involves a distinction between General Relativity and Special Relativity, in the former time dilation is superimposed with a contracted spatial component due to gravitational attraction and space curvature whereas in the latter time dilatation is associated with an unconstrained spatial boundary. Furthermore, the superposition of spatial and temporal features simultaneously is achieved in a relativistic quantum mechanics formulation. Its functioning occurs instantaneously and simultaneously, hence this implies removing the limitations of space and time. A better understanding of quantum biology allows us to see the whole, so to paraphrase a well-worn adage, we will be able to see the forest for the trees. *Viewing the living system as a whole, including the energy supply to the system, one has to incorporate the entire sum of the cells into a functioning organism including their interactions and the fluxes of energy flowing through the system.*

Explaining the mechanism of coordination of the organism as a whole is a major challenge in the field of biology. Methods of communication in the human body are commonly thought to include electrical signaling by the central and peripheral nervous systems, chemical signaling through blood flow with hormones and cytokines. Moreover, inter- and intra-cellular connections in the body, which form a structurally and mechanically connected link between the cytoskeleton, the nuclei of cells, and the extracellular matrix indicate the possibility for mechanical transduction with much faster speeds than traditional chemical signaling. Intra-cellular polymeric structures involving microtubules, AFs, and collagen have been considered as bio-nanowires that form a mechanical tensegrity matrix throughout the body; in addition to their structural roles, they are potentially capable of high-speed electrical, protonic, ionic, and possibly electronic signaling. Bioelectrical signaling is known to be involved in the regeneration and functional ordering of the organism.

Since chemical signaling systems have very limited speeds, and there is ample evidence that an organism takes advantage of these bio-nanowires to attain faster message passing speeds, it is expected that electromagnetic signaling will be found to play more and more significant roles in explaining biological coherence.

The main theories for a body-wide information network based upon bio-nanowires come from the tensegrity theory that envisages the existence of sub-molecular communication systems in the body [85]. Within this model, the cell is viewed as a micrometer scale tent-like structure with the membrane representing the canvass, the microtubules representing the posts, and AFs seen as cable-like tension resistant elements (Figure 3.15). The cytoskeleton is acting to provide pre-stress, meaning internal mechanical integrity for this tiny tent while its membrane receptors that adhere to the substrate function like pegs that plant this living tent firmly to the ground.

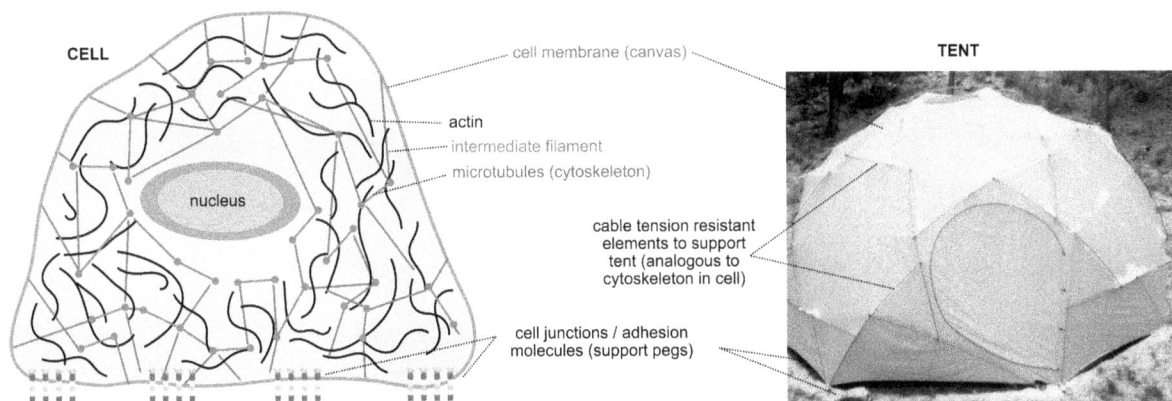

FIGURE 3.15 Cell structure compared to tent tensegrity model. Cell cytoskeleton act as bio nanowires that provide mechanical support and shape to cells. Cell adhesion/ junctions molecules adhere the cell to an extra-cellular matrix (ECM) similar to pegs that help in binding the tent to the ground. Source: cell diagram adapted from Gan, B.S. (2020) "Tensegrity in Biological Application: Cellular Tensegrity". In: *Computational Modeling of Tensegrity Structures*. Springer, Cham. https://doi.org/10.1007/978-3-030-17836-9_8. Tent photo from Juggler, used under Creative Commons license https://creativecommons.org/licenses/by-sa/4.0/legalcode.

However, there is more to the cytoskeleton than just its mechanical properties. The Cytoskeleton Information Processing Model describes the role of MTs and AFs in information processing in the brain as a fully interconnected sub-molecular messaging system providing signals, information, and instructions to the various parts of the body as a synchronized whole. As described earlier, this messaging system is capable of several parallel channels of communication utilizing coherent integration of electric and magnetic field sensitivities and the interaction with chemical transmission modes involving synaptic connections. These modes of communication within microtubules can be further mediated by electronic oscillations and excitonic degrees of freedom within tubulin dimers. Outside microtubules but interacting with them and the cytoplasm are the counterion waves (Figure 3.16), namely positive ions such as potassium, sodium, and calcium that are most directly engaged with the tubulin's flexible and electrically charged C-terminal tails.

This picture goes significantly beyond the conventional information transfer methods that include chemical signaling (for example neurotransmitters, autocrine, paracrine, and endocrine signaling) and electrical signals (for example ion channels) which pass through synapses and gap junctions and can spread in parallel fashion involving neuronal depolarization. Non-chemical information transfer mechanisms may include not only electrons, ions, and protons (as well as magnetic fields generated by these currents), but also excitons and other types of quantum quasiparticles such as polarons, phonons, and magnons (spin waves).

The coherent integration of magnetic fields, electric conduction with synaptic chemical transmissions is an attractive but challenging proposition since the time scales of these interactions are vastly different.

Electromagnetic signals propagate with the speed of light (c = 300,000 km/s), electric conduction is not as fast (at a small fraction of c) but still much faster than ionic waves (on the order tens of m/s) while chemical signaling is the slowest since it largely depends on diffusion and drift.

Therefore, integration of these signals must involve a hierarchical organization where the faster processes affect the slower ones.

For example, in microtubules, tubulin assembly could be the slowest process, followed by GTP hydrolysis, which is slower than C-termini oscillations that cause ionic waves, which still slower than electromagnetic triggers that could be initiated by mitochondrial proton pump action.

However, it is becoming apparent that the entire organism is organized in an interconnected matrix that allows for long-range information storage and processing. The extracellular matrix, which defines the space between cells, forms connections to the cell's interior through integrins, which connect to the intracellular cytoskeleton. The cytoskeleton, in turn, connects to the nuclei of cells forming a continuous mechanical linkage. This connective matrix suggests the possibility of high-speed communication systems at work in the organism. Since most of the proteins are negatively charged, especially tubulin and actin, which form the key filaments of the cytoskeleton, they attract cytoplasmic ions with positive electric charge (cations), especially potassium, sodium, and calcium ions. These ions partially condense on these protein filaments and partially localize in the vicinity of these proteins and can propagate along their lengths when stimulated by voltage changes in their vicinity. These ionic waves can serve as signals for cellular processes, e.g. cell division, differentiation, or motility.

The Dendritic Cytoskeleton Information Processing Model (Figure 3.17) describes the role of MTs and actin in information processing in the brain as a fully interconnected sub-molecular messaging system in the body as a whole. This implies that

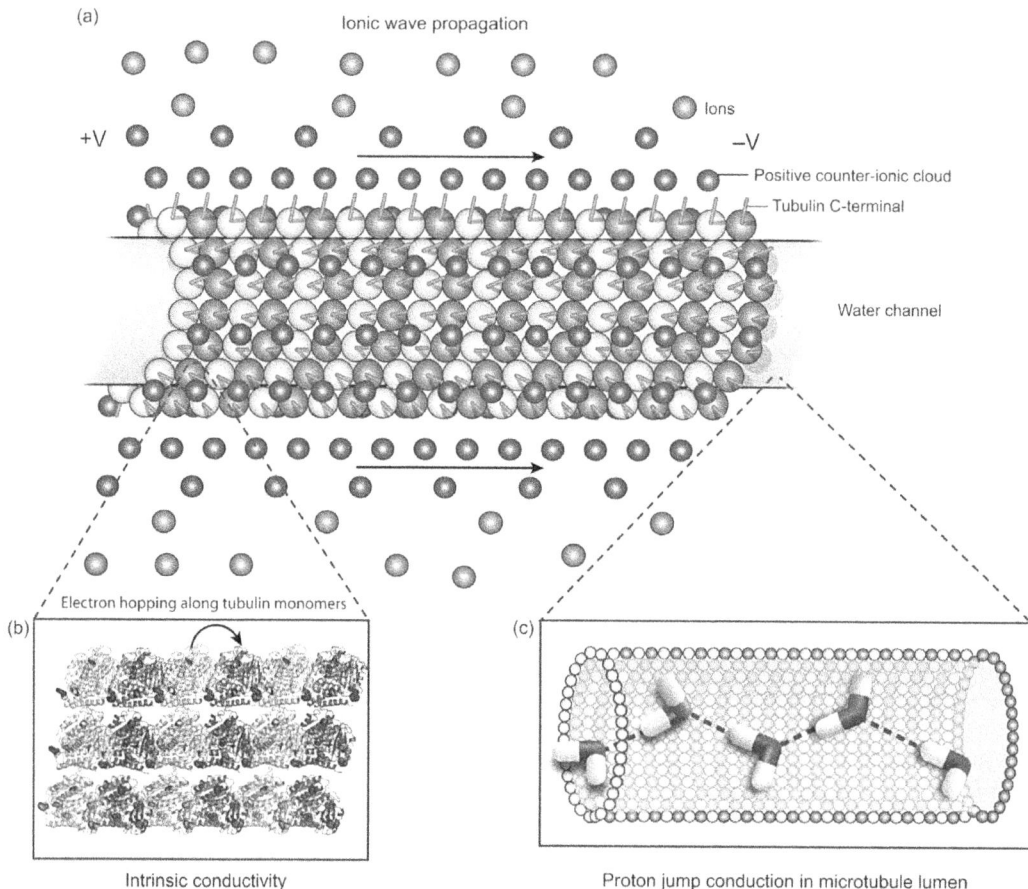

FIGURE 3.16 An illustration of a microtubule surrounded by ions that form counterion waves propagating along the surface of a microtubule. Additionally, intrinsic conductivity due to electronic motion is hypothesized. Source: from Friesen, Douglas E., Craddock, Travis J.A., Kalra, Aarat P., and Tuszynski, Jack A. (2015) "Biological Wires, Communication Systems, and Implications for Disease". *BioSystems* 127: 14–27, used with permission.

intracellular signals can be transmitted to neighboring cells in the organ or tissue and then across tissues and organs within the organism. Developments in the fields of bioelectricity, electromagnetic medicine, and biophotonics have largely been slow due to the challenge of probing electromagnetic phenomena, and other nanoscale events in biological systems, both in the classical and especially quantum regimes. The above processes are physical because they do not directly elicit chemical reactions. Chemical signals include hormones and neurotransmitters since they involve chemical reactions. The advent of nanotechnology, more precise instruments, and the interest in developing biological circuits promise to lead to rapid developments in this field. The domain of electromagnetic effects in biology is vast and exceeds the scope of the present chapter.

As noted in section 3.18, Albert Szent-Györgyi (1941) pioneered the notion that proteins can act as semiconductors and today we see that this is indeed true, especially applied to protein polymers such as actin filaments and microtubules. Semiconductors are solid materials that are neither insulating like rubber nor conducting like metals but somewhere in between due to their ability to allow some of the valence electrons to be excited into the conducting state. The idea was that proteins can also have a small number of electrons that can be excited and promoted to a conduction band.

Instead of electric currents flowing as in metals, in proteins, electrons could hop from one site with an attractive potential to another leading to a small electric current, especially if they could be assembled into linear filaments, such as an actin filament or a microtubule.

MT's have also been theorized to support kink-like ionic soliton waves (see Figure 3.18), which theoretically travel at a speed of 2-100 m/s. Soliton waves can be distinguished from extended waves by comparison between tsunamis and regular ocean waves. Tsunamis are an example of a soliton, and they are spatially localized, large-amplitude nonlinear waves that propagate with no loss of energy. In regard to the roles of these signal propagation effects, perhaps the most interesting and well developed is the possible connection between MTs and memory storage and retrieval in neurons, which we briefly discussed above. These structures could form a veritable network of bioelectronic conducting wires within living cells. However, their conductive properties could also include proton and ion conduction leading to a multi-channel signaling ability.

While organisms conduct electricity, as evidenced by medical equipment such as electrocardiograms,

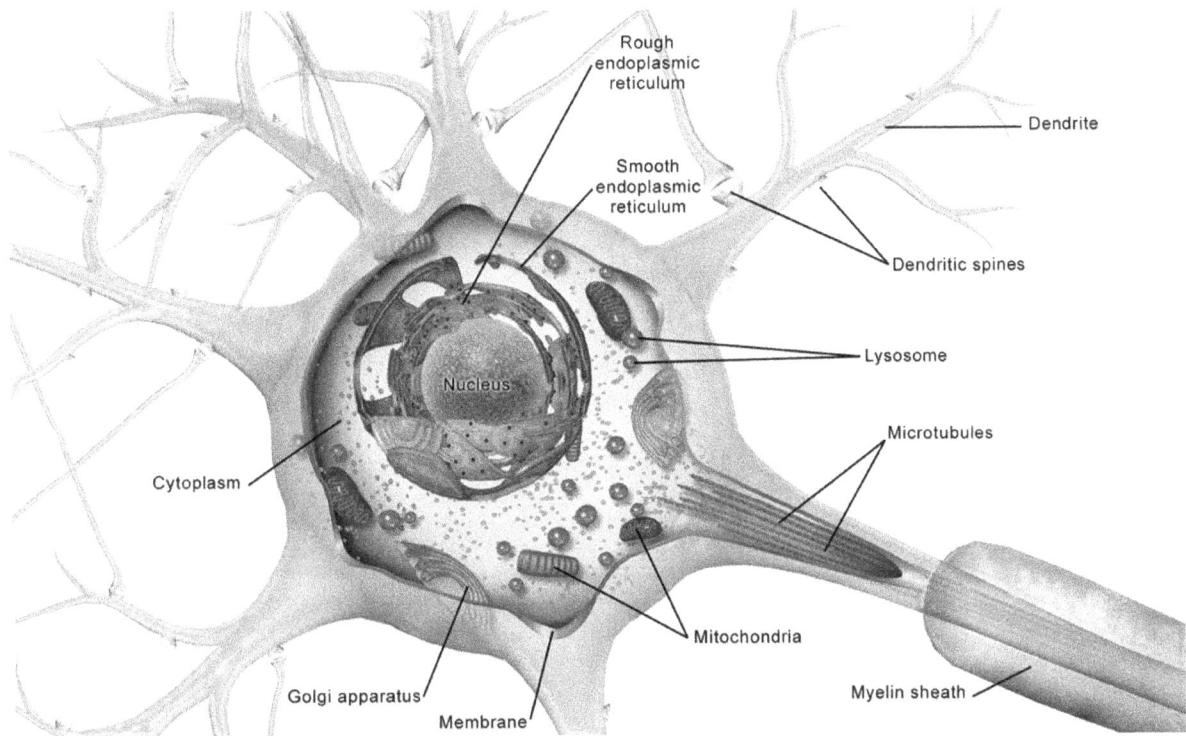

FIGURE 3.17 Pictorial representation of a typical neuron. Dendrites have been shown to be actively involved in information processing in patch-clamp electrophysiology experiments. Source: by Bruce Blaus. Used under Creative Commons license https://creativecommons.org/licenses/by-sa/4.0/legalcode.

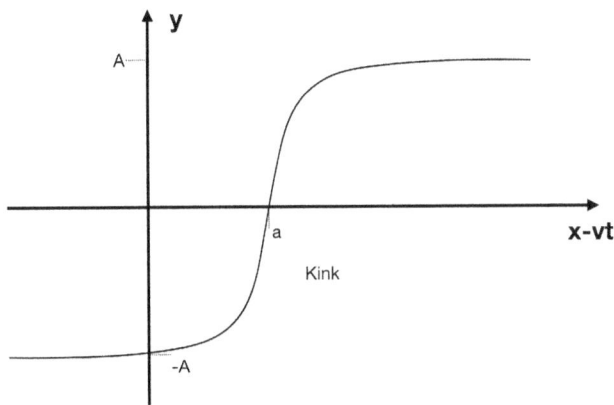

FIGURE 3.18 A typical profile of the kink soliton wave traveling along a propagation direction with velocity versus its amplitude is along the vertical axis.

electroencephalograms, and transcutaneous electrical nerve stimulation, questions remain as to what the charge carriers in living systems are, and what all the functions of these electrical signals are.

Well known are the action potentials that allow for neural firing and the control of the nervous system.

A closer examination of the body reveals the exquisite role of membrane potentials, ionic currents, and endogenous electromagnetic fields in cellular proliferation, morphogenesis (embryonic

development), neurogeneration and neurite growth (neuronal cell body projection), and regeneration (e.g. wound healing).

Also emerging in investigation is the potentially crucial role of water in the proliferation of biological signals due to the structure imparted by hydrophilic surfaces as extensively elaborated on in the work of Gerald Pollack. In particular, Dr. Pollack identified the physical state of water in biological systems as a fourth phase, which is neither liquid, nor ice, nor vapor, but a structured phase of water somewhat similar to liquid crystal.

The Jesus Christ lizard derived its name because it is able to walk on water (Figure 3.19) due to the surface tension which prevents it from sinking as it steps on it.

This phase is characterized by long-range spatial correlations, which is why it would lower the number of thermal fluctuations and could also enhance electric signals due to a lower dielectric constant and dipole ordering. Gerald Pollack demonstrated an ice-like ordering of water molecules in biological systems. Charge carriers under investigation in relation to protein semiconduction have largely been electrons and protons. Electron tunneling involved in enzymes has also been well documented in the literature. Evidence for protonic conduction in proteins (analogous to electron tunneling) has been found for collagen, keratin, cytochrome c, and hemoglobin. *Protonic conduction involves hydronium water structures, which surround proteins, and is different from electronic tunneling due to the shorter distances traveled by individual protons* (being much heavier than electrons) and a tightly organized water environment as hypothesized by Mae Wan Ho [86] for proton

FIGURE 3.19 Jesus Christ or basilisk lizard running on water. Source: licensed from Science Source Images.

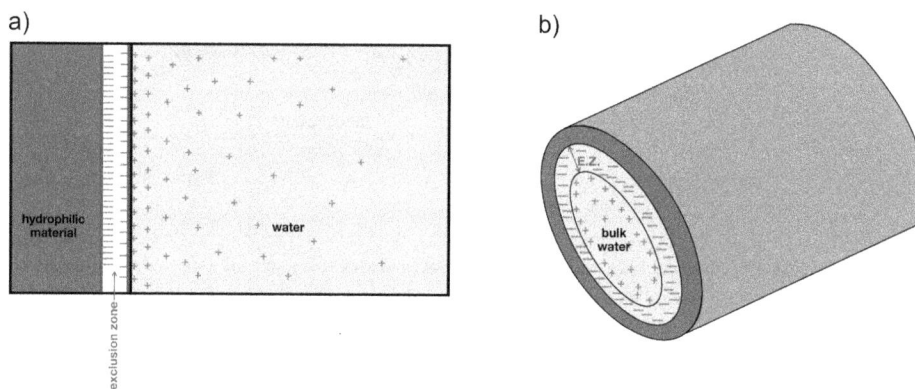

FIGURE 3.20 Graphic illustration of structured EZ water versus bulk water. Source: adapted from Pollack, Gerald H. (2013) *The Fourth Phase of Water: Beyond Solid, Liquid, and Vapor.* Ebner & Sons Publishing, Seattle.

conduction and by Gerald Pollack for EZ water (Figure 3.20) organization.

In these contexts, electric potential gradients are known to guide the movement of material in a specific direction, as for example in neurite growth or wound healing phenomena. Electron tunneling involved in enzymes has also been well documented in the literature and it is remarkable that this growing body of literature indicating the importance of electronic transitions in biology has not been able to draw major attention of the medical profession. MTs are theorized to be involved in numerous forms of information processing, such as transferring ionic, mechanical, and possibly quantum communication messages involving photons, excitons, and other degrees of freedom (electron tunneling).

The question that one might ask is why this body of evidence has not caused a major paradigm shift in biology and medicine yet. The answer probably involves both resistance to change and friction between different disciplines in science that impedes research translation. Perhaps the most elucidating and explosive connection will be between electromagnetic and mechanical signaling interactions, which in turn may have critical implications for mind-body insights as well as insights ultimately for clinical therapeutic interventions that could fundamentally transform standards and expectations into beautiful realities. Obviously, therapeutic successes with

neurodegenerative diseases such as Alzheimer's should be enabled using electromagnetic methods since pharmacology has so far failed miserably in this regard.

Being able to use this knowledge to augment or enhance our memory storage capabilities as well as cognitive processing powers sounds like science fiction but it can soon become reality due to the possible connection between MTs and memory storage and retrieval in neurons.

Understanding the relativity of time is elemental to space and what it means to evolve a positive vitality by engaging in life's challenges and recognizing the things you can as opposed to things you cannot control. The same applies to much of quantum biology, for example to quantum metabolism that can be interpreted as maximally efficient spatially unconstrained synchronistic energy production with relative time dilatation.

The convergence of the laws of thermodynamics, quantum biology, electromagnetism, and physical chemistry, when properly integrated will evolve to describe the emergent collective properties of entire organisms and their behavior such as that characteristic of human beings viewed as complex adaptive systems.

One critically important distinction between living and non-living systems is the presence and absence of metabolism, respectively. At the most fundamental level, metabolism means the continual generation of metabolic energy quanta in the form of ATP molecules. The main manifestations of energy transfer from ATP hydrolysis include 1) motor proteins motion and 2) ion/molecule active transport. Both of these represent kinetic energy. A third manifestation of energy transfer from ATP hydrolysis is catabolism, i.e. energy within the bonds of synthesized molecules. This involves potential energy. The fourth fundamental manifestation of energy transformation from ATP hydrolysis is enzymatic catalysis, which involves both kinetic and potential energy. Enzymatic catalysis may include the phosphorylation of an enzyme or its effects upon a substrate. The energy is harnessed in the bonds of the phosphorylated product. Alternatively, the phosphorylation releases energy, i.e. is exothermic and hence it derives the kinetic energy molecules around it. Enzyme catalysis also involves the kinetic energy of the molecular configuration of both the enzyme and the substrate it acts on. In the context of explicitly complex biology, ATP hydrolysis (including for example the phosphorylation of the enzyme, which turns it into a negatively charged particle and in addition sets it in motion) promotes enzyme conformational changes that allow the precise integration of magnetic fields that promote a subsequent electrostatic attraction between the enzyme and the substrate. Perhaps enzyme catalysis may involve the role of a motor protein moving the enzymes themselves or the substrates as charged particles creating current and hence magnetic fields, which mediate the interactions of substrates into their products. This may help explain the signal transduction processes of molecular biology inside the cell.

By understanding the notion and implications of a hormetic stress response provides a reason for a wide diversity of stressors, which are necessary for optimal physical, psychological, and emotional health. In particular, at one dose a given stressor is therapeutic whereas at a higher dose it becomes toxic. For example, arsenic at a very low dose is actually listed to have a recommended daily allowance. At higher doses, it is the number one major public health problem. However, modest doses are tolerated in some, yet lethal in others. Running marathons, even by seemingly healthy individuals but with undiagnosed underlying health issues, may lead, as it is now known, to chronic physical disease, such as cardiovascular complications. The optimal healthy dose of exercise is highly individual. It is an extraordinarily low percentage of the population, even probably among marathon runners, in whom this level of activity is optimally healthy while for the vast majority of the population running a marathon would cause physical collapse.

The relevance of the insights and information of this book to clinical medicine and health care is at a number of levels. It allows people to help themselves by understanding the conceptual underpinnings of the various perspectives of energy flow in biological systems. It allows doctors to act in the role of a "teacher" (which is what "doctor" really means) to in turn empower the patient by transferring the

useful understanding, perspectives and insights. It provides a deeper appreciation to understanding the ecological symmetries and in this context, the relationship to healthy and unhealthy epigenetics and microbiome influences on the sophisticated genetic complexity of our rudimentary number of genes, rapidly and unavoidably becoming mainstream topics in health care and clinical medicine.

Other crucial elements promising to advance the diagnostic and therapeutic efficiency and potential include the electronic medical records and computer simulations capable of ingesting huge sets of data that find bottom-up and top-down solutions to complex problems not possible by the traditional establishment Newtonian linear compartmentalized model of health care that focuses on fixing the broken part.

Understanding conceptually the different types of thermodynamic engines creates a foundation for understanding complex adaptive systems as biological engines. Although in some ways analogous, the physiology of organismic and human processes does not function like a heat engine. Accordingly, physicians are typically not mechanics that fix most chronic problems by finding the broken part. This is in contrast to acute problems such as infection, arrhythmia, seizure, or respiratory failure, for example, which fortunately lend themselves well to algorithmic protocols. Understanding the distinction between the various thermodynamic engines is an essential perspective to the application of clinical health care.

Assimilation of Gibbs free energy, redox potential, and acid-base status when healthy allows optimal informational energy flow over four dimensions and has a clear relationship to healthy biological function. However, when things tend to go awry with oxidative stress, energy is dissipated in the form of heat, which is responsible for both acute and chronic inflammation.

When chronic pathology of inflammatory disease states emerges, degenerative diseases occur. Senescence itself is a similar albeit more insidious process underpinned by the chronic progression of oxidative stress, disturbances in Gibbs free energy, redox potential, and acid-base status with the associated deterioration of informational energy to low-grade heat and associated thermal entropy. Cancer, Alzheimer's disease, and accelerated cognitive decline are some chronic diseases with a well-recognized relationship to the aging process to which there are emerging novel therapeutic implications.

The progeny of thermodynamic entropy notions of information entropy and related evolutionary entropy are each closely related to the notion of informational energy. There are major differences in thermodynamic entropy from information and evolutionary entropies, however, recognizing their contrasting distinctions and similarities illuminates a deeper understanding and meaning to the integrated perspective of energy flow through biological systems and its relationship to health and disease.

There are many problems with the current explanation of what is going on inside the cell. One of the primary problems is the efficiency of finding partners for biological reactions. How efficiently and rapidly do molecules find their way and position to interact with the right partner? The current *Brownian theory is insufficient to explain the necessary speed and efficiency for biological processes.* It would probably be the greatest elucidation of biology if it were directly demonstrated that systems at the level of proteins or enzymes actually use quantum recognition algorithms to avoid accidental collisions and interactions with incompatible partner molecules that are otherwise almost inevitable in a crowded cellular environment. Living systems are actually quantum coherent and therefore disturbing the delicate balance may have consequences, such as diseases including cancer. Hameroff's paper [38] provides support to the notion that cancer is a disturbance of quantum coherence at the cell level.

It is important to remember that classical physics was not thrown out when quantum mechanics was discovered. It remains very useful, although its role has been greatly diminished in light of the onset of electronic communication theory that predominantly drives modern technology. However, classical physics is still very important at the level of macroscopic activities such as designing trains, planes, and automobiles, infrastructure construction (roads, bridges, tunnels). Furthermore, even at a molecular level, modeling structures of macromolecules such as proteins and nucleic acids largely employ energy functions in the classical physics realm. However, considering the fundamental nature of matter, one cannot do without quantum mechanics. The same is true of biology. The cartoon image of how cells operate has a classical physics component at the larger scale, however, the proper understanding of how mitochondria work, how various enzymes operate, or how recognition operates at the subcellular level, may require quantum mechanics. Vlatko Vedral suggested that it would be illuminating to put oneself in the perspective of people in the early 1920s when the laws of quantum mechanics were first discovered. It was said at the time that it is extremely difficult to apply this to the simplest of all atoms, the hydrogen atom. Then to recognize quantum mechanics not from the perspective of a single atom, but from that of a piece of solid that consists of 10^{24} (1 followed by 24 zeros) atoms; how would we ever come to understand that? Actually, this happened very shortly thereafter as the field of solid-state physics was launched, which is the very basis of all modern technology! Therefore, we should anticipate an exponential speed of scientific discovery in the years ahead and we may be optimistic about the future of quantum biology and perhaps quantum medicine.

Certainly, the field of quantum biology has potential for explosive advances in applied science including medicine.

Finally, an important aspect of quantum biology, already emphasized strongly in the past by Sir Roger Penrose in the context of the functioning of the brain, *is the non-algorithmic nature of quantum science.* The intersection between physics and biology via mechanisms of quantum phenomena and synchronized physiology plausibly translates to the domains of athletic performance and success in general in life's pursuits ("if you believe in yourself you can make it happen") in the context of the quantum nature of free will and consciousness.

The field of modern medicine should be further developed to include the totality of a human person, both the physical aspects including the latest discoveries in the areas of systems biology, molecular biology, quantum biology, etc. but also the intangible aspects of human behavior that continue to inspire, surprise and amaze us all to achieve our full human potential.

In the following chapter, we discuss how the rapidly developing field of systems biology can help medicine change its focus from independent parts to integrated wholes.

REFERENCES

1. Bell, J.S. (2004). *Speakable and Unspeakable in Quantum Mechanics: Collected Papers on Quantum Philosophy.* Cambridge University Press. Cambridge, England.
2. Aspect, A., Dalibard, J. and Roger, G. (1982). "Experimental Test of Bell's Inequalities Using Time-Varying Analyzers". *Physical Review Letters* 49(25):1804.
3. Griffin, A., Snoke, D.W. and Stringari, S. (1996). *Bose-Einstein Condensation.* Cambridge University Press. Cambridge, England.
4. Schrödinger and Erwin (1992). *What Is Life?: With Mind and Matter and Autobiographical Sketches.* Cambridge University Press. Cambridge, England.
5. Beck, F. and Eccles, J.C. (1992). "Quantum Aspect of the Brain Activity and the Role of Consciousness". *PNAS* 89:11357–11361.
6. Vattay, G., Kauffman, S. and Niiranen, S. (2014). "Quantum Biology on the Edge of Quantum Chaos". *PLOS ONE* 9(3):e89017.
7. Pollack, G.H. (2013). *The Fourth Phase of Water: Beyond Solid, Liquid, and Vapor.* Ebner and Sons Publishing. Seattle, WA.
8. Ho, M.-W., et al. (2006). "The Liquid Crystalline Organism and Biological Water". In: *Water and the Cell* (pp. 219–234) Gerald H Pollack, Ivan L. Cameron, and Denys N. Wheatley (eds). Springer Netherlands. Dordrecht, the Netherlands.
9. (a) Engel, G.S., et al. (2007). "Evidence for Wavelike Energy Transfer Through Quantum Coherence in Photosynthetic Systems". *Nature* 446(7137):782–786. (b) Karafyllidis, I.G. (2017). "Quantum Transport in the FMO Photosynthetic Light-Harvesting Complex". *Journal of Biological Physics* 43(2):239–245.
10. Popp, F.-A. and Yan, Y. (2002). "Delayed Luminescence of Biological Systems in Terms of Coherent States". *Physics Letters, Section A* 293(1):93–97.
11. Albrecht-Buehler, G. (1992). "Rudimentary form of Cellular "Vision"". *Proceedings of the National Academy of Sciences of the United States of America* 89(17):8288–8292.
12. Kumar, S., et al. (2016). "Possible Existence of Optical Communication Channels in the Brain". *Scientific Reports* 6.
13. (a) Gutteridge, J.M. (1985). "Superoxide Dismutase Inhibits the Superoxide-Driven Fenton Reaction at Two Different Levels. Implications for a Wider Protective Role". *FEBS Letters* 185(1):19–23. (b) Gutteridge, J.M., Maidt, L. and

Poyer, L. (1990 July 1). "Superoxide Dismutase and Fenton Chemistry. Reaction of ferric-EDTA Complex and Ferric-Bipyridyl Complex with Hydrogen Peroxide without the Apparent Formation of Iron(II)". *Biochemical Journal* 269(1):169–174. (c) Mao, G.D., Thomas, P.D., Lopaschuk, G.D. and Poznansky, M.J. (1993 January 5). "Superoxide Dismutase (SOD)-catalase Conjugates. Role of Hydrogen Peroxide and the Fenton Reaction in SOD Toxicity". *Journal of Biological Chemistry* 268(1):416–420.

14. Priel, A., Tuszynski, J.A. and Woolf, N.J. (2010). "Neural Cytoskeleton Capabilities for Learning and Memory". *Journal of Biological Physics* 36(1):3–21.

15. Mayburov, S.N. (2011). "Photonic Communication and Information Encoding in Biological Systems". *Quant. Com. Com* 11 arXiv:1205.4134.

16. Berman, M.H., et al. (2017). "Photobiomodulation with Near Infrared Light Helmet in a Pilot, Placebo Controlled Clinical Trial in Dementia Patients Testing Memory and Cognition". *Journal of Neurology and Neuroscience*. 8(1).

17. Chen, H., et al. (2017). "Quantum Dot Light Emitting Devices for Photomedical Applications". *Journal of the Society for Information Display*. 25(3), 177–184.

18. Yu, W., et al. (1997). "Photomodulation of Oxidative Metabolism and Electron Chain Enzymes in Rat Liver Mitochondria". *Photochemistry and Photobiology* 66(6):866–871.

19. Rieke, F. and Baylor, D.A. (1998). "Single-Photon Detection by Rod Cells of the Retina". *Reviews of Modern Physics* 70(3):1027.

20. Franco, M.I., Turin, L., Mershin, A. and Skoulakis, E.M. (2011). "Molecular Vibration-Sensing Component in Drosophila Melanogaster Olfaction". *Proceedings of the National Academy of Sciences of the United States of America* 108(9):3797–3802.

21. Pauls, J.A., Zhang, Y., Berman, G.P. and Kais, S. (2013). "Quantum Coherence and Entanglement in the Avian Compass". *Physical Review. Part E* 87(6):062704.

22. Scholes, G.D., Fleming, G.R., Olaya-Castro, A. and van Grondelle, R. (2011). "Lessons from Nature About Solar Light Harvesting". *Nature Chemistry* 3(10):763–774.

23. Dörnemann, D. and Senger, H. (1986). "The Structure of Chlorophyll RC I, a Chromophore of the Reaction Center of Photosystem I". *Photochemistry and Photobiology* 43(5):573–581.

24. Sension, R.J. (2007). "Biophysics: Quantum Path to Photosynthesis". *Nature* 446(7137):740–741.

25. Craddock, T.J.A., et al. (2014). "The Feasibility of Coherent Energy Transfer in Microtubules". *Journal of the Royal Society. Interface / The Royal Society* 11(100):20140677.

26. Tonello, L., et al. (2015). "On the Possible Quantum Role of Serotonin in Consciousness". *Journal of Integrative Neuroscience* 14(03):295–308.

27. Demetrius, L. (2006). "The Origin of Allometric Scaling Laws in Biology". *Journal of Theoretical Biology* 243(4):455–467.

28. Demetrius, L. and Tuszynski, J.A. (2009). "Quantum Metabolism Explains the Allometric Scaling of Metabolic Rates". *Journal of the Royal Society Interface*: rsif20090310.

29. Demetrius, L.A. Kafatos, M.C., and Tuszynski, J.A. "The Quantization Paradigm in Physics and Biology: An Analogue of the Planck Constant in Cellular Metabolism" (unpublished preprint).

30. Demetrius, L.A. and Simon, D.K. (2012). "An inverse-Warburg Effect and the Origin of Alzheimer's Disease". *Biogerontology* 13(6):583–594.

31. Fröhlich, H. (1986). "Coherent Excitation in Active Biological Systems". In: *Modern Bioelectrochemistry* (pp. 241–261). Springer, US.

32. Paul, R., Chatterjee, R., Tuszyński, J.A. and Fritz, O.G. (1983). "Theory of Long-Range Coherence in Biological Systems. I. The Anomalous Behaviour of Human Erythrocytes". *Journal of Theoretical Biology* 104(2):169–185.

33. Nardecchia, I., et al. (2014). "Experimental Detection of Long-Distance Interactions between Biomolecules through Their Diffusion Behavior: Numerical Study". *Physical Review. Part E* 90(2):022703.

34. Lundholm, I.V., et al. "Terahertz Radiation Induces Non-Thermal Structural Changes Associated with Fröhlich Condensation in a Protein Crystal". *Structural Dynamics* 2(5) (2015):054702.

35. Titova, L.V., et al. (2013). "Intense THz Pulses Down-Regulate Genes Associated with Skin Cancer and Psoriasis: A New Therapeutic Avenue?". *Scientific Reports* 3:2363.

36. Kirson, E.D., et al. (2007). "Alternating Electric Fields Arrest Cell Proliferation in Animal Tumor Models and Human Brain Tumors". *Proceedings of the National Academy of Sciences of the United States of America* 104(24):10152–10157.

37. Sahu, S., Ghosh, S., Fujita, D. and Bandyopadhyay, A. (2011). "Computational Myths and Mysteries That Have Grown Around Microtubule in the Last Half a Century and Their Possible Verification". *Journal of Computational and Theoretical Nanoscience* 8(3):509–515.

38. Krebs, H.A. and Kornberg, H.L. (1957). "Energy Trans-formations in Living Matter". In: *Energy Transformations in Living Matter* (pp. 212–298). Springer, Berlin Heidelberg.

39. Hameroff, S.R. (2004). "A New Theory of the Origin of Cancer: Quantum Coherent Entanglement, Centrioles, Mitosis, and Differentiation". *Biosystems* 77(1):119–136.

40. Shannon, C.E. (1949). "Communication Theory of Secrecy Systems". *Bell Labs Technical Journal* 28(4):656–715.

41. Casagrande, D.G. (1999). "Information as Verb: Re-conceptualizing Information for Cognitive and Ecological Models". *Journal of Ecological Anthropology* 3(1):4.

42. Bateson, G. (1979). *Mind and Nature: A Necessary Unity.* Dutton, New York.

43. Stonier, T., (1990). Information and the Internal Structure of the Universe. *An Exploration into Information Physics.* Springer-Verlag, Berlin.

44. Ingber, D.E. (1997). "Tensegrity: The Architectural Basis of Cellular Mechanotransduction". *Annual Review of Physiology* 59(1):575–599. (b) Ingber, D.E. (2008). Tensegrity and Mechanotransduction. *Journal of Bodywork and Movement Therapies* 12(3):198–200. (c) Ingber, D.E., Wang, N. and Stamenovic, D. (2014). "Tensegrity, Cellular Biophysics, and the Mechanics of Living Systems". *Reports on Progress in Physics. Physical Society* 77(4):046603. doi:10.1088/0034-4885/77/4/046603. PMID:24695087. PMCID:PMC4112545.

45. Oschman, J.L. (2015). *Energy Medicine: The Scientific Basis.* Elsevier Health Sciences. Amsterdam, the Netherlands.

46. Ho, M-W (2008). *The Rainbow and the Worm: The Physics of Organisms*. World Scientific. New Jersey.

47. Szent-Györgyi, A. (1977). "The Living State and Cancer". *Proceedings of the National Academy of Sciences of the United States of America* 74(7):2844–2847.

48. Friesen, D.E., Craddock, T.J., Kalra, A.P. and Tuszynski, J.A. (2015). "Biological Wires, Communication Systems, and Implications for Disease". *Biosystems* 127:14–27.

49. Landauer, R. (1991). "Information is Physical". *Physics Today* 44(5):23–29.

50. Bennett, C.H. (1982). "The Thermodynamics of Computation—A Review". *International Journal of Theoretical Physics* 21(12):905–940.

51. Vedral, V. (2006). *Introduction to Quantum Information Science*. Oxford University Press on Demand.

52. Pothos, E.M. and Busemeyer, J.R. (2013). "Can Quantum Probability Provide a New Direction for Cognitive Modeling?". *Behavioral and Brain Sciences* 36(3):255–274.

53. Feynman, R.P., Robert B Leighton, R.B. and Sands, M.L. *The Feynman Lectures on Physics*. Addison-Wesley Publishing Co., Reading, Massachusetts : ©1963-1965. See also https://en.wikiquote.org/wiki/Richard_Feynman.

54. Pauling, L. (1946). "Molecular Architecture and Biological Reactions". *Chemical and Engineering News* 24(10):1375–1377.

55. Kanigel, R. (2016). *The man who knew infinity: A life of the genius Ramanujan*. Simon and Schuster, USA.

56. Avula, S., et al. (2014). "Treatment of Mitochondrial Disorders". *Current Treatment Options in Neurology* 16(6):292. doi:10.1007/s11940-014-0292-7.

57. Sud, N., et al. (2008). "Asymmetric Dimethylarginine Inhibits HSP90 Activity in Pulmonary Arterial Endothelial Cells: Role of Mitochondrial Dysfunction". *American Journal of Physiology. Cell Physiology* 294(6):C1407–C1418.

58. Spiegelman, Bruce M. (2013). "Banting Lecture 2012: Regulation of Adipogenesis: Toward New Therapeutics for Metabolic Disease". *Diabetes* 62(6):1774–1782.

59. Keys, A. et al. (1950). *The Biology of Human Starvation (2 Volumes)*. University of Minnesota Press. Minneapolis, Minnesota.

60. Albrecht-Buehler, G. "Cell Intelligence". http://www.basic.northwestern.edu/g-buehler/FRAME.HTM.

61. Ley, R.E. (2010). "Obesity and the Human Microbiome". *Current Opinion in Gastroenterology* 26(1): 5–11.

62. Cawley, N.X., Li, Z. and Loh, Y.P. (2016). "60 years of POMC: Biosynthesis, Trafficking, and Secretion of Pro-opiomelanocortin-derived Peptides". *Journal of Molecular Endocrinology* 56(4):T77–T97. doi:10.1530/JME-15-0323. Epub 2016 Feb 15. PMID:26880796. PMCID:PMC4899099.

63. Tononi, G. and Koch, C. (2015). "Consciousness: Here, There and Everywhere?". *Philosophical Transactions of the Royal Society of London Series B* 370(1668). doi:10.1098/rstb.2014.0167. PMID:20140167. (b) Koch, C. (2018). "What is Consciousness?". *Nature* 557(7704):S8–S12. doi:10.1038/d41586-018-05097-x.

64. Player, T.C. and Hore, P.J. (2018). "Posner Qubits: Spin Dynamics of Entangled Ca9(PO4)6 Molecules and their Role in Neural Processing". *Journal of the Royal Society, Interface / the Royal Society* 15(147). doi:10.1098/rsif.2018.0494. PMID:20180494.

65. Schwartz, J.M., Stapp, H.P. and Beauregard, M. (2005). "Quantum Physics in Neuroscience and Psychology: A Neurophysical Model of Mind-brain Interaction". *Philosophical Transactions of the Royal Society of London Series B* 360(1458):1309–1327. doi:10.1098/rstb.2004.1598.

66. Hameroff, S. and Penrose, R. (1996). "Orchestrated Reduction of Quantum Coherence in Brain Microtubules: A Model for Consciousness". *Mathematics and Computers in Simulation* 40(3–4):453–480.

67. Hameroff, S. and Penrose, R. (2014). "Consciousness in the Universe: A Review of the 'Orch OR' Theory". *Physics of Life Reviews* 11(1):39–78.

68. Tuszynski, J.A., Cocchi, M., and Bernroider, G. "Energy, Information and Time Scales in Human Brain Dynamics: Can There be Quantum Computation?" (Submitted to Biosystems).

69. Libet, B. (1993). "Unconscious Cerebral Initiative and the Role of Conscious Will in Voluntary Action". In: *Neurophysiology of Consciousness* (pp. 269–306). Birkhäuser, Boston.

70. Craddock, T.J.A., Tuszynski, J.A. and Hameroff, S. (2012). "Cytoskeletal Signaling: Is Memory Encoded in Microtubule Lattices by CaMKII Phosphorylation?". *PLOS Computational Biology* 8(3).

71. Hodgkin, A.L. and Huxley, A.F. (1952). "A Quantitative Description of Membrane Current and Its Application to Conduction and Excitation in Nerve". *The Journal of Physiology* 117(4):500.

72. Soon, C., Brass, M., Heinze, H.J. and Haynes, J.D. (2008). "Unconscious Determinants of Free Decisions in the Human Brain". *Nature Neuroscience* 11(5):543–545. doi:10.1038/nn.2112.

73. Aellen, M., Burkart, J.M. and Bshary, R. (2021). *No Evidence for General Intelligence in a Fish*. BioRxiv.

74. Carls-Diamante, S. (2017). "The Octopus and the Unity of Consciousness". *Biology and Philosophy* 32(6):1269–1287. doi:10.1007/s10539-017-9604-0.

75. Jean, D.C. and Baas, P.W. (2013). "It Cuts Two Ways: Microtubule Loss during Alzheimer Disease". *EMBO Journal* 32(22):2900–2902.

76. (a) Nogales, E. (2001). "Structural Insight into Microtubule Function". *Annual Review of Biophysics and Biomolecular Structure* 30:397–420. (b) Goodson, H.V. and Jonasson, E.M. (2018). "Microtubules and microtubule-associated proteins". *Cold Spring Harbor Perspectives in Biology* 10(6):a022608. doi:10.1101/cshperspect.a022608. PMID:29858272. PMCID:PMC5983186.

77. Priel, A., Ramos, A.J., Tuszynski, J.A. and Cantiello, H.F. (2006). "A Biopolymer Transistor: Electrical Amplification by Microtubules". *Biophysical Journal* 90(12):4639–4643.

78. Förster von, Th. (1948). "Zwischenmolekulare Energiewanderung und Fluoreszenz". *Annalen der Physik* 2(6):55–75. (b) Donaldson, L. (2020). "Autofluorescence in Plants". *Molecules* 25(10):2393. doi:10.3390/molecules25102393. PMID:32455605. PMCID:PMC7288016.

79. Poznanski, R.R., Tuszynski, J.A. and Feinberg, T.E. eds. (2016). *Biophysics of Consciousness: A Foundational Approach*. World Scientific. New Jersey.

80. Craddock, T.J.A., et al. (2014). "The Feasibility of Coherent Energy Transfer in Microtubules". *Journal of the Royal Society. Interface / the Royal Society* 11(100):20140677.

81. (a) Holick, M.F. (2007). "Vitamin D Deficiency". *New England Journal of Medicine* 357(3):266–281. (b) Kim, H. and Giovannucci, E. (2020). "Vitamin D Status and Cancer Incidence, Survival, and Mortality". *Advances in Experimental Medicine and Biology* 1268: 39–52. doi:10.1007/978-3-030-46227-7_3. PMID:32918213. (c) Holick, M.F. (2014). "Cancer, Sunlight and Vitamin D". *Journal of Clinical and Translational Endocrinology* 1(4):179–186. doi:10.1016/j.jcte.2014.10.001. PMID:29159099. PMCID:PMC5685053.

82. Fels, D. (2018). "The Double-aspect of Life". *Biology (Basel)* 7(2):28. doi:10.3390/biology7020028. PMID:29735890. PMCID:PMC6023002.

83. Tucker, J.B. (2005). *Life Before Life: A Scientific Investigation of Children's Memories of Previous Lives* (256pp). St. Martin's Press, New York. ISBN 0-312-32137-6.

84. Wolf, F.A. (1994). "The Dreaming Universe". *Psychological Perspectives* 30(1):36–41.

85. Ingber, D.E. (2003). "Tensegrity I. Cell Structure and Hierarchical Systems Biology". *Journal of Cell Science* 116(7):1157–1173. (b) Turvey, M.T. and Fonseca, S.T. (2014). "The Medium of Haptic Perception: A Tensegrity Hypothesis". *Journal of Motor Behavior* 46(3): 143–187.

86. Ho, Mae-Wan. (2014). "Illuminating water and life." Entropy 16, no. 9: 4874–4891.

4

From Systems Biology to Systems Medicine

Chapter Overview

Has the reductionist approach to science run its course and if so, why? Can we replace it with something more powerful to better understand biology and medicine? What is systems biology and is there a need for systems medicine? Is chaos necessary for life and can it be beneficial? How do we connect complexity science with medicine?

now, most of this research is limited to either meta-analyses of population studies or to model animal systems (e.g. yeast, fruit flies, *E. coli* bacteria, or *C. elegans* worms). However, with these model systems used as validation tools, in the future systems medicine approach will be applied to personalized precision medicine. This chapter provides a broad stroke–style painted canvas, showing the reader the present-day foundation (e.g. chaos, fractals, solitons, cellular automata) for this type of approach and directions for explorations (e.g. artificial intelligence, big data analytics) and applications in the future.

As argued in the previous chapter, reductionism in science has run its course. Complex systems, such as the various life forms, are nonlinear by design. They are, in fact, in almost all cases, constructed from systems operating based on the principles of nonlinear response to external perturbations. One of the most interesting properties of these systems is the possibility of symmetry breaking, which allows for a change of state (a dynamical state). This is a powerful concept, fully exploited in the physics of phase transitions where the same physical system can exist in different phases, say liquid water or solid ice. The same applies to the structure of elementary particles that are composed of quarks and quantum fields. We propose to use these ideas in the context of medicine where the same system can exist in different states, e.g. a state of health and various types of disease or pre-diseases states. Transitions between these states represent analogs of physical phase transitions and differences between these states highlight the associated broken symmetries. The ultimate phase transition for a living system is that from being alive to the state of death. The march toward this ultimate destiny of all living things is called aging and it is characterized by a gradual increase of entropy, hence, loss of information and heat generation (manifested by inflammatory state of the body). All of the above can be quantified and parameterized enabling the construction of a personalized physiological fitness landscape. In the future, such a fitness landscape will be used to navigate the individual's lifestyle and, if necessary, to design therapeutic interventions in order to avoid treacherous valleys of pathological attractor states or to climb a "mountain" range separating a pathological valley from an area characterized by well-being. Luckily, the task of accomplishing such algorithmic analyses is not as daunting as it was even a decade ago. This is because of the explosive growth of the field of systems biology and now, systems medicine. These areas of biomedical research are building not only a methodological framework for an engineering approach to the living system of systems, but also amassing reams of parametric data that feed the construction of predictive models—not just retrospective but also prospective. For

4.1 Problem Solving: Reductionism versus Simplifying Complexity

Methodological reductionism is a philosophical position based on precise definitions of subjects and objects that exist in reality. These definitions are premised on fundamental laws or rules of nature, rigorously studied in isolation, and analytically confirmed scientifically. In the natural sciences, the application of methodological reductionism attempts to explain entire systems in terms of their individual constituent parts and their interactions. For example, the temperature of a gas is reduced to nothing else but the average kinetic energy of its molecules in motion. By definition, methodological reductionism is an approach that informs or educates. It is rooted in investigation at the lowest level in terms of biochemistry and molecular biology. Accordingly, it studies linear mechanistic relationships providing insights, explanations, and information. Although often motivated by ontological reductionism, it does not integrate molecular-level discovery with the investigation of higher-level features [1]. This highlights the inherent flaw of reductionist thinking in medicine because methodological reductionism is often erroneously extrapolated to ontological reductionism (the fundamental theoretical flaw of biological reductionism of nonlinear systems) across hierarchical disciplinary scales of biological structures and functions.

By distinction, epistemic reductionism reflects the knowledge about a safe system in terms of the idea that one scientific domain or discipline, a higher-level feature of the system, can be reduced to and explained by lower-level features. This highlights the same logical deductive reasoning that underscores the concept of ontological reductionism but is applicable beyond the relationship of biology to physical chemistry. The premise of reductionism is that these mechanistic laws or rules derive from lower scales of organization and can explain higher-level, more macroscopic, organizational behavior. Hence, higher scales of behavior in a system (such as biological systems) can be reduced to the lower scales or its constituent parts, such as cells and biomolecules. This brings the notion of

DOI: 10.1201/9781003149873-4

symmetry into the picture, which is a feature of reductionism and a central philosophical scientific principle of physics that looks for commonalities and extracts similar behaviors.

> *A puritanical form of reductionism indeed implies that a system is nothing more than the sum of its parts.*

A variation of this understanding of reductionism is that the whole is composed entirely of its parts although the system as a whole will have features that none of the parts contain.

> *Accordingly, reductionism holds that biology is really biochemistry. Biochemistry can be reduced to chemistry. Chemistry is really physics. Physics is really math. Math is really logic. Logic is really philosophy. Philosophy is really psychology. Psychology has really emerged from biology.*

Hence, we are back to where we started whereby each discipline is simply an application of lower hierarchical scales. Some argue that physics is the lowest hierarchical scale, while others argue that math, logic, or even psychology are the most fundamental hierarchical scales to which biology is simply an application.

However, despite the remarkable achievements of the application of methodological reductionism to science and the amassed amount of knowledge from its discoveries, it has limitations in terms of its power in studying systems, which demonstrate emergent properties such as those in biological systems. Attempts are made to rigorously study such systems from various perspectives. In this context, it seems necessary to employ the notion of heuristics, such that any approach to problem solving, learning, or discovery, which engages a practical method is not guaranteed to be perfect or optimal but sufficient for immediate goals. The most promising approach to such non-reductionist problem solving is the use of computers to discern relatively few "simple rules" which represent empirical relationships that can be used to predict the emergent behavior of a system. Although such approaches to computationally analyze huge volumes of data, made possible by the "omics" revolution germinated from the human genome project, are indeed promising, deterministic models of human disease remain daunting. One obvious barrier is posed by an immense number of combinatorial possibilities. The term immense set refers to a set with more than 10^{110} elements, which represents the so-called computational barrier. This is a real impossibility of even inspecting such a set, which is easily exceeded in biological systems. Many examples of immense sets in biology can be given, such as the number of peptide chains or the number of specific variants of a protein. In the latter case, this is because of the stochastic ability of cancer cells and microbes to mutate and to additionally produce post-translational modifications leading to an astronomically large number of such possibilities.

Further application of the idea of heuristics recognizes that the most practical approach is an assimilation of both reductionist and non-reductionist thinking, that is to fully integrate reduction but constrain its applicability by empirical

information. For example, a mechanistic therapeutic strategy can be successfully incorporated into big data analytics of risk and benefit for a given drug.

Many of the problems that mankind seeks to solve are often daunting and critical to our survival at an individual or even global scale. Taking information and harnessing from it a deeper understanding that facilitates scientific problem solving overall can largely be ascribed to alternative basic philosophical positions, reductionist thinking, on the one hand, versus a less mechanistic broader scope of embracing and simplifying complexity on the other hand. The latter position concerns the essential thought patterns that organize into systems of parts and wholes that this book endeavors to promote.

Reductionism was the crown jewel of scientific discovery and the legacy of the 20th century. It is based on the notion that everything can be broken down to its smallest constituent parts and that the whole is a simple sum of the parts. Notable examples it taught us include discovering that molecules are made up of atoms, that the universe of galaxies is expanding from the moment of the Big Bang, and that our genes are made up of a double helix of DNA containing complementary base pairs that spell out the instructions for all our physical characteristics. It has given us insight into how to build lasers and computers. Reductionist thought has fueled some of the most technological awe-inspiring advances and broken barriers to our understanding of biology and of the universe. Idealist proponents of this philosophical position argue that any complex phenomenon, whether it be a physical system, a biological system, any process, even a thought, can be broken down into its component parts, and knowing the rules of interactions *between these component parts can reconstruct the entire system. This may work in many cases but with the advent of systems biology and nonlinear phenomena, limitations to this reductionist approach are becoming more and more apparent.* In this chapter, we argue that the physiological system of the human body is too complex to be functionally reducible to its constituent parts (cells, molecules, atoms).

4.2 Symmetries, Conservation Laws, and Symmetry Breaking

Science provides us with an understanding of the far reaches of the universe while at the same time teaches us about a singularity to which all biological systems are connected as a biosphere and to the four forces of nature. While the fundamental forces of nature highlight the notion of inherent symmetry or a unifying origin, it also illustrates the central relevance of breaking symmetry as a natural phenomenon. Notably, it has been shown that the strong force, the weak force, and the electromagnetic force all derive from a common single force that separated in a broken symmetrical event as a consequence of the expansion of the universe. Only gravitational force remains to be integrated into a unified theory of everything.

> *The notion of breaking symmetry underscores the deeply interconnected foundation of the universe. This conceptual unification should ultimately*

include biological systems within themselves and with the universe, in which they are embedded. Symmetry breaking leads to the breakdown of the ability to recreate a system whole in an additive linear fashion from the individual parts, which underscores the reductionist philosophy.

The understanding of symmetry involves a number of different perspectives. The tangible geometrical symmetries are the simplest. Other perspectives of symmetry or symmetry breaking are conceptually more challenging but are foundational to modern physics. These concepts require careful thinking, even when one is a physicist. If one is not a traditional student of physics, as I am not, some of these concepts require revisiting until they eventually make perfect sense.

From the perspective of a physician practitioner, the notion of symmetry breaking is critical to the premise of the nonlinearity of systems at many hierarchical scales that underpin human physiology. These scales include the allostatic pathways and networks that maintain the homeostasis of free energy, redox status, and pH, which sustain life. Symmetry breaking at any of these scales has consequences on all scales and to the human body as an integrated organism.

The spherically symmetric shape of the universe is related to general relativity, which describes how space–time is curved and bent by mass and energy. Most things in nature are shaped like a sphere. Stars, planets, and moons are spheres. All the atoms in an object pull towards the center of gravity. They are resisted outward by forces pulling them apart. The result is often a geometrical sphere. Consider water droplets dripping from a faucet or raindrops falling on the ground. The water molecules from all sides pull towards the center of mass (gravity and surface tension pulling it in) and molecular forces push it outward. The perfect balance is called "hydrostatic equilibrium". The spherical shape minimizes the surface area and has the lowest surface area to volume ratio, thus, the most stable shape with the least potential energy.

Symmetry carries explanatory power for many phenomena observed around us. The breaking of symmetry has given rise to our universe as it exploded from a singularity in space–time. A deeper understanding of symmetry gives us greater reasoning skills. The symmetry of an object is a measure of its invariance. The geometrical invariance of a circle shows complete rotational symmetry, it appears the same rotated by any angle. A square has rotational symmetry at 90-degree rotations or turns. A brick wall shows translational symmetry in that it appears the same with the translation (or displacement by unit length) of each brick. A symmetrical space or object is useful in reducing the number of calculations. For example, a cube has six identical sides, thus, knowing the dimension of any one of them simply needs to be multiplied by six to know the surface area of the cube.

However, symmetry has much deeper significance beyond that of geometry. Emmy Noether, in the early 1900s, demonstrated that symmetry is a manifestation of a physical property in a statement known as Noether's theorem. This theorem claims that *any continuous symmetry in a physical system is associated with a corresponding conserved quantity.* The time invariance (unchanging with the passing of time) of the physical universe (the classical or macroscopic and quantum atomic and subatomic level) is associated with the conservation of energy.

The conservation of energy that is neither created nor destroyed is part and parcel of time invariance of the laws of physics. Similarly, space invariance means that the same laws of physics do not change at different points in space. Furthermore, analogous to the conservation of energy arising from time invariance of the universe, the conservation of momentum arises from translational space invariance of the universe.

The most profound significance of the notion of symmetry is seen when symmetries in nature are broken. At the simplest conceptual level, this may be thought of as taking a square and cutting out a corner or cutting a slice out of a circle, hence, breaking the rotational symmetry. A classic example in nature is the spherical geometry of water droplets, when they freeze to ice it breaks symmetry forming a less symmetric crystal configuration (described below). Magnets are another example, they generate a magnetic field by picking out a particular direction in space. The electromagnetic, strong, and weak forces are another example of symmetry breaking, which at very high temperatures give different perspectives of the same force. However, early on in the expansion of the universe, the strong force separated, breaking symmetry, and subsequently, the weak force broke symmetry from the electromagnetic force. While these forces were part of the same unified force, their interactions could not be distinguished. As separate forces, they all are now known to have distinct interactions. Nonetheless, the weak force and electromagnetic forces have a clear unifying origin and at very high energy levels merge again into the same force.

The electro-weak force is endowed with a conserved quantity, which is the so-called gauge invariance. "Gauge" refers to redundant degrees of freedom (the number of independent parameters) of classical field theory. One such degree of freedom is the phase of the wave function, which has no classical analog and only appears when a quantum field is introduced as a complex function. All symmetry, in one respect or another, derives from a common origin. Gauge symmetries are best understood as being of a common origin, parts of a whole, the functions of which could not be distinguished until the symmetries are broken. Imagine if the laws of physics, which are inherently time-invariant, did in fact lose this feature of time invariance. In this case, Newton's force (equal to the product of mass and acceleration) would only apply in a given instance in time but not in another. Physics not only would be more complicated and utterly chaotic but the implications for the universe and life itself would be profound. The Higgs boson is rooted in the breaking of quantum mechanical symmetry, but its major significance is due to the fact that it explains the origin of mass, one of the most fundamental properties of matter. The Higgs field is not only a unifying force in the standard model but also an explanation for the short range of weak forces.

The deepest relevance of symmetry for the life sciences, biology, and medicine concerns many levels.

> *The existence of organismic life itself was made possible by the breaking of the electroweak force symmetry allowing electromagnetic solar rays to promote photosynthesis whereby Einstein's celebrated formula $E = mc^2$ eventually translates into the physiological source of the calorie.*

Furthermore, as Nobel Laureate Philip Anderson described, the hierarchical scales of science from the most microscopic subatomic level to the molecular, cellular, and ultimately more macroscopic physiological levels require the breaking of symmetries to form new symmetries (which occur at ever hierarchical scale).

> *Without symmetry breaking there would be no complexity and no emergence.*

Furthermore, what symmetries nature "decides" to break—such as those of time or space invariance and the inextricable conservation of energy or momentum, respectively—always ends up having profound influences on the modern laws of physics.

Just like the philosophical mainstream of reductionism brought us the insights of analytical science, Einstein's relativity was pivotal to framing new philosophical perspectives. The most important of such perspectives include nonlinear systems and the notion of symmetry, which dominate modern fundamental physics. Quantum theory further shaped these foundations. Moreover, and of the highest concern to this discussion, is the relevance to biological function. Indeed, it is the geometry of space–time on which the laws of physics depend. In the absence of gravity, time and space are one and the same, they cannot be extricated. In the presence of gravity, space–time symmetry is broken. The greater the gravity, the greater the mass and *vice versa*, and the greater the asymmetry and curvature of space–time. Accordingly, the constraining of both space and time is the patterned asymmetry of time moving faster in a smaller amount of space. This is what is meant by the loss of time–space invariance.

Conceptually added to the breaking of time–space symmetry is the symmetry perspective of the origin of life, the transition of inanimate to organic matter, and the evolution of organism species.

> *The origin of life is the transformation from thermodynamic equilibrium reversible chemical states at the lowest energy state to biological systems as dissipative structures. Such structures are self-organizing and self-adapting inherently complex systems as a function of irreversible steady states, which are far-from-equilibrium lowest energy states.*

Hence, *the formation of dissipative structures accompanies the spontaneous appearance of symmetry breaking from the* symmetrical reversible thermodynamically lowest energy state. The steady states of these complex systems are created by a dynamic interplay of energy flow through pathways of networks that plateau to a time-independent state of conserved variables. This may be exemplified by homeostatic vital sign parameters or allostatic neuronal activity of the binary autonomic nervous system, or allostatic hormonal activity of insulin and glucagon or of neuroendocrine pituitary hormones. Each allostatic steady state of the given parameters (e.g. glucagon, insulin, growth hormone, cortisol, T3, testosterone, estradiol, neuronal sympathetic or parasympathetic activity, and so on) represents an oscillatory bifurcation of a double pendulum.

An explanatory distinction between entropy and dissipation of energy is crucial; this understanding helps to elucidate these frequently misunderstood abstract concepts. The dissipation of energy is a process in which energy is transformed from an initial to a final form whereby the latter is less capable of doing mechanical work. The rate of energy dissipation can be viewed as being similar to entropy which in accordance with the second law of thermodynamics represents heat that is dissipated unavailable for doing useful work. *Heat, which is energy in transit, moves from energy that is responsible for the configuration of particles such as atoms and molecules in ordered more constrained fashion, i.e. with fewer degrees of freedom, and which is responsible for the work of creating and sustaining complex adaptive systems, to a spatially less constrained dissipation.* Thus, dissipated heat is the transfer of internal energy from a hotter body to a cooler one, such as what occurs in the aging and mortality process of animals and humans.

> *Once a dissipative structure, i.e. a complex adaptive organism, is created, the maintenance of the steady state of the organism as a system or of the many subsystems of the organismic system requires many transformations of internal energy.*

The loss of heat with each transformation contributes to the rate of entropy production. This loss or dissipation of heat is energy that can no longer be used for useful purposes. The high pent-up energy of electromagnetic rays that create dissipative structures is energy that moves "into the cool" as it forms complex adaptive organisms, which in relative terms represent dissipation of heat. The Gibbs free energy generated as a parallel to the ideologically rejected term negative entropy does not violate the second law but rather delays the ultimate rise in entropy. Because complex adaptive systems as organisms do not represent positive entropy *per se*, the rate of dissipation of energy may be a more useful property to accommodate this intermediate phase in the unidirectional arrow of entropy production.

> *The breaking of symmetry also occurs in parallel with biological evolution. Conceptually, this may be described by an original cluster of indistinguishable organisms that are split into distinguishable sub-clusters, corresponding to speciation. This highlights the symmetry-breaking characteristic of pattern formation in biological space.*

Similarly, the appearance of a litter of pups prior to the pattern distinguishing characteristics of the individual pups that evolves through the ontogeny of their lifetime, shows a degree of symmetry to one another which is subsequently lost (i.e. broken symmetry).

"Spin" is one of the characteristics of elementary particle symmetry, for example, of electrons and other subatomic particles. Spin is not actually what it was originally thought to be, rotation about the particle's axis, like the rotation of planets. To do so, the particle, like an electron, would have to move faster than the speed of light. Rather, *spin is considered to be an inherent magnetic property. Importantly, electricity and magnetism are really just two sides of the same coin, hence called electromagnetism.* In general, symmetry applies meaning to a relationship of proportional equality of harmonizing different elements of a geometric structure or function into a unitary whole. Accordingly, magnetism is a force created by electric currents, themselves caused by moving electrons.

The symmetry of electromagnetism derives from its unitary nature and invariance that is independent of the measurement frame of reference. For example, the electromagnetic frequency may be equally measured by determining either the magnetic force or the electric current. The property of electromagnetic frequency remains unchanged—that is it is invariant—under the transformation of the system from an electrical current to a magnetic field. It is a form of conservation of energy as a function of the invariance of time. The idea of breaking symmetry occurs also at the transition of the reversible thermodynamic equilibrium state to the macroscopic ordered far-from-equilibrium state. In addition to applying symmetry breaking to the origin and evolution of life as described above, it may be applicable to quantum spins of the electron pairs in magnets. At higher temperatures of a magnet, the electron spins are oriented in randomly different directions. Lowering the temperature of the magnet below the threshold Curie temperature results in the alignment of the spin directions leading to a macroscopic magnetic state with a net magnetic field exhibited by the material called a ferromagnet. Other materials, called paramagnets, also have spins but they do not spontaneously align, hence their net magnetic field is almost zero unless they are exposed to external fields that are strong enough to force their alignment.

Analogously, the motions of the atoms of a crystal below a certain temperature (Debye temperature) are quantum correlated in the form of phonon modes, which is also a symmetry-breaking phenomenon. Above the Debye temperature, the total kinetic energy of the atoms prevents their orderly quantum synchronized motions otherwise characteristic of crystals. The Debye temperature of a crystal, just like the Curie temperature of a magnet, represents a critical point whereby infinitesimally small fluctuations induced by the injection of heat affect the system's fate (presence or absence of quantized magnetic correlations; classical disorderly regime versus orderly quantum regime of phonon modes of a crystal) by determining which branch of a bifurcation is taken. *This process is called symmetry breaking because it moves the system from a symmetric disorderly state, typically a minimum energy state, into a more orderly but multistable state with more than one possible representation.*

The creation of far-from-equilibrium dissipative structures itself requires symmetry breaking from the systems of inorganic physical chemistry that try to achieve a metastable state of free energy and/or reduction of thermodynamic entropy. These creations of complex living dissipative structures are characterized by energy, which becomes "de-homogenized", a form of broken symmetry allowing the formation of regular-solid configurations as new symmetries. Each hierarchical scale, for example, molecular biology—cell biology—macroscopic physiology, is associated with breaking symmetries of the scale below it with the formation of new symmetries in order to accommodate the global structural and functional stability of the organism as a complex adaptive system. This is often (but not always) a reversible and dynamic process.

The breaking of space–time symmetry is particularly intriguing. The interwoven continuum of the four dimensions of time–space symmetry is a singularity of the two quantities. Time breaks spatial symmetry, and conversely, space breaks temporal symmetry. *Gravity breaks the time–space symmetry by creating the asymmetries of both space as three dimensions, and time as a fourth dimension.* Einstein's special relativity describes a moving object that approaches the speed of light such that its length contracts relative to time that dilates. That is, time and space become more symmetrical as time moves slower and space becomes less constrained. The breaking of symmetry with the separation of the quantities of time and space as a consequence of gravity essentially creates the physical reality of space and time as at least partially independent entities. This is one of the essential components required for the origin and existence of life, that is the viability structures that dissipate free energy forming non-linear self-organized patterns (cells, tissues, organisms—including humans—societies). This type of symmetry-breaking phenomenon results in a profound effect on the second law of thermodynamics, which is locally violated by entropy reduction in a living system. This effect, however, is not only spatially constrained but also limited in time because all living systems age and eventually die. Hence, *the second law of thermodynamics eventually always prevails, and this leads to the introduction of the arrow of time that gauges the aging process.* Organisms, as examples of dissipative structures of free energy, generating negative entropy, are complex systems described by non-equilibrium and nonlinear dynamics.

> *The more pathways within networks of systems biology that fall to their lowest energetic states, that is thermodynamic equilibrium, by the processes of redox disturbance with excessive free radical formation, oxidative stress, and inflammation, the more accelerated is the entropic process and the physiologic arrow of time, or senescence.*

4.3 Systems: Open and Closed, Simple and Complex

A set of connected parts forms a complex or unitary whole. A complex whole is greater than and hence different from the sum of its parts. A unitary whole, such as that composed of simple or complicated systems, is the linear sum of its parts. A system may be defined arbitrarily, it may be confined to a reaction in a beaker, may consist of as few as two interacting parts or it may consist of the entire universe.

> *The notion of an open system has its application in the natural sciences integrating the organism, thermodynamics, and evolutionary theory as well as information theory and systems theory.*

Its borders are permeable to both energy and mass. For example, an organism such as a plant, animal, or human is permeable to energy and mass such as the solar electromagnetic energy in the form of light provides the energy and information crossing the boundaries of the organism. *In the case of humans and animals, mass is exchanged with the surroundings, which may be exemplified by the plants that they consume. Mass also permeates the boundaries of plants as marine algae became subsumed as chloroplasts and bacteria as mitochondria in higher organisms.*

> *The basic idea of an open system is that it interacts with its environment through giving and receiving information, energy, and matter, whereas closed systems are closed off from the outside environment and all interactions occur only within the closed system.*

An emergent property of a system is the higher-scale property, a static or dynamic pattern or behavior that results from the interactions of elements that cannot be predicted from the simple, lower-scale entities or components of the system. That is, it is a feature of a nonlinear system.

> *The behavior of an entire nonlinear system of interacting parts cannot be predicted on the basis of the structure and function of the parts per se.*

Emergence is also characteristic of biological processes and the phenomenon of life itself. There are many examples of this in medicine. Consider the complexity of cytokines, which are small molecular components of the overall inflammatory response. They convey information from one cell type to another, having receptors on most, if not all, types of immune cells. They exhibit classical features of complexity including interdependency, pleiotropy (multiple effects), redundancy (multiple cytokines having the same effect) and demonstrate both positive feedback (can amplify the effect of the other cytokines) and negative feedback (can dampen the effect of other cytokines) [2]. Given the complexity of these interactions, it follows that the pattern of their behavior in a relatively simple system consisting of the cytokines along with effector and regulatory immune cells is dynamically nonlinear. For example, typical cytokine dose response curves show an initial threshold (no effect below a certain concentration) followed by an exponential response prior to reaching a plateau [2]. In healthy individuals, one type of cytokine, TNF, promotes high circulating neutrophils at low concentrations but low circulating levels of neutrophils at high concentrations [3]. Analogous

to the flock of birds, one cannot understand and behavior of the system by studying an isolated cytokine or cell type. Therefore, *data analysis using linear models based on proportionality between two or more elements and/or relationships described by linear differential equations would not be able to predict dynamic behavior.*

The overall pattern of the geometrical shape of the cytokine response or the promoted behavior of the immune response is the emergent property of the system. The component interacting parts of a system, the various cytokine and cell types collectively interact to affect the emergence of the system's behavior as a whole at the highest hierarchical scale, in this case, the cell scale. The collective properties refer to the individual entities attributed equally to different organization (or hierarchical) levels of a system. For example, the cytokine and cell components of a system are collectively considered equal components. However, the activities of the molecular scale of the system, the cytokines levels, produce the emergent properties of the next hierarchical scale, the cellular scale. The next level of the hierarchical scale may include physiological vital sign parameters, blood pressure, pulse, respiratory rate, and temperature, which would be equally attributed to as collective components. Conversely, emergent properties result from the activities of lower-level entities on the next higher organizational level. These emergent properties are not present on the lower levels.

> *A simple system is predictable, deterministic, and linear. A complicated system is similarly linear, deterministic, and predictable although in the latter case there are bifurcations. Simple systems may evolve into complicated systems due to environmental perturbations. Complex systems are nonlinear, deterministic but unpredictable. Quantum systems are nondeterministic and unpredictable. Biological processes and life itself are an intersection of all of these systems.*
>
> *A linear system is one that can be reduced to its component parts and by understanding the parts in isolation, the system as a whole can be understood. The analysis of a linear system is based on the output either being proportional to the input or can be determined by the application of linear differential equations to the input.*

A linear system has mechanical periodicity, like a simple pendulum that in an orderly fashion oscillates (Figure 4.1) or tends toward a particular fixed point (attractor). Over the course of time and iterations an element of a system, for example, endogenous insulin levels in a healthy individual will oscillate toward a physiological level that balances the broader system of blood glucose. The postprandial state drives the oscillations to another fixed state or attractor. Hence, this exemplifies a simple system with multiple equilibrium points, each representing a vertical state, the fixed state, or attractor of a simple pendulum. These points may also be described as homeostatic set points of a physiological system. However, in the case of perturbations of the system that are external to the simple system but only indirectly external to the organismic system (the body) such as due to excessive intake of food (i.e.

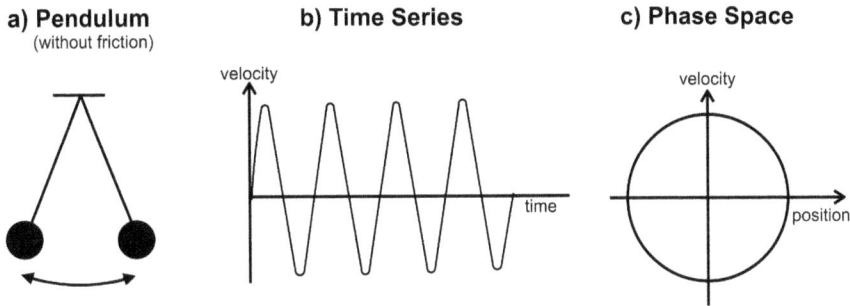

FIGURE 4.1 Oscillations of a simple pendulum a) in terms of its time series for velocity as a function of time b) and a so-called phase portrait where velocity is plotted in terms of position c).

is mediated more directly by other hormonal and metabolic molecular signaling), allostatic changes manifest so to maintain homeostasis of higher scale hierarchical systems, the highest of which is ultimately vital signs and vital organ system function. Allostatic changes in this example may include a host of hormonal integrations whereby the simple equilibrium relationship between blood glucose and insulin evolves into interactions between heterogeneous elements, for example, free fatty acids, hormones such as cortisol, growth hormone, and the sex hormones. These elements collectively interact as a complex system that self-organizes, reaching steady states that cannot be predicted solely on the basis of the structure and function of these interacting parts.

There are a number of theories stating that insulin resistance is an adaptive allostatic regulation. One such theory posits the downregulation of insulin receptors of hypertrophic adipocytes (and accompanying increase in circulating insulin levels) and release of free fatty acids from the breakdown of triglycerides in the fat cell by removing the lipolytic suppressive action of insulin that occurs to protect against obesity. This allostatic strategy would only work if the excessive intake of food were transient. If persistent, overindulgence of energy intake occurs, unremitting insulin resistance cascades into associated leptin resistance, deleterious ectopic fat deposition, and a hormonal profile that is intended to prioritize energy resources away from anabolism and reproductive function. This process becomes pathogenic in feed-forward fashion as impaired social vigor ensues along with depressed mood, motivation, and endurance accompanying obesity and often type 2 diabetes, the manifestation of allostatic overload. The allostatic regulations of the elements of the broader system that may appear to exhibit random behavior actually tend toward and manifest a likelihood of certain states, i.e. attractors or new steady states. This leads to the phenomenon of the emergence of adaptive behavior. *In physiology, there are simple and complex systems that are dynamic over time.* Physiologic allostasis is exemplified by the evolution of the simple system of glucose and insulin equilibrium to a complex system of multiple steady states designed to be a temporary adaptation to prevent the broader system (e.g. an organism) against obesity. When chronic, the complexity of allostasis degenerates to allostatic overload, and hence instead of organized complexity of emergence, it is characterized by disorganized complexity, pathological rhythms, and pathogenic disease.

A simple system can be linear, like a periodic (self-repeating) simple pendulum. However, it may start out as a simple system and dynamically evolve as a result of external perturbations to a more complicated nonlinear system, as with multiple equilibrium points.

This occurs in the case of pre- and postprandial endogenous insulin levels in a healthy individual. Simple linear systems may evolve into nonlinear chaotic systems. The allostatic adjustments as discussed are probably best considered as an example of chaotic behavior. In contrast to linear systems portrayed by a simple pendulum, chaotic behavior is portrayed by a double pendulum. Starting the pendulum from a slightly different initial condition would dynamically result in an entirely different trajectory. In fact, the double rod pendulum is one of the simplest dynamical systems that has chaotic solutions (Figure 4.2).

In addition to hormonal allostatic patterns, there are many classical manifestations of chaotic behavior in medicine. The example of cytokine behavior exhibiting emergent patterns was previously discussed under emergent collective properties, see above. There are many other examples. Heart rate variability is a particularly clinically relevant example because of the increasing use of autonomic nervous system testing in endocrinologist and cardiologist offices. Heart rate variability involves the analysis of spontaneously varying inter-beat intervals (R-R intervals). Sinoatrial node intrinsic rhythm generated heartbeat is modulated by many feedback circuits of the autonomic nervous system integrated with various systems. A

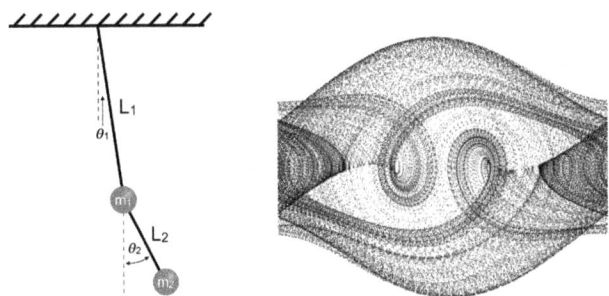

FIGURE 4.2 A double pendulum with two masses connected by two limbs of arbitrary lengths a) and its rich dynamical phase portrait b).

normal heart rate displays complex fluctuations in response to breathing and changes in position, that is, going from sitting to standing to lying position. This is the basis for autonomic nervous system clinical testing protocols.

Emotion and exercise also promote fluctuations in inter-beat intervals. Loss of such variability, such as with taking a deep breath in relative to exhaling, or changes in position, suggests a loss of the healthy integrated modulatory feedback circuits. *Loss of such nimble flexibility does not refer to the inverse correlation with robustness as described by Demetrius in characterizing the genetic evolution of microbes relative to their host or of cancer. Rather, it refers to a component of robustness characterizing subsystems that maintain the overall robustness of the larger system of the organism (e.g. a human being) as a whole.* Accordingly, loss of healthy heart rate variability is associated with illness [4–6]. Individuals with congestive heart failure are known to have reduced heart variability [7, 8]. Furthermore, congestive heart failure complicated by autonomic nervous system dysfunction portends higher arrhythmic and total mortality following acute myocardial infarction [9,10]. The heart rate variability illustrates nonlinear dynamics related to the control mechanisms of the electrical system of the heart. An important feature of chaotic behavior, as it describes nonlinear behavior and physiology is fractal organization.

> *The application of fractal analysis appears promising as an approach that predicts cardiac risk including sudden death.*

In such important cases, the nonlinear interactions displayed on phase portraits and their fractal organizations break down [11]. It appears that this is a consequence of impaired autonomic neuronal modulation of the sinoatrial node in additional sympathovagal imbalance [12, 13]. In healthy individuals, there is a predominance of the vagal component of the autonomic nervous system to maintain variability [14]. Also important is the interdependency and interconnectedness of the organ systems within the total organismic system. Thus, when some systems become isolated or uncoupled from other systems, or "biological oscillators", such isolation correlates to greater regularity with loss of healthy variability, flexibility, or allostasis, among the organismic subsystems and associated loss of robustness at the organismic system level [15–17]. Reduced vagal tone and/or sympathetic dominance that correlates to impaired heart rate variability makes the heart less adaptable to the hemodynamic demands of the environment of perturbations [18–19].

> *Chaos theory helps understand both normal and abnormal behavior of the subsystems of systems. The application of fractals that characterize phase portraits of these systems and subsystems has real value in the early warning stages such as cardiac arrhythmic death.*

Principles of systems, the science of complexity, and chaos theory also seek to invoke this knowledge in managing illness such as to establish "recoupling", i.e. normal connections and variability and hence complexity between systems. Such recoupling re-establishes the "edge of chaos" without overt chaotic behavior (e.g. cardiac arrhythmias).

Simple systems follow basic rules such that with the knowledge of the elements that make up the systems and the rules that govern them, the future of that system is determined (i.e. deterministic), and is accurately predictable under various conditions. A different type of system that is deterministic but not predictable, as discussed above, is the nonlinear chaotic system. Initial recorded conditions of the system, or other perturbations from outside the system, result in unpredictable behavior. In this type of simple but nonlinear system, the trajectory of the system may be determined by the knowledge and the elements of the system and the rules that govern them. However, the equations which incorporate these deterministic factors are very sensitive to the initial conditions such that imprecise recording makes the model unpredictable to the dynamic behavior of the system relative to the perturbations coming from outside the system. This type of system has been referred to as predictably unpredictable.

The alternative nonlinear system to a chaotic one has been referred to as unpredictably unpredictable. The notion of complexity from complexity theory evolved from chaos theory and has attempted to capture what happens beyond what chaos theory describes into the realm of creating new systems from existing ones, and the emergence of order from disorder [20–21].

> *In medicine, nonlinear systems are modeled according to complexity theory, which are distinct from chaos models. Such systems are both nondeterministic (that is there are no equations that incorporate all the potential elements and governing rules) and unpredictable.*

The endothelial cell is an example of a complex system. Its function may be altered by the dynamic interactions with a very large number of heterogeneous stimulating elements, both internal and external, to the endothelial cell. Such stimuli include cytokines, drugs, oxygen content, physical stress, neural behavioral changes, temperature, organ dysfunction, circadian rhythm, host-pathogen interactions, thrombus, and all of the cardiovascular risk factors [21]. Although in vitro studies of the individual list of exposures on the endothelial cell have framed our mechanistic understanding, applications of linear models to such complex nonlinear systems have been unsuccessful. It was shown that cytokines interact with many transcriptional activators and inflammatory pathways suggesting complex nonlinear function [22]. In addition to the distinction of complex and chaotic nonlinear systems, the broader distinction between linear and nonlinear systems may be contrasted by the analogy of individual musicians and their instruments in isolation to predict the tremendous interdependency and harmony of an orchestra.

4.4 Stability, Biological Complexity, and Energy Flows

The best example of regularity, predictability, and equilibrium is a simple pendulum. *Linear, deterministic equations accurately predict the dynamic evolution of linear systems, such as a simple pendulum, which predictably oscillates in a periodic (regular) fashion. This is in contrast to a complex system, which must be cast in terms of probabilistic measures.* A predictable and deterministic system is inherently stable. Once the initial conditions are known, the future is reliably predictable. The trajectory of a simple pendulum is typically predictable and deterministic whereby displacement away from the equilibrium point, the vertical position, like in any stable system, tends to asymptotically die down taking the system back to its equilibrium position. The perturbation is transient if we account for friction. A simple system, however, like a simple pendulum, may behave unpredictably if the perturbation is large enough because its representation becomes nonlinear. The system taken from the stable zone of order to the inner zone of chaos is searching for a new equilibrium point and ultimately it either falls back to the original balance point or it spirals into the outer zone of the "edge of chaos" and ultimately becomes permanently detached from regulatory control. A chaotic nonlinear system is analogous to a double pendulum that is inherently unstable, searching for new steady states, and that will never recover the original position. In other words, the original attractor or equilibrium balance point is no longer the goal to which the system seeks to settle.

In human physiology, for example, both stable and unstable systems may be healthy or unhealthy depending on the system.

> *Equilibrium points are characteristic of homeostasis, resistance to change.*

Vital signs are homeostatic equilibrium points characterized conceptually as periodic oscillations with a predictable trajectory. Allostasis means stability through change.

> *The allostatic hormonal adjustments and other molecular and autonomic nervous system oscillations are examples of systems at the "edge of chaos" searching for new balance points that have the compensatory capacity to maintain homeostasis of vital signs and hence vital organ system function.*

The simple pendulum of the vital sign parameters has a narrower displacement than do the oscillations of the allostatic hormones such as cortisol, insulin, and the sex hormones because the latter systems have a lower calibrated priority for energy requirements necessary for survival. The inner zone of the "edge of chaos" is an active space or boundary of self-organization and self-adaptation and characterizes flexibility of the allostatic subsystems and organismic robustness (e.g. the system of the human body). This is a principle of organized complexity.

> *Prolonged or severe stress prevents the allostatic mechanisms from re-establishing their equilibrium balance at the zone or order from the inner zone of the "edge of chaos".*

The inner zone of the "edge of chaos" is analogous to the double pendulum that searches for new equilibrium steady states until the underlying stress or perturbation to the system is restored to the baseline state at which point the conceptual double pendulum reverts to the conceptual single pendulum. If the prolonged or severe stress is not returned to baseline, ultimately, the energetic capacity of the system is drained as it deteriorates into the outer zone of the "edge of chaos", whereby the conceptual double pendulum swings more erratically and becomes progressively detached from the orchestra of regulatory control of other systems. At this point, insulin resistance and glucose intolerance become metabolically irreversible type 2 diabetes with higher systemic blood glucose levels and circulating insulin levels that are no longer relatively depleted to the level of high blood glucose but are depleted in absolute terms. The double pendulum at this point is no longer capable of reverting back to a single pendulum and re-establishing the prior healthy balance equilibrium set point of a simple subsystem. This is a manifestation of the subsystem-in-failing. This phase is referred to as allostatic overload whereby the allostatic flexibility is lost, pathogenic disease ensues and the protection of the homeostatic minimally displacing simple pendulum becomes threatened. As the energetic flow of these allostatic systems is lost, the organized complexity of our physiology becomes progressively compromised. Hence, the self-organizing and self-adapting potential is increasingly lost in a feed-forward fashion as the informational energy of the system's pathways and networks deteriorates into the "edge of chaos" with associated increased thermodynamic entropy. Ultimately, the system's complexity becomes so compromised and so simplified to represent a single metabolic tendril, which represents life support and the ominous portent of mortality.

> *The outer "edge of chaos" is represented by disorganized complexity, loss of functionality, and entropy heading for the end state. The outer "edge of chaos" typically refers to random disorder as does to a greater extent the end state.*

Predictable regularity represents the territory of the minimally displaced pendulum from its vertical state in terms of the oscillatory swings from the equilibrium set points of any given system or subsystem. In a healthy individual, predictable regularity is exemplified by the pendulum swings of the system's vital organ system function such as that of heart rate, blood pressure, respiratory rate, and temperature. The inner zone of the "edge of chaos" is neither random disorder nor predictable regularity. Rather it is characterized as "order masquerading as randomness" (Edward Lorenz) designed to restore order.

> *A hallmark of healthy biological complexity is the metabolic flexibility of pathways that form networks or systems, to form new steady states allostatically with built-in redundancies for the purpose of maintaining resilience and homeostasis of higher priority systems.*

These new steady states can also be viewed as symmetry-breaking phenomena because they correspond to the emergence of new properties in analogy to the new quantum numbers arising in the standard model. Allostasis is the ability to maintain stability through change while homeostasis is resistance to change. When newly emerged allostatic steady states do not maintain homeostasis of the next hierarchical scale of metabolic priority, those systems themselves allostatically form new steady states with the goal of maintaining homeostasis of the next hierarchical scale of metabolic priority.

> *The most fundamental properties required to be homeostatically regulated are free energy, redox, and acid-base balance.*

Allostatic capacity, precipitously, in the setting of severe acute metabolic challenges or, insidiously, in the setting of more indolent but persistent metabolic challenges, ultimately loses adaptability and functionality of safeguarding those sacred metabolic parameters essential for life. *Consequently, death ensues after a variable duration of infirmity characterized by allostatic overload.*

The complexity and robustness of energy flow in the maximally healthy state are manifested by homogenized free energy within the boundaries of the organism as a system of systems. This may be thought of as symmetry in a similar way that the reversal of health is seen via its free energy loss due to heat and associated entropy increase, ultimately progressing to a maximum entropy state under constraints, i.e. thermodynamic equilibrium. The notion of symmetry of free energy within the boundaries of an organism, however, should only be considered within a narrow context as an abstraction. It should be noted that it is not symmetry in the conventional sense that entropy in systems in thermodynamic equilibrium represents. *Maximum entropy corresponds to the symmetry of homogeneously distributed heat in a closed system.*

Gravity, felt by us all, is created by the massive size of Earth. According to Einstein's theory of general relativity, gravity increases the curvature of space–time whereby it constrains space and constricts time. That is, the greater the asymmetry of space–time, the faster time moves within a more constrained boundary of space. However, this increase is not felt by us all. *Pathways that form networks of systems biology interactions become degraded by the processes of disturbed redox physiology and loss of free energy to heat.* Consequently, the ensuing loss of complexity that represents the interconnectedness between the subsystems that comprise the whole organism correlates to the segregation of the subsystems. This overall process corresponds to advancing allostatic overload and loss of metabolic flexibility necessary for adaptation and resilience for survival. By extrapolation, the implications of the curvature of space–time suggest an increasing loss of free energy and biological complexity corresponding to progressive functional compartmentalization of subsystems. This amounts to boundary constraints of energy flow and time constriction with accelerated aging. This metaphor connecting the theory of relativity and aging may not be intuitively obvious, but it will be explained in more detail in Volume 2. Next, we expand on this metaphor to make it clearer to the reader.

The larger the amount of free energy that is engaged biologically in a functional role, that is the faster and more robust the flow of free energy is, and hence the less spatially constrained that flow, the slower is the physiological march of time (aging) relative to the total energy utilized and lost. In other words, for any given amount of effort and energy put into a process or task, less of it is lost to heat and entropy when that energy flow is less spatially constrained, and hence the rate of flow is faster. In this case, time is relatively dilated and biological aging relatively slower. A greater robustness of energy flow across the pathways of networks and systems correlates to the adaptability of allostatic pathways to find new steady states maintaining homeostasis in the face of environmentally induced perturbations. *Steady states are time independent, i.e. cyclical.* This time invariance and the nimble capacity to find new steady states, that parallels the robustness and complexity of energy flow through systems biology, illustrates the adaptive nature of symmetry breaking and finding new symmetries.

Another symmetry that correlates to healthy biological function is the symmetrically coupled redox cycling of metabolic enzymes across the mitochondrial biomembrane. Moreover, this symmetry correlates to the synchronization of metabolic cycles through gene regulation within and between cells [23]. This synchronization involves nonlocal coherent communication of pathways and systems underpinned by this notion of quantum metabolism described by Demetrius and Tuszynski [24]. *The optimal biological functioning breaks down when conditions become unfavorable such as in the setting of nutritional excess and oxidative stress.* The quantum de-coherence is associated with constraining of time and space, isolation and compartmentalization of energy flow, and related downstream effects. Furthermore, chronic disease and/or accelerated senescence ensue.

Ants, bees, or bats in a colony or birds of a flock represent the individual components of elements of a system whole (the colony or flock). The ecosystem to which a colony or flock belongs represents a system on a larger or more macroscopic scale. It is an inherent design of nature that a system is comprised of parts to that system whereby the system as a whole becomes a part of the next organizational more macroscopic scale system in a hierarchical order. In turn, that system whole becomes a part of the next even larger system as a continuum of functional integration of systems of parts and wholes. As such, a colony of organisms becomes a part or an element within many other colonies and organism species of predators and prey that comprise a larger organizational scale. This connectedness includes an extraordinary complexity of energy flow through the global ecosystems. Complexity means having many (sometimes different) parts that interact in a specific fashion. The rules of interactions may be simple and still lead to complex behavior such as in the famous "Game of Life" or cellular automata (Figure 4.3).

This complex behavior that results from simple rules may be exemplified by real-life systems of networks, such as an ant colony, flock of birds, or school of fish, each of which is composed of seemingly identical interacting organisms. In this case, the complexity is a simple complexity involving systems whose component parts are similar and whose interactions occur within the same hierarchical scale. Complex adaptive systems in this scenario appear to act according to

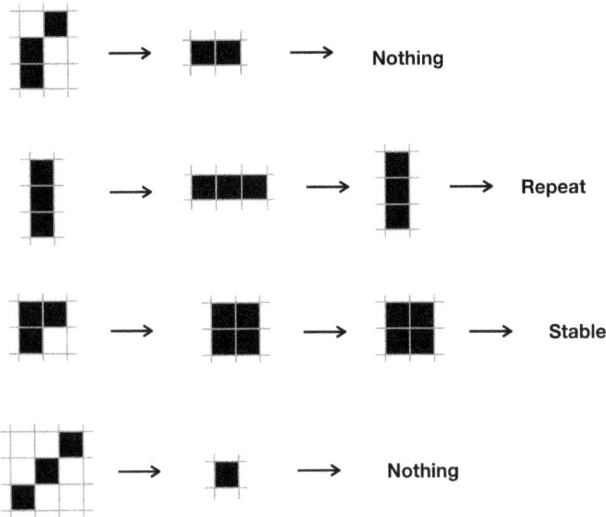

FIGURE. 4.3 Example patterns and their evolution in Conway's Game of Life. Source: adapted from http://pi.math.cornell.edu/~lipa/mec/lesson6.html.

relatively few rules, which produce emergent properties and globally coherent behaviors of many types. Examples here may include the signaling patterns responsible for a biological immune response, defined by a pattern of antibodies, cytokines, and other mediators in response to a pathogenic exposure. Additionally, the behavioral properties of animal species—such as the honeybee that does "the waggle dance" to identify the locality of a nest, or ants that forage by following chemical pheromones dropped by other ants after finding a food source—determine the complexity of the emergent behavior. All of these systems represent extraordinary adaptive collective behavior mediated by simple rules of recognition and response. Computer simulations may help determine

or elucidate simple empirical rules governing these phenomena. An example of a system defined this way is the flocking patterns of migratory birds (Figure 4.4).

Computerized simulations of flocking behavior create a model whereby the flock moves according to three basic rules: 1) separation (each bird avoids crowding neighbors, or short-range repulsion), 2) alignment (each bird steers toward the mean heading of neighbors), and 3) cohesion (each bird steers the mean position of neighbors, or long-range attraction). With these three simple rules, the computerized flock moves in an extremely realistic fashion with complex motion and interactions. The actual fundamental laws of nature, which these birds obey are unknown, but these simple or basic rules succeed in creating a realistic simulation of their behavior. In conjunction with mechanistic insights, the so-called simple rules provide a heuristic model for the future of problem solving in medicine. There is a functional eloquence of all natural complex systems that provide adaptive value. In the case of flocking of birds, the adaptive value includes an effective and safe journey of migration south during the winter. However, because the mechanical rules are not the same as the actual rules devised by nature, predictability is not always accurate, as seen in cases of statistical anomalies for computer models. Such statistical anomalies provide the impetus to self-correct the computer model to improve its accuracy and predictability. *Bottom-up approaches to learning in general are characterized by self-correcting over time.* In fact, Wikipedia although is widely criticized for not being accurate, in the hard sciences, physical sciences, and life sciences is quite accurate. The long-term input from many fragmented truths is self-correcting and ultimately has the potential to be highly accurate over time. As a matter of fact, in the physical sciences and life sciences, Wikipedia outcompetes *Encyclopedia Britannica*, which is a classical top-down approach to acquiring knowledge. In the case of *Encyclopedia Britannica*, a select number of scholars

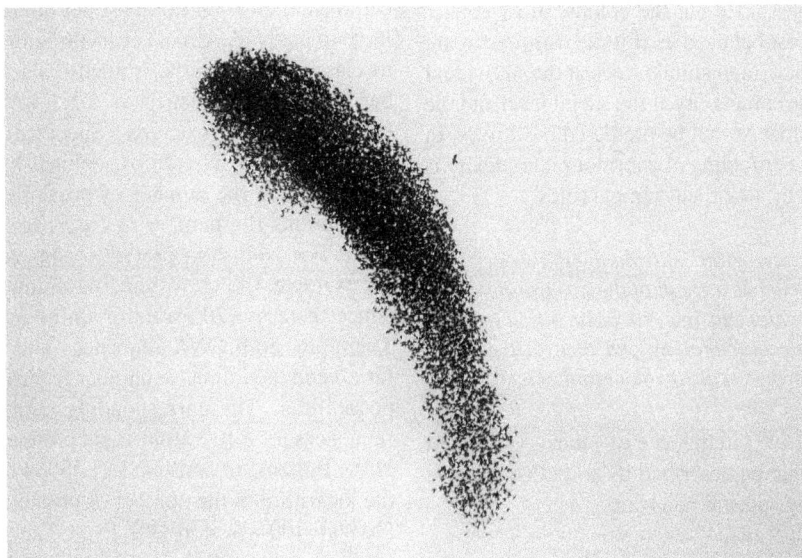

FIGURE 4.4 An example of a flock of birds forming a pattern in the sky. Source: by Mostafameraji, used under Creative Commons license https://creativecommons.org/licenses/by-sa/4.0/legalcode.

dictate a blueprint of the text from which others derive their information.

> *The collective behavior of a flock of individual birds, just like the collective behavior of the individual components of multi-systems biology represents the unpredictable emergence of higher-order parameters or hierarchical scales from lower-order parameters.*

The different fundamental laws of nature that in reality exist at each of the organizational levels of nature are officially replicated, albeit with less accuracy, by more simplistic or basic rules enlisting the convenient strategy of "simplifying complexity".

Another example of simplifying complexity analogous to the simple rules of migratory birds can be illustrated from the work done by Nicholas Perony and coworkers [25] studying the bonding between networks of bats. During the winter, the bats hibernate in isolation. However, they leave isolation in the early spring joining up again with those individual bats they previously interacted with during the previous year. Social bonding has been known to occur among animals such as primates, elephants, and dolphins with much larger brains than the peanut-sized brains of bats. It turns out that by computer remodeling analysis, a few simple individual rules explain these social patterns and complexities, and how the bats roost and forage together. Further observation of the bats underscored how these simple rules guiding social complexity provide adaptability and resilience in the setting of environmental perturbations. For example, a particularly cold winter resulted in the loss of approximately one-third of the bat colony members. Interestingly, the bats did not form separate communities like it had the previous winter, which would have been too small for them to survive. Rather, they formed a single cohesive colony unit emerging a behavioral pattern of social complexity. Obviously, the individual bats were not aware of what was occurring on a larger scale but the colony, in its collective wisdom did, or it least behaved as if it did. Similar to ant colonies, extreme simplicity and simple rules at the individual level result in remarkable complexity at the social level and the evolution of emergence that cannot be predicted intuitively. In these models, synergistic unfolding of enormous complexity is guided at the local level by so-called "simple rules".

> *In biological systems of an individual organism, emergence also occurs as a result of the integration of many different scales starting with basic physics, progressing to molecular biology and then to the phenotypic macroscopic scale of the organism.*

Analogous to the collective intelligence of models discussed above, emergence also can be described by so-called "simple rules" of humans guiding societal behavior.

> *The emergence of complex systems involving many non-similar interactions also occurs as a result of crossing different scales anywhere in the spectrum from elementary particle physics to atomic physics,*

> *to molecular physics, to condensed matter physics, to chemistry, to biochemistry, to molecular biology, to cell biology, all the way to the organismic level (and beyond to societal behaviors and ultimately to the broader universe). This notion of emergence highlights the limits to reductionism in biological systems.*

Perhaps the best description of emergence as a modern science is the one described by Alwyn Scott. In his 1995 book, *Stairway to The Mind* [26], he argues that *reductionism cannot account for the emergence of cognition as a representation of not only any macro-biological manifestations of living systems or organisms, but also of inorganic systems such as water or the flame of a candle.* He contends that the power of reductionism breaks down very rapidly in view of the so-called combinatorial barrier derived from the concept of an immense set or immense number. This is the number beyond which it is impossible to even inspect, let alone analyze, a set composed of that many elements and is rooted in the notion of biological entropy. A discussion of biological entropy and evolutionary entropy as a formal similarity to the concept of entropy in physics is included elsewhere in the book.

The definition of the immense number is 10 to the power of 110 (10^{110}). This number is essentially the product of 1) the number of particles in the universe and 2) the age of the universe in picoseconds. The number of particles in the universe ($\sim10^{80}$) is derived by dividing the total mass of the universe (1.46×10^{53} kg) by the mass of a hydrogen atom (1.67×10^{-27} kg). The age of the universe is around 14 billion years ($\sim4.5 \times 10^{29}$ picoseconds). Therefore, if the entire universe were a supercomputer functional at the speed of quantum transitions, it would be able to perform only up to $\sim10^{110}$ operations during its entire existence. Such a number is obviously much smaller than all the combinatorial possibilities of DNA and protein structures in all the living organisms. This is a hard fact, and it poses an insurmountable barrier to reductionism, because even if one could enumerate an immense set of elements, it would be too large to analyze. Almost everything in biology that one wants to classify exceeds the immense set starting with the number of possible sequences of DNA to the number of possible encoded protein structures. Specifically, the DNA sequence is described by its entropy, defined by four to the power of n. Four (4) is the number of possibilities of nucleotides and n represents the number of base pairs forming a DNA molecule. For even short proteins composed of 100 amino acids, for example, the corresponding number of possibilities is 20 (since there are 20 essential amino acids) to the power 100. Therefore, both DNA sequences and amino acid sequences far exceed the immense number regarding the number of their possibilities. The corresponding entropy of the amino acid sequences for a 100-amino acid protein is equal to the product of the Boltzmann constant k (1.380649×10^{-23} Joules·K^{-1}) and the logarithm of the number of possible sequences: $S = k \times \ln (20^{100}) = 100 \times k \times \ln (20)$. Even if this computational barrier could be achieved, the relationship between this vast number of elements of a system cannot possibly be analyzable in any meaningful way using methodological reductionism due to an even larger number of combinations and permutations of pairs, triplets, etc. of the interacting elements that form an immense set. Alternatively, such enormous complexity would only be possible to analyze by process of simplifying complexity that requires finding "simple rules". Such simple rules are not based on any fundamental laws or principles of nature and hence, are far from an exact science. It involves taking shortcuts and coming up with observations or some empirical rules or relationships that constitute an empirical model. Such a model is subsequently applied to other similar data sets of complex systems. Statistical anomalies promote an understanding of the "rules behind the rules" so to speak. These refined simple rules create new models moving it closer to the actual truth but with a theoretically infinite number of trials, this process asymptotically approaches the truth but never reaches it. The process of simplifying complexity utilizes computers such as analyzing big data, may also consider the method invoked by IBM's Watson super-computer that works by pattern recognition using artificial intelligence algorithms. *This analytical process too has limitations but is nonetheless a powerful tool.*

Aside from the quantitative barrier that limits the viability of reductionist models of analysis the essential argument qualitatively against the usefulness of the reductionist philosophy of scientific inquiry, as Alwyn Scott contends, is that there exist different fundamental laws or rules of nature across the different hierarchical scales. Moving from the most microscopic scale of biological organization from quarks and leptons to protons and neutrons to atoms to molecules and macromolecules to cells to tissues and organs to organisms to ecosystems and even increasingly greater hierarchical scales of science, each scale obeys its own rules and laws designed by nature.

The rules at higher scales of organization must not contradict the rules at lower scales, hence biology cannot be in violation of physical laws.

The relationships that span across these organizational levels underpin the critical phenomenon of emergence that cannot be understood or inferred by methodological reductionism. For example, one cannot use the rules of particle physics to explain, let alone deduce the rules of biochemistry or molecular biology. Alwyn Scott examines the phenomenon of consciousness in terms of cognitions and the human mind whereby the ultimate manifestation of thought and knowledge is an emergence that cannot be reduced or broken down to lower levels of understanding in terms of space or time. That is, the laws and rules at the organization scale of atomic physics or cell biology cannot be simply extrapolated to explain the eventual expression of thought, or any other macroscopic manifestation of organismic biology. More simplistic examples of emergence outside of biology, as Scott points out, include the flame of a candle or the behavior of water which are inherently indescribable and incomprehensible from lower levels of description in terms of the first law; that is, explaining mechanistically how energy flows from one level of organization to the next. Nobel Laureate Philip W. Anderson [27] explains the notion of transcending across hierarchical scales of biological organization as a "breaking of symmetry" (see discussion of symmetry) and therefore more macroscopic scales of biology are not reducible to lower or microscopic scales of organization. In his highly cited article, "More is Different" [27], he outlines the heart of the understanding of interdisciplinary symmetry breaking. The synergistic unfolding of enormous complexity in each realm is guided at the local level by simplicity of interactions and fundamental rules. Anderson describes the sequence in the linear hierarchy of sciences as not simply the application of one science to another. He emphasizes that each scientific discipline has its own symmetry rules, and their functional integration requires breaking symmetries and forming new ones, each tier having new laws and generalizations that explain behavior. These interdisciplinary relationships speak both to the system's complexity, the many similar parts that interact at each hierarchical scale, and to the complicated interactions of the system. Something that is complicated has many non-similar interacting parts such as what especially occurs between hierarchical scales. The characteristic distinction here is that in "simple complexity" the parts are similar in contrast to "complicated complexity" where the parts are different. These are two very different but commonly misunderstood terms. *Both interactive processes typically exhibit nonlinear emergent behavior and require a bottom-up approach to understand them. Complex patterns, which are complicated typically entail greater uncertainty than the complex patterns which are not complicated. Something that is complex but not complicated does not generally involve breaking and forming new symmetries. The interacting parts of such a system within the same plane or manifold are similar.*

Reductionist thinking is particularly difficult to apply to biological systems because they are not only complex but complicated.

In biology, there have been a lot of naive attempts to reduce everything to mutations or genes and basically jump several levels of organization. Conversely, it is necessary to come up with some kind of analogy or empirical language in mathematics, i.e. basic or "simple rules" that are specific for a given level.

Radical organizations that use the internet to recruit the sympathetic support of young people have provided a blueprint top-down approach. Once a threshold number of youths have become engaged, it becomes a self-organizing phenomenon detached from regulatory control from the outside world. This self-organizing phenomenon is a bottom-up process whereby each individual component of the group does not see or understand the whole picture from any objective or large-scale perspective. Rather, they act according to local interactions with others in the group that together form a collective intelligence and outcome. This exemplifies nonlinear complexity, the notion that the whole is greater than the simple sum of its parts. It is a fundamental property of the bottom-up autonomously regulated network systems of biology; the structural boundaries of the global system may be an individual organism, or it may be a group of organisms. The latter may be a group of migratory birds. Alternatively, it may be a colony of ants capable of producing spectacular accomplishments such as forming large bridges using their own bodies for the rest of the colony to travel across.

In the insect world, part of the swarm mentality is the notion of personal sacrifice for the greater good of the colony. Male honeybees seek out the opportunity to die by mating with the queen bee, in a so-called sexual suicide ritual. Energy-rich stores are too limited to provide for all the honeybees during the winter months, so the male bees sacrifice themselves in the mating process. Larger animals, mammals, and even humans also act in nature according to local rules that form a collective intelligence. Humans, in fact, share a surprising number of genes with very distant relatives, even insects.

The important issue here is that a swarm mentality, or collective intelligence, was designed to be an adaptive healthy metabolic phenomenon deeply rooted in our evolution. Unfortunately, however, the consciousness of free will, debatably unique to humans, allows cynical motivations with the potential to manipulate this phenomenon. Political despots and religious cults provide a catalyst for driving misplaced veneration to ideologies. Similarly, the modern availability of social media has allowed the potential for a top-down blueprint that galvanizes a very destructive cause by co-opting what has in human history turned out to be a genetic evolutionary vulnerability. The purpose of this analogy is to underscore the phenomena of top-down (having a central command) and bottom-up (having no central command, other than a permissive catalyst) processes that promote learning and problem-solving skills in clinical medicine.

4.5 Implications for Clinical Practice

The applied science in which clinicians practice medicine largely involves guidelines. However, an understanding of physics provides the physician the tools and flexibility to not be bounded by the constraints of such guidelines. Guidelines are broad-stroke standards of care that do not address the uniqueness of individuals. Biology is applied chemistry, chemistry is applied physics, and physics is applied mathematics. Galileo has famously been quoted as saying, "Mathematics is the universal language of nature". No matter what part of the world we live in, everyone understands the same math. It is not the intent of this writing to explicate the mathematical and statistical formalisms that underpin concepts in physics. This is in part because such an approach would lose the interest of most readers interested in the broader picture and in part because it is beyond my own understanding. Ironically, most scientists who study physics as a career, particularly in certain cultures steeped in such an "ivory tower" tradition, are led to believe they are isolated from the rest of the world. What they ultimately find, however, is that they are wonderfully poised to study any discipline and topic of science.

The essential question to be answered here is why physicians should care about the basic sciences, whose results they apply to their patients, but which represent only the raw materials that are not directly applicable as an algorithm to patient care. Crude oil must go to the refinery capable of creating a useable product by processes that remove irrelevant materials not necessary for the combustion of petroleum such as in an automobile heat engine. Guidelines of care represent an analogous but more subjective top-down process that filters available knowledge into what is deemed most appropriately applicable to the largest subset of patients that present with a defined constellation of signs and symptoms. However, unlike automobile engines, which are man-made machines and accordingly follow reductive principles, human biology is highly complex and variable in terms of presentations and therapeutic responses. That is, patient care approached categorically with empirical reductionism is often ineffectual and may be fraught with dangerous untoward effects. Furthermore, people inherently want to know the source of their problem in order to deal with it and possibly remedy it if properly understood. Patients have very different fears about their illness as well as expectations for how it is managed and to what outcome. Some of this may relate to the impact symptoms have had on their life. *The relationship between the physician and patient must be collaborative and dynamic.* It should be a relationship that sets goals and priorities that are adjusted in response to feedback. Similarly, there should be a set of relationships that also involves collaboration between physicians involved in the patient's care as well as sometimes between the physician and non-clinician scientists. The latter may be computer bioinformatics specialists who may likely become the interface between the physician and the basic sciences of physics and biological chemistry. The clinician would then effectively assume his or her intended role as both healer and teacher, functions, which are often inextricably linked. This not only empowers the patient for self-care but improves feelings of self-control, reduces anxiety, and often helps patients cope with pain. The physician can then appropriately echo what Isaac Newton famously quoted, "If I can see far, it is because I can see standing on the shoulders of giants".

The creative individual applied scientist (clinical care provider), as well as basic scientist, with the greatest energy, drive, knowledge, and expertise escape the silo problem. Our system must seek to reward people according to the adage "your success is my success" that encourages interactive participation involved in a patient's care including for example bench researchers, computer bioinformatics scientists, and clinical care providers. *The creators and the discoverers are the people who best know the intricacies of a particular line of scientific inquiry and may offer unparalleled value to clinical care providers as applied scientists.* Unfortunately, the basic scientists do not often tend to seek to avail themselves to clinicians nor do clinicians routinely tend to seek to enlist their insights for individual application. To an extent, there seems to be an arrogance of superiority that supports the barriers and silos of the basic and applied scientists [28]. These attitudes are overdue for a change.

An important concept of medical education in general is integration. This includes integration of basic science knowledge into clinical teaching and of clinical cases into basic science curriculum (The International Association of Medical Science Educators [IAMSE]). It also includes integration of systems biology and of the interdisciplinary sciences interfacing with clinical pathological disease. This chapter posits that a panoramic perspective of interdisciplinary and multidisciplinary insights engages a deeper understanding of the concepts of clinical medicine. This is underpinned by an integration not only of the medical subspecialties but of molecular biological science, biological chemistry, and even physical chemistry into the spectrum of clinical medicine. This improves biological thinking and flexibility of the clinical decision problem-solving process. This integrative conceptual approach in addition to strengthening clinical reasoning skills provides psychological benefit in a number of ways. It amplifies confidence that the physician is both capable and worthy of taking care of the patient. It also elevates the intellectual joy of the task that is often lost by the monotonous task of diagnostic pattern recognition routinely tethered to a top-down set of instructions or guidelines, the value of which for any given patient may be questioned. These factors are critical for avoiding the reduced sense of personal accomplishment with which physicians are frequently confronted, tending to see work negatively and as meaningless in the context of losing resilience that is characteristic of physician burnout. This devitalizing loss of fulfillment promotes a downward spiral of physical and emotional exhaustion. It accompanies a demoralizing attitude and a diminishing sense of moral obligation. Accordingly, there is a cynicism toward patients and their concerns, and an unwillingness to be grounded in their circle of pain and suffering.

4.6 Framing Energy by the Creation of Time and Life, and by the Breaking of Symmetry

The panoramic perspective of energy includes its origin from solar nuclear fusion and fission as was amply described above. Its transformation is mediated by the forces of nature into ultimately the self-organizing biological complex adaptive structures of which we as human beings are the apex predator.

> *Scholarship in the sense of knowledge and learning in the sciences and in clinical medicine fundamentally requires integration at all levels to satisfy deeper understanding and meaning.*

Where does energy come from? How is it used to drive our life processes as living healthy beings? Where does it go when we die and how is it responsible for our senescence and chronic disease?

The physical laws of nature and the energy derived from the subtraction of atomic mass in the Sun perpetuates the notion of cyclicity borrowed from phenomena of nature to germinate and sustain the rhythm of life. Our biological systems on an individual scale dance in cycles described as circadian (e.g. 24-hour sleep–wake patterns), infradian (e.g. greater than 24-hour menstrual patterns), and ultradian (e.g. less than 24-hour hormonal releasing and heart rate patterns). On a population scale, life-death cycles describe the generations.

It is in fact more mundane and practical than just cultural enlightenment for physicians to appreciate the flow of energy occurring both internal to our cutaneous boundaries and external as well, the latter (not addressed in this chapter) with sociological and spiritual implications. In addition, the concept of symmetry breaking, including reductionism, the crowned jewel of scientific-analytical thought, is inextricably linked to the flow of energy that forms the structures and functions of life.

Energy is created by mass, initially in the Sun by nuclear fusion. It ultimately creates the mass and energy (structure

and function) of all of us. We then create bio-photons that are the electromagnetic field with us. The electrical currents that create electromagnetic fields, amplify, and synergistically entrain our cells, tissues, and organs as an organism. Hence, the mass of our subatomic particles is transformed into energy that optimizes our own structure, function, and energy and accordingly controls our available time (and space) to live and have influence on the world. Perhaps human beings are indeed still evolving, not so much to enhance survival as a species by ensuring reaching the age of reproduction, but for longevity determination *per se*. The more energy we have while we are here, the longer it allows us to live. That is, higher productive energy enhances a more coherent synergistic structure and function as an organism, happier, adaptive, and interactive in a synergistic fashion physically, psychologically, and interpersonally. This is in addition to the time dilation that occurs relative to greater energy expenditure.

> *The four fundamental forces and twelve particles provide the energy and chemistry respectively, and accordingly the flow of information that is responsible for the physical universe and the resplendent complexity and beauty of biology and the systems of nature, which have evolved here on Earth.*

The universe constructs time by the presence of cycles exemplified by electrons orbiting around atoms as well as planets and astronomical bodies circling around the Sun and other suns, respectively. The Earth orbiting the Sun and the moon orbiting the Earth are responsible for the light-dark cycles on Earth and ocean tides, respectively.

> *Light-dark cycles are responsible for the origin of life as well as the circadian clocks of many biological functions (the cycles of the body) that sustain life.*
>
> *The first two laws of thermodynamics frame the laws of energy and entropy of the physical universe. The first law frames the time invariant nature of energy-matter as the axis of time; the second law, which defines systems as always moving unidirectionally towards increasing entropy, frames the arrow of time.*

Energy flow is the flow of information as the energy interfaces with our body's molecular chemistry and biochemistry. If constructive, free energy is positive and structure and function is created, increased, or maintained. If destructive, free energy is lost to heat and associated entropy, to use the lexicon of thermodynamic physics, or to oxidative stress and inflammation, to use the lexicon of pathobiology and pathophysiology of clinical disease.

> *The first law of thermodynamics underpins time in an absolute sense of energy preservation. The second law underscores time in a relative or temporary sense in terms of the loss or dissipation of information energy of our biological structures and functions as heat randomizes away from the complex molecular organization of our systems biology*

*to the entropy of the universe. The combination of
two laws and our circadian clocks creates the foundation of biological time and the process of aging
relative to energy flow within the constraints of the
boundaries of a biological system.*

The transformation of inanimate matter into living systems
may largely be described in terms of systems going from
thermodynamic equilibrium states to far-from-equilibrium
steady states. This breaks the symmetry of reductionism.
That is, time invariance of the physical laws of the universe—
equations that are deterministic and reductive, for example,
Newton's laws or the first law of thermodynamics. Biology,
which cannot be described, let alone predicted, by a mathematical equation, does not violate the first law but rather, as
a subsystem of the universe, is encompassed within the first
law. This process forms new symmetries of biological stability anchored by time-variant steady states (steady states that
vary over time). *The arrow and the cycles of time describe the
biology of time, which breaks the symmetry of the axis of time
and underscores the nonlinearity of systems biology and the
limitations of reductionism.*

4.7 Steady States, Attractor States, Strange Attractors, and Chaos

From the perspective of clinical practitioners, fundamental to
the understanding of nonlinear biological systems is the breaking of the time-invariant symmetry of scientific reductionism. Genomics, which is the study of the expression patterns
of our genes, may be described as the generation of systems
that are attracted to dynamical states of stability, or steady
states. Following a perturbation arising from environmental
circumstances, these systems will eventually predictably settle
back to the point of these stable or steady states. These are
points of attraction within the space of states and are known
as attractors.

*The attractors characteristically represent the emergence
of the system's dynamics in the form of spiral loops or circles.*
Each spiral loop is composed of fractals of self-similar recursive iterations across different scales. Each recursive iteration
within a given loop tends toward a center point. The recursive iterations are never self-repeating, that is overlapping but
instead represent an infinite dimensionality in the abstract
concept of phase space (*albeit* their image is projected on a
two or three-dimensional screen).

When the system becomes increasingly unstable, the attractors draw the stress. That is, an attractor of a given spiral loop
bifurcates or splits and changes direction forming new spiral
loops as a representation of a qualitatively new behavior of
the system searching for stability (or order) with a new steady
state. Each spiral loop represents a different attractor of the
same system. New spiral loops continue to form as the system
searches for solutions to the stress put on the system. These
geometric configurations of phase space viewed on the computer screen represent a compelling modern physical model
describing how biological systems (as well as inorganic systems of nature) generate new behaviors in response to stress

from outside the system as a geometric manifestation of
allostasis. Hence, each spiral loop of a system represents a
bifurcation from one attractor to another resulting in new possibilities that keep the system alive and dynamic. Each of these
attractors in a chaotic system is characterized by fractals as
described above. *When multiple spiral loops are present in a
chaotic system, they are called strange attractors.*

Chaos theory teaches us to expect the unexpected as
described by fractal mathematics, which captures the infinite
complexity of nature. Recognizing and understanding this
behavior provides powerful insights into the infinite wisdom
of nature and biological systems. Such insight should set the
bar for standards of care in medicine. What nature teaches us
ultimately is how the microscopic and macroscopic components of our systems biology are connected not only within
the boundaries of our body per se (our cutaneous borders or
interface with the environment) but how we are connected to
the surrounding on many levels such as ecosystems, social
systems and so on. Hopefully, the powerful insight of understanding nature's wisdom of biological systems will avoid
healthcare practices, which end up being detrimental to our
long-term well-being.

*It is easy for DNA to specify a single repeating process
of bifurcation and development of the various anatomical
branching systems of the body such as the gastrointestinal,
pulmonary, vascular, excretory and more. These self-repeating processes of anatomical bifurcation are structural fractals, which are the simplest and most efficient possible ways
for genetic programming.* If a separate gene were required for
each neuron, cell of the vascular tree or of the pulmonary tree,
etc., there would not be enough genes in our DNA [29]. Hence,
nature has evolved the concept of organismic biology using
the efficient space-filling of anatomical fractals. However,
as described above, the spiral loops of chaotic systems are
representations of fractal biology albeit it in a more abstract
dynamic or temporal form of fractals as systems evolve and
emerge over time manifested as attractors that draw stress to
the organism as a whole and search for new steady states or
allostatic steady states and maintain survival and health of the
larger organism and doing so enlisting the efficiency of fractal
systems.

To use the anatomical vascular system as an example of the
amazing efficiency of space filling by fractals, the vascular
system is as compact as a single line entwining itself into a
perfect sphere yet retaining the surface area of the entire line.
There is no wasted surface area. In the conventional sense one
dimension represents a line, two dimensions a plane, three
dimensions a cube while a fractal represents a fraction of a
dimension. The vascular system for example comes within the
reach of five cells of every cell of the body delivering oxygen and nutrients yet only represents roughly 5% of total body
mass. This is an incredibly efficient system that is said to have
a fractal dimension of 2.7.

Fractal patterns were first recognized by computer simulations of nonlinear mathematical equations as descriptions
of many natural phenomena (Figure 4.5). Such phenomena
include coastlines, mountains, and parts of living organisms. Important properties of fractals are self-similarity and
non-integer (fractal) dimensions. Self-similarity may refer to

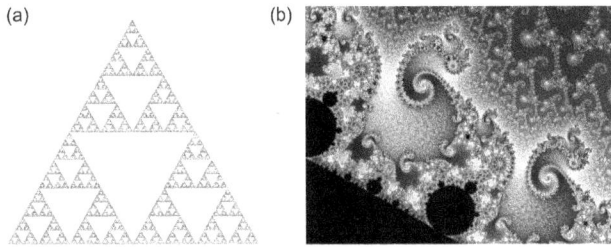

FIGURE 4.5 Illustrations of examples of geometric fractals. a) Sierpinski triangle. b) Partial view of the Mandelbrot set. Sources: a) by Vahram Mekhitarian, used under Creative Commons license https://creativecommons.org/licenses/by-sa/3.0/legalcode; b) created by Wolfgang Beyer with the program Ultra Fractal 3. and used under Creative Commons license https://creativecommons.org/licenses/by-sa/3.0/legalcode.

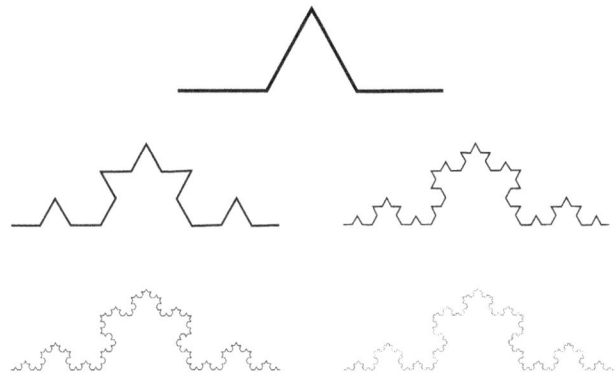

FIGURE 4.6 Illustration of the Koch curve, a simple geometric fractal, generated by subsequent iterations. Source: adapted from Falconer, Kenneth. (2004) *Fractal Geometry: Mathematical Foundations and Applications*. John Wiley & Sons, Chichester.

functional characteristics, which are even more interesting, however anatomical ones are more classically recognized. An example may be a fern leaf whereby every small leaf is part of a bigger one, each level being geometrically similar to the whole fern leaf. Hence, fractals are replications at different scales of a motif of the same pattern.

Fractal scaling is also a characteristic property, which is the presence of the same level of detail across all scales within the fractal. Non-integer dimensions of fractals represent a property that is an aberration from the classical geometrical dimensions. A point in classical geometry has 0 dimensions. Lines and curves have a single dimension, squares and circles are two-dimensional figures whereas cubes and spheres are three-dimensional. The fractal geometry of many natural phenomena is somewhat between these integer dimensions. For example, a string is a one-dimensional object. However, if it is twisted into the shape of a square in self-similar iterations, it has neither one nor two dimensions but rather has approximately 1.26 dimensions [30].

Benoit Mandelbrot, a Polish-born French mathematician coined the term fractal and described fractal dimensions as a way of measuring the system's property (e.g. surface area or length using yardsticks of increasingly shorter length). Fractal dimensions illustrate purpose for chaotic systems designed to maximize surface without increasing volume [30]. The vascular system for example has an effective fractal dimension of 3, bent and packed, entwining itself into a perfect sphere while being as compact as a single line. *The efficiency of fractal distribution is reflected in the fact that although a whale weighs ten million times that of a mouse, it requires only 70% more branches in its circulatory tree* [31]. The branching system described by the mathematics of fractals is manifested in all systems of the body, not only vascular, lymphatic, tracheal, bronchial but also the Purkinje fibers of the heart, bile ducts in the liver, calyceal system in the kidney and so on. Importantly, only a few simple rules or bits of information explain their emergence using a mathematical algorithm [32].

The core idea behind fractals, like in all nonlinear systems, is a feedback in iteration. The creation of most fractals involves applying simple rules to a set of geometric forms or numbers and then repeating the process from the perspective of the new magnitude using the same set of rules. These simple iterative

functions can produce an infinite variety of patterns of a given geometric form. Consider a standard geometric structure like a circle or a square. The more we can zoom in on it with higher and higher resolution, the more it appears like a simple line, with less and less variety. Conversely, if the same were done to a fractal structure, we see an infinite variety of structural detail. Even if we zoom in with a magnification a millionfold greater, we will continue to see infinitely detailed self-similar structures within the structure as a whole. Another feature of fractals, in addition to an infinite form existing within a finite form, is the existence of an infinite length within a finite boundary. This is exemplified by a fractal called the Koch curve (Figure 4.6) with the classic analogy of an ocean coastline, an infinite length within a finite boundary created by self-iterations whereby if we followed each iteration, it would prevent us from ever reaching the end [33].

There are many examples of nonbiological fractal systems in nature such as coastlines, crystals, snowflakes, river networks, lightning patterns, electrical discharge effects, hurricanes and many more. Many biological systems are also fractal chaotic in nature or non-chaotic complex systems, which in contrast to machines are nonlinear. There are many examples of fractals in biology. Suffice it to recognize that branches of a tree, just like branches of the pulmonary or vascular tree represent geometric fractals in addition to there being many dynamic fractals (non-static over time that characterize much of our physiological systems, such as the conduction system of the heart or electrical transmission pathways of the brain. The self-similar outpouching network of alveoli, the circulatory arterial tree, and the Purkinje network of the heart conduction system are among the many fractal geometric structures of the body [34–35] (Figure 4.7). Additionally, *there are many fractal structures observed in processes in time such as heart rate variability, electrical patterns in the brain, and hormonal secretory patterns.* Hence, fractals are not only observed in physical structures but in dynamic biological processes. Accordingly, the fractal fluctuations exhibited at a given time scale parallel the same process observed at larger or smaller time scales. That is, there are often both spatial and temporal components to the self-similar fractal properties of systems. For example, recordings of ion channel currents display

(a)

Windpipe (trachea)
Right upper lobe
Right bronchus
Right middle lobe
Left lower lobe

(b)

(c)

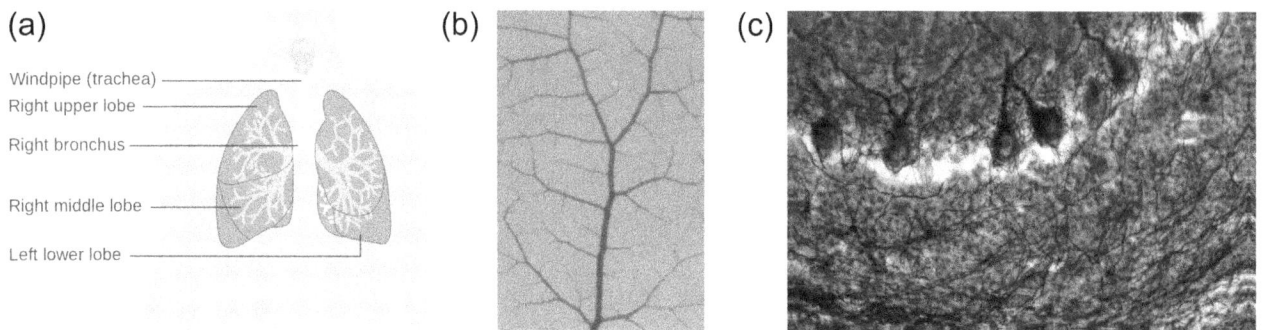

FIGURE 4.7 Illustration of fractals in biological systems: a) respiratory system, b) circulatory system, and c) Purkinje network in the cerebellum. Sources: a) by Cancer Research UK, used under Creative Commons license https://creativecommons.org/licenses/by-sa/4.0/legalcode; b) by Michael Jeltsch, used under Creative Commons license https://creativecommons.org/licenses/by-sa/4.0/legalcode; c) by Lazar.zenit, used under Creative Commons license https://creativecommons.org/licenses/by-sa/4.0/legalcode.

self-similar patterns of ion channel opening and closing over different time scales [36]. A *sine qua non* of a chaotic system is the fractal nature of the attractor.

A geometrical fractal is a shape that has symmetry of scale, which means that we can zoom in on the shape of a fractal, any part of it an infinite number of times and it would look the same on every scale. This again captures the features of scale-free, self-similarity, and fractal components of chaotic systems. Fractals patterns may be recognized in the snowflake, the branches of a tree or the human circulatory vasculature, or in the temporal fractals of a phase portrait of a chaotic system over time.

Chaotic systems are characterized as deterministic and unpredictable. Unlike deterministic Newtonian systems, which also describe the philosophical position that for every event there exist conditions that can cause no other event (that is, preconditions are determinative of a subsequent event). The chaotic system is unpredictable (in counter-distinction to the Newtonian system, which is predictable).

Deterministic chaos with strange attractors describes a system, which, unlike equilibrium systems, periodic systems, and quasiperiodic systems, is no longer confined to repeating a particular rhythm and is free to respond and adapt. It is constrained only by boundary conditions imposed to prevent it from collapsing.

Chaos is found in a wide range of physiological systems most characteristically described for the cardiovascular system and brain. A chaotic system has a high level of flexibility and efficiency for maintaining homeostasis through allostasis as geometrically characterized by strange attractors and the phenomenon of emergence. This type of system, despite being referred to as chaotic or even random, does not represent randomness or chaos per se in the colloquial sense. On the contrary, the fractal nature of these patterns and this behavior represent and amazing complexity, beauty, and eloquence of biological systems in finding adaptive solutions to stress from the environment. To the extent that chaotic systems parallel allostatic adaptive responses, allostatic overload underscores maladaptive behavior and consequently disease states. This underscores a breakdown of the system whereby the strange attractor of a chaotic system deteriorates into the behavior of true randomness or true chaos that is detached from regulatory control and as such, the fractal nature of the dynamic behavior of these systems is lost. Aside from the deterministic non-predictable chaotic systems with strange attractors, which represent one important type of system, there are four additional types of biological systems. An equilibrium system describes a classical physiological system at an equilibrium steady state. An equilibrium state of living systems is a point attractor held within a very narrow physiological range such as physiological pH (7.35 to 7.45). This narrow physiological range of a system represents a stable constant condition of homeostasis.

Homeostasis is a biological advantage in that it allows an organism to function effectively in a broad range of environmental conditions.

Such homeostasis, which represents physiological equilibrium is regulated allostatically by other systems such as hormones and neurotransmitters regulated by the other four types of systems including chaotical systems. The physiological equilibrium type of system such as that of acid-base balance is a far-from-equilibrium steady state. A chemical equilibrium state is not a far-from-equilibrium state (one defined by requiring energy input into the system) but rather an equilibrium state defined by being at the lowest energy of the system whereby the rates of forward and reverse reactions are equal, and the concentrations of the reactants and products remain constant. Therefore, a reversible equilibrium steady state of biological systems, particularly an equilibrium system, which must be held within a narrow physiological range, would be incompatible with life. This is a critical distinction between a physiological equilibrium system at steady state and a chemical equilibrium system at a reversible steady state. Again, allostatic regulation in biology maintains the homeostasis of equilibrium systems. The next type of biological system is a periodic system.

Periodicity describes a system with a simple rhythm in which a cycle is repeated with a set frequency.

This is seen in many circadian rhythms such as those of the central and peripheral body clocks, which in turn control many other circadian rhythms. A periodic system is characterized by a periodic attractor, which may be a simple oscillator or a multi-oscillator system that is perfectly synchronized such as in the case of the central and peripheral (molecular) biological clocks which regulate many other circadian rhythms.

> *A quasiperiodic system is slightly more complex and is characterized by systems that oscillate with at least two cycle frequencies.*

This is a common type of physiological system and may be exemplified by the normal heart rate variability physiological pattern. Quasiperiodic behavior is displayed by multi-oscillator systems, such as when water is required during times of low demand. Many of these systems when stressed can qualitatively change or bifurcate into a chaotic system with strange attractors, as discussed above. This bifurcation increases biological flexibility and adaptation to stress from outside the system. The fifth and final type of biological system studied in the context of chaotic systems is the truly random behavior that represents the breakdown of order in the system whereby behavior is uncoordinated [35]. It is important to note that equilibrium, periodic and quasiperiodic systems in biology can be studied using a reductionist approach whereas chaotic systems with strange attractors cannot be. The key determinant for whether reductionism is a useful tool in studying a given type of system is dependent not on an underlying deterministic nature of a system but rather on the predictability nature of the system.

> *Many systems exhibit so-called strange attractors, which typically represent two attractor states between which the system may oscillate in a chaotic fashion.*

Each of these attractor states has a region of relative stability, a so-called basin of attraction (Figure 4.8). However, there is

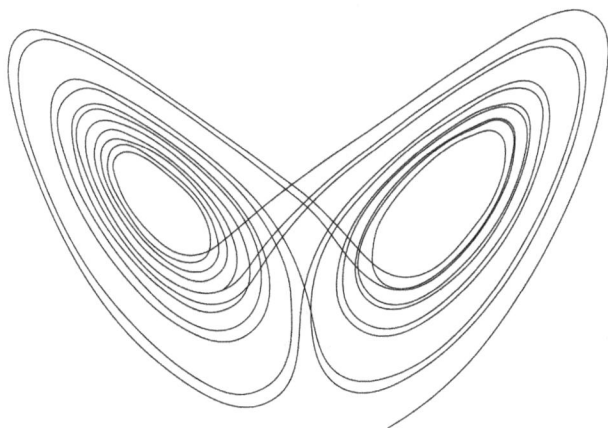

FIGURE 4.8 Illustrations of the concept of a three-dimensional strange attractor. Source: adapted from Ogunjo, Samuel. (2013). "Increased and Reduced Order Synchronization of 2D and 3D Dynamical Systems". *International Journal of Nonlinear Science* 16: 105–112.

also a region separating the two states where even the smallest perturbations throw the system off balance and cause a transition from one basin of attraction to the other. This variation in steady states is not noise in the system, or imprecision, for example by imprecise measurements of parameters of the system that have been perturbed. Rather, variability and unpredictability are inherent in the system. On a two-dimensional dynamic (over time) computerized portrait of the system's trajectories in its phase space (called a phase portrait), it can be seen that the system's dynamics represent a fractal pattern with its characteristic property of self-similarity. This is exemplified by the iconic Butterfly Effect, the original Lorenz attractor describing the system of weather patterns where a slight perturbation causes the folding over of the trajectory of the attractor to form the opposite wing of the phase-space trajectory in the shape of a butterfly (Figure 4.8).

The Butterfly Effect was so named because of the appearance of a butterfly on the computer-simulated model. Its characteristic chaotic behavior was due to the sensitivity to rounding off at three digits to the right of the decimal. The metaphorical effect of the flapping of a butterfly's wings in one area of the world can result in a dramatic effect on weather patterns in faraway areas. Small changes in the parameters of weather patterns may result in huge positive feedback loops resulting in qualitative changes in resultant behavior. This is an example of a system, which is infinitely sensitive to the initial conditions.

The metaphor used to describe it is the case of a butterfly that flaps its wings in the Amazon, which then later causes by a snowball effect a storm in Texas. The phase diagram of such a system represents the composite information regarding the values of the relevant control parameters (e.g. weather patterns, temperature, humidity, and pressure). Depending on the range of values of these control parameters, the systems may be characterized by a single attractor state or multiple attractor states such as the strange attractor situation. As pointed out above, an important characteristic of these chaotic systems is their fractal nature. The computerized portrait of the individual states of the system in its phase space, rather than densely filling a two-dimensional structure, creates a pattern that has voids on all length scales. It can be characterized by a measure called the fractal dimension.

The fractal dimension (determined by the scaling law) relates the coverage of the phase space area as a function of the change in dimensional scale, a measure of the space-filling capacity of a pattern. This is the mathematical definition of a fractal dimension which is fractional, unlike a Euclidean dimension which is an integer (1, 2, 3, etc.). In Euclidean space with dimension d, an object increases in size in proportion to its length raised to the power d. For example, the volume V of a sphere whose radius is r, is proportional to the third power of r, namely, $V = (4/3)\pi r^3$. The volume of a fractal object with fractal dimension d_f measured within a radius r is proportional to the radius raised to the power d_f, i.e. r^{d_f}. A consequence of the fractal's property of self-similarity is its appearance being the same independent of the magnification. It looks the same at all magnifications independent of complexity or variability that is independent of how dense the fractal pattern is at the different scales of magnification. These fractal patterns of strange attractors take on beautiful shapes, which may resemble forms not only such as butterfly wings, but as strolling clouds, spider's web, leaves, and other shapes.

4.8 Nonlinear Interactions: Positive and Negative Feedback Loops

It is important to understand that complexity science explains how order arises from complexity masquerading as disorder or even chaos when looking superficially at nonlinear systems. In fact, the field of complexity science evolved by the seemingly unnatural likelihood, as explained by the traditional linearity of reductionism, of the organization of systems unfolding over time with such order.

> *Complex adaptive systems manifest macroscopically as a collective result of nonlinear interactions dynamically over time as self-organizing systems.*

What defines the adaptability of complex adaptive systems are not the elements, or agents, that comprise the systems per se but rather the interactions between these agents. Essentially, the agents of a complex adaptive system are information processors, which can process information and react to changes contained in that information. These agents continuously act and react to what other agents are doing. Although the range of interactions may be short, each agent only paying attention to its local environment, the range of influence is typically much wider due to trickle-down effects. Information is carried throughout the system through feedback loops, both positive and negative whereby qualitative changes occur in the system as it evolves. Feedback processes affect the interacting elements driving the rate of change to either put on the brakes or to amplify the new emergent behavior.

> *Negative feedback seeks stability and homeostasis at a given far-from-equilibrium state. Conversely, positive feedback drives further qualitative change and hence is destabilizing away from homeostasis. The end game is for these networks to establish adaptation to the surrounding environment.*

Each new hierarchical or macroscopic scale forms a new complex adaptive system from the agents that form it, each of which typically is a complex adaptive system in and of itself. Each new organizational scale is the creation of novelty and adaptability for the larger complex adaptive system, which also imparts adaptation to the constituent elements or agents that form the system. For instance, the cells of an eye tissue of an evolving species improve the survival advantage of the organism stalking prey while in the process ensuring their own viability as individual cells. The cells, themselves complex adaptive systems, within the biological environment, may be considered agents to the next hierarchical scale complex adaptive system, the eyes, which together with the other tissues and organ systems form ultimately the organism-wide complex adaptive system within the ecological environment.

The family unit may be considered the next organizational level complex adaptive system to best optimize the survival of species by providing for and protecting the young. Analogously, a single bird of a flock is a complex adaptive system in and of itself. The flocking pattern is a larger scale complex adaptive system. The element of surprise and unpredictability is fundamental to the creativity of living systems that seek superiority over their competitors. Diversity is critical to the functioning of complex adaptive systems. The absence of heterogeneity of agents in terms of how agents process information is essential for change and/or growth.

A "simple" complex system is distinct from a "complicated" complex system in that in the former case the interacting agents are similar whereas in the latter case they are dissimilar. That is not to say, however, that similar interacting agents within a given hierarchical scale are identical (e.g. birds of a flock). This is analogous to the individuals of a human population who are not identical. Individual birds of a flock are exposed to different external information, such as potential threats. The processing of that information may result in a particular response by a single bird followed by local interactions with neighboring birds, which has far-reaching effects on changing the qualitative behavior of the entire flock. It is possible that an individual bird of a particular flock may process the same external or even local information and react in a novel way. Although it may be only one bird of a large flock, an aberrant behavior of processing and reacting to information by that single bird may have a ripple effect of local interactions with neighboring birds that can have a far-reaching and adaptive influence on the behavior of the overall flock. This unpredictable behavior may have imparted survival adaptation to larger predatory birds, which have been evolving their own successful strategies for penetrating the flock. This new pattern of behavior is associated with new rules. The application of medicine to complicated complex systems highlights interacting agents, which are qualitatively more disparate. In both simple and complicated complex systems, the critical nature of diversity or heterogeneity to the functioning of complex adaptive systems is central. However, the notion of symmetry breaking when moving from one, more basic, scientific scale of organization, (e.g. molecular biology) to a more macroscopic scale, such as cell biology, involves new fundamental laws or rules of nature.

Reductionism represents a periodic pattern. That is, it is inherently time-invariant or time-independent, such that over time the pattern is exactly the same. Environmentally induced perturbations are only transient displacements from the preexisting steady state to which it will continue to be attracted as a low-energy state of the system. These clockwork-resembling cyclical patterns are representations of symmetry with a corresponding conservation law of the system as a manifestation of the laws of physics. The cyclicity around the axis of time protects or conserves the system at its existing steady state. Following perturbations, the system seeks to re-establish that steady state. Again, time invariance defines a form of symmetry that conserves energy in matter (first law of thermodynamics), hence, information does not change over time. Biological steady states are transient representations of time invariance in living systems (second law of thermodynamics).

Chaotic nonlinear and hence unpredictable systems represent an example of breaking the symmetry of reductionism. Chaotic systems describe the anatomical branching of the circulatory, bronchial, renal, and calyceal trees. Chaotic dynamics also describes the dynamic functionality of systems such as neuronal behavior in the brain or the immune system that exemplify nimble biological adaptive flexibility to find solutions in the face of perturbations. Cardiac arrhythmias are examples of chaotic behavior that in contrast to neuronal behavior in the brain, and perhaps cognition, exemplify behavior at the "edge of chaos" that may be detrimental to health. Rather than occurring at the inner "edge of chaos", as the case of a healthy immune system or of normal signaling brain activity (i.e. a normal EEG), cardiac arrhythmias occur at the outer "edge of chaos", that is entropic. Therefore, it is not surprising that chaotic behavior of the cardiac conduction system is a common cause of sudden death.

4.9 Why Life Exists: A Chaos Theory Perspective

The purpose of existence likely lies at the interface of philosophy and material science. Science describes an inherent order in the form of rules, principles, and laws. Laws and rules exist because regularities exist, which is because of molecular anatomy enabling the interactions ultimately promoting higher-level purpose or biology.

Optimal biological decisions are environmentally context dependent. The complexity of biological organization is underscored by the fact that no universal law connects all the hierarchical scales from elements to ecosystems. Rather, generalizations at each scale are sought instead.

In other words, at each scale new symmetries are formed. The rules that find the generalizations capture the dynamic behavior of biological systems and hence emergent properties from the assimilated behavior of interacting components. The development of novel measurement technologies and date collection is facilitating this. Biology exists in the purpose and not just in the physical interactions *per se.* Feedback loops may be considered the physical equivalent of the purpose.

4.10 A Pedestrian Overview of Systems Biology

The problems emerging from the previous analysis are both fascinating and bewildering. For complex biological systems such as the human being, comprehensive approaches are needed to explore roles, relationships, and interactions of the various types of biomolecules that comprise the cells of an organism. At present, both experimental techniques and computational methodologies are being used to understand the behavior of cells, tissues, and organisms by quantifying individual characteristics of their genes, proteins, or small metabolites. This effort has spawned a plethora of disciplines that have been termed using the suffix "-omics" when

classifications and taxonomy are involved while the suffix "-metrics" distinguishes the creation of quantitative databases in this regard. They include the following fields:

- *Genomics* (the study of genes and their function).
- *Proteomics* (the study of proteins).
- *Metabolomics* (the study of molecules involved in cellular metabolism).
- *Glycomics* (the study of cellular carbohydrates).
- *Interactomics* (the study of interactions between and among proteins, and other molecules within a cell).
- *Lipomics* (the study of cellular lipid).
- *Transcriptomics* (the study of transcriptomes [the sum total of all the messenger RNA molecules expressed from the genes of an organism] and their functions).

These areas of investigation provide the tools needed to explore the cellular behavior, produce enormous amounts of data regarding the functional and structural characteristics of the cells but also develop analytical approaches to understand, quantify and integrate the data. Their integration is accelerated by the progress in computational hardware, software, and analytical algorithms that adequately represent the networks of interacting biological molecules in terms of their properties and functions in living systems. Systems biology deals with the complexity of data (rather than managing many specific context-dependent types of data) without breaking it down to its smallest component parts. Systems biology is a major methodological and epistemological departure from reductionism.

This area has been termed "Systems Biology" due to its use of mathematical models that describe signals and circuits in a way, which is very similar to those developed in electrical, chemical, and mechanical engineering. This analysis represents active elements of a biological system as interconnected functional modules with well-defined individual properties and mostly binary interactions. The approach also accounts for the sequences of chemical reactions, which control the network's behavior just like signal processing circuits are used in electrical systems. Consequently, *living cells exchange information with their surroundings via receptors embedded in their membranes and passing this information through signaling pathways into the cytoplasm and the nucleus, which results in changes in the cell state, or gene expression. Intracellular signaling via cell-specific pathways has been modeled by sophisticated computer simulations of the dynamical information transfer within cells. Likewise, intercellular signaling provides dynamical communication and exchange of information between cells within a tissue and between functional biological modules.* Modeling of these processes is based on graph theory and network theory adapted for the purpose of complex biological networks that describe dynamical interactions within an organism at the biomolecular level. The challenge ahead is the comprehensive integration of both the intracellular and intercellular components, calibrated with data obtained in laboratory measurements and incorporated into a dynamical computer-based simulation.

The concept of attractors describes order emerging from chaos within the theory of complex systems. It describes the circumstances that can result from the random interaction of multiple agents. They can apply to a steady state of physiological behaviors as a function mediated by the many binary systems of biology including the parasympathetic autonomic nervous system balance and many molecular biological systems. Many steady states or attractors describe the balance of hormonal functions

> *The fitness landscape is an important property for any complex system and can involve many dimensions or corresponding variables. It is a geometrical representation of a dynamical system's stability against perturbations.*

This means, that a valley in the fitness landscape describes the area of local stability since the system tends to stay in the valley when pushed uphill by an occasional perturbation. On the other hand, a peak in the fitness landscape represents a very unstable state, which will be driven downhill by even the slightest perturbation. For simplicity, we are using a three-dimensional metaphor where there are two independent directions (variable parameters) and one coordinate representing the system's state (altitude above the surface). For example, a dimension representing a controllable variable can be environmental temperature and another could refer to a type of cell chemistry, such as insulin receptor status. Human metabolism works best within a very narrow range of temperature, so we have an optimal fitness value at a point in phase space corresponding to insulin receptors and the optimal temperature. Other examples of metabolic chemistry may work best at a different temperature so the landscape will have other peaks and valleys corresponding to a combination of temperature and metabolic receptor activity (e.g. glucagon receptors). The valleys correspond to evolutionary stable systems.

Organisms optimize themselves not only against the environment per se, but even more so against other organisms within the same ecosystem or its niche.

> *The corresponding fitness landscapes evolve in tandem with the evolution of organisms that both compete and cooperate.*

A predator will evolve better eyesight to see its prey; however, the prey will simultaneously evolve a disguise negating the eyesight advantage. That is, the hill the predator attempted to climb to gain a competitive evolutionary advantage has moved. The fitness landscape peak has changed as a result of co-evolutionary changes in the prey. Hence, the inhabitants of any ecosystem evolve in a closely coupled dynamic and high nonlinear system over time.

4.11 Relevance of Chaos Theory to Human Biology

Chaos theory can be and has been applied to better understand the architecture and complex functions of human (and other

organic) beings, something which is greater than the sum of its parts. However, we often refer to our hearts as being like clockwork along with other metaphorical analogies, such traditional reductionism is an oversimplified portrayal of the nonlinear complexity of biological systems as machines. Homeostasis of our far-from-equilibrium chemical biology is maintained within narrow boundaries by feedback regulation. Our systems, to the extent that they are physical systems guided by the laws of physical chemistry, albeit complicated, are predictable and explainable by the language of reductionism. To this extent, biology is reducible to the laws of physics. If this were applicable to all of biology, it would suggest that the mind is nothing more than the physical *brain, free will cannot exist and all behavioral phenomena are reducible by classical reductionism. The centerpiece of traditional reductionism of biology is the genetic sequence of DNA that directly correlates to phenotypic traits directed by the interactions of molecules such that each single gene ultimately translates to independent gene products.* It turns out, however, that genomics, in contrast to genetics, is the expression of genes that considers the bidirectionality of the effects between genes and the genetic environment. The effect of the environment on genetic expression, or epigenetics, may account for well over 70% of the hundreds, or potentially even thousands, of final protein products per gene. This underscores that nongenetic factors account for the majority of chronic disease incidence.

Systems biology embraces the nonlinear, non-predictable, and non-machinelike behavior that breaks from the traditional reductionist philosophy of scientific thought. This accounts for the majority of biological behavior conceptualized by chaos and complexity theories. Chaos theory is interesting because it classically describes non-organic behaviors such as weather patterns and a dripping faucet. However, organismic systems also display chaotic behavior. This has potentially high value because it provides an important research tool for predictive modeling of an impending organ failure that cannot be done easily on living systems. For example, a dripping faucet as depicted on a phase portrait has a very analogous geometric and statistical signature as an electrocardiogram (ECG) plot as the two very different systems evolve from regularity to one that is random and uncontrolled. The random chaotic behavior of ventricular fibrillation is lethal and often unpredictable. By applying the same rules of studying the dripping faucet to systems such as cardiac conduction, data may be gathered and manipulated in a time series that may allow timely life-saving interventions for patients at risk for life-threatening arrhythmias.

> *Chaotic systems are classically characterized by extraordinary sensitivity to initial conditions whereby minuscule differences in the starting parameters (initial conditions) of a system result in grossly amplified changes in the trajectory of the system.*

For example, subtle changes in the pressure of interacting particles in the atmosphere may be the difference between a calm, clear sunny day, and a hurricane. When the parameters are plotted in a three-dimensional phase portrait, the trajectory of recursive loops of interactions may resemble the wings of a butterfly, as in the case of Lorentz's attractor as was discussed above. If chaos is taken to be random patterns, chaotic actually do not display chaos *per se* but rather patterns of the interface between both stability and order and unstable random disorder, which is often referred to as the "edge of chaos". *As stress parameters cause a system to become increasingly unstable and chaotic, driven by the destabilizing influence of positive feedback, the system maintains bounded constraints under the stabilizing influence of negative feedback.* Accordingly, the three-dimensional phase portrait illustrates qualitative changes as folding patterns or bifurcations. If this were not to occur, the system would escape the spiraling bounded "edge of chaos" becoming overtly unstable and ceasing to remain extant, i.e. become extinct.

The real beauty of chaotic systems is its intriguing character of fractal geometry. A single DNA sequence of a gene can code for the repeating process of folding bifurcations, each one regulated by negative feedback to constrain the system, keeping it within the bound of the "edge of chaos". Accordingly, the balance between unstable chaos and predictable regularity is the optimal zone whereby the fractal branching patterns maintain optimal vitality and longevity of the living system. *The unpredictable quality of chaotic systems is a fundamental characteristic and requirement of natural selection.* The rhythm of cardiac conduction is governed by the sinoatrial node electrical excitation that is responsible for atrial depolarization and P wave formation corresponding to expansion (spreading out on the three-dimensional phase portrait) followed by conduction through the atrioventricular node, Purkinje fibers, and, ultimately, ventricular contraction (ventricular depolarization), corresponding to the folding pattern on phase portrait that constrains the system, i.e. keeps it from becoming unbounded random chaos. This is the QRS complex on the ECG printout.

Finally, ventricular repolarization is the T wave on an ECG representing the snapback, beginning to spread back out preceding the expansion phase on the phase portrait. This is the critical part that corresponds to the development of fatal arrhythmias that occur when the T wave (or snap back) is superimposed by the P wave (expansion on the phase portrait). The statistical signature of stress parameters in a time series is comparatively like the analogous phase portrait responsive to stress parameters of a dripping faucet. When the stress parameter, the faucet pressure, is increased the system destabilizes into chaos. Using phase portraits to study biologically similar models is a promising research strategy aiming to find rules of statistical correlation that can be translated over to clinical heath care and medicine capable of saving lives by accurately predicting fatal arrhythmias before they occur. The ratio of sympathetic to parasympathetic nervous system activity with a pattern of sympathetic dominance regulatory balance on the heart is the best thus far known predictor for fatal arrhythmias. However, it remains an imprecise predictor with limited power for targeting what drives this aberrant pattern. The goal of studying phase portraits of inorganic systems that are statistically similar to organic systems is ultimately defined by finely grained rules behind the rules used to predict the pathogenic behavior, which can be targeted to prevent morbidity and mortality. Nonetheless, the binary balance between the

sympathetic and parasympathetic autonomic nervous system branches is a complex interplay representing together a chaotic system. Importantly, in the setting of sympathetic dominance of the autonomic nervous system, the heart rate variability is reduced, and such regularity moves the system away from the optimal zone of the "edge of chaos" towards excessive predictability that is discouraged in the process of natural selection. *The relative complexity of positive feedback reduces the optimal variability.*

The functional fractal design of cardiac electrophysiology, analogous to the anatomical branching system of vasculature, is illustrated by the self-similarity, non-integer fractal dimensions and self-similar scaling demonstrated on phase portraits. There is also a rich interplay between the chaotic cardiovascular system mediated by the autonomic system and the respiratory system. There is an important relationship between respirations and heart rate variability whereby increased respiratory rate increases heart rate and heart rate variability. This represents a feed-forward regulation of enhanced variability as a result of an interdependency of the two chaotic systems. A plausible advantage of heart rate variability is to limit fatigue because continuous repetition of the same movement promotes fatigue, just as it does skeletal muscle activity. Additionally, heart rate variability is better able to adapt to rigid changes in cardiac demands because it allows a smoother transition [29a, 31].

The brain function is inherently the most chaotic process of the human body in the sense that the "edge of chaos" in the brain has the most expensive positive feedback and its behavior is the least predictable. The computerized phase portrait image is the least bounded of all the organ systems. By comparison to the electrical conduction of the heart, the EEG of a normal brain is analogous to the pattern of the ECG of the heart in ventricular fibrillation. Conversely, the EEG of a brain during a seizure has a regular pattern analogous to the relatively regular ECG of a normal heart. The reason for this is thought to be related to the concept of search required of the brain to find optimal solutions critical to learning and rapid transitions of thought. The goal is not to find a solution but rather the best one, requiring significant positive feedback and relatively unbounded space. This has a hidden order that when plotted on a phase portrait is attracted to a certain region. There are multiple fractal attractors in the brain that change as thinking processes change. The geometry of EEG activity varies relative to human cognitive and creative processes. It is interesting that the zone of the "edge of chaos" in the brain is breeched typically into unregulated positive feedback and chaotic randomness that can be fatal; conversely in the brain it is classically breeched into pathological disease due to overregulated negative feedback and predictable regularity. Arnold Mandell, psychiatrist and proponent of chaos theory, stated the brain has "more than 50 transmitters, thousands of cell types, complex electromagnetic phenomenology, and continuous instability based on autonomous activity on all levels" [32]. This highlights the intricate neuronal interconnectedness. Characteristic of chaotic systems, such as weather patterns, small internal uncertainties are amplified over time making it impossible to give long-term predictions of brain activity [37].

Chaotic systems are pervasive in the healthy physiology of human beings. It even appears they represent the basis for the origin of life. In contrast to disorder being somewhat synonymous with disease, the disorder represented by chaos or perhaps more aptly, the "edge of chaos", is characteristic of resilience and health. Chaotic systems are a form of complex adaptive systems and in counter-distinction to linear processes that, when slightly pushed off track, tend to remain off-track whereas complex systems tend to get back on track.

> *Complex adaptive systems have feedback and feedforward mechanisms, allostatic and homeostatic attractors that drive the process to getting back on track, maintaining its stability in the subtlety of external changes and perturbations. The composite parameters of the systems evolve in a dynamical fashion towards homeostatic set points, which are the basins of attraction.*

The potential for practical application of chaotic systems in medicine and medical research is robust and itself represents an active area of investigations.

Examples of applying chaotic theory to medicine include developing diagnostic tools to distinguish between healthy and unhealthy tissues based on fractal dimensions and looking at inorganic chaotic systems such as a dripping faucet to find correlating predictive signs for fatal cardiac arrhythmias or myocardial infarctions. *Studying phase portraits of human cognition and creativity may lead to developing applicable forms of artificial health.* Identifying the chaotic attractor in the spread of disease may lead to the prediction of epidemics. We are at the dawn of this discipline and its vast potential has unlikely been realized, even remotely.

Fractals in anatomy and histology (microanatomy) are organized to maximize fitness that facilitates the rate at which energy and/or material resources are taken up from the environment (or wastes eliminated from the body) for maximizing metabolic capacity, allocated to some combination of survival and reproduction. For example, the leaves of a tree absorb carbon dioxide and solar rays for photosynthesis. The vascular tree delivers oxygen (from the lungs) and organic nutrients (from the gut) to peripheral tissues of the body. The dynamic trajectory of fractal attractors of chaotic systems in physiology is organized analogously to maximize fitness and metabolic capacity.

The important perspective of symmetry in reductionism is based on the time invariance, or time-independent biological systems of the information of matter/energy in terms of the conservation of their steady states. Breaking the symmetry of reductionism with the formation of complex chaotic and non-chaotic systems promotes an adaptive responsiveness to perturbations that serves to maximize fitness in terms of allocation to some combination of reproduction and survival. Breaking symmetries inherently implies nonlinear systems. It forms new symmetries in complex and chaotic systems as allostatically adjusted new steady states form. This encompasses Andersen's notion of breaking symmetries across hierarchical scales of biological science. The molecular genetic scale, for example, dictated by fractal genes in response to minor

perturbations, results in ultimately more macroscopic behavior of structure and functional physiology. In chaotic systems, as described above, the fractal nature of steady states represents variation, which in contrast to reductive systems is an inherent design as part of the system. The threshold of criticality reached by any insignificant or minor perturbation highlights the feature of unstable steady states of chaotic fractal systems. In this case, the double pendulum metaphorical comparison is most appropriate.

Although chaotic systems are an example of complexity, many biological systems are complex but non-chaotic. All complex systems are distinguished by the characteristic of feedback relation and interconnectedness. Non-chaotic but otherwise complex systems do not share the feature of fractal designs. Further, *unlike chaotic systems, non-chaotic complex systems do have the capacity to move from an allostatic new steady state in response to a perturbation than back to the baseline resting steady state following the remission of the perturbation and correction of the metabolic stress.*

Breaking the symmetry of space–time implicates the relationship between mass and energy by the equation $E = mc^2$. Space offers the room for matter to exist and move in. Time offers the means of keeping track of what matter does and in what order. Time invariance is the conservation of the mass/energy of the physical universe over time. Space invariance is the conservation of momentum of the physical universe over space. This includes linear momentum such as solar photons of light and angular momentum of the electrons of atoms. The concept of time invariance as the conservation of the quantity of mass and energy takes the analogy of time being the axis of mass and energy. The universe helps to construct time by the presence of cycles, repeating patterns of relationships between the conserved particle-like quantities of mass/energy. These repeating patterns are exemplified by electrons around atoms in orbits of planets and astral bodies, the latter creating the repeating pattern of light and dark cycles. The virtue of not changing with time as a function of this cyclicity may be conceptualized as a protective and self-preservation effect of the conserved quantity.

> *Light–dark cycles sustain life within the context of the first law of thermodynamics but within the limits of the arrow of time in accordance with the second law of thermodynamics.*

While time invariance reflects the absolute conservation of mass/energy of the physical universe, its implications to humans and other living organisms are relative. This is heralded by the steady states of free energy within the boundaries of every life form. Light–dark cycles contribute to the circadian rhythms of our biological systems.

The central clock, the suprachiasmatic nucleus that is located in the visual sensory receptor pathway of the brain receiving neuronal projections from the retina recognizes the light-dark cycles. This suprachiasmatic nucleus located at the base of the hypothalamus (and it actually is not a single body but multiple discrete bodies) contains tens of thousands of neurons that project to a number of structures in the brain. One of these structures is the pineal gland, which releases melatonin, important in regulating the suprachiasmatic nucleus or central biological clock. This central biological clock has melatonin receptors that suggest a feedback regulatory function. There are many behaviors and functions of the body (such as sleep/wake cycles, various hormones, hunger, body temperature, and metabolic rate) which fluctuate in circadian fashion. In addition to the central biological clock, there are molecular clocks that are genetically expressed. These molecular genetically expressed clocks are referred to as peripheral clocks. There is a coordinate oscillatory regulation between the central and peripheral clocks. Biological circadian patterns are adjusted, or entrained, by local external cues of light-dark cycles. Intriguingly, as discussed elsewhere, the synchronization of mitochondria and of metabolic cycles as reported by Demetrius, et al, occurs through gene regulation within and between cells that invoke collective and coordinated transcriptional cycles.

> *Chronobiology is a field of biology that examines cyclicity patterns of biological activity in living organisms, which may be circadian, ultradian, or infradian.*

The latter may be monthly, annually, or longer. Species of organisms are even characterized by life-death cycles, or average longevity that may differ at baseline and following environmental perturbations. There appears to be a central role for the circadian system in fractal regulatory networks in the time organization of neurophysiological processes. The time-invariant nature of fractals in contrast to linear periodic systems, which have absolute time invariance, is only relative. In other words, the time-independent steady state of a fractal system has inherent variability and thus the steady state may be considered wider or looser until it reaches a point of criticality and changes qualitative behavior leading to a new steady state.

An example of broken symmetry would be a biological system. An exception would be a biological system whereby a point of therapeutic intervention may be at the plane of molecular biology with a predicted response occurring at a physiologically more macroscopic scale. In contrast, the flying patterns of flocking birds are not characterized per se by breaking symmetry across hierarchical scales of organization. The flocking behavior is nonetheless complex and adaptive characterized by internal feedback loops, diversity of interactions, and nonlinear behavior. The interactions in this case occur between agents, which are relatively similar and within the same hierarchical plane. In either the case of the complicated complexity of systems biology or simple complexity of a collective intelligence, the patterns of these self-organizing systems are unpredictable. *That is, the spontaneous emergence of new structures and new forms of behavior in open far-from-equilibrium systems are characterized by internal feedback loops and mathematically by nonlinear equations. In the complicated complexity of biology there is a symmetry in the sense of unity and linearity between genetic coding in DNA and our physical characteristics.* However, this example may serve both to highlight where the reductionist model is a successful predictor, and

because of the complexity of genetic expression, it has predictive limitations. At present, it can be safely stated that our genetic code represents only a minority of disease incidence. Conversely, epigenetics, the environment that mediates the expression of genes, is responsible for a majority of disease incidence due to the process of oxidative stress. This process is nonlinear and poorly predictable.

4.12 Self-Organization and Self-Regulation

Biological systems by definition are open and require a supply of energy, information, and matter from the environment. The challenge of understanding their behavior involves the presence of many different constituents involved in nonlinear interactions leading to the emergence of self-organization and self-regulation, i.e. homeostasis. Emergent properties can arise from a response to environmental pressures, to which the system tends to adapt if these pressures are limited. Away from thermodynamic equilibrium, when energy flows through a collection of many interacting molecules, the emergence of new patterns is facilitated so the energy can be more easily dissipated.

> *Self-organization typically refers to a class of systems, which are able to change their internal structure and their function in response to external perturbations. Self-organization cannot occur in isolated systems, because it requires a source of instability and non-equilibrium conditions giving rise to a phase transition.*

Abrupt changes in temperature, radiation, pressure, flow of matter, mechanical stress represent typical external influences. In biological systems, processes leading to self-organization often are due to the search for thermodynamic stability, which can result in the formation of lipid membranes. Adaptation and evolution are processes where the changes in the structure and function of a biological system depend on the mutual exchange of information between it and a dynamic environment.

Gottfried von Leibniz developed a simple yet profound idea that a theory has to be simpler than the data it explains, otherwise, it does not explain anything, only describes it. This idea is compatible with the approach to measure complexity proposed by the Russian mathematician A.N. Kolmogorov and G. Chaitin an Argentinian-American mathematician and computer scientist. This approach is referred to as "algorithmic complexity" and is defined as the length in bits of the shortest program, which can describe an entity such as a data set, an image, a material object, or a living form. Accordingly, an object without internal structure cannot be described in any meaningful way but only by storing all its features. A random object has maximum complexity, since the shortest program able to reconstruct it needs to store the object itself. Within this approach, *a measure of complexity is correlated with the amount of information needed to describe a system or a process, which offers the computability required to describe the evolution of the systems.*

4.13 Playing Simple Games with Profound Implications: Cellular Automata

Researchers try to model complex adaptive systems by capturing these local rules and using computational science like cellular automata and agent-based modeling to try to simulate how these systems are shaped by the iterations between agents (elements of the system) and evolutionary forces. The enormous complexity of life forms becomes purposefully complex because it increases the chance of survival. The automata theory has a different perspective than the Darwinian theory of natural selection to explain biological evolution. Rather than relying on the random process of optimizing mutations, automata theory shows that simplicity can naturally generate complexity. Cellular automata are a great way to see the principles of both simple rules as building blocks and very small differences causing emergent phenomena. Additionally, it demonstrates fractal behavior across many different scales such that the emergence is scale interdependent. Cellular automata are best exemplified by the Game of Life designed by John Conway in the 1970s [38]. The Game of Life, like other cellular automata, starts off with a set of initial conditions and a simple set of local rules for how one can metaphorically reproduce into the next generation with the emergence of intriguing patterns. The rules proposed by Conway in this example are as follows:

1. Any live cell with fewer than two live neighbors dies, as if caused by under-population.
2. Any live cell with two or three live neighbors lives on to the next generation.
3. Any live cell with more than three live neighbors dies, as if by overpopulation.
4. Any dead cell with exactly three live neighbours becomes a live cell, as if by reproduction.

Starting from an initial configuration, by means of a series of subsequent transitions, the system undergoes an automatic evolution by growing and generating a sequence of configurations giving rise to numerous complex patterns in space and time.

The Game of Life shows how a complex thing like the mind may come about from a basic set of rules. The simulation consists of a grid, like a chessboard, extending infinitely in all directions. Each square of the grid can be either lit up, which equals life, or turned off, which equals death. Whether a square is dead or alive depends on what is happening with the 8 other squares that surround it. For example, if a living square has no other living squares around it, it will die of loneliness. If a living square is surrounded by three or more living squares it will die of overcrowding. However, if a dead square is surrounded by three living squares, it becomes lit, or is born. Once we set an initial state of living squares and let the simulation run, these simple rules determine what will happen in the future. The results are surprising. As the program progresses, shapes appear and disappear spontaneously. Collections of shapes move across the grid bouncing off one another. There are whole kinds of objects, species that interact, some can even

reproduce just as life does in the real world. These complex properties emerge from simple rules though contain no concept like movement or reproduction. It is possible to imagine that something like the Game of Life with only a few simple rules might product highly complex species, perhaps even intelligence. It might take a grid of billions of squares but we have hundreds of billions of cells in our brain.

In the overwhelming majority of cases the pattern ceases after a number of sequences, that is, they fail or go extinct. In addition to the emergence of intriguing patterns, another critical property is that of convergence. Convergence refers to the situation where of all the diverse potential starting points, only a very limited few emergent patterns (or biological phenotypes of the analogous evolution of wildlife) successfully over time beat the odds of going extinct. Another important feature of convergence is that roughly half will look remarkably similar, sharing variations of a stereotypical pattern or analogously inhabiting similar adaptive phenotypes. There are examples of biological species evolving from entirely different ancestor lineages and in very different parts of the world, but with similar environmental pressures that result in strikingly similar phenotypes. However, the most classical characteristic of chaotic systems is that minor differences in starting states extend into very different consequences, portraying the Butterfly Effect. The phenomena of the Butterfly Effect and convergence are alternative strategies of emergence that encapsulate the nonlinearity of chaotic systems. It is impossible to predict over a long range of generations either the starting state or the mature state. From the starting state of cellular automata, there is no way to predict a priori whether the state will emerge, go extinct or persist unchanged over time as a non-animated static pattern.

Agent-based modeling has its origin in two-dimensional cellular automata. Many of the earlier agent-based systems were modeled using extraordinarily simple rules that led to highly complex emergent behaviors. Over the past decade or two, agent-based modeling software tools and development of environments (the agent's location dynamically tracked as it moves across the landscape, acquires resources, and encounters other situations) have broadly expanded. Both cellular automata and agent-based modeling provided explanatory insights into the collective behavior of systems following simple rules with interactions among their components. These two computer models are very similar with a primary difference however being the agents of agent-based modeling are free to move around within the bounds of the model whereas in cellular automata they are spatially fixed. Agent-based modeling for example typically studies natural systems of animal and organism behavior such as flocks of migratory birds and foraging of ants described above. Cellular automata may be useful in studying cancer behavior, for example by reconstructing a prostate tumor compared to an observed tumor (the environment of each cell consisting of four or eight neighbor cells depending on the specific model being a variant of the Game of Life) and the model can simulate the progression of cancer. Alternatively, agent-based modeling may be useful to study the metastatic behavior of cancer and the immune responses to tumor formation or viral infection. Stimulating the latter also uses cellular automata modeling techniques and it is not unusual to find hybrid forms from both agent-based modeling and cellular automata methods used, each providing complementary advantages.

The best cellular automata and agent-based modeling simulations are bottom-up designs that use simple rules of interactions and provide an understanding of the emergent behavior from initial conditions to determine failure points. The simple rules of interactions between agents lead to coupled behavior.

The usefulness of defining simple rules is not to understand the intricacies of the system's complexity but to predict behavior for the next complex adaptive system.

Whether it be for the next flock of birds faced with predatory attack or the next patient at risk for a myocardial infarction or arrhythmia (at a future time when computer simulations are used for clinical management of patient care). The predictive value also pertains to the relationship to what is driving the rule. The rules may be analytical mathematical relationships of the interactions of the agents of a system at an individual level analogous to the relationship between two objects described by Newton's second law F=ma. That is, force exerted on one body allows the calculation of mass or acceleration of the body, so long that one of the two variables is known. Another example is the neuronal connections in the brain. They are simple at the individual neuron level, however, the pathways and networks of connections responsible for adaptive complex functioning are awe-inspiring due to the emergent complexity.

Cellular automata were originally introduced by two great mathematicians of the 20th century, John von Neuman and Stan Ulam. Cellular automata consist of a regular grid of geometrical cells, each of which can be found in one of a finite number of states, for example, an on or off state. An initial state of the systems is selected by assigning a state for each cell. New generations of states of the system are created advancing in time steps, according to some fixed rules of state transitions that determine new states for each cell in terms of the previous state and on the states of the immediate neighborhood cells.

4.14 Biological Networks

Systems of elements (nodes) connected by interactions (edges) are called networks and can be found at every scale. For example, biological cells are networks of molecules connected by biochemical reactions involved in complex metabolic reactions (Figure 4.9). At higher scale societies, too, are networks of people linked by friendship, family, or professional ties, while ecosystems can be represented as networks of species competing or cooperating depending on their relationships. Networks can also be found in transportation and telecommunication systems.

Network theory studies such topics as measures of network complexity, connectedness and subnetwork structures as well the emergence and structural evolution of complex systems represented by networks. For simple systems, their behavior can

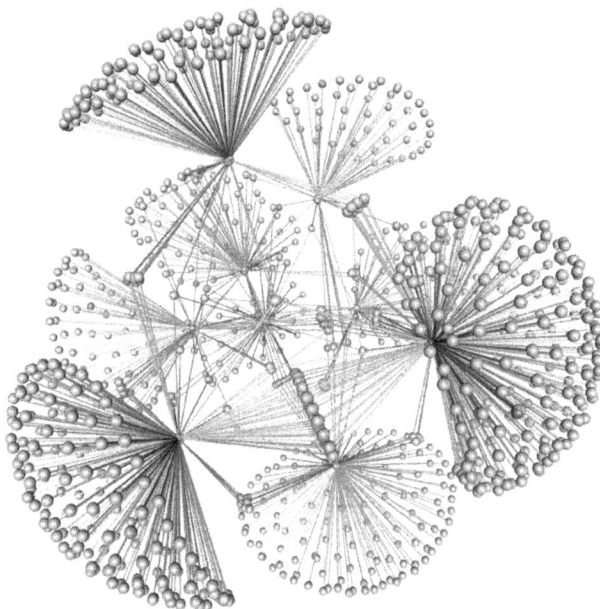

FIGURE 4.9 Schematic illustration of an example of a protein-protein interaction network. Source: by Dedalus, used under Creative Commons license https://creativecommons.org/licenses/by-sa/4.0/legalcode.

be understood intuitively, but for more complex networks, sophisticated tools based on graph theory are required for describing and visualizing them as sets of nodes with connections between pairs of nodes shown as edges.

Historically, the concept of a network was first introduced by Immanuel Kant in connection with a topological problem involving crossing the bridges on the river Pregel in Koenigsberg, East Prussia. In the 20th century, two Hungarian mathematicians, Paul Erdos and Alfréd Rényi introduced a general model in which random networks were composed of N interconnected nodes, with k interconnections. If N does not change in time and if the nodes are all equivalent, so that the tendency of each one to link to another is the same, the probability P(k) of a node to have k interconnections follows a Gaussian distribution with a bell shape and a maximum. When k is small, a Gaussian probability distribution becomes Poissonian. However, most real-life networks such as biological networks are not random but scale-free networks, and their probability distribution is expressed by the following exponential law:

$$P(k) \propto k^{-\chi} \tag{4.1}$$

where χ is a characteristic empirical parameter. The term scale-free network has been coined by A.L. Barabasi in connection with a model of the growth of a network that exhibits a preferential attachment such that new nodes tend to connect to nodes that are already connected.

In general, all networks can be grouped into three classes: 1) regular, such as crystal lattices, 2) random, which has a well-defined average and noise around it, and 3) scale-free, which best describes a spontaneous organization of complex systems at all scales forming hierarchies.

In the latter case, the corresponding hierarchical models are based on the replication of small clusters linked to the central node of the original cluster. The resulting network is scale-free because it has a power-law distribution.

In a living cell, the processes characterized by mass, energy, and information transfer are connected through a complex network of cellular constituents and biochemical reactions (Figure 4.10). A living cell is built and maintained based on thousands of interconnected enzymatic reactions that control the energy fluxes and the production of biomolecules involved in cell metabolism. As shown in the pioneering research by Barabási and coworkers [39], involving the metabolism of *Escherichia coli*, cellular metabolic processes form scale-free networks. These networks include thousands of interconnected enzymatic reactions, which control the energy fluxes and the production of cellular metabolites.

Scale-free networks are ideally suited for the creation of an increasing level of complexity that emerges from hierarchical interactions between interacting subsystems, where the formation of structures and autocatalytic processes is used to attain the energy necessary for the maintenance, growth, and development of a biological system of systems.

However, although scale-free networks can be well-described graphically and characterized mathematically regarding both global and local properties, as nonlinear complex systems, they cannot be predictable based on a few mechanistic laws or rules because the behavior of such a system cannot be defined by an integrable equation. Inherent in complex biological systems is the presence of multiple hierarchical planes of organization each defined by a separate set of mechanistic laws or rules. Accordingly, each plane represents a closed set of elements and their interactions manifesting inherent unity and symmetries.

In medicine, a point of therapeutic intervention may be at the plane of molecular biology with a predicted response occurring at the physiological macroscopic scale. In this case, drug therapy that targets a molecular scale may have different mechanistic relationships than the series or hierarchical scales above it and that contributes to the clinical effect of the drug. For example, a vasodilating drug given in order to lower blood pressure by targeting a calcium channel in the vascular smooth muscle often promotes a feedback rise in heart rate serving to mitigate the reduction in blood pressure hence confounding the predictability of the system in terms of clinical response and untoward side effect profile. Accordingly, it is hoped that complex systems may be better predicted by a strategy that defines the system by simple rules involving the relationships between the elements of a biological system rather than reliant solely on more scientifically based mechanistic fundamental or natural laws or rules for such relationships.

Computer simulations may help determine or elucidate simple empirical rules. An example of a system defined this way is the flocking pattern of migratory birds as mentioned above. Analogous basic rules to describe other complex systems such

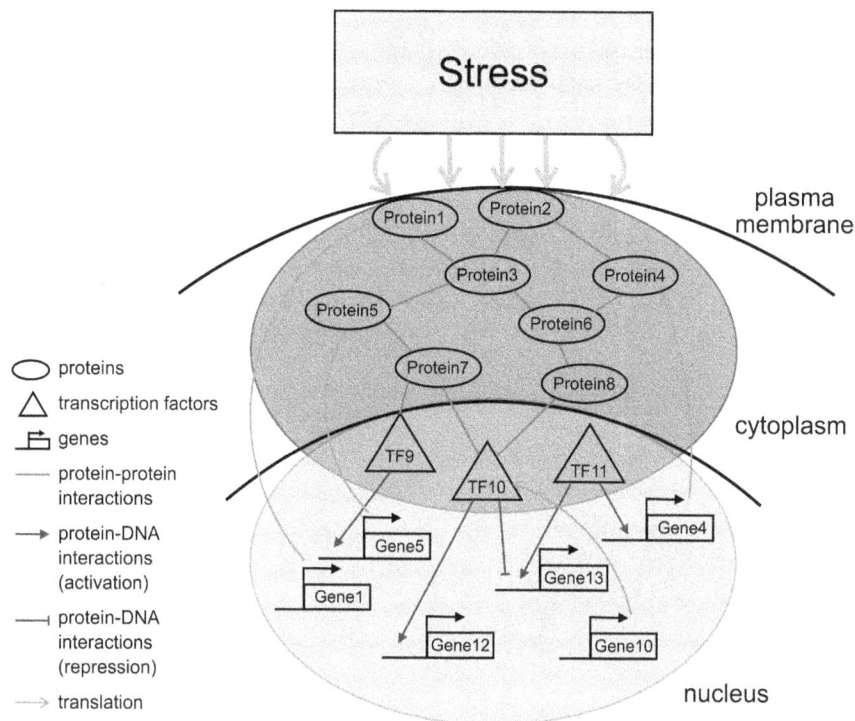

FIGURE 4.10   Schematic diagram of the integrated cellular network. The integrated cellular network consists of two subnetworks. The signaling regulatory pathway (upper) contains protein–protein interaction (PPIs). The gene regulatory network (lower) contains transcription regulations. The transcription factors serve as the interface between the two subnetworks. Source: adapted from Chen, B. and Wu, Chia-Chou. (2013). "Systems Biology as an Integrated Platform for Bioinformatics, Systems Synthetic Biology, and Systems Metabolic Engineering". *Cells* 2: 635–688. 10.3390/cells2040635.

as biological systems have the potential to impart valuable predictive capacity. There is a functional eloquence of all-natural complex systems that exhibit adaptive qualities. In the case of the flocking behavior of birds, the adaptive value includes, for example, an effective and safe journey of migration south during the winter. However, because the mechanistic rules are not the same as the actual rules devised by nature, predictability is not always accurate, as seen in cases of statistical anomalies to computer models. Such statistical anomalies provide the impetus to self-correct the computer model to improve its accuracy and predictability. *The collective behavior of a flock of individual birds, just like the collective behavior of the individual components of multi-systems biology represents the unpredictable emergence of higher-order parameters or hierarchical scales from lower-order parameters.* The different fundamental laws of nature that in reality exist at each of the organizational levels of nature are officially replicated, albeit with less accuracy, by more simplistic or basic rules enlisting the convenient strategy of "simplifying complexity".

The linear relationship between genetic coding in DNA and our physical characteristics is classical reductionist biology whereby our qualitative traits are reducible to our DNA. However, reductionism in biology breaks down because of a number of factors. One, combinatorial space of genetic entropy (discussed above and below) for the outcome of a particular phenotype is such a large number that it is not only non-analyzable but is not even computable. Second is the issue of symmetry breaking as we transcend across planes from molecular biology to cell biology and ultimately physiological expression. Different rules at each organizational level pose an obstacle to reductionist analysis. Further, there is the interaction at the molecular level between genetic information and its environment. This environmental impact on genetic DNA is epigenetics whereby histone separation from the DNA strands allows methylation or acetylation providing another layer of information derived not from the biological system but from its interaction with the environment it inhabits. The result in terms of evolution is the formation of further subspecies as a feedback loop to the natural selection process. This is a positive feedback loop if it favors further subspecies formation yet a negative feedback loop if it favors species elimination. In humans, who do not seem to be further evolving in terms of subspecies, increasing toxic environmental and dietary exposure promotes disease pathogenesis by modifying the DNA. This additional layer of information further weakens the reductionism of genetic biology. Our genetic code represents roughly 30% of disease incidence. The environment that mediates the expression of genes is responsible for roughly 70% of disease incidence due to the process of inflammatory and oxidative stress that opens up the protective shield of histones around DNA and promotes punitive DNA oxidation as well as acetylation or methylation to occur.

Finally, and not mutually exclusive of epigenetics is *the existence of noncoding DNA, or micro-RNA, previously naively called "junk DNA". Such DNA underscores the incredible complexity of a seemingly rudimentary genetic code by regulatory factors, adaptive but also potentially maladaptive, by harmful environmental exposures.* Hence, counter to the traditional establishment philosophy of reductionism, the cornerstone of analytical thought in medicine, even genetic biology, the prototype for reductionism, is profoundly nonlinear, non-reducible, and hence non-predictable. It exemplifies the complexity of many interacting parts represented by regulatory genes turned on or off by the altered environment. A worm, or a grain of rice, has at least three-fold the number of genes that humans have in their entire genetic code (90,000 to 200,000 versus less than 30,000 in humans). So, what makes us so complex and sophisticated if we have so few genes relative to these other simple organismic systems? The answer is the complexity of our gene expression.

Interestingly, there is a molecule called Zonulin that is released by the epithelial cells that line the gastrointestinal tract. This occurs when these cells are exposed to cytotoxins, for example, gluten, as a result of disturbed tracking of molecules in the lumen of the gut. Zonulin is responsible for causing disruption and disassembly of the tight junctions that link the epithelial cells together. Consequently, the immune system gut-associated lymphoid tissue or GALT, becomes chronically activated by the immune-stimulating antigenic component from the gut. The resulting systemic auto-inflammatory process completes antioxidant molecules in relevant cells exposing their DNA through oxidative stress disrupting the finely tuned regulation of genetic coding by noncoding DNA. Hence, DNA expression in predisposed individuals will result in any given disease process determined by the weakest genetic links. That is, those genes which are affected first are the least robust in terms of interactions with noncoding DNA. Under healthy conditions, noncoding DNA forms intricate and finely tuned regulatory networks. However, in the fortunate 10% to 15% of the population who are not either genetically heterozygous or homozygous for carrying the Zonulin protein are significantly protected from expressing their otherwise predisposed disease states, for example cardiovascular, neurodegenerative, autoimmune diseases, or cancer, because the pernicious inflammatory state from the disassembly of tight junctions does not occur.

4.15 Simplifying Complexity

Complexity, defined by many interacting parts, or nodes, but also by the emergence of properties that are unpredictable, by definition defies the capacity to be understood by reductionist thinking.

> *Simplifying complexity is a term applied to the process of finding the simple rules from which complexity emerges.*

The process is a collaborative interdisciplinary endeavor that itself is a nonlinear but reductionist approach in the sense of predictability independent of mechanistic insight. Such a process may be based on observation but typically utilizes computer simulations for finding the so-called "simple rules". Fundamental laws of physics are reductionist tools for predicting the behavior of linear systems such as macroscopic bodies (prototypically, Newton's second law: $F = ma$) or the laws of thermodynamics (an ideal gas equation: $PV = nRT$). Although these laws may provide the ability to predict reductively patterns of behavior, defining the principles themselves is far from reductive because further simplification of the laws is not feasible. Importantly, the understanding of complex systems by rules requires analysis of relationships between component parts of the system and cannot be achieved by studying an isolated component part, such as a single bird in a flock or a single particle of matter. One molecule of water does not have a temperature, only a group of interacting molecules does since temperature is a statistical property Indeed, the molecular explanation of temperature is the average kinetic energy of all the molecules. The faster-moving particles collide with slower-moving particles transferring heat from the former to the latter such that the average of the kinetic energy of all the molecules determines the temperature of a system (notably, heat describes energy in transit). The temperature and pressure of their composite interactions determine whether the molecules will form a vapor, ice, or liquid water, which is illustrated schematically as a so-called phase diagram (Figure 4.11). This in fact represents an emergent quality of a physical system.

Phase transitions represent changes in the equilibrium phase of the system as a result of changing external conditions such as temperature and pressure. They occur in nonlinear systems that are understood not so much by laws per se but by observations and quantification of behavior using mathematical modeling. This provides an important distinction between reductionism and non-reductionism. The laws of electromagnetism that describe the response to electric fields of a single water molecule cannot be simply extended to explain such behavior when the interactions among water molecules whether in the solid, liquid, or gaseous state are included. Water exemplifies a simple but excellent example of emergence in a physical system. Thus, water represents a complex system and as such cannot be understood by equations describing only its constituent molecules without including interactions between them. This is in contrast to linear or even complicated nonlinear systems, which are not complex. An example of a very simple yet highly unpredictable

nonlinear system is a damped-driven anharmonic oscillator, which exhibits transition to chaos when a periodic driving force is applied to it since it can have nonlinear response and friction, a fairly common occurrence in real mechanical systems. Although the fundamental emergent properties of animals may be described by "simple rules" they are not easy and almost always nonlinear, often requiring the collaboration of physicists, biologists, mathematicians, and computer scientists to develop realistic mathematical models of their behavior.

The origin of reductionism in the sense of modern science dates back to Newton's laws. Impressively, Newton in the 1700 s did not have the availability of multidisciplinary collaboration in arriving at the relationship between force, mass and acceleration. Furthermore, he did not have the benefit of arriving at his laws by reductive reasoning because the laws themselves are fundamental, hence there is nothing for them to be reduced to. However, the Newtonian equation $F = ma$ is not flawless. It applies within the context of a relatively symmetric relationship of space and time (i.e. "particles" moving much below the speed of light). Newton's laws provide the tools to interpret the motion caused by gravitational force between two masses. The force of gravity is defined as "every particle of matter in the universe attracts every other particle with a force that is directly proportional to the product of masses of the particles and inversely proportional to the square of a distance between them". This as it turns out does successfully apply to the physical behavior of objects, masses, and particles anywhere in the universe despite the lack of time–space symmetry as pointed out in Einstein's special relativity. In the case where speeds of objects or particles are much slower than the speed of light, Newtonian mechanics is useful as an approximation.

Newtonian mechanics is applicable to physical systems that are predictable and by definition do not display emergent behavior. Every system can and should be understood as systems of parts and wholes. The ecosystem to which a colony of bats or ants belongs represents a system on a larger or macroscopic scale. As such, a colony of organisms becomes a part or element within many other colonies and organism species of predators and prey that comprise a larger organizational scale.

4.16 The Limitations in Molecular Biology and Reductionism in Explaining the Living World

The main ideas of biology were influenced by classical physics and chemistry. Molecular biology germinated from the Watson and Crick description in 1953 [40] of the double-helical structure of DNA, which comprises the genetic material. It is from this that the reductionist model of medical research and clinical care was developed. The basis of this model is that DNA and RNA encode all genetic information necessary for the structure and function of biological organisms. In effect, the contention of the model is that animals, including human beings, are simply the sum of their parts. The essential question arises, are human beings comparable to a machine like a car motor or a clock that can be taken apart entirely and rebuilt? It is important to underscore that this pervasive thinking is only a theory, one that has been widely adopted and

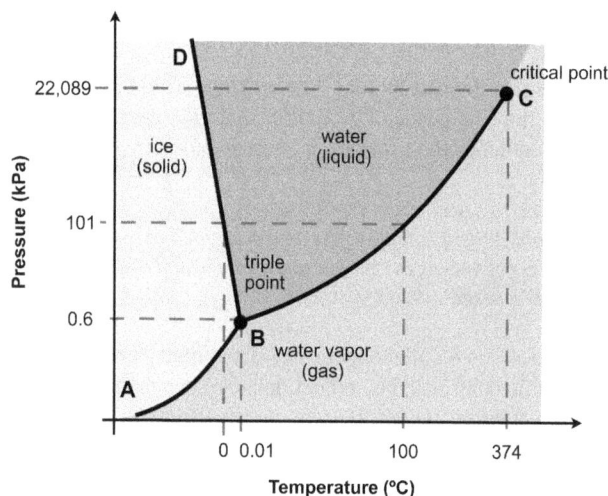

FIGURE 4.11 Phase diagram of water illustrating its three phases (ice, water, and water vapor). Source: adapted from https://chem.libretexts.org/Courses/Lumen_Learning/

one that has great value, but a theory, nonetheless. The central dogma [41] raised by the theory posits the following. The notion that all information is contained in DNA and RNA; that mutations are purely random; that it is only the environment that selects mutations; and that there is no feedback regulation to the information coded in DNA.

> *The major ideas of molecular biology theorize that biological phenomena can be reduced to the following behavior: 1) that information is stored in only certain molecules; 2) only short-ranged specific interactions occur; 3) its characteristics are essentially those of classical equilibrium; 4) water is mainly a passive solvent; and 5) the cell is a bag filled with a solution of molecules that collide randomly to form complexes or interactions leading to a higher level of function.*

There are many problems with this understanding to adequately explain the complexity and the integral nature of humans and other living organisms. For example, phenotypic plasticity and self-organization are phenomena that illuminate the limits of the central dogma of molecular biology. That is, DNA coding for RNA that in turn codes for proteins and protein enzymes that create and maintain all the complexities and the integral nature of biological organisms and beings is untenable. This model is not capable of explaining the nonlinear and self-organizing systems with long-range electromagnetic interactions. It does not adequately explain the origin of life. The probability of molecular biology explaining the development of the first living cell has been estimated at one chance to 10 to the power of 40,000 (1 in 10^{40000}). The chance of this model giving rise to man is roughly 10 to the power of 24 million over four billion years. As quoted by Nobel Laureate François Jacob, a French biologist who together with Jacques Monod conceptualized the idea that control of enzyme levels in all cells occurs through transcription regulation [42]: "Chance is not enough". Furthermore, the metabolic coordination of such an enormous number of specific chemical reactions occurring in a cell at precisely the right place, conformation, and time would be of an astonishingly low probability in the absence of cells utilizing quantum recognition algorithms. Hence, both the origin and maintenance of life on the basis of molecular biology alone appear impractical and difficult to support. *The information content in the cell is much greater than the potential provided by what is coded for in the DNA.* Notably, the human genome has 25,000 genes, which is very few in comparison to a worm (90,000 genes) or rice (200,000 genes). The question then arises, what makes this so complex and sophisticated. Clearly, it goes beyond the Watson and Crick model of DNA structure. There are other phenomena that are also not explained (or even addressed) by molecular biology, such as embryonal development, brain activity, and consciousness.

4.17 Systems of Wholes and Parts

As elaborated earlier in this chapter, relationships exist between sub-particles of an atom or between atoms of molecules,

molecules of a macromolecular complex, between molecules of a cell, between cells and the organ system, between organ systems of an organism or individual, between individuals of a species, between species of an ecosystem, between ecosystems of a planet, and even between planets of a galaxy or between galaxies. Therefore, relationships exist between members of a family or between families whereby the family may be an atom, a molecule, a planet, or the universe. There is virtually an infinite number of relationships in the universe that can be studied, and there are many new and different ways of looking at old relationships. A system is comprised of parts and a whole and may be defined by how the parts are related to one another in a meaningful way.

Understanding new or unique relationships requires new perspectives. This, in turn, requires thinking that forms new mental models and, hence, new knowledge that extends information into an organization that can improve problem solving.

> *Thinking is a way to organize information to give it meaning. Information is not enough to solve problems. Thinking solves problems by organizing information as a mental model that imparts meaning from that information and hence transforms that information into knowledge which in turn is vital and critical for problem solving.*

Therefore, seeing relationships between the component parts of a system that give the system as a whole meaning requires new perspectives. The greater the number of perspectives capable of organizing the information of the parts of a given system or between systems into wholes, the more robust the number of mental models and hence the greater the mental agility for problem solving. A system may be a physical inorganic chemical system (defined by tangible physical properties such as including component particles or objects as mass, velocity, momentum, position, shape, etc.) or it may be a chemical-biological system defined as a group of parts that interact to make up a whole. The parts and wholes of a biological system may be defined across many hierarchical scales with a fractal-like nature. Accordingly, elementary particles are components of an atom. The atoms are parts that make up a molecule which in turn comprise a supramolecular structure. Supramolecular structures in turn form the parts of the cell. Cells represent component parts of an organ system which in turn form the parts of the whole organism or individual, which in turn represent the parts of an ecosystem, and so on.

The ontological definition of reductionism refers to a biological system that is constituted by nothing but molecules and their interactions such that biological properties supervene upon physical properties. That is, each biological process is identical to a type of physical and chemical process. There is often symmetry at each of these scales in the sense of the system being logically reduced to its parts to explain the behavior of the systems as a whole. However, the assimilation of multiple scales breaks symmetry, and hence the systems that form cannot be understood by simply reducing higher-level behavior to their lowest level component parts.

Ontological reductionism implies that a biological system is constituted by nothing but molecules and their interactions, such that biological properties supervene upon physical and chemical properties.

This notion of ontological reductionism applies higher-level features of a system to explain lower-level features. This thinking is consistent with the euphemism that biology is applied chemistry, chemistry is applied physics, which in turn is applied math. Accordingly, this reductionist thinking is Newtonian-like whereby equations such as F = ma can be applied to understand the natural universe. By extension, biological thinking that invokes reductionist thinking can solve challenges in medicine by applying linear equations. In contrast to the argument that symmetry is broken across hierarchical scales, ontological reductionism premises that biological structures and functions supervene upon physical chemistry. That is, *biology takes what physics and chemistry might break down into smaller units and builds it up over larger systems into a superstructure applying physics and chemistry.* Contrarily, the notion of symmetry breaking is consistent with the finding that qualitatively different superstructures emerge across each hierarchical scale. These emergent structures may be considered phase transitions or bifurcations occurring along the trajectory of a nonlinear system. Accordingly, the model of punctuated equilibrium underscores temporal and spatial zones of stability, tethered by negative feedback mechanisms to an equilibrium attractor. Energy input into the system results in consonantly incremental linear responses within any given hierarchical scale (for example of molecular chemistry) in proportion to the component parts. This time invariant symmetry of systems holds as a steady state only within the given hierarchical scale.

As the system extends into a different level symmetry is broken; it transitions to a new phase or bifurcates as a manifestation of positive feedback.

The system is now an emergent qualitatively different superstructure and higher hierarchical scale.

Various physical systems will be described as well with the purpose of providing meaningful perspectives to the writer in terms of the interaction of the parts of the system to produce the whole. The big picture or overshoot of this writing is to develop an understanding of how organic systems self-organize and integrate to produce higher-level biological systems. This intriguingly draws comparison to material inorganic systems in ways that should be anticipated to become increasingly invoked by some of the greatest healthcare challenges we face. Such perspectives also open the curtain to new insights into the relationship among parts of whole systems that have credible and explosive potential relevance to clinical healthcare. The distinctions, systems, relationships, and perspectives discussed in this writing aim to improve the problem-solving skill set of the reader by imparting a more flexible and nimble capacity for imparting meaning to biological systems. This is intentionally a provocative discussion that challenges not as a replacement for but rather as an extension of and integration with much of evolving traditional establishment thinking in clinical medicine.

4.18 Complexity and Information

Information is an abstraction inherently devoid of meaning. Claude Shannon, the father of information theory, introduced the notion of information [43] as something counterintuitive to the conventional understanding of the term. Rather than information being an objectively useful fact or description, Shannon conveyed it as something abstract. He described information in terms of bits whereby a string of bits is information independent of whether they are true or not, meaningful or not. In fact, ironically a random (in a relative rather than absolute sense) string of bits carries more information than an orderly one. Orderliness implies predictability and hence has low information content. The element of surprise in a system has greater inherent value and is fundamental to the emergence of complexity in systems that is fundamental to their survival. Unpredictability is an essential characteristic of complex systems finding solutions for the general notion of survival in the context of otherwise adversarial surroundings/environment.

The highest amount of information in a system lies in the balance between predictability and total randomness.

It is in this zone that the evolution of a dynamic self-organizing system is maximally adaptive. Stated another way, the purpose for organization of information of any complex system is to serve an adaptive value that promotes its own survival.

Information in biology lies at the interface of energy and chemistry.

It is present in our DNA and in the proteins coded from DNA in our cells, cell metabolites, microbiome, pathways, and networks. Energy exists as potential energy in the bonds and interactions of our chemistry and as kinetic energy of the bonded elements as well as translational movements of ions, molecules, supramolecular complexes, or cells.

The energy of structure and function is informational energy.

However, finding solutions is the capacity of this information to adaptively organize and reorganize in response to the challenges of the environment. This represents its metabolic flexibility and resiliency that equates to a problem-solving potential enabling its own survival. The greater the complexity of interactions between the elements or component parts of systems and subsystems, the greater the energy-efficient adaptive flexibility and resilience or robustness.

According to Shannon, information is a statistical property that is considered on the same level as entropy in the sense that each concept represents the number of possible ways the components of a system can be rearranged.

The fewer possible states in which it can be rearranged, the lower the entropy and the information, the more predictable it is. A system that is predictable is tethered to a limited number of ways it can be arranged and, accordingly, the less flexible or resilient the system is in terms of its ability to reconfigure itself for a workable solution that allows it to survive in a challenging environment. Information alone is not enough to solve problems. It must have an interactive capacity or complexity of component parts dynamically conformable to its surrounding demands. Each individual bird in a flock behaves autonomously according to simple rules of local interactions with neighboring birds. There is no central control. Growing out of the simple rules of local interactions are "emergent properties which have nothing to do with the original rules" (Frank Heppner, University of Rhode Island) [44]. The complexity of interactions is highlighted by how much switching and shifting places of individual birds takes place demonstrated by computer tracking systems. This occurs within the synchronous fluid-like motions of the flock that changes shape like a jellyfish. It is assumed that the purpose of birds flocking is to defend against predators. However, the purpose is likely to be far more extensive. Flocking behavior occurs in wave-like patterns that are found in many biological species and systems. Such patterns are often evolutionary bursts of arousal and quiescence that evolve to conserve energy. This is an evolutionary purpose to find the most efficient solution, which is fundamental and found throughout the physical universe. Whatever the system, the goal is to rest until it receives its cue.

Whether it is the information of a system of birds (flock) that reconfigures into a given shape at intervals of excitation cues, or neurons that coordinately fire to consolidate information into an organizational pattern of knowledge that can be stored as memory, the same principle applies. In fact, the same laws apply to how photons are governed. In each case, a signal originating within the system can spread widely in a fraction of a second with virtually no distortion or diminution [45].

> *The greater the capacity for components of the system to connect and interact, the greater the complexity of the system and its potential to reconfigure the information of the system into emergent energy-efficient solutions. The greater the number of potential patterns available to it, the greater the flexibility or robustness to the circumstances and chances of ultimate survival.*

The greater the information or entropy of the flock of birds, the greater the potential for the emergence of the optimal organizational pattern. Conversely, if the birds are relatively immobile with a limited capacity for rearranging themselves into an optimal flying pattern as a flock, the more limited their dynamic adaptive self-organizing capacity for survival as a group.

> *Thinking is a specific type of biological process that involves cognitive interpretation of information bits into something meaningful and of tangible value. Cognitive problem-solving transforms information into knowledge, which is part of the broader strategy for a system as an organism, such as a human being, to maintain vitality or even survival itself.*

Seeing relationships between the component parts of a system that give the system as a whole meaning requires new perspectives. The greater the number of perspectives capable of organizing the information of the parts of a given system or between systems into wholes, the more robust the number of mental models and hence the greater the mental agility for problem solving.

The study of dynamical systems looks at how the state of a system changes over time. One model of doing this illustrates a given system in phase space, which is a graphical representation on a two- or three-dimensional screen. In mathematics and physics, a phase space is a model for studying dynamic systems. *Phase space is a space in which all possible states of a system are depicted with each possible state of a system corresponding to a unique point.* Each state is an assimilation of the parameters of the system at a given point in time to provide a multi-dimensional coordinate. The system's evolving state over time traces a path of the system's parameters along usually two or three coordinates. The dynamic trajectory of an equilibrium system is a transient motion that gravitates to a stable equilibrium state. In a physical system, it may be the point of the lowest potential gravitational energy, but in a biological system, it is a steady state within a narrow range of homeostasis or relative point of equilibrium. Paragon examples of biological periodic systems are the circadian clocks as well as the circadian rhythms, which they control. For example, there is a central suprachiasmatic nucleus in the hypothalamus of the brain, which senses light–dark cycles visually. Peripheral molecular clocks are present in most cells of the body and are synchronized to the central suprachiasmatic nucleus in the brain. These biological periodic systems control circadian rhythms of the body such as the neuroendocrine axes, which originate in the hypothalamus. Classical examples are the circadian rhythm of cortisol secretion, thyroid stimulation, and catecholamine neurotransmitters. Rather than being a point of equilibrium, periodic systems are more like cycling around an equilibrium point. In phase space, these trajectories are represented by circles.

> *Unlike equilibrium and periodic systems, which are simple systems with a single attractor state, complex systems involve multiple interactive forces acting on a system with multiple attractor states over time.*

A classic example of such a complex system is the double pendulum. Specifically, this is a chaotic system whereby small changes in the initial conditions of the system can lead to very large changes in the long-term trajectory. Chaotic systems, in part, are defined by their high sensitivity to the initial conditions such that minor differences result in widely divergent long-term outcomes. Accordingly, long-term predictability of the system becomes impossible. In contrast to non-chaotic complex systems, or nonlinear complex systems, which have many components that are nondeterministic, chaotic systems are nonlinear simple systems with very few elements or interacting parts. While complex and seemingly chaotic unpredictable behavior would be expected of a nonlinear complex system, simple deterministic systems would not be expected to behave in this fashion. In chaotic systems, future behavior is deterministic. That is, the future behavior is fully determined by the initial conditions. It is counterintuitive that their deterministic nature would not make them predictable. The two limbs of a double pendulum are strictly deterministic and predictable when taken in isolation. However, when we joint the two limbs, the simple linear systems exhibit nonlinear chaotic behavior. This unpredictable behavior emerges out of the interactions between components.

Chaotic systems are a fundamental and inescapable scientific reality as laws of nature of nonlinear systems. Inherent in these systems is the effect of exponentially changing trajectories leading to an unpredictable and counterintuitive output. Nonlinear dynamic systems, through feedback loops, grow exponentially. What this means is that negligible errors, effects, or differences within nonlinear systems can grow in an exponential fashion where small effects or errors feed back onto the system at each stage of its development causing the size of the error or difference to exponentially grow. This was the case in the seminal Lorentz weather prediction models. The values of parameters of the weather prediction equations were thought to be the same ones used in a prior experiment. He was simply repeating the experiment to confirm his findings and validate his equations. However, a tiny rounding error of a millionth of a decimal point of the parameters resulted in wildly divergent weather patterns. Through the iterations, the same differences in the parameters would grow nonlinearly but exponentially making the resultant output weather pattern unpredictable within a relatively short period of time. This highlights the phenomenon of sensitivity to initial conditions, which has popularly been called the "Butterfly Effect" after a 1972 presentation by Lorentz [46] who metaphorically likened the flapping of a butterfly's wings in Brazil to setting off a tornado in Texas a month later. The flapping of the wings represents the change in initial conditions leading to a chain of events and, ultimately, a large-scale phenomenon. Importantly, the flapping of the wings does not cause the tornado but rather defines the initial conditions. It sets off a chain reaction through feedback loops that result in the long-term unpredictable trajectory.

4.19 Nonlinearity, Bifurcations, and Phase Transitions

Nonlinear systems, again, are defined by positive and negative feedback loops. Positive feedback amplifies change in a supra-linear fashion in contrast to negative feedback, which stabilizes or dampens the trajectory of the system toward linear or equilibrium behavior. Positive feedback growth is an unstable trajectory that must be constrained by negative feedback loops. Positive feedback is the key driver of phase transitions characterized by bifurcations and multistability. A phase transition may be defined by some smooth small change in a quantitative input variable (control parameter) that results in a qualitative change in the system's stable state. An example of a phase transition may be the transition from ice to water due to a small change in the control or stress parameter of the system's environmental temperature (raising it from 0 °C to

+1 °C). At the critical temperature, the system is governed by a new set of parameters with a new set of properties (liquid rather than solid). Another example of a phase transition may be the change within a colony of bacteria. When we change the heat and nutrient input to a bacterial culture, which may be a microbiome for example of our gut, we change the local interactions between the bacteria and get a new emergent structure and dynamics to the colony. Although this change in input value may only be a linear progression in terms of temperature or nutrient composition (small differences at the local level), it results in an emergent qualitatively different pattern of the microbial colony on a more macro scale. Not only has a new bacterial colony structure emerged but the actual rules that govern the system changed as well. That is, due to some small change in a parameter a regime shift or qualitative change occurs in the bacterial colony structure. This, in turn, feeds back into the colony system by new interactions reflecting the qualitative shift and the so-called rules or interactions between elements under which the system operates. Another way to talk about this is in the context of bifurcation theory that describes branching or bifurcations analogous to the phase transitions that represent a change in the qualitative properties of a system.

Bifurcation theory describes that a small change in a control parameter can result in a structural or dynamical change in the system. This results in new attractors emerging with a new basin of attraction and a new regime or equilibrium steady state.

Essentially, phase transitions are physical examples of bifurcation phenomena whereby the physical system can at some point on the phase diagram coexist in one of two equilibrium states. As a real-world example of bifurcation, a chief medical resident may have plans to pursue a career in cardiology, representing his basin of attraction despite his multi-specialty exposures cycling through many different domains along the course of training. However, a mentor, whom he could have met locally, representing a local interaction, convinced him that the opportunities in the field of endocrinology offered a more attractive horizon. Consequently, the long-term trajectory in that physician's career bifurcated. Finally, nonlinear systems are capable of exponential growth in contrast to the incremental growth of linear systems, which are characterized by the model of linear transformations with punctuated equilibria. This model is highlighted by a marked dynamic between positive and negative feedback. The negative feedback holds the system within a basin of attraction that represents a period of stable development, however, punctuated by periods of positive feedbacks that move the system to areas far from its equilibrium (that is from its baseline far-from-equilibrium state) to a phase transition characterized by a new attractor state or bifurcation. An example of punctuated equilibrium may be seen in the development of a human being. As we go from childhood to adulthood to old age, each period represents a stable basin of attraction, which changes between each period, marked by rapid and defining physiological change.

The notions of phase transitions, bifurcations, and punctuated equilibria of nonlinear systems provide an applicable

framework for some counterintuitive findings in clinical medicine that otherwise rely on a deductive model of reasoning and problem solving. Consider the phenomenon of metabolic memory. This is a concept that may explain the ACCORD trial outcome findings that tight glycemic control in diabetic individuals with historically chronic hyperglycemia paradoxically provokes increased mortality. This heightened mortality importantly is independent of hypoglycemia. How can this possibly make sense? It does so if we consider the qualitative emergence of new steady states of a system governed by new sets of rules or interactions between constituent elements of the system. The thermodynamic parameters of a physical system such as a solid crystal lattice that absorbs a certain amount of surrounding heat will raise the temperature of the lattice by one degree Celsius. However, at a threshold of the melting temperature, physical and chemical properties change.

A physical system characterized by linear dynamics may be exemplified by physical entities such as planetary bodies. A planet may obey the properties of an equilibrium system if studied in isolation or if negating "negligible effects". As such, a planet being cycled remains in its lowest gravitational energy state. A periodic system may be exemplified as the orbiting planet. As described above, both equilibrium systems and periodic systems are linear, hence they are both deterministic and predictable. The linearity relates to a proportional or incremental matching between the energy of the system's output with the amount of energy input into the system or a proportionality between cause and effect. For example, the gravitational force is an energy input onto the orbiting planet that matches the product of the planet's mass and acceleration. Strictly speaking, it is Newton's second law applied to rotational motion that should be used here, namely that the torque, which is the product of gravitational force and the level arm equals the moment of inertia times angular acceleration of the planet. *In linear systems, the larger the perturbation applied to the system, the proportionately greater the change with respect to its original state. In nonlinear systems there is no proportionality between the cause and effect.*

Another type of linear system may be illustrated by a mechanical internal combustion heat engine. It represents a continuously linear system that is not characterized by a time interval between measurements of the system's parameters. Continuous systems are understood using the language of differential equations of calculus. The thermodynamic variables, temperature, and pressure are exploited in the mechanical energy produced by a combustion heat engine. That is, the input of heat from the potential energy in the bonds of the molecules of petroleum fuel is linearly titratable and proportional to the work performed by the engine. The free energy liberated into the combustion chamber is the Helmholtz free energy characterized by increased pressure of water and carbon dioxide vapor molecules and associated increased molecular collisions that result in increased temperature and subsequent translational movements of the molecules that cause volume expansion that allows the pressure to fall. The volume expansion inside the combustion chamber promotes the work of compressing the piston cylinder that rotates the crankshaft and ultimately the wheels of the motor vehicle. A gallon of gasoline will predictably provide a number of locomotive miles for an automobile, according to the engine's efficiency.

Thermodynamic parameters of temperature and pressure also describe the Lorentz attractor for weather patterns. These thermodynamic parameters are plotted in phase space assimilated into a unique point along three axes as the attractor describes the evolution of this system over time. The nonlinear nature of a weather system is rooted in its sensitivity to initial conditions due to positive feedback loops that amplify change in a supra-linear (that is, non-incremental) fashion. Some small change in a quantitative input of temperature or pressure results in a qualitative change in the weather pattern beyond the short-term as represented in phase space as bifurcations.

In addition to the nonlinear systems of weather patterns, chaos systems, biological systems are commonly and characteristically nonlinear. Nonlinear dynamical systems in the world, such as the chaos systems of weather patterns and many chaos systems and complex systems of biology, are significantly more complex than the linear equilibrium or periodic systems of nature or of machines. It is worth noting that machine engines convert energy into mechanical work whereas motors are a subset of engines that produce motion as the mechanical work. A distinction should also be made clear between machine engines and biological engines. Indeed, it is the latter metaphor that most powerfully assimilates an understanding of systems biology and improved biological thinking and problem solving. *Complexity and unpredictability of biological systems relate to the interactions between component parts of the systems.* Accordingly, the emergent output of a nonlinear system is greater than the sum of its parts. In other words, a nonlinear system as a whole represents a greater output than would be expected on the basis of input. Nonlinear systems are by definition a function of interactions of positive and negative feedback loops.

The interactions of the elemental parts of biological subsystems according to rules of local interactions evolve the subsystems adaptively in response to stressors that allostatically find solutions for stability of the larger organismic whole (that is, the living system). These local interactions such that computer simulations have demonstrated for flocks of birds or colonies of ants, bees, or bats, represent the goal of Big Data and bioinformatics (using genomics, proteomics, metabolomics, and microbiomics) to elucidate the understanding of human biological systems. The complexity of interactions is enormous but only a relative and definable few are responsible for the outcome behavior, and so it is not necessary to understand the mechanistic underpinnings of all the relationships. Rather, the goal is to delineate the "simple rules" and to elucidate the mechanisms of interactions using these "simple rules" that explain statistical aberrations necessary to understand pathogenesis and treatment or prevention of a disease state.

Punctuated equilibrium is a model that moves a system away from existing equilibrium steady state relationships between elements of a system into new interactions that protect the system from otherwise toxic conditions imposed on it. The punctuated equilibrium detaches the system from its pre-existing constraining forces of negative feedback pushing it into a bifurcation and causes a phase transition driven by the forces of positive feedback.

The pathogenesis of the apparent majority of obesity is promoted by the feedforward phenomena of catecholamine resistance, insulin resistance, and leptin resistance. Bioinformatics and Big Data analytics will need to ultimately elaborate upon and refine this framework, which will make it applicable for individualized healthcare. However, it suffices for now to highlight how current challenges in medicine illustrate the concepts of phase transitions, bifurcations, and punctuated equilibrium as they relate to biological systems.

4.20 A Biological Example: Metabolic Memory

Genomic insights have identified SNPs (Single Nucleotide Polymorphism) of catecholamine receptors such as β_2 and β_3 receptors as well as of hormone-sensitive lipase in subcutaneous adipose tissue that is genetically predisposing for obesity. Accordingly, there are differential rates of lipolysis between the different fat depots in the body whereby lipolysis is reduced in subcutaneous adipose tissue and increased in visceral adipose tissue. These gene polymorphisms are the underlying mechanism for catecholamine resistance in subcutaneous adipose tissue, which is an antecedent factor in the obesity triumvirate of resistance to catecholamines followed by resistances to insulin and subsequently leptin.

Intriguingly, amlexanox, a benign topical cold sore medicine (brand name Aphthasol) inhibits phosphodiesterase activity [47, 48] resulting in the reduction of catecholamine resistance in obese tissue. Insulin resistance occurs consequent to fat cell hypertrophy related to outgrowing the blood supply resulting in hypoxia and hypoxia-inducible factor (HIF) induced inflammatory cytokine expression. This is hypothesized to be a protective mechanism against obesity and further fat cell hypertrophy.

Resistance to insulin prevents the uptake of glucose while up-regulating the inactivating pyruvate dehydrogenase kinase (PDK) enzyme that phosphorylates pyruvate dehydrogenase enzyme complex responsible for decarboxylating pyruvate to acetyl Coenzyme A (CoA) to be utilized in the Tricarboxylic acid cycle (TCA) cycle for energetic needs of the cell. Conversely, mitochondrial bioenergetics relies on long-chain fatty acid b-oxidation for its source of acetyl CoA. Additionally, insulin resistance at the level of the adipocyte prevents the suppressive effect of insulin on lipolysis that in effect decompresses the "overstuffed" adipocytes. Furthermore, this accelerates the lipolytic activity of the visceral adipose tissue depot (which at baseline, unlike subcutaneous adipose tissue, is not insensitive to the lipolytic effect of catecholamines).

The insulin resistance response to obesity widens from adipose tissue to include the liver due to the flux of free fatty acids from the portal circulation. This excess flux of free fatty acids, due to peripheral insulin resistance, is likely to be responsible for hepatic steatosis. The excess fat deposition in the liver may render hepatocytes less sensitive to insulin action in turn leading to hepatic insulin resistance and related dyslipidemia. The antecedent genomic predisposing factors for obesity and dietary excess lead to the feedforward process of insulin resistance as an emergent evolutionary adaptation to obesity as a solution to abrogate the worsening hypoxia and inflammatory state of hypertrophied adipose cells. However, *failure to attenuate the caloric excess dietary consumption relative to energy expenditure results in progressive fat cell hypertrophy and essentially the failure of adipose insulin resistance to provide physiologic adaptations.*

Accordingly, the overstuffed fat cells overflow the storage capacity for neutral lipid reserves [49]. The result is an increase in circulating free fatty acids that deposit ectopically into tissues including the liver and skeletal muscle. The deposits in liver cells exacerbate hepatic steatosis and hepatic insulin resistance, whereas in skeletal muscle, such deposits cause or contribute to skeletal muscle insulin resistance. Notably, at least in a subset of insulin resistant individuals, skeletal muscle ectopic fat may precede hepatic steatosis due to genetic factors, brain, pancreas, myocardium, and other tissues [50]. Important to this discussion is that *the development of adipose insulin resistance illustrates the concept of punctuated equilibrium characterizing a qualitative bifurcation and phase transition of biological behavior searching for a new equilibrium steady state.*

Insulin and leptin are both satiety hormones and are considered the two most important adiposity signals. Accordingly, the circulating levels of both hormones rise in proportion to fat mass to suppress hunger by acting on receptors in the arcuate nucleus of the hypothalamus. There is a complex interdependent relationship between insulin resistance and leptin resistance. One such relationship is insulin resistance as an antecedent cause of leptin resistance due to the high circulating free fatty acids and ectopic deposition in the neurons of the arcuate nucleus. Lipid metabolites, for example acyl CoA and diacylglycerol within the neurons interfere with leptin signaling as well as insulin signaling. Plausibly, the leptin response to obesity and consequently insulin resistance is initially an adaptive response intended to reduce the underlying obesogenic state by quelling hunger as well as increasing energy expenditure. However, unless the root contributory factors of dietary excess relative to physical activity are adjusted, leptin resistance ultimately ensues as a function of insulin resistance with intracellular lipid metabolites and inflammatory signaling in addition to downregulation of leptin receptors due to prolonged exposure to high circulating levels of leptin. There is an ensuing inextricably linked disruption shared between insulin and leptin signaling. Thus, the leptin response, like insulin resistance, represents attempted negative feedback mechanisms to maintain body weight set points as an equilibrium steady state. However, these physiological responses as subsystems (biological systems) of the larger system as a whole (the living system) exemplify processes of positive feedback and allostatic new steady states seeking to maintain stability or homeostasis of the greater whole. Thus, the local interactions between parameters of a biological system of an individual living system, an individual being, parallel those of a material system and the principles of physics to describe the concepts of bifurcation, phase transition, and punctuated equilibrium of nonlinear systems.

The phenomenon of metabolic memory also illustrates the concepts of bifurcation, phase transition, and punctuated equilibrium of nonlinear biological systems. While highlighting the fact that essential interactions between parts of the systems responsible for the evolution to a new steady state remain undiscovered, again, bioinformatics and Big Data analytics will be expected to facilitate these insights coupled to randomized control trials and other mechanistic studies. Bioinformatics and Big Data should help to "simplify complexity" of biological systems by finding the "simple rules" underlying emergent behavior analogous to those that describe flocking behavior of birds. The phenomenon of metabolic memory has been recognized since the mid-1980s at the experimental level and has been further highlighted by the large-scale clinical trials, Diabetes Control and Complication Trial (DCCT) and United Kingdom Perspective Diabetes Study (UKPDS). These studies reveal that initial tight glycemic control in the early years of the diagnosis of diabetes provided a so-called "legacy effect" whereby these individuals were protected from the vascular complications even if subsequent glycemic control was poor. Conversely, the Accord Trial, which included previously poorly controlled diabetics who were subsequently brought under tight control, was prematurely terminated due to increased cardiovascular mortality. Although it is not precisely clear which are the specific interactions that explain this phenomenon, certain elements and interactions are clear. Hyperglycemia results in oxidative stress through a number of pathways including via the activations of NADPH oxidase and xanthene oxidase but mainly via the surplus of glucose in the mitochondria resulting in electron leak along the electron transport chain promoting superoxide species. The initial events, classically described by Brownlee, that lead to chronic vascular complications include the polyol pathway, the hexosamine pathway, the protein kinase C pathway, and the production of advanced glycation end products (AGEs). AGEs activate the receptor for AGE (RAGE) which further promotes, in feedforward positive feedback fashion, generation of intracellular reactive oxygen species expression. This positive feedback mechanism moves the system to a new steady state, and as such, prior AGEs can maintain RAGE overexpression even in the absence of persistent elevated glycemia. This is a major pathway in the oxidative damage to mitochondrial DNA and proteins and resulting impairment of mitochondrial function.

Reduced activity of mitochondrial electron transport chain complexes in the heart occurs following about ten days of hyperglycemia whereas in skeletal muscle, the liver, and kidney, this occurs after about 30 days of hyperglycemia. An inflammatory cascade and inextricably linked oxidative stress in the cell occur underpinned by high levels of NFkB, protein kinase C, AGEs, and RAGEs that pull protective histones away from nuclear DNA and results in epigenetic altered gene expression. These epigenetic alterations are the manifestation of the positive feedback mechanisms of oxidative stress and inflammation. Accordingly, this subsystem of the organism moves into an allostatic steady-state equilibrium that stabilizes and protects the organism, maintaining a survival homeostasis. Again, all the mechanistic underpinnings responsible for this are not clear. However, improving glycemic quality ironically

moves subsystems of the organism into new equilibria such that, on balance, the systems of epigenetic changes destabilize the survival homeostasis of the organism. Bifurcation, phase transition, and punctuated equilibrium of given subsystems invoke the notion of search for solutions to challenges that threaten the homeostasis of the living system, in this case, the individual. However, such allostatic changes consecutively superimposed may ultimately break down the capacity to maintain the order of the greater whole, i.e. the living system. This is a biological term considered allostatic overload, that is, pathogenic disease or even death. *Computational bioinformatics and Big Data will hopefully elucidate the biological elements and their interactions that will enable therapeutic stratification of patient subsets for targeted guidelines for the optimal range of glycemic control tailored to the individual. Further, this may ultimately lead to relevant molecular level mechanistic understanding to broaden the scope of patient-specific therapeutic intervention or preventive measures.*

4.21 *Plus ça Change, Plus C'est la Même Chose*

The change in physical and chemical properties that are due to heat transfer causing a rise in temperature of the solid by 1 degree Celsius is known as the specific heat of this solid. Temperature is a macro-level property proportional to the average kinetic energy of the particles of the solid. Because collisions represent the interactions between the elements or parts of the system, temperature is a manifestation of the system as a whole. Each temperature elevation of one degree Celsius represents a process of positive feedback following a period of negative feedback and underscores the model of punctuated equilibrium with each one degree Celsius rise representing a new equilibrium steady state. Water has a relatively high specific heat because strong hydrogen bonds connecting water molecules are capable of absorbing a considerable amount of heat before breaking. The broken hydrogen bonds allow increased translational kinetic energy of the water molecules with resultant collisions and hence increased temperature. The broken hydrogen bond is, therefore, the harbinger of a bifurcation or phase transition from one macro-state to another. At a threshold water temperature of 100°C, a particular qualitative transformation takes place underpinned by a positive feedback mechanism whereby the liquid phase begins transforming into a vapor phase. Each period of positive feedback depicts a nonlinear phase of the system whereby the output behavior may be an exponential diversion from the incremental input. This complex set of interactions defines a nonlinear system that uniquely behaves in a predictable fashion. Most nonlinear systems, however, are not predictable because their interactions are not defined precisely enough to be able to predict the evolution of their behavior. *Predictability of emergent behavior requires an understanding of all the relevant interacting parts of the system and the local rules of their interactions responsible for the emergent qualitative behavior of the system as a whole. Particularly unpredictable are many complex biological systems relevant to medicine.*

Fractals are often found as beautifully symmetric pictures of nature. Chaos theory describes non-predictable aspects of nonlinear systems due to their sensitivity to initial conditions, while fractals represent the order behind the apparent chaos. Symmetry is a form of order. There is symmetry in the periodic activity of systems of self-repeating cycles, reflecting the notion of sameness.

The two sides of a butterfly are virtually mirrors of one another. A snowflake has geometric forms, which are almost replicas of one another. Symmetry defined mathematically is an object or process, which is invariant through a transformation. For example, the first law of thermodynamics reflects the time invariance of energy. That is, energy cannot be created or destroyed over the course of time, it can only be transformed from one form to another. Another example is the gauge invariance of the electromagnetic and weak forces of nature, that is, the electroweak force, which underscores that these forces of nature are neither created nor destroyed but just transformed from one form to another through gauge operations. Simply stated, symmetry is that which stays the same despite a change. However, symmetry is more than that. It is the secret of science. It is one of the most powerful concepts in mathematics and science. Science is largely about finding patterns in the world and symmetries in these patterns. Finding these patterns and encoding them in models allows us to describe a wide variety of phenomena with simple equations. The absence of symmetries would mean we would be unable to create a compact representation or understanding of the world. Symmetry indeed allows the basis for deductive logic and the philosophy of reductionism. Our understanding of DNA and molecular biology is based on reductionism and symmetry, so is Einstein's mass-energy equivalence principle by, and so is the periodic nature of time itself and the usefulness of clocks. Without symmetry, our scientific map of the world would have to be the same size as the world itself and hence useless. The concepts of symmetry and asymmetry help us understand the abstract concepts of order and in a distinctly geometric and visual form [33].

In the phenomenon of chaos, we see the breaking of symmetry as two things that start out similar become increasingly dissimilar as the symmetry between them becomes broken, ultimately resulting in asymmetry. Fractals have scale invariance, meaning they exhibit symmetry across scales. That is, despite a changing scale, the geometric configuration will repeat itself across the different scales of magnitude. This scale variance, analogous to time variance or gauge invariance, is a self-similarity reflecting the inherent notion of sameness characteristic of symmetry. It is the scale-invariant form of symmetry that defines fractals. Fractal structures are abundantly present in the real world, however in chaos theory on computer screens of phase space they represent mathematical models and provide the resplendent and eloquent order behind otherwise apparent chaotic processes.

Symmetry is our way of scientifically understanding physical, biological, and living systems of the universe. The powerful concept of symmetry allows us to reduce the expansive phenomena of the universe to linear equations that in turn make problem solving possible. Without symmetry, again, our map of the scientific world would be the size of the world itself. This map would be useless in terms of a capacity for translating science into efficient problem solving. It would mean the inability to predict future events or the inability to extrapolate understanding of one phenomenon to another. The perspective of symmetry in science, including the life sciences and medicine, equates to philosophical reductionism and reductionist reasoning using analytic and deductive methods. This in a sense requires only one perspective. Uniformity (or invariance) from different perspectives is symmetry. Symmetry breaking is a process by which such uniformity is broken or the number of points to view uniformity is reduced with the outcome of generating greater stability. Symmetry breaking is a prevalent process in biology. Organismal survival depends critically on structural and functional stability and diversification at both microscopic and macroscopic scales.

Symmetry breaking allows the generation of cells with different fates in the process of differentiation and underpins the complex arrangement of tissues and organs during embryogenesis [51]. Symmetry breaking is responsible for the complex rearrangement of tissues and organs, and it stems from molecular assemblies to subcellular structures and subsequently to cell types and tissues [51].

Dynamic actin filaments form contractile bundles resulting in the asymmetric generation of force that drives cell motility in a given direction [52, 53]. Apical and basal polarity of epithelial cells that is generated in part by dynamic actin filaments promotes different components and functions of these epithelial cells. Apical polarity involves the core proteins, which act to amplify and reinforce asymmetry within cells by a mechanism that involves contact and cooperation with neighboring cells. This is critical for the organizational emergent patterns at the level of tissues and organs [54, 55]. In essence, the roots of symmetry breaking can be found at a lower hierarchical scale such as the molecular organizational level that results in emergent qualities at higher hierarchical scales of organelles, cells, tissues, and organs.

Physical and biochemical positive feedback amplification can break symmetries and promote system stability through the emergence of qualitatively different phenotypic structures and functions. This process evolves from systems of lesser complexity (of interactions) to those of greater complexity as they move across higher (increasingly macro) scales of biological underpinnings. The engine that drives symmetry breaking incorporates the dynamic interplay between internal and external cues onto the system to drive the nonlinear emergent outcome. This prevalent process in biology is critically responsible for organismal survival. "Symmetry breaking is the dynamic consequence of systems of molecular interactions operating in the physical context of living cells and their immediate environments" [51a]. For example, key elements (agents) that control symmetry breaking include the directional assembly of cytoskeletal polymers, the molecular motors that drive the various types of motions in biological and living systems from cell division or fusion to chemotaxis, to ciliary motility to muscle contraction. The regulatory actions of key enzymes such as GTPases, kinases, and proteases also represent key agents in mediating the dynamic interplay between cues and the cytoskeletal motors. *Quantitative* biomedicine and bioinformatics are fields that use computerized modeling to understand the complex molecular and biochemical interactions that underlie symmetry breaking as statistical, mathematical, and quantitative explanations for systems that transition to new qualitative states. These computational fields of science including medicine are also increasingly fundamental to understanding the role of symmetry breaking in downstream physiological and pathophysiological consequences. Symmetry breaking is essentially the conversion of linear to nonlinear processes that accordingly cannot be deductively anticipated or understood. Symmetry is as rich and diverse in biology as it is in the inorganic universe. Consequently, these fields of science will continue to grow becoming fundamentally important to biological and medical research and increasingly so to clinical patient care as well.

Reductionist reasoning is typically not applicable to nonlinear systems whereby symmetry is broken. Therefore, in the absence of invoking the powerful tool of reductionism based on underlying symmetries requires viewing and understanding systems through a number of perspectives that provide insights into the asymmetries. Since solutions for such systems cannot be provided in a reductionist manner from equations, they must be inferred instead from creative thinking. Creativity can be powerful, however its potential parallels the number of perspectives from which a system is scientifically appreciated and understood. Creativity is not something that can be memorized and regurgitated. In the management of diabetes, it is productive to approach problem solving from the perspectives of the many asymmetries of the disease process. This may be highly dependent on the individual patient perspectives that should be considered, which for example include those of the microbiome; glycemic index and gut absorption patterns; degrees of insulin resistance and efficiency, hormonal patterns for example the common coexistence of estrogen dominance; body composition of muscle mass; fat and the topography of fat; sleep and stress. Stress, in turn, importantly should be approached from the perspectives of its type, whether it be physical such as dramatic, microbial, toxicant, allergic or dietary. Conversely, psychogenic stress may be social (the most toxic form of stress) or a non-social form of stress such as work-related, financial, or health. The concept of positive feedback as discussed earlier may be understood in the context of physical and psychogenic forms of stress. Physical stress induces the expression of cytokines in the body, which are often inflammatory in nature. These inflammatory cytokines have the potential to act in the region of the prefrontal cortex of the brain and in the limbic regions of the brain to reduce the threshold for which environmental pressures are perceived as threatening and furthermore out of our control. The inability to control threatening challenges, in fact, defines psychogenic stress. When inflammatory cytokines interfere with the inhibitory actions of the prefrontal cortex on the emotion centers of the brain (the limbic system), it inherently detaches rationality from the circumstance. The degree of threat is exaggerated from an imagination perspective and so is the inability to separate those things that we can and cannot control.

Those things that we can control should be viewed as inherently vitalizing, that is something with

potential to evolve a positive vitality. This distinction is important because those things that cannot be controlled should not consume conscious concerns and hence should allow a sense of tranquility and some very otherwise cage-rattling circumstance that is often an ineluctable circumstance of life.

Nonetheless, the exaggerated psychogenic perception of stress activates the neuroendocrine axis and catecholamine levels accompanying gut motility and absorption patterns, gut microbiome composition and so on which further potentiates the physical component of stress and elevation of circulating inflammatory cytokines which further exacerbate in feedforward fashion the psychogenic perception of stress. Feedforward processes are positive feedback that pulls systems out of stable zones of equilibrium tethered to negative feedback. A balance of positive and negative feedback is often healthy for the body to maintain homeostasis (resistance to change by allostasis [stability through change]) ultimately re-establishing new stead states transiently until the underlying stress remits is the healthy evolutionary purpose of these dynamics.

4.22 The Physics of Heat and the Biology of Inflammation: Are They Related?

Another useful example to illustrate the phenomenon of positive feedback takes a very different perspective of thermal physics. Collisions between molecules are the molecular definition of temperature. As the temperature of a body or system is reduced, the number of collisions between the component molecules is reduced. This reduction in molecular collisions further reduces the temperature in a feedforward or positive feedback mechanism. This example of considerations in the management of diabetes is intended to exemplify a disease state, which is highly asymmetric and is a common clinical scenario and challenge in clinical healthcare. Understanding the disease state from a number of perspectives indeed helps problem-solving capacity on an individual level. Moreover, the perspectives highlighted above can be further understood not only from classical thermal physics points of view but emerging practical applications of nonlinear systems with fractal dynamics such as the neuro-autonomic mechanisms regulating heart rate variability and highly complicated signal transduction cascades involved in, for example, neuro-endocrine and immune dynamics in which interactions and crosstalk occur over a wide range of temporal and spatial scales [56]. Other perspectives of chaos and its offshoot complexity theory will undoubtedly provide useful perspectives for the clinical management of diabetes on an individual scale. *The perspectives of bioinformatics and Big Data involving computer simulations and electronic medical records will increasingly become more relevant to clinical healthcare. The phenomenon of quantum metabolism and other phenomena of quantum biology are likely to contribute, perhaps even in an explosive fashion, to therapeutic implications for many disease states. The* perspective of biological engines such as fuel cell, far-from-equilibrium and molecular quantum mechanics will all likely enhance the creative potential for biological thinking and solutions in clinical medicine.

Thus, science, particularly medicine, which often cannot be approached or understood by reductive thinking, implies a breaking of symmetries, and hence the systems become nonlinear in their interactions or complexity. Inherently they become unpredictable and noncompressible, that is, unable to be generalized in a way that symmetrical systems may. In essence the implication for required creative thinking is the art of science with individualized application, one patient at a time. Implicitly this requires more than a single perspective. The more perspectives, the more flexible and creative the problem-solving capacity.

Reductionist reasoning is the traditionally established way of thinking in both research and clinical medicine. This thinking is compatible with a fractal-like nature of the parts and wholes of biological systems with scale invariance and hence a continuous symmetry across hierarchical scales. The implications of this are that a whole biological system of a lower biological scale becomes a part of the next higher organizational level. Accordingly, subatomic particles are components of an atom. The atoms are parts that make up the molecule which in turn comprise a supramolecular structure. Supramolecular structures in turn form the parts of the cell. Cells represent component parts of an organ system which in turn form parts of a whole organism of individual, which in turn represent the parts of an ecosystem, and so on. *Medicine typically considers the living system, or individual, as the highest hierarchical level.* However, as argued by Nobel Laureate P.W. Anderson, and others, reductionism is often flawed biological thinking in medicine because symmetry is actually broken between biological systems that cross hierarchical scales.

Because there is much we do not know, the practice of clinical medicine is as much art as it is science *per se*. Stated another way the practice of clinical medicine is an art of science. It requires creative thinking which in turn demands an understanding of the science from different perspectives.

Biological symmetry may be exemplified by tissue cell invariance. That is, the systems of a given type of cell do not change as processes extend from a single cell to those involving all the cells across the tissue. However, visceral adipose tissue, comprised of myeloid which developed from myeloid progenitor cells are distinct from subcutaneous adipose tissue, the latter cell types store neutral fat. Visceral adipose tissue, which is derived from the bone marrow in contrast to subcutaneous adipose tissue cells, which are derived from mesenchymal cells, are immune cells and hence evolved with an immune cell function rather than a function to store neutral fat. Consequently, the loss of one type of fat cell (for example surgically), may portend different effects as the removal of subcutaneous adipose tissue may have deleterious effects on the metabolic health of an individual whereas the removal of visceral adipose tissue may have favorable effects. Of course, this is a simplified interpretation because our understanding of these different types of fat cells is rudimentary, and much heterogeneity and complexity undoubtedly exists beyond what we know about these tissues. With that said, one study [57] that looked at this did not show differences in the inflammatory markers between the subgroups of cohorts, but there were

significant limitations to the study. It is useful to consider the concept of linear systems and the notion of symmetry in terms of reductionist reasoning by geometric representations of computer graphics. Accordingly, a linear system is represented as a straight line reflecting the usefulness of linear deductive reasoning that is predictive of the future behavior of a given system based on simple equations. A periodic system in biology is often represented as a circle. This is because the cyclicity of periodic behavior by definition shows repeating behavior whereby the trajectory behavior of the system lines up with past and future behavior just as a straight line connects prior and future behavior. A conceptual characteristic of a linear system is one whereby the past predicts the future. This underscores a fundamental difference to nonlinear systems and nonpredictability characteristic of many biological systems that form the basis for the issues of clinical medicine.

Reductionist reasoning applied to biological systems inherently requires symmetry defined as uniformity or invariance of a system independent of perspective. Reductionist reasoning may aptly apply to systems when they are linear or in a linear phase bounded by negative feedback between interacting parts. In linear systems with equilibrium and periodic basins of attractions, the sum of the parts is equal to the whole, and an understanding of the whole can be achieved by breaking it down to its smallest component parts. However, as described above, biological systems are characteristically complex assimilations of multiple scales that break symmetry in the process, and hence the systems that form cannot be understood by simply reducing higher-level behavior to their lowest level component parts.

A further distinction should be made between the closely related concepts of ontological reductionism, which has to do with what exists, and epistemological reductionism, which has to do with what we know. Ontological reductionism contends that the lowest and smallest part of any system of reality can account in causal and linear fashion for all higher levels of the system.

> *Epistemic reductionism contends that the lowest and smallest part of a system of reality is sufficient to reconstruct knowledge of the larger system, no matter how large, complex or sophisticated that system may be.*

In both ontological and epistemological reductionism, whether it be an existence of a system or the knowledge of a system, respectively, the central premise is that local deduction can reduce any system to its component parts. This is classical Newtonian thinking.

In summary, *Newtonian thinking is difficult to apply to biological systems, which are not only complex but also complicated, that is involving many layers or hierarchical planes of organization.* Each scientific discipline has its own symmetry of laws or rules, and their functional integration requires breaking symmetries and forming new ones. These interdisciplinary relationships speak both to the system's complexity, the many similar parts that interact at each hierarchical scale, and to the complicated interactions between the organization scales of the overall system whole. Something that is complicated has many non-similar interacting parts such as what especially occurs between hierarchical scales of biological systems. The characteristic distinction here is that in "simple complexity" the parts are similar in contrast to "complicated complexity" where the parts are different. These are two very different but misunderstood terms. Both interactive processes typically exhibit nonlinear emergent behavior and require a bottom-up approach that studies the interaction between the agents, to understand. *Complicated complex problems as exemplified in the challenges of medicine, typically entail greater uncertainty than simple complex problems. Something that is complex but not complicated does not generally involve breaking and forming new symmetries, for example, the flocking behaviors of birds and other displays of collective intelligence.*

The strategy of simplifying complexity focuses not on mechanistic relationships between the elements or agents of a system, but rather, using computational strategies, looks for key interactions between agents independent of mechanism that forecast the global behavior of the system. Although this strategy as well as computational problem solving in medicine in general face many challenges, it does abrogate the limitations of reductionism in approaching nonlinear problems of biological systems. Self-organizing properties of complex adaptive systems are studied using computer simulations and a bottom-up approach with the notion for example in the flocking of birds or foraging of ants there is no "smart" bird or ant "that gets things organized". Rather, a self-organizing system evolves from local interactions among agents following "simple rules". These observations are augmented by studies of human physiological systems as examples of "self-organizing systems" [58].

Reductionist reasoning applied to biological systems requires symmetry. However, the concept of symmetry is rooted in physics and its understanding may be more complex than a common colloquial interpretation may infer. Phrases such as time invariance, space invariance, gauge invariance, or scale invariance to most people sound confusing. Hence, the extension of the concept to biological systems must be clear. Let us try to simplify this without breaching its fundamental intended meaning.

Symmetry is conceptually a uniformity or invariance of a system independent of perspective. It is often tied to a conserved quantity as in Noether's theorem. Accordingly, energy is the conserved quantity in time invariance—it is neither created nor destroyed over time. In other words, a number of analogous descriptive terms, symmetry, uniformity, conservation, or invariance are all like terms that apply to energy over the dimension of time. This is what is meant by the term time invariance. The quantity of energy is often not explicit, but implied. Similarly, space invariance implies conservation of momentum. A moving car that rams into the back of a stationary vehicle demonstrates a so-called inelastic collision whereby the momentum of the moving car is transferred in part to the car it impacted. An elastic collision may be the full transfer of momentum (mass times velocity) from an object like a cue ball to another ball in the game of pool. All of the kinetic energy (velocity) from the first ball is transferred to the second ball. In the case of the motor vehicle accident (an inelastic collision) the difference in the velocity component of

momentum between the two cars after the first car comes to a full halt is reflected in the damage in the mass of the two vehicles. Bumper cars are fun to ride because the collisions are elastic causing no damage to the vehicles. The conservation, symmetry, or invariance here involves momentum, which is neither created nor destroyed across the dimension of space. Space is the medium in which momentum is conserved analogously to time being the medium in which energy is conserved. It is important to note that in "time invariance" or "space invariance", it is not time or space that are implied to be the symmetrical or invariant quantity, but rather the conserved quantities of energy and momentum, respectively.

Gauge invariance is a more complex type of symmetry that is rooted particularly in quantum mechanics and the quantum mechanical behavior of electromagnetism, the weak force, and the strong force. Gauge theories in physics provided a unified framework for describing these three of the four fundamental forces of the universe known as the Standard Model. The Theory of Everything is an elusive model that seeks to incorporate the fourth fundamental force of nature, gravity, into the Standard Model. Although the Standard Model shows that the strong, weak and electromagnetic forces evolve from a single force (the weak and electromagnetic together known as the electroweak force separated from the strong force seconds after the Big Bang (see separate discussion of the fundamental forces and particles of nature), the separation of forces resulted in a breaking of symmetry. Accordingly, different conserved quantities apply according to the particular gauge variance. A gauge is a field transformation whereby the field describes forces, or fundamental interactions between elementary particles, and gauge invariance implies a conserved quantity of phases. For example, the conservation of electric charge applies to the electromagnetic field transformation between the wave-like properties of an electron (of classical electromagnetism) to the corresponding quantum particle, the photon. Hence, the negative electrical charge of the electron is neither created nor destroyed as it moves in various forms or transforms from one form to another along field transformations in the classical and quantum realms of electromagnetism. As in all conserved quantities in the concept of symmetry, there is no measurable change.

Scale invariance of fractals seems to be an exception whereby the invariance or symmetry does not imply a corresponding conservation principle in accordance with Noether's theorem. Rather, it implies an iterative process of self-similarity. The invariance may reflect structural symmetry in space or a process, which is symmetrical over time, but which may be displayed by computer models of a process in phase space as an abstraction with geometrical representation.

The symmetrical or reductionist thinking in medicine often relies on the self-repeating nature of systems. Dynamical periodic systems repeat their behavior with oscillatory cyclicity like a simple pendulum. These systems in biology are at a far-from-equilibrium steady state that may be considered to display a time-invariant symmetry. That is, the energy output and associated behavior of the system are conserved in the sense that it does not change over time. There is predictability in the future behavior of the system. It is analogous to the time invariance of the first law of thermodynamics in that energy in the system does not change quantitatively, it just transforms from one form to another. The first law, however, considers the universe as an isolated system in contrast to biological systems, which are connected to a web of interactive interdependence with other systems. The flow of energy between chemical bonds and the cells of the central clock in the suprachiasmatic nucleus of the hypothalamus as well as of the molecular clocks in other cells of the body coordinately synchronized to the pulsatile circadian release of hypothalamic releasing factors that control patterns of circadian hormones such as thyroid hormone and cortisol. So long that the living system, or individual, remains in an energy robust healthy state, the energy fluxes and quantitative behavior of circadian systems will be predictable and unchanged over time. Thus, these systems lend themselves to reductionist logic and analytical thinking for problem solving based on the predictability of their future behavior. Equilibrium systems that maintain function within a narrow physiological range, for example, pH, also apply to this same notion of steady-state time invariance. Physiological parameters of heart rate, blood pressure, and respiratory rate, may also be considered equilibrium systems, however, there are circadian components to them. For example, there are increases in each of these in the early morning. Nonetheless, these parameters that are required for vital organ system function are resistant to change and maintain within a relatively narrow homeostatic range. With that said, within their homeostatic scope these systems maintain vital organ system functions by an element of allostatic capacity (the capacity to maintain stability through allostasis). This is probably best exemplified by the quasi-periodicity pattern of heart rate variability.

This leads to a brief discussion of chaotic systems and their fractal symmetry. The focused intent here is to distinguish the nature of the systems from the time invariance symmetry of biological systems, the latter, which are both analogous to the symmetry of the first law of thermodynamics and may be exploited by traditional reductionist thinking. The fractal self-similarity of a chaotic system dynamically over time is characteristically unpredictable. In fact, the value of these systems to adaptively contributing to the survival of the living system is critically linked to its unpredictability. The unpredictability of systems is due to their interactions, a measure of complexity. Chaotic systems are a type of "simple" complexity because they only consist of few interacting parts. Nonetheless, the output behavior of the system, despite being deterministic (due to the simplicity of determinant component elements), is virtually unpredictable because of those interactions. Conversely, non-simple complex systems, which have many interacting components are nondeterministic because the elements of the systems are not all known, that is, they are indeterminate. However, as a distinction of nonlinear systems, simple, chaotic, and non-simple complex systems are non-predictable. In the short-term, chaotic systems do have a degree of predictability, highlighted by the orderliness of the system's fractal pattern that may be depicted graphically as a single recursively iterating loop on a computer screen of phase space. The single fractal loop illustrated in phase space represents a qualitative steady-state behavior constrained by negative feedback interactions so long that this behavior

is an efficient calibration of energy to maintain homeostasis of the organism. Whereas equilibrium and periodic systems may break their symmetries, this occurs later in the overall breakdown and compromise in energy flow through the organismic systems biology. Much earlier, as initial lines of defense against threats to energy balance and healthy physiology, allostatic adjustments occur. The flow of energy through systems such as the cardiac conduction system is reflected in heart rate variability, a quasiperiodic system, insulin secretory patterns, and other compensatory hormonal and catecholamine changes accompanying chaotic systems highlight such allostatic adjustments. Chaotic systems include the immune system, neuronal firing patterns in the brain, and perhaps cardiac conduction when this quasiperiodic system may undergo a phase transition to chaotic behavior, which may be either an adaptive or a pathological phase transition. Neuronal firing patterns in the brain, for example, are considered to be an adaptive healthy phenomenon when displaying chaotic behavior whereas in the case of cardiac conduction may be life threatening. Many nonlinear complex systems are nondeterministic and hence not chaotic systems *per se*. Many adjustments in hormonal secretory patterns including pancreatic insulin secretion may reflect this type of allostatic response. In the case of chaotic systems that are engaged as allostatic means of maintaining homeostasis, the single fractal loop representing the baseline steady-state attractor, a simple attractor, breaks symmetry forming multiple chaotic-appearing loops, now referred to as a strange attractor. Despite the overt chaotic appearance of the computerized graphical representation of a system, the continued fractal nature of the attractor underscores the orderliness behind the apparent disorder. By distinction, the fractal symmetry of the system has not been breached. Rather, symmetry in terms of the linear periodic dynamic of the simple attractor was broken as the system evolved to a strange attractor in search of the most energy-efficient steady state for maintaining homeostasis of the larger system (in the case of this discussion, the individual). It is the nonlinearity of chaotic systems in this context that would escape the success of reductionist thinking for intervention and problem solving in research and clinical medicine. Importantly, the orderliness imposed by fractal symmetry requires high-quality informational energy, that is, robust availability of free energy that is available to do the work required of systems biology in times of stress. When the fractal composition of a strange attractor breaks down, that means there is no longer adequate energy to sustain the process integral to the search for solutions capable of maintaining overall homeostasis. Accordingly, the system of highly ordered apparent chaos deteriorates to disordered true chaos. Unfortunately, this qualitative state of infirmity, catalyzed by an imbalance of positive to negative feedback, becomes an increasingly feedforward process of widespread interconnected disease states.

True maximum entropy demonstrated on a computer screen of phase space, a mathematical abstraction of all the possible places energy would in principle be given by unbounded data points occupying the entirety of the screen. Conversely, a chaotic system would show the appearance of chaos but not occupy the entire screen. It would be bounded to a region on the screen. This may be referred to as constrained randomness. Furthermore, careful inspection at increasing magnifications reveals highly ordered patterns of self-similar fractals. The notions that *information of biology is the product of the interface of energy and chemistry and that the level of information is largely gauged by the number of ways building block component parts of systems can be reconfigured adaptively into system wholes underscores the critical value of chaotic systems and organismic biology.* Bits of information that are highly ordered may have inherent meaning such as a given pattern of vowel and consonant letters as a word, however, this meaning is limited if the letters cannot be reconfigured in various ways to provide different meanings. This is analogous to the periodic predictable patterns nature provides as certain biological systems. These stable systems have steady-state basins of attraction that are energy efficient representing states of homeostasis. The greater the capacity, the greater the dynamic flow through the biological chemistry of systems that is capable of adaptively and allostatically reconfiguring that chemistry into systems with new steady states. This has value for maintaining homeostatic systems at their baseline steady states and hence organism survival. The greater the capacity for this, the greater the information content of that system. The versatility in meaning of any biological system, like the capacity to shuffle a string of consonants omitting vowels to make new words with different meanings (that provide communication value in different contexts), parallels the amount of inter-convertible informational components of energy and chemistry that provide survival equivalents. The omission of vowels in a word provides on the surface the appearance of randomness or entropy analogous to the constrained randomness that one sees as the representation of a chaotic system in phase space on a computer screen. The loss of this apparent disorderliness may be a sign of pathology such as in the case of neuronal firing patterns in the brain during a seizure. In this case an EEG goes from the normal pattern of remarkable irregularity to one of periodicity/regularity and predictability.

Any allostatic system, such as the nonlinear complexity of insulin secretory pattern, as described above, highlights initially an adaptive value. *That allostatic behavior, like nonlinear chaotic behaviors, requires the availability of energy input coupled to the chemistry substrates to drive the systems even further from their baseline equilibrium steady states.* The lack of informational energy is responsible for the decline in the structure and function of biological systems. For example, the conformational folding of proinsulin and the breaking of disulfide bonds in the secretory granules of pancreatic b cells becomes impaired due to insufficient energy flow through the system. The lack of energy flow may be due to, for example, energy diverted to the high energy process of auto-inflammation in the region and systemically in addition to unavailable energy deposited in adipose reserves due to poor quality energy intake (lack of phytonutrients, minerals, and vitamins necessary for antioxidant and anti-inflammatory responses, and enzyme cofactors). Consequently, relative insulinopenia results and type 2 diabetes ensues. Thus, *allostatic nonlinear systems are critical to the maintenance of linear homeostatic biological systems and ultimately the survival of*

an individual. Nonlinear dynamical systems require a high degree of informational energy and its sine qua non *quality of unpredictability and apparent randomness.*

4.23 Distinctions Between Homeostasis, Dynamic Equilibrium, and Steady States

Equilibrium of a chemical system is established when the system is most stable, which occurs at its lowest free energy state. Net free energy neither enters nor escapes the system. Over time, the difference in entropy between the system and the external environment tends to disappear. Equilibrium applies typically to a chemical equilibrium of ionization.

A dynamic equilibrium is different from a reversible chemical reaction, which represents the chemical equilibrium of an inorganic system. A dynamic equilibrium is exemplified by the concepts of homeostasis as well as a steady state. Homeostasis refers to the entire internal environment. Steady states may refer to specific mechanisms. A cell is in homeostasis because every mechanism that keeps it alive is in a steady state. For example, sodium/potassium ATPase uses energy from adenosine triphosphate (ATP) hydrolysis to maintain the ion pump. This maintains the internal potassium ion concentration of a cell in a normal steady state. A steady state requires free energy to be continuously put into the system. The conditions of a steady state are stable. Finally, over time, the system is maintained at a higher state of order than its surroundings.

How nature works is how we should seek to understand nature. *Fundamental to life is to find symmetries characteristic of periodic and equilibrium systems that maintain homeostasis of vital organ system functions.* However, nonlinear behavior requires breaking these symmetries from the most energy-efficient adaptive perspectives of allostatic solutions. A greater number of perspectives provide a more extensive search for solutions in terms of the flow of energy through pathways and networks of systems. The greater the robustness of this capacity the more malleable or flexible and hence successful nature is in achieving these solutions. This requires a high degree of informational energy enlisting principles of physics and chemistry into systems biology. *Transformation of information into valuable adaptive self-organizing systems that translate into improved survival appears to be nature's strategy.* Analogously, transformation of information of systems biology into deeper meaning guided by an increasing number of perspectives for understanding it is the compass for this exposition.

4.24 Classes of Systems: Man-Made and Biological

Machines are man-made systems, which may be simple or complicated, however in either case they are linear and predictable. Natural systems are often chaotic fractal in nature. The parallel concepts of information and entropy, ideas for mathematical physics, shed light on complex living systems.

A migratory flock of a cloud of thousands of starlings maintains beautiful fluid formations in the absence of a conductor. Each individual bird does not see the big picture or watch the movement of the entire flock as a whole. The model that best accounts for this behavior assumes maximal entropy of the entire flock. That is, as a statistical group, the behavior of the individual birds is as random as possible. In contrast to a model of lower entropy whereby the birds fly in a relatively fixed relationship to most or all of the other birds (the more fixed that relationship amongst the birds, the less flexible their potential for rearranging themselves relative to one another and hence the less their entropy as a group). The birds are highly fluid whereby each bird in the flock matches the flying pattern of a surprising few other neighboring birds flying on average equidistant from one another. The homogeneous dispersion between the locally interacting birds (and in turn an entire flock of birds) holds true whether the distance between the birds is mere inches in the case of a vast black cloud of birds, or farther apart in the case of a thinner formation of a smaller flock. Accordingly, there is an amazing number of ways the birds can rearrange themselves to form the flock. Analogous to the entropy of a thermodynamic system whereby it is impossible to predict where in a gas system of atoms the position of a particular atom may be, the same can be said of an individual bird within the flock. What is predictable, however, is that the birds' flock as a whole will form a model of maximal entropy because this is the interaction of the component parts that are most nimble at forming the flock. This exemplifies how the notion of *constrained or bounded randomness (maximal information or entropy short of unfettered randomness) optimizes the chances of success for the system.* This concept of information as entropy also fits the model of chaos theory, another model for complex systems displaying the phenomenon of emergence.

The information of birds flying in a flock with maximal entropy formation recognized on a computer model is only part of creating an effective model that can predict its behavior as a whole. A predictive model also requires understanding the rules of local interactions between the neighboring birds, the component parts of a system, responsible for the emergent behavior. Similarly, cognitive information is a string of bits that by itself is not enough to solve problems. It requires thinking that understands how the bits, or parts, relate to one another to create the greater whole. Further, these perspectives allow an understanding of how systems in nature form solutions. This promotes a capacity to reproduce these solutions and to resolve barriers that lead to their breakdown.

4.25 Application of Molecular Biology of Insulin Resistance and Type 2 Diabetes to Clinical Enigmas

In this section we provide examples illustrating the uncertainty of a mechanistic reductionist model of using molecular biology to understanding clinical outcomes. The study of insulin signaling from the work of Ron Kahn's group at Mass General Hospital has noted over 1,700 different insulin-signaling pathways [59]. This highlights complexity at the molecular scale. An example of reductionism in medicine uses the mechanistic example of molecular biology whereby insulin resistance is associated with down-regulation of the insulin receptor (a tyrosine kinase receptor) such that the signaling cascade that ultimately promotes a translocation of GLUT4, the glucose transport molecule, to the cell membrane responsible for promoting the uptake of glucose into the cell is impaired. Classical treatments for type 2 diabetes include the use of sulfonylureas to promote the secretion of endogenous insulin from the pancreatic b cells. Alternatively, the use of exogenous insulin injections is widely used. In either case whether endogenous insulin or exogenously administered insulin the idea is to increase the saturation of insulin receptors such as in the muscle so to ultimately increase the translocation of glucose transport molecules to the cell membrane surface driving glucose into the cell, hence lowering the circulating blood glucose and improving metabolic control. However, there are many unexplained nuances to both orally administered sulfonylurea agents and exogenously administered insulin.

Sulfonylurea therapy was the mainstay and only oral treatment for type 2 diabetes in the 1970s. Even in the 1990s it was routine that patients in coronary care units were being prescribed sulfonylurea therapy to treat their high blood sugar. However, the University Group Diabetes Program in the 1970s demonstrated increased incidence and severity of heart attacks due to these agents. Although these agents were effective at lowering the circulating blood glucose, they also bind to receptors on the myocardial cells that block potassium channels and promote the uptake of calcium into the cell, hence potentiating myocardial contraction. This accentuated myocardial contractility worsens the ratio of myocardial supply and demand (for oxygen) hence driving a higher incidence and severity of myocardial infarction. This phenomenon is referred to as ischemic preconditioning. It is analogous to transient ischemic attacks as premonitory events to strokes. In the case of the heart, episodes of angina are associated with the opening of potassium channels and closing of calcium channels as a protective mechanism that gets abrogated in a setting of sulfonylurea therapy. So, this represents an example of where mechanistic reductionism at the molecular level cannot always be used as a safe predictor of outcome in emergent systems whereby the complexity at both the molecular and even more so the higher scale of cellular biology (hence complicated complexity across hierarchical scales) cannot be predicted.

The predictability of outcomes using exogenous insulin is similarly plagued by clinical uncertainty in terms of potentially pernicious effects of cancer and even sudden death. Again, while insulin therapy exogenously administered whether subcutaneously or intravenously, often effectively improves metabolic control in terms of circulating blood glucose levels other unanticipated effects occur. Not only does insulin help to bring glucose into the cell (by both saturating insulin receptors as well as up-regulating the insulin receptors as a result of the improved or lower blood glucose levels), but it brings the end product of glucose metabolism of the metabolic pathway of glycolysis that occurs in the cytoplasm of the cell, pyruvate, into the mitochondria capable of more efficient energy production. That is, glucose taken into the cell is metabolized glycolytically to the end-product of pyruvate. Pyruvate is taken into the mitochondria by the enzyme complex pyruvate dehydrogenase complex. Insulin promotes the dephosphorylation of pyruvate dehydrogenase complex, which activates the enzyme promoting the uptake of the decarboxylated form of pyruvate, acetyl CoA, into the mitochondria (see Figure 7.15 in Chapter 7 of Volume 2). The acetyl CoA combines with another molecule inside the mitochondria called oxaloacetate to produce citrate. This is the beginning of the citric acid cycle or TCA cycle within the mitochondria which feed into the process of oxidative phosphorylation and the ultimate production of anywhere from 32 to 36 molecules of ATP per molecule of glucose. This is a very efficient form of energy production of ATP in contrast to only two net molecules of ATP produced outside the mitochondria enlisting only the process of glycolysis.

However, *a fundamental pathogenesis of diabetes and insulin resistance is the loss of mitochondrial structure and function*. Therefore, rather than the typical 1,000 mitochondria which exist per cell in the absence of insulin resistance or diabetes, there are far fewer mitochondria functioning normally. Hence, the uptake of the breakdown products of glucose into the mitochondria in the setting of insulin therapy and high circulating glucose levels means an extraordinary burden imposed on the limited mitochondrial capacity. Accordingly, high levels of electron leakage occur leading to the formation of superoxide-free radical species. This imposed oxidative stress damages mitochondrial DNA as well as its structural proteins and lipids further impairing the structural and functional capacity of mitochondria. The implications here are numerous.

In terms of cancer, the competition between cancer cells, which are always present but are typically outcompeted by native cells, is actually given a distinct advantage of outcompeting the native cells. Cancer cells are generally non-functioning (the less differentiated the less functioning) and whose primary goal is for cell replication. It requires more energy for highly functioning cells so less differentiated cancer cells are less dependent on mitochondria for their energy requirements. However, because they are reliant on the significantly less efficient process of glycolysis the uptake of glucose in the setting of hyperglycemia is advantageous to their goal of self-replication. The preferential use of the pathway of glycolysis over oxidative metabolism inside mitochondria by cancer cells is called the Warburg Effect. Although insulin does not transform cells, it is a growth promoter working through signaling pathways, which promote the proliferation of existing cancer cells. Estrogen in fact has its growth-promoting effects of estrogen-dependent tumors rooted in its cross-reacting through insulin signaling pathways. In any event, increased oxidative stress and inflammation potentiate the effects of gene mutation and cancer cell initiation while the hyperinsulinemia promotes growth promotion of cancer cells through insulin signaling cascades which are extra-mitochondrial. This relationship of exogenous insulin to cancer cell promotion is not only plausible by the above brief molecular and cell biological mechanisms but epidemiological analyses for this relationship are quite strong. The very few interventional studies that negate the relationship between cancer and exogenously administered

insulin are both limited and flawed. Overall, the purpose of highlighting this important potential relationship is to underscore *unrecognized and unpredictable clinical consequences at the organism level (in this case humans) of medical mainstay Holy Grail management of disease states based on the extrapolation and hence reductionist application of isolated molecular level mechanisms across the hierarchical scales of the organism as a whole*.

A final example of an intriguing and totally unanticipated finding is demonstrated by the international NICE SUGAR study [60] that demonstrated increased mortality among intensive care unit patients treated with intravenous insulin with associated tight glucose control but even in the absence of hypoglycemia. This outcome was totally surprising and has led to an entirely new standard of care in terms of targeted glucose levels in critically ill patients on intravenous insulin. This is another powerful example of how molecular level reductionist mechanistic philosophy applied to standards of care in medicine is not only limited but this form of empirical therapy can be a dangerous game costing human lives. This unpredicted outcome of mortality may relate to the sudden uptake of a high burden of glucose into the cells and ultimately imposing an enhanced burden to the mitochondria resulting in a deleterious loss of much of the remaining mitochondrial structural and functioning capacity. Although glycemia is controlled in the acute intensive care unit setting, its destructive effect at the mitochondrial level and resulting oxidative stress extends to the system's biology level whereby impaired mitochondrial capacity in the autonomic nervous system consequently resulting in a sympathetic dominance and pro-arrhythmogenic potential leading to sudden death. This is a speculation as to a plausible mechanistic explanation to the unfortunate potential outcome of intravenous insulin uncovered by the NICE SUGAR study [60]. However, it is plausible and underscores the implications that *reductionism applied to biological systems in medicine has limitations and can be potentially dangerous*.

Some other excellent examples of this include some unexpected outcomes involving other hormonal interventions such as growth hormone in the intensive care unit setting leading to increased mortality demonstrated by Gert Van der Berg's group in Copenhagen [25].

4.26　Integrated Complexity of Systems Biology into an Optimally Functioning Whole

The superposition of an electron and photon of energy and entanglement of these wave particles across cells, tissues, and organ systems throughout the body underpins instantaneous and simultaneous transfer of energy into the work of biological functions synchronically as an organismic whole. The mechanism for how this energy is transferred may be independent of the universal biological energy mediated by ATP.

The transfer of electrons extracted from the breakdown components of fatty acids and glucose entering the mitochondrial TCA cycle to the electron transport chain is responsible for this phenomenon. Traditional molecular biochemistry describes how the transfer of electrons from these metabolized nutrients to the electron transport chain along the inner mitochondrial membrane leads to the subsequent transfer of electrons along an electron gradient produced by the coupling of redox cycling enzymes. The final acceptor of electrons is oxygen, which combines with hydrogen protons to form water molecules. The energy made available in this process is captured in the high-energy phosphate bonds with the conversion of ATP from ADP (adenosine siphosphate). However, independent of this molecular biological currency of energy in the form of ATP there is a quantum mechanical phenomenon that occurs along the electron transport chain. The transfer of electrons exhibits a tunneling phenomenon, a characteristic quantum mechanical effect. The other important characteristics of quantum mechanics are the phenomena of superposition, entanglement, and delocalization, which are typically interdependent processes. The duality properties of electron wave-particles highlight the notion of superposition. That is, the electron is both a particle and a wave simultaneously. The electron transfer along the electron transport chain (ETC) has plausibly been considered to exhibit the superposition phenomenon of wave-particle duality. The quantum photons in this case have been referred to as bio-photons. The delocalized phenomenon of quantum mechanics may be exhibited by the process of entanglement. Entanglement refers to a communication between electrons of opposite spin that may occur simultaneously at different places at the same time (superposition of entangled particles) and may be separated by considerable distance at the same time (delocalized superposition of entangled particles).

The energy imparted by this superposition of electron wave duality in a delocalized fashion across cells, tissue, and organ systems of the body accordingly escapes the constraints of space and time. In the following we expand on the potential consequences of this phenomenon for biology. In the absence of spatial constraints, time is maximally dilated. In the context of biological structures and function, spatial constraints are relatively minimized, and as such biological time moves more slowly. In this case, optimal biological complexity and maximal informational energy (the interface of energy with the body's chemistry) exist across the organization of systems biology. In this case, there is maximal and optimal physical, cognitive, emotional, and psychological health and capacity of the body. In order for such quantum phenomena to persist, it requires the virtual absence of noise in the system. In the case of a quantum computer, the slightest amount of heat in the system results in quantum de-coherence. In the case of biological systems, heat refers to free energy that cannot be captured as work or harnessed for useful purposes. Such heat dissipates as entropy and can lead to a collapse or decoherence of a quantum state.

> *With advancing age and with disease, oxidative stress and the inextricable process of inflammation become a feed-forward (that is, positive feedback) process and functions as noise that prevents the capacity for delocalized biological function at increasingly more constrained regions of space of systems biology.*

The integrated complexity of structure and function of an organism becomes increasingly compartmentalized and compromised. Accordingly, as the flow of energy within a given tissue or organ system becomes progressively spatially constrained and in tandem time constrained, resulting in accelerated biological aging.

The classical counterpart to the quantum photon is the electromagnetic wave. The movement of electrons along the electron transport chain is an electrical current that creates an electromagnetic field. Metaphorical to the fuel cell analogy, the entrainment of electron transfer along electron transport chains across mitochondria of the same cell, across cells, and even across tissue types underscores a synchronization of bio-energetics and potentially as an organismic whole analogous to the electromagnetic phenomenon responsible for the coordinated motion of an electrical car.

4.27 Systems Theory: A Perspective

The reductionist model has been the traditional establishment way of scientific thinking since the onset of the Newtonian era in the late 1600s. The basic premise of reductionist thought is the predictability of cause-and-effect interactions between objects based on fundamental laws of those interactions. These objects are the determinant elements or variables in the system. The outcome of the system is linear, deterministic, and predictable. Characteristically this model assumes that complex phenomena can be broken down into simple parts and reassembled to reconstruct the whole. These features are in counter-distinction to systems theory that critically emphasizes the whole as being greater than the sum of the component parts of the system.

> *The interactions between constituent components of systems models are guided by simple rules and patterns analogous to that of physical laws and equations that describe reductionist systems. What are unique to systems theory are the qualities of nonlinearity and non-predictability underlying emergent patterns.*

Inherent in this distinction is that in contrast to reductionist systems there is an inability to fully understand the intricacies of the determinant elements of a system and their interactions. Nonetheless, simple rules have predictive power, independent of having a full knowledge of the mechanistic underpinnings, the latter that is fundamental to reductionist models. It was the theory of relativity published in 1905 (special theory) and 1916 (general theory) (61) that exposed an apparent flaw in the mechanical reductionist view. The predictions of quantum physics and relativity showed that some of the most basic assumptions about time, space, and causality of reductionism are incorrect. This void catalyzed the evolution of theories and understanding of nonlinear systems. Network theory recognizes complex systems in terms of how energy flows through interconnecting pathways of the networks of systems. Complex adaptive systems are the product of patterns of self-organization and rules that guide the interactions of physical and chemical processes. A lot of systems theory comes out of the area of computation derived from computer science, engineering, and information theory.

Information theory is the branch of computer science that involves the quantification of information developed to find the fundamental limits of signal processing operations involving the reliable storage and communication of data. Concepts of information theory include a "message", which conveys a quantifiable amount of information and a "channel", which conveys a quantifiable information capacity.

These include the notions of data compression and robustness. For example, the more common a word is, the fewer are the bits required to store or communicate it. Information entropy applies the concept of entropy to information theory.

The greater the average number of bits needed to store or communicate one symbol in a message the greater the entropy of the information, that is the greater the quantity of uncertainty.

An infrequently used word requiring more bits to store or communicate it hence has less information entropy or more uncertainty in its prediction. Analogously, a coin flip has a predictability of 50% versus a die roll, which has a predictability of only slightly greater than 15%. Thus, the die roll would require more bits if this example were extrapolated to the frequency of how often a given word is used. Conceptually fundamental to information theory is that maximum entropy, symbolized by a string of random letters, contains no information. Likewise, a strain of identical letters also contains no information. The former scenario has no predictive capacity in contrast to the latter scenario, which has 100% predictive capacity and zero entropy. Maximum information is stored and communicated midway between maximum and zero entropy. Claude Shannon introduced the concept of information entropy in 1948 [43]. He is considered the founding father of the electronic communication age.

The Shannon Index is a measure of biodiversity which is a popular measure of entropy, broadening the analogy between thermodynamic and information entropy. Therefore, the Shannon Index is a mathematical analogy to the second law of thermodynamics. Accordingly, the greater the number of categories and the more equal the number of individuals in each category the greater the biodiversity, hence the mathematical equivalency of entropy. In the most diverse ecosystem, the rainforest, it is unpredictable which species of plants we will find next as we walk across the rainforest. Theoretical physicist Roderick Dewar used information theory to reformulate the laws of thermodynamics showing that *maximum entropy production is the most likely behavior of an open complex non-equilibrium system*, for example organismic species wherever conditions of resource availability are unbounded and external pressures are not constrained [62]. The rich biodiversity of a rainforest exemplifies this. This contrasts with the counterpart of classical thermodynamic entropy that is described only for closed or isolated systems (energy cannot enter or exit to or from the system). Hence, information theory extends the concept of entropy and the second law of thermodynamics to the natural selection of biological organismic open systems (open to the energy from sunlight).

The reductionist perspective of modern-day biology is epitomized by the mechanistic model of DNA coding for proteins. Francis Crick famously quoted, "The ultimate aim of biology is to be explained in terms of chemistry and physics" [63]. This view is consistent with the extension of a concept of thermodynamic entropy to biology captured by the term evolutionary entropy coined by Lloyd Demetrius [64]. Demetrius equates organisms to the molecules of a gas whereby the quantity evolutionary entropy is the mathematical equivalent of thermodynamic entropy presenting another interdisciplinary mathematical application of entropy. Evolutionary entropy, rather than representing physical randomness as is the case in thermodynamic entropy, describes the randomness in terms of the age range over which organisms reproduce. Over long periods of evolution, Demetrius predicts that *natural selection increases evolutionary entropy because organisms that reproduce over long periods of time are more adaptable environments and limited resources.* His prediction is analogously consistent with Dewar's finding using information theory that maximum entropy production is the most likely behavior of an, also analogously, open complex non-equilibrium system such as populations of organismic species. Similar to Dewar's findings, Demetrius's predictions also appear to hold in the absence of bounded conditions of resource availability. He nonetheless argues that albeit analogous there are fundamental differences between the entropy of natural selection to that of thermodynamic entropy. He recognizes that the evolutionary pressures beyond the molecular level of biological systems are very different from that of simple physical systems such as the addition of heat to a gas in a closed system. He states, "As you go from molecules to cells and higher organisms, (natural) selection involves replication, and there is no replication in physics. It is what distinguishes the living from the dead".

An important distinction between reductionism of simple linear systems from that of nonlinear ones such as chaotic or complex adaptive systems is that in the latter cases, they have a purpose with the directed goal of avoiding extinction. While it is farfetched that all biology would not be explainable by the underlying principles of chemistry and physics, the linearity of that relationship in terms of it being a simple application is just as farfetched. The predictability or likelihood of a physical gas to progress towards maximal dynamic entropy is 100%. However, the likelihood of predicting the position or velocity of a given particle or element within the gas is infinitely negligible. Furthermore, the essential distinction between linear systems, modeled by reductionism in nonlinear systems, is the lack of predictability of the latter. One particular irony to the analogous applications of the concept of entropy to biology described by information theory and natural selection (information entropy and evolutionary entropy, respectively) is the aim of thermodynamic entropy is an equilibrium state incompatible with life. Systems at thermodynamic equilibrium with the surroundings represent a state characterized by the lowest energy gradient and maximum useless energy. Conversely, *the unpredictability of the natural selection process creates the best opportunity for survival, while the equilibrium state of maximum biodiversity showcases the greatest distribution of survival underpinned by the concepts of both information entropy and evolutionary entropy.*

It was in the 1940s that marks the origin of systems theory proposed by biologist Ludwig von Bertalanffy [65]. The emphasis was that real systems are open to and interact with their environments and evolve qualitatively new emergent collective properties in counter-distinction to the compartmentalized concepts of reductionism. Cybernetics formed the foundation of complexity theory.

> *Cybernetics is a trans-disciplinary approach for exploring regulatory systems of how people, animals, and machines control and communicate information.*

For example, it compares the automatic control systems of natural selection and the brain with the electrical mechanical communications systems of machines. It focuses on circular feedback mechanisms of complex systems that are goal oriented. This was a revolutionary contrast to the linear mathematical and deterministic models of traditional Newtonian science. In the mid to late 1900s, the study of control systems underpinned much of the theoretical background to modern computing. Cybernetics grew out of information theory, with counterpart major contributions to the science of complexity.

4.28 Chaos Theory

Chaos theory is a part of cybernetics because chaotic systems that generate order do so because of the positive feed-forward energy driving the system and the negative feedback loops that tame the system. Together the feedback mechanisms result in the organized collective activity. Chaos theory is a form of cybernetic theory of regulatory systems that specifies how order emerges from chaos [66]. In contrast to the common implication of "chaos" meaning "a state of disorder", in chaos theory it is mathematically defined by sensitivity to initial conditions as well as the property of topological mixing. The latter means that the system will evolve over time so that any given region of measurements of the system's variables will overlap (or mix) with any other given region. In a weather system an example is the mixing of parameters or air and water molecules. Another example may be the mixing of food dye in a glass or water. Chaotic systems are predictable for a while, i.e. in the short-term, before becoming apparently random. Edward Lorenz, the father of Chaos Theory, in 1963 [46, 66], when performing weather calculations, rounded off one of the parameters as he entered it into a computer from 0.506127 to 0.506. As a result of the rounding off, running the program a second time did not reproduce the results. The uncertainty in a weather forecast increases exponentially with time; doubling the forecast time more than squares the proportional uncertainty in the forecast. Characteristically, chaotic systems are deterministic and goal-oriented although they are unpredictable. The apparent randomness and disorder of these systems are deceptive as order actually emerges from such chaos. *Chaotic behavior is driven by uncontrolled bursts of energy in a given region. Such regions are vortices in which information due to the interactions between component elements that flow through a system highlights a pattern of unpredictability.*

Vortex indicates a process such as a hurricane or tornado in which there is a strong circular motion of air or water. This represents a metaphor for chaotic behavior whereby things are spinning out of control. However, the chaotic behavior of chaos theory involves the motion of search for stability rather than order or disorder per se. Ultimately, however, the dynamical interactions over time in pursuit of optimal solutions result in the optimal state of stability. A hurricane weather pattern created by the Butterfly Effect represents the stability of a system rather than the system "out of control. The metaphorical relationship of chaos to vortices may be extended to the interaction of parts of any physical or chemical system whereby the emergent properties of water arise out of chaos. In addition to weather patterns such as a hurricane representing a vortex, galaxies by the same criteria or vortices as are organisms such as human beings. In contrast to the apparently random nature of chaotic behavior, the priority for seeking stability, as in self-organizing complex organisms, is underscored by its precedence for creating a negative change in local entropy. The chemical species that flow through biological systems exemplify order created from chaos whereby every cell of the body is a living vortex. Hence, the human body has about 50 trillion living vortices, which increases to roughly ten-fold that number if the symbiont microbiota is included. Nature creates life across many scales of complexity including physical chemistry, chemical, molecular, cellular, and macroscopic biology. Norbert Weiner, the father of cybernetics, quoted, "We are nothing but whirlpools in a river of water that flows constantly. We are not habitable substance, but self-perpetuating patterns". Rather than chaos being the traditional interpretation of randomness and disorder, chaotic behavior of chaos theory highlights the capacity of systems to generate sublime order.

Chaos theory began to open up a new world of complexity science as an emerging post-Newtonian paradigm, hence extending the Newtonian framework into the world of nonlinear systems. Like special and general relativity that came before it, chaos theory further proved that some of the basic assumptions of Newtonian theory are flawed. The most central and critical of such faulty assumptions are that systems are linear such that the whole is no greater than the sum of its parts.

4.29 Complicated Systems and Complex Systems

Some important distinctions about the commonly misunderstood terms simple, complicated, and complex warrant clarification. *Simple is easy and knowable. Complicated is not easy but knowable. Complex is not fully knowable but is reasonably predictable. A system is complicated in the absence of complexity when there is a high level of difficulty in terms of problem solving. The complicated nature of a system is independent of the number of interactive component parts.* Computing power is useful in solving complicated problems when the determinate nodes (component parts, elements, or agents) are known. There are typically many component parts in a complicated system but the parts are dissimilar in contrast to complex systems in which the interconnecting parts are similar. Complicated is difficult to understand because it has a lot of different parts and detail. Problems involving complicated systems are solvable when the component parts can all be separated and coordinated back together to reconstruct the whole, for example a clock. *Complicated systems are engineered, fully predictable, and understandable by analyzing the components of a deconstructed system.* It often requires specialized expertise. An example of a complicated problem requiring high levels of expertise in a variety of fields is sending a rocket to the moon. Once a complicated problem is solved, such as successfully landing a rocket on the moon, the mission can typically be reliably duplicated. A critical feature is the lack of emergent behavior. Computational simulations analyzed designs of physical processes before committing to a constructed product such as a rocket. Similarly, computer simulations play out the implications of altering or adjusting the engineering design of for example the tilting control by the flow of air over an airplane wing. Formulas are critical and necessary for managing complicated systems. For example, Newton's second law, F = ma can be employed to solve an unknown if two of the three variables of the equation are known. If a car that weighs 1,000 kilograms runs out of gas and we push it with an acceleration of one-half meter per second squared, the force applied to the car can be readily calculated. Reconfiguring the equation as force F = Δ(MV)/Δt, the change (Δ) in momentum, p = MV, of a body is directly proportional to the acting force (F) and inversely proportional to its mass. It is equations such as these that are necessary to solve truly complicated problems.

Complexity of a system refers to many similar interacting parts with the hallmark feature of nonlinearity. That is again, the whole is more than the simple sum of its parts. Accordingly, we cannot separate the parts from the whole because the interactions of the parts create the emergent behavior that cannot be understood by analyzing the individual components. For example, we cannot understand the flocking behavior of migratory birds by studying the individual birds independent of the flock. A contrast to an engineered complicated system, complex adaptive systems, cannot be constructed with predictable and reproducible outcomes. Complex adaptive systems are adaptive in that they have the capacity to change and learn from experience. Examples of complex adaptive systems include living organisms, especially human beings, immune systems, and ecosystems. The networks of interactions of a complex adaptive system result in rich collected behavior that feeds back onto the individual parts. Complex adaptive systems are reasonably but not precisely predictable because they are typically goal-oriented and follow certain simple rules. We can achieve a level of understanding by studying how the whole system operates. Nonetheless, predictability is often the exception rather than the rule.

Understanding the distinction between complicated and complex is crucial to successfully manage problems. In healthcare, treating the systems of a disease process as complicated that is often complex in nature, tries to influence it mechanistically. This again highlights the flaws of the Newtonian era enlisting approaches of problem-solving treatments that rely on blueprints and predictability rather than recognizing the inherent indeterminacy of not knowing all the elements and their interactions. Further, by not appreciating the process as one that is complex, and hence not understanding how the whole system operates, predictability is even lower. In other words, we are treating by a blueprint that does not match what we are treating. Furthermore, we are in effect treating the pathology of the complex system blindly because not only is it not fully knowable, but unpredictable as well. This is commonly, unfortunately, so the case in clinical healthcare enlisting mechanistic paradigms more suitable for fixing machines than for problem solving of chronic disease. Alternatively, if we achieve an understanding by studying how the system operates as a whole, we can more effectually implement well thought out and constructive interventions. Rather than working from a set plan, management of complex systems requires the adaptive nature of management just as the complex systems themselves, as complex adaptive systems, by necessity, learn from experience to maintain adaptability. For example, working from a range of small innovative empirical interventions with continuous evaluation and adjustments, re-evaluations and readjustments is an art of science. In counter-distinction to intervening decision making for complicated systems, for which there is often a best choice, for complex systems it involves a collective interpretation. Unlike complicated systems, which are often known with a clear set of guidelines to both follow and instruct, complex systems entail the tasks of learning from the responses to empirical interventions while simultaneously planning the next step and building on what works according to emergent directions of the system. *The task of managing complex adaptive systems requires relationship building, working with patterns of interactions. In contrast to looking for interventions that will fix a broken or pathologic complicated system, the goal for managing complex systems is more appropriate to find leverage points that can be adjusted to improve the system.*

> *One-size-fits-all approaches are rarely effective in complex adaptive systems and solutions are highly context-dependent. Modeling complex adaptive systems applied to managing treatable but incurable chronic disease states is an ongoing process that is in a sense an art form, often engaging creativity, open discussions and collaborations while carefully observing for clinical response, tolerance, and adverse effects.*

Molecular biology-derived pharmaceutical interventions have been suggested to provoke cancer signals with an increasing number of agents. This has been the cause of some well-deserved anxiety in the medical community amongst patients and prescribing physicians alike. Examples include the GLP-1 agonist drugs with regard to pancreatic cancer, and the osteoporosis drug teriparatide linked to sarcomas. How may this be mediated? It is an unanticipated consequence of targeting a lower hierarchical scale of science (i.e. molecular biology) coupled with insufficient understanding of the complexity of parameters (interacting nodes or components) and may result in the detached regulatory control of cell division. Complicated complexity and uncertainty are inextricably intertwined in the study of systems biology. Multi- and interdisciplinary efforts to expanding the armamentarium of top-down and bottom-up non-reductionist (in the sense of mechanistic and predictable) approaches to finding diagnostic and therapeutic solutions for human chronic disease states with maximally efficient reduction of risk profiles indeed deserves center stage attention.

Einstein stated that, "We cannot solve problems by the same kind of thinking that created them". It is quite possible that he was referring to our tendency to view problems in isolation analogous to the oversimplification of reductionist thinking. Treating chickens and farm animals with subclinical doses of antibiotics to make them fatter overlooks the very real potential for detrimental effects to human health. For example, antibiotic resistance, especially to quinolones results in antibiotic-resistance microbes such as salmonella harbored by the animals that humans later ingest. Similarly, blue-dye-containing arsenic as a growth promoter given to chickens contributes to the potential for arsenic poisoning those who eat the chicken. This is especially true if the individual has genetic polymorphisms of the relevant bio-detoxification enzymes in the liver. Genetic modification of widely consumed crops of fruits and vegetables is considered to contribute to the rising autoimmunity and autoinflammatory conditions including gluten intolerance, another example of the over-simplicity applicable to Einstein's statement. Genetically modified crops may be associated with gene transfer of the genetically modified plant genes into the human body or bacteria in the human body. This consequently may promote adverse effects such as antibiotic-resistant genes transferred to disease-causing microbes in the gut of humans or animals consuming genetically modified foods. *E. coli* is a microbe in the gut of cows that consume the crops and subsequently, when ingested by humans, causes a devastating illness [67]. This "contributes to the growing health problems of antibiotic resistance" [68]. Genetically modified foods may also be contributing to the obesity and diabetic epidemic, which parallels a similar timeframe. The unanticipated disorders that arise from these practices are exactly analogous to the reductionist mindset that treats individual symptoms or molecular targets of a disease without recognizing the complexity of the body's systems biology. Einstein also was quoted as saying, "Make everything as simple as possible, but not simpler". Reductionist medicine often oversimplifies the behavior of our systems biology. Alternatively, the computer-generated approach of simplifying complexity by finding "simple rules" is typically a non-mechanistic form of reductionism that is often a useful tool for recognizing predictable patterns of complex processes. Many TED talks explain how complexity leads to simplicity. If we want to know the effect of one animal species on another and focus only on that link and blackbox the remaining complexity of interacting influences, we are actually less likely to understand the specific effects with a level of predictability. Rather, it is more productive to step back and look at all the links, and from that place home in on the sphere of influence that matters most within three degrees of influence of the nodes of interest. The more we step back and embrace the complexity, the better the chance we have of finding the simple answers, which are more often accurate. This is a general theoretic framework of physics. In contrast to biologists' approaches to complex systems, which look at complexity and differences, physicists try to look at what's common and extract behaviors from their

commonness. This is compatible with the notion that science is the discovery of symmetries.

Treating a complex problem as if it were complicated with a blueprint using computer simulations needs to be recognized as not only inefficient and suboptimal but misdirected. Computer simulations evaluate known parameters of a system to look for patterns, cycles, and fundamental rules that predict those patterns even if many (even up to 50% [69]) of the parameter values are missing.

Another important model for managing chronic disease states is the institute of functional medicine model that looks at antecedent, triggering, and mediating factors (the ATM model) used to determine the likelihood of a therapeutic intervention with the highest safety profile. This model is an individualized form of patient care and stresses a notion of treating the individual person, the whole system, rather than the component parts of the individual, which sees biological systems as Newtonian. The focus of this ATM model is to treat the whole individual as a top-down approach without getting lost in the above mechanisms. Although it recognizes mechanisms, the understanding of such mechanisms is assimilated in an approach to patient care, which relies on balance more on intuitive insights rather than as systems that can be predicted based on component parts alone. However, it is probably not best to classify this type of model as bottom-up or top-down in a strict sense. Certainly, the creators of this model prioritize the significance of treating the individual rather than its component parts. It does this by focusing on often external antecedent, triggering, and mediating factors driving the genesis of a disease. In this way the ATM model is considered a top-down approach.

It should be possible to inspire not only the multi-disciplinary expertise between medical subspecialists, but interdisciplinary cooperation of physicists, mathematicians, and computer scientists to develop computerized programs in pursuit of solutions of complex disease states to either find the diagnostic "simple rules" to guide therapeutic interventional efficacy in a top-down approach where possible and whereby providing the optimal solutions to problem solving, or alternatively diagnostically stratify and treat in highly individualized fashion in a model such as the ATM model where appropriate size fits all or set of specific instructions on how for example to raise a child. There is no formula or algorithm such as can typically be applied to complicated problem solving. This is certainly also true in managing chronic disease states in contrast to typical acute care management of hospitalized patients. In the latter case for example cardiopulmonary resuscitation protocols are followed in cases of cardiac arrest. It is impossible to know every wire and the gazillion combinations of how the circuitry interacts to understand and predict the emergent behavior and personality of a young child as he becomes a young adult. Computer simulations are now being used to study complex solutions (such as at the Santa Fe Institute) to find salutations for treating complex system disorders. These models are observed to understand and find the patterns and fundamental rules that predict emergent behavior. However, patterns and simple rules of child development have become understood over many hundreds of generations of empirical observation showing that parenting and home values provide the best predictive value for the emergent behavior and personality of the child transitioning into a young adult. Such values hence represent fundamental rules that need not rely on a comprehensive understanding of the unknowable list of nodes and infinite detail of interactions of the regulating network parameters that are mechanistically responsible for outcome behavior.

The parameters of a system at an instant of time describe the system's state. A state often expresses a point in a multi-dimensional geometrical space (phase space) with each axis corresponding to a parameter that mirrors a dimension. The number of parameters defines the number of dimensions of both the space and the system. Each point in phase space represents a way in which the system can be at a given instant. Plotting by computation the geometric configurations of the parameters of a system over time may also be described as a phase portrait. The phase portrait represents a dynamical expression of the system in terms of the integrated parameters in the theoretical dimensions of phase space. The overall health of an individual may include the parameters of blood sugar, heart rate, blood pressure, lung capacity and so on. The composite of all these parameters forms a single point with the dimensionality corresponding again to the number of parameters. The path of the integrated composite point over time is represented by a line that forms recursive iterations forming the system's phase portrait as it traverses through phase space following a trajectory of navigating attractors. Chaotic systems are classically described by computer simulations providing phase portraits. They are characterized by features that include self-similarity and scale free properties, which means that it looks alike at different scales. The other characteristic of chaotic systems is that they are fractal, which is a never-ending pattern of infinite complexity. Together with the features of scale-free self-similarity chaotic systems are driven by repeating simple processes over and over in an ongoing feedback loop with a characteristic geometry driven by recursive iterations along a trajectory that approximates an attractor that is different for every qualitative behavior. Chaotic behavior describes many forms of nature as strategies of the system to survive. There are many examples of chaotic systems in systems biology. Chaotic systems are exemplified by the flocking behavior of migratory birds or the foraging behavior of ants. It is also exemplified by every system of human physiological behavior. Chaotic systems are a form of complex adaptive behavior. In human biology, complex chaotic and complex non-chaotic behaviors are inextricably linked.

The study of the foraging behavior of ants using computer simulation has determined cyclical patterns from which rules are extracted. These rules include ants finding food and drop pheromones along the path back to the colony. Other ants follow the scent of the pheromone and then perpetuate the same pattern of dropping pheromones along the path back to the colony after finding food. As increasing number of ants contribute to increasing the scent of pheromones this serves as a positive feedback mechanism that amplifies the behavior. Notably, positive feedback is an important component to the behavior of chaotic systems. The mechanistic highly context-dependent bottom-up approach, unless the system is limited to three or four component parts of less, is generally too complex

to translate into an ability to find patterns because we become lost in the weeds or to put it differently, we cannot see the forest for the trees. *It becomes essential to stand back and embrace the complexity, without knowing all the intricate details, and observe the patterns from which simple rules may be applied. In other words, solutions are found by so-called simplifying complexity. Clinical medicine standards of care as well as clinical research designs paradoxically are rooted in reductionist models that are not fitting to broad applications. This mismatch of the scientific approach to the intended target population underpins frequent unexpected absence of a beneficial response.* Moreover, outcomes are far too often unforeseeably adverse or even tragic. There are prodigious examples in all fields of medicine, whether it be hospital-, oce-, or clinic-based, seen virtually every day by physicians and other healthcare providers. In the field of clinical endocrinology there are some classical examples of this recently presented by Greet Van den Berghe from Belgium involving the outcomes of a host of clinical research designs.

Examples of emergent behavior are endless and include the origin of life itself and the natural selection of any given species of organism. It includes solutions that are arrived at by cognitive creativity or the hemodynamic response such as blood pressure of an individual in response to the individual facing an environmental challenge. In the case of the foraging ants that ultimately create a trail of scent that becomes concentrated, the initial ants stumbled randomly upon the food source that subsequently paved the way for later ants to coordinate the effort that became increasingly less random. Hence, the random nature of the emerging process actually appears as if it were orchestrated from a centralized control such as a queen aunt laying out a set of instructions. Such a set of instructions represents the blueprint characteristic of reductionism. Computerized input of information data of what is known about the ants includes video of ant colonies and foraging makes it possible to both recognize the behavioral patterns and to synthesize the simple rules. Relevant information compiled and input into computer simulations about the insects that allowed or facilitated this process of simplifying complexity may have been derived from mechanistic studies. In isolation, however, such reductionist modeling cannot be extrapolated to the general behavior. The parts (either the individual insects or components of the insects) cannot contain the whole (the emergent behavior) because the complexity is not present in the elements but rather created by the relationships between the elements. In other words, no element in the system can control the system. We cannot understand group behavior by studying an individual within the group. Such mechanistic pieces to the broader puzzle may for example include food signals messaging to DNA that codes for the process or secreting pheromones; the scent of the pheromones signal messages to DNA that code for the biological coordination of locomotion toward the scent. Context-dependent mechanistic understanding derived from specific bottom-up approaches allows the extraction (from computer simulations) of fundamental patterns that provides bridges to apply to a broader context of top-down approaches. That is, a graphical representation of an upside-down triangle with the vertex at the base illustrates the broader top emerging into the specific behavior at the base. *Reductionist models that start from the bottom are typically too narrow to understand the broader context, however, many contextual simulated perspectives provide or help provide useful understanding to help elucidate the patterns and simple rules of behavior.* Another example of well-documented interactive local rules that explain complex global behavior involves the flocking of migratory birds. In this case, the majority of birds the migrate on social cues can be simulated on the basis of three fundamental rules: separation or short-range repulsion; cohesion or long-range attraction; and alignment or steering towards the average heading of local flockmates. Although it is not known if the birds actually follow these rules in nature, the framework used in computer graphics provides useful realistic-looking representations and complex flocking behavior that determine points of transition to new qualitative behaviors such as when confronted with obstacles or when reacting to an external threat. Mechanistic factors the mediate the interactive local emergent flocking behavior involves magneto-receptors visual and acoustic sensory channels. Chemical communication in birds involving pheromones appears also to be contributory to social behavior, however, the physiological role remains largely undefined as birds are either anosmic (partial or complete loss of the sense of smell) or microsmatic.

The key value to understanding fundamental rules for complex or chaotic systems is that we allow the ability to influence systems empowered by a central command. That is, rather than relying on bottom-up, highly context-dependent strategies that are just as unpredictable as the spontaneous emergent patterns themselves, simpler rules are more capable of predictably influencing the complex global behavior while embracing the temporally dynamic interconnectedness and interdependence of the system of top-down approaches. This is achieved by very many computer simulations that manipulate parameters of a system to find the simple rules that can be used to determine failure points. The computer entry required for computer simulation includes all the knowable components of the system and their interactions, that is as much bottom-up mechanistic data as possible. *The daunting challenges of unraveling the biological complexity of human disease are now entering a new era of metabolic mapping and computer simulations. This model uses personalized genetics, epigenetics, microbiomics, proteomics, lipomics, and metabolomics to provide an exhaustive detail of constituent parts possible to allow the most individualized accuracy of the simulations.*

Chaos theory, also a derivative of cybernetics and information theory, is another important contributor to the genesis of complexity theory. Chaotic systems, like complex ones, are chemical and physical systems that evolve self-organizing and far-from-equilibrium behavior by exporting entropy. These adaptive systems are heterogenous networks of interacting elements. Like complicated and complex systems, it is important to understand the small-scale behavior, the simple or fundamental rules, and the low-level interactions. For example, to best predict the long-range personality qualities of decency of a future adult raised from childhood, it is the small-scale, low-level consistent interactions that instill the proper values. The best long-range perspective in this example is to focus on short-range interactions rather than long-range ones such as, is

he going to go to college one day? and ultimately what will he do for a living?

Chaotic systems are a type of complex system, but with distinct characteristics. Complex non-chaotic systems are characterized by repeated interactions between many slightly heterogenous but similar parts that result in a richness of collective behavior. This emergence that feeds back into the behavior of the individual parts continually shapes asymptotically the emergent properties. Chaotic systems by distinction involve few interacting parts. However, analogous to other complex systems, the interactions have very intricate dynamics. Complex non-chaotic systems theoretically could be represented in the same fashion, however, there are typically too many interacting component parameters and hence dimensions, many of which are often indeterminate, that would form the attractor's trajectory. *A critical quality of chaotic systems is the notion that tiny differences can have consequences that magnify and amplify into a butterfly effect. Chaotic systems start with simple rules from which emerging patterns of complexity evolve.* In fact, it is largely from this discipline of chaotic systems that complexity theory evolved and the significance of finding simple rules of complex systems, i.e. simplifying complexity.

The top-down "blueprint" or "rules" approach to finding solutions for deterministic systems germinates from Reductionism. Such top-down "blueprints" are rooted in Reductionism because mechanistic understanding frames the basis for such a top-down approach. It is considered top-down because it is not directed at any particular component element of a system but rather directed at the system as a whole. However, the mechanistic insights to drive this "blueprint" in the first place require the understanding of the mechanistic relationships between the parts that form the system. That is, bottom-up approaches were required for understanding how the system works. This bottom-up form of analysis of a system is reductionism. That is, based on mechanistic relationships between parts of a whole that can be used to predictably understand the whole system. These systems whose understanding is rooted in Reductionism have an outcome behavior, which is predictable from the component parts that make the system. This predictability may be linear or nonlinear. In either case, the relationship between the parts to the output or outcome behavior may be described as an equation.

A system, as briefly described above, may be defined as simple, complicated, or complex. A complex system involves many interactions between component parts with many constituent components of the system. When these component parts are all the same it is referred to as a simple complex system. When the component parts are different it is described as a complicated complex system. Complicated in this context is different qualitatively from a complicated noncomplex system, as briefly described above. Biological systems are characteristically complicated complex systems. Although reductionism applied to biological systems has enjoyed great success in the 20th century, however, due to this inherent non-predictability of complicated complex biological systems the fruits of this philosophical approach are limited. These types of complex systems cannot be defined by the sum of their parts and hence cannot be described by an equation. Linear systems such as simple systems may of course be described by simple equations. Nonlinear systems such as complicated systems may be described by equations and simple complex systems may be described by equations. Complicated and simple complex systems are nonlinear but nonetheless definable by equations. Newton's second law of motion equation is an example of a simple linear equation. An example of a simple complex system is seen when many billiard balls interact with one another, described also by Newton's laws. *In any event, reductionism is the fundamental philosophical basis for understanding both bottom-up and top-down approaches to understanding and manipulating systems when mechanistic insights are the basis for this understanding or manipulation of the involved system. This is in counter-distinction to the notion of simplifying complexity by finding "simple rules" that can be used to understand the behavior of a system independent of the mechanistic relationships between the component parts of the system.* This idea of simplifying complexity is a strategy for understanding complex systems that cannot be done so by reductionist strategies. This approach is largely facilitated by computer simulations of systems under study.

For the purposes of medicine, it can be used to understand biological systems for diagnostic and therapeutic purposes. For example, invoking the success of Watson's IBM computer has great promise in utilizing the enormous potential for generating biological input data fueled by the human genome project. There are estimated goals of being able to feed such sophisticated computer models with a million data bits including genomic, proteomic, metabolomic, microbiomic, and other data. This enormous potential for information, when interfaced with an equally spectacular computer processing capability, underscores a capacity to provide diagnostic and treatment potential which is dizzying and mind-numbing. Medicine from this approach will be transformative away from the current traditional establishment paradigms. It will represent an entirely new era and cultural change. It is hard to predict at this inchoate period in the evolution of this cultural shift many of the pros and the cons of the overall impact on the healthcare industry. However, the overarching value is the ability to optimize diagnostic and treatment actions on a highly individualized scale. The assessment of genomic and other data, because it involves looking at the component parts of a system, the biological system representing the organism or individual person, represents a bottom-up approach. The computer programming is apparently so sophisticated it is capable of analyzing the interactions of all these component parts, independent of mechanism, and finding the so-called "simple rules" that can then be used as a top-down approach to treating this individual as well as subsequent individuals who fall into the same pattern of simple rules. This top-down approach because it does not rely on mechanistic underpinnings, is non-reductionist. However, for the purpose of understanding the distinction of top-down and reductionist in this context, if the computer analysis were to find for example a protein that may be an enzyme that was not functionally expressed, treating that individual by up-regulating its function would be considered mechanistic. Therefore, a top-down approach in this case would not be in accordance with so-called "simple rules" or "blueprint" but rather would be a bottom-up reductionist approach to treatment. In this case it is reductionist because it is mechanistic, and it is bottom-up because it targets a component part of the system. Conversely, an individual with respiratory failure in an emergency room would be managed according to a blueprint or "simple rules" of pulmonary resuscitation independent of whether the respiratory failure is due to congestive heart failure, pneumonia, Chronic Obstructive Pulmonary Disease (COPD), asthma, or whatever the mechanism may be. In this case the management is independent of mechanism between the component parts that make up the respiratory system and intervention is directed at the system whole rather than any of its constituent parts. Hence, it is top-down and non-reductionist.

A typical example of routine management of chronic disease in medicine, which is based on reductionism and a bottom-up approach is the management of elevated cholesterol. The therapeutic target for the cholesterol-lowering statin drug such as Lipitor is a specific enzyme (HMG-CoA reductase), whose inhibition leads to lower cholesterol biosynthesis. This is reductionist because it targets a mechanistic component of the system, which is based on understanding the mechanism for the interaction between the elements of the system. Empirical drug treatment targets specific molecules, often a statin drug for elevated cholesterol for example, without clear knowledge that these specific molecules are the root problem for a given individual being treated.

Rather, it is an assumption based on statistical likelihood that blocking this particular enzyme will fix the problem. Theoretically blocking this particular enzyme, HMG reductase, does lower cholesterol, however, it may not work on a given individual due to many other variables involving component parts of the biological system. Furthermore, the assumption of a statistical likelihood that this particular class of drug or more specifically this particular drug can be given safely without adverse events. However, this is again a statistical probability assumption that many unknown variables involved in this complicated complex biological system will not come into play to result in toxicity for the individual patient.

4.30 Bottom-Up and Top-Down Approaches

When the parameters and their interactions within a system are known they are said to be determinate. Often, we are dealing with indeterminate systems in medicine because control parameters to the system are not all known. Furthermore, inherently we do not know what we don't know about complex systems with many interacting component parts. In contrast to indeterminate, which means not knowing all the component elements of a system, nondeterministic means not knowing how the elements of a system interact to create the emergent outcome (see discussion on emergence). Both indeterminacy and non-determinacy contribute to the non-predictability of complex systems. Additionally, the more complicated the complexity of a system the more substantial the uncertainties, i.e. the un-predictabilities.

If the parameters are known, such that responses to input interventions are predictable, the system is said to be deterministic. In this case, a "blueprint" or "cookbook" reductionist approach to understanding or managing the system is most appropriate. The parameters are known and the interactions or rules (mechanistic or not) between these parameters are understood. In this case, the systems may be linear or non-linear. They may be simple, complicated, or simple complex systems but not typically complicated complex systems such as biological systems. The exception however in the latter case, as described just above is when these complex systems are computer analyzed to find "simple rules". However, in contrast to management of chronic disease states in medicine the management of acute care, typically inpatient hospital emergencies or critical care does follow sets of rules as algorithms of care which may be reductionist, i.e. mechanistic, or non-reductionist. When they are reductionist, it typically involves a bottom-up approach because we are addressing a component part of a system and it is based on mechanistic insights. If it is independent of mechanistic insights, it is non-reductionist, as described earlier, and when it targets the individual rather than any of the component parts of the system of the individual it is a top-down approach. An example of the above is intubating a respiratory failure patient may be considered a top-down non-mechanistic approach as described above. However, perhaps for the purposes of simplicity of understanding it may be considered also a bottom-up reductionist approach that indeed does work in healthcare similar to other acute care emergency settings. Intubating a respiratory failure patient may be considered bottom-up by recognizing the respiratory system as a component part of the individual and reductionist in the sense that treating this component part mechanistically keeps the individual alive. Likewise, treating a grand mal seizure or administering CPR (cardiopulmonary resuscitation) to an individual with cardiac arrest represent fine examples in acute care settings where reductionism and a bottom-up approach works nicely. *Top-down approaches to patient care may for purposes of understanding be best considered as largely non-reductionist such as an example of finding the "simple rules" by a Watson IBM computer using genomic and other data.*

In a bottom-up approach, if we were to visualize an upside-down triangle with the vertex at the bottom, the simple rules apply to interactions at the individual scale between agents that represent the reductionist model of breaking complexity to its simplest component parts. Finding these rules but without seeking to understand further intricacies between the agents as increasing emerging complexity evolves into high levels of incomprehensible interactions that are the beautiful and elegant dynamic portrait of a flock of migratory birds or the metabolic adaptability, flexibility and robustness that facilitates an Olympian individual gymnastic performance. In both cases, the emergent behavior represents the complex interactions of component parts functioning in unison (in an impressive fashion), representing the top, broadest part of the upside-down triangle. Hence, emergent behavior is a bottom-up process as is the process of finding the simple rules, i.e. from the bottom-up. The purpose of finding the simple rules for a complicated linear system is to understand the system, as in the laws of physics, for example Newton's laws of motion. These laws may be applied to the broadest context (top-down) to accurately predict context-specific solutions or outcomes. That is, any range of the equation's variables exemplifies the

broadest context that may be applied to determine specific values of the interactions or the parameters because of the linearity of their mathematical connections. The predictability of applying top-down linear equations is 10%. Conversely, complex systems are nonlinear and simple rules are not intended to understand the entire more complex system because that would be impossibly incomprehensible. Rather, it is to predict with statistical probability (hence less than 100% predictability) the behavior of a similar *albeit* not identical system, such as another patient with the same or similar parameters from which the rules apply. Accordingly, it is useful for determining failure points. This is a bottom-up approach that juxtaposes the rules to the data collection. Rules create or describe patterns such as those of cellular automata, for example, the Game of Life described previously; attractor trajectory forming fractals found in probably all biological systems; or any variety of healthy or unhealthy patterns of simple, complicated, and complex behavior. The rules may be as simple as patterns of exercise or of an immunization schedule. They may be described as the patterns of interactions between components the gut microbiome in the setting of a particular diet that gives rise to certain patterns of metabolic phenotypes including type 2 diabetes, organ fat deposition, and obesity. These patterns are independent of further mechanistic understanding of the interactions of components of a system at higher levels of complexity. They are recognized only by standing back and observing three or more generations of nodes.

Another purpose for simple rules of bottom-up approaches is to find out what is driving the rule, for example determining the basic etiology of a pathogenic process, or failure point. Ultimately, the rules and patterns uncovered in a bottom-up approach may be to extract the fundamental patterns that provide bridges to apply to a broader context as top-down diagnostic and therapeutic approaches. In other words, recognizing a particular pattern of metabolic disturbance correlating to a particular abnormal microbiome may be established from bottom-up studies. Subsequently, this pattern may be recognized when applied to an individual patient or group of patients (i.e. top-down). A therapeutic top-down approach would then lead to strategies that target the pathogenic microbes and replacing them with healthy ones.

The next question becomes: can simplifying complexity analogous to the simple rules described above be extended to the level of complexity inherent in systems biology to study complex diseases like cancer? It should be regarded that even in the above apparent simple models, it often takes hundreds of computer simulations studying patterns of colony behavior to arrive at the so-called simple rules on a local scale that are critical or correlate with the emergent properties of the system as a whole. However, can an analogous approach be applied to the complexity of human biology for the purposes of advancing medicine by solving in a non-reductionist way that reductionism cannot? *Big Data invokes systems in networks biology utilizing "omics" approaches (e.g. genomics, proteomics, lipidomics, microbiomics, metabolomics, etc.) to look for patterns of interactions at the local scale to understand effects on the global scale.* Such analysis involves hundreds, thousands, hundreds of thousands, or even more of different biological molecules simultaneously.

A system in the sense of systems biology is the integration of networks that exist at each hierarchical scale such as molecular, cellular, organ, and individual (organismic) into which can be referred to as the "network of networks".

Notably, a system can be defined arbitrarily and, hence, can be a subset of molecular networks, a particular grouping of cells, or a single organ. Alternatively, the organism as a whole consisting of many networks can be viewed as a system. Two or more organs work together to form organ systems, e.g. the nervous system and the digestive system that are responsible for complex tasks. Together, all the organ systems work together coherently and sometimes synchronously as quantum biological systems that allow optimal health.

Thinking in the context of systems biology should always be thought of in terms of systems and parts and wholes. For example, ants or bats of a colony represent the individual components or elements of the system whole (the colony). Every system can and should be understood as systems of parts and wholes. The ecosystem to which a colony of bats or ants belongs represents a system on a larger or macroscopic scale. As such, a colony of organisms becomes a part or element within many other colonies and organism species of predators and prey that comprise a larger organizational scale.

Although the quantification of deterministic modeling for human health and disease may be outside our reach, as will be further discussed ahead, there remains important value both in terms of maximum utilization of increasing availability of data as well as creative advances that may not be in a predicted sense. Clearly, predictability will be either prevented or reduced by mutational events in cancer or microbes, or by the computational barrier characteristic of biological systems. However, despite the indeterminacy of many component nodes of a system, insights, hypotheses and computational algorithms may be created on the basis of any select number of control and order parameters. Subsequent usage of the algorithm may demonstrate statistical anomalies, which help to refine and self-correct newer generations of the algorithm. These algorithms are non-mechanistic *per se* and thus are intended to approximate rather than represent the "truth". Although not mechanistic or reductionist *per se*, such algorithms of simplifying complexity may incorporate elements of reductionism.

The way people think about problems from a complex system's perspective is that we have a number of given "simple rules" and we end up with some form of emergent property. However, that would be very difficult to predict just by knowing these rules alone. For example, in the case of the weather or other chaotic systems, minuscule deviations of recorded initial conditions can lead to wildly different endpoints. Accordingly, short-term predictability is reasonable, however, beyond which the trajectory becomes exponentially separated from the anticipated path. In the case of a complex system, as it relates to health and disease, is that we don't know really what a lot of these simple rules should be and there is no reason to suspect that these simple rules would be constant across time. So, *the problems with disease are even more difficult to predict than other complex emergent systems such as the* weather or of emergent swarm intelligence of insects or the colony behavior of bats.

4.31 Algorithmic Medicine?

The role of computers in medicine is met with both enthusiasm and pessimism. The goal of the following few paragraphs is to provide a short synopsis underscoring the likely false expectations of using simple rules to predict outcomes, while still highlighting their significant value in advancing medicine. There is an important distinction between pessimism about the prospects of computer models working contrasted to the enthusiasm of many that a widely adopted explosion in terms of the application of computer models in medicine to understand and treat complex systems and diseases. These things are not necessarily antagonistic.

Watson-like software could have predicted the Ebola outbreak in West Africa much sooner than was realized by the public health community. There are many applications for these adaptive machine learning approaches to move the power of medicine forward by leaps and bounds. In medical school, we are taught to be pattern recognition machines—in as much as we input findings and we output diagnoses and plan for interventions. Symptoms or other findings may be considered the "simple rules", analogous to those attached to colony of flocking behavior of relatively simple species of animal in terms of pattern recognition. For example, shortness of breath and wheezing gives an empirical diagnosis of asthma, despite the "noise" of many other less diagnostic signs and symptoms such as an accompanying cough, chest pain, anxiety, nausea, palpitations, and so on. Simple rules for an acute abdomen would be abdominal pain and rebound tenderness. For pyelonephritis (kidney infection) it would be flank pain plus exaggerated punch tenderness. However, using the "simple rules" it may be more difficult to discern cause–effect relationships, such as perioral swelling due to a particular toothpaste, generalized pruritus due to a fungal dysbiosis in the gut, or weight gain and fatigue due to insulinoma. The example of perioral swelling related to toothpaste is especially relevant because it highlights the challenges of predicting emergent behavior. In the cases of symptoms that may be rooted in fungal dysbiosis or insulinoma, the challenges lie in identifying the emergence of these manifestations from their root causes. Another example is identifying (or pattern recognition) the precise profile of benefits and adverse effects of receiving a statin drug for elevated cholesterol on an individual level. Indeed, computerized algorithms with sufficient data may realistically be able to output these kinds of answers. Another example is the use of computers to determine in whom the benefit of tight blood glucose control for a diabetic, even independent of hypoglycemia, promotes the risk of sudden death. One more example is using computerized algorithms for determining the best therapeutic regimen for a given patient with colon cancer or thyroid cancer.

The greater the availability of data input to the computer the greater the potential or pattern of "simple rules" to emerge.

Clearly, on this level, computers will always be better at pattern recognition than humans. However, there are other important things that go into being a physician that define it, from the way we process data into the way we are able to make connections. A computer has the ability to store an incomprehensible amount of data and process it in a rapid amount of time—which is vastly superior to what physicians as human beings will ever have the capacity to do. Within the complexity of biological systems in medicine there is seemingly an insurmountable number of rules, and some computer programs, for example Watson's IBM software, are capable not only of utilizing an algorithm containing the "simple rules" but in fact have an additional capacity for finding the rules. It does this by using the data it ingests about a given patient interfaced against the huge data bank of information available from the world literature about interactions of nodes resulting in emergent patterns. Computer software programs in the future will be increasingly more sophisticated in terms of a capacity for pattern recognition among huge volumes of data.

Nonetheless, there are other things we can do better than computers, such as make connections between things in a way that would be very hard to codify in terms of an adaptive learning computer system. For example, a face-to-face interview with a patient whose affect, mannerisms, and personality may be well known to a physician who is able to perceive subtle changes indicating that something is wrong. Another example is the equivalent of creativity. Consider that one of the things that the Watson IBM computer was doing that was failing is they would ask on *Jeopardy*, who was the first woman in space? The computer's answer would be Wonder Woman because Wonder Woman was the first recorded piece of information about a woman being in space. However, what was implied (i.e. not explicit) in that question is that they were talking about a real woman who actually went to space. The information was not in the question, it was all implied. Although the computer's algorithm had been corrected to discern fictional characters, there remain plenty of inferences from unspoken interactions to which a computer does not have the capacity. *Empiricism is an expression of creativity. There are limits to predictability capacity of computers using simple rules and biological systems in health and disease. There are only approximations to the truth because these "simple rules" are not fundamental laws, and the mechanistic basis that some of these rules incorporate is far from having a linear relationship with emergent outcome behavior.* Consider that an indeterminacy of a system may be 50%, i.e. 50% of the nodes or elements of a system are unknown. That may, in fact, be sufficient for statistical likelihood for a pattern between a select number of these elements to be predictive of emergent behavior. However, that predictability is far from 100% because it is highly nonlinear and not based on any predictability of a fundamental law. Therefore, empiricism is the role of the physician and will best be a conjugate value along with diagnostic and therapeutic guidelines of computer programs. The role of a physician is clinical judgment of whether to accept computer guides as well as to artistically titrate therapeutic trials in the framework of an optimal risk to benefit profile ratio. Meanwhile, statistical anomalies whereby the computer has it wrong can be used in a self-correcting fashion

to help understand the "rules behind the rules" which in turn helps to refine computer algorithms.

Sequencing the human genome was thought to change medicine forever. To a certain extent, it has, but not to the extent of expectations. The hope, of course, will be to advance computer algorithms capable of incorporating perhaps in excess of a million data points derivable from genomics and other "omics" big data. Analogously, the stock market approach accesses an enormous amount of data and practically unlimited financial resources, and again has made inroads but not inroads in the predictable sense. Trading rates between buyers and sellers represent the many thousands of companies on the New York Stock Exchange. While the rules of the trading system are simple, the resulting patterns may be very complex, for example a global glut of oil may result in the increase of the price of gold. This results in remarkable complexity at the group level and the evolution of emergence that cannot be predicted intuitively. They figured out that one can speed up trade rates to gain a competitive advantage, but those competitive advantages have not been in the predicted sense. Thus, we will likely come up with creative ways to improve health through computers, but they may not always be in a predictive sense.

There is an enormous dimensionality inclusive of the many scales of information flowing through the biological system that makes it difficult to separate the effects of a drug from all the "noise" in the system. Every drug entity has multiple effects, both by design and accidentally, which lead to targeted and off-target molecular interactions. In addition, these molecular interactions are heterogeneously distributed throughout the human body compounding the complexity of the problem. Moreover, drugs become metabolized in the gastrointestinal system, giving rise to both known and unknown metabolites that independently and differently interact with the molecular targets within the various organs and tissues. The resultant complexity is truly daunting, and it makes predictive algorithms in drug discovery very error prone.

4.32 An Added Layer of Complexity: Gut Microbiome

Biomolecular organization—from genes through proteins to metabolites—provides information (or data). Accordingly, important insights into biological processes, including aging and disease, offer the hope of augmenting the development of new, highly predictive treatments at the individual patient level. Crucial is the relationship involving the gut microbiome inhabiting all animals including humans.

> *The gut microbiota co-evolves with the human host from birth and throughout life.*

The composition of the microbiome is influenced by many factors including the mode of birth, diet, the immune system, severe or chronic stress, antibiotics, steroids, and other environmental stimuli. The interaction between the microbiome and host is mediated by more than four million gene products from the microbiome that can interface with the immune system. Two-thirds of the immune system is located in the wall of the gut. Inflammatory signaling from the microbial flora induces the infectious immune response. This may be for example the molecular origin of the low-grade inflammation that characterizes the onset of obesity and type 2 diabetes. Different gut flora seems to create correspondingly heterogeneous metabolic profiles in the host (70). Martin, et al. showed how human gut microbiomes transplanted into mice resulted in dramatic metabolic effects in the transplant-receiving mice. This was reflected in the metabolic profiles of their body fluids (urine and plasma), as well as in the metabolic profiles of organs such as the liver. These differences correlated, for example, to changes in bile acids in the ileum and short-chain fatty acids in the cecum. The mouse recipients of transplanted human microbiota showed a shift in bile acid composition away from cholic acid to a predominance of taurine conjugated bile acids that correlates to the microbial composition. This shift is due to the inability of intestinal bacteria to de-conjugate bile acids (in this case taurine conjugated) consequently enhancing the absorption of dietary fat. Changes in plasma lipid profile also correlate to the type of gut microflora. Hence, the application of metabolomics ("systemic study of the unique chemical fingerprints that specific cellular processes leave behind their small molecular metabolic profiles") illustrates that different microflora result in different metabolic phenotypes.

Nicholson, et al. [71] showed interesting results in mice recipients of microbiome transplants from discordant Malawian infant twins for kwashiorkor malnutrition. The mice that received the transplanted flora from the kwashiorkor twin but not the normal twin demonstrated impaired growth and less weight gain. These mice also demonstrated disturbances in amino acid, carbohydrates, and TCA cycle metabolism while the mice were maintained on a traditional Malawian diet that only resolved when changes to a protein-enriched diet. This highlights microbiome processing of nutrients and extraction of energy from the diet. Similarly, the transfer of microbes from Japanese sushi is an example of horizontal gene transfer—one that provides seaweed-digesting enzymes that are valuable and necessary to the Japanese population [72]. It was shown that both systemwide, i.e. the organism as a whole, and organ specific changes in the metabolic profile are a consequence of populations of gut microflora. The implications are that microflora represent a significant source of metabolic variability in the host. Hence, host genetics alone cannot adequately explain metabolic profiling [71]. That is, it appears that for example the gut flora constitutes a significant influence on the degree of intestinal fat absorption as well as where it is distributed; unhealthy gut microbial flora, independent of diet and genetics, is contributory to non-alcoholic liver disease, obesity, and type 2 diabetes. *Metabolic phenotypes and profiles are modulated by interactions of genes, diet, and symbiotic gut microorganisms that favorably or unfavorably influence response to therapeutic interventions.* Data-rich molecular signatures provide valuable information to aid in clinical decision-making.

Burcelin [73] showed that the epidemic of obesity and type 2 diabetes and metabolic syndrome are related to nongenetic causes. It highlights the significance of metabolic diversity and the need for personalized medicine. For example, similar hemoglobin A1c and different body weight, plasma insulin, and fasting or red blood sugars have different pathophysiology. Therefore, the parameters that determine therapy vary according to the metabolic diversity. What is needed are biomarkers that stratify this metabolic diversity and classify subgroups of metabolic similarities. Accordingly, *therapeutic strategies would have enhanced predictability to fundamentally changing the course of the disease, and if early or pre-disease phases are targeted, the chronic disease itself abrogated. Therapeutic strategies may be based on vaccines, pharmacology, prebiotics and probiotics* for example.

Metabolic stratification of a population with type 2 diabetes ir prediabetes may be based on bottom-up computational predictability models that determine etiologic failure points. Such failure points may turn out to be altered microbial flora causing increased fatty acid absorption due to microbial elements that reduce conversion of amino acid conjugated bile salts to primary and secondary bile acids. Additionally, certain microbes produce short-chain fatty acids. Consequently, abnormally increased fat absorption from the small bowel may promote hepatic steatosis, insulin resistance, and/or type 2 diabetes consistent with what was demonstrated by Martin and Nicholson, et al. [71, 73] Bottom-up research such as this establishes rules and patterns (in this case the microbial composition and corresponding hepatic fat predicting insulin resistance and/or type 2 diabetes) that can be subsequently applied at the time of visit to individual patients in "top-down" fashion guided by computer simulation analysis that finds these patterns to optimize diagnostic and potentially early intervention. Notably, the patterns may be provided to the computer, or as will be illustrated ahead, computers may be constructed that are capable of finding the patterns *de novo*.

4.33 Electronic Medical Records

Electronic medical records may be used to harness an extraordinary amount of research association data into top-down solutions. It may do so with the aid of inputting simple rules or in some models, it finds its own rules. There is great technological capacity for electronic medical records (EMR) behind the nonlinear processes of complex systems. This role is underscored by its general intended use to stratify patient populations in order to take a finite number of resources and use them as wisely as possible. When a patient goes to a doctor's oce or to a hospital, electronic medical records form the information management between the triad of the patient, the provider, and the computer to optimally effect outcome (Figure 4.12). The EMR looks to be the bridge to an anticipated explosive potential of clinical healthcare by the integration of an unfathomable amount of raw data. The computer draws conclusions by the association of data. Such data may be simple, such as body weight, or may be quite sophisticated including genetic, genomic, proteomic, metabolomic, or metagenomic. The history of EMR—with very modest available data—gives an estimate of how great the impact of this may be. The following is a synopsis from a personal correspondence with Greg Shorr, who along with his group of colleagues built the first EMR.

Greg Shorr was a medical student in the late 1960s when he had the opportunity to do a rotation on an Indigenous

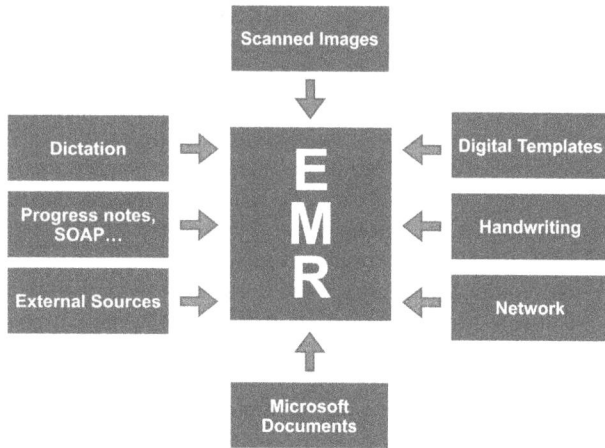

FIGURE 4.12 Schematic illustration of a generic structure of the Electronic Medical Record (EMR) Database. Source: adapted from Aldosari, B. et al. (2018) "Assessment of Factors Influencing Nurses Acceptance of Electronic Medical Record in a Saudi Arabia Hospital". *Informatics in Medicine Unlocked* 10: 82–88. ISSN 2352-9148, https://doi.org/10.1016/j.imu.2017.12.007

reservation. He had a background in computer science and electrical engineering prior to going to medical school. He went to the Papago Indian reservation where he and a group began building an electronic medical record, a new concept at the time. The group he joined largely consisted of ex-NASA scientists. Together, they built the first EMR. Since then, this large-scale EMR has been remodeled many times and has now served more than two million Native Americans at 400 sites. There are a lot of uses of the system, not only for one-on-one care but for healthcare population overviews. Shorr was very interested in the population views because, more than anything else, it allows one to turn information into good results. At the time of writing, Shorr is at the Santa Fe Institute where he is using machine learning to collate huge datasets on interface designs, and is turning that data into analytical requirements that allow for the use of a variety of tools in standard types of statistical analysis. Machine learning is where the computer looks for patterns and programs itself to recognize those patterns.

Thomas J. Watson was the chairman and CEO of IBM between 1914 and 1956. Many years later, IBM transformed from a manufacturer of typewriters to a global leader in computational hardware and software solutions that won the television quiz show *Jeopardy*. On this show, an answer is presented, and each contestant is asked to come up with a relevant question. For example, a picture of George Washington could be shown, and the contestant could respond with "Who is George Washington?" It involves a lot of semantic nuances. Watson claimed that IBM could build a computer that, in real time, would compete with the best *Jeopardy* contestant ever. He indeed built a system that not only competed but knocked those brilliant people, long-time *Jeopardy* winners, right off their pins. The fundamental mechanism through which it did that was machine learning, where the machine ingested and digitized huge volumes of information on topics that were commonly presented on *Jeopardy*. It also learned English and

learned patterns from the information basis that allowed the computer to recognize the answers that in the case of *Jeopardy* were questions. There is an enormous amount of inference used in this game, and it was almost unimaginable that a machine could do this.

Machine learning is commonly used in the investment world, banking, and marketing research. It finds patterns that often would not even be asked for or considered. This is an interesting way to identify the clinical parameters that hold the solutions for critical diagnostic and therapeutic dilemmas in healthcare. The interesting thing here is we are not feeding it rules. The machine is writing its own rules based on a large set of data. A group was contracted with the Watson team to do cancer research. They were feeding in raw information from the literature—absolutely understood at a certain level by the machine itself—and they were feeding it data by antecedents and outcomes. The machines (the computers) were building up associations between data elements, which is the level of learning it predicts. Every date element has a network of related data elements and the big picture of this network structures the information in a way that it is intelligible to computation. Watson was both the founder of IBM and Sherlock Holmes, who was intended to be a sidekick in many ways. Traditionally the computers were given the rules, as with the EMRs in the late 1960s, 1970s, and 1980s. The computers were presented with a set of facts, they would then apply the rule and come up with a conclusion. In the case of the IBM model for *Jeopardy*, for the large part it made its own inferences. There are inference engines in use today, for example, Google's search engine is based on the same thing. One can ask a fairly vague question and, in the vast majority of times, it will provide something that will at least get one started on answering that question. The reason it can do that is through the huge history of questions it has been asked. It finds relationships between certain words, building up associations between these data elements. Nobody is sitting in Google offices programming a bunch of rules that can govern every kind of question. Rather, it is a framework that finds these relationships and applies them. There is currently work going on applying this type of technology to healthcare that is extraordinarily powerful. It is based on huge sums of associations between data that trains models. The inherent nonlinearity of biological systems presents a challenge, however, to even the most sophisticated machines. Therefore, *the nimble capacity of models for finding their own patterns to diagnostic and therapeutic problem solving is coupled with the information data that it is fed, both in terms of training the model with a long history of reductionist research, and superimposed personalized data.*

4.34 An Anecdote Shared by Greg Shorr Regarding the Use of His Electronic Medical Record on the Native American Reservation

Shorr and his group again invented the EMR and people from all over the world came to see it. The day it went live, however, it was obsolete. The reason for this is that it was designed to support one-on-one clinical care in the oce, where one gets an integrated health summary, which is what is out in clinical practices today where information from various places the patients were seen is integrated. Accordingly, we can summarize all the information we use to get to know the patient very quickly. It also synthesizes information regarding such things as whether the patient is due for things like immunization, breast exams, etc. What it did not tell us was what physicians wanted to know right away, for example identifying all the patients with diabetes or hypertension so that resources can be directed in the right way. It was this population overview that the EMR did not have an answer for early on.

The number one health problem on the reservation at the time was gastroenteritis and the mortality rate from diarrhea in babies was at a third-world rate in the 1970s. For every 1,000 live births, a devastating 75 babies died in the first year of life from gastroenteritis. There are no pediatricians on the reservation. So, Shorr and his partners went to the World Health Organization for help and were put in touch with a pediatrician who described his experience with a similar epidemic in Egypt. In Egypt, they dealt with the problem by developing a rehydration solution where salt and sugar were reconstituted with water, allowing them to change the dynamic balance between input and output of fluids. This kept the babies alive long enough to get them to the hospital where they could be put on intravenous hydration. Secondly, they incorporated village health workers into the program. These are people that are trained to talk about things like sanitation, isolation, hydration, and so on. The combination of these two interventions allowed the group in Egypt to take control of the epidemic. In the United States, in this case Arizona, the epidemic only occurred during the summer when it was hot, but nonetheless the same interventions were applicable—though there were probably not enough village health workers to tackle the problem. However, there were some available health workers that could be trained in sanitation and breastfeeding, how to mix up the solution, and so on. They met with a group of village health researchers and realized they did not really have enough health workers, but considered the enormous value in being able to predict ahead of time who is going to get sick. This would allow the ability to take their meager resources and focus on the high-risk individuals. Hence, they needed to be stratified for high risk. There were three-to-four years of information in the EMR database which could be used to identify risk factors that could be tested retrospectively. Shorr and his group came up with about 40 different risk factors that they examined and tested retrospectively. They found seven of them seemed to be predictive of subsequent morbidity and mortality. The types of data that the system had included were things such as gestational age and weight, overcrowding in the home, alcoholism, and so on. They built a few statistical prediction models, for example, Bayesian and multiple regression that seemed to be predictive of subsequent morbidity and mortality. They, arbitrarily, said that anybody above a certain threshold of risk would have a village health worker sent to their home. The threshold was necessary to match the limited resources available. In April, they trained the workers and applied the model. In August, a priest came to the reservation from one of the districts the model was being applied to

(it was only applied to the remote western districts) and met with Shorr saying he was not sure what was going on. Shorr responded, "What do you mean?" The priest said, "We haven't buried any infants this summer". Sure enough, the model was highly predictive. The next they got NIH funding to care for the entire reservation (not just the remote districts) and for the first time there were no deaths. In fact, for the next ten years there were no deaths. Medical informatics, hence, can be clinically powerful, as seen in this case of a dramatic change to a restricted population.

The impact of medical informatics demonstrated in this narrow context the extraordinary and amazing potential of predictive computation for reducing morbidity and mortality.

This, coupled with the IBM *Jeopardy*-type model of innovation built around healthcare, hints at an imminent explosive armamentarium for healthcare delivery. Over the next decade we should experience the application of computational simulations with an enormous pool of accessible data from the worldwide research derived from complex and chaotic systems. This will provide powerful interventional tools at the bedside with a level of predictability for chronic diseases not only like we have never seen before, but never even imagined.

4.35 The Neuroendocrine and Immune System Hormonal Stress Responses: Adaptive Versus Pathologic and the Role of the Fitness Landscape Model

The understanding of the hypothalamic pituitary relationship in 1990 marked the origin of the field of neuroendocrinology. These essential hormonal regulating systems of metabolism, growth, and reproduction opened the curtain to the communication between the brain and the body. What we have witnessed beyond the final decade of the 20th century has truly and remarkably contrasted with the limitations of the theater that existed prior to that time and which may be likened to the dark ages of medicine. The complexity of integration between the brain and the body has come to be appreciated as mediated by the intersection of the central nervous system with the endocrine system. This has evolved into a fuller understanding of the stress response, not only in a physiological sense through the notion of allostasis [74], but in a pathophysiological sense brought about by the same mediators of allostasis but through the notion of allostatic overload [75]. Thus, these mediators, primarily hormonal, catecholamine, and immune responses, paradoxically, are both protective—in the case of allostasis— and harmful—in the case of allostatic overload. The purpose of physiological allostasis is a mechanism of adaptive resilience of the living system involving the endocrine, nervous, and immune systems. Allostasis means stability through change. Homeostasis means resistance to change. The purpose of physiological allostasis is to maintain homeostasis of vital organ system function. The most fundamental parameters of homeostasis are the parallel indices of free energy (Gibbs' free energy equation), redox (Nernst equation), and the acid–base

(Henderson Hasselbalch equation [76]) status. The inextricably linked inflammatory NFkB and disturbed redox impair both free energy flow and acid–base chemistry. The NFkB regulated generation of damaging oxidative free radical molecules alter the structure and function of protein, lipid, and nucleic acid components of cells leading to tissue injury. Like the neuroendocrine system, peripherally secreted proinflammatory cytokine "hormones" of the immune system and their interaction with the brain and relationship to chronic disease only germinated not far before the 20th century, in the mid to late 1980s. *Chronic immune system activation was recognized itself to be pathogenic as a function of the secreted circulating proinflammatory protein "hormones" which ushered the manifestations of disease.*

> *Allostatic overload occurs as a result of chronically elevated hormonal, catecholamine and inflammatory mediators of allostasis detached from regulatory control.*

It is often accompanied by phenomena of cortisol and catecholamine resistance, or a compensatory rise in the proinflammatory responses not counter-regulated by glucocorticoids, depending on the type of allostatic overload. Resistance phenomena such as resistance to cortisol are contextual dependent on the cell type. For example, despite immune cell resistance, chronic elevation of cortisol levels in the brain is shown to cause hippocampal atrophy. These processes leading to chronic disease is further contributed to by secondary mediators of allostasis, for example, markers of insulin resistance including low HDL, high triglycerides, and, significantly, endogenous hyperinsulinemia. (Primary mediators of allostasis include cortisol, catecholamines, and inflammatory markers.) Hence, protracted central nervous system neuroendocrine and autonomic stress responses mediated by endogenous cortisol have been reported in the literature to be responsible for pathological states of obesity, hypertension, dyslipidemia, diabetes, urinary tract infections, and osteoporosis [77, 78]. Moreover, it may be cogently argued that the same pathogenic foundation, in addition to insulin resistance and related type 2 diabetes, impaired satiety, obesity, and other features of the metabolic syndrome, also serves as the origin or contributor to virtually all chronic disease states that occur with aging. These states include cancers, cardiovascular disease, Alzheimer's disease, and even accelerated cognitive decline with aging. This opens a new chapter in the rapid evolution of modern medicine.

An appreciation of the relationship of the neuroendocrine system with the immune and gastrointestinal systems is proving to be critical in the overall construct of the body's systems biology. *The immune system was designed to keep us safe from pathogens and to promote wound healing in the setting of injury. The astonishing numbers of gastrointestinal microbiota, including pathogenic, commensal, and symbiotic, underscores the central importance of the immune and gastrointestinal systems.*

Furthermore, the neuroendocrine, immune, and gastrointestinal systems form a framework for understanding the richness of the tapestry of complexity between the body and the brain. However, the human gene pool is surprisingly rudimentary. This exploitation of vulnerable genes accounts for an estimated 70% of chronic disease. Such expression is ultimately caused by the inextricable processes of disturbed redox and inflammation and accompanying breakdown of free energy and acid–base homeostasis.

4.36 Concluding Remarks

A theme of this book is the stress response state of arousal to external factors, both real and imagined. They have the dual capacity to vitalize psychophysiology and optimally shape genetic potential or, conversely, to cause psychological and physiological pathology. The former scenario is achieved largely by top-down regulation of thought, emotion, and behavior, mediated by the strengthening of neural and synaptic plasticity of the prefrontal cortex and hippocampus. Alternatively, chronic excessive and uncontrollable stress responses dictate bottom-up processes characterized by the overload of allostatic systems including decreased parasympathetic vagal tone, increased sympathetic activity with reduced heart rate variability, a chronically activated hypothalamic pituitary adrenal axis with increased time-integrated free cortisol, a proinflammatory cytokine dominance with increased acute phase reactants, insulin resistance with impaired glucose tolerance, chronic hyperinsulinemia, and other susceptibility states for chronic metabolic diseases including neurodegenerative and cardiovascular disease and cancers.

The quantum manifestations of thought which occur at the interface of mind–body in the prefrontal cortex and limbic regions of the brain is the highest-level intrinsic control parameter in terms of psychophysiology and metabolic health. The exaggerated and prolonged stress response is a fundamental component and often the origin of virtually all chronic disease. Structural plasticity of the prefrontal cortex and hippocampus is critical for high-level consciousness, or conscious cognitions, responsible for goal-directed behavior of free will. Consciousness is quantifiable as a consciousness quotient (CQ), which is analogous to the intelligence quotient (IQ). By effectively controlling the emotional centers of the brain, specifically the amygdala and related structures, subconscious chronic feelings of fear and anger, as well as emotionally driven motivational behavior for food and other hedonic rewards, are prevented. This blocks the driving force of systemic proinflammatory and out of control bottom-up processes that accelerate biological aging and chronological diseases of aging. An interesting and important interdisciplinary metaphor is the notion that biological inflammation represents heat, and the incineration of information contained in the amazing and beautiful complexity of the human body at all scales of organization is lost to entropy. This represents an acceleration of biological aging subject to the unavoidable force of the second law of thermodynamics. The premise of accelerated biological aging invokes another metaphor that connects biological and physical systems to the same laws of physics. If time does not exist as an absolute quantity consistent with the theory of special relativity, biological aging indeed likely does progress at varying speeds. Accordingly, the optimal and most fundamental means of dilating time in the aging process, theoretically, should be by strengthening structural plasticity of the prefrontal cortex and hippocampal neurons in the brain. This represents a critical control parameter in the fitness landscape model and is a promising target for clinical intervention. The notion of quantum metabolism is a firmly supported scientific phenomenon. It highlights that our metabolism involves the quantum nature of energy production. *This correlated bioenergetics makes possible the extraordinary complexity and efficiency of human physiology. Top-down regulation of psychophysiology attenuates allostatic overload and associated inextricable processes of insulin resistance and mitochondrial dysfunction.* This can be understood as slowing the degradation of the biological fabric of space–time, relatively preserved by the phenomenon of quantum metabolism.

The model of fitness landscape is borrowed from physics, comparatively a far more naturally scientific discipline, to understand biological systems in the context of the field of medicine, where the impact of these insights has the greatest potential on reducing the toll of human suffering. This model is a Nobel Prize–worthy concept and its applicability to physiology has been demonstrated.

> *The compartmentalization of scientific disciplines, particularly clinical medicine from the branches of physics, should be recognized as a pathological isolation.*

Moreover, it should be described as an extraordinary and even tragic oversight that this model has not yet been invoked into the standards of clinical medicine. By homing in on various aspects of the stress response, in the context of the metaphorical fitness landscape, metabolic susceptibility states for disease can be identified as the control and order parameters to phase transitions from a healthy state. Further, the trajectory to chronic disease may be predicted on the basis of mathematical models of these parameters and their attractors. This has profound implications for both diagnostic and therapeutic purposes. Points of criticality may represent the threshold of a phase transition from normal to the disease state as well as an onset of irreversibility.

> *Thus, the model of fitness landscape serves as a critical tool and strategy to the development of therapeutic interventions capable of targeting control parameters and changing the trajectory of order parameters from reversible stages of disease or susceptibility states to normal states.*

Additionally, the neuroendocrine system, and the brain as the principal organ of allostasis along with the primary and secondary mediators of allostasis deserve center stage attention for the opening curtain to the future of medicine model of fitness landscape.

We close this chapter with a recent medical case study that illustrates the complexities of the human physiology that can be best considered using the systems biology framework.

4.37 An Unsuspected Trigger of Mental Status Change

This case involves an 84-year-old woman with a history of autoimmune thyroid disease with a baseline thyroglobulin peroxidase antibody (TPO) titer from a few years previously in the 40s range. Approximately eight weeks after receiving the COVID-19 vaccine, she developed worsening memory causing her to forget to take her thyroid replacement pills. Within a span of about three weeks, her memory lapses progressed to intermittent overt changes in her mental status with episodes of confusion. The patient's daughter noted a full supply of the thyroid medication in the medicine cabinet, which should have been near empty. An immediate biochemical profile revealed TSH of 100, total T4 of 4.2, total T3 42, and a TPO titer 4,000 (normal range less than ten).

A likely assessment of this presentation is a vaccine-mediated amplification of the thyroid autoimmune response. Mechanistically, this likely is rooted in a viral origin of the autoimmune state, which is widely reported in other autoimmune disorders. This author proposes that the worsening memory of this patient predated her not taking her thyroid supplement with the worsening inflammatory infiltrate of the thyroid gland promoting a decline in her thyroid function. This effect will dictate a higher thyroid hormone replacement dose independent of requiring a higher temporary dose due to the three weeks of missed dosing.

Notably, the half-life of thyroxine, or T4, is seven days. Since a new steady-state thyroid blood level takes 4–6 half-lives, this patient was advised to take liothyronine, or T3, which has a half-life of less than one day with an onset of action in only a few hours. Thus, the strategy was not immediately to increase the dose of thyroxine, an inactive prohormone, but rather to start liothyronine, the active form of the hormone, to induce an expedited return to a euthyroid state and a baseline mental status. Within one day the patient reported feeling subjectively better, and her cognition was notably improved.

A significant inference relates to the remarkable complexity of biological systems with inherent unpredictability. In medicine, we do not have yet organizing principles as does the discipline of physics with laws and rules such as the allometric scaling laws of metabolic rate across species. Consequentially, it is a dangerous game we sometimes play in medicine. *People with all forms of underlying autoimmune disorders should be carefully observed for potential exacerbation of the disorder in the setting of viral infections or vaccines.* Nonetheless, the unpredictable likelihood requires careful calibration of potential risks versus benefits, which is the *modus operandi* of proper clinical decision making in medicine with subsequent empirical observation.

REFERENCES

1. Wimsatt, W.C. (2006). "Reductionism and Its Heuristics: Making Methodological Reductionism Honest". *Synthese* 151(3):445–475. doi:10.1007/s11229-006-9017-0.
2. Seely, A.J. and Christou, N.V. (2000). "Multiple Organ Dysfunction Syndrome: Exploring the Paradigm of Complex Nonlinear Systems". *Crit. Care Med.* 28(7):2193–2200.
3. van der Poll, T., van Deventer, S.J., Hack, C.E., Wolbink, G.J., Aarden, L.A., Büller, H.R., ten Cate, J.W. (1992). "Effects on Leukocytes After Injection of Tumor Necrosis Factor into Healthy Humans". *Blood* 79(3):693–698. (b) Varfolomeev, E. and Vucic, D. (2018). "Intracellular Regulation of TNF Activity in Health and Disease". *Cytokine* 101:26–32.

4. Lombardi, F. and Mortara, A. (1998). "Heart Rate Variability and Cardiac Failure". *Heart* 80(3):213–214.

5. Braun, C., Kowallik, P., Freking, A., Hadeler, D., Kniffki, K.D. and Meesmann, M. (1998). "Demonstration of Nonlinear Components in Heart Rate Variability of Healthy Persons". *Am. J. Physiol.* 275(5):H1577–H1584.

6. Glass, L. (1999). "Chaos and Heart Rate Variability". *J. Cardiovasc. Electrophysiol.* 10(10):1358–1360. (b) Glass, L. (2009). "Introduction to Controversial Topics in Nonlinear Science: Is the Normal Heart Rate Chaotic?" *Chaos* 19(2):028501. doi:10.1063/1.3156832. PMID:19566276.

7. Butler, G.C., Ando, S. and Floras, J.S. (1997 June) Fractal Component of Variability of Heart Rate and Systolic Blood Pressure in Congestive Heart Failure. *Clin. Sci. (Lond.)* 92(6):543–550. doi:10.1042/cs0920543. PMID:9205413.

8. Hedman, A.E., Poloniecki, J.D., Camm AJ and Malik, M. (1999). "Relation of Mean Heart Rate and Heart Rate Variability in Patients with Left Ventricular Dysfunction". *Am. J. Cardiol.* 84(2):225–228.

9. Tapanainen, J.M., Thomsen, P.E., Køber, L., Torp-Pedersen, C., Mäkikallio, T.H., Still, A.M., Lindgren, K.S. and Huikuri, H.V. (2002). "Fractal Analysis of Heart Rate Variability and Mortality after an Acute Myocardial Infarction". *Am. J. Cardiol.* 90(4):347–352.

10. Lin, L.Y., Lai, L.P., Lin, J.L., Du, C.C., Shau, W.Y., Chan, H.L., Tseng, Y.Z. and Huang, S.K. (2002). "Tight Mechanism Correlation between Heart Rate Turbulence and Baroreflex Sensitivity: Sequential Autonomic Blockade Analysis. *J. Cardiovasc. Electrophysiol.* 13(5):427–431.

11. Goldberger, A. L., Amaral, L. A. N., Hausdorff, J. M., Ivanov, P. C., Peng, C. K., & Stanley, H. E. (2002). Fractal dynamics in physiology: alterations with disease and aging. *PNAS* 99, Suppl. 1, 2466–2472.

12. Schwartz, P.J., Vanoli, E., Stramba-Badiale, M., De Ferrari, G.M., Billman, G.E. and Foreman, R.D. (1988). "Autonomic Mechanisms and Sudden Death. New Insights from Analysis of Baroreceptor Reflexes in Conscious Dogs with and without a Myocardial Infarction". *Circulation* 78(4):969–979. doi:10.1161/01.cir.78.4.969. PMID:3168199.

13. Guzzetti, S., Signorini, M.G., Cogliati, C., Mezzetti, S., Porta, A., Cerutti, S. and Malliani, A. (1996). "Non-linear Dynamics and Chaotic Indices in Heart Rate Variability of Normal Subjects and Heart-Transplanted Patients". *Cardiovasc. Res.* 31(3):441–446.

14. Wagner, C.D. and Persson, P.B. (1998). "Chaos in the Cardiovascular System: An Update". *Cardiovasc. Res.* 40(2):257–264.

15. Pincus, S.M. (1994). "Greater Signal Regularity May Indicate Increased System Isolation". *Math. Biosci.* 122(2):161–181. doi:10.1016/0025-5564(94)90056-6. PMID:7919665.

16. Godin, P.J. and Buchman, T.G. (1996). "Uncoupling of Biological Oscillators: A Complementary Hypothesis Concerning The Pathogenesis Of Multiple Organ Dysfunction Syndrome". *Crit. Care Med.* 24(7):1107–1116. doi:10.1097/00003246-199607000-00008. PMID:8674321.

17. Ellenby, M.S., McNames, J., Lai, S., McDonald, B.A., Krieger, D., Sclabassi, R.J. and Goldstein, B. (2001). "Uncoupling and Recoupling of Autonomic Regulation of the Heartbeat in Pediatric Septic Shock". *Shock* 16(4):274–277. doi:10.1097/00024382-200116040-00007. PMID:11580109.

18. Goldberger, A.L. (1996). "Non-Linear Dynamics for Clinicians: Chaos Theory, Fractals, and Complexity at the Bedside". *Lancet* 347(9011):1312–1314.

19. Lombardi, F. (2000). "Chaos Theory, Heart Rate Variability, and Arrhythmic Mortality". *Circulation* 101(1):8–10.

20. Gallagher, R. and Appenzeller, T. (1999). "Beyond Reductionism". *Science* 284(5411):79.

21. Higgins, J.P. (2002). "Nonlinear Systems in Medicine". *Yale J. Biol. Med.* 75(5–6):247–260.

22. Ling, P.R., Smith, R.J., Mueller, C., Mao, Y. and Bistrian, B.R. (2002). "Inhibition of Interleukin-6-Activated Janus Kinases/Signal Transducers and Activators of Transcription But Not Mitogen-Activated Protein Kinase Signaling in Liver of Endotoxin-Treated Rats". *Crit. Care Med.* 30(1):202–211.

23. Bianchi. (2008). "Collective Behavior in Gene Regulation: Metabolic Clocks and Cross-Talking". *FEBS Journal* 275(10):2356–2363.

24. Demetrius, L. and Tuszynski, J.A. (2010). "Quantum Metabolism and the Allometric Scaling of the Metabolic Rates". *J. R. Soc. Interface* 7(44):507–514.

25. Kerth, G., Perony, N. and Schweitzer, F. (2011). "Bats Are Able to Maintain Long-Term Social Relationships Despite the High Fission-Fusion Dynamics of their Groups". *Proc. R. Soc. Lond. B* 278(1719):2761–2767.

26. Scott, A. (1999). *Stairway to the Mind: The Controversial New Science of Consciousness.* Springer Science & Business Media. Berlin/Heidelberg, Germany.

27. Anderson, P.W. (1972). "More is Different". *Science* 177(4047):393–396.

28. Emmert, M., Sander, U. and Pisch, F. (2013 February 1). "Eight Questions About Physician-Rating Websites: A Systematic Review". *J. Med. Internet Res.* 15(2):e24. doi:10.2196/jmir.2360. PMID:23372115. PMCID:PMC363 6311.

29. Ives, C. (2004). "Human Beings as Chaotic Systems". *Life Science Technology*: 1–7. (b) Ogle, W.O. and Sapolsky, R.M. (2001 September 30). "Gene Therapy and the Aging Nervous System". *Mech. Ageing Dev.* 122(14):1555–1563. doi:10.1016/s0047-6374(01)00286-x. PMID: 11511396. (c) Dumas, T.C. and Sapolsky, R.M. (2001 December) Gene Therapy Against Neurological Insults: Sparing Neurons versus Sparing Function. *Trends Neurosci.* 24(12):695–700. doi:10.1016/s0166-2236(00)01956-1. PMID:11718873.

30. Briggs, J. (2002). *Fractals: The Pattern of Chaos.* Touchstone, Simon and Schuster, Inc, New York, NY.

31. Ward, M. (2001). *Beyond Chaos. The Underlying Theory Behind Life, the Universe and Everything.* St. Martins Press, New York, NY.

32. Gleick, J. (1987). *Chaos: Making A New Science.* Penguin Books, New York, NY.

33. Ding, J, Fan, L., Zhang, S.Y., Zhang, H. and Yu, W.W. (2018). "Simultaneous Realization of Slow and Fast Acoustic Waves Using a Fractal Structure of Koch Curve". *Sci. Rep.* 8(1):1481. (b) Addison, P.S. (1997). *Fractals and Chaos: An Illustrated Course* (p. 19). Institute of Physics. ISBN 0-7503-0400-6. (c) Lauwerier, H (1991). *Fractals: Endlessly Repeated Geometrical Figures* (p. 36). Translated by S. Gill-Hoffstädt. Princeton University Press. ISBN 0-691-02445-6.

34. Sahli Costabal, F., Hurtado, D.E. and Kuhl, E. (2016). "Generating Purkinje Networks in the Human Heart". *J. Biomech.* 49(12):2455–2465.

35. Sharma, V. (2009). "Deterministic Chaos and Fractal Complexity in the Dynamics of Cardiovascular Behavior: Perspectives on a New Frontier". *Open Cardiovasc. Med. J.* 3:110–123.

36. Liebovitch, L.S. and Toth, T.I. (1991). "A Model of Ion Channel Kinetics Using Deterministic Chaotic Rather Than Stochastic Processes". *J. Theor. Biol.* 148(2):243–267.

37. Skarda, C.A. and Freeman, W.J. (1990). "Chaos and the New Science of the Brain". *Concepts Neurosci.* 1(2):275.

38. Gardner, M. (1970). "MATHEMATICAL GAMES The Fantastic Combinations of John Conway's New Solitaire Game "Life"". *Sci. Amer.* 223(4):120–123. (b) Izhikevich, E.M., Conway, J.H. and Seth, A. (2015). "Game of Life". *Scholarpedia* 10(6):1816.

39. Ravasz, E., Somera, A.L., Mongru, D.A., Oltvai, Z.N. and Barabási, A.L. (2002). "Hierarchical Organization of Modularity in Metabolic Networks". *Science* 297(5586):1551–1555. (b) Almaas, E., Kovács, B., Vicsek, T., Oltvai, Z.N. and Barabási, A.L. (2004). "Global Organization of Metabolic Fluxes in the Bacterium Escherichia Coli". *Nature* 427(6977):839–843. (c) Shen, Y., Liu, J., Estiu, G., Isin, B., Ahn, Y.Y., Lee, D.S., Barabási, A.L., Kapatral, V., Wiest, O and Oltvai, Z.N. (2010). "Blueprint for Antimicrobial Hit Discovery Targeting Metabolic Networks". *Proc. Natl Acad. Sci. USA* 107(3):1082–1087.

40. Watson, J.D. and Crick, F.H.C. (1953). "Molecular Structure of Nucleic Acids; A Structure for Deoxyribose Nucleic Acid". *Nature* 171(4356):737–738.

41. Crick, F. (1970). "Central Dogma of Molecular Biology". *Nature* 227(5258):561–563.

42. Jacob, F. (1976). *The Logic of Life*. Translation from French edition by Princeton University Press, 1993. Princeton, New Jersey. ISBN 0394472462.

43. Schneider, T.D. (2006). "Claude Shannon: Biologist [information theory used in biology]". *IEEE Eng. Med. Biol. Mag.* 25(1):30–33, Jan.-Feb. 2006. doi: 10.1109/MEMB.2006.1578661. (b) Shannon, C.E. (1997 July-August). "The Mathematical Theory of Communication. 1963". *M D Comput.* 14(4):306–317. PMID: 9230594.

44. Heppner, F.L., Ransohoff, R.M. and Becher, B. (2015). "Immune Attack: The Role of Inflammation in Alzheimer Disease". *Nat. Rev. Neurosci.* 16(6):358–372.

45. Adler, J. (2012). "Erasing Painful Memories". *Sci. Am.* 306(5):56–61. (b) Black, I.B., Adler, J.E., Dreyfus, C.F., Friedman, W.F., LaGamma, E.F. and Roach, A.H. (1987). "Biochemistry of Information Storage in the Nervous System". *Science* 236(4806):1263–1268.

46. Lorenz EN (1963). "Deterministic Nonperiodic Flow". *J. Atmos. Sci.* 20(2):130–141.

47. Saltiel, A.R. (1987). "Insulin Generates an Enzyme Modulator from Hepatic Plasma Membranes: Regulation of Adenosine 3', 5'-Monophosphate Phosphodiesterase, Pyruvate Dehydrogenase, and Adenylate Cyclase". *Endocrinology* 120(3):967–972.

48. Makino, N., Dhalla, K.S., Elimban, V. and Dhalla, N.S. (1987). "Sarcolemmal Ca2+ Transport in Streptozotocin-Induced Diabetic Cardiomyopathy in Rats". *Am. J. Physiol.* 253(2 Pt 1):E202–E207. doi:10.1152/ajpendo.1987.253.2.E202. PMID:2956889.

49. Gustafson, B., Hedjazifar, S., Gogg, S., Hammarstedt, A. and Smith, U. (2015). "Insulin Resistance and Impaired Adipogenesis". *Trends Endocrinol. Metab.* 26(4):193–200.

50. Samuel, V.T. and Shulman, G.I. (2016). "The Pathogenesis of Insulin Resistance: Integrating Signaling Pathways and Substrate Flux". *J. Clin. Invest.* 126(1):12–22. doi: 10.1172/JCI77812.

51. Li, R. and Bowerman, B. (2010). "Symmetry Breaking in Biology". *Cold Spring Harb. Perspect. Biol.* 2(3):a003475. doi:10.1101/cshperspect.a003475. (b) Bowerman, B. and Sugioka, K. (2019). "Breaking Symmetry: Worm Cue Finally Found". *Dev. Cell* 48(5):593–594.

52. Wang, Z., Gerstein, M. and Snyder, M. (2009). "RNA-Seq: A Revolutionary Tool for Transcriptomics". *Nat. Rev. Genet.* 10(1):57–63.

53. Mullins, F.M., Park, C.Y., Dolmetsch, R.E. and Lewis, R.S. (2009). "STIM1 and Calmodulin Interact with Orai1 to Induce Ca2+-dependent Inactivation of CRAC Channels". *Proc. Natl Acad. Sci. USA* 106(36):15495–15500.

54. Nelson, W.J. (2009). "Remodeling Epithelial Cell Organization: Transitions between Front-Rear and Apical-Basal Polarity". *Cold Spring Harb. Perspect. Biol.* 1(1):a000513. doi:10.1101/cshperspect.a000513. PMID:20066074. PMCID:PMC2742086. (b) Nejsum, L.N. and Nelson, W.J. (2009). "Epithelial Cell Surface Polarity: The Early Steps". *Front. Biosci. (Landmark Ed.)* 14:1088–1098.

55. Vladar, E.K., Antic, D. and Axelrod, J.D. (2009). "Planar Cell Polarity Signaling: The Developing Cell's Compass". *Cold Spring Harb. Perspect. Biol.* 1(3):a002964. doi:10.1101/cshperspect.a002964. PMID:20066108. PMCID:PMC2773631.

56. Goldberger, A.L. (1996). "Non-linear Dynamics for Clinicians: Chaos Theory, Fractals, and Complexity at the Bedside". *Lancet* 347(9011):1312–1314.

57. Meydani, S.N., Das, S.K., Pieper, C.F., Lewis, M.R., Klein, S., Dixit, V.D., Gupta, A.K., Villareal, D.T., Bhapkar, M., Huang, M., Fuss, P.J., Roberts, S.B., Holloszy, J.O. and Fontana, L. (2016). "Long-term Moderate Calorie Restriction Inhibits Inflammation without Impairing Cell-Mediated Immunity: A Randomized Controlled Trial in Non-Obese Humans". *Aging* (Albany NY) 8(7):1416–1431.

58. Singer, W. (1986). "The Brain as a Self-Organizing System". *Eur. Arch. Psychiatry Neurol. Sci.* 236(1):4–9. (b) Dioguardi, N. (1989). "The Liver as a Self-Organizing System. I. Theoretics of Its Representation". *Ric. Clin. Lab.* 19(4):281–299.

59. Boucher, J., Kleinridders, A. and Kahn, C.R. (2014). "Insulin Receptor Signaling in Normal and Insulin-Resistant States". *Cold Spring Harb. Perspect. Biol.* 6(1):a009191. doi:10.1101/cshperspect.a009191. PMID:24384568. PMCID:PMC3941218. (b) Rask-Madsen, C. and Kahn, C.R. (2012). "Tissue-Specific Insulin Signaling, Metabolic Syndrome, and Cardiovascular Disease". *Arterioscler. Thromb. Vasc. Biol.* 32(9):2052–2059. (c) Katic, M. and Kahn, C.R. (2005). "The Role of Insulin and IGF-1 Signaling in Longevity". *Cell. Mol. Life Sci.* 62(3):320–343.

60. Nice-sugar Study Investigators. (2009). "Intensive versus Conventional Glucose Control in Critically Ill Patients". *N. Engl. J. Med.* 2009(360):1283–1297.

61. Einstein, A. (1915). "Explanation of the Perihelion Motion of Mercury from the General Theory of Relativity". *Sitzungsber. Preuss. Akad. Wiss. Berl. (Math. Phys.)* 1915:831–839.

62. Dewar, R.C. (2005). "Maximum Entropy Production and the Fluctuation Theorem". *J. Phys. A Math. Gen.* 38(21):L371.

63. Crick, F. (1966). "Of Molecules and Men". 10. University of Washington Press, Seattle, WA.

64. Demetrius, L., Legendre, S. and Harremöes, P. (2009). "Evolutionary Entropy: A Predictor of Body Size, Metabolic Rate and Maximal Life Span". *Bull. Math. Biol.* 71(4):800–818.

65. Von Bertalanffy, L. (1968). "General System Theory". *New York* 41973(1968):40.

66. Lorenz, E.N. (1995). *The Essence of Chaos.* University of Washington Press, Seattle, WA State, USA.

67. Bawa, A.S. and Anilakumar, K.R. (2013). "Genetically Modified Foods: Safety, Risks and Public Concerns-A Review". *J. Food Sci. Technol.* 50(6):1035–1046. doi:10.1007/s13197-012-0899-1. (b) Karalis, D.T., Karalis, T., Karalis, S. and Kleisiari, A.S. (2020). "Genetically Modified Products, Perspectives and Challenges". *Cureus* 12(3):e7306. doi: 10.7759/cureus.7306.

68. Uzogara, S.G. (2000). "The Impact of Genetic Modification of Human Foods in the 21st Century: A Review". *Biotechnol. Adv.* 18(3):179–206. (b) Paitan, Y. (2018). "Current Trends in Antimicrobial Resistance of Escherichia Coli". *Curr. Top. Microbiol. Immunol.* 416:181–211.

69. Costenbader, E. and Valente, T.W. (2003). "The Stability of Centrality Measures When Networks are Sampled". *Soc. Netw.* 25(4):283–307.

70. Martin, F.P., Dumas, M.E., Wang, Y., Legido-Quigley, C., Yap, I.K., Tang, H., Zirah, S., Murphy, G.M., Cloarec, O., Lindon, J.C., Sprenger, N., Fay, L.B., Kochhar, S., van Bladeren, P., Holmes, E. and Nicholson, J.K. (2007). "A Top-Down Systems Biology View of Microbiome-Mammalian Metabolic Interactions in a Mouse Model". *Mol. Syst. Biol.* 3:112.

71. Smith, M.I., Yatsunenko, T., Manary, M.J., Trehan, I., Mkakosya, R., Cheng, J., Kau, A.L., Rich, S.S., Concannon, P., Mychaleckyj, J.C., Liu, J., Houpt, E., Li, J.V., Holmes, E., Nicholson, J., Knights, D., Ursell, L.K., Knight, R. and Gordon, J.I. (2013). "Gut Microbiomes of Malawian Twin Pairs Discordant For Kwashiorkor". *Science* 339(6119):548–554.

72. Hehemann, J.H., Kelly, A.G., Pudlo, N.A., Martens, E.C. and Boraston, A.B. (2012). "Bacteria of the Human Gut Microbiome Catabolize Red Seaweed Glycans with Carbohydrate-Active Enzyme Updates from Extrinsic Microbes". *Proc. Natl Acad. Sci. USA* 109(48):19786–19791. doi:10.1073/pnas.1211002109. Epub 2012 Nov 12. PMID:23150581. PMCID: PMC3511707. (b) Koyanagi, T., Nakagawa, A., Kiyohara, M., Matsui, H., Yamamoto, K., Barla, F., Take, H., Katsuyama, Y., Tsuji, A., Shijimaya, M., Nakamura, S., Minami, H., Enomoto, T., Katayama, T. and Kumagai, H. (2013). "Pyrosequencing Analysis of Microbiota in Kaburazushi, A Traditional Medieval Sushi in Japan". *Biosci. Biotechnol. Biochem.* 77(10):2125–2130. doi:10.1271/bbb.130550. Epub 2013 Oct 7. PMID:24096680.

73. Grasset, E. and Burcelin, R. (2019). "The Gut Microbiota to the Brain Axis in the Metabolic Control". *Rev. Endocr. Metab. Disord.* 20(4):427–438. (b) Burcelin, R. (2012). "Regulation of Metabolism: A Cross Talk between Gut Microbiota and Its Human Host". *Physiol. (Bethesda)* 27(5):300–307.

74. Sterling, P. (2012). "Allostasis: A Model of Predictive Regulation". *Physiol. Behav.* 106(1):5–15. (b) Sterling, P. (2014). "Homeostasis vs Allostasis: Implications for Brain Function and Mental Disorders". *JAMA Psychiatry* 71(10):1192–1193.

75. McEwen, B.S. (1998). "Protective and Damaging Effects of Stress Mediators". *N. Engl. J. Med.* 338(3):171–179.

76. Po, H.N. and Senozan, N.M. (2001). "The Henderson–Hasselbalch Equation: Its History and Limitations". *J. Chem. Educ.* 78(11):1499–1503.

77. Kleiman, A. and Tuckermann, J.P. (2007). "Glucocorticoid Receptor Action in Beneficial and Side Effects of Steroid Therapy: Lessons from Conditional Knockout Mice". *Mol. Cell. Endocrinol.* 275(1–2):98–108. (b) Kolber, B.J., Roberts, M.S., Howell, M.P., Wozniak, D.F., Sands, M.S. and Muglia, L.J. (2008). "Central Amygdala Glucocorticoid Receptor Action Promotes Fear-Associated CRH Activation and Conditioning". *Proc. Natl Acad. Sci. USA* 105(33):12004–12009. (c) Howell, M.P. and Muglia, L.J. (2006). "Effects of Genetically Altered Brain Glucocorticoid Receptor Action on Behavior and Adrenal Axis Regulation in Mice". *Front. Neuroendocrinol.* 27(3):275–284.

78. Tandogdu, Z. and Wagenlehner, F.M. (2016). "Global Epidemiology of Urinary Tract Infections". *Curr. Opin. Infect. Dis.* 29(1):73–79. (b) McLellan, L.K. and Hunstad, D.A. (2016). "Urinary Tract Infection: Pathogenesis and Outlook". *Trends Mol. Med.* 22(11):946–957.

5

Introduction to the Roadmap of Future Medicine: The Physiological Fitness Landscape

Chapter Overview

Is medicine en route to become a branch of applied science? Why is the concept of the physiological fitness landscape such a potentially powerful diagnostic and therapeutic tool? What is the link between insulin resistance, diabetes, and cancer? Symmetry breaking is a fundamental concept in physics, what meaning does it have in biology and medicine? What is hormesis and why is important? Can we handle the huge amount of medical data being generated and how? How can the physiological fitness landscape methodology help us navigate to optimize health, cure diseases and delay aging?

We have attempted to construct a rich tapestry of elements, interconnections, and relationships involving the most important aspects of metabolic health at all scales. It is tempting to say that the emerging picture shows a bottom-up hierarchy starting at the molecular level and concluding at the whole-organism representation of the metabolism's hierarchy of coherence and synchrony. However, such a sweeping generalization is too simplistic since the hierarchy involves a flow of signals in both directions, from molecules to cells to tissues and the entire body as well as a reverse flow of signaling which, in fact, begins with our interactions with the outside world. As it turns out, the outside world cannot be cleanly separated from ourselves, as we are social "animals" whole lives are tightly enmeshed with those of others close to us and with society in general. This is a source of both vitalizing stress and toxic stress, depending on the situation we are in, our individual sensitivities, and the severity as well as the duration of stresses. Another part of our "greater" selves is the microbiota, which is comprised of more genes and more cells than those that are an integral part of the human body. In a sense, we are indeed, what we eat. Another important relationship between our body and the external world concerns the biology of time and how we synchronize our internal biological clocks with circadian cycles of the Earth revolving and orbiting around the Sun. It is becoming increasingly clear that not respecting the cycles of a day, year, and lifetime results in negative consequences for our health. How then do we integrate the many factors, both internal and external to our bodies in an internally consistent, logical, and quantifiable framework that can guide us through life and also provide important information to the physicians in their diagnostic and therapeutic approaches aimed at maintaining optimal health and combating

disease? In this chapter, we have provided an outline for a future methodology that does exactly this and does it at an individual level. This framework is inspired by physics, and it is based on the identification of order and control parameters that are respectively response functions to external perturbations to the homeostatic equilibrium that living organisms tend to preserve. The response of our body to external perturbations, best exemplified by the stress test used to assess our cardiac health, is a measure of the system's flexibility or resistance to change. In terms of a mathematical formulation of the resultant picture of our state of health (or disease), we proposed to use the so-called physiological fitness landscape. This is a close analogy with the free energy function commonly used in thermodynamics and in the physics of phase transitions. This general formula is indeed a function of many parameters and forms a multi-dimensional manifold that allows navigation, akin to the use of GPS when we travel far and wide, but in this case the journey involves lifestyle choices and pharmacological interventions aimed at health-risk avoidance and maximum possible fitness. The input data for the construction of the personal physiological fitness landscape, of course, depend on our access to genomic, proteomic, metabolomics as well as physiological information about the construction and functioning of our body. Hence, in the future, the entire battery of big data analytics can be brought to bear on the resultant construction of precision medicine algorithms that the fitness landscape platform will enable. An additional important element in the development of this methodology is the aspect of time, more specifically aging processes. The landscape will be periodically updated, and its projections refined as new data become available. As always, the devil is in the detail, and we hope that this book has provided a sufficient amount of details to spawn future studies in the area of metabolism as both a branch of science and medicine. In Volume 2 of this book, we will delve deeply into the detailed descriptions of the key physiological aspects of human metabolism and its relationships to the state of optimum health, the onset of disease, disease progression, aging, and eventual demise.

5.1 Models Inspired by Physics Can Help Understand Biological Systems

Biological systems evolved from physical systems in the sense that negative entropy of self-organizing complex adaptive systems evolved from the molecular building blocks of physical systems as a result of "inventing" metabolism.

DOI: 10.1201/9781003149873-5

Metabolism results in biological systems being in a state removed from thermodynamic equilibrium for as long as they are alive. The death of biological systems brings them back to a thermodynamic equilibrium as soon as metabolic processes cease to exist. Furthermore, the driving force of the evolution of living systems is, ironically, the dissipation of pent-up solar heat "into the cool" of the organism with transiently reduced entropy. Ultimately, the process is faithful to the second law of thermodynamics returning greater heat and associated entropy to the universe than would have been possible in the absence of the formation of the negative entropic complex adaptive systems. The driving force of solar heat for the evolution of organized complexity is analogous to the coupling of exothermic reactions in biological systems to endothermic reactions requiring work. Biological systems do this continuously and cyclically for as long as they are fueled by metabolic energy. In this context, it makes sense to assume that the behavior of order parameters of biological systems in response to control parameters is analogous to physical systems undergoing a phase transition in terms of forming structural and functional order versus disorder. For example, heat applied to a permanent magnet raising its temperature above Curie temperature destroys the ordered symmetric state of magnetization resulting in a disordered structure of magnetic dipoles in a paramagnetic state. Analogously, any disease can be viewed as resulting from a phase transition from the state of health (order, synchrony) to a pathological state (disorder, dyssynchrony). This phase transition can always be identified with an order parameter (potentially with several components) that corresponds to a physiological marker of that particular disease (e.g. blood glucose level, creatinine concentration, blood pressure, etc.).

> *Dietary fat excess and inadequate fiber may break the ordered symmetry state of gut microbial diversity and diversity of communications with the human host leading to a disordered symmetry state with loss of such diversity. Similarly, the loss of gut microbial diversity in response to the external control parameter of an unhealthy diet may be viewed as inducing a phase transition breaking a metabolic symmetry ordered state of insulin sensitivity forming a disordered pathological new symmetry state of insulin resistance.*

The concepts of phase transitions taking place in physical systems, such as vapor to water and water to ice or a paramagnetic state to a ferromagnetic state, may be applied in principle to physiology. In fact, it is somewhat surprising that this has not been done on a systematic level before. In all these examples of physical system phase transitions as in physiology, there are no rules that can be applied in a reductionistic manner that predict the behavior of these complex systems composed of a multitude of molecular components. As in statistical physics, physiological systems require too many parameters to describe them with arbitrary precision, which may be upon the order of the number of constituent atoms, i.e. 10^{23} or higher making this problem intractable deterministically without making major simplifications. Additionally, the vast majority of these

parameters are not known (in fact, not even knowable), that is they are indeterminate, not to mention the complexity of their interactions. Taken together, the possibility for a full understanding of the elements of a complex biological system and predicting its trajectory (e.g. disease progression) is insuperable. However, all is not lost.

5.1.1 A Free-Energy Landscape Model

A similar conundrum plagued physics in the 19th and early 20th centuries, as it gathered reams of data on the thermodynamics of diverse physical systems composed of many particles. First-principles approaches to describe these systems were not possible but very insightful observations by theoretical physicists eventually resulted in yielding a practical solution. Some of the most ground-breaking advances made by the brilliant scientists of that time are due to Lev D. Landau. *He proposed a vast but powerful simplification that involves constructing a model free-energy landscape that hugely reduces the complexity, indeterminacy, and complicated nature of systems undergoing phase transitions. This landscape invokes metaphorical peaks and valleys where each valley is a parabola that represents a state of stability of a master (also called primary) order parameter. The peaks represent unstable states of the system from which it eventually transitions to the nearest valley. The deepest valley is an absolutely stable state, and any other valley represents a metastable state, i.e. a state with limited stability.*

In addition, the plot of this free-energy landscape is typically made in terms of a single variable, namely the primary order parameter, a parameter that crucially determines the state of the system. For a ferromagnetic-to-paramagnetic phase transition this order parameter is naturally chosen as the net magnetization, which is non-zero in the ferromagnetic phase and zero in the paramagnetic phase, hence it easily discriminates between these two states. In addition to these primary order parameters, there may be secondary order parameters when other physical characteristics of the system play a major (but not dominant) role. These secondary order parameters are, for obvious reasons, sometimes called slave modes. For example, some magnetic systems may also change their shape when their magnetization changes as a result of so-called magnetostrictive properties. In such cases, a secondary order parameter in the form of an elastic strain can be added to this description with attendant coupling terms in the free energy between magnetic and elastic order parameters. This can then describe both magnetization and shape deformation that can be associated with a phase transition.

It is plausible that such a free-energy landscape with several order parameters (each of which relates to one type of morbidity) can be introduced in the area of physiology as pertains to health and disease. *For example, obesity measured by the BMI (Body Mass Index) may be determined to be the primary order parameter while high blood pressure is a secondary order parameter. There may be several additional secondary order parameters describing co-morbidities or associated physiological abnormalities.*

Moreover, Landau simplified this free-energy landscape for the region in the vicinity of a phase transition with the fundamental notion of bifurcation in the free energy function. There, the free energy changes from a single parabola to a double parabola (with a double-well shape) when the system's stability changes on going from one phase (e.g. paramagnetic) to another (ferromagnetic). The paramagnetic phase corresponds to a single well with a stable equilibrium at the zero-magnetization point. The ferromagnetic phase corresponds to the presence of two minima (hence a double-well), each of which corresponds to an opposite orientation of the magnetization vector (up or down), due to an arbitrary selection of the magnetization axis. As the system changes stability from a single minimum to two minima (two valleys on the landscape), it undergoes a phase transition that is characterized by a transiently flat landscape. When this happens, the system is said to be infinitely susceptible to perturbations because there is no confining slope of the landscape allowing it to easily explore the vast expanse around it.

This free-energy description changing from a single- to a double-well metaphorically describes an entire system simplified to its "bare bones" by focusing only on its dependence on the essential (primary) order parameter and represents a practical (albeit approximate) way to describe the system. Of course, this can be made in a more complicated and accurate way by adding more features such as several secondary order parameters or more intricate free-energy landscapes with more valleys (all of which has been done by the scientists who applied these notions to various branches of physics).

Landau was awarded a Nobel Prize based on this work applied to superconductors and superfluids. Important consequences of the Landau theory of phase transitions include the derivation of scaling laws for the order parameter and the generalized susceptibility of the systems as well as the notion of hysteresis, which represents the memory of a physical system. Hysteresis means that when a control parameter directly interacting with the order parameter is varied, the response of the order parameter changes depending on the direction in which the control parameter is varied (upward or downward). This is well known in magnetic systems where magnetization is the order parameter and an externally applied magnetic field is a control parameter. The resultant magnetic hysteresis loop has been known to physicists and engineers for centuries. We have illustrated graphically these properties in Figure 5.1. This is a critically important concept because it allows predictability and control of a system that is very complex and nonlinear. We believe that similar behaviors can be found in physiology and medicine where both reversible and irreversible transitions in the status of the patient's health are well known to exist. In addition, a phase transition in the context of physiology can be a point in time when the person succumbs to a disease, or conversely, when that person fully recovers from that disease.

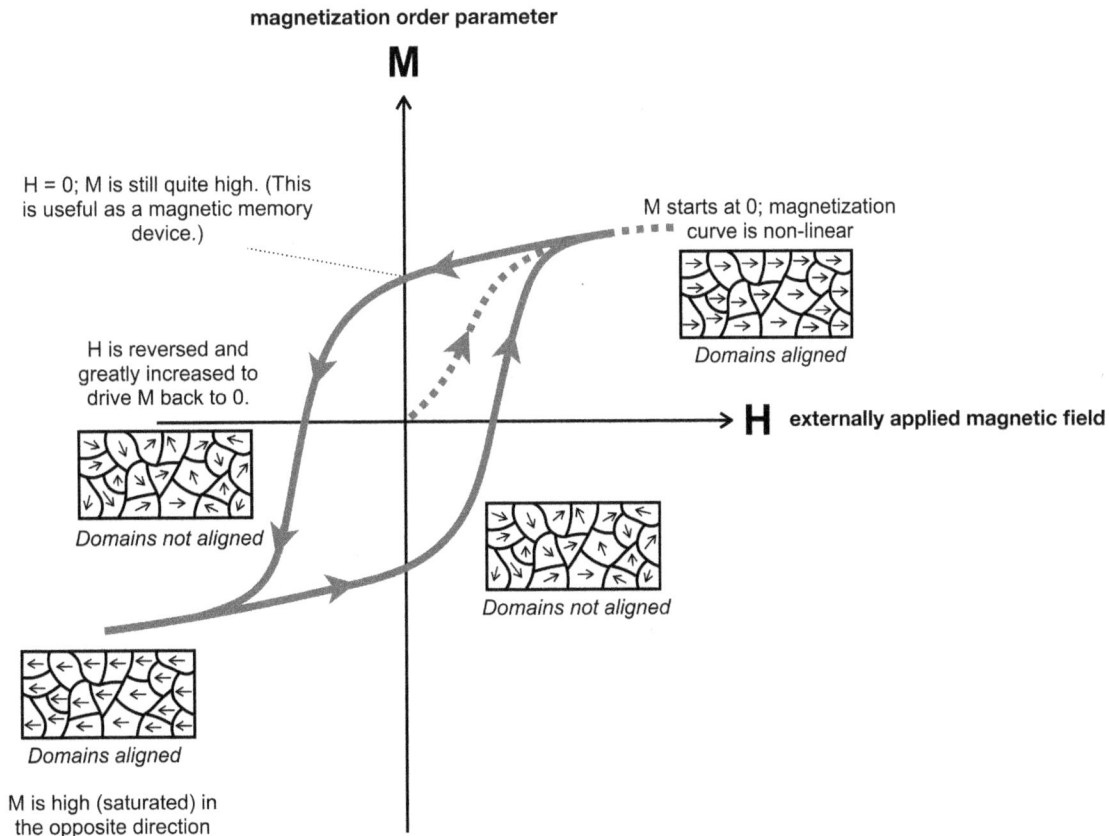

magnetization order parameter

M

H = 0; M is still quite high. (This is useful as a magnetic memory device.)

M starts at 0; magnetization curve is non-linear

Domains aligned

H is reversed and greatly increased to drive M back to 0.

Domains not aligned

Domains not aligned

H **externally applied magnetic field**

Domains aligned

M is high (saturated) in the opposite direction

FIGURE 5.1 Hysteresis loop showing that for a ferromagnetic material, its magnetization is "history dependent". Once the ferromagnetic material is saturated, the magnetic field H can be removed, and the material will keep most of its magnetization (it "remembers" the magnetic field). Source: adapted from Young, Hugh D. (1992). *University Physics*, 8th Ed. Addison-Wesley, Reading, Massachusetts.

The most important implication of Landau's work for medicine could prove to be its applicability to physiology. It should be appreciated that physics is a far more mature science than is biology and medicine and so intuitively its conceptual advances could prove invaluable if adapted as models with appropriate modifications in medicine both at the research and clinical levels.

> *The concept of a free-energy landscape described by Landau can, in the opinion of this writer, be successfully translated as a "physiological fitness landscape" and be directly applicable to physiology.*

It may be useful in understanding many unexpected outcomes in human experimentation, such as the increased mortality found in the international NICE SUGAR study [1] that provided tight glycemic control to critically ill patients using intravenous insulin, or in critical care patients who received growth hormone. Further extensions of the Landau methodology have been developed in physics as warranted by the complexity of the system studied. An external control parameter such as the application of a magnetic field can induce a dual ordered state phase transition with electron spins of every other atom aligned in alternating opposite directions from a completely randomly oriented electron spins of atoms in a paramagnetic state. Interestingly, this is known as an anti-ferromagnetic state whose spins form two intercalated sublattices, with sublattice magnetizations that are oriented in opposite directions adding up to no net magnetization overall. This intricate symmetry is hidden within this seemingly non-magnetized system. If an additional small magnetic field is applied, electron spins of one of the subsystems can flip over resulting in a field-induced phase transition and a net magnetization of the entire system that is now in a ferromagnetic state.

A metaphor of such a phase transition in physiology may be the coexistence in a group of cells of a tissue whereby some cells operate in a glycolytic fashion and others in the mode of oxidative phosphorylation. Triggering one of the domains to switch its metabolic state into the other by applying a pharmacological agent represents a phase transition, which in this case is not spontaneous but field-induced using the terminology of the physics of phase transitions. For example, a drug such as dichloroacetate (DCA) activates the enzyme complex pyruvate dehydrogenase by promoting its dephosphorylation. This results in the enzymatic decarboxylation of pyruvate to acetyl coenzyme A (CoA) in the mitochondria converting glycolytic bioenergetic domain to oxidative phosphorylation bioenergetics. *This may prove useful in reversing an insulin-resistant susceptibility state of cancer as a phase transition or even inducing a phase transition that reverses the cancer state itself, which is probably in many cases, especially in early stages, close to the criticality threshold and hence still reversible.*

Alternatively, a metaphor to the induction of a two-component order parameter phase transition in physiology analogous to the induction of an anti-ferromagnetic state from a paramagnetic state by the application of a magnetic field may be exemplified by applying an agent such as pioglitazone or metformin. These drugs may both increase the size and distensibility of subcutaneous adipocytes as well as promote subcutaneous adipocyte replication. Each phenotypic phase coexists in an analogous fashion to that of the coexisting subsystems in the anti-ferromagnet. Pioglitazone and metformin may change the topography of adipose tissue that increases the capacity for neutral fat storage in subcutaneous adipose tissue while reducing the potentially more deleterious storage of fat in the liver, reducing hepatic steatosis, and in the visceral adipose depot. This profile of fat storage improves metabolic health despite an overall increase in adiposity.

In terms of the example above involving two coexisting bioenergetic pathways in a group of cells of a tissue, some cells working through the glycolytic domain and others through the oxidative phosphorylation domain, this may represent a susceptibility state for chronic diseases of aging, the particular disease to which is susceptible depends on genetic predisposition. In the case of cancer, the cell membrane potential represents at least an epiphenomenon of bioenergetic potential and the correlated nature of ATP production and coherent work processes of maintaining self-differentiation and coherent intercellular behavior.

> *Insulin resistance and mitochondrial dysfunction impair this process of quantum metabolism, which is responsible for intra- and inter-cellular coherence and related work. Impaired bioenergetics result in a shift from oxidative phosphorylation to glycolysis.*

The Warburg effect mediates cancer proliferation giving cancer cells a competitive advantage over normal cells in a hypoxic environment. Extrinsic control parameters that may be used therapeutically include insulin-sensitizing measures. There are a number of these including pharmacologic agents and lifestyle changes. Lifestyle changes may include caloric restriction and intermittent fasting, among other measures. Resonance biophoton therapy or vitamin and mineral supplements may represent additional potential control parameters as is mitochondrial nutritive support.

5.1.2 Biological Motors as Mechanical Engines

> *In this book, we argue that metabolic impairments are central to the development of numerous diseases. Cell metabolism utilizes molecular motors as discussed earlier and there was a parallel drawn between biological motors and mechanical engines. Here, we discuss in more detail the similarities and differences between biological and physical engines.*

Both types of engines utilize and require hydrocarbon fuel. In machine engines, this fuel is in the form of petrochemicals or fossil fuel. In the case of biological engines, it requires nutrient dietary hydrocarbons as fuel. In a biological engine, enzymes of glycolysis, particularly hexokinase or glucokinase in skeletal muscle and liver respectively, and phosphofructokinase (PFK) spark the flow of energy and represent the analog of the spark plug of an internal combustion engine. The hydrocarbon fuel undergoes an explosive process of combustion in the presence of oxygen in the combustion chamber. Analogously, in a biological engine, this oxidative combustion occurs in mitochondria.

The products of this combustion process of fuel in the presence of oxygen are carbon dioxide and water vapors. The accompanying heat that is liberated from the reconfiguration of the bonds of the hydrocarbons of fossil fuel or dietary nutrients into the lower potential energy of more stable bonds of carbon dioxide and water vapor molecules in what is called an exothermic reaction (heat-liberating). The heat-induced expansion of carbon dioxide and water vapors that cause the compression of the cylinder piston corresponds in a biological engine to the analogous expansion of heat that is transferred into the bonds of ATP that subsequently following hydrolysis transfers that energy as work into the motion of muscle contraction as the clearest example within the biological engine. Thus, the compression of the cylinder piston that represents the work (force x distance) translated into motion, described in mechanical engines as power (work per unit time), is an easy comparative metaphor to the work or power, respectively, of muscle contraction of the biological engine.

It should be recognized that the information content of the biological engine (biological chemistry held together by the glue or potential energy of its bonds) when combined with additional energy, in the form of either heat or work, becomes transformed into something of greater value. The example above involves muscle contraction, which because it involves work or power, energy in this case is transferred into mechanical movement. The body has many types of molecular motors, one type of such motor protein, myosin, mediates muscle contraction when organized into thin and thick filaments with actin. The other major type of work, also mediated by the hydrolysis of ATP, in biological engines, is involved in active ion transport. The energy that is transferred as heat in biological engines also typically mediated by ATP hydrolysis, drives anabolism.

Although the above description highlights the metaphorical comparison between a machine and a biological engine, the component of molecular motors underscores another important, even more important similarity. *The universal currency of energy in biological systems, ATP, which is predominantly produced in the metaphorical combustion chamber is mediated by the molecular motor ATP synthase which is represented as Complex V of the electron transport chain. This motor bears striking and even stunning similarity to a mechanical electric motor in each case constructed with a stator and a rotor as its main components. The implication for this is how designs of nature should be carefully studied not only for mimicking in the field of biotechnology, but to understand them in order to facilitate better outcomes in the practice of medicine.*

Fuel may be in the form of gasoline hydrocarbons in the case of a machine engine or glucose in the case of a biological engine. The metaphorical sparkplug of a biological engine is the hydrolysis of ATP that generates the flow of energy and the process of combustion of hydrocarbons. Specifically, the catalyst of hexokinase or glucokinase which phosphorylates glucose trapping it inside the cell is the initial catalyst followed by a major regulatory enzyme of glycolysis, phosphofructokinase.

The combustion chamber of a mechanical engine is where fuel is oxidized with the addition of heat from the spark plug resulting in an explosive process whereby the combustion of petroleum fuel leads to the heat-induced expansion of carbon dioxide and water vapor molecules as products of this combustion. Analogously, in the biological engine the combustion chamber is the mitochondria and particularly the electron transport chain whereby the combustion of nutrient hydrocarbons occurs in the presence of oxygen to produce ATP.

The hydrolysis of ATP and the biological engine equates to compression of the cylinder piston by the heat-induced expansion of the carbon dioxide and water vapor molecules of the mechanical engine. The transfer of work from ATP to muscle contraction systems (or other molecular motors or active ion transport that translates to microscopic and macroscopic scales of movement) equates to the transfer of work from the expansion of the heat-induced vapors of a mechanical engine due to the compressing cylinder piston. The power of a mechanical engine relates to the movement of the cylinder piston and ultimately the crankshaft rotations. The engine is what drives the motion of the machine and consists of the combustion chamber and the cylinder piston. This engine is the hardware as well as the information (information representing a substrate that when energy is added to it, it is transferred as work). Hence the combustion chamber and cylinder piston may be considered information content of a mechanical engine whereby the addition of energy as heat produces the work of moving the cylinder piston.

Metaphorically, the biological engine includes the mitochondria (combustion chamber), the ATP (and its hydrolysis), molecular motors, and the macroscopic scales of all the moving parts to which the component parts of the biological engine are formed at micro- and macroscopic scales. For example, the transfer of work from the potential energy of ATP to muscle contraction is analogous to the transfer of work from the heat-induced expansion of carbon dioxide and water vapor products of combustion to the compression of the cylinder piston whereas the engine in a mechanical device is confined to the combustion chamber and the cylinder piston. *In a biological engine, the hydrolysis of ATP promotes muscle contraction mediated by molecular motors as well as virtually every other work-related process of the body is driven by the hydrolysis of ATP. Thus, the totality of the human body may be considered a biological engine.* Notably, information that is transformed from one form to another as a result of the transfer of energy as either heat or as work allows the transformation of that information to something of greater value or usefulness.

5.1.3 Biological Thermodynamic Engines

Biological motors are central force generation elements to the machinery of living systems just as mechanical motors are to a car engine. In both cases the motor is responsible for converting one form of energy into a force that generates motion involving either translational or rotational kinetic energy. In a limited but most common notion of a motor, it implies an electrical device as an arrangement of a stator (stationary device) containing a permanent magnet with a rotor (rotating device) containing wire coil windings as an electromagnet connected to a power source (a battery or fuel cell). The rotor rotates hence turning the shaft as a result of the force generated by the interaction between the electromagnet of the rotor with the permanent magnet of the stator.

Similarly, an engine in a limited sense refers to a piston cylinder device. However, in a broader sense, a motor is a subtype of engine that produces motion. By distinction, an engine takes in energy per unit time and possibly raw materials and converts those into useful outputs. It is a more or less efficient transformation of one form of energy into another (e.g. chemical into mechanical). A biological system as an organism may be considered not only a biological engine but a complex structure composed of several types of biological engines as well as motors intricately coordinated into a single larger system. In general, *biological motors at a subcellular level can be divided into: 1) linear motor proteins that move in a processive fashion along protein filaments, e.g. kinesin along microtubules or myosin along actin filaments, 2) rotational motors such as F-ATPases, 3) beating complexes of motor proteins and filaments such as the microtubule-dynein arrangements in cilia and flagella, and 4) contractile arrangements of interconnected long and short filaments involving actin-myosin complexes.*

One molecular motor, ATP synthase, is particularly structurally and functionally analogous to the electrical motor. It contains a stationary stator with which the rotor, the rotating component is associated. The stationary stator anchors the complex within the bi-lipid inner mitochondrial membrane including the knob, the site for catalyzing ADP to ATP, and the hydrogen ion channel for protons to be transported through the channel to participate in the conversion of ADP to ATP. Notably, Complex V of the electron transport chain, or ATP synthase, is known to be involved in iron metabolism. It is intriguing to consider if iron is mechanistically a controlling factor in the rotational component of ATP synthase. Iron, having a net magnetic moment, is polarized in a magnetic field, which is produced by the electric current generated in the electron transport chain (from the moving electrons as well as the molecules containing iron) whereby the electrons ultimately combine with oxygen to produce water at Complex IV. Therefore, it is conceivable that the interaction of the moving electrons, the electromagnet, and iron, the permanent magnet expected to be present in Complex V, ATP synthase, may be responsible for the torque applied to the rotor component of Complex V. Regardless, however, of whether this rotational mechanism is due to the interaction of iron (a permanent magnet) and electron transport (a conductor analogous to a copper wire producing a magnetic field around it), a spectacularly similar engineering design is seen in how nature has devised a molecular biological motor and how a man-made electrical motor is built. This observation highlights that studying nature provides the best blueprint for designing machines.

Biological thermodynamics concerns energy transformation in living matter, which in fact is the title of the German-British medical doctor and biochemist Hans Krebs' 1957 book [2]. Famously, the mitochondrial TCA cycle was named after and described by Hans Krebs. *The focus of biological thermodynamics is on principles of chemical thermodynamics applied to biology and biochemistry. The concepts employed here include the first two laws of thermodynamics, Gibbs free energy and the origin of life. Accordingly, thermodynamic biological engines may be considered to include far-from-equilibrium thermodynamic states in which the concept Gibbs free energy is invoked to create entropy reduction or, equivalently, an increase in information.* Such information gain in biology is an essential characteristic of living systems where it is generated across many hierarchical organizational scales ranging all the way from elementary particle physics (to accommodate physical principles) to the scales of molecules and macromolecules subsequently leading to the transition to life's building blocks such as organelles, cells, tissues, and organs.

Finally, on the scale of biological organization, information is further produced by entropy reduction due to structure formation enabling information processing powered by energy input from nutrients. It should be noted that all prokaryotic and eukaryotic organisms, either directly or indirectly, are dependent on the energetic input from the sun. The fundamental outcome of the theory of special relativity as mentioned above is the equation $E = mc^2$. This relationship may be interpreted to be a quantum mechanical relationship highlighting the equivalence between mass and energy. Energy, which was initially produced from mass by the process of nuclear fusion in the Sun is initially reconverted from the form of electromagnetic rays back to mass in prokaryotes and plants by the quantum mechanical process of photosynthesis. Subsequently, the spectacular phenomenon of quantum mechanical transduction again appears to manifest itself in specialized cells or organelles, mitochondria, the powerhouse of cells. *The mitochondria are the cell's most efficient energy-producing factory whereby nutrient hydrocarbons are stripped of electrons with the ultimate transformation to the universal biological currency of energy, ATP, sometimes referred to as the biological energy quantum. In this sense, the dependence of biological systems on mitochondria and hence on quantum mechanics provides a fundamental basis for considering organismic life as a quantum mechanical–biological engine. Furthermore, there are additional contexts in which higher-level organisms possess quantum mechanical thermodynamic engines.*

Important is the notion of quantum metabolism described by Lloyd Demetrius and advanced by Jack Tuszynski. Quantum metabolism is a function of the coherent synchronized ATP production as a macroscopic orchestration of mitochondrial metabolism. Structured water, first described by Mae Wan Ho has quantum mechanical properties in biological systems and coordinates intra-cellular activities by providing a medium for long-range order, which has been well-documented by Gerald Pollack. Hence, *this so-called fourth phase of water discovered and described by Pollack may transfer information between disparate regions of the body, such as from the microtubules in the neurons of the brain to the metabolic enzymes in the cells of visceral organs or of skeletal muscles of the extremities or perhaps even to the motor proteins myosin present in skeletal or cardiac muscle.*

Wan Ho proposed that water aligns in a highly constrained fashion within the nanotubes of the triple helices of collagen that span throughout the entirety of the extracellular space of the body. It transfers the information of specific resonance frequencies from where it originated (e.g. the neuronal microtubules in the central nervous system, to intracellular domains elsewhere in the body). Upon transfer of information from the extracellular to the intracellular space the water molecules align as the conduction of electricity *sine qua non* propagation of hydronium ions constrained within the nanotubes of double-stranded cytoskeletal filaments of actin (in the intracellular space) until it reaches and transfers its information to specific targets. The targeted structures may be those with the same frequency *via* a phenomenon called resonance. The amplification of that frequency may be what is required to impact the function of those structures. In this fashion the so-called *positive electricity represents a flow of information approaching the speed of light so that it underpins a potential synchronized and coherent organismic level integration of function from the brain (controlled by the mind) to all parts of the body. The concept of biological water can hence be recognized as part and parcel of a quantum mechanical engine.* Perhaps this is a manifestation of a well-orchestrated symphony integrated with the metabolic activity of mitochondria and the process of oxidative phosphorylation and quantum metabolism on a macroscopic scale.

Critically, structurally organized and stabilized microtubules in neurons of the brain appear to be key players in the formation of consciousness, definable as an awareness of being aware, or as higher-level cognitions. Microtubules, as the mediating structures of sensory input, are information processors, translating information into an output response. Information comes in from our sensory neurons of the optic, auditory, olfactory, and somatosensory nerves, which ultimately synapse with neurons in higher-level centers of the brain within which microtubules integrate this sensory input processing and coupling it to mental cognition. These cognitions are the manifestations of awareness. The depth of that awareness may be considered a non-intellectual or emotional IQ, a form of perceptiveness. *Simple awareness of any sentient organism is what is derived from the senses independent of cognition. Awareness of being aware is a proposed minimum criterion defining consciousness.* An awareness of being aware may be exemplified by the pride that one may feel of an intellectual achievement or the anxiety and depression one perceives how others may think negatively of him or her. Higher-level degrees of consciousness may include an awareness that one is aware of being aware, and so on, as may be exemplified by transcendental meditation.

Although intellectual cognition may also be mediated at the level of neuronal microtubules in the brain as information processors as well as storage and retrieval of memory such as from the hippocampal neurons, this is not a defining feature of consciousness *per se*. The notion of consciousness equates to many on an intuitive level to the "soul" or identity of an individual. It is this soul or identity of consciousness, which is imponderable in terms of where it comes from that defines us and where it goes when we die. One interpretative description of consciousness includes both an awareness of being aware, particularly but not exclusively limited to human beings, and is likely fundamentally rooted in quantum mechanics. The physical substrate for this quantum phenomenon appears to be the microtubules of neurons in the brain. However, there is a daunting impossibility of determining the origin of a quantum process of cognition beyond the microtubule as well as of predicting the destiny of our identity beyond our mortal lifespan.

Quantum neurobiological processes are characteristically coherent, i.e. information is shared between sites effectively. Additionally, the sites have waveforms that are synchronous, that is, are in phase whereby the peaks and valleys happen at the same time in both waveforms. There is an in-phase timing relationship also referred to as spatio-temporal correlation. The flow of energy and information in quantum systems may be unconstrained spatially and temporally such that manifestations of quantum phenomena may occur instantaneously and simultaneously between spatially distant areas. Mae-Wan Ho argues this in the case of information carried by electrically activated water through the nanotubes of collagen throughout the extracellular space of the body. In this sense, time dilation may be coincident with the unconstrained spatial domain to the flow of information and/or energy. This, in a very intriguing way, parallels the length contraction-time dilation aspect of the theory of special relativity.

The concept of quantum metabolism is similarly rooted in a mathematical formalism invoking Debye's specific heat of a solid to explain the metabolic rate of a living organism as a manifestation of quantum behavior. The formal analogy between thermodynamic variables and metabolic parameters also includes the parallel between the absolute temperature of an inanimate solid and a metabolic cycle time of an organism. *The basic premise is that characteristic time does not exist in nonliving systems while it is a critically important parameter in living systems as they span a range of time scales in a hierarchical fashion from protein- to cell- to tissue- to organism-levels. Conversely, most living systems are isothermal, and temperature must be maintained within a narrow range to sustain the living metabolic state, hence temperature as a nearly constant physical variable is not relevant in biology. While time is a characteristic of living systems on each scale or organization, the greater the quantum nature of that living system defined by quantum metabolism, the more dilated time becomes, i.e. the more slowly it moves. Systems defined by quantum metabolism have a correlated energy production in terms of coherence and synchronicity on a macroscopic scale of ATP energy to biological work and entropy reduction. This amounts to optimal efficiency of energetics and transformation of Gibbs free energy to do useful work.*

However, with natural senescence, the maximal entropy reduction, and metabolic stability that was achieved at a young age, it gradually gets degraded by the inevitable accumulation of reactive oxygen species (ROS) and inflammation. These destructive processes are inextricably linked. Reactive oxygen species and reactive nitrogen species bind to DNA, protein, and lipid components that contain biological information (negative entropy) disrupting its structural and functional integrity, and in the case of DNA, potentially causing mutations. Oxidative stress elicits an inflammation response by up-regulating the expression of NF-κB, a central hub transcription factor with a role to amplify the open processes of inflammation and reactive oxygen species. Hence, the "regulation" is actually a feed-forward dysregulation. *With aging, mitochondrial structure and function decline as a hallmark of the reduction of metabolic stability and loss of quantum metabolism as well as other likely integrated quantum biological processes.* Thus, from the perspective of classical physics, the question of where our consciousness goes when we die, may be answered in physical terms by the second law of thermodynamics. That is, it is simply dissipated as heat that increases the entropy of the universe.

Moreover, *as our consciousness, or individuality is lost, along with the other biological manifestations of quantum metabolism, the loss of metabolic stability and the passage of time become more accelerated.* Analogous to a quantum computer, which has vast cryogenic requirements to prevent quantum decoherence, perhaps the heat associated with the natural inflammatory states of aging can be thought of in the same sense. However, it is understandable that this explanation will be unsatisfying and disquieting to those who believe the soul does not die. It is quite possible that it does not, and the alternative views do not necessarily need to be mutually exclusive. Indeed, there are plausible theories to suggest that the universe is deeply entangled at scales far more macroscopic than the boundaries of an individual. There are views such as the many-worlds theory which posits essentially that an individual's "soul" or "consciousness" is not destroyed by the mortal processes of inflammation and oxidative stress, but rather transferred to the entanglement of other dimensions to which our awareness in the current existence does not extend, i.e. is 'blind'.

Another, more mechanistic, natural question arises regarding how may the integration of quantum metabolism, microtubules, and consciousness or quantum cognitions be assimilated? Although the electric fields generated by the tubulin arrangements comprising microtubules may not be explained by classical science, the neural sensory input into the brain from the surrounding environment in the presence of accompanying neuronal influences such as emotional memories all contribute to the formation of cognition. *Classical components of neuronal activity include in part, the release of neurotransmitters from the terminals of neurons into synaptic clefts resulting in the opening of cation channels (e.g. sodium and potassium channels).* This promotes a change in the transmembrane potential with depolarization and firing of action potentials. The influx of cations generates ionic waves, which propagate along both actin filaments and microtubules of the neuronal cytoskeleton, although the mechanism of wave propagation is different, and they play different roles in directing neuronal behavior.

The underlying mechanisms may be even more complex. Microtubules or indeed other biomolecules such as the DNA, may have a large number of degrees of freedom, some of which can operate in the classical realm, e.g. vibrational dynamics, while others, e.g. tryptophan excitations may operate in the quantum domain. These two types of degrees of freedom (classical and quantum) may even interact with each other such that classical dynamics may affect quantum behavior. Cations are counterions attracted by the negatively charged amino acids of tubulin or to the surface of actin double-helical filaments. Ionic waves of these cations have been shown experimentally and described theoretically to form and propagate as soliton-like waves as ionic "clouds" along actin filaments virtually without loss of wave amplitude. When the wave reaches the end of the actin filament it can trigger a polarization event, for example causing neurotransmitter release from a synaptic button. Similarly, cations as counterions are attracted by the negatively charged amino acids of tubulins on the surface of microtubules. However, the conduction of ionic waves along microtubules may give rise to spiraling signals, which may involve an electric dipole type phenomenon (discussed elsewhere in this book) that acts like a transistor in the sense that it significantly amplifies the input wave signal.

Microtubules can also propagate ionic waves into their contact with actin filaments, which can then relay these waves all the way to ion channels. It is also known that microtubules interact with some ion channels directly. Therefore, these bioelectric elements are unlikely to act in isolation but instead, can form integrated bioelectric networks or indeed logical circuits with sophisticated and complex functional characteristics. Moreover, an integrated model of the quantum nature of consciousness, i.e. of cognitions, with the notion of quantum metabolism mediated by coherent mitochondrial production of ATP on a macroscopic scale could be considered in this connection. Interestingly, in the context of cellular metabolism, microtubules associate with mitochondria which move within the cell along microtubule tracks using the energy they produce in the form of ATP molecules as fuel for motor proteins such as kinesin and dynein that propel them along these microtubule tracks. There may also be additional forms of interactions between microtubules and mitochondria since mitochondria produce electric fields and emit weak electromagnetic radiation during their activities. *Electric fields are produced by mitochondria established by the proton gradient across the inner mitochondrial membrane.* Additionally, weak electromagnetic radiation is emitted in the form of biophotons produced by the recombination of reactive oxygen species (see discussion of biophotons in this volume). How these complex intra-cellular structures orchestrate their activities using physical and chemical signaling is largely unknown but future research is bound to discover an amazingly complex and coordinated world of molecular interactions, most likely showing a great amount of purpose and determination in their dynamics far different from a jumble of randomly colliding Brownian particles.

Microtubules, in the immediate proximity of mitochondria, may absorb electric signals, transducing them into bits of information stored in the subtle structural changes such as C-termini states or electronic transitions. This information can possibly affect the way microtubules conduct ionic waves and hence may influence the whole cell's response to stimuli. In terms of ionic waves, albeit they exhibit classical behavior *per se*, they may explicitly play a role in positively or negatively influencing quantum effects. This would be of special significance in neurons where a change in the cell's threshold for firing an action potential can have an effect on the synaptic connection between neurons.

Ionic waves may, for example, exert a number of differential effects that may potentially enhance or inhibit quantum regime, which consonantly improves or impairs respectively the quality of mental and emotional cognition. For example, ionic waves may enhance motor protein binding to microtubules. *Motor proteins are responsible for moving mitochondria along microtubules, due to the notion of quantum metabolism motor traffic may be closely integrated with neuronal cell quantum processes. Alternatively, ionic waves may promote microtubule disassembly and accordingly discourage microtubule-mediated neuronal quantum effects. Microtubules whether as part of a quantum-mediated function or a classical function, the latter possibly related to ionic waves independent of microtubule function per se, may influence response to neurotransmitters and hence affect depolarization and neuronal firing. These mechanisms, in turn, can connect groups of neurons and make them more or less active, eventually affecting larger-scale cognitive processes in either a positive or negative way depending on the particular mechanism and the specifics of these interactions.*

5.1.4 Framing Energy by the Creation of Time and Life

The panoramic perspective of energy transformations includes its origin from solar nuclear fusion and nuclear fission. Its transformation is mediated by the forces of nature, which ultimately direct it into the self-organizing biologically complex adaptive structures of which we as human beings are the apex predator. Scholarship in the sense of knowledge and learning in the sciences and in clinical medicine fundamentally requires integration at all levels to satisfy deeper understanding and meaning. This also includes the knowledge of energy, its sources, its transfer routes, and its dissipation. Where does energy come from? How is it used to drive our life processes as living healthy beings? Where does it go when we die and how is it responsible for our senescence and chronic disease?

The physical laws of nature treat energy as a central property of matter. On our planet, most of energy supply is derived from the reduction of atomic mass in the sun, which perpetuates the notion of cyclicity borrowed from natural phenomena that germinate and sustain the rhythm of life. *Our biological systems on an individual scale dance in cycles described as circadian (e.g. 24-hour sleep–wake patterns), infradian (e.g. greater than 24-hour menstrual patterns,) and ultradian (e.g. less than 24 hour hormonal releasing and heart rate patterns). On a population scale, life–death cycles describe the ebb and flow of generations. It is in fact more mundane and practical than just a cultural enlightenment for physicians to appreciate the flow of energy occurring both internal to our cutaneous boundaries and external as well, the latter (not addressed in this book) with sociological and spiritual implications.* In addition, the concept of symmetry breaking, including reductionism, the crown jewel of modern scientific analytical thought, is inextricably linked to the flow of energy that forms the structures and functions of life.

The four fundamental forces and 12 particles of subatomic physics provide the matter and energy framework for structure formation and their combinations give rise to the almost infinite complexity of chemistry. This then leads to the flow of information that is responsible for the physical universe and the resplendent complexity and beauty of biology and the systems of nature, which have evolved here on Earth. The universe constructs time by the presence of cycles exemplified by the motion of electrons around atomic nuclei as well as orbits of planets around the Sun. The Earth orbiting the Sun and the Moon orbiting the Earth is responsible for the daily light–dark cycles on Earth and the longer lunar cycles, respectively. Light–dark cycles are responsible for the origin of life as well as the circadian clocks of many biological functions (the cycles of the body) that sustain life. *The first two laws of thermodynamics frame the laws of energy and entropy of the physical universe. The first law frames the time-invariant nature of energy-matter as the axis for the cycle of time; the second law, which defines systems as always moving unidirectionally towards increasing entropy, frames the arrow of time.*

Energy flow also affects the flow of information as the energy interfaces with our body's molecular chemistry and biochemistry. Free energy can be seen as constructive when structure and function are created, increased, or maintained. If destructive, free energy is lost to heat and associated entropy, to use the lexicon of physical thermodynamics, or to oxidative stress and inflammation, to use the lexicon of pathobiology and pathophysiology of clinical disease.

The first law of thermodynamics underpins time in an absolute sense of energy preservation. The second law underscores time in a relative or temporary sense in terms of the loss or dissipation of information energy of our biological structures and functions as heat randomizes away from the complex molecular organization of our systems biology to the entropy of the universe. The combination of these two laws of thermodynamics and our circadian clocks creates the foundation of biological time and the process of aging relative to energy flow within the physical constraints of the boundaries of a biological system.

The transition of inanimate matter to living systems may largely be described by systems going from thermodynamic equilibrium states to far-from-equilibrium steady states. This breaks the symmetry of reductionism (that is, time invariance of the physical laws of the universe—equations that are deterministic and reductive, e.g. Newton's laws) imposed by the first law of thermodynamics, obeyed by biological systems. All life forms represent a subsystem of the universe, which internally reduces entropy by being alive but taken together with their surroundings uphold the second law of thermodynamics since the universe moves towards increasing entropy. *In the process of being alive, new symmetries arise from adaptations of biological stability to the physical constraints within which biology functions. This is anchored by time-variant steady states (steady states that vary over time, hence are quasi-steady in a strict sense). The arrow and the cycles of time describe the biology of time, which breaks the symmetry of time invariance in physical systems and underscores the nonlinearity of systems biology and the limitations of reductionism.*

From the perspective of clinical practitioners, fundamental to the understanding of nonlinear biological systems is breaking the time-invariant symmetry of reductionism. Genomics, which is a branch of science that studies the expression of our genes, may be described as the generation of systems that are attracted to states of local stability, or steady states. Following a perturbation incurred from environmental conditions, these systems will eventually predictably settle back to the point of these stable or steady states. These are stable states known as attractors. Many systems however have so-called strange attractors, that is, following an outside perturbation the system is pulled in an unpredictable (stochastic) way toward more than one attractor state and exhibits chaotic transitions between these two or more attractors. This variation in steady states is a true property of this chaotic system. It is not due to noise in the system, or imprecision, for example by imprecise measurements of parameters of the system that have been perturbed. Similarly, this variation does not reflect that the system has been perturbed and is on its way back to its stable attractor steady-state but not yet reached it, as would be the case in a periodic predictable steady state of reductionist systems. Rather, the variation is inherent in the system. *Each new steady state following a perturbation approximates but never becomes the same as the previous or original steady state. Furthermore, once a new steady state reaches a critical threshold, a point of criticality, the system changes behavior qualitatively with an entirely different steady state in response to any, even minimal, subsequent perturbation.*

On a two-dimensional dynamic (over time) computerized portraits (of phase space) illustrating the trajectory of the fractal system (characteristic of chaotic systems) will demonstrate a folding pattern. This is exemplified by the iconic Butterfly Effect, the original Lorenz attractor describing the system of weather patterns as a result of slight perturbation causing the folding over of the trajectory of the attractor to form the opposite wing of the butterfly (see the section on chaotic systems). Over time, the trajectory of the system may have reached its stable steady state attractor, but because of the inherent variation of the fractal pattern of chaotic systems this state may approach a threshold of criticality whereby the slightest perturbation causes it to qualitatively change behavior, folding back over to the other wing of the butterfly which itself is another steady state. The attractor thereafter will flip back and forth unpredictably in the case of this system forming the appearance of a butterfly.

The phase portrait depiction of a system represents the composite information of control parameters such as in the case of the butterfly representing weather patterns, the control parameters include such properties as temperature and pressure. These parameters code for a pattern that creates the phase portrait represented by a trajectory that is infinitely long in a finite space. As pointed out above, an important characteristic of these chaotic systems is their fractal (or self-similar) nature. The infinitely long trajectory represents the computerized phase portrait of the system, which is very different from a regular orbit in a two-dimensional phase space in a linear system, such as a circle for a harmonic oscillator. The corresponding dimension has a numerical value which is a fraction of the integer. Hence it is referred to as a fractal dimension which is a measure of the space-filling capacity of a pattern that tells how a fractal scales differently from the space it is embedded in (3). *The other distinction of a fractal pattern is that it is scale-invariant. That is, its appearance is the same independent of the magnification.* It looks the same at all magnifications independent of complexity or variability that is independent of how dense the fractal pattern is at the different scales of magnification. These fractal patterns of strange attractors take on beautiful shapes, which may resemble forms not only such as butterfly wings, but as strolling clouds, spider's web, fern leaves, tree branches, blood vessels in the liver, and other shapes.

Reductionism can be viewed as a periodic pattern. That is, it is inherently time-invariant or independent such that over time the pattern is exactly the same. Environmentally induced perturbations are only transient displacements from the preexisting steady state to which it will continue to be attracted as a low energy state of the system. This periodicity or clocklike cyclical patterns are representations of symmetry with conservation of the system as a manifestation of the laws of physics. *The cyclicity around the axis of time maintains the system at its existing steady state. Following perturbations, the system seeks to re-establish that steady state. Again, time invariance defines a form of symmetry that conserves energy in matter, hence, the dynamic behavior of the energy in matter (inextricably linked); information does not change over time. Biological steady states are transient representations of time invariance in living systems. However, unlike the symmetry time invariance, which conserves energy/matter, the time-invariant steady states in biological systems are only transient because they are subject not only to the first but also to the second law of thermodynamics that introduces the arrow of time.*

However, biological reductionism breaks down as there are not nearly enough genes to code for each bifurcation point in the biological tree of life.

It turns out that many biological systems are so-called chaotic systems rather than reductionistic linear systems. Chaotic nonlinear and hence unpredictable systems represent an example of breaking the symmetry of reductionism. Chaotic systems describe the anatomical branching of, for example, the circulatory, bronchial, renal, calyceal trees. It also describes the dynamic functionality of systems such as, for example, neuronal behavior in the brain or the immune system that exemplify nimble biological adaptive flexibility needed to find solutions in the face of continuous external perturbations. *Cardiac arrhythmias are examples of chaotic behavior that in contrast to neuronal behavior in the brain, and perhaps cognition, illustrate a dynamical system's behavior at the "edge of chaos".* Rather than occurring at the inner "edge of chaos", as is the case of a healthy immune system or of normal signaling brain activity (i.e. a normal EEG), cardiac arrhythmias occur at the outer "edge of chaos", that is entropically. Therefore, it is not surprising that chaotic behavior of the cardiac conduction system is a common cause of sudden death.

Fractals found in anatomy and histology (microanatomy) are organized to maximize fitness that facilitates the rate at which energy and/or material resources are taken up from the environment (or wastes eliminated from the body) for maximizing metabolic capacity, allocated to some combination of survival and reproduction. For example, the leaves of a tree absorb carbon dioxide and solar rays for photosynthesis. The vascular tree delivers oxygen (from the lungs) and organic nutrients (from the gut) to peripheral tissues of the body. The dynamic trajectory of fractal attractors of chaotic systems in physiology is organized analogously to maximize fitness and metabolic capacity.

The important perspective of symmetry in reductionism is based on the time invariance, or time-independent biological systems of the information of matter/energy in terms of the conservation of their steady states. Breaking the symmetry of reductionism with the formation of complex chaotic and non-chaotic systems promotes an adaptive responsiveness to perturbations that serves to maximize fitness in terms of allocation to some combination of reproduction and survival. Breaking symmetries inherently implies nonlinear systems. It forms new symmetries in complex and chaotic systems as allostatically adjusted new steady states form. This encompasses Nobel Prize–winning physicist P.W. Anderson's notion of breaking symmetries across hierarchical scales of biological science (4). The molecular genetic scale, for example, dictated by fractal genes in response to minor perturbations, results in ultimately more macroscopic behavior of structure and functional physiology. In chaotic systems, as described above, the fractal nature of steady states represents variation, which in contrast to reductive systems is an inherent design as part of the system. The threshold of criticality reached by any insignificant or minor perturbation highlights the feature of unstable steady states of chaotic fractal systems or systems close to bifurcation points. In the latter case, this is illustrated by a double pendulum as a metaphorical physical comparison, which is most appropriate. Although chaotic systems are an example of complexity, many biological systems are complex but non-chaotic.

All complex systems are distinguished by the characteristic of feedback relation and interconnectedness.

Non-chaotic but otherwise complex systems do not share the feature of fractal designs. Further, unlike chaotic systems, non-chaotic complex systems do have the capacity to move from an allostatic new steady state in response to a perturbation then back to the baseline resting steady state following the remission of the perturbation and correction of the metabolic stress.

Breaking the symmetry of space–time implies the relationship between mass and energy according to Einstein's famous equation $E = mc^2$. Space offers the room for matter to exist and move in. Time offers the means of keeping track of what matter does and in what order events occur. Time invariance is the conservation of the mass/energy of the physical universe over time. Space invariance is the conservation of momentum of the physical universe over space. This includes linear momentum such as solar photons of light and angular momentum of the electrons of atoms. The concept of time invariance as the conservation of the quantity of mass and energy takes the analogy of time being the axis of mass and energy. Again, notably mass and energy are interchangeable since $E = mc^2$.

The universe helps to construct time by the presence of cycles, repeating patterns of relationships between the conserved particle-like quantities of mass/energy. These repeating patterns are exemplified by electrons circling around atomic nuclei and the orbiting of planets and Celestial bodies, the latter creating the repeating pattern of light and dark cycles. The virtue of not changing with time as a function of this cyclicity may be conceptualized as a protective and preservation effect of the conserved quantity, such as life forms that exist on earth including humans. Light–dark cycles sustain life within the context of the first law of thermodynamics but within the limits of the arrow of time in accordance with the second law of thermodynamics. While time invariance is the absolute conservation of mass/energy of the physical universe, its implications to humans and other living organisms are relative and heralded by the steady states of free energy within the boundaries of every life form. Light–dark cycles contribute to the circadian rhythm of all biological systems. Below we briefly discuss the way living cells keep time at a molecular level. A detailed exposition of chronobiology will be given in Volume 2 of this book.

The central clock, the suprachiasmatic nucleus that is located in the visual sensory receptor pathway of the brain receives neuronal projections from the retina that recognizes the light–dark cycles (Figure 5.2). This suprachiasmatic nucleus is located at the base of the hypothalamus (composed of not just a single body but multiple discrete bodies) contains tens of thousands of neurons that project to a number of structures in the brain. One of these structures is the pineal gland, which releases melatonin, important in regulating the suprachiasmatic nucleus or central biological clock. This central biological clock has melatonin receptors that suggest a feedback regulatory function. There are many behaviors and functions of the body, which fluctuate in circadian fashion, for example sleep/wake cycles, hormone cycles, hunger/satiety,

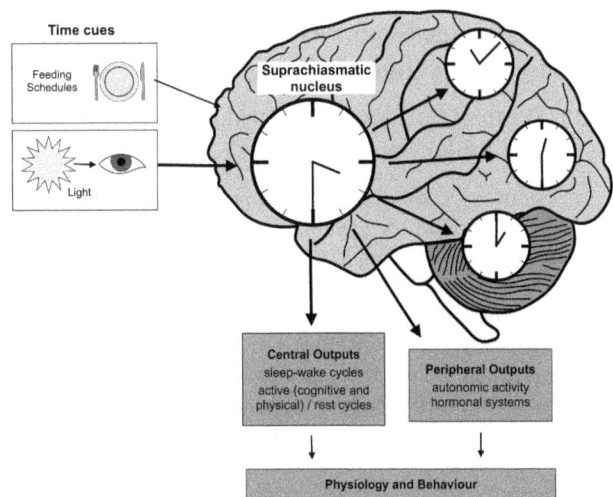

FIGURE 5.2 An illustration of the master clock—the suprachiasmatic nucleus (SCN) controls circadian physiology and behavior. It receives input from the visual cortex that perceive light/dark cycles. It also recognizes other visual cues that maintain behaviors such as feeding schedules. The SCN sends outputs to multiple brain regions that govern sleep-wake cycles, autonomic activity, and hormonal regulation. Source: adapted from Videnovic, Aleksandar, Alpar S. Lazar, Alpar, S., Barker, Roger A., and Overeem, Sebastiaan. (2014). "'The Clocks That Time Us'— Circadian Rhythms in Neurodegenerative Disorders". *Nature Reviews Neurology* 10(12): 683.

body temperature and metabolic rate. In addition to the central biological clock, there are molecular clocks that are genetically expressed. These molecular genetically expressed clocks are referred to as peripheral clocks. *There is a coordinate oscillatory regulation between the central and peripheral clocks. Biological circadian patterns are adjusted, or entrained, by local external cues of light–dark cycles.* Intriguingly, as discussed elsewhere, the synchronization of mitochondria and of metabolic cycles as reported by Demetrius, et al, occurs through gene regulation within and between cells that invoke collective and coordinated transcriptional cycles.

> *Chronobiology is a field of biology that examines cyclicity patterns of biological activity in living organisms, which may be circadian (24-hour daily cycles), ultradian (cycles shorter than 24 hours) or infradian (cycles longer than 24 hours).*

The latter may be monthly, annually, or longer. Species of organisms are even characterized by life-death cycles, or average longevity that may differ at baseline and following environmental perturbations. There appears to be a central role for the circadian system in fractal regulatory networks in the time organization of neurophysiological processes. The time invariant nature of fractals in contrast to linear periodic systems, which have an absolute time invariance, is only relative. In other words, the time-independent steady state of a fractal system has inherent variability and thus the steady state may be considered wider or looser, until it reaches a point of criticality and changes qualitative behavior. This signifies a transition to a new steady state.

5.2 The Bridge from Physics to Physiology and Medicine

5.2.1 Symmetry, Symmetry Breaking and Reductionism

> *Philosophy provides the building blocks to logic. Logic provides the building blocks for math. Math provides the building blocks for physics. Molecular-level physics is the application of atomic-level physics, which in turn is the application of elementary particle physics. Chemistry is an application of molecular physics to complexes of atoms connected by chemical bonds. The chemical macromolecules of life form the structural elements of biochemistry, which then provide the basis for cell biology, etc. all the way to physiology and medicine. In this manner all sciences can be seen as hierarchically organized with an increasing level of complexity as we move up the length scale and coincidentally also up the time scale as larger physical objects take longer to respond to stimuli.*

However, it is important to stress that these are not straightforward applications of physical principles to the building blocks of the universe in which we live because at each increasing hierarchical scale, both the building blocks and the rules "of the game" change. Chemistry is rooted in physics as well as in the aforementioned more elementary disciplines. Biochemistry is rooted in physical chemistry. However, biochemistry represents symmetry breaking from physical chemistry as it follows a different set of rules. For example, a nucleic acid has specific binding rules to other particular nucleic acids whereas when viewed from the perspective of general physical chemistry such a nucleic acid would only be considered an acid *per se*. Accordingly, such an acid would not be guided according to the rules of biochemistry but rather by physical chemistry. *As we move up the scale of scientific knowledge the rules applied at a higher level are not in contradiction with those at a lower level. However, they cannot be derived from the lower-level rules due to the emerging complexity of the higher-level systems they describe.* For example, stereochemistry provides 3D descriptions of how molecular bonds guide the formation of molecules from atomic constituents. However, these rules cannot be derived from the Standard Model of subatomic physics.

This is even more strikingly obvious when considering the genetic code where the 64 possible triplets of the four nucleic acids code for the 20 naturally occurring amino acids plus a stop instruction. This type of information cannot possibly be deduced from the equations of physics such as Newton's or Schrödinger's equation. However, reassuringly the rules at a higher-level scale (e.g. Hodgkin-Huxley equations for action potentials in neurons) do not contradict those below but provide shortcuts and simplifications often arrived at by empirical observations. For this reason, condensed matter physics is not in contradiction to the Standard Model of elementary particle physics, but it is only concerned with its consequences regarding electrons, protons and neutrons with no regard for muons or the Higgs boson. Without such simplifications and shortcuts, it would be virtually impossible to make progress along the hierarchies of knowledge. This is a good example of how a more fundamental discipline is suitable for a direct application to a higher hierarchical plane such as the evolution of physical chemistry to biochemistry.

The same is true in each of the above-mentioned disciplinary hierarchical scales going from most fundamental to further downstream and more applied ones. From biochemistry evolves molecular biology again as a result of breaking its inherent symmetry and forming new rules. From molecular biology evolves cell biology and subsequent tissue biology, further yet multi-organ systems biology and subsequently physiology at each hierarchical scale representing new disciplines with new rules in a breaking of the symmetry or symmetries present at a more fundamental level. Further yet psychology represents a higher hierarchical scale of organismic disciplinary science that examines the way humans interact with other humans through cognitive abilities, emotional states and introspection. While psychology should not contradict the rules of physics (unless it ventures into the territory of parapsychology), physics is incapable of formulating the rules of human behavior.

Each symmetry-breaking transformation represents the creation of a new system, each can correspond to a different discipline distinct with a separate set of rules and a new symmetry that assimilates the antecedent more fundamental disciplines that have to break symmetry prior to assimilating into a new symmetry at a higher hierarchical scale. Within the realm of physical chemistry and biochemistry, each new system evolves as a result of energy transformation within each hierarchical scale. Energy is transferred as heat in tandem with reconfiguration of information. Symmetry breaking here describes the breaking of old rules and forming new rules within each hierarchical scale of science. While symmetry breaking necessarily invokes nonlinearity (lack of proportionality between cause and effect), the new order that emerges from a broken symmetry produces new rules resulting in locally applicable approximations in the form of linear equations with predictability of patterns of information. These linear equations are approximations that work well up to a limit, that is up to the emergence of nonlinearities leading to the next symmetry breaking event. *Each symmetry breaking results in new organizing principles.* The formation of new rules accompanies each hierarchical manifold of broken symmetry that resulted from the inadequacy of linear equations that were only capable of describing the symmetry of relationships that existed within the discipline before it. This subsequently leads to emergence and the associated nonlinearities that are required for symmetry breaking to take place.

Symmetry in the realm of physics may be described in terms of invariance, which is the unchanging nature of some physical quantity pertaining to the defined invariance. For example, time invariance is the unchanging nature of that quantity over time. Rotational invariance is the unchanging nature of another quantity with each rotation or cycle. With each invariance there is a physical quantity associated that is conserved which represents the unchanging nature of that quantity. Thus, time invariance represents conservation of energy and is represented in the first law of thermodynamics.

Rotational invariance represents a cyclical invariance in space that conserves angular momentum. Linear momentum is the product of mass and velocity and its change over time defines a force. Conservation of linear momentum is a fundamental property of space–time and it can be proven to represent translational invariance of physical laws in space. For example, when an electromagnetic solar wave within the physical spectrum collides with chlorophyll of plants that momentum (momentum p= mass x velocity or in the case of an electromagnetic wave it equals Planck's constant/wavelength of the wave or $p = h/\lambda$). In the case of this collision, momentum changes as the moving object decelerates, and hence the change of momentum over time represents a force (force = mass x acceleration but more generally force = dp/dt so in this case force = hdλ/dt) exerted on the chlorophyll. Hence, the energy of the solar ray gets quite efficiently transformed into an electronic excitation energy in chlorophyll that is then used to make biochemical bonds. In any event, the notion of translational symmetry with respect to time and space leads to the conservation laws of energy and momentum, respectively.

Furthermore, all of the laws of physics, whether it be Newton's laws of mechanics, Maxwell's laws of electromagnetism, the laws of thermodynamics, and the laws of quantum mechanics, all these laws of physics conform with the conservation laws of energy, momentum, angular momentum, charge, spin, etc.; all of which are so-called conserved quantities. Conserved quantities cannot be created or destroyed; they remain constant.

Hence, the laws of physics are fundamentally framed by the conservation laws, which are applicable to all material systems governed by physics, and they represent the organizing principles of reality and are generated by hypotheses. Notably, symmetry is defined, as described earlier, as rules capable of forming linear equations with invariance of patterns of information is inherently a representation of conservation, and hence it makes sense that the notion of symmetry within the field of physics ultimately leads to the conservation laws to which all of physics applies. This statement has been formally proven by the so-called Noether theorem. According to this theorem, *every conservation law is associated with a symmetry in the underlying physical system that is subjected to a conservation law.* In accordance with Noether's theorem a one-to-one correspondence exists between every conserved quantity and a differentiable system symmetry. As a case in point, energy conservation is implied from the time-invariance of physical systems. The fact that physical systems behave the same irrespective of their rotational coordinates gives rise to the conservation of angular momentum. This is one of the most powerful insights into the way the Universe works.

A law of conservation states that a measurable property of an isolated physical system does not change with time. Properties covered by such conservation laws include energy, linear momentum, angular momentum, and electric charge. Approximate conservation laws also apply to lepton number, baryon number, mass, parity, lepton number, baryon number, hypercharge, and strangeness, in certain classes of physics processes. A local conservation law, by contrast with a universal conservation law, is expressed as a continuity equation, a partial differential equation, which relates the amount of a quantity and the "transport" of that quantity over some distance in space or time. Any change can take place only by the amount of the quantity, which flows in or out of the system. Conservation laws, mostly absolute and exact, are deemed fundamental to nature, with broad applications in physics, biology, chemistry, engineering and geology.

Conservation speaks to the uniformity of space and the inseparably linked manner in time. Symmetry is a manifestation of patterns, pattern recognition is a means of understanding the universe and has been a characteristic trait of intelligent species such as humans so that by recognizing patterns of danger or of opportunity, these intelligent members of the species could take appropriate steps to adapt to the changing environmental conditions, survive and thrive. Thinking involves critical analysis aimed at recognizing patterns, which means finding a symmetry in a seemingly chaotic universe. Moreover, there is an underlying dynamic.

> *Biology is largely empirical. There are no fundamental laws of biology and, in fact, the notion of hypothesis in the original sense used by physicists intended to uncover fundamental laws of physics by which the Universe is governed, does not (yet) exist in biology.*

Gibbs free energy may be the first such organizing principle that leads to laws of biology, for example Demetrius's directionality entropy and the concept of evolutionary entropy. This may be the beginning of uncovering some universal laws of biology.

The development of disease or pathology may be described in terms of increased entropy, loss of information and inextricably in parallel, a new minimum free energy state corresponding to pathologies, hence increased disorganization and loss of complexity of connections between systems.

We need to find organizing principles or patterns that in a nonlinear fashion, inclusive of breaking of symmetries, can be coordinated into fundamental principles and ultimately universal laws applicable to living systems. Such laws may not allow predictability *per se* of nonlinear biological systems in medicine but may facilitate understanding the interfaces of classical physics with biological chemistry and ultimately the manifestations of health versus disease of biological systems on a macroscopic scale. Accordingly, this will direct areas of research and clinical aspects of medicine. In the setting of systemic insulin sensitivity, for example, the transfer of heat across the connecting pathways that promote and coordinate bioenergetic information with anabolism and other information such as cell cycle and hemodynamic (for example vasodilatation) function promote optimal function of the biological engine. This parallels a conservation of energy and patterns within and even across hierarchical scales of science. Insulin resistance, as a result of changes in the values of important control parameters such as nutrient depletion, energy excess, or chemically polluted diet, results in increased oxidative stress and inflammation. This works in tandem with decoherence of quantum metabolism and other quantum processes and hence amplifies and accelerates entropy increase associated with disorganization of patterns and hence loss of symmetry. Entropy, as a function of state invoked by the second law, modifies the first law in the sense that energy while not destroyed is dissipated as heat, or becomes energy in transit, that cannot be used to do work. Hence, symmetry breaking may be described in biological terms as the loss of healthy function and the development of disease, for example accelerated cognitive decline, Alzheimer's disease, cancer, and cardiovascular disease. Physics, therefore, offers an important insight into biology and medicine, which is one of the recurring themes of this book.

Contrary to the case in physics, the theoretical framework of symmetries is neither associated with invariance nor with transformations preserving invariance, in biology. Rather, it addresses permanent changes of symmetries.

In case of neural networks and cell proliferation (where nonlinear effects are very important), hierarchical integration and regulation of cellular contributions to a nonlinear framework at the tissue level often need to be sharp. In the case of cell proliferation, regulation by the tissue and the organism seems to prevent pathological changes (e.g. cancer) from taking place. Each mitosis represents a bifurcation in the cellular system and a symmetry change because the resultant two new cells are not identical. This is essential for specialization ontologically (evolution of the organism) and phylogenetically (evolution of species). This specialization and permanent symmetry change apply to whether the system is a cell, an organism, or a species.

It should be noted that this is in contrast to the case of cells that secrete a given protein (for example a hormone) where regulation does not need to be sharp because the amount of protein that is sufficient depends on the tissue level rather than the cell level. The process of cell culture growth is nonlinear, it follows a logistic map with saturation due to spatial and nutritional constraint. The allometric laws of physiology are fractal with an exponent close to three-quarters, not linear where the corresponding exponent is one. Accordingly, the evolution of these systems is nonlinear, indeterministic and hence nonpredictable. This nonlinearity is a function of the complex relationships between the elements of the system, in particular the positive and negative feedback regulation as well as likely random or even quantum effects. In contrast to physical systems whereby symmetry implies a predictability defined by linear equations and conservation of quantities, for example energy and angular momentum, invariance over time and cycles, respectively (i.e. no change over the course of time or the course of cycles), biological systems do continually change symmetries with respect to all control parameters including time. Possibly even angular momentum may be related to temporal cycles such as reproductive cycles and life/death cycles, etc.) as it involves rotational symmetry and periodicity with a characteristic angular frequency $\omega = 2\pi/T$ where T is the period, so in this sense any periodic process has a characteristic angular frequency and hence angular momentum (which is the product of angular frequency and the moment of inertia, related to mass distribution) associated with it [5].

5.2.2 Biological Mechanisms of Survival and Stress

Despite adapting well to premodern environments, humans have been experiencing modern industrialized urban existence, which resulted in dramatically different evolutionary pressures. Although human life expectancy has been increasing steadily over the past century or more, the prevalence of cardiovascular disease, cancers, and other chronic diseases imposes a daunting toll in terms of human suffering and spiraling cost of long-term care. The irony that people on average are living longer is explainable by the development of effective acute care paradigms of healthcare (antibiotics, emergency medicine, non-invasive surgery) that make it far less likely to die from accidents or acute complications of chronic diseases than used to be the case in the past. Indeed, chronic disease is a difficult road to travel and requires an improved navigational system. The author of this book also believes that a number of new paradigms must be integrated into medical practice that are based on recent advances in exact and quantitative sciences.

One of the trade-offs of the modern urban lifestyle that brings so much positive change is stress with its unrelenting negative effect on human physiology. Stress represents any challenge to the normal balance of biological systems of the body.

Interestingly, the original context of stress was an engineering term to describe mechanical stress on a support beam or any such solid object. *Stress causes material strain and the relationship between these two physical properties, known as the stress-strain relationship*, is one of the earliest quantitative formulas found by modern science dating back to Sir Robert Hooke and his description of elastic materials known as Hooke's law. Its equivalent in biology is called the dose-response. However, Hooke's law predicts a linear response of strain to stress while the dose-response is typically characterized by an exponential or sigmoidal formula. Hans Selye in the 1970s [6] first adopted the term stress to biology, referring to alarm signals that trigger the acute physiological responses of the body. Indeed, biological stress within the body is fundamentally rooted in endogenous reactions to external factors.

In its extreme form stress can become the predominant or even the only source of a plexiform of pathways that cause pathological transformations. The pathogenic nature of the stress response processes is inextricably feed-forward. Surprisingly, it was not until the 1980s that a new historical perspective unveiled the curtain to the realization that immune responses to pathogens or injury are not uniformly beneficial. Proteins produced by macrophages and other immune cells, i.e. cytokines, are quite capable of causing tissue damage. It is intriguing to note that history seems to have come full circle back to the ancient Greek concepts of disease that look inward to an imbalanced production of body substances as the root cause of disease.

Walter Cannon, in 1929, recognized *medicatrix naturae*, the concept that the body heals itself when its normal state is upset [7]. Cannon first described his model of homeostasis, which stated that through complex coordinated physiological adjustments the body maintains a state of constancy, or homeostasis. However, the stage for the conceptual framework had been initially set by physiologist Claude Bernard in 1878 when he stated the following:

> *Living organisms seek to preserve constant their internal environment conditions of life despite shifts in the circumstances of the outer world.*

In fact, he declared that "the preservation of constancy is the sole object of all vital mechanisms". Classical examples of homeostasis regulation highlighted by Cannon include energy production by substrate molecules of glucose, fat, and protein, also required for tissue growth and repair. Other constituents for the cells need an activity that requires constancy within a narrow range include water, salt, calcium, oxygen, osmotic pressure, pH, and temperature [7]. Using the previously discussed thermodynamic perspective, we can state that the preservation of constancy acts to oppose entropy increase in living systems since dissipation of heat in an irreversible process.

> *Homeostasis is a process whereby physiological "set points" exist, and any deviation from their values results in a feedback error correction mechanism.*

Both Bernard and Cannon deserve credit for their intellectual pioneering and channeling the concept that health is the maintenance of physiological homeostasis in the face of changes in the external environment. This was one of the watershed seminal departures from prevailing medical doctrines that provide an intriguing, valuable, and historical perspective. Prior to the work of Bernard, in the late 1800s, and later Cannon, in the early 1900s, dating all the way back to before 200 AD. as posited by Galen (Claudius Galen), disease was believed to be the result of an imbalance in the body's humors. In fact, by some accounts, this widely held view can be traced back to Hippocrates around 400 to 300BC. His theory of the makeup and the workings of the human body posited a relationship between the four humors: black bile; yellow bile; phlegm; and blood. These substances were said to be produced by internal organs, the imbalance of which was the cause of disease. It was believed that each humor was associated with a characteristic disposition. Black bile, if overproduced, resulted in an individual who was despondent and melancholic. Alternatively, too much yellow bile caused a person to be choleric or bad tempered and easily angered. Phlegmatic unemotional persons had a preponderance of phlegm. The optimal state of the body occurred when blood was most prevalent, not overproduced, allowing a sanguine or confident and hopeful disposition.

While Bernard held that the body seeks constancy in response to external factors, Cannon based his theory on the "fight or flight" system. "Fight or flight" systems, as described by Cannon involved the sympathetic nerves and the adrenal medulla sine qua non the release of catecholamines, considered to be responsible for priming the body for action [8] as an adjustment response mechanism to achieve this [7]..

As stated above, Hans Selye, an early Hungarian-Canadian endocrinologist, employed the concept of stress response to the signals that threaten the body's state of equilibrium [5], which is in fact analogous to the "fight or flight" response described by Cannon. The Selye stress system involved the hypothalamic-pituitary adrenal (HPA) axis as the mechanism by which the body copes with stress. This is the origin from which cortisol became recognized as the "stress hormone". However, the term stress had come to be an ambiguous one and had many diverse connotations over the years. For this reason, in 1988, Sterling and Eyer coined the term allostasis [9, 10] to refer to the active physiological changes of the body to stressors from the outside world that challenge the normal state, or homeostasis of the body.

> *Allostasis literally means stability through change while homeostasis means resistance to change.*

While the model of allostasis has since largely supplanted the model of homeostasis as the core theory of regulation under acute stress, the shift in thinking maintains the model of homeostasis, but that homeostasis is preserved by allostasis. The purpose of the stress response throughout evolution has been to escape mortal danger such as that of a predator or some other threat to life and limb.

> *Allostasis conceptually describes mechanisms that abrogate constancy by overriding local negative feedback to affect a coordinated variation in physiological parameters that most efficiently reacts to the challenge.*

In contrast to the Bernard model that "constancy" is the sole object of all vital functions, the model of allostasis holds that the true object of all vital functions is chronological survival, that is survival in order to reproduce [11]. Inherent in this model is the body's need to calibrate a hierarchy of priorities so that in escaping a predator or fighting a threat that engages the acute stress response, the purpose of the neuro-endocrine-immune system signals must be to defer energy-expensive processes of reproduction and anabolism in favor of the immediate need for catabolism, fuel production and wound healing. Also central in optimizing efficiency of the allostatic stress response is predictive regulation, for example such as the one that occurs with stress hyperglycemia in nondiabetic individuals anticipatory to competitive sporting events [7]. Insulin and a host of hormones, which regulate fuel supply, are modulated by neuronal output from the brain based on anticipatory mechanisms [12, 13].

Clinically relevant to efficient management of the type 1 diabetic is the importance of administering insulin prior to a meal based on predicting food intake to prevent exaggerated blood glucose swings postprandially.

Even more intriguing is the plausible conceptual model that insulin resistance developed as an evolutionary adaptation to prevent obesity and cancer, the latter either directly or indirectly as a result of preventing obesity. Here, the predator of modern times is cancer, and the allostatic response to the stress caused by a pathologic excess energy uptake into cells is a physiologic reaction responsible for the sequelae, for example involving hyperglycemia, dyslipidemia and hypertension. This is the cost incurred to maintain homeostasis of vital physiological functions and hence preservation of life. Reduction of chronic nutrient excess is necessary to reduce excess nutrient storage so as to alleviate adipose cell stress.

Persistence of nutrient excess leads to the development of new set points, which is characterized by failure to establish basal equilibrium, increasing threshold of fat mass storage accompanying

inflammatory infiltration and progressive degrees of insulin resistance. This condition further exposes the patient to other modern day predators such as cardiovascular disease.

While allostasis is the concept of divergent physiological mediators that serve the body's protective coping responses to acute threatening or unpredictable environmental stimuli, chronically these stress responses become dysfunctional and ultimately exact declines in cognitive and physical functioning. Indeed, biological stress responses are endogenous reactions to external factors.

These endogenous responses, when prolonged or severe, ultimately become the cause of virtually all chronic diseases.

The terms allostatic load and overload were first coined by Bruce McEwen in 2000 [14] as descriptive terms for the physiological cost incurred when the person is exposed chronically to the stress response. That is, as acute stressors cycle into chronic ones, the stress response progresses from adaptive and protective to maladaptive reactive changes of the central and autonomic nervous systems as well as the endocrine/hormonal and immune systems (allostatic load) and ultimately to chronic disease states (allostatic overload). An example of allostatic load may be highlighted by the evolution of the body's ability to calibrate the priority of metabolic resources to optimize survival in the context of acute stress. In this case, a functional hypogonadism develops because the energy-expensive process of reproduction would be counterproductive. However, in prolonged stress, the persistence of a male hypogonadal state with reduced androgens leads to not only inability to reproduce, but to abdominal obesity, muscle loss, depressed mood and social vigor, and increased irritability. These changes represent allostatic load because they reflect the cost of the stress response, which in the acute phase is adaptive, however when prolonged is pathogenic. The subsequent development of type 2 diabetes or cardiovascular disease would represent their progression to the additional cost of allostatic overload. Figure 5.3

Allostatic Overload

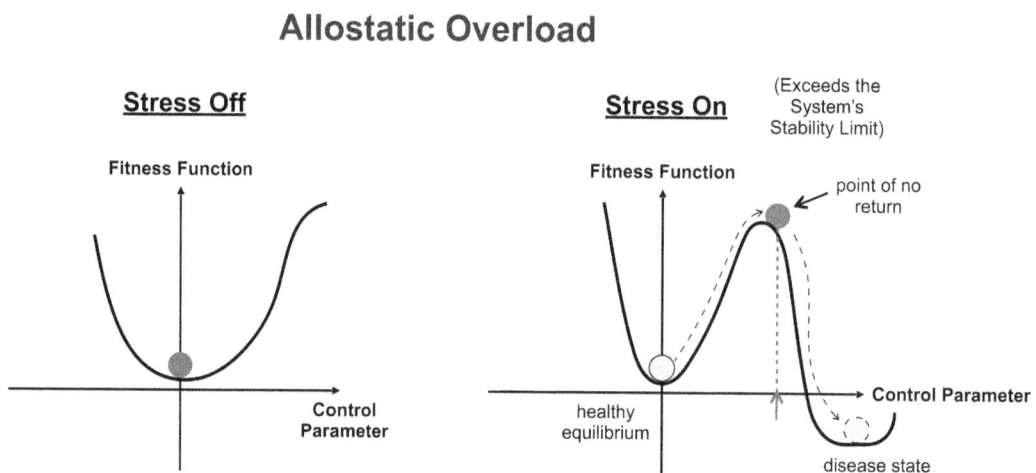

FIGURE 5.3 Illustration of allostatic overload under conditions of prolonged stress where a suprathreshold ("point of no return") is reached and the system's stability limits are exceeded (right). Here, allostasis can no longer maintain homeostasis resulting in disease states.

illustrates the concept of allostatic overload graphically using the physiological fitness function as a description to the system's response to a control parameter with and without the presence of stress.

The cost of allostasis is mediated by the coordinated series responses of the nervous system, endocrine/hormonal and immune systems to protect survival (of species). When these core supersystems become dysregulated in the setting of chronic stress, allostatic load and overload supervene. The following questions now emerge: why does the stress response become maladaptively prolonged? Where anatomically does it originate? *Why Zebras Don't Get Ulcers* is the title of the classic award-winning book written more than a decade ago by animal biologist Robert Sapolski [15]. The overarching concept expanded on this book is that zebras, like all prey animals in the wild, are faced with the intermittent acute life-threatening stress of escaping predatory attack. Despite the full engagement of the stress response in a dramatic fashion, to be successful in eluding the predator, their neural, hormonal, and immune systems need to be restored in an integrated manner to their baseline homeostatic balance. Adrenaline and cortisol levels are no longer elevated and immune system activation recoils.

This is adaptive behavior, which is necessary for individual survival and also for survival of species. Maladaptive human behavior, although considered an individual phenomenon, in many ways, is becoming endemic in our society. This behavior manifests the stress response assuming chronic character, thus overextending the very mechanisms and systems that promote survival in the wild.

Ironically, when this occurs survival is threatened. The integrative physiology of the core supersystems has gone awry leading to a dysregulated state of alarm that ultimately results in chronic disease in a feed-forward circuitry of self-amplification detached from regulatory feedback control. If the root causes are not mitigated, a wide range of functions and cascades fuel and stoke an inflammatory process that is systemic rather than anatomically confined. The resulting disease states relate more to the individual's weakest link of genetic predisposition than to the etiology and origin of the problem. Intra- and extra-cellular communication and message signaling at the molecular level is referred to as signal transduction.

Whether the stress is generated due to the presence of a predator, a toxin, a microbe, an allergen, obesity, ischemia, arthritis, or diabetes, at the cell level, the body does not know the difference, i.e. the transduced signals despite disparate etiologies lead to the same stress response. The difference is in the extent or more importantly, the duration of the stress or allostatic response.

The allostatic mediators of the chronic stress response promote inflammation. Inflammation causes a disturbed redox environment, i.e. oxidative (sometimes nitrosative) stress, which in turn is recognized at the tissue and cell level and this allows signal transduction to occur. The host of chronic disease states not only threaten our longevity and quality of life by incineration of the body's tissues, but additional harm results from delay in extinguishing the origins of the inflammatory process.

In order to extinguish, rather than sometimes unwittingly stoke the flames, by looking for diagnoses and treatments we must look for answers and for an understanding of what is really the root cause of the problem. Although this is sometimes (perhaps most of the time) not easy, it is vital to recognize that such origins occur at the body/environmental interfaces, especially the brain (where the senses of sight, hearing, and smell intersect), and the gastrointestinal tract. In each case, they feed into the core neuro-endocrine immune supersystems, which in turn ultimately mediate disease by inflammation and disturbed redox processes that cause chemical imbalances.

The remarkable and complex brain tissue, the hypothalamus, continuously and simultaneously integrates cascaded input arriving not only from the major afferent limbs, the brain, and the gastrointestinal tract, but also the skin and every organ system of the body.

It carries amazing functions of nearly incomprehensible evolutionary engineering capable of measuring the totality of sensory input and spontaneously recognizes alteration in some or all of these parameters. Further, it instantaneously elicits the compensatory and adaptive coordinated series of neural, endocrine, and immune system allostatic responses that serve to optimize survival of the individual. What the hypothalamus is not able to discern is the chronic nature of signal transduction inputs. In baseline and acute stress circumstances, the hypothalamic output of allostatic changes maintains homeostasis and protects survival (and health) of the individual or animal. However, the prolonged psychogenic perception of threat or of the persistent presence of gut-derived toxins, allergens, or pathogenic microbes, all pose serious challenges to which man did not evolve.

5.2.3 Why Do We Need a New Medicine?

In college pre-med studies, we learned physics including principles of thermodynamics, organic and inorganic chemistry, calculus, botany, biology, physiology, genetics, evolutionary biology, and so on. The first two years of medical school teach us some of the same. Somehow, at the height of our basic science knowledge, all of that went out the window at the start of clinical rotations, in the third year of medical school. We subsequently pursue internship, residency, fellowship, and private practice of clinical medicine never looking back to those initial informative years of studying basic science underpinning human physiology and anatomy. We do not use it, we do not think about it, and in fact, we see it as clinically irrelevant. The next question becomes, do we need it? Does it have the potential to facilitate independent, original, and creative thinking? If not, are we replaceable by nurse practitioners, physician assistants, or other nonphysician practitioners who are more than capable of following guidelines? The purpose here is not to diminish the important value of these clinicians. Rather, the purpose is to answer some fundamental questions.

Should medical school be reduced to a two-year training for those who wish to pursue the privilege of a career in patient care, whether it be a surgical or medical specialty? If this would be the case then the college premedical academic course work would conceivably provide a discerning curriculum capable of directing those students who may wish to pursue a career in basic science bench research. It is a relevant question therefore to ask why do we need to essentially duplicate two years of course work that serves to delay the process and provide unnecessary expense for the majority of students who ultimately go on to clinical practice in one form or another? On the other hand, does understanding the basic science of energy flow from a conceptual qualitative perspective improve biological thinking and problem-solving skills that translate into patient care? I believe it does. If it does, this supports the value of the current four-year medical school course curriculum. Moreover, this assimilation of basic and clinical science at the most sophisticated level rises to the height of intellectual gratification. There is inherent joy in learning, which is evidenced at the most basic level by the pleasure we derive from our senses.

> *What is meant by energy flow here is the associated informational flow as energy interfaces with molecular chemistry and biochemistry. If constructive, free energy is positive and structure-function relationships are created, increased, or maintained. If destructive, free energy is lost to heat and associated entropy to use the lexicon of thermodynamics, or to oxidative stress and inflammation to use the lexicon of pathobiology and pathophysiology of disease.*

The essential issues are as follows. Does an understanding of the basic sciences truly promote independent and creative flexible thinking that is relevant to clinical healthcare and would this be an improved model for evolving the standard of clinical care? The essential ingredients here are the mechanistic understanding of the component parts of the system, the nonlinearity in how these parts interact to produce emergent outcomes, and finally how we collect knowledge and data and apply it. It is proposed here that both mechanistic basic science and non-mechanistic pattern recognition are necessary for biological thinking and problem solving.

When solving routine or simple cases, expert endocrinologists, similar to cardiologists rely on quickly processed non-analytical reasoning and pattern recognition, accounting for roughly 75% of patient encounters. However, the other approximately 25% of patient encounters are more difficult cases where pattern recognition is not readily apparent. In these cases, mechanistic understanding is known to be helpful. Physicians are in fact superior in this capacity to students and non-physician healthcare providers. The scientific nature of medicine has been traditionally considered in terms of data-driven recognition and understanding cause-effect relationships. However, cause and effect are often manifested experimentally by therapeutic intervention to discern apparent responses (favorably and unfavorable) to highlight the art component of medicine. Hence, mechanistic Reductionism

that brought analytical methodology to medicine is also ironically responsible for medicine being a formulation of both art and science. This is because of the intrinsic unpredictability of therapeutic empiricism to top-down reductionism. This uncertainty in turn is rooted in the inherent nonlinear complexity of our systems biology to which we are inefficiently and at times even dangerously imposing reductionistic assumptions.

Reductionistic "evidence-based medicine" is useful as algorithms for treating conditions in a population on average, however, it is incapable of predicting the response in an individual patient. Thus, such "evidence-based medicine" is ill-suited to address the individuality of a situation that includes co-morbid diseases, genetics, and environmental

circumstances that underpin antecedent, triggering, and mediating factors of a presenting process. Algorithm guidelines and large-scale multicenter clinical trials are only capable of giving a rough approximation of a course of action. Physicians must have the skill set of basic science to handle the uncertainty of the complexity of a given patient-centered decision making [16]. The teaching of basic science is important because clinical medicine itself is changing. What separates physicians from other healthcare providers is their ability to manage more difficult problems [17].

> *The medical education of physicians must not only enable them to follow practice guidelines but to write such guidelines.*

This requires a deep fundamental understanding of clinical science at the basic science level, hence branding physician experts the authority to violate practice guidelines [17, 18].

Mechanistic understanding derives from formal basic science training that is largely a top-down (having a central command) method by students learning what scholarly instructors profess. Furthermore, students evolve their understanding by their distinctions (interactions) amongst themselves, a bottom-up process (no central command). In this example, the top-down set of instructions is a linear blueprint that can be broken down into its component parts whereas the bottom-up collaborative process of learning is nonlinear. It is the latter process that is the basis of the most effective subsequent strategy for acquiring general knowledge and insights as a clinical practitioner. An analogous understanding applies to the presentation of a health problem by a patient, himself or herself considered a "physiological system". An individual is a system of the biological interactions involving complex subsystems at many hierarchical scales. These subsystems may be understood at an isolated level mechanistically, hence reductively. However, their interactions are largely nonlinear, non-reductive, and non-predictable, and accordingly, our strategy for understanding them must be a bottom-up approach.

> *The ultimate goal is to apply the gestalt knowledge, both of underlying principles and of the circumstances of the individual patient being studied, to a top-down therapeutic approach directed with as much precision as possible to maximize the benefit-to-risk ratio.*

The distinction obtained from various perspectives of why bottom-up non-reductionism is advantageous becomes fundamental to successful differential diagnosis. The premise here is rooted in the logic that parallels the nonlinear self-organizing complexity of all living systems. That is, the process by which biological solutions in nature are derived is one that is characteristically bottom-up. The converse is true of building or fixing a clock, which may be complicated, involving many links, but the system is linear. In the case of chronic disease states involving biological systems, processes are distinctly nonlinear, which is a central theme of this body of work. *The current paradigm of traditional establishment medicine essentially abrogates any real encouragement to assimilate an understanding of basic and interdisciplinary sciences into patient care at the "bedside".* Instead, there exist clinical guidelines that on certain levels supplant the expertise of the putative experts, the practicing physicians.

I believe this model is at best therapeutically and intellectually constraining. Moreover, this focus implies a priority for medical-legal protection. While not intended to be overly derisive, the reliance on practice guidelines, distinctly a top-down approach, may be satirized as a few gray-bearded erudite committee members determining in their infinite wisdom what is the correct set of rules for everyone. This shall be the Holy Grail blueprint set of instructions for every doctor to follow for every patient with any given chronic disease. Since most chronic pathogenic biological processes are complex and hence nonlinear, their solutions, with the lowest risk-to-benefit ratio at the individual level, are unlikely to represent a top-down linear blueprint. A colony of thousands of ants displays collective emergent intelligence as do many species of insects and animals capable of performing extraordinary feats of collective behavior. A common denominator of these behaviors is the absence of a central control, no managing director, or top-down control-and-command set of instructions. Rather, this emergent behavior evolves from local interactions between elements (e.g. organisms) of the system (e.g. an ant colony), or so-called "simple rules".

Computers allow collaboration between people beyond the traditional organization boundaries, which has been shown to increase the collective intelligence of the larger group. A study published in the journal "Nature" a number of years ago compared Wikipedia to the Encyclopedia Britannica with regard to hard-nosed facts in the physical and life sciences. Although there are many things in Wikipedia that are ludicrously incorrect, when it comes to hard-nosed scientific facts, this has not been found to be the case. In this study, the two sources of knowledge were virtually identical at a time when Wikipedia was still in its infancy. It has since had more than five years of self-organizing and self-correcting. This is an amazing study outcome since the *Encyclopedia Britannica* was written by a group of scientific law-giving scholars including 110 Nobel Prize winners and five United States presidents. In fact, as a result of plummeting sales in the face of competition with Wikipedia, after 244 years of annual publication, 2012 became its last printed edition.

This demonstrates the power of the bottom-up approach of finding solutions through a self-organized effort. In the bottom-up example of Wikipedia, at the time more than three-quarters

of one million ordinary people have contributed to it, which amounts to input from a diverse set of people with varying opinions. Its outcome becomes an accurate self-correcting adaptive system with no blueprint, just very simple local rules looking for patterns shared between individuals with self-correcting abilities. This is collective intelligence analogous on a different scale to thousands of ants forming a colony capable of finding extraordinary and genius solutions to their own challenges. An even more efficient version of Wikipedia is achieved when people not only put in their own opinion but the individual opinions have different ratings with some given greater weight than others. *This self-correcting self-organizing system highlights the wisdom of the crowd with its incredible accuracy and predictive power. It is important however that the system must stay unbiased toward conformity.*

Other examples of bottom-up approaches facilitated by machines are electronic medical records and computational analyses in the field of bioinformatics, or quantitative biomedicine that invokes the collaboration of interdisciplinary scientists. In each case, the analysis of large cohorts of data represents a bottom-up approach that searches for patterns, or rules, to find nonlinear solutions to complex challenges in healthcare. Just like the consumer internet of Wikipedia, crowdsourcing in healthcare is the direction of technology, building a global brain by sharing knowledge. This will move in the direction towards collective action at the lay public scale. At the physician and scientific expert scale, the cultivation of deeper and more diverse realms of knowledge will promote access to vast resources in real time, enabling close interactions between physicians and scientists in different specialties and between basic science and clinical care. *This bottom-up process will expand the frontiers of knowledge and evolve problem-solving capacity analogous to the evolutionary process in the animal kingdom itself.*

IBM's Watson computer is a cognitive system that partners people with computers and promises to transform medicine. It competed in popular game shows like *Jeopardy* (hosted by the late Alex Trebek on American Broadcasting Corporation) against former champions and defeated them. It is now being redesigned for medical decision-making and represents a future new generation model of electronic medical records. It uses computer-generated algorithms based on simple rules, which are associations between input data and diagnoses (discussed below). It takes all of the data and the diagnoses available to it to give differentials with assigned probabilities. It is based on all available medical knowledge including peer-reviewed publications and the results of clinical trials. It is accurate, consistent, bias-free, and capable of generating hypotheses and evaluating the strength of those hypotheses. The database already available to IBM's Watson computer includes a multitude of textbooks, massive databases of medical journals and thousands of patient records from Memorial Sloan Kettering Cancer Center.

Together, IBM's Watson has analyzed 605,000 pieces of medical evidence, two million pages of text, 25,000 training cases, and had the assist of 14,700 clinical hours of fine-tuning for its decision making [19]. IBM has partnered with MD Anderson Cancer Center where the Watson model helps recommend leukemia treatments. The physician and nurse might input symptoms and other factors into the system. The Watson computer then identifies key pieces of information and scours the patient's data to find relevant facts about family history, current medications, other existing conditions combined with results from testing for genomic, proteomic, and metabolomic data if available. It takes this information, forms hypotheses by data association, and tests those hypotheses relative to existing treatment guidelines, electronic medical record data, and peer-reviewed research and clinical studies.

> *Fundamentally, healthcare going forward will be increasingly computer-centric, not only in terms of electronic medical records and big data bioinformatics, but hopefully will maintain the engagement of active physicians and include scientific symbiotic participation required to edit information in real time.*

It will be crucial that the experts in all disciplines of the physical and life sciences work collectively to find healthcare solutions. This applies to evolving and increasingly interdisciplinary culture not just in the case of physicians but all branches of scientific expertise and must be rooted in the training of young experts. Finding successful solutions requires asking the right questions. None of us have infinite knowledge, however, knowing enough to ask the right questions founded on a well-rounded knowledge base has the potential to assimilate our experience and judgment with an almost infinite collective intelligence. Ultimately, for personal growth, for the benefit of our patients, and for the greater good, we all need to speak each other's language and to interact.

This body of work provides a model for how seemingly disparate branches of science may interconnect in unconventional but logical fashion. This emulates a type of interdisciplinary thinking that removes the compartmentalizing barriers that constrain the potential for asking the right questions. Interdisciplinary insights also facilitate improved understanding of concepts by using metaphors. For example, understanding the functioning of a combustion heat engine process using thermodynamics may help gain insights into the bioenergetic combustion of nutrients in cellular mitochondria.

There are many daunting challenges to changing the paradigm of biomedical education of both clinicians and researchers who can work across traditional disciplinary boundaries. *One obvious challenge is the explosion of medical and biological information referred to as Big Data.* However, there are a number of models recently promoted from the perspective of the physician researcher. These models can help better analyze and interpret the data. What I am proposing is to extend that perspective to include the clinical physician.

It is important to recognize that learning is not limited to the years we spend in formal training. We do learn the basics in formal training but in the scope of everything that is known, the robust amount of knowledge that we accumulate by the time of graduation prior to going into clinical practice, in relative terms amounts to only a scintilla of knowledge. Much more important is to understand the value of continued learning and the focus on how that is best achieved. Competence should be assessed in the various stages of training in terms of

demonstrating a nimble capacity to reason. The crucial priority at the training stage is to recognize emergencies that are too commonly overlooked. For example, a 35-year-old hospitalized patient with a pituitary adenoma and a cortisol level that is low-normal and now has recurrent hypoglycemia and mild orthostatic vitals, should not be discharged home with instructions to drink plenty of fluids and intake frequent carbohydrates. Similarly, a patient with diarrhea and a low bicarbonate level on the chem-screen (CO_2 on the basic metabolic profile) should not be given Imodium along with a bolus of digoxin for her atrial fibrillation and discharged home from the emergency room.

> *In addition to a multidisciplinary integration, a skill set should also be encouraged to include translational and transdisciplinary reasoning and problem solving from multiple perspectives of understanding systems in terms of their parts and wholes and recognizing distinctions between linear and nonlinear systems as differentiated by bottom-up and top-down approaches.*

This latter distinction is critical. Top-down mechanistic insights exemplify reductionism. Conversely, bottom-up self-organizing systems as complex adaptive systems taken as a whole span multiple scales of microbiology, cell biology and organismic biology and therefore are fundamentally nonlinear. Understanding any part of a whole biological system invokes reductionistic reasoning. However, to understand the system as a whole both reductionism and symmetries break down, and the reasoning process must be nonlinear. Understanding the complexity of nonlinear systems requires finding the so-called simple local rules that underpin the emergent phenomenon.

This book concerns the multidisciplinary and interdisciplinary perspectives of energy flow that connects seemingly disparate fields of science with the ultimate goal of intersecting with clinical medicine in addition to serving clinical value to college science students and medical students. It seeks to integrate an understanding of energy from the perspectives of physics, chemistry and biology to include molecular, organismal and ecosystem scales of biology. The aim is to promote a deeper and usable understanding, knowledge or scientific literacy that improves biological thinking and reasoning skills. One strategic theme is to absorb the thought process of the audience of readers by framing concepts from alternative perspectives and wherever applicable, understanding systems as parts or wholes, recognizing relationships as linear, complex or casual, and perceiving distinctions and similarities of closely related concepts.

There are many wonderful researchers and teachers admittedly far in the weeds at the subcellular level who are not able to connect their understanding to the macroscopic scale with tangible relevance, especially to biological systems and health or healthcare. "The small hydrogen atom, what does that mean?" "How will understanding the hydrogen atom change or affect my life?" These are the type of questions that college professors sometimes are asked by students. There is real value for understanding that couples the insights of chemistry and physics at the atomic and subatomic realms to the characteristics of a healthy biological state; to the macroscopic manifestations of the pathology of chronic diseases; and to the most effective diagnostic and therapeutic strategies for which there is emerging and widespread enthusiasm. This versatility is advantageous at the student level to apply the symmetries and integration of the various scientific disciplines as well as the links that make it all relatable. On the other side of the spectrum, it fastens healthcare practitioners to an acumen that enhances problem solving.

5.3 Creative Thinking, Information Transfer, and the Physiological Fitness Landscape

The execution of the medical profession is fundamentally rooted in making diagnoses based on signs and symptoms and treating those signs and symptoms. The field of medicine is bravely vulnerable to being replaced by technicians and computers armed with artificial intelligence software. Diagnosis and management are algorithmic and their connection to basic science is fragmented. Furthermore, it is largely non-intuitive with limited consideration of patient fears, expectations, biases, and belief systems.

This necessitates integration of information flow in the body and brain, which can be translated into an improved execution of the medical profession by invoking the physics models of Physiological Fitness Landscapes that will be described in subsequent chapters. *The Physiological Fitness Landscape (PFL) is generated by a map that describes a temporal trajectory of disease and disease-susceptible states and how to change that trajectory as long as potential reversibility exists.* The need for precision personalized medicine invokes computer science capable of integrating large volumes of bioinformatics and deductive problem solving, which also must involve sophisticated expertise of physicians and other health care providers connected both to the basic science and to the patient from a non-algorithmic intuitive perspective of problem solving that is irreplaceable by computers.

This perspective highlights the need for the adaptation of the Physiological Fitness Landscape model in clinical practice. *The major extrinsic control parameters include those describing diet (quantity, quality, and timing), circadian behavior (synchronized versus desynchronized) and the stress response (toxic, tolerable or vitalizing). The major intrinsic control parameter (secondary order parameter) is the microbiota. The role of reduced heart rate variability is a fundamental and promiscuous (non-discriminating) parameter for chronic disease (without putting a label on the disease) as a function of the loss of free energy from the richness of the connecting interactions that define complexity and of human biological health at the molecular level.* The loss of heart rate variability (HRV) reflects the underlying metabolic dysfunction and a deterioration of the biology of time in its correlation to chronological time. This is a central and clinically useful marker for a compromised temporal state of synchronized physiology, a portent of not only fatal cardiac arrhythmias but of the onset of premature chronic disease as well as a cause of mortality. The loss of heart rate variability is a manifestation of the impairment of chronobiology and a self-amplifying feed-forward inextricable web of all the defined extrinsic and intrinsic control parameters that define human diseases.

The body dynamically interacts with the environment and responds to stress by the phenomenon of allostasis, i.e. stability through change, primarily in autonomic, hormonal, and immune system parameters. *Homeostasis is resistance to change. The most fundamental parameters of homeostasis are Gibbs free energy, redox, and acid base.* When homeostasis is disturbed, these parameters move in tandem. This chapter concludes with a perspective of mitochondrial health and dysfunction and its role in the chronic states of cancers and dementias as metabolic diseases. This perspective offers a promising potential to transform the practice of medicine. In all that we show that the PFL approach focuses on stressing a system to determine a healthy, disease (e.g. using cosyntropin stimulation test to assess adrenal sufficiency; a cardiac stress test) or susceptibility state (e.g. a postprandial lipid profile, a postprandial glucose assessment; a post fluid restricted serum creatinine or creatinine clearance). It can also be used to determine stable versus unstable states (for example by determining a dietary excess BMI). Furthermore, it is theoretically possible to assess disease reversibility by introducing order parameters such as the metabolic pathways of oxidative phosphorylation and glycolysis or other markers of cell differentiation to which oxidative phosphorylation is typically linked.

This methodology may be relevant for applications to any of the chronic diseases of aging. We need to understand the mechanism behind the reverse Warburg effect because oxidative phosphorylation is necessary for cognition and is the primary mode of energy production of neurons in the brain. Microglial cells, on the other hand, utilize glycolysis as the predominant mode of energy production. The inverse relationship between the prevalence of Alzheimer's disease and cancer is intriguing. Hopefully, the reverse Warburg effect is not irreconcilable with the idea that oxidative phosphorylation is the mode of energy production in the brain under healthy circumstances and that the breakdown of this process is not only inefficient but represents the de-coherence of the correlated nature of energy production, which likely accompanies the metabolism. Thus, *the breakdown of this process leads to Alzheimer's disease or accelerated cognitive decline in the same way as it does in the body, or the brain, in cancer states.* Possibly, perhaps the mathematical model of reverse Warburg effect relates to the fact that the brain requires both robust capability of glycolytic metabolism for energy production (in the microglia) as well as oxidative phosphorylation (for neurons). If oxidative phosphorylation is shut down or significantly reduced, cancer becomes a significant risk. In these cases, glycolytic metabolism is upregulated with downstream dysregulation of numerous cellular processes including cell cycle maintenance. Alternatively, if glycolytic metabolism is upregulated in the brain, it does not compete with oxidative phosphorylation in the sense that it predisposes to cancer. Rather, it works in concert with oxidative phosphorylation, which may be the essential distinction accounting for the model of reverse Warburg effect.

5.3.1 Physiological Fitness Landscape

The notion of the stress test, for example the prototype cardiac stress test, invokes changing control parameters to assess the response of the order parameters in the system as a measure of the state of health or disease. There should be allostasis present in order to maintain homeostasis. Exercise-induced ischemia on a treadmill is an order parameter of homeostasis that indicates a state of disease. Exercise-induced tachycardia is an allostatic response that occurs to maintain stability of homeostatic parameters of nutrient and oxygen supply to the rest of the body to keep up with metabolic demand. Ischemia signifies allostasis unable to occur without taxing homeostatic parameters. The chronic psychogenic stress response leading to anxiety from the adrenergic excess of the overactive autonomic system and to depression from the lack of dopamine and serotonin normally produced by the gut symbiota, that is perturbed as a function of changes in extrinsic control parameters and their downstream effects on intrinsic parameters of physiology (e.g. immune compromise and desynchronized metabolism). Adding the corresponding Physiological Fitness Landscape to this model provides not only a quantitative representation of these physiological states and their respective stability but also

describes the dynamics of the system in terms of a trajectory over time. Thus, *a time course of disease progression can be mapped in terms of extrinsic and intrinsic parameter changes and their transitions from the values in a stability zone to those in an unstable zone (disease state). This also provides information on how steep the slope is in both the stability and instability areas, and how high is the relative amplitude between optimal health and end stage disease or death.*

Figure 5.4 shows a typical schematic representation of a Physiological Fitness Landscape for human progression from health to disease with characteristic potential barriers and areas of stability (homeostasis) separated by them. Figure 5.5 represents a corresponding trajectory in the abstract parameter space of order and control parameters over time. *The stable trajectory around the state of health shows an almost perfect periodicity of physiological characteristics. The unstable trajectory of transitioning into a disease state manifests an unhealthy diet, chronic exaggerated stress response, loss of circadian behaviors with a disturbed microbiota composition including loss of diversity.* All of this is happening in tandem resulting in the loss of coherence and synchronized bioenergetics (quantum metabolism), of other metabolic transformations that connect the endogenous cell clocks to nuclear hormone

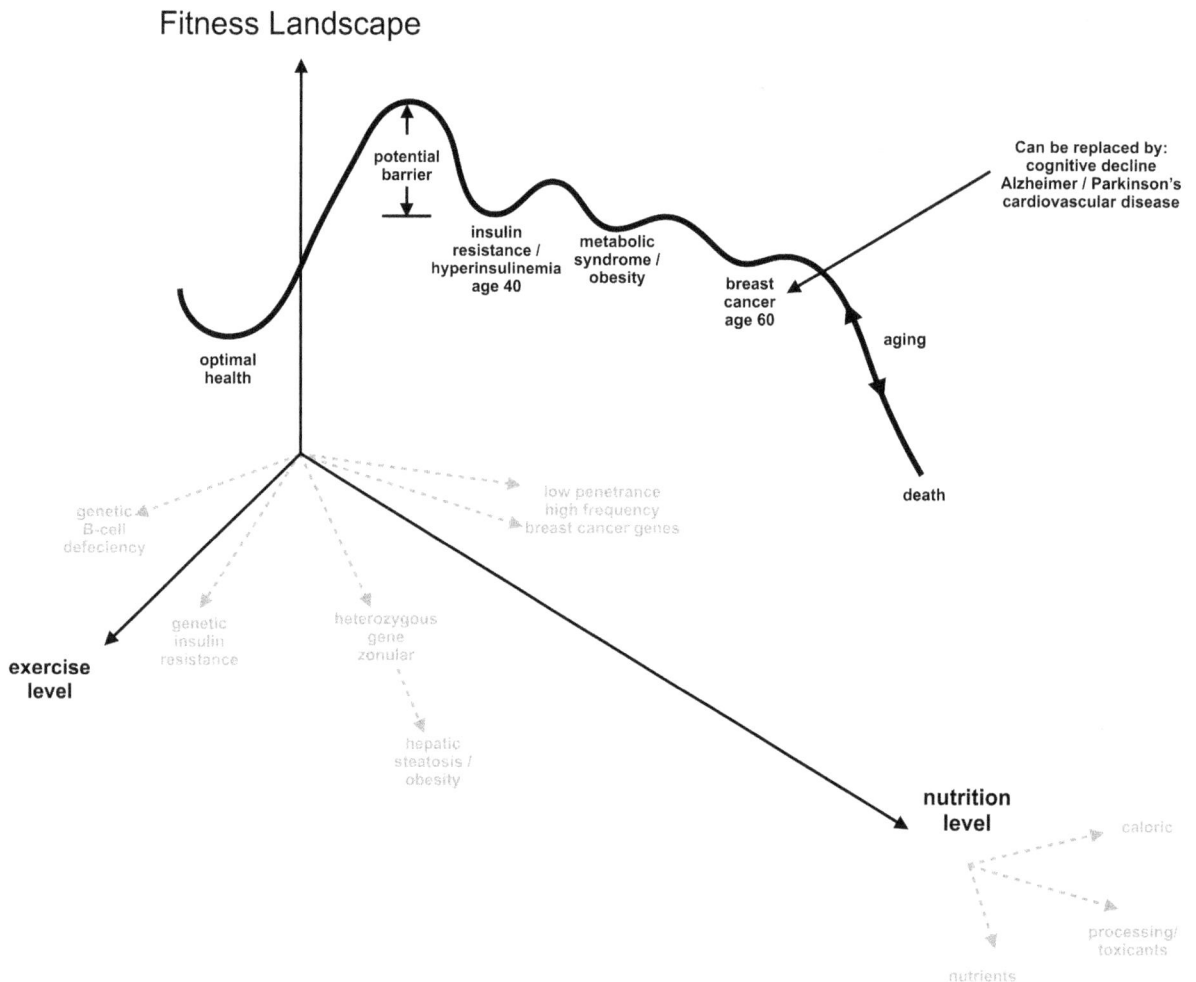

FIGURE 5.4 Illustration of the Physiological Fitness Landscape for human progression from health to disease. Optimal health depends on a variety of parameters including exercise level and nutrition level.

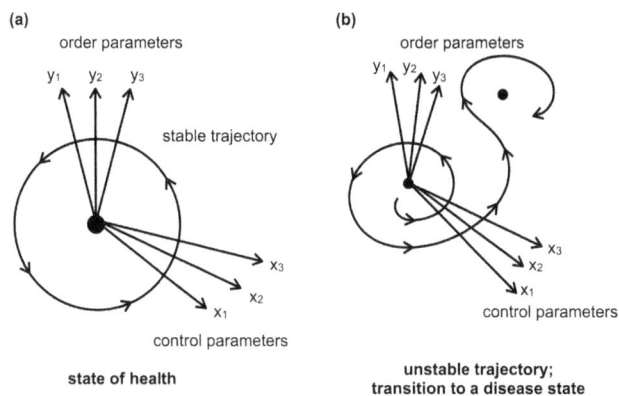

FIGURE 5.5 Illustration of the Physiological Fitness Landscape trajectory in abstract parameter space over time. A stable trajectory with perfect periodicity represents states of health (left), while an unstable trajectory represents states of disease due to loss of synchrony and coherence (right).

receptors, bile acids, and other transcriptional regulators that mediate clock-controlled gene output and ultimately physiology. Accompanying this is free radical generation, redox and inflammatory stress, and disturbed homeostatic parameters of pH and Gibbs free energy. This describes the mechanistic basis of disease, such as the chronic diseases of aging, which ultimately manifest as cardiovascular disease, dementias, or cancers depending on genetic predispositions.

The inextricable web spanned by the triumvirate of extrinsic control parameter relationships characterizing human diseases (diet, circadian behaviors, and the stress response) and the microbiota, which is an intrinsic control parameter, activates the allostatic responses of the autonomic nervous system and neuroendocrine hormonal branches of the stress response. When chronic, it causes a positive feedback on the control parameters that stimulate it, promoting a self-amplifying and perpetuating instability of the control parameters. Consequently, allostatic parameters such as proinflammatory cytokines, cortisol, DHEAS, and catecholamines eventually reach the point of allostatic load. *Allostatic load is a threshold of criticality, or irreversibility, beyond which restoration to normal control parameter values is unable to reverse the homeostatic parameters of the Gibbs free energy, redox, and pH homeostasis. Hence, this represents a point of no return for the human body succumbing to a chronic disease.* Once homeostasis of the vital organ system parameters is moved beyond this point, the state of organized complexity represented in the Physiological Fitness Landscape has been lowered in amplitude, bringing it closer to biological thermodynamic equilibrium with increasing tendency for instability and disease progression. *A state of health is characterized by a property similar to thermodynamic stability, namely robustness and resilience that allows it to return to its stable trajectory following an external perturbation such as exposure to toxicants, physical exertion, or temporary hunger. Beyond the critical point, however, the state of allostatic overload ensues, which is a manifestation of disease.* Parameters of any given manifestation of disease, for example neurodegenerative disorders, cancers, cardiovascular or autoimmune diseases, may be defined along specific trajectories of the Physiological Fitness Landscape.

The autoimmune manifestation of Celiac disease is one in which gluten may be defined as a trigger. However, some people who have been exposed to gluten for even 50 years or longer and have tolerated it, for some reason, may develop an immune intolerance, with allostasis that progresses to the autoimmune and auto-inflammatory exhibition of allostatic overload. The microbiota appears to be the core disturbance that causes a change from gluten tolerance to intolerance. For example, the relative depletion of strains of *Firmicutes lactobacillus* and *Actinomyces bifidobacteria* may largely be responsible for such deterioration of the state of health. These microbial strains ferment host non-digestible fibers to the short-chain fatty acid butyrate, which in turn promotes epithelial cell differentiation and counteracts actions of zonulin. In addition, this may directly inhibit the pathological trafficking of gluten and its interaction with the intestinal epithelial cell that instructs zonulin-mediated disassembly of tight junctional complexes.

What are the control parameter stressors that perturb the system capable of altering the microbiota and lead to the change from gluten tolerance to intolerance? I believe much has to do with a changing world of demands and associated toxic stressors. A toxic or devitalizing stress may be considered to be a life challenge, which cannot be controlled because it does not match the person's skill set relative to the type and dose (intensity and duration) nor does the person have the necessary support system to control the outcome of the stress. The psychogenic, circadian, and dietary stressors of an individual are an entangled set of control parameters that disturb the microbiota. The magnitude of stress affecting the health of any given bodily system can be measured in terms of a healthy or unhealthy response, and this is definitely patient-specific as various individuals can tolerate different amounts of stress depending on their physical, psychological, and social makeup. Examples of systems experiencing stress are the cardiovascular, renal, or other extra-intestinal organ systems, and hence stress involves multiple systems as they respond to it. These include prolonged psychogenic stress, as well as the extrinsic control parameters circadian and eating habits, intestinal dysbiosis, genetic predisposition, and in the case of autoimmune disease, it may require a specific trigger. In Celiac disease, the trigger is gluten. It is really not known yet what the trigger may be for type 1 Diabetes, Multiple Sclerosis (MS) and other autoimmune disorders. Furthermore, there may be unforeseen specific triggers for cancers or cardiovascular disease. For example, gluten in addition to promoting the actions and release of zonulin, instructs transcriptional regulation of predisposing genes to manifest the development of Celiac disease. Analogously, the same may be true for another trigger that not only stimulates the actions and release of zonulin but also contributes to transcriptionally expressing predisposing genes that manifest as type 2 diabetes, obesity, cancers, or cardiovascular disease. Superimposing a multidimensional set of control and order parameters on a generic Physiological Fitness Landscape can help elucidate responsible triggers for chronic disease states, autoimmune diseases, and other pathologies. *Monitoring stressors and responses to these stressors in the case of an individual patient can, in the future, provide valuable insights into disease progression as well as recommendations for optimal therapeutic regimens.*

The conceptual centerpiece of this book is the adaptation from physics into medicine of the methodology developed for phase transitions and critical phenomena as well as more recent advances in systems biology and quantum biology. *Concepts such as the Physiological Fitness Landscape with the attendant control and order parameters as well as symmetry breaking (which defines transitions between stability zones), can provide profound insights into the physiology of aging and development of chronic diseases.* Much of human intellectual activity is occupied by the search for patterns in space and time. These are usually repetitive arrangements of objects or signals. Geometrical symmetries such as tiles are primary examples of patterns. More complex patterns, such as branched structures of the cardiovascular system are examples of geometric fractals. Analogous temporal patterns can be found in musical compositions or the fractal self-similarity of New York Stock Exchange stock postings. There are also spatiotemporal patterns where repetitiveness can be found both along the time axis and in space, for example in the so-called Conway game of life. There, physical systems persist over time by forming recursive patterns that may change trajectory making pattern formation quite complex and interesting to follow. Similar to a Lorentz attractor of a weather pattern in response to a perturbation, the mechanism of allostasis may result in a transition to a disease state when the human body is subjected to environmental pressures whose external control parameters exceed their critical values. This form of a dynamical phase transition from health to disease may exhibit the same or similar pattern across spatial and possibly temporal scales.

Part of this temporal ontogenesis is in the life course of a human who is proceeding through stages from infant to child to a healthy adult. Subsequent changes biologically represent insidious aging punctuated by the more dramatic phase transitions imposed by disease. The inability to restore previous fractal patterns, such as for example found with heart rate variability, as it decreases over time, also applies to the fidelity of its fractal nature. These gradual and sometimes sudden deteriorations are a function of the free energy lost from the interactions and connectedness that create the complexity of the organismic whole. The breakdown of the temporal and spatial synchronicity and coherence of living systems makes them vulnerable to punctuated deteriorations in health (e.g. cancers, infections, and cognitive decline). All of these processes are self-perpetuating and self-amplifying. The brain, for example, has been hypothesized to have evolved for the purpose of the acquisition of food. However, in states of illness such as cancer or infection, associated anorexia is further debilitating, perhaps as an attempt of allostasis to starve the cancer cells or infecting microbes. As free energy is progressively lost from the organized complexity of the individual as a functioning whole, a threshold of allostatic load is reached and allostatic overload, or a state of chronic disease is unavoidable. At this point in time the reversal of the allostatic changes back to the previous state of homeostasis is no longer possible.

As mentioned above, the central extrinsic control parameters of diet, circadian behaviors, and the stress response represent a triumvirate, which forms an inextricable web that may define pathologically self-amplifying perpetrators of chronic disease. The symbiotic nature of the intestinal microbiota is emblematic of coherent singularity that determines biological complexity and healthy physiology. The native cells of the host are themselves symbiotic in the sense of their synchronized and coherent interactions at all scales of physiology, including the subcellular scale with mitochondrial synchronization, as well as within and across the tissues and organ systems of the whole body. *The endogenous clocks are extraordinary molecular timepieces that evolved as biological incarnates of the physical bodies of the universe responsible for the chronological framework of time created by the Earth's rotation around its own axis and rotation with respect to the Sun.* These endogenous clocks form the biological nature of time superimposed on physical chronology. The exquisite and beautiful organizational design of Nature enlists the biological clocks as orchestrators of total body physiology as an organismic whole. This is metaphorically similar to a well-conducted orchestra playing a beautiful symphony. The biological clocks are mediated by crucial elements of transcriptional regulators such as the microbiota, bile acids, and the thyroid, and steroid superfamily of nuclear receptors. The gut lumen and the intestinal microbiota represent such a fundamental part of what we are, as the major interface with the environment, and an ambassador to the outside world, which mediates the expression of the host genome.

The symbiotic and mutualistic relationship between the human host and its inhabiting microbiota is a function of the coevolution of these bodies of cells that co-opt each other's molecular machinery for the common good of the interwoven system as a whole. There are assimilated coherence and synchronized clock-controlled transcriptional output and metabolic physiology. *The temporal and spatial coordination of the intestinal microbiota uniquely positions it as an intermediate between the extrinsic behavioral control parameters and the intrinsic order parameters of host human physiology. The gut microbiota strongly interfaces with the host immune system.* Inflammatory and redox stress and associated loss of free energy from the organized connectedness and complexity that defines aging and chronic disease are most fundamentally generated by the breakdown in the microbiota superorganism as a secondary order parameter, or intrinsic control parameter, of human health.

Robustness of human physiology against environmental perturbations is compromised by the vulnerable central positioning of microbiota at the interface of the battlefield between friends and foes, exposed to many thousands of ingested toxicants. Furthermore, chronic psychogenic stress responses with immune compromise and inextricably linked disturbances in circadian behaviors allow pathogenic microbes an opportunity to outcompete and infiltrate the commensal and symbiotic microbiota within the intestinal lumen and along the mucosal surface. Ancient cyanobacteria eventually became subsumed into the cells of eukaryotic organisms to provide the efficient bioenergetic mitochondria organelle that is a central parameter of human health. Conversely, mitochondrial dysfunction is a central parameter of disease. Similarly, a synchronized symbiota is likely in the midst of a transitionary phase that is evolving with the directionality and purpose of providing greater physiological robustness and stability by becoming fully subsumed and insulated within the internal milieu of the human host.

The adaptive nature of organized complexity of interactions between independently living organisms providing functional stability of the system as a whole may be exemplified by the majestic and extraordinary dynamic of many thousands of starlings flying in seemingly acrobatic fashion intended to entertain. However, the interactions of these birds with one another are guided by local rules that cannot be appreciated by viewing the large flock from afar. This is analogous to observing the exquisite totality of human behavior. In the case of the flock, each of the individual birds is protected from a predatory falcon, which is not the case when it would fly away from the flock. Just like the loss of the synchronized nature of the parts of human physiology that comprise an adapted system whole, the increasing compartmentalization and loss of coordination between the birds in the flock represent an inherent loss in survival advantage. Consequently, escalating predator attacks on the starlings by falcons that degrade the flock is metaphorically equivalent to the action of microbial pathogens, inflammatory and oxidative stress, all of which is responsible for the loss of Gibbs free energy required for local interactions and organized complexity that parallels the progression of human disease.

It is important to understand the concept of Physiological Fitness Landscape as applied to the field of medicine and frame it tangibly in terms of parameters that lose connections in the context of disease. *Biological health is characterized by resilience to environmental perturbations. It is the capacity to adaptively respond to stress in the sense of maintaining a stable quasi-steady state of physiological parameters as a result of allostasis, i.e. stability through change.* Allostasis maintains homeostasis of vital organ system parameters pH, Gibbs free energy and redox states. Allostasis may take the form of a classical fight-or-flight stress response activation of the central and peripheral autonomic and hormonal systems. Alternatively, it may be defined as a nutrient stress such as an overnight fast whereby metabolic flexibility is manifested by peripheral insulin resistance and reduced pancreatic secretion of insulin. The latter case may be considered a specialized one of the general fight-or-flight stress response. No matter what the specific stress response is, it reflects a flexibility or responsiveness to changes in environmental conditions, or perturbations, capable of maintaining resilience or homeostasis (Figure 5.6).

The metabolic switch to fatty acid metabolism in skeletal muscle, away from glucose metabolism as an anticipatory response to overnight energetic stress, constitutes an intricate interplay of circadian patterns of physiology including the interwoven nighttime rise in melatonin that induces slow-wave sleep. This form of sleep is linked to nocturnal rise in growth hormone, which represents an exquisite interlocking coherent network of systems biology. It promotes the energy-requiring anabolic activity of building antioxidant systems and other stress resistance programs of cell and DNA repair, precisely timed so that it does not compete with the physical and cognitive energy-demanding daytime activities. The scale of resilience of allostatic anticipatory and reactive changes parallels the level of coherence and synchronized interconnectedness of the many parameters that define the biologically organized complexity over time and space.

Allostasis is a biological phenomenon, which is analogous to the physical effects of symmetry breaking that maintain stability by finding new symmetries of a system that is experiencing criticality due to external control parameter changes. The potential for allostasis to maintain homeostasis and thus for a system to break its "old" symmetry to find "new" symmetries may be framed by the available energy held in the interactions and bonds between molecular parameters of the system. Symmetry breaking is applicable within a context of the forces of Nature leading to new phases of matter (e.g. ferromagnetism, liquid crystals, or elementary particles) as well as to physiological states of a living system.

Homeostasis

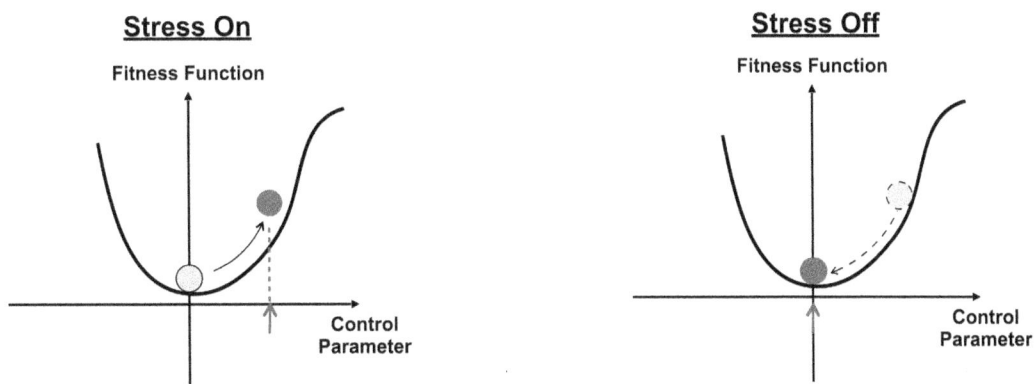

return to the original state and
the original fitness level after
stress application

FIGURE 5.6 Schematic illustration of the fitness function for homeostasis, which represents a tendency toward a stable equilibrium. The system can be perturbed during circumstances of acute stress; however, it will return to the same previous stable state once the stressor is removed. This flexibility and responsiveness help maintain resilience.

As an example of symmetry breaking in a living system, consider the international NICE SUGAR study [1]. It showed increased mortality in critically ill patients treated with intravenous insulin that establishes euglycemia (even in the absence of hypoglycemia) in the context of a history of chronic hyperglycemia following the diagnosis of diabetes. The results of the NICE SUGAR study highlight a phenomenon of metabolic memory, which is present after a number of years whereby an ontological history of chronic poorly-controlled glycemia following the diagnosis of diabetes portends increased mortality in the setting of critical illness and intensive care unit intervention with intravenous insulin to maintain euglycemia. The Physiological Fitness Landscape model applied to this situation identifies the susceptibility state of euglycemia as an unstable state with a risk of mortality. *It is seemingly paradoxical that hyperglycemia represents a stable zone within the Physiological Fitness Landscape, and "therapeutic" euglycemia breaks the symmetry but is unable to find a new symmetry state within a stability zone.* This importantly underscores that a normal blood glucose level is contextual. That is, a normal blood glucose level is not tantamount to a healthy blood glucose level in the circumstance of critical illness with a history of chronic hyperglycemia.

Immune system dysfunction and impaired HPA axis functions may be consequences of IV insulin driving glucose into skeletal muscle. The history of chronic disease and current critical illness results in the loss of available free energy that is prohibitive for new coherently integrated connections across systems biology such that establishing euglycemia itself represents a perturbation. The system as a whole, in this case, over time has lost complexity whereby interactions between parts of the system are reduced. The lost interactions are needed for allostasis to bridge the restoration to normal parameters such as euglycemia without compromising the stability of other parameters, even vital homeostatic parameters of redox, free energy, and acid base. That is, hyperglycemia may be necessary to supply the energetically expensive immune system function of the critically ill state. This in turn may be pathologically linked to the phenomenon of relative adrenal insufficiency because the intricately responsive and trans-tissue coherence is compromised due to the loss of robustness or redundancies in the framework of integrated systems biology. *In a more resilient healthy system, establishing euglycemia in the state of critical illness may have local or other alternative mechanisms for providing sufficient energy to immune cells to maintain a vigorous immune response.* The loss of free energy in the system may underpin the loss of redundancies. This may be plausibly hypothesized as responsible for relative adrenal insufficiency and associated mortality. Although the specific mechanisms are not known, the overshoot of biological complexity of interactions being lost as a fundamental manifestation of disease may be better understood in the PFL context.

In a nutshell, the Physiological Fitness Landscape is an analytical map to our understanding of a temporal trajectory of disease and disease-susceptible states and how to change that trajectory by both pharmacological and non-pharmacological interventions as long as there is potential for its reversibility. The need for precision personalized medicine that involves computer science capable of investigating large volumes of bioinformatic data and deductive problem solving also must involve sophisticated clinical expertise of physicians and other health care providers committed to both the basic science and to the patient from a non-algorithmic non-intuitive perspective of problem solving that is irreplaceable by computers.

5.3.2 Order Parameters, Control Parameters, and Physiological Fitness Landscape for Disease State

One of the most important take-home messages of this book has been linked to the involvement of thermodynamic concepts in medical practice. We have argued that metabolism, which is similar to power generation in internal combustion engines, can be analogously characterized by thermodynamic functions and framed in terms of mathematical equations governing thermodynamic systems, especially those undergoing phase transitions. The connection to phase transitions comes naturally due to our interest in health and disease, two different physiological states that can be thought of as thermodynamic phases. In this vein, we have proposed to simplify and systematize our approach to these physiological states using such concepts as extrinsic control parameters of health and disease, order parameters, and fitness functions. The reference point for this description can be either the state of health if we wish to investigate the onset of disease, or *vice versa*, the state of disease, if our attention is focused on the healing process.

> *The number of both control and order parameters is in principle unlimited and the only limitation is our ability to quantify them by laboratory tests or other means of measuring their values.*

For example, we could use the following factors as order parameters relevant for cancer: 1) molecular markers such as the rates of over-expression of biomarker proteins or the prevalence of mutations in various genes; 2) cellular-level pathologies such as aneuploidy, membrane morphology aberrant insulin signaling, cortisol levels, immune suppression, impaired nuclear hormone receptor signaling, change in the transmembrane potential, pH, lactic acid production, upregulated signaling pathways (e.g. PI3K, KRAS) and the extent of Warburg effect; 3) organ-level characteristics such as tumor size, grade and type; 4) organism-level quantification by vital data such as blood pressure, weight, oxygen saturation, heart rate, etc. Each of these measures can provide a single data point for these variables seen as axes for a multi-dimensional space which, when constructed, could give rise to a Physiological Fitness Landscape as a manifold in this parameter space.

To do this, we need to also define corresponding extrinsic control parameters, which trigger a physiological response of the human physiology system. As described above, one of the most powerful order parameters in cancer initiation and progression is linked to hyperinsulinemia, which triggers cellular signaling pathways such as the ERK/MAPK pathway that includes the signaling cascade of Ras-Raf-MEK-ERK. These pathways are stimulated by proteins that activate cell membrane epidermal growth factor receptors (EGFRs). Insulin resistance and endogenous hyperinsulinemia are major contributors to a large spectrum of epithelial-derived cancers such as pancreatic, colorectal, stomach, esophageal, hepatocellular, and others which show a strong relationship between cancer and obesity and type 2 diabetes. This relationship is mediated by the endogenous hyperinsulinemia. As examples of control parameters, we could list those associated with pharmacological interventions, i.e. the drug type, its dose and scheduling, radiotherapy described by the treatment plan, diet including nutritional supplements, vitamin D levels, microbiota, exercise regimen as well as potentially negative effects on the state of health such as addictions, stressors, sleep disturbances, etc.

Having gathered all this information, we can then start building a multidimensional map that not only places the patient's data in the context of normal values but also follows the patient's response to treatment over time. This can be used as a trajectory in the course of the disease progression or remission depending on the outcome. In view of the huge number of data points that can eventually be introduced into the model, bioinformatic resources and algorithms such as IBM's Watson can be brought to bear on the description of the state of health/disease of each patient. Moreover, artificial intelligence platforms that will no doubt enter into the field of medicine in a big way, can then predict and optimize the treatment regimen based on the emerging Physiological Fitness Landscape. This is envisaged as an iterative process as the patient's history and the disease progression can be represented as a trajectory on this multidimensional Physiological Fitness Landscape.

Below, we have schematically described the process as a flow chart (Figure 5.7).

In a very simplistic sense order parameters can be related to disease symptoms at all levels of physiological organization: from DNA mutations to vital sign changes outside the normal range of values. On the other hand, control parameters represent interventions that affect order parameter changes. Therefore, these can include gene therapy, pharmacological agents including antibodies, immunotherapy as well as lifestyle changes that can be controlled such as the amount and content of the person's diet, its timing, exercise, and physical activity, sleep patterns, presence, and the intensity of stressors of various kinds (relationship-related, work-related, financial, etc.).

The process described in Figure 5.7 critically depends on the availability of physiological data at all levels of patient analysis, from molecular biomarkers to blood work to vital signs. This data integration requires models, both mechanistic and artificial intelligence type. If the data are collected from an individual patient, then we can truly talk about precision medicine, which is patient specific. This is, of course, not always available, so surrogate generic data inputs can be provided from statistical and epidemiological databases as estimates for a given patient, in which case the term precision medicine is not entirely warranted but the process can still be of value. As a result, one creates a Physiological Fitness Landscape, which is always an approximation but the level of precision in this approximate diagnostic process depends on the availability of relevant data for analysis (Figure 5.8).

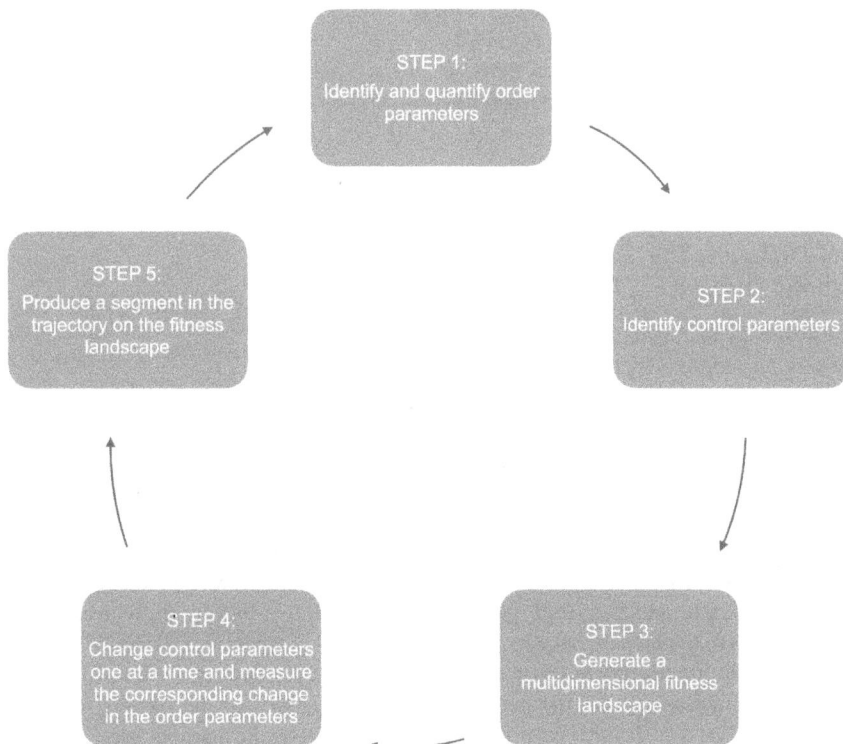

FIGURE 5.7 A Physiological Fitness Landscape trajectory flow chart describing the proposed steps involved in diagnosis and treatment utilizing this method in precision personalized medicine.

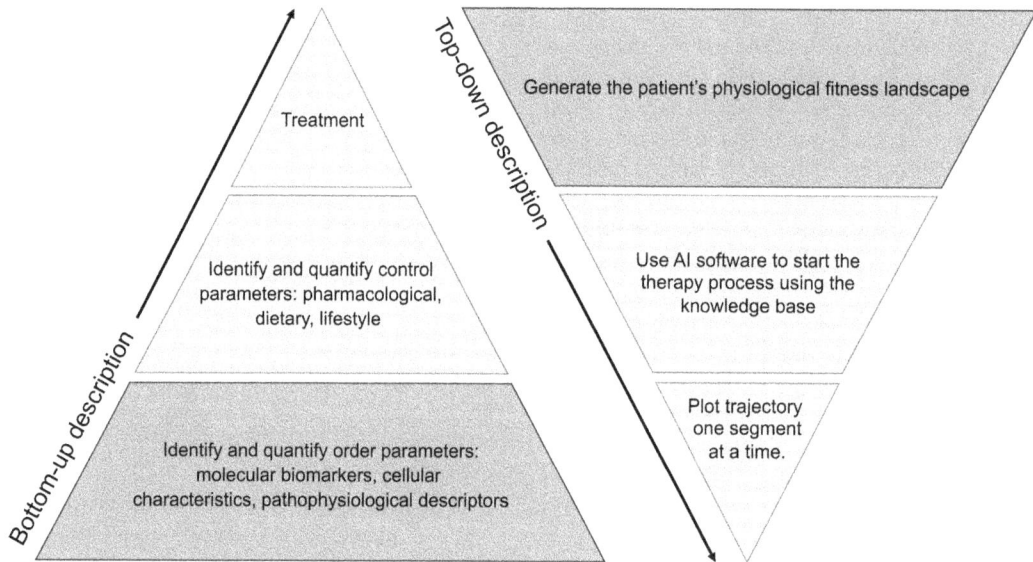

FIGURE 5.8 Bottom-up and top-down approaches to treatment utilizing the Physiological Fitness Landscape in the diagnosis and treatment of disease.

Moreover, as the arrows indicate, *this is an iterative process whereby new patient data should be used over time to revise the construction of the corresponding Physiological Fitness Landscape, which would then inform about the need for a change in the course of therapy.* In other words, this concept is not static but dynamic and its determination changes in the course of therapy with the available updates to the patient's status. The frequency of such updates should be implemented depends on the practicalities of performing required tests and physical examinations but in the case of chronic diseases it is expected to occur on a monthly or semi-annual basis while for acute conditions, daily or even hourly updates may be required. In critical emergencies, minute-by-minute updates may need to be taken.

5.3.3 An Example of Order and Control Parameters of Diabetes, The Classic Metabolic Disease

Here, the mechanisms involved in metabolic dysregulation leading to chronic diseases, including diabetes, are described in detail. Glucagon-like peptide 1 (GLP-1) is a hormone produced by intestinal epithelial endocrine L-cells in the ileum. The increases in GLP-1 production that occur following bariatric surgery contribute to a phase transition of type 2 diabetes into remission *via* effects in the pancreas, in the brain, as well as in the gut regarding postprandial gastric emptying, making GLP-1 a governing control parameter.

Neuropeptide Y (NPY) and the agouti-related peptide (AgRP) are produced by neurons in the arcuate nucleus of the hypothalamus. NPY/AgRP-containing neurons activate orexigenic responses mediated by connections with the paraventricular nucleus (PVN), projections which otherwise inhibit the nucleus tractus solitarius (NTS) satiety center and lateral hypothalamic area (LHA) orexigenic neuropeptide orexins and melanin-concentrating hormone (MCH). Hormones

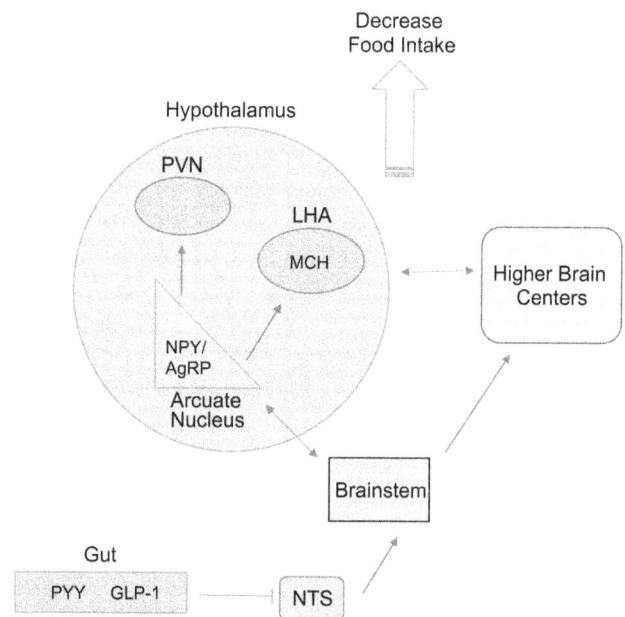

FIGURE 5.9 Schematic representation of hormone and neuropeptide pathways involved in satiety.

produced by epithelial cells of the distal gut, primarily peptide YY (PYY) and GLP-1, inhibit these orexigenic responses to promote a satiety effect (Figure 5.9). A satiety effect has metabolic benefits related to a reduction in caloric intake, particularly for carbohydrates, with a lowering of the body weight set point. Furthermore, some of the benefits may be due to ketogenic effects, which offer a multifactorial range of mechanisms discussed in the Calorie Restriction Chapter (Volume 2). This serves to improve peripheral insulin sensitivity, ultimately lightening the load on pancreatic β cells.

In the theoretical setting of a complete reversal of peripheral insulin resistance, pancreatic β ccell hypersecretion of insulin would not be required to restore normal fasting and postprandial glycemic levels and other associated metabolic parameters unless the pre-existing diabetes was advanced with an absolute level of insulinopenia. The changes in GLP-1 production that occur following surgery result in their own quartet of effects. These include effects on the pancreatic α and β cells, effects in the brain, as well as on the gut regarding postprandial gastric emptying. GLP-1 also has beneficial effects on the heart, although in terms of the post-bariatric effects of GLP-1 contributing to a phase transition of type 2 diabetes into remission, or improving the attractor trajectory of metabolic control, it is the quintet of effects on the pancreatic α and β cells, gut, and brain to which GLP-1 is the governing control parameter. The effects of GLP-1 on the pancreatic β cell in rodent models include an improvement of insulin secretory dynamics, reduced apoptosis, increased proliferation and even neogenesis, that is, de novo formation of new islet β cells. This regulation of β cell mass includes signaling pathways of cyclic AMP (cAMP)/ protein kinase A (PKA) response element-binding protein (CREB), and phosphoinositide-3 kinase (PI3K)/protein kinase B (AKt)/mammalian target of rapamycin (mTOR), in the case of antiapoptotic and mitogenic effects. Neogenic effects are reported to be regulated by cAMP/mitogen-associated protein kinase (MAPK)-extracellular signal-regulated kinases (ERKs) along with phospholipase C (PLC)/protein kinase C (PKC)/ERKs signaling pathways (Figure 5.10). Despite an impressive and growing signaling roadmap in rodents, with GLP-1 and other stimulants of β cell mass leading to up to a compensatory 30-fold increase to counter insulin resistance, the capacity for enhancing adult human β cell mass is very limited. Furthermore, the paucity of available mitogenic signals has hampered the development of a clear roadmap or even the ability to hypothesize testable models [20, 21].

Because of this, complete reversal of type 2 diabetes and increase of β cell mass in humans has not been possible. However, following bariatric surgery, a gestalt of improvements in the human metabolic state is often sufficient to restore normal glucose tolerance and fasting glucose. These improvements include a reduced threshold of satiety, lower body weight setpoint, enhanced peripheral insulin sensitivity, and favorable effects on the be β cell, particularly the dynamic of insulin secretory function. Nonetheless, this will prove to be non-durable if there is a return to the pathogenic extrinsic environmental control parameters of diet, toxicant exposures, activity level or stressors (i.e. psychosocial, work, financial, personal health, or health of loved ones, etc.) that mediated neuroendocrine and autonomic pathological output in the first place. While there are genetic predispositions at many levels as will be discussed ahead, environmental factors account for approximately 70% of the genetic expression of disease risk.

The composite pathological chronic stress response promotes emotional top-down processes of hedonic appetitive consumption that include excessive quantity and/or quality of dietary intake. Furthermore, the immune suppressive effects of the sympathetic autonomic and neuroendocrine systems lead to changes in the gut microbiota composition. It is also likely to cause consequent destruction in the normal bile acid metabolism and the enteroendocrine system of the gut responsible for insulin resistance and pancreatic β cell dysfunction, in addition to direct antagonistic effects of the hormonal neuroendocrine and sympathetic autonomic nervous system components of the stress response. Moreover, the gut-brain axis is disturbed in part due to reduced PYY and GLP-1 hormonal and vagus nerve sensory fibers. The vagus nerve afferent fibers originating in the gut have dendritic branches in the lamina propria, in close proximity to the enteroendocrine cells that release satiety factor hormones. In addition to hormones such as PYY and GLP-1 having access to the brain *via* an endocrine effect carried through the systemic circulation to specialized fenestrated areas of the blood-brain barrier, the vagus carries the sensory information to the NTS satiety center in the brain stem. The signaling of GLP-1 carried by sensory afferent vagal fibers and transmitted to the NTS has broader effects on the brain. Further synaptic interactions occur within the NTS, with GLP-1-neurotransmitter-containing neurons that are separate from the anorexigenic signal of GLP-1, from the periphery.

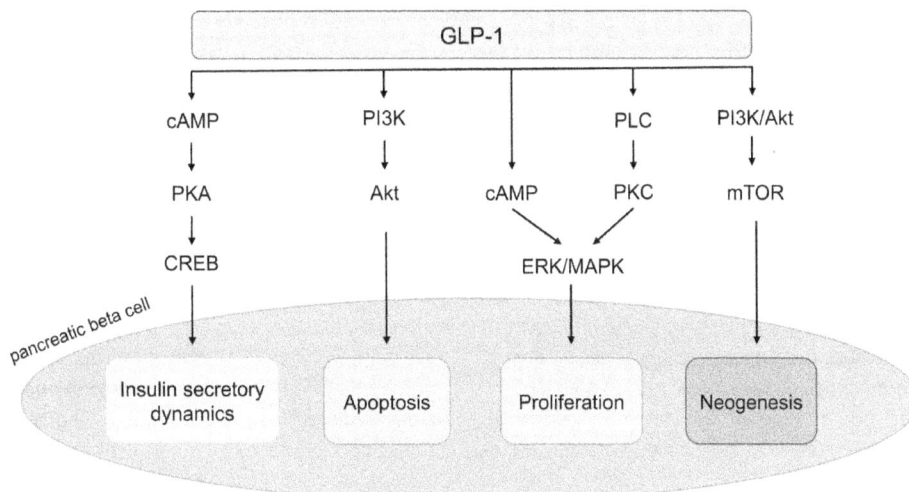

FIGURE 5.10 Illustration of GLP-1 signaling cascades and resulting downstream effects regulating insulin signaling, apoptosis, cell proliferation, and neogenesis in pancreatic β cells.

These neurons project widely throughout the central nervous system. This may have explosive implications for GLP-1 neurotransmitters in the brain, not only in terms of energy homeostasis regulated through the hypothalamus, but also in terms of its ability to govern the stress response. It appears to promote cognition memory and learning as well as hedonic reward-motivated behavior—a healthy goal-oriented motivational reinforcement that builds emotional resilience against an exaggerated stress response. This serves to counter an emotional top-down control parameter physiological fitness function over cognition, and in general, the downstream effects of pathological psychophysiology and chronic disease. Hence, the gut-brain axis, when disturbed, exaggerates the stress response interceded by circulating inflammatory cytokines as well as impaired GLP-1 signaling. Furthermore, healthy gut microbiota produces a host of neurotransmitters including GABA, serotonin, dopamine, norepinephrine, and acetylcholine, in addition to short-chain fatty acids. These neurotransmitters favorably influence resilience of the stress response, and they all may have antidepressant or antianxiety effects. Accordingly, altered gut microbial composition compromises the secretion of these chemicals, which is another component of the gut-brain axis.

Moreover, the GLP-1 neurotransmitter-producing neurons of the NTS promote biogenic amine neurotransmitters serotonin, dopamine, and norepinephrine in neurons of the ventral tegmental area, mesolimbic (nucleus accumbens) and hypothalamic nuclei containing GLP-1 receptors. These signals not only promote satiety in the hypothalamic regions but reduce food motivation as well as motivation for nicotine, alcohol, and psychostimulants. Bariatric surgery, by increasing GLP-1 and vagally transmitted signals to the NTS, may induce a broad salutary effect on addictive behaviors not limited to food (22). These same monoamine neurotransmitters appear to be responsible for improvement of Parkinson's disease movement, meaning that neuropsychiatric manifestations of GLP-1 agonist therapy may represent a potential mechanism for the common coexistence of type 2 diabetes and Parkinson's disease [23, 24]. The GLP-1 neurotransmitter neuronal system with cell body in the NTS also has effects on lipids and glucose levels, mediated through hormonal and vagal efferent fibers from the brain innervating the liver and skeletal muscle. One mechanism implicates endogenous brain-derived GLP-1 neurons extending from the NTS to the paraventricular nucleus of the hypothalamus, activating corticotropic releasing hormone (CRH) and Nesfatin-1 neurons, that are responsible in part for inducing satiety and may have an effect on lowering blood glucose by regulating skeletal muscle glycogen. GLP-1 neurons in the arcuate nucleus, PVN, and LHA contribute to an anorexigenic effect, as mentioned above. Brown adipose tissue-stimulated thermogenesis is also implicated in the weight loss effects of the brain-derived neuronal system of GLP-1eurotransmitters. This is suggested to be mediated by AMP kinase upregulation of uncoupling protein-1 (UCP) [25].

If brown adipose tissue were included along with the skeletal muscle and hepatic effects of GLP-1, the tissue or cell types affected by GLP-1 would be a septet. However, these effects may be included with the brain effects since it is from there that the neuronal or humoral signals of the GLP-1 system derive.

Along these lines, in addition to the pancreatic and extra-pancreatic effects of GLP-1 already mentioned, we include effects on kidney function. It inhibits the proximal tubule Na^+/H^+ ion exchange pump along with other mechanisms that regulate glomerular filtration rate, underpinning in part the antihypertensive effects of GLP-1. Accordingly, the entirety of Ralph DeFronzo's iconic ominous octet—proposed in his 2009 Banting Lecture regarding a pathophysiological approach to the core defects of type 2 diabetes—is either directly or indirectly impacted by the impairments of GLP-1 [26]. Central to the emphasis of the Stress Chapter of this book is the role of the stress response to metabolic health and chronic disease, both hormonal and neurotransmitter GLP-1, which represent an important intrinsic control parameter impacting the fitness function of an individual.

5.3.4 Physiological Fitness Landscape as a Guiding Concept in Medical Diagnosis

Medicine is an art that sits on top of the science of human anatomy and physiology. Reductionism, the crown jewel of 19th-century science is rooted in the fundamentals of physics and in equations capable of predicting outcomes of physical processes. It is also applicable to fields such as chemistry, logic, and philosophy. Fundamentally, it is tethered to mathematics. While biology as a scientific discipline is a less mature science, roughly a century behind its physics counterpart, the greatest achievement of its interactions with physics is the unlikely germination of living systems from a primordial "soup", which ultimately led to human life. However, it is almost shocking how naive and nascent is the interwoven field of biology and medicine. In fact, it was not until late in the 19th century that the most basic element of chronic disease, pro-inflammatory protein cytokine molecules, even began to be discovered. These cytokines and related immunologic molecules highlighted the inflammatory basis of chronic disease. For example, in the 1980s it was revolutionary for cardiovascular disease to be considered an inflammatory process. The major implication was that it is not the control parameters of the environment extrinsic to the body that causes chronic disease but rather intrinsic factors within the body.

The health of a stable state correlates with three parameters: 1) the height of its altitude; 2) the slope of the subsequent mountains; and 3) the height of the energy barrier on the upslope of the mountain. Dietary overconsumption is a classic stress that beyond a threshold overcomes an energy barrier to an unstable state of a metaphorical mountaintop. Accordingly, chronic overeating moves an ideal or healthy body weight set point to an unstable state of weight gain that does not correlate with the quantity of calorie consumption. It follows that weight gain may progress despite dietary restraint. This is likely a consequence of diet-induced disturbance in gut microbiota diversity and composition that results in abnormally increased energy harvesting. The excess body weight is a major contributor to an insulin-resistant state, including metabolic syndrome. This complicates an increased body weight set point characterizing a new stable state positioned lower in amplitude within the topological terrain of the Physiological Fitness Landscape. Additionally, as depicted in Figure 5.11 the

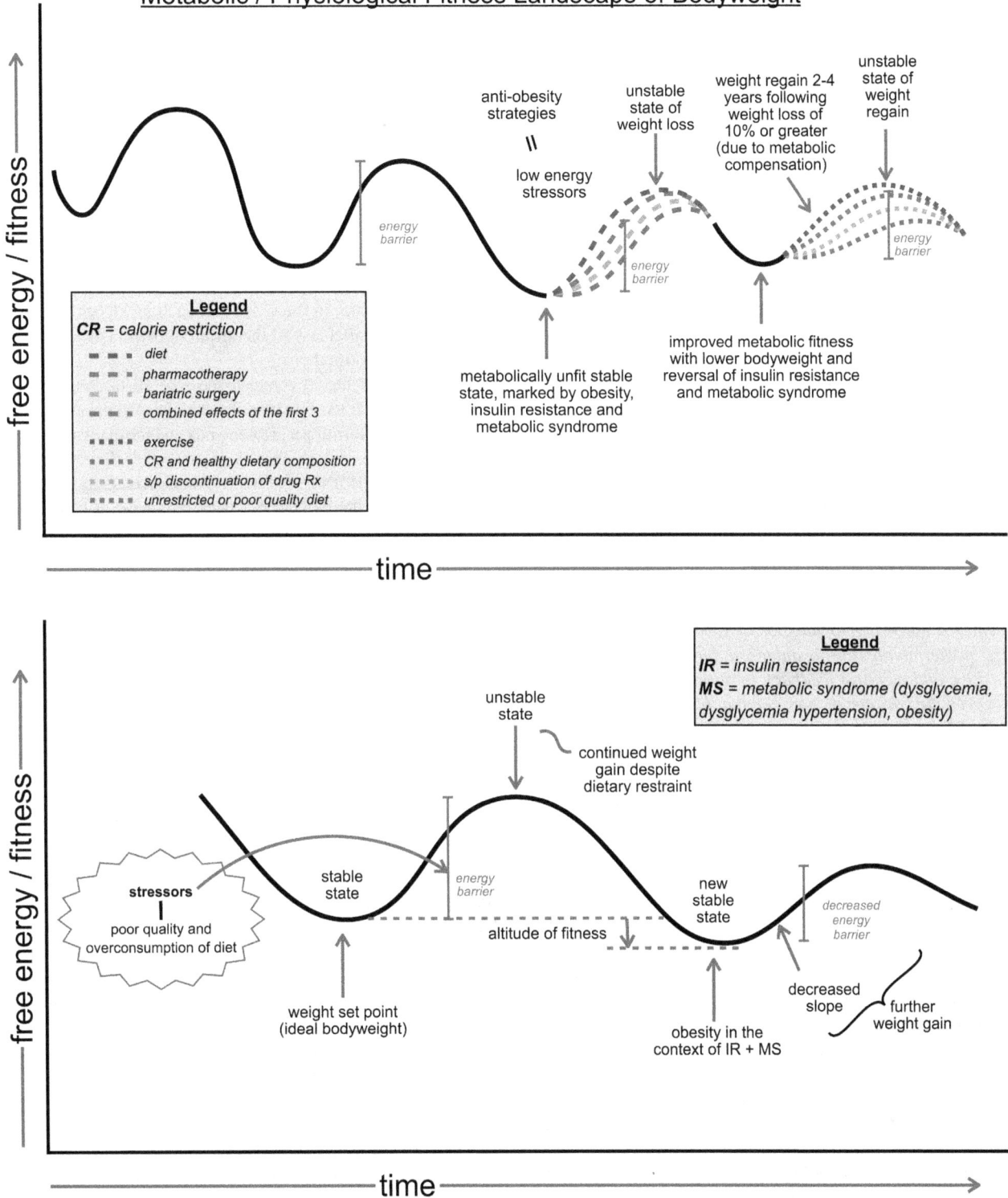

FIGURE 5.11 Illustration of variation of metabolic/Physiological Fitness Landscape (depicted as a fitness function on the Y-axis) with time (X-axis) through states of health and disease.

steepness of the slope to the next mountain top is typically shallower and the energy barrier is lower and easier to overcome than was the case for the preceding mountain. The slope up the next mountain and the position of the energy barrier represents resistance to weight loss against various weight loss strategies. In this schematic, successful weight loss is greatest for those strategies with the shallowest slope. Once the energy barrier is overcome, reaching the unstable state of the mountain top, weight loss becomes effortless, until falling down the other side of the mountain into a new trough. This trough becomes the new stable state characterized by an improvement in parameters of metabolic fitness. The journey up the next mountain is the metaphorical representation of weight regain over time following significant weight loss. Figure 5.11 shows the relationship of alternative behaviors to that of metabolic compensation and slowing of metabolic rate in the direction of restoring the previous body weight set point.

The rich tapestry and complexity of interconnecting elements of the integrated macroscopic biological system comprising many subsystems is the consummate display of resplendence and beauty. It is the stress of molecular oxidation and inflammatory activity that degrades and incinerates the information of biological complexity, which is the foundation of chronic disease for which medicine seeks to find solutions. Despite the fact that reductionism provides us with scientific methodology as a framework for studying biological systems in medicine, it is imperfect and often misleading when the change of the output of a system is not proportionate to the change of the input. This nonlinearity and non-predictability is the hallmark of all living systems, particularly those systems with the highest level of complexity such as human beings. Accordingly, this accounts for often unexpected and sometimes tragic outcomes of interventional studies and clinical practice. Unlike physics, for example using the Newtonian equation F=ma, and knowing two of the three variables in the equation reliably predicts the value of the unknown variable, *biological systems not only involve a much higher number of variables, many of which are indeterminate, that is unknown, but the presence of both positive and negative feedback loops makes the relationship between the parameters non-computable.*

It will be a great advantage if such interventional investigations and clinical practice recognize not only parameters of fitness and free energy in a physiological sense but also susceptibility states and pathological chronic disease states. Free energy parameters of a fitness landscape are quantifiable measures of a system under investigation, hence are personalized characteristics of this system. *This "landscape" of physiological fitness is how this fitness (or relative fitness in the case of susceptibility or disease states) depends on variables that can be applied to control it. Thus, control parameters are external to the system, typically in medicine or clinical studies this would be external to the individual.* These could comprise both pharmacological and non-pharmacological interventions into the state of the human body. The parameters that change in response to control parameters are both primary and secondary order parameters. Secondary order parameters are intrinsic to the system but control downstream primary order parameters within the same system.

The complexity of biological systems is reflected in the multidimensional network of control, primary and secondary order parameters. The fitness and free-energy landscape are metaphorically conceptualized by a terrain of mountainous peaks and valleys. Crucially important are several notions; one is that the trough or bottom of the valley represents a metastable state resistant to minor perturbations, and another is the idea of histories. Histories signify the memory of a system that may be rooted phylogenetically or oncogenetically. This memory is the recall of the evolutionary or individual history of the person or species, respectively. In the latter case it is an extension of the process of natural selection. The parameters of a landscape may be described in terms of either free energy or a fitness function as discussed ahead. The most efficient or stable pathway of the system metaphorically described in the framework of a landscape looks for the most stable free energy or peak performance, which is the most efficient and optimal choice of the available possibilities. In the case of declining parameters of fitness, the system breaks the existing symmetry and finds a new symmetry, which is obtained in order to balance the interaction of all of the order and control parameters, intrinsic and extrinsic to the system as a relative state of stability or metastability. The Physiological Fitness Landscape methodology could be extended to exploit the concept of a quantum algorithm, in the sense that such an algorithm could efficiently in parallel test all the different pathways, ultimately settling on the optimal one.

The deeply entangled connectedness of all elements of life within the boundaries of an organism and beyond must be recognized as systems forming a greater whole, which in turn becomes part of yet a larger system, an even greater whole. Intuitive thinking is critical in medicine. Strict interpretation of study outcomes often leads to faulty clinical practices. The level of complexity of biological systems is striking and is only paralleled by the nothing-short-of-amazing scale of wisdom in nature's evolutionary designs. Accordingly, this brings us to the notion of purpose that lies at the intersection of material science and philosophy. *Purpose is found in the goal-oriented positive and negative feedback loops, which are the hallmarks of biological complex adaptive systems providing them with directionality. The symbiotic microbiota in the human gut is a co-evolutionary product that was driven towards maximizing physiological health and survival of both the microbes and the host.* Each of the symbiotic co-evolving bodies co-opts one another's molecular machinery for the common good of the body as a whole. This is an example of a mutualistic interplay whereby each of the trillions of bacterial cells are free living in addition to the human host in the absence of the microbiota being free living. Thus, each microbial organism as well as the human organism represents the highest hierarchical scale system whole.

A human being has many component parts to his or her biology, which form systems within the body. These systems are actually subsystems of the larger system, the organismic individual or the person. So, many systems across many hierarchical scales form the human being. The largest system that evolves within the complex self-organizing biology is the individual, in this example the human being. This is the ultimate largest system whole from which many systems assimilate to create. The single human being, however, becomes a part of a greater system as he or she joins a network within society. So, this is an example of the hierarchical scales of a system becoming a part of a greater whole. This idea will be further explained from the most fundamental systems from which humans evolved.

The compartmentalization of clinical medicine treated as insulated specialties contributes significantly to the limitations of applied science that inherently gets lost and cannot see the proverbial forest for the trees.

This is underscored by the reality that approximately half of the well-accepted standards of care over the years are determined to actually be wrong, and only roughly a third of which have been validated by repeated and reproducible proof-of-concept studies. The very nature of hypothesis in medicine and in biology is radically different from that in physics. Physics is a more mature older science that strives for disproving a hypothesis rather than confirming it. Rather than longing to prove the hypothesis correct, its priority is focused on the longer term with a series of repeated adjustments to the hypothesis until it approximates the ultimate objective truth about a natural phenomenon.

The Physiological Fitness Landscape is an organizing framework based on models in physics that can be used to study the relationships between order and control parameters of a susceptibility state or a disease state manipulated to induce a phase transition of the susceptibility or disease state. This Physiological Fitness Landscape is a free energy landscape that may be invoked to study the susceptibility and disease states of all chronic diseases of aging, which should be considered fundamentally metabolic disease states. Accordingly, disturbances in correlated energy production are rooted in impaired mitochondrial health with reduced capacity to transform energy into the universal biological currency of energy, ATP, through which all the work of physiology is generated. The building blocks of work mediated by ATP include active transport, molecular motor function, and enzymatic activation that drive signal transduction kinases and biosynthetic reactions. The notion of insulin resistance is inextricably linked to mitochondrial disease in a bidirectional manner. Furthermore, *insulin resistance lies at the core of virtually all, if not all, chronic diseases of aging including cardiovascular disease, cancers, Alzheimer's disease, and even accelerated cognitive decline.* Research aims to find treatments and cures capable of inducing phase transitions and healthy attractor states. While type 2 diabetes and obesity are each considered disease states *per se*, for the purpose of this organizing model of a fitness or free-energy landscape, these conditions may be treated as risk factors. In other words, they are susceptibility states.

Another way that this landscape can be framed is from the perspective of order and control parameters of the overall state of the individual or human organism. That is, cancer or heart disease are morbidity states or susceptibility states, and secondary order parameters to downstream mortality of the individual. In this case, each secondary order parameter "susceptibility state" contributes to the overall burden of redox and acid-base disturbance and loss of free energy modeled by the Nernst, Henderson Hesselbach, and Gibbs free energy equations, respectively. Ultimately, the severity of a given chronic disease secondary order parameter or cumulative severity of the multiplicity of disease states compromises the metabolic homeostasis of the individual to result in death. The metabolic health is quantitatively indexed by the Nernst, Henderson Hesselbach, and Gibbs free energy equations, which have a striking correlation to one another. Intriguingly, these quantitative parameters correlate in parallel the systemic inflammatory process driver of clinical disease and senescence, which represents the biological equivalent to heat dissipation and associated entropy increase that drives the physical arrow of time, in compliance with the second law of thermodynamics.

Truly, *the intersection of physics and biology is rich and powerful but as yet very sketchily and inadequately explored for the benefit of medicine.* Moreover, this is more than an intellectual playground of entertainment and research. This free energy Physiological Fitness Landscape model has real value for applied medicine. The clinical realm of patient care is an art on top of the science that relies on intuitive instincts for which there is no definitive truth or predictable outcome. A connection to the basic sciences and to the available data strengthens the nimble biological problem-solving skills of patient care at the individual scale, particularly when it is coupled to an understanding of the biases, fears, expectations as well as knowledge of the psychosocial support systems of a given patient.

The weaknesses of clinical medicine mentioned above can be addressed by a skillful integration with physics. The assimilation of physics into clinical medicine allows for models of assessment and decision-making that are critically consistent with the most basic ways in which critical thinking makes sense of the world.

Aside from understanding how some material or ideas are distinct from one another, that is, for example how two closely related ideas are the same and how they are different, in science the world should be understood in terms of systems. This includes understanding any system, defined arbitrarily, as systems of parts and wholes and the relationships between the parts and the wholes. Furthermore, the greater the number of perspectives for viewing the same system(s), the deeper the understanding of that system(s). *Thus, understanding chronic disease states from perspectives of control parameters that are extrinsic to the system itself, i.e. the environment, is crucial to understanding how that system works. If we change the environment a cell is grown in, we change the cell.* This change is a phase transition. If a system is an individual person, the environmental control parameters include the person's diet, psychosocial support systems, career, and other activities. These important, often critical modifiable, control parameters for metabolic disturbances of insulin resistance, obesity, metabolic syndrome, type 2 diabetes as well as downstream chronic disease states are very often, even typically, overlooked in routine clinical practice. This highlights the value of using a PFL-based model. It provides a broader and more organized perspective on the patient's state of health or disease.

Furthermore, order parameters intrinsic to the individual are similarly structured as secondary and primary order parameters. Therefore, the former or upstream control are causally related parameters to the latter, albeit they are more proximal to the order parameter and often less fundamental to the genesis of a pathological order parameter than is an extrinsic control parameter. For example, *insulin resistance is a secondary order parameter to type 2 diabetes. While genetic coding is an intrinsic secondary order parameter to insulin resistance, lifestyle factors represent the extrinsic control parameters for the development of insulin resistance.* It can be argued, based on work by Gerald Schulman and others, at least for some individuals, their genetic factors are more fundamental to the development of insulin resistance, which may occur in

Second Order Parameters as a Self-Amplifying Circuit

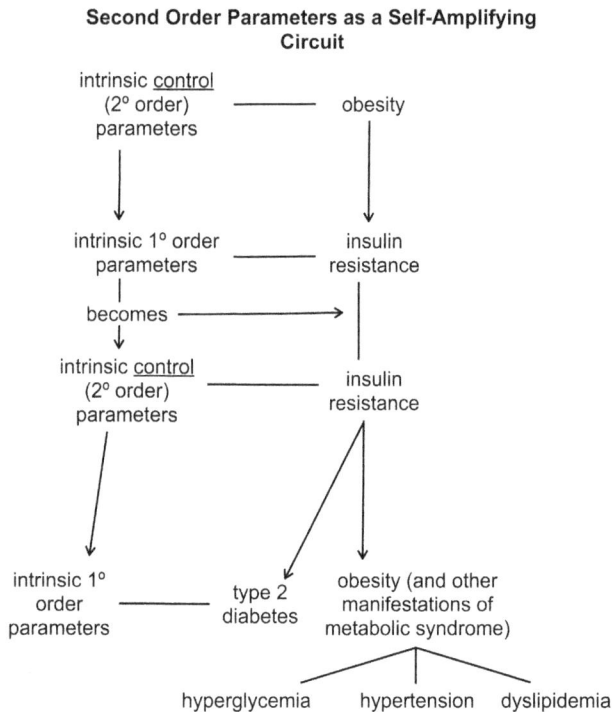

FIGURE 5.12 Illustration of secondary order parameters as a self-amplifying circuit in the pathogenesis of disease.

the absence of external lifestyle contributing causes. In any event, diabetes when uncontrolled may be considered a secondary order parameter to insulin resistance. Obesity should be recognized as a secondary order parameter to insulin resistance as well as to the sequelae of insulin resistance including hypertension, dyslipidemia, elevated small dense LDL cholesterol, and type 2 diabetes (Figure 5.12).

However, *conversely insulin resistance is also a secondary order parameter to obesity. This reciprocal relationship is the basis for the positive feedback with a circuitous feed-forward activity.* Accordingly, treatment that targets either one is likely to have a more favorable effect on the other. This increases the number of potential strategic therapeutic approaches that can vary depending on the details (the risk-benefit ratio profile) of the case and the preferences of the individual. Insulin resistance may be, in 50% of cases, a secondary order parameter to hypertension, again that is, having a causal relationship to the development of hypertension. The presence of insulin resistance increases the further downstream order parameters of cardiovascular event risk. Thus, it makes sense according to this relationship that treatment of the hypertension with an insulin-sensitizing agent, for example, an angiotensin-converting inhibitor or angiotensin receptor blocker, rather than a β-blocker that will promote weight gain, fatigue, or worsened insulin resistance, would be preferred. Many of these decisions are made reflexively and properly.

Problem solving in medicine is typically a function of pattern recognition. However, mental training to use a PFL model forces the routine consideration of the complexities of biological systems. This may help not only sometimes find a superior treatment choice but in the frequent scenario of a reciprocal feed-forward relationship between secondary

order parameters, increasing the number of treatment options may allow an approach that has lower risk, less undesirable side effects, and is more palatable relative to the subjective fears or preferences of the individual patient. An obese type 2 diabetic has two secondary order parameters to many forms of cancer as well as to Alzheimer's disease. However, intriguingly Alzheimer's disease and cancer seem to be inversely related to one another by a statistical phenomenon known as the Reverse Warburg effect. This relationship further underscores the metabolic nature of these chronic disease states because the Warburg effect of cancer is rooted in mitochondrial dysfunction of native ancestor cells, which are outcompeted by cancer cells for nutrient resources. The less efficient glycolytic bioenergetic pathway is sufficient to maintain the simple needs of cancer cells for replication not requiring the energetic more expensive process of cell differentiation and coupled to the anaplerotic ability of cancer cells to use TCA cycle intermediates as building blocks for cell replication. Another reciprocal bidirectional positive feedback phenomenon analogous to that of the two secondary order parameters of obesity and insulin resistance is that of mitochondrial dysfunction and insulin resistance. This relationship will be described ahead. However, this analogous relationship to metabolic disease or susceptibility states such as obesity and insulin resistance is useful to differentiate the often interchangeable terms of Physiological Fitness Landscape and free-energy landscape.

Free energy refers to the amount of energy that can be used to do work, and in the case of Gibbs free energy, it refers to energy used to do biological work. The framework of the free-energy landscape *per se* invokes the bioenergetic pathways of, for example, the mitochondrial pathways with ultimate oxidative phosphorylation mode of ATP production. *With mitochondrial dysfunction, which causes insulin resistance, and insulin resistance, which causes mitochondrial dysfunction, the compromise capacity of the production of ATP by oxidative phosphorylation requires the cells to attain energy from a much less efficient process of glycolysis which occurs outside of the mitochondria.* The significance of this is profound. It is the fundamental metabolic basis for all chronic disease states of aging. Furthermore, as discussed at length and in detail in Chapter 4, mitochondrial dysfunction leads to a loss of correlated ATP coherent or synchronous energy production, which is required for the coherence of macroscopic physiological processes and optimal health.

Indeed, *the breakdown of this potential, rooted in mitochondrial dysfunction, corresponds to an acceleration of the arrow of time with the generation of heat and associated entropy increase and the biological equivalent of redox disturbance and systemic inflammation that accelerates biological aging itself.* It can be argued what comes first, mitochondrial dysfunction or insulin resistance. Although it is likely case-specific, this debate is akin to that of the chicken or the egg. In either event, it becomes a positive feedback deleterious cycle. Both mitochondrial dysfunction and insulin resistance or susceptibility states are secondary order parameters to one another, each of them having a causal relationship to the pathogenesis of chronic diseases of aging including cardiovascular disease, Alzheimer's disease, and cancers.

The most upstream secondary order parameter and susceptibility state that lies at the intersection between the environment and internal milieu of the body is an alteration in the composition of the microbiota. There is a rich interplay between the metabolites of the microbiota and the human host. It is interesting to recognize that the human genome consists of less than 20,000 genes. Consider this relative to the example of a kernel of rice, which consists of roughly 250,000 genes.

> *What makes humans so complex and sophisticated is that we are each the product of two gene pools, that of our own and that of the microbiome which involves more than two million genes. The metabolites of the microbiota control the expression of the human genome.*

For example, *the microbiome is responsible for producing enzymes including methylases and histone deacetylases that mediate epigenetic expression of the human host.* Furthermore, the microbiota produce enzymes that convert primary bile acids to secondary bile acids as well as produce short-chain fatty acids, which in turn mediate metabolic effects such as the release of gut satiety hormones from enterochromaffin cells and a host of related insulin-sensitizing effects in the case of a healthy microbiota and insulin-resistant effects in metabolic disturbances, including obesity, metabolic syndrome, type 2 diabetes, cognitive decline, cancers, and cardiovascular disease, in the case of dysbiosis. This is discussed at length in this current chapter. Again, these relationships can perhaps be best understood in the context of the Physiological Fitness Landscape because the apparent favorable risk-benefit profile of simple fiber or prebiotics and certain strains of Bifidobacterium and/or Lactobacillus probiotics may offer robust responses and hence compelling strategies for inducing phase transitions of metabolic disease susceptibility states. This again relates fundamentally to the notion that the applied science of clinical medicine is more of an art than an actual science per se with strong connection to the basic sciences and inchoate yet blossoming fields such as metagenomics that should be included in the armamentarium of patient care in the context of the Physiological Fitness Landscape of susceptibility and disease states.

> *A framework model of control and order parameters should be invoked for clinical decision making that can lead to promoting phase transitions to healthy attractor states from a pathological state with the most favorable risk-to-benefit ratio profile, which is also sensitive to subjective preferences of the patient.*

Thus, a dysbiotic microbiota is proposed here as a secondary order parameter to the primary order parameters of insulin resistance and obesity as well as other downstream metabolic disturbances. Furthermore, these primary order parameters are described as such relative to the secondary order parameters of the microbiota. However, secondary pathological order parameters are defined this way because their relationship to primary order parameters is one of causation. They represent an etiology of a pathological susceptibility state or a disease state. That is, the secondary order parameter is a special type of control parameter, one that is intrinsic to the system, in this case defined as the human body. In this context again, insulin resistance and obesity are secondary order parameters to one another. In the context of glucotoxicity, type 2 diabetes may be considered a secondary order parameter to insulin resistance and consequently to diabetes as well in the sense that severe hyperglycemia causes downregulation of insulin receptors. The complexity of interactions of the Physiological Fitness Landscape of metabolic susceptibility and disease states can be further developed into elaborate schemata to highlight casual and causal relationships in the evaluation of their many manifestations.

The free-energy landscape of physical systems is a concept that we can use as a mental metaphor for the Physiological Fitness Landscape translated to physiology. In describing systems that develop instability, we basically use a free energy function. In general, this is a function that quantifies the stability of a system under a specific set of physical conditions such as temperature, pressure and constant volume. This is equivalent to finding the most probable state of the physical system, which corresponds to the highest entropy of a system. The free energy is plotted as a function of the order parameters. This is a simplification because this system has a huge number of variables but we choose one as an order parameter to describe the state of the system in the simplest way, e.g. magnetization in a physical system or BMI in a living system. Normally, in healthy circumstances the free energy looks like a parabola with a single minimum. If we perturb the parabola to the left or to the right by changing the order parameter externally or internally, the system will eventually tend to return to the middle, its equilibrium or normal set point, so to speak. For example, in physiological terms, if we overeat today, we tend to eat less and increase energy expenditure tomorrow in order to maintain our body weight set point over a longer term. This is a healthy normal situation such that our system gravitates to a normal set point.

When things go awry, the organism moves to metastable states, which although not favored typically, may become favored as states of relative instability. For example, when normal eating patterns and activity routines are disrupted for whatever reason, the free energy climbs over a barrier into a new parabola. Free energy minimum of the new parabola is higher and hence less stable than the more stable original equilibrium point. This new less stable state is known as a metastable state. The area between the two parabolas is of course higher and represents an unstable state with a higher free energy. It should be clarified that the free energy minimum is always the most stable state occurring at the bottom of the parabola, because it equates to the greatest entropy state of the system parameter or set of parameters. The new parabola defines a metastable free energy state. The phase transition from a single valley to the generation of a second valley, a metastable free energy state, occurs only after the free energy of the system parameter or set of parameters exceeds the barrier and traverses the mountain between the two troughs. The land on top of the mountain between the troughs is reasonably flat, with no minimum free energy until the system finds a valley. Eventually, it arrives at one, the metastable state. A critical state of free energy is represented by a relatively flat plain region at the bottom. This critical state is different from the meaning of critical state in medicine, which means near death. In physics, a free energy landscape critical state just means that there are large fluctuations in the person's physiological parameters. It represents the position between two stable states. *The initial "set point" stable state is represented as a trough and accordingly is the most stable state with the lowest free energy minimum at the bottom of the trough.* Again, consistent with the second law, the lowest free energy corresponds with the highest entropy and hence metaphorically the most stable state.

The subsequent stable states have a free energy minimum, which is higher because the overall landscape rises as the free energy overcomes the energy barrier of its pre-existing trough as a result of a perturbation. A perturbation in the case of body weight set points is whatever external or internal control parameters (see separate discussion) pushes the system out of the trough into a less stable free-energy landscape. This perturbation represents symmetry breaking. Ultimately, a system parameter or set of parameters finds a new metastable state that represents the formation of a new symmetry. The area in between the stable and metastable states on the relatively flat terrain is unstable and the physiological parameters, for example body weight, fluctuate widely. The new metastable free energy state is a susceptibility disease state and the unstable free energy on the flat terrain prior to reaching a valley with relative free energy stability, again is the point of criticality, where a phase transition occurs resulting in a disease state. New metastable states may represent increasing body weight set points, however, without yet crossing a point of criticality. If the external or internal control parameters persist, this will push the system order parameters, for example body weight set point, into a juggernaut of increasingly less stable, higher free energy minimums, or troughs, i.e. metastable states. Ultimately, with an associated genetic predisposition the criticality state will be reached crossing into the susceptibility disease state of insulin resistance with hyperinsulinemia (Figure 5.13). Hence, the primary order parameter of body weight becomes a secondary order parameter, i.e. an internal control parameter for insulin resistance. Insulin resistance in turn is both a primary order parameter and a secondary internal control parameter for further downstream order parameters for example the additional susceptibility states of prediabetes as well as escalating obesity and chronic diseases of aging such as cancer, cardiovascular disease and Alzheimer's disease. *Another internal control parameter, mitochondrial dysfunction, has a reciprocal secondary order parameter relationship with insulin resistance that evolves in parallel as well as representing powerful and fundamental internal control parameters for the chronic disease states of aging.* An additional concept of significance here is the notion of a reversible versus irreversible phase transition.

A reversible phase transition into a disease state is one that is consistent with the disease state being a metastable state capable of being converted back to a health state by external control parameters such as lifestyle, pharmacologic and surgical measures for example targeting body weight as an internal control parameter for the weight management of the disease state order parameters of insulin resistance, prediabetes and/or diabetes. An irreversible phase transition is one in which the metastable state becomes absolutely stable with the loss of potential for returning to the normal healthy state of the initial parabola.

It should be understood that the parabolas, the metaphorical valleys between the mountains, are necessary for stability. When there is enough "free energy" to climb over the barriers of the troughs there is an area of instability with wide fluctuations in parameters. This was illustrated above in terms of well-described physical systems such as a ferromagnet undergoing a phase transition to a paramagnet when heated above its Curie temperature, a situation also applicable to physiology. Importantly, *the hallmark of a biological system*

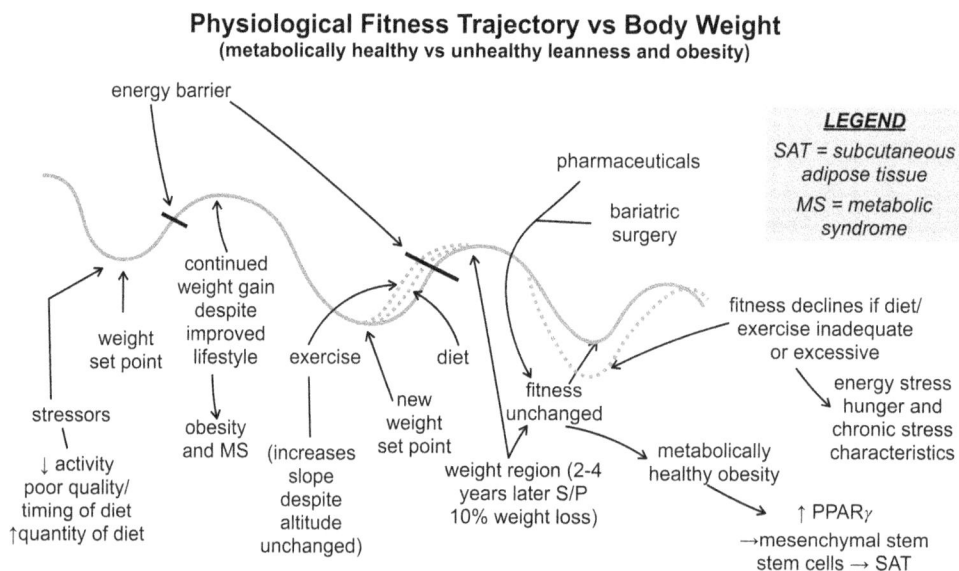

Physiological Fitness Trajectory vs Body Weight
(metabolically healthy vs unhealthy leanness and obesity)

FIGURE 5.13   Schematic illustration of the Physiological Fitness Landscape trajectory variation with body weight.

is a far-from-equilibrium state whereby in the optimal state of health there is stability, maximally reduced entropy of the living system. There is also a maximum amount of Gibbs free energy available to perform useful work of biological and physiological function. This is a state of homeostasis and in parallel, of redox and acid-base chemistry.

We should be thinking about living systems in terms of parts and wholes whereby the complexity of interactions between the component parts is intrinsic to the system forming a whole and in turn becoming a part of another system on a more macro hierarchical scale, which is extrinsic to the initial system. The significance here is that extrinsic control parameters for system parts become intrinsic order parameters for the system as defined on a larger scale highlighting the deeply entangled nature of all biological systems. The most elementary plane of science is particle physics, which is rooted at the subatomic scale. Defined as a system this may be considered to be a control parameter to the next hierarchical scale of a defined system in terms of solid-state physics. A system of solid-state physics that contains the primary order parameters of subatomic particles serves as a control parameter to a chemical system, which becomes a control parameter to a biochemical system, which becomes a control parameter to a molecular biological system and subsequently to a system of cell biology, physiology and so on. *The components of each successive scale transition from an intrinsic order parameter at the lower hierarchical plane to a control parameter that influences the development of the next higher scale to which it is extrinsic* until becoming incorporated as a part, or order parameter of that greater whole. This conceptual framework of a free-energy landscape describes the normal healthy evolution of macroscopic biological systems. However, the understanding of disease in medicine considers this same conceptual framework in the context of pathological susceptibility and disease states on a reduced Physiological Fitness Landscape.

As an organismic human individual, we are an order parameter of our social network or of our community. As we assimilate into this broader network the extent to which we influence the behavior of that larger system as a whole determines the extent to which we represent a control parameter. In this sense, being a control parameter is defined as extrinsic to that system but affecting its response to change. This distinction is critical to appreciating the role of environment (e.g. psychosocial interactions) and support systems of a patient in the assessment of a Physiological Fitness Landscape and the control parameters for metabolic susceptibility and disease states. The stress response is a fundamental factor as a secondary order parameter to all metabolic susceptibility and disease states. The stress *per se* is a control parameter but typically is a less important control parameter than the prolonged irrational perception of that stress. Mark Twain famously stated: "I have suffered a great many misfortunes in my life, most of which never happened". That stress response represents physiological allostasis, stability through change. What changes is an upregulated hypothalamic pituitary adrenal system axis, other neuroendocrine responses to a lesser extent, and the autonomic nervous systems responses. When prolonged, these responses perturb intestinal motility, alter the gut microbiota and bile acid metabolism as well as a number of hormonal

signals that adversely affect metabolic health. Ultimately, the impairment of the homeostasis of redox and acid-base status and of Gibbs free energy results in a continuum that promulgates a cascade of chronic metabolic diseases. The inflammatory processes mediated by metabolic endotoxemia, low-grade bacteremia, and other promoting causes of a proinflammatory cytokinemia, such as short-chain fatty acids, further reduce the threshold of the stress response in a feed-forward pernicious circuit.

The Physiological Fitness Landscape implies a fluid and dynamically changing landscape, which is a topography of stable and unstable regions (valleys and peaks) whose coordinates are given in terms of control and order parameters. In this discussion, these parameters involve susceptibility and disease states. Insulin resistance is a susceptibility state that is downstream to genetic coding as a secondary order parameter. *The worsening of insulin resistance may be a function of aging in the context of genetic predisposition but is more dramatically a function of environmental control parameters including a poor quality or high caloric density diet and/or reduced energy expenditure.* Social support systems that compromise the optimal or vitalizing stress response are examples of the control parameters of stress which typically relate less to the actual stress than to the environmental factors. That is, the notion of hormesis is crucial to whether a stress, physical or emotional, is adaptive/beneficial or harmful (Figure 5.14). Note that the notion of hormesis explains that the dose of stress is of critical importance. Too little or too much is harmful, but some optimal amount is beneficial because it builds resilience.

The factors that determine the outcome of this situation in the case of psychogenic stress are often control parameters qualitatively distinct from the stress itself. Furthermore, *an exaggerated stress response causes the emergence of additional control and secondary order parameters that amplify the stress response.* For example, dietary intake under stress is of poor quality virtually 100% of the time independent of whether the amount of macronutrient is inadequate or excessive. A poor-quality diet is defined as one, which is usually both pro-inflammatory and lacks adequate micronutrients. Dietary intake, because it is extrinsic to the body, represents a control parameter. Moreover, the prolonged stress response and HPA activity with hypercortisolemia represents another secondary order parameter to the encouragement of obesity as well as to the exacerbation of insulin resistance. The insulin resistance and obesity are again reciprocally secondary order parameters of one another. Additionally, as mentioned before, the prolonged stress response is itself a secondary order parameter to a change in gut microbiota composition underpinning an inflammatory response, which is in turn an additional secondary order parameter to exacerbating insulin resistance.

All of these secondary order parameters of systemic inflammation in addition to serving as secondary order parameters to insulin resistance also perpetuate the psychogenic as well as physical components of the stress response in a feed-forward fashion with inputs from a multiplicity of control and secondary order parameters. For example, diet is a control parameter. There may be alcohol and substance abuse associated with stress, which characteristically upregulates the reward motivation limbic system of the brain. The HPA axis and sympathetic

Hormesis

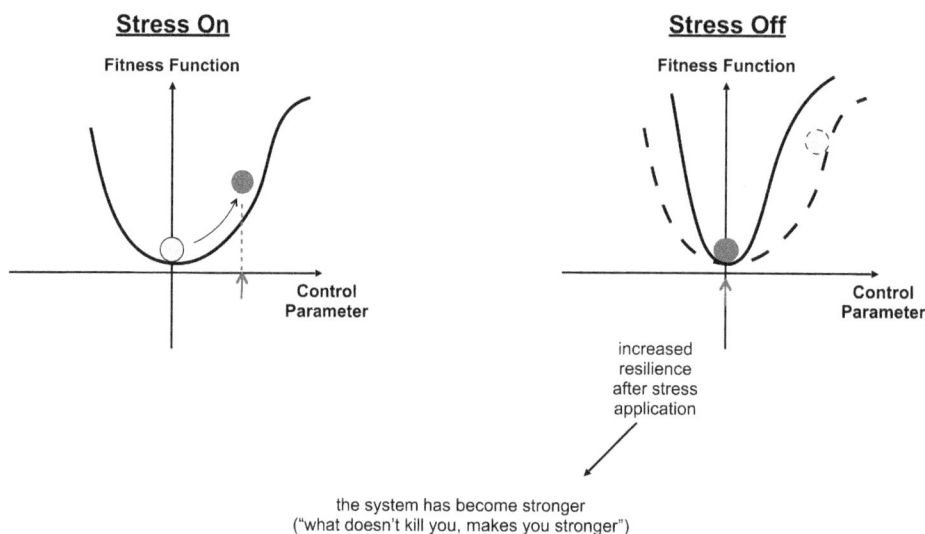

FIGURE 5.14 Illustration of the notion of hormesis in terms of relationship between the control parameter of stress and fitness function. Stress can be either adaptive (as in the case of acute stress that builds resilience—right) or harmful (as in the case of chronic stress that reaches a state of allostatic overload—left).

nervous system efferent branches of the stress response are secondary order parameters, which are both counter-regulatory to insulin sensitivity directly as well as indirectly by promoting a disturbed composition of gut microbiota and associated inflammatory mediator's consequent to gut microbial dysbiosis. Furthermore, a disturbed microbial composition as described at length in the following discussion results in insulin resistance and disturbances in energy balance mediated by altered bile acid metabolism as well as enteroendocrine impairments.

Healthy physiological perturbations to the system from the environment, for example routine dietary intake, physiological mechanisms of allostasis respond to control the challenge to the system with an increase in insulin secretion to control glycemic excursions and regulate energy balance, satiety, and availability of energy to skeletal muscle. Leptin levels rise to enhance energy expenditure and satiety and insulin sensitivity. There is an increase in bile acid delivery to the gut, which aids not only in the absorption of fats and fat-soluble vitamins, but actively participates in the overall energy homeostasis of the body by potentiating the effects of insulin. Hormones such as GLP-1 and PYY are released from the enteroendocrine cells of the gastrointestinal tract that complement the effects of insulin on satiety and other energy homeostatic effects. GLP-1 enhances the first phase of insulin release while inhibiting the counter-regulatory glucagon release from the α cells of the pancreas. There is far more complexity in the regulation of energy homeostasis in response to food, but this gives the overall idea of the free energy landscape in a healthy state of physiology whereby each of the parameters mentioned is in its maximally stable state of the parabola. Accordingly, homeostasis of Gibbs free energy, redox, and acid-base status and vital organ systems function is preserved.

In the case of dietary excess as an extrinsic control parameter, the system parameters of energy homeostasis adapt

allostatically beyond the limits of normal healthy physiology. We call this allostatic load. The baseline is furthest from the equilibrium stable state of symmetry of parameters and the system undergoes a bifurcation forming a new state of symmetry represented metaphorically by a parabola, which initially represents a metastable state. As such, it is an intermediate state between the high informational maximal state of negative entropy that is the hallmark of optimal fitness and physiological health, and the equilibrium state, which is the most stable state of nonliving inanimate systems corresponding to maximum entropy. *The rich and exquisite complexity of living systems corresponds to the negative entropy state of the system, which is a high informational state.* That is, it is capable of transducing the informational energy within the bonds of the biochemistry of the system to create new bonds that represent the transformation of energy and information, all part and parcel of the same process. An equilibrium non-living state is not capable of making these transformations that require continuous energy inputs.

> *The metastable state of disease is characterized by higher entropy due to inflammation and dysregulation resulting in the loss of efficiency and entropy production.*

The processes involved in entropy reduction in living systems, i.e. healthy metabolism, underscore the connectedness of interactions within the living systems whereby this hallmark of biological complexity parallels physiological health. In any event, a metastable state of allostatic load is a susceptibility state for breaking down that connectedness of complexity, that is, reducing some of the negative entropy (equivalent to information) to heat generating associated randomness making it incapable of being transferred into useful work of biophysiology. It has been suggested that the allostatic load state of insulin resistance is initially protective against obesity, that is, it prevents the excessive adipose lipid accumulation. Independent of whether that is the intended teleological significance of insulin resistance, it can only be temporarily effective if the extrinsic control parameter of dietary excess persists. However, in the short term, it may be posited that insulin resistance promoting hyperinsulinemia for some finite duration acts on neurons of the hypothalamic arcuate nucleus in the brain with a differential sensitivity that is increased relative to the lipoproteins' lipase activity in the capillaries of adipose tissue. Although this is theoretically possible it is the opposite differential sensitivity that is typical of the insulin-resistant obese state whereby the hypothalamus is more sensitive than the lipoprotein lipase present in the capillaries of adipose tissue.

A hallmark of biological complexity is the context-dependent action of hormones, which differ dependent on the physiological circumstances, for example in this case the early obese versus the prolonged obese state, which sometimes is tissue-dependent. Furthermore, there are interactions with other satiety signals, such as the gut peptides. For example, GLP-1 and PYY hormones do interact with leptin at the level of the arcuate nucleus of the hypothalamus to reduce the threshold to satiety even in the setting of leptin resistance. Further, it is hard to discern the effects of leptin from insulin because of their crosstalk with leptin co-opting insulin signaling pathways. Teleologically, it certainly makes sense that the adaptive benefit of insulin resistance enters into maintained inter-meal glucose availability when external resources are unavailable. *This adaptive allostasis explains differential insulin relative increased resistance in skeletal muscle as well as in the hunger and satiety centers in the hypothalamus to encourage food-seeking behavior because deep in human evolution food availability was not constant.*

Additionally, and importantly, adipose tissue insulin resistance including adipose tissue capillary lipoprotein lipase is relatively reduced consistent with a priority for allowing fat storage. While it is interesting to discuss the teleological significance of insulin resistance, the reality is that insulin resistance, although protective in a different historical context, is a susceptibility state in the current environment. In the free energy landscape this early insulin-resistant state is a metastable state displaced from the most stable equilibrium state. A classic example of this being an adaptive allostatic state is the hibernating bear during the winter in which the phenomenon is cyclical. Strictly speaking this state is defined as allostatic load [14], exceeding the limits of normal physiological parameter fluctuations, however with no pathological consequences because of the cyclical nature of this phenomenon. However, in humans the process is not usually cyclical and accordingly describes a susceptibility state for allostatic overload [14], or disease states. To distinguish the allostatic load state of insulin resistance in a bear from that in a human, it is useful to illustrate how the two systems compare in regard to the free-energy landscape. In the former case the metastable state of the parabola has greater stability, and it is easier for the parameters of, for example, body weight of adiposity, to bounce back and undergo a phase transition back to the lean healthy state. In the human, the parabola does not give the same stability, and it is not as steep making it easier to overcome the free

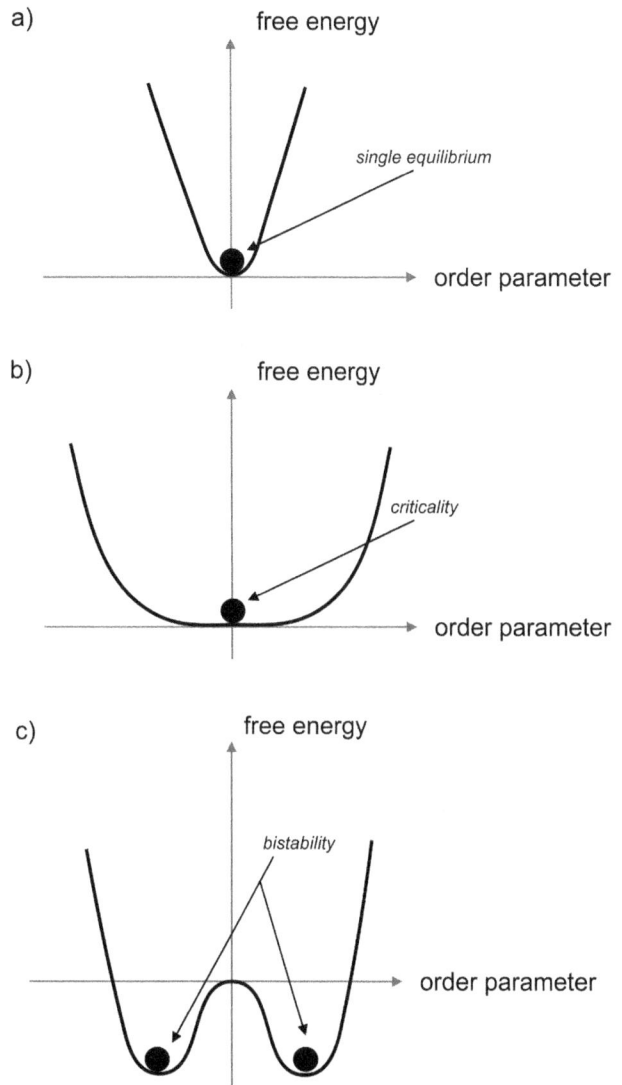

FIGURE 5.15 Illustration of the system's stability based on free energy as a function of order parameter in three canonical cases: a) a single equilibrium point exists, b) the system is at criticality, and c) the system has bifurcated and has two stable equilibria.

energy barrier that moves the parameter, in this case adiposity, further away from maximum complexity of the negative entropy state (i.e. farthest from equilibrium most stable and healthy state with maximum information; see Figure 5.15).

Once the change of the control parameter causes the system to escape from the parabola of the susceptibility state or allostatic load state, it is free to roam on a flat plain with wide swings for its greater adiposity until it becomes anchored to the next metastable valley even further to the right of the maximum stability point at the threshold of criticality. The system now reaches a point whereby it is a disease state or a state of allostatic overload.

The next question is what causes the flattening of the fitness landscape? In physical systems, phase transitions often occur in response to the extrinsic control parameters such as temperature, pressure, or magnetic field intensity. When systems are isolated from the environment, they easily find a stable state due to the second law of thermodynamics. However, when the system's temperature approaches an instability point (also called the critical point), then the order parameter of the system fluctuates widely giving rise to not only infinite fluctuations in its response to external perturbations but also to long-range correlations such as local fluctuations, which propagate over macroscopic distances. This is called structural instability. In a classic example of a magnet, at cool temperatures below the Curie temperature, the magnet is in a ferromagnetic state with all the outer most unpaired electrons having their spins (magnetons) aligned in the same direction and exerting a magnetic field due to a large net magnetization. Conversely, at higher temperatures, above the Curie temperature, the magnet is in a paramagnetic state and as such, the outermost unpaired electrons' spins are randomly oriented, which does not result in a net magnetic field, and is a state of thermodynamic equilibrium with no organizational stability. This is analogous to the equilibrium state that is incompatible with life, and only occurs in non-living systems. It is a state of maximum entropy, the absence of negative entropy that characterizes living systems. However, the arrow of time is an inevitable consequence of the second law of thermodynamics to which all humans (and all living systems) eventually succumb by achieving a maximum entropy state.

The changes in the order parameter values as a result of control parameter variations (e.g. increasing the pressure on a mechanical system) may lead to both reversible and irreversible transitions depending on the range of these control parameters vis a vis the system's stability. This is illustrated in Figure 5.16. Since systems undergoing phase transitions are by definition non-linear, they have a limited range of stability. For moderate changes in the control parameter values, the resultant order parameter changes are reversible. However, beyond a certain value, there is permanent loss of stability, and the system cannot be returned to the former equilibrium state. This resonates with some disease states. For example, mild infections cause reversible changes, and the person returns to health over

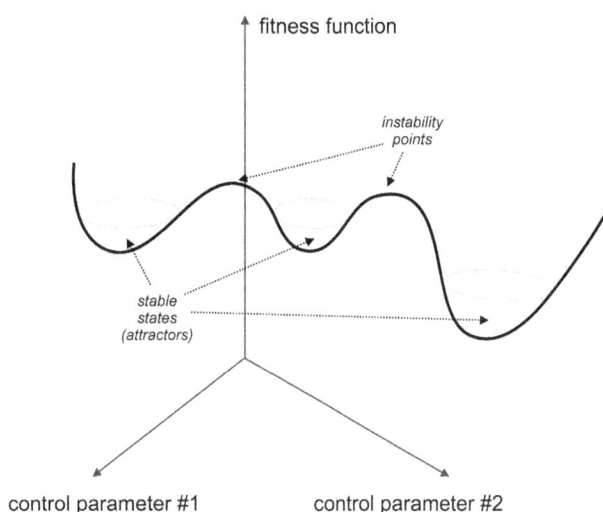

FIGURE 5.16 Stability in various states of fitness. The stable states are three-dimensional bowls, and the instability points are rounded hilltops. Note the presence of two control parameters spanning a three-dimensional space in which the fitness function is plotted.

a period of time with the use of antibiotics. However, beyond a certain point an infection can overcome the body and result in an irreversible transition to either permanent loss of some function (e.g. organ failure) or death. In the former case, these extrinsic primary order parameters become intrinsic secondary order parameters that take the larger system whole of which all intrinsic order parameters are part, the human individual, closer to a thermodynamic equilibrium state which is mortality.

When blood sugars are historically high, particularly in the initial months following diagnosis of diabetes, a phenomenon of "metabolic memory" occurs (Figure 5.17). This signifies that once (metabolic) symmetry is broken and a new symmetry is formed in terms of interactions and bonds between molecules across many hierarchical scales, there is a change in the Physiological Fitness Landscape around the new stable attractor state. It has shallower stability zones around its dynamic equilibrium within the Physiological Fitness Landscape. It is also separated by a high energy barrier making return to the previous equilibrium state of health virtually impossible even though the blood sugars may have subsequently improved. When a critically ill person is in the intensive care unit on IV insulin with normal blood sugars, Van den Berg has demonstrated a 50% reduction in mortality potential occurs. In this scenario, unfortunately this benefit is lost and mortality is ironically increased, even in the absence of hypoglycemia. This seeming paradox has never been explained properly. However, the Physiological Fitness Landscape does explain it in the sense that *in the setting of critical illness, establishing "normal healthy" metabolic control does not result in the availability for the necessary integrated complexity of total body vital organ system interactions across the many hierarchical scales. Accordingly, when medical interventions attempt to achieve something artificially against the natural ontological evolution of the person, there is no available stability zone within the Physiological Fitness Landscape, not even a shallow one to maintain this induced state.* The result is manifested by falling off the metaphorical outcome and away from the beautiful far-from-equilibrium state that defines life. Return to this stable equilibrium state is not viable.

The cure of a disease manifestation, e.g. a cancer, albeit not a reversal underlying disease in terms of free energy that has been irreversibly lost from the system, may nonetheless stop the accelerated aging that is tissue-specific and otherwise would have resulted in a more premature mortality. The accumulated toll of the disease process may determine the ultimate age of mortality of the individual. This will depend, however, on the extent of separation of biological age of the oldest vital organ system at the time the disease (e.g. cancer) was cured relative to chronological age. Accordingly, this biological age will dictate the extent of the adaptive resilience of the individual to stressors as determined by the location, slope, and depth in the stability zone of the new symmetry state within the Physiological Fitness Landscape (Figure 5.18).

New-onset diabetes with prolonged uncontrolled hyperglycemia and fatty acidemia results in progressive mitochondrial dysfunction in tissues systemically due to glucotoxicity and lipotoxicity. This is associated with a feedforward cascade of increasing inflammatory and redox stress interwoven with progressive insulin resistance. Intravenous (IV) insulin in the setting of critical illness had been a standard of care because

PFL vs Glycemia - Metabolic Memory

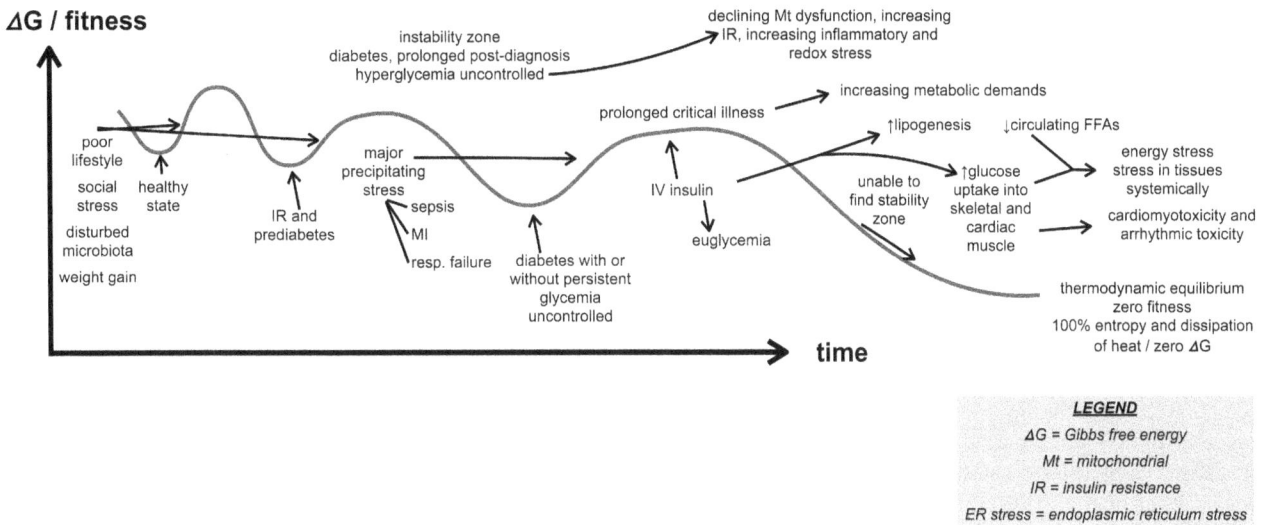

FIGURE 5.17 Illustration of the Physiological Fitness Landscape *versus* glycemia as an example of metabolic memory. Note the time dependence of the fitness function in this graphical representation.

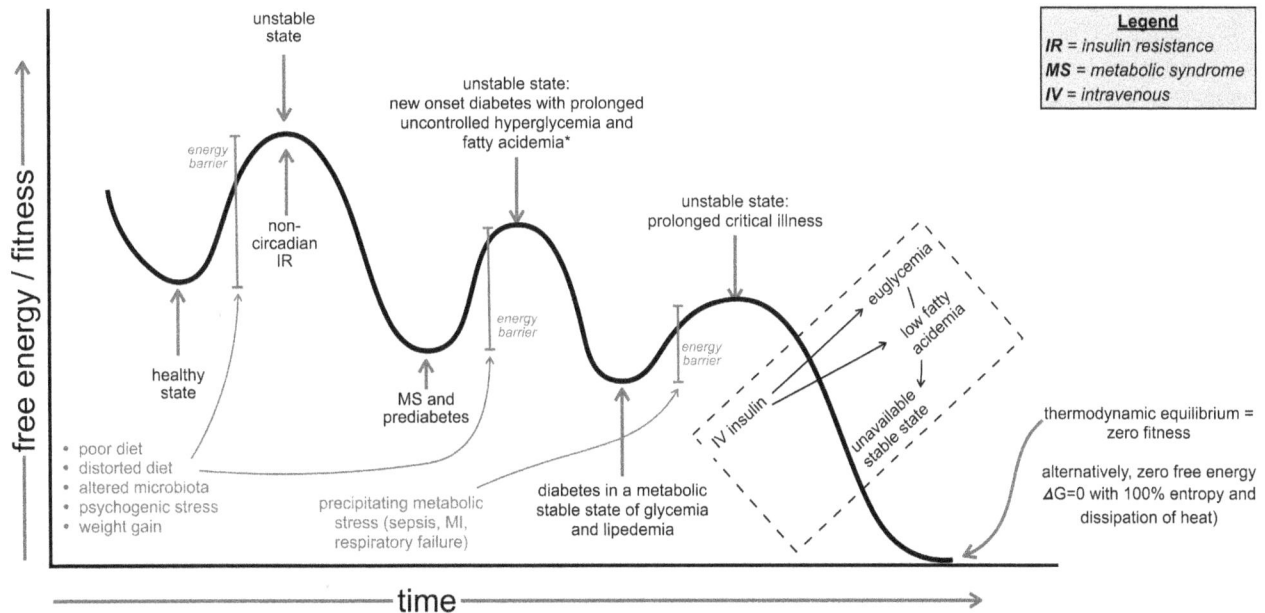

FIGURE 5.18 Visualization of the Physiological Fitness Landscape as a function of time illustrating metabolic memory of diabetes in critical illness.

it had been demonstrated to reduce intensive care unit (ICU) mortality by 50%, from an overall 20% to 10% [27]. However, more recently the Nice Sugar Study showed that IV insulin in this setting may paradoxically promote mortality despite the absence of iatrogenic hypoglycemia. This phenomenon of metabolic memory may be rooted in the levels of glycemia and lipemia. Accordingly, a provocative metabolic stress that induces critical illness by pushing the metaphorical landscape over an energy barrier, generates another instability zone. This zone is characterized by a cascade of deteriorating bioenergetic capacity in the context of very high metabolic demands. It follows that medical management with IV insulin suppresses adipose tissue lipolysis while enhancing lipogenesis.

This compromises the availability of fatty acids as an energy source for cells and tissues systemically. Additionally, glucose uptake into skeletal and cardiac muscle is promoted, further depriving energy substrate availability for immune cells and tissues such as the brain, the intestine, and liver. Consequently, there is insufficient energy to maintain the functional physiology of these cell types and tissues. Furthermore, the increased glucose uptake into cardiac muscle exceeds the capacity of already impaired mitochondrial function. This leads to redox and inflammatory stress that together induce cardiomyotoxicity and arrhythmogenicity. Taken together, these events provide a plausible basis for an otherwise unexpected causative relationship connecting IV Insulin mediated euglycemia to mortality in the critical care setting. Moreover, the PFL model could be predictive on a personalized precision scale for whom IV insulin would be a life-saving strategy or a dangerous one.

Crucially important in this connection is the notion that *a "normal" value for a parameter such as blood sugar or body weight is not necessarily a healthy one in the sense that such values are predicated upon a foundation of sub-parameters and interconnecting parameters that exist in the healthy state.* Moreover, these values should always be compared to the patient's recent past in order to see a trajectory, not just a point. As argued throughout this book, the values of the order parameters are themselves less important than the rate of change and also the response of these values to external stresses. In this context, the cardiological stress test is a primary example where the patient's response to controlled perturbation of the external conditions informs the physician about the state of health or disease, not a single point, such as the heart rate or blood pressure. Heart rate variability and blood pressure changes between lying down, sitting, and standing up contain more valuable information about the human system's response. Such stress tests should be commonplace in all areas of medical practice, not only in cardiology, as they provide information on metaphorical steepness of the hills around the valley, not just the altitude of the valley above sea level (i.e. baseline). The implications for medical practice are vast and even minor changes in this direction in routine procedures, in terms of algorithmic guidelines and intuitiveness of this approach, can make a major difference in the quality of healthcare practices and patient outcomes.

> *The relationships of extrinsic control parameters and intrinsic primary and secondary order parameters with each other can be described in the map of the free energy or Physiological Fitness Landscape.*

"Free energy" of a free-energy landscape represents a metaphorical sense of energy. In fact, the entirety of the model of free-energy and Physiological Fitness Landscape is a metaphorical abstraction. Physiological Fitness Landscape refers to the degree of fitness. While the maximum stability of a free energy landscape occurs at the bottom of the valleys, which are most favored by the parameters of a system, the maximum stability range of the Physiological Fitness Landscape occurs at the peak of performance (that may occur under extreme external stress) which corresponds to the top of the valley, the mountains. A susceptibility state for example obesity may have a free energy zone at the top of a lower trough.

5.4 Physiological Fitness Landscape as an Organizing Principle for Understanding Health and Disease

5.4.1 Survival and Design Principles for Its Achievement

An integral feature of disease is the loss of spatial and temporal synchronized capacity of an organism to function as a singular whole. A state of health, on the other hand, can be characterized by symbiotic interactions between all of the parts of organismic physiology including the microbiota and the host. This highly organized synchrony can also be seen in animal and human societies, and it takes an effort to achieve this state of coordination. Moreover, this is an example of an emergent phenomena where the collective system is greater than the sum of its parts. In many cases, this emergent superorganism develops a greater intelligence beyond the sum of its individual members. Collective intelligence such as the flocking behavior of starlings or a school of herring is superior not in the sense of intellectual capabilities of an individual, but rather within the notion that a collective whole is smarter by being capable of adaptations not possible individually. This is clearly seen in highly developed human societies or even when we compare the range of possibilities accorded by forming an orchestra as opposed to a solo artist.

Synchronized or coherent behavior with both temporal and spatial orchestration among the parts of a system whole often characterizes collective intelligence. It is this author's strongly held belief that this quality is a necessary feature of human health. In the case of human physiology this also includes its inhabiting intestinal microbiota. *Moreover, social support networks form an extension of this orchestrated organization by allowing for additional adaptability when an individual is faced with challenges that life brings with it.* Converse is true of disease states where organizational breakdown can be seen at the level of subcellular structures (e.g. impaired mitochondrial function in diabetes), cells (e.g. epithelial-to-mesenchymal transition as a precursor to cancer), organs and tissues (highly inflamed cells in an infected organ) and, finally, the entire organism (e.g. the lack of vitalizing energy in a patient presenting with major depression).

Individual and species survival defines purpose in biological systems, which is intriguingly analogous to Conway's Game of Life that exemplifies the concept of cellular automata applicable to biological and physical systems in the sense of creating order and self-organization from chaos or randomness. The notion of reproduction and survival may be strangely applied to physical systems that perpetuate patterns. A research area within computational physics was spawned that studies such systems and it is called "artificial life". Both artificial and biological life systems are characterized by intrinsic symmetries such as fractals, with their infinite recursive geometric or temporal patterns at progressively smaller hierarchical scales. To characterize fractal patterns in space and time one applies rules for recursive generation of a given motif. However, there is always eventually a break in the symmetry required to maintain the pattern. When that occurs, a new symmetry is generated, that can be seen as a phase transition that forms a new governing pattern. The new symmetry is actually another iteration on a more macroscopic scale. Hence, at various scales there may be different rules that apply and to connect between the scales we must construct a new mathematical object called a multi-fractal.

As we build an organism from cells of various types to tissues and organs performing different functions, we encounter a multi-fractal object, which is much more complicated than physical fractals. A frazzled coastline of a country like Norway is fractal-like due to its self-similar scaling geometry. The length of such a coastline is irregular and counterintuitively immeasurable. The smaller the yardstick we use to measure it, the longer the measured value of its length, eventually reaching infinity if we were to truly use atomistic scale in the measurement process. Biological systems are also replete with structural self-similar symmetries, including the vascular and bronchial trees. This illustrates the infinite wisdom of nature and evolution, which is metaphorical to the relative infinite surface area provided by this design. However, behind this beautiful design there is a very pragmatic functional outcome. For example, it allows for enormous capacity for gas exchange in the lung maximizing surface area in a given volume with each respiratory cycle. In fact, every cell of the body is within a distance of no more than five cells from a capillary. *The cardiac conduction system exemplifies a temporal fractal whereby as the time scale becomes shorter and shorter, the heart rate variability pattern repeats recursively the patterns from longer time scales.*

Importantly, the parallel between patterns of biological and physical systems may be exploited in the sense that if the signature pattern of a given physical system matches that of the biological system, it is far easier and more convenient to study the former that may be applicable to physiology. For example, the temporal pattern of a dripping faucet resembles the conduction system of a beating heart. There is ventricular expansion as a result of venous return from the atrium coinciding with atrial contraction and the PR interval, analogously to the expanding bead of the dripping water. This is followed by ventricular contraction coinciding with ventricular depolarization and the QRS complex, which is analogous to the tapering of the water droplet. Finally, if the temporal space between the T-wave of ventricular repolarization and the subsequent P-wave of atrial contraction is lost, ventricular fibrillation occurs. Similarly, if the space between the contraction of a dripping water droplet and the subsequent expansion of the next droplet is lost such that the two consecutive droplets are partially superimposed, a chaotic pattern is triggered in a phase transition. Ventricular fibrillation is a fatal arrhythmia that accounts for many lost human lives with no antecedent rhythm patterns capable of predicting these tragic events. Astonishingly, by studying computerized simulations of the physical system of a dripping faucet, it may be possible to discover life-saving predictive patterns that portend the onset of chaotic events of fatal arrhythmias.

Patterns representing a physical entity or living state are the manifestations of their enduring nature. Stability and symmetry captured in these patterns are the hallmarks of durability. Phase transitions represent breaking of internal symmetries, both in space and time, to form new symmetries or stability zones. In this process of discovering new symmetries and their stability ranges, we can uncover larger-scale spatial or temporal patterns. In essence, we are searching for an organizing principle that would describe the evolving state of health and its transitions to various stages of disease. An important aspect of biological systems, as opposed to physical systems, is the aging of the former and virtual immortality of the latter. The ontogenic course of transformations across the aging process of a human lifetime represents an insidious sequence of breaking and forming new symmetries as a consequence of losing free energy from the system and dissipating

it into the environment with a concomitant entropy production. While all living systems reduce entropy in order to be alive, as aptly and profoundly discussed by the great Austrian physicist, Erwin Schrödinger in his book *What is Life?,* the rate of entropy reduction determines the system's robustness [28]. As aging sets in, this rate of entropy reduction diminishes as a result of accumulated damage, and the effects of a multitude of stressors. It is important to emphasize that not all stressors are toxic and detrimental to the vitality of an individual. There exist numerous vitalizing stressors that energize the person and also successful allostatic responses that build resilience against toxic stress. This makes it not only complicated but first and foremost individual because what is toxic stress to some, may in fact be vitalizing stress to others.

We discuss this in the framework of the Physiological Fitness Landscape, which is a mathematical concept applied to human health and disease that should be thought of within the context of personalized medicine. Its origin is in the theory of phase transitions and it additionally requires the definition of (extrinsic) control parameters and (intrinsic) order parameters, i.e. those factors that applied externally or internally may affect our internal physiological response functions, respectively. The Physiological Fitness Landscape is a functional representation of the metaphorical vital energy level (fitness) as a function of these control and order parameters. As will be discussed in more detail below, the topography of this *Physiological Fitness Landscape involves three crucial properties: the altitude (energy level), the steepness of the terrain around the energy valley (stability zone), and the height difference between the valley and the neighboring peak (energy barrier).* The latter property quantifies how easy or hard it is to traverse the landscape from one stability zone to the next, which could represent different states of health or indeed could represent the difficulty related to a return to the state of health from that of a chronic disease.

5.4.2 The Various Types of Stress and the Physiological Fitness Landscape

Psychogenic stress is the most under-recognized element of public health. Crucially, however, this type of stress can be either the most toxic and debilitating extrinsic control parameter to human disease or the most vitalizing control parameter to human health and well-being. What determines these diametrically opposite responses to stress is often how life's challenges match the skillset and the experience of the individual experiencing the stress. The dose, duration, and intensity of stress are also important factors. *Critically important is the perception of whether the outcome of the stressful challenge can or cannot be controlled.* The outcome of a stressful challenge defines the intrinsic control parameter, or secondary order parameter, referred to as stress response. Emotional resilience allows one to make a metaphorical lemonade out of lemons.

For example, in 2008 I watched in horror as my beautiful car was being engulfed and eventually destroyed by a fire, which was caused by an electrical problem. The visually displayed structure and function of this fabulous piece of engineering being lost to entropy and heat, created a major emotional distress for me. How high my jugular vein distended, and hence whether I could make "lemonade" out of this "lemon", depended on who would pay for this mishap! The essence of who we are, in terms of our talents and assets, our core values and what we are not may be defined by which stressors are vitalizing and which are debilitating. What makes the practice of medicine such a great privilege is to connect with other human beings in such a way that can powerfully help people reduce their emotional stress due to concerns about their physical health. When things go well, this can fundamentally broaden the opportunities for the patient. The opportunity of time given to the patient as a result of the physician's diagnostic and therapeutic recommendations allows the patient to achieve his/her goals and to enjoy life experiences with the loved ones. *When we can reduce the stress of others, this action becomes deeply entangled with our own stress response and is the root of human connection that cannot be replaced by even the most powerful computational algorithm or technological advance.*

Prolonged toxic stress literally steals time away from the life and health span of a human living system because it is associated with a faster metabolic rate in a classical sense of ATP production per unit time. A higher metabolic rate generates increased reactive oxygen species in the bioenergetic machinery of the mitochondria and specifically in the electron transport chain. This leads to inflammation, which is the biological equivalent of entropy production with the associated heat loss, which is predicted by the second law of thermodynamics. The aging process represents the time element of the metabolic rate. *Accelerated aging equates to a higher rate of entropy production and associated greater liberation of heat. It parallels the chronic inflammatory diseases of aging mediated by the bidirectional relationship of mitochondrial dysfunction and reactive oxygen species generated along the electron transport chain. This pathogenesis is linked to another bidirectional relationship of mitochondrial dysfunction and insulin resistance.* Insulin resistance promotes obesity and other components of the metabolic syndrome, cardiovascular disease, dementias including Alzheimer's disease, and cancers. In fact, the understanding of disease from the perspective of metabolism provides a more fundamental armamentarium in the approach to managing these diseases and their evolution from susceptibility states. Insulin resistance and mitochondrial dysfunction represent intrinsic control parameters of a self-perpetuating and amplifying interwoven web that underpins the chronic disease states of aging that are manifestations of the accelerated aging process itself. This is perhaps best described by the Physiological Fitness Landscape model.

The Physiological Fitness Landscape describes susceptibility or disease states in the context of net interactions of a constellation of a high-dimensional system of control and order parameters that define the state of health or disease of the human physiological system. It is a function of stability versus acceleration (instability) of the entropy production rate, which is tantamount to the pace of aging and in parallel to the metabolic rate. This mathematical model is Nobel Prize–worthy in physics (as attested by Lev Landau's 1962 Nobel Prize for his theory of the superconducting phase transition) and has been shown to be applicable not only to physical systems and even social systems, but eminently so to physiology as elaborated on in this book. Vexingly, it has never been invoked for its invaluable transformative potential for medicine, let alone applied to concrete disease states. I have been in clinical practice for 30 years, and as an endocrinologist for 25 years. I have never seen two diabetics who behave and respond to therapy in exactly the same way.

Of central and crucial significance, it is not the disease *per se* as a diagnostic entity we are treating, as treatable promiscuously by a standard-of-care therapy. Rather, *it is the uniqueness of the relationships between control and order parameters of the individual patient that directs the evolution of the disease and potentially the optimal choice of therapeutic interventions that should be patient specific. The Physiological Fitness Landscape shows the composite view of these parameters as an attractor that can undergo phase transitions from normal (healthy) to abnormal (unhealthy) state, or vice versa, in dynamic fashion that can further change its trajectory over time on its own or as a result of therapeutic interventions.* Nonetheless, the trajectory within the basin of attraction of the nearest attractor state provides at least short-term predictability beyond what is otherwise possible given the daunting complexity and unpredictability that is characteristic of all living systems, especially human beings.

5.4.3 Main Features of the Physiological Fitness Landscape

Relative to the PFL model, a disease state represents the loss of available free energy, or fitness level, with instability and stability zones for the presence or absence, respectively, in the progressive loss of free energy or fitness. One guiding principle for the behavior of a physiological system, a composite attractor state of order parameters within the metaphorical Physiological Fitness Landscape, is the notion of history, both ontogenetic and phylogenetic, the latter case being an extension of natural selection. Variables in the basic structure of the Physiological Fitness Landscape include the amplitude (or elevation) of the valley, steepness of the slopes of the hills surrounding the valleys, and the elevation difference between the valley and the neighboring peak (see Figure 5.19).

The amplitude represents the fitness level, which invariably drops with the passage of time as we succumb to the loss of vitality due to aging. Nonetheless, *one can maintain optimal health at any age although that optimum moves down with the passage of time starting in early adulthood.* Disease and unhealthy lifestyle can further force the fitness level to drop from the optimum level. The good news is that improving the level of health and adjusting the lifestyle of an individual can raise the level of fitness toward its optimum for a given age (and person). Steepness of the landscape around a stable point that corresponds to a particular state of health and disease represents resistance to change as a result of external control parameters being changed along a given axis. The many external control parameters correspond to the various stress factors, both physical and psychological. A steep landscape around a metaphorical valley means internal stability and resistance to change. For example, a resilient, experienced and healthy individual will maintain both physical and mental stability when faced with the stress of public performance. An inexperienced,

A) non-equilibrium level attractor states

B) non-equilibrium level of metabolism

C) barrier to disease reversal

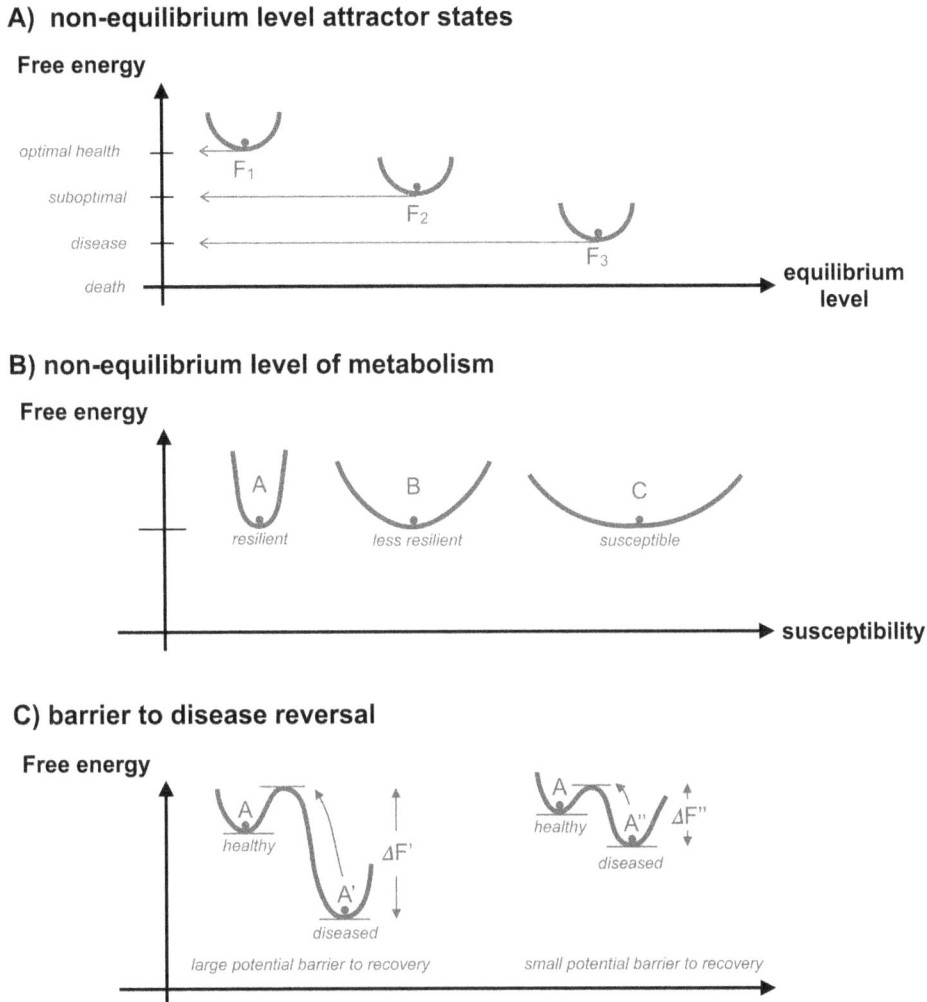

FIGURE 5.19 Illustration of variables that influence the basic structure of the Physiological Fitness Landscape.

fragile, and insecure person can be driven to the verge of a mental breakdown when faced with the same challenge.

Numerous other challenges represent stresses of life and all of these situations can, in principle, be modeled mathematically in terms of the Physiological Fitness Landscape. A classic example is, of course, the cardiological stress test to determine the heart and cardiovascular system's health level. Both the response to a stressor and the rate of return to normal values when stress is removed represent variables that determine the metaphorical steepness of the Physiological Fitness Landscape. In a steep landscape situation, the system rapidly returns to its normal values when the stress factor is removed. This also resembles a mass connected to a spring. When the spring is rigid, the mass rapidly returns from the stretched-out position to equilibrium when a force that was applied to it is removed. A soft spring will take much longer to return to equilibrium than a rigid spring. This analogy illustrates the dynamics of a restoration of the physical equilibrium in physical systems which can be translated into physiology by representing the spring as our homeostatic response. Hence, the speed with which the spring contracts back to its original length illustrates the rate of return to a state of health, for example following a bout with an acute case of influenza or

infection. Young, healthy individuals will, on average, regain their state of health much faster than elderly people with additional morbidities who may develop a chronic condition resulting from the same health hazard. In fact, in many cases returning to physiological equilibrium of health may not be possible at all, very much like a worn-out spring or an old elastic band that will stop mid-way on its way back to equilibrium. This is worth discussing in more detail.

As described above, *homeostasis represents resistance to change physiologically represented by external control parameters such as stress.* The equilibrium state of health is illustrated as a valley surrounded by hills. Applying a stressor is equivalent to pushing the system out of the valley and up the hill. When the stressor is removed, in most cases the system will return to the bottom of the valley, i.e. the state of health. However, there is a critical point at which so-called allostatic load is applied and it means the stress level is such that the system reaches the neighboring peak and then slides down to the valley beyond the hills. Reaching the peak is called in the medical literature "allostatic load" and sliding over to the next stability state (neighboring valley) is referred to as the "allostatic overload". *Allostatic overload means the system has adopted a new set of equilibrium parameters (set points) and is unable to return to the former state of equilibrium.* These parameters may represent a chronically elevated blood pressure, substantial weight gain, or respiratory difficulties. We show an illustration of homeostasis, allostatic load, and allostatic overload in Figure 5.20.

Finally, the third key characteristic of the Physiological Fitness Landscape is the difference in elevation between the valley in which the person's state of health is located at present and the peak separating it from the neighboring valley, which may represent another stability region such as a more advanced stage of disease or, conversely, an improved state of health. When a tall mountain needs to be climbed to be able to reach the neighboring valley, this means a major energy boost is required. This energy boost may require both mental and physical effort such as positive thinking, healthy lifestyle, sustained exercise regimen, highly nutritious food or indeed pharmacological agents. Hence, it is expected that a return to a state of greater health requires a "tall hill to climb on the part of the patient. On the other hand, little "effort" is required to neglect one's health and slide into the neighboring state of chronic disease such as diabetes or heart disease as a result of poor nutrition, disturbed sleep patterns, lack of exercise, etc. This metaphorical "slide" into a disease state virtually lacks a free energy barrier, or an obstacle, and can sneak up on the person insidiously like a thief at night. *In this state resilience to stress is lost, allowing minor environmental perturbations, that is stressors, to provide enough external energy on the human systems to push it over the energy barrier of a stability zone to an instability zone,* across the mountain top before falling to a more advanced form of disease. In this case, intrinsically applied resilience is insufficient to resist pressures of time and challenges applied to it (Figure 5.21).

Extrinsic stressors, including psychogenic, behavioral (diet, activity and circadian), infectious and traumatic, have the potential to move a physiologic and metabolic state of fitness out of a healthy stability zone. In the acute setting of stress, homeostasis of vital organ system parameters is maintained by autonomic, hormonal, and immune system responses. However, these allostatic responses, when prolonged or recurrent, impose wear and tear cost on metabolic efficiency and physiological fitness. Such costs are referred to as allostatic load and overload, whereby the former may be considered

intermediate between health and disease, while the latter is itself the manifestation of disease. These concepts may be applied to the PFL model. Accordingly, allostatic load ensues when stressors push parameters of metabolic homeostasis across a metaphorical energy barrier as shown in Figure 5.21.

All chronic diseases of aging may be considered metabolic disorders in the sense that they represent the loss of core energy and redox homeostasis. Moreover, metabolic diseases, typically rooted in mitochondrial dysfunction and often inseparably interwoven insulin resistance, are appropriately modeled in the topological terrain of a metabolic, or Physiological Fitness Landscape. While circadian insulin resistance is physiological, the loss of circadian cycling (see Allostatic Load (AL) sitting on the first mountain peak beyond the initial energy barrier, symbolic of an "instability zone") represents the transition from health to a susceptibility state for subsequent metabolic and chronic diseases.

Each successive mountain top and ensuing trough in the topological terrain of the Physiological Fitness Landscape delineate instability and stability zones, respectively. There is a progression over time of declining altitude and correlated fitness level that ultimately manifests as the metabolic and chronic disease states of aging. The initial trough is a stability zone with metabolic homeostasis that represents a state of physiological health. The subsequent troughs between mountains are the stability zones characterized in Figure 5.21 as states of Allostatic overload (AO); these are metabolic syndrome, diabetes type 2, and other downstream chronic disease sequelae of pathogenic insulin resistance, respectively. The mountaintops preceding these troughs symbolize pathophysiological changes that evolve into these metabolic disease states. Importantly, in reality, there are a huge number of dimensions representing independent physiologically relevant variables, with associated parameter values, that define a PFL. For simplification purposes of illustration, Figure 5.21 shows only a small representation of dimensions, and composites of dimensions drawn as a single quantitative characterization. Nonetheless, it should be recognized that this model is the product of massive multidimensionality reduction easily compressing thousands of parameters into few critically important ones. Moreover, it is a computerized mathematical model that is capable of incorporating thousands of metabolomic bits of

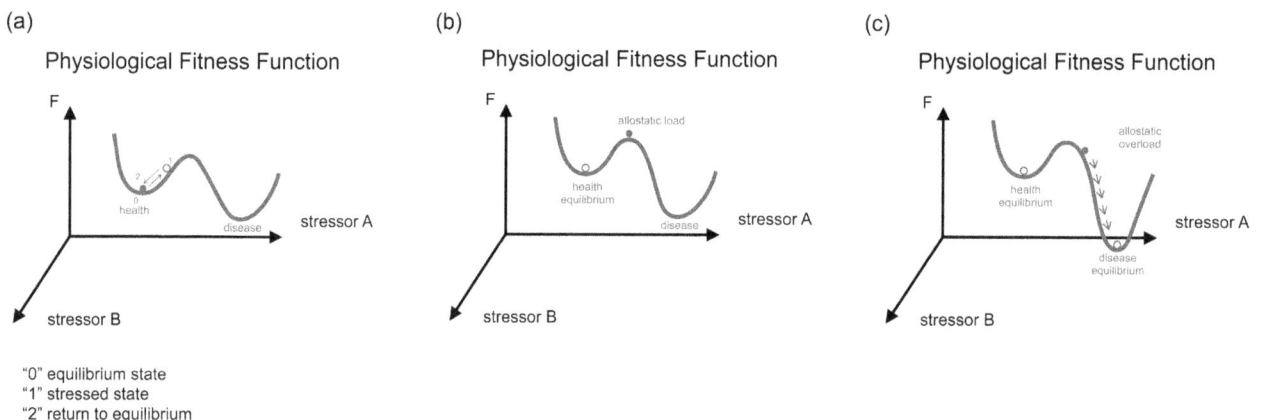

(a)

Physiological Fitness Function

(b)

Physiological Fitness Function

(c)

Physiological Fitness Function

"0" equilibrium state
"1" stressed state
"2" return to equilibrium

FIGURE 5.20 The Physiological Fitness Landscape during conditions of homeostasis (a), allostatic load (b), and allostatic overload (c).

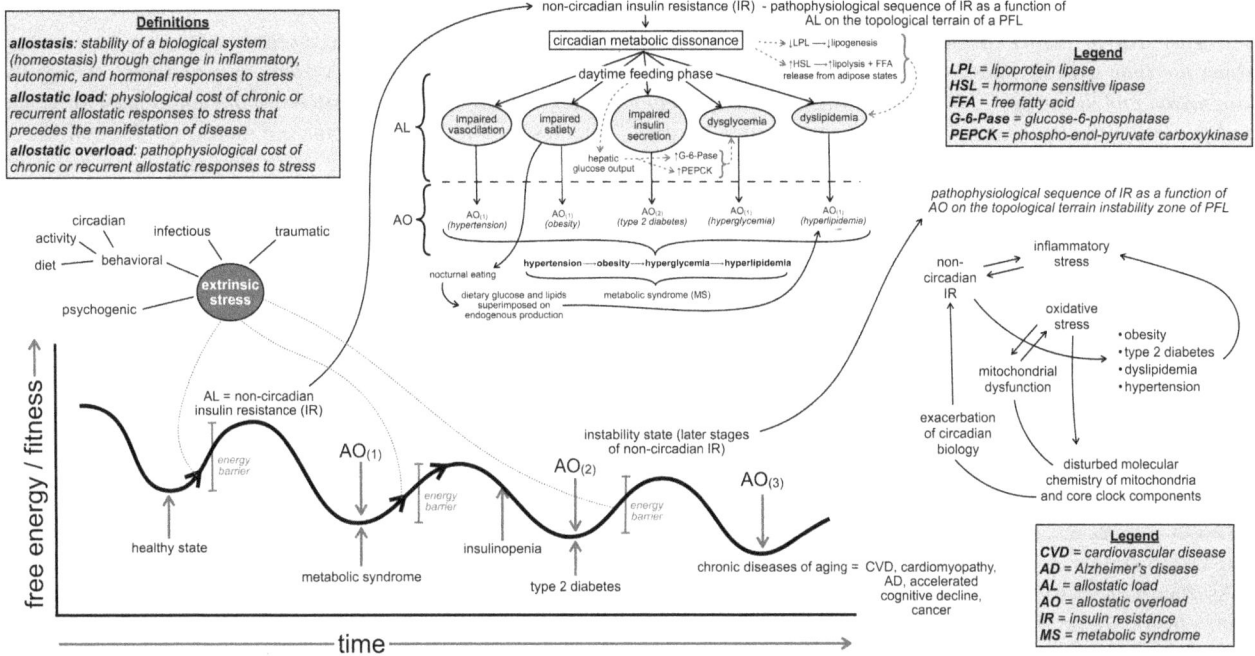

FIGURE 5.21 Illustration of the impact of allostatic load and allostatic overload of insulin resistance mapping on the Physiological Fitness Landscape as a function of time.

data for each parameter. *Critically, the trajectory of the topological fitness terrain changes with time, and in response to therapeutic interventions or lifestyle changes.* It is also important to recognize that parameters of fitness improve in response to some extrinsic conditions but decline in response to others. The aging process itself is a major impetus of the deterioration of metabolic fitness, overall physiological health, and consequently health span and longevity. The PFL model proposed aims to be a functional scaffold for the incorporation of metabolomic and other big data. It can also incorporate artificial intelligence algorithms and guide the physician through changing trajectories of the patient's fitness over time, for the purpose of optimizing human health at a personalized level.

Another conceptual linchpin to understanding the notion of allostatic load is being extrinsically pushed out of the stability zone of optimal health to an instability zone at the mountaintop before falling to a subsequent trough that has lost amplitude in the Physiological Fitness Landscape signifying movement of homeostatic organ systems parameters. Subsequent peaks and valleys within the terrain that accompany increasing loss of altitude signify stages of allostatic overload, of progressive disease. The parameters of the stress response represent allostatic markers. Resilience is analogous to a brisk spring that causes the return of these parameters to the base line powering resolutions of the stress without losing amplitude in the terrain of the Physiological Fitness Landscape. The recently recognized phenomenon whereby pro-inflammatory immune system cytokines provoke or reduce the threshold for eliciting cytogenic stress response is novel and almost counterintuitive. However, the qualitative framework of the Physiological Fitness Landscape explains this as well. The stress response at all hierarchical scales of a living system is a hallmark of allostasis. *The resilience of a system is best defined as the ability to resist change of vital organ system homeostatic parameters. This is the most fundamental evolutionary design of a biological system.* The psychogenic pre-autonomic, autonomic, and hormonal-rooted stress response is a top-down control system that is regulated in a positive feed-forward self-amplifying fashion from bottom-up processes such as the allostatic immune system responses.

Having a support system of family, friends, colleagues, and social workers can act as an additional barrier to allostatic load when a person is experiencing challenging stresses. This can be seen as externally increasing the height of the hill that separates the valley representing the equilibrium state of health from the neighboring valley representing chronic disease. Conversely, an accumulated toll of the life's challenges may effectively lower this barrier and act to facilitate an allostatic load even under a moderate amount of stress. In such cases the immune system is activated into pro-inflammatory and immune dysfunction, which causes a change in the microbiota and further promotes changes in the allostatic pro-inflammatory cytokines, enhancing prolonged response to stress. As a result, emotional pathological stress response ensues and all of these factors disturb dietary behaviors and other circadian behaviors, as well as the behavior of appropriate quantity and quality of food. *Stress response disturbs circadian behavior and then circadian behavior disturbs microbiota.* Extrinsic control parameters can be classified broadly as: social support and social pressure, circadian behaviors, extrinsic stresses, and diet. Intrinsic order parameters include microbiota, and the parameters of allostasis, namely immune, hormonal, and autonomic system parameters.

5.4.4 Entropy Increase along the Time Axis and Aging

The process of traversing the Physiological Fitness Landscape along the time axis signifies an accelerated pace of entropy production and hence aging, which is consistent with the second law of thermodynamics. The lowest energy state on the Physiological Fitness Landscape can be compared to the sea level and it signifies the end of life. Therefore, in allegorical terms our life's journey can be compared to an initial ascent to the highest peak, our own Mount Everest, when we reach maximum health in young adulthood, followed by gradual descent to the sea level. The trajectory we traverse both to our own personal peak and the subsequent ascent may have its unique peaks, ridges, and valleys, sometimes even canyons and cliffs representing both opportunities to improve our state of health and sometimes deadly challenges to it. If we could map the terrain at every step of the trek through these Himalayas of health fitness, we could not only avoid falling down some treacherous cliffs but also prolong and smooth out the journey making it as long and as close to being level as possible.

As energy is lost from the system, such as that resulting from a history of poor metabolic control following diagnosis, amplitude is lost and consequently intrinsic biological aging advances. This brings the state of the system of organismic physiology further away from maximum organized complexity and further from the state of equilibrium, which is death. Allostasis, i.e. stability through change, is now occurring in the setting of disease. This state of allostatic overload is not capable of finding a new healthy and stable steady state. Claude Bernard's concept of homeostasis is maintained in response to stressors to the system whereby that homeostatic resistance to change is accomplished by the allostatic stress response. The most basic homeostatic parameters of vital organ system function are the inseparably linked Gibbs free energy, redox state, and acid-base status, definable by the Gibbs free energy, Nernst and Henderson-Hasselbalch equations, respectively.

However, *when physiological allostasis becomes pathological, reaching allostatic overload, homeostasis set points become irrevocably moved.* These changes in homeostasis set points define disease, underscored in the Physiological Fitness Landscape as a decline in the overall amplitude of free energy, or fitness. Consonantly, allostatic parameters of physiology, including for example blood glucose, have new "normal" healthy ranges that parallel a reduction in health. Nonetheless, in this context the allostatic attractor state, its interacting control and order parameters continue to respond to extrinsic stressors seeking to break and find new symmetries in zones of local stability. In the unstable state of critical illness, in the context of a given ontogenetic history, the greater the loss of Gibbs free energy from the system, the larger the difference in what is the new "normal" range relative to the absolute healthy normal physiological parameters of the attractor state. Accordingly, this attractor state in the Physiological Fitness Landscape finds new symmetries at lower amplitudes, and hence more displaced ranges of homeostasis. Progressively, the steepness of the valleys declines as do the slopes from stable troughs to unstable peaks.

5.4.5 Curing a Disease Is Not Reversing Aging

The cure of a disease manifestation, e.g. a cancer, albeit not a reversal of the underlying disease in terms of free energy that has been lost from the system, may nonetheless stop the accelerated aging that is tissue-specific and otherwise would have resulted in a more premature mortality. The toll of the disease process may determine the ultimate age of mortality of the individual. This will depend, however, on the extent of separation of biological age of the oldest vital organ system at the time the disease, for example cancer, was cured relative to chronological age. We have earlier discussed the issue of accelerated aging within the context of Special Relativity applied to biological systems where the synchronization of biological processes is crucial for the most efficient energy flow through this system of systems. Furthermore, in a somewhat controversial way, we can say that everyone who dies from a chronic disease dies at the maximum human life expectancy, e.g. at the age of 113 or 120, albeit the chronological age of that person may be 60, 82, 95, or indeed 113 or 120 depending on the rate and stage of disease progression over the years leading up to death. Accordingly, this biological age will dictate the extent of the adaptive resilience of the individual to stressors as determined by the location, slope, and depth on the stability zone of the new symmetry state within the Physiological Fitness Landscape.

It is also worth stressing that in the opinion of this writer and consistently with the message of this book, *curing a disease is not equivalent to the return of the state of optimal health* and it cannot be. This would effectively amount to time reversal and would contradict the second law of thermodynamics. Energy lost from the system in the form of dissipated heat in the course of disease is an irreversible process and hence this loss can be translated into years of life lost. However, this does not mean that there is no point in reaching for the optimal level of health following (or indeed even during) the disease. This effort will result in a slower deceleration of energy loss and hence slower aging than when compared to the same person's decline without lifestyle and diet modifications. Finally, in this connection, it is worth mentioning the hypothesis put forward by Lloyd Demetrius who stated that the formal equivalence between physical and biological systems leads to a direct relationship between temperature and time, respectively. Therefore, inflammatory states of human physiology that produce heat and increase temperature are formally equivalent to the acceleration of time and hence faster aging.

Additionally, it is crucially important to realize that the notion of *a "normal" value for a physiological parameter such as blood sugar or body weight is not necessarily a healthy one in the sense that such values are predicated upon a foundation of subsidiary parameters and interconnected parameters that exist in the healthy state. It is more important to follow the trajectory of parameter values*, especially when the patient is subjected to stressors, and based on this a response of the system can be analyzed informing about the direction of the trajectory and the resilience or lack thereof with respect to stressors. The implications of this new way of thinking for medical practice are vast. In the future, they should become routine in terms of both algorithmic guidelines and intuitive understanding, which in many situations could be significantly different from what has been traditionally used as the standard of care.

Certain individuals, given their ontogenesis and the location of their stable state on the Physiological Fitness Landscape, may have very limited options for finding a new symmetry state corresponding to available molecular interactions in this increasingly less complex multi-dimensional attractor state of interacting parameters and disturbed set points of homeostasis. Thus, for example a healthy blood glucose range in a critically sick patient is incorrectly assumed to be the same as the glucose range in healthy individuals. This faulty reasoning may be responsible for the composite attractor state to fall further in amplitude on the Physiological Fitness Landscape, further pushing the free energy and fitness in the system towards zero accompanying the movement of the other parameters of vital organ system homeostasis in the direction of thermodynamic equilibrium. Ultimately, the system is pushed off the cliff to an equilibrium state of mortality and thermodynamic equilibrium, unable to tether itself to some level of organizationally successful stable zone, without even a metabolic tendril of life support.

Outside of the critical care setting there remain other examples of the sometimes-precarious nature of present-day standards of care, for example plausible and legitimate concern for the risks of both cancer and cardiovascular disease. The Physiological Fitness Landscape provides a framework for a precision personalized scale model that can be argued should be invoked and has the potential to transform the power of medicine in the future to guide these and countless other challenges.

Thomas Kuhn provided a historical framework for the progress in science and introduced the famous concept of "paradigm shift" meaning that major advances in science occur through revolutions in our understanding of the universe around us. *The new insights cannot be understood from the old perspective since the two paradigms are incompatible.* He also famously stated that science further evolves by first incorporating new ideas that make things more complex until it shapes them into a model that is much simpler and hence stronger empirically. It is this author's belief that the current state of complexity in the field of medicine and physiology is characteristic of an early stage of an evolution of a scientific paradigm and we should soon see great simplifications leading to deeper understanding of human health and disease. It is indeed the hope of the author and the goal of this book to bring about an accelerated transformation toward a new paradigm shift in the field of medicine.

5.4.6 Summary

This chapter emphasizes the role of physical concepts in shaping the future of medicine as converging the other fields of science and becoming personalized. The key concepts around which new approaches to disease prevention and therapeutic intervention are developed involve homeostasis, allostasis and allostatic load. One of the most important aspects that needs to be quantitatively incorporated into these approaches, not only in the diagnostic but also therapeutic framework, is

stress. Another problem area is diet and nutrition. *Persistence of nutrient excess leads to the development of new set points, which is characterized by failure to establish basal equilibrium, increasing threshold of fat mass storage accompanying inflammatory infiltration, and progressive degrees of insulin resistance.* This condition further exposes the patient to other modern-day predators such as cardiovascular disease. It should be kept in mind that stress can be generated by various sources. For example, due to the presence of a predator, a toxin, a microbe, an allergen, obesity, ischemia, arthritis, or diabetes. Stress can occur at a cellular, tissue or organism level. However, irrespective of the source of stress and the level at which it occurs, the body does not know the difference, i.e. the transduced signals despite disparate etiologies lead to the same stress response. The difference is in the extent or more importantly, the duration of the stress or allostatic response. However, it is also important to note *that stress is a part of life and it can be vitalizing. In fact, chronic lifelong low-intensity stressors counter disease and favor healthy aging and longevity* (Figure 5.22).

1) (a) High altitude, (b) steep slope (c) high energy barrier.
2) Unstable state created by entwined series of interlocking positive feedback pathogenic loops ultimately responsible for a decline in the altitude, slope. and energy barrier of the next stability zone over time.
3) Changing composition and diversity of gut microbiota.
4) Altered composition and diversity of gut microbiota ultimately promote a breach in the mucosal barrier which is the initial manifestation triggered by overcoming the metaphorical energy barrier. This is followed by the state of subclinical endotoxicosis, hepatic and visceral adipose tissue, insulin resistance, and the feed-forward interlocking loops depicted above (2 $\longrightarrow$ 4) within the instability zone marked "5".
5) Unstable state of 2 $\longrightarrow$ 4,.
6) Intercurrent stress $\longrightarrow$ exceeds energy barrier.
7) Potentiated unstable state of 2 $\longrightarrow$ 4.
8) Major life crisis.
9) Inflammatory redox stress $\longrightarrow$ oncogenic mutation.
10) Profound metabolic inefficiency, mitochondrial dysfunction.
11) Warburg effect and cancer cell proliferation.
12) Mortality $\longrightarrow$ free energy and fitness of the human host biological system is zero (i.e. thermodynamic equilibrium).

As stated earlier, metabolism plays a major role in both health and disease. *Metabolic inflexibility is characterized by the inability of cells to optimally utilize lipid and carbohydrate food and to transition between them using fatty acid in the resting state and glucose in the fed state.* Both insufficient and excessive calorie sources of energy, as well as poisoning

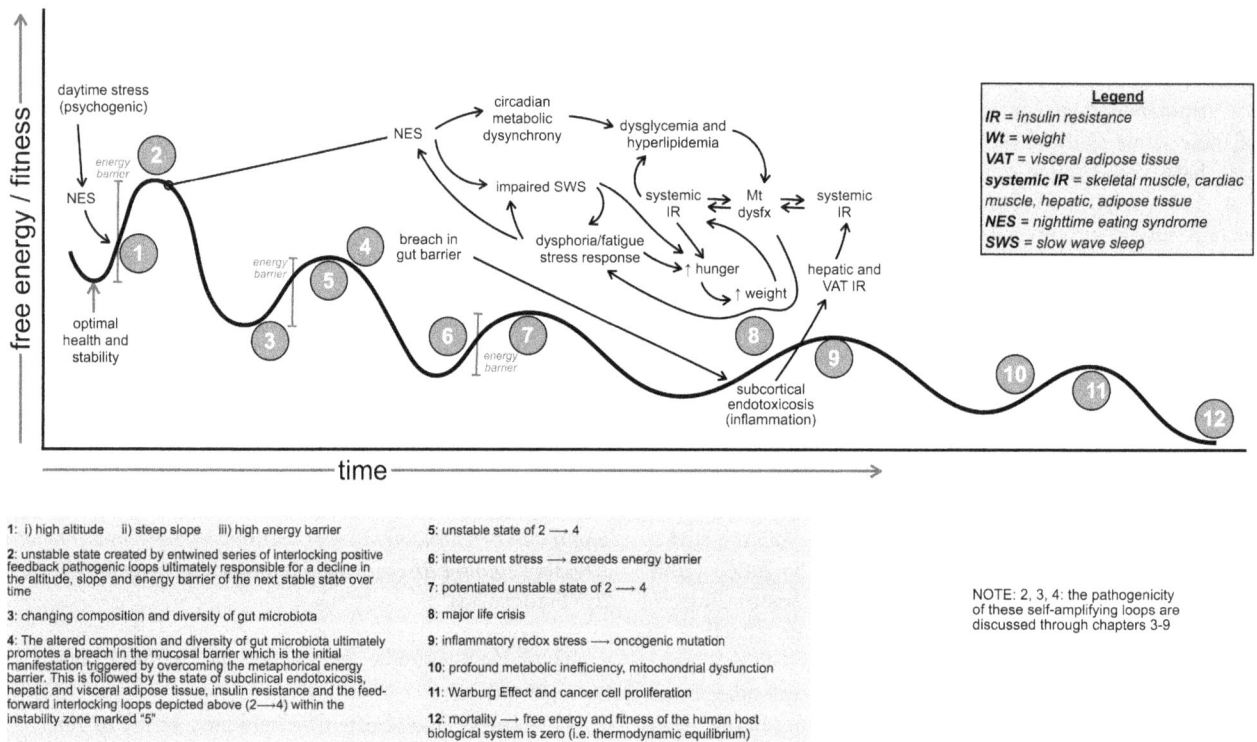

FIGURE 5.22 Illustration of various characteristics (one through 12) of the Physiological Fitness Landscape as it evolves over time in the context of states of health and progressive disease.

or interference of electrochemical conversion within mitochondria are sources of oxidative stress, disturbed redox and inflammation. These disturbances are the equivalent of abnormal metabolism, which sits at the interface of metabolic disease and thus tissue damage and dysfunction. The change in the composite microbiome, for example, due to significant stress, infection, a course of antibiotics or in pregnancy, may result in a switch from immune tolerance to immune activation and trigger autoimmune diseases including type 1 diabetes. In order to develop quantitative precision approaches to medical decision-making, we propose the implementation of the Physiological Fitness Landscape idea borrowed from physics.

A Physiological Fitness Landscape is an individual patient's road map to healthy living and can provide a trajectory to optimized recovery from a disease state (Figure 5.23). This requires a proper introduction of both order and control parameters with which, changing controllably the latter and measuring the response of the former, we obtain a multidimensional Physiological Fitness Landscape manifold. *The development of a pathological state may be described in terms of increased entropy, loss of information, and a new minimum free energy state highlighting increased disorganization.* The human body, being a complex system organized into interlocking hierarchies, requires an application of the laws of complexity of connections between systems in order to understand the underlying causes of pathological transformations. A message that is often lost is that there is always an optimal range for everything. On the one hand, if something is good for our health, it does not mean that more will be better. Conversely, completely eliminating something bad is not

recommended either due to a phenomenon called hormesis, which basically strengthens the body's preparedness against a deleterious effect, somewhat akin to immunization. Linear thinking is oblivious to the existence of optimum ranges of parameters because these values arise from a balancing act between beneficial and deleterious effects in a nonlinear system such as the human body. Every linear extrapolation in real systems has a limit and this message is often lost on the pundits and pontificating prophets of every new fad, be it in the area of physical fitness, nutrition, prevention, pharmacological intervention or diet.

5.5 A Look at the Elements of the Metabolism Story

5.5.1 The Stress Response

Colloquially speaking, stress is often used to describe an unpleasant situation (e.g. an overbearing boss), the reaction to the situation (i.e. headache, chest pain, heartburn), or the cumulative response to these reactions (i.e. an ulcer or a heart attack). Historically, stress was often perceived as negative and synonymous with distress—a physical, mental, or emotional strain. However, stress has both positive (health-promoting) and negative (health-damaging) functions. *Stress is any challenge to the normal balance of biological systems of the body. Stressors may include work or school stress, social conflict and isolation, financial stress, adjustment stress, bereavement stress, competition stress, and health stress. The type,*

a) Healthy aging (optimal case)

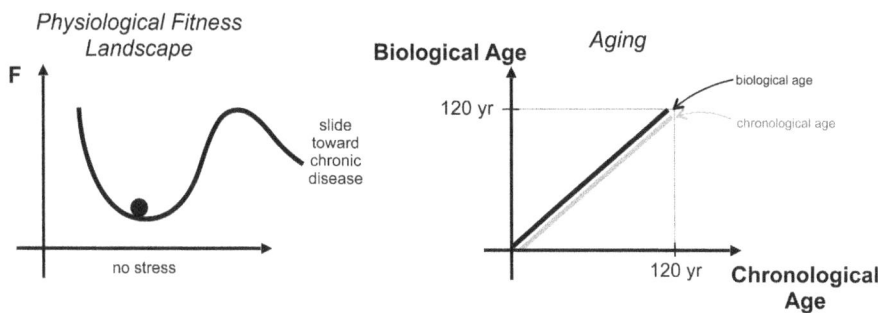

b) Toxic stress followed by vitalizing stress

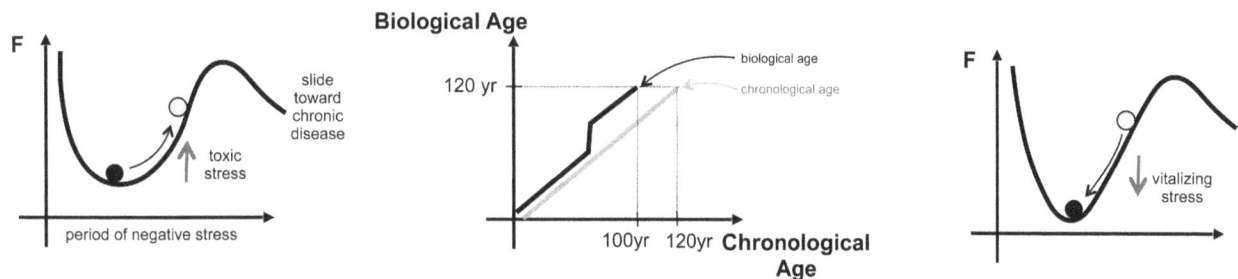

c) Toxic stress followed by chronic stress

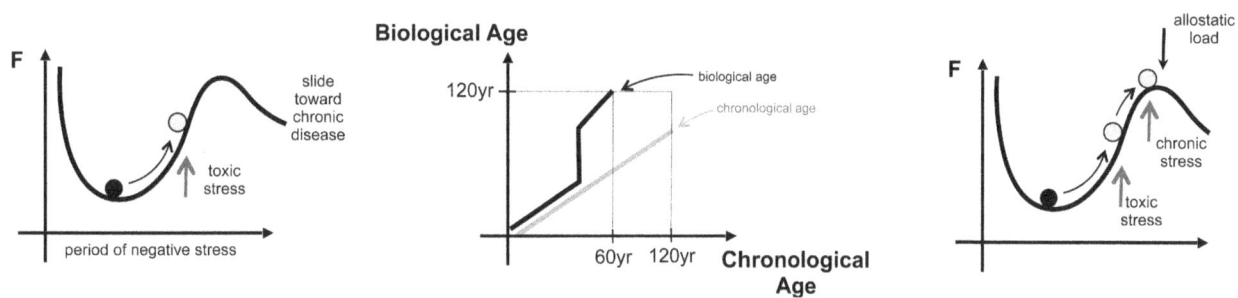

FIGURE 5.23 An illustrative summary of the properties of the Physiological Fitness Landscapes under conditions of a) optimal health, b) toxic stress following vitalizing stress, and c) toxic stress followed by chronic stress. Note the effects of stress on the disparity between chronological and biological age.

amount, and effect of stress on the body can be quite subjective depending on physical, psychological and social makeup. However, resilience to stress is required for a healthy state just as debilitating stress is required for a diseased state (Figure 5.24).

Stress, in the context of this volume, is defined as any disruption or a threat of disruption to homeostasis, the body's dynamic equilibrium. Homeostasis is the maintenance of morphological, physiological, and behavioral daily routines of the life cycle through allostasis. Allostasis is the healthy, adaptive response to maintaining homeostasis through the hormonal, autonomic, and immune systems. Allostatic load is the cumulative result of an allostatic state and a subthreshold critical point whereby homeostasis of vital organ systems is maintained. Beyond this threshold is allostatic overload, which is tantamount to the onset of chronic disease. This is represented when allostasis can no longer maintain homeostasis of vital organ systems. Disharmony in this exquisite, organizationally complex system can drive the body from health to disease.

When the brain senses a stressful situation, it activates the autonomic nervous system or the body's "fight-or-flight" response, triggering a cascade of stress hormones that produce well-orchestrated physiological changes. The brain, particularly the hypothalamus, signals the adrenal glands on the kidneys to release stress hormones such as adrenaline, cortisol, and norepinephrine. As these hormones travel through the bloodstream, they increase heart rate, increase blood pressure and dilate the air passageways of the lungs to bring in more oxygen with each breath. Extra oxygen delivered to the brain increases alertness and heightens the body's senses, priming the body for instant action. *Following this fast-acting surge of hormones, the secondary stress response system activates what is known as the hypothalamic pituitary adrenal axis, continuing to release hormones into the bloodstream if the brain continues to perceive a threat. Acute and short-lived stress promotes enhanced cognition and emotion regulation whereas prolonged and chronic stress deteriorates learning and memory and accelerates the trajectory to mental illness and biological diseases of aging.*

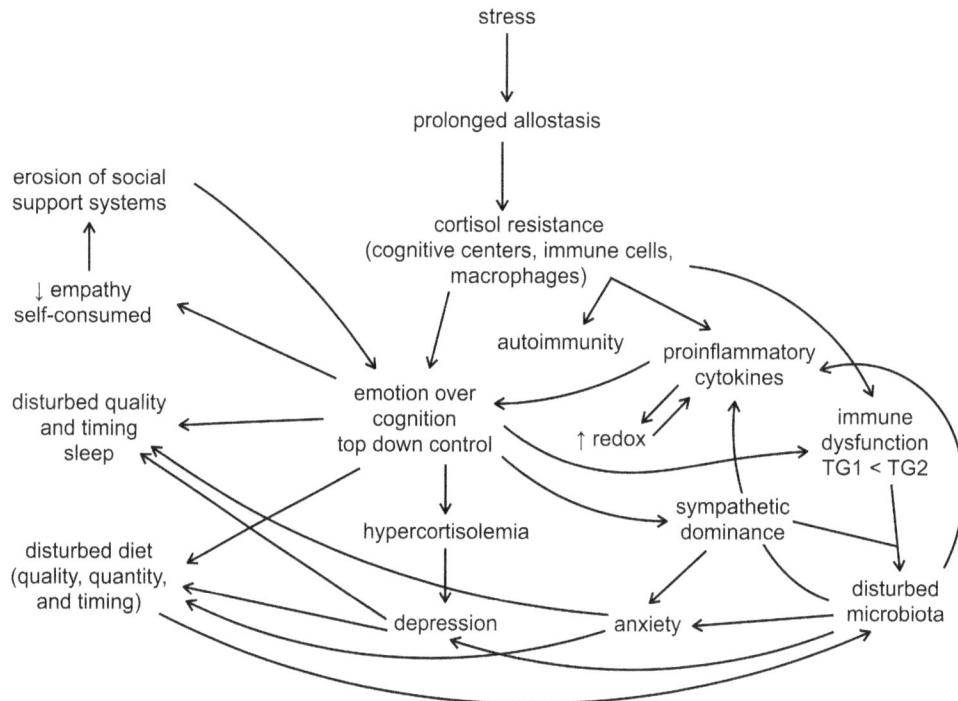

FIGURE 5.24 An illustrative summary of the stress response phenomenon. Prolonged or chronic stress leads to a variety of downstream physiological outcomes including increased inflammation, immune system dysfunction, emotional and cognitive disturbances, anxiety, depression, disturbed microbial composition, and sleep disruption.

Importantly, the perception of stress underscores its effect on the body. Overall, people who report experiencing high levels of stress increase their risk of death by 43%. However, people who report experiencing high levels of stress but do not perceive the stress as harmful to their health have a considerably low mortality rate, comparable to those who report low levels of stress. So, can the body's perception of stress alter life expectancy? New research on the effects of stress on the body, presented in great detail in Volume 2 of this book, suggests that this may be possible. For example, when the brain senses a stressful situation, it signals increases in heart rate, blood pressure, and oxygen intake. These physiological changes are often interpreted as signs of anxiety or the body's incapability of coping with stress. However, if individuals perceive these physiological changes as the body's preparation to successfully meet the challenge it is facing, psychological and physiological effects of stress are dampened, and the body is able to return to homeostasis.

The intention of the chapter on stress response in Volume 2 is to give the stress response the recognition it deserves. This chapter hopes to shed light on how the stress response provides a unifying explanation for virtually all chronic diseases of aging, which is often the most underappreciated element in medicine and public health. The perception of stress can alter life expectancy, which is the single greatest challenge to the profession of medicine—predicting death and intervening appropriately to delay it. The notion of stress can be extremely useful in clinical medicine as a model for testing susceptibility states of disease that cannot be assessed reliably in the baseline state, e.g. using a cardiac stress test to assess adrenal sufficiency. Monitoring responses to stressors and adaptability

once the stressors are removed on a patient-by-patient basis can provide more valuable insights into disease prediction and progression as well as recommendations for optimal therapeutic interventions.

5.5.2 Metabolism and the NHR Superfamily

In humans, the regulation of growth, metabolic homeostasis, and development processes involve extensive intercellular communication. This is achieved by various endocrine signals that often communicate with intracellular receptors that regulate gene expression. In this latter process, transcription factors, specifically nuclear hormone receptors (NHRs), participate in the up- or down-regulation of gene expression. Nuclear hormone receptors are ligand-inducible transcription factors that mediate changes to whole-body metabolic pathways (Figure 5.25).

Nuclear receptors include steroid ligand nuclear receptors such as: androgen receptors (AR), estrogen receptors (ER), progesterone receptors (PR), glucocorticoid receptors (GR), and mineralocorticoid receptors (MR). These classical hormone nuclear receptors bind to DNA as homodimers inducing transcription. They have evolved to regulate carbohydrate and lipid metabolism, development, reproduction, and electrolyte balance.

The regulation of ligands that bind to these hormone receptors takes place *via* the classical hypothalamic-pituitary axis negative feedback mechanisms. The extended family of steroid nuclear hormone receptors such as GR and peroxisome proliferator-activated receptors (PPARs) utilize energy and are involved in sensing. For example, the hormone glucocorticoid binds to the GR, as it regulates hepatic and systemic glucose

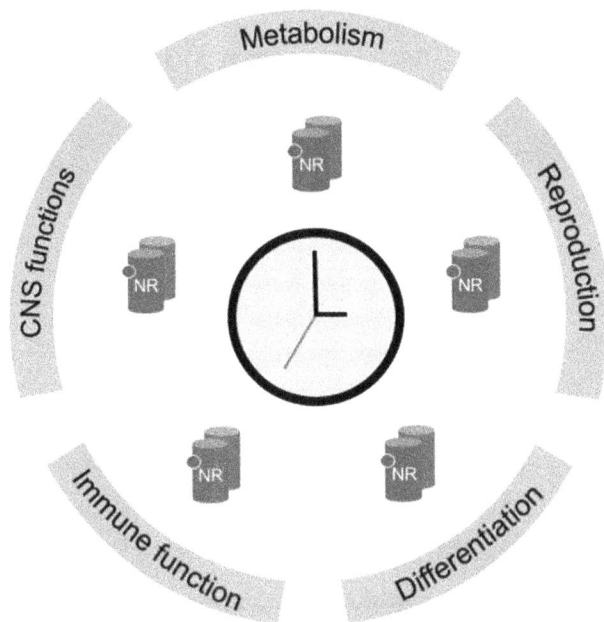

FIGURE 5.25 An illustrative nuclear hormone receptor summary. NHRs are involved in numerous physiological processes including metabolism, reproduction, cell growth and differentiation, immune function, and CNS functions. Source: adapted from Yang, X. (2010). "A Wheel of Time: The Circadian Clock, Nuclear Receptors, and Physiology". *Genes & Development* 24(8): 741–747.

metabolism. PPARs regulate whole-body glucose and lipid metabolism. Thus, under normal physiological conditions, GR and PPARs maintain systemic energy homeostasis.

Additionally, recent studies have suggested that NHRs are tractable targets for cardiovascular disease and diabetes therapy, especially Farnesoid X receptors (FXRs) and liver X receptors (LXRs), which regulate multiple metabolic pathways. As metabolic regulators, FXR and LXR play a major role in glucose and lipid metabolism. Upon activation by bile acids, FXR regulates various aspects of lipid and glucose metabolism. LXRs play a crucial role in regulating the reverse cholesterol transport pathway in lipogenesis and in the maintenance of whole-body glucose homeostasis.

Current studies have indicated that both LXR and FXR are associated with the development of metabolic diseases. Therefore, FXR and LXR might have therapeutic implications for the treatment of metabolic diseases such as type 2 diabetes and cardiovascular disease. In addition to this, ~25% of the adipose tissue genes are regulated by the circadian clock to maintain lipid energy metabolism. Thus, *the nuclear receptor family plays a pivotal role in mediating communication between circadian rhythms and metabolic functions to maintain whole-body energy homeostasis.*

5.5.3 The Biology of Time

Biological cycles are categorized as circadian (24 hours), infradian (longer than 24 hours), and ultradian (shorter than 24 hours) systems, which together are intricately interactive and interconnected. The aging process is a manifestation of the desynchronization away from the exquisitely orchestrated

beauty of biological perfection of cyclical processes at all levels of organization that operate as a singular whole in youthful states of optimal health.

The overarching message of this section is the notion of a biological cycle. We also underline the *strong connection between biological cycles and the superfamily of thyroid and steroid nuclear hormone receptors*. These receptors act as transcriptional regulators of hormone- and lipid-derived ligands to maintain the acid-base, redox, and energy branches of metabolic homeostasis organism-wide.

Time is an essential variable in biological systems, measured in terms of cycles. The constant influx of energy in a living system maintains it in a far from equilibrium state. Time is measured externally with mechanical clocks composed of physical oscillators. Biological processes are measured and conducted by internal endogenous clocks, molecular machinery present in all cells that have DNA. These molecular clocks are the astonishing evolutionary incarnate of the earth's rotation around its own axis.

The cell's endogenous timepieces include central (SCN) and peripheral (non-SCN) circadian oscillators that link metabolic pathways of physiology and behavior to 50% of the human genome. This temporal organizing strategy is essential for maintaining energy and redox homeostasis, and hence life itself. The SCN receives primary sensory cues in the form of visible light and thus synchronizes to the diurnal light–dark cycle of the external environment. The circadian pattern of neurotransmission originating from the master pacemaker reaches many other regions of the brain such as the hypothalamus and the pineal gland. Central circadian regulation of these areas of the brain from the SCN in turn provides neural and hormonal cues to different tissues and organ systems throughout the body, thus temporally coordinating many aspects of physiology. These cues include autonomic fibers, hypothalamic-pituitary-adrenal gland axis (HPA axis), and melatonin. Accordingly, while the autonomic nervous system and HPA axis mediate the body's stress response, and melatonin promotes sleep induction and slow-wave sleep, these are also central conductors of systems biology within and between tissues of the body. This allows the time organizational coherence of physiology to function as a system whole.

It follows that diurnal external light cues coming in through the retina of the eyes ultimately entrain the cell-autonomous circadian peripheral clocks that drive the timing of organ system physiology. Entrainment is the stable synchronization of external cycling events of the environment to the internal cycling pathways that conduct physiology. Teleologically, this orchestration between the central and peripheral clocks allows the anticipation and accordingly the adaptation to environmental changes, promoting maximum metabolic efficiency and homeostasis. There are many inputs received by the body that can entrain physiology, with the two strongest cues being light (for the master clock) and food for peripheral clocks.

Coordination with the major cycling external cues of light and food includes but is not limited to behavioral cycles of fasting/feeding, sleep/wake, and rest/activity. Additionally, aligned to external cues and to behavioral cycles are physiological cycles of core body temperature, neuroendocrine function, and autonomic function (Figure 5.26).

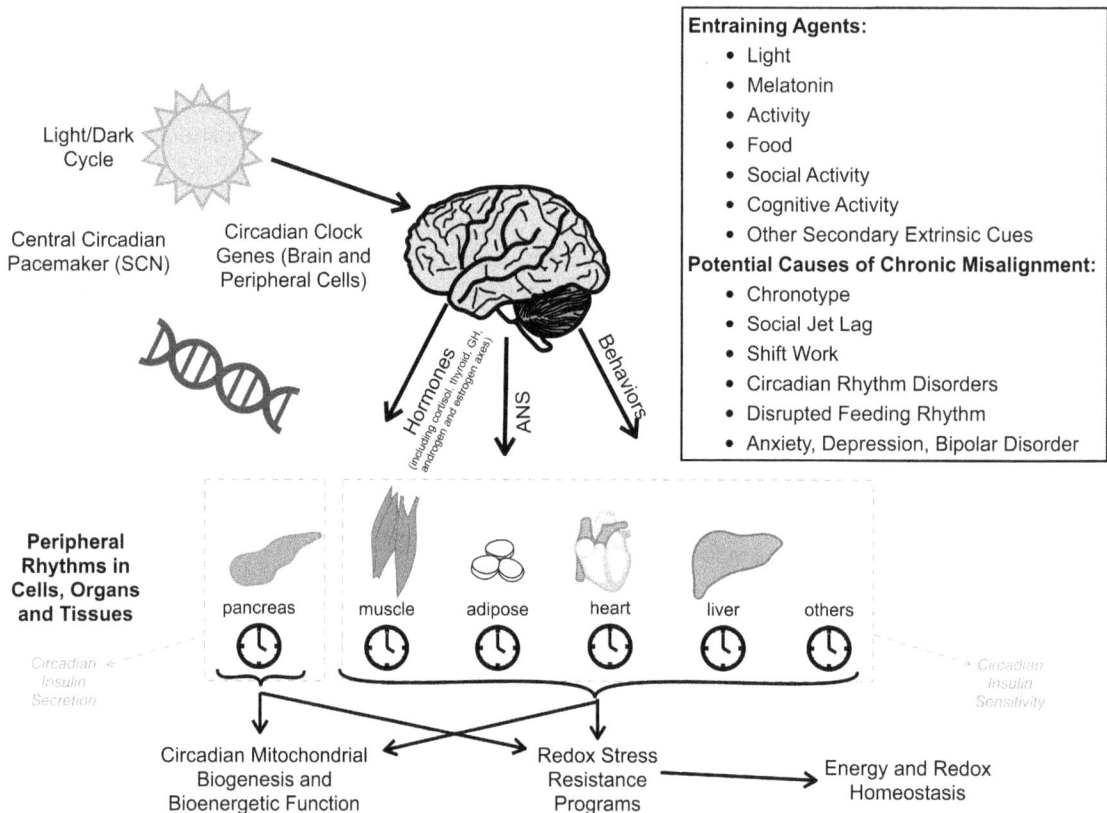

FIGURE 5.26 Schematic illustration on the role of sunlight as the primary visual cue which transmits information about the environment to the master clock, the suprachiasmatic nucleus (SCN). The SCN in turn signals to numerous brain regions and tissues that regulate hormone systems, the autonomic nervous system, and behavior. Source: adapted from Gooley, Joshua J., and Chua, Eric, C.-P. (2014). "Diurnal Regulation of Lipid Metabolism and Applications of Circadian Lipidomics". *Journal of genetics and genomics* 41(5): 231–250.

The master metabolic regulator, AMP activated protein kinase (AMPK), and sirtuin 1 (SIRT1) are energy sensors with a complex and strong bidirectional interplay between themselves and with clock function in several tissues, most notably in the hypothalamus and the liver. AMPK and SIRT1 are intricately coupled to the steroid and thyroid superfamily of nuclear hormone receptors (NHRs) and other transcriptional regulators that govern the circadian cycles of the energy and redox programs of metabolic homeostasis. Accordingly, these programs in states of excellent physiological health, align with circadian behaviors of the sleep-wake and fasting-feeding cycles. During nocturnal fasting, energy-consuming pathways are inhibited while energy (ATP) producing pathways are stimulated. Thus, for example, NHR peroxisome proliferator activated receptor γ (PPARγ), which promotes anabolic pathways of adipogenesis and in parallel adipocyte filling lipogenesis, is upregulated during the daytime feeding phase of the circadian cycle. Conversely, NHR PPARα, which promotes fatty acid oxidation, is diurnally activated during the nocturnal and fasting phase.

In the metabolically healthy obese state (Figure 5.27a) persistent dietary intake in the context of a pear-shaped body type, or pharmacologic drug prescription with a PPARγ agonist, promotes weight gain in the absence of developing pathogenic metabolic parameters. In these individuals there is greater adaptive storage capacity for excess lipids that correlates with the predominance of subcutaneous adipose tissue.

Accordingly, there is resilience to the evolution of insulin resistance features (such as dysglycemia, hypertension, and dyslipidemia), and correspondingly to a decline in the amplitude of the metaphorical Physiological Fitness Landscape stable state. In the metabolically unhealthy obese state (Figure 5.27b), persistent dietary overconsumption in the context of an apple-shaped body type has less adipose tissue storage capacity. Consequently, there is less resilience to the loss of metabolic fitness, represented in a decline in the altitude of the metaphorical stable state within the topological terrain of the fitness landscape.

The timing of eating plays a key role in the timing, coordination, and efficiency of metabolic pathways. It appears evident that 12 hours, and even up to 18 hours of consistent fasting on a daily basis of time-restricted eating (TRE), particularly during the night, is a powerful if not crucial method for achieving optimal physiological body weight and health. This is rooted in the circadian rhythm of the synchronized symbiotic intestinal microbiota with human host metabolism. When we eat at a time that our body anticipates it, as is the case of TRE, the feeding cue reinforces and amplifies the circadian rhythms. However, when we eat at irregular times, when our body is not prepared for it, it provides conflicting cues to the circadian pathways that guide physiology and behavior.

Nocturnal eating induces impaired fasting glucose, fasting hyperlipidemia, insulin resistance, and weight gain through

a) Metabolically Healthy Obese

b) Metabolically Unhealthy Obese

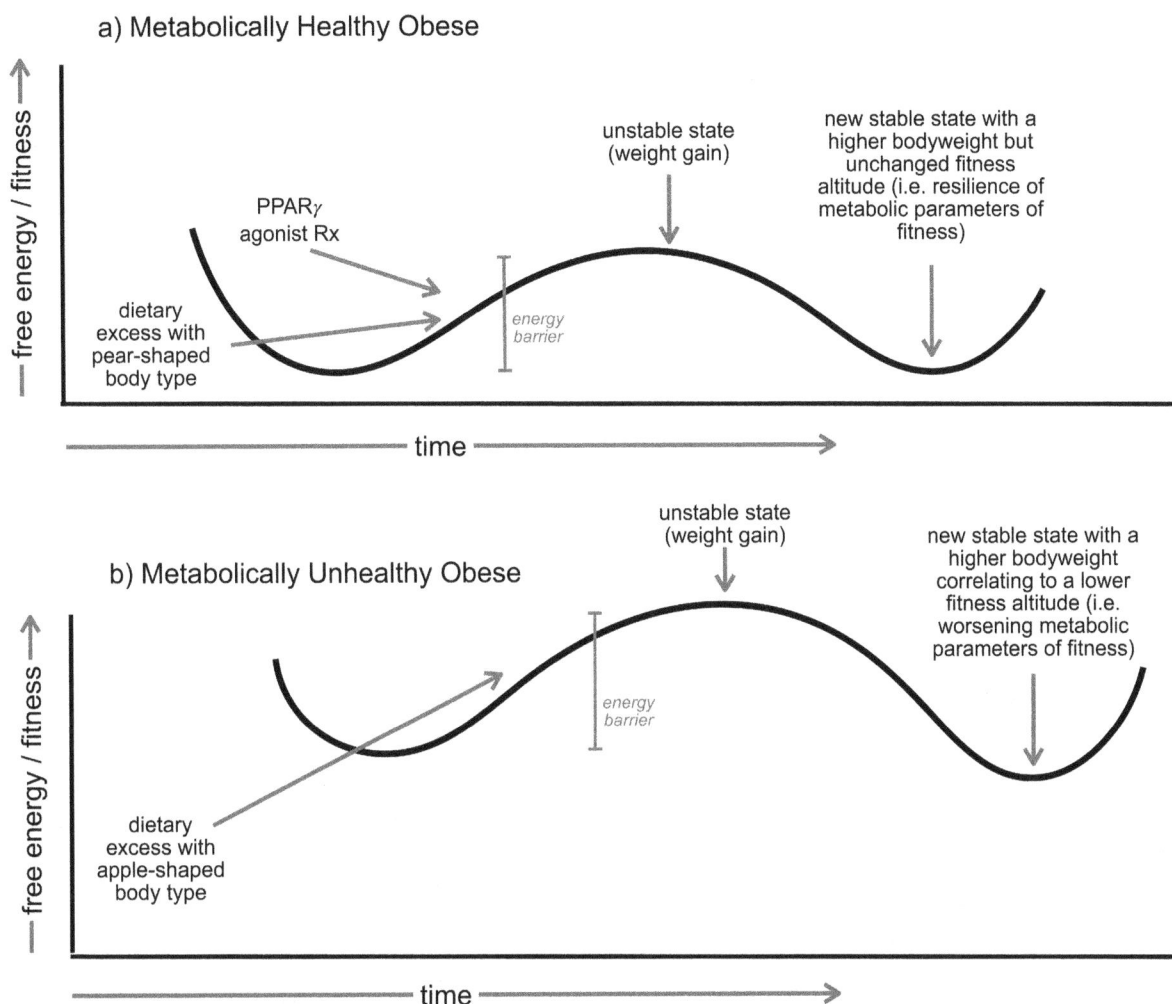

FIGURE 5.27 Illustration of the metabolic/Physiological Fitness Landscape of changing bodyweight as a function of time for: a) a metabolically healthy obese individual and b) a metabolically unhealthy obese individual.

a number of mechanisms. These include carbohydrate and fat consumption superimposed on a malalignment of nocturnal processes of hepatic glucose output and adipose tissue release of fatty acids, respectively. These events may lead to glucose toxicity and lipotoxicity which induce inflammatory and redox stress, and insulin resistance that disrupt metabolic homeostasis. An additional contribution is the diurnal releasing pattern of melatonin at night. Melatonin inhibits pancreatic release of insulin and thus accordingly, this is another exacerbating factor that potentiates not only hyperglycemia with nocturnal eating, but also other manifestations of insulin resistance.

Eating patterns at times that conflict with other cues to the circadian clocks, and hence at a time when the body is not prepared, compromise metabolic function. Summarily, in addition to diet and other circadian behaviors, control parameters of biological time in the form of cycles include the stress response and the microbiome. Spatial-temporal desynchronization of metabolic pathways is consequent to a conflict in behavioral and physiological circadian cues. What ensues is a feedforward self-amplifying matrix of reverberating cascades whereby the control parameters and the primary markers of disease itself, i.e. redox, energy, and inflammatory stress, cannot be extricated. Taken together, human disease may be defined from the perspective of a breakdown in the temporal organization of physiological processes and control parameters.

5.5.4 Calorie Restriction, Intermittent Fasting, and Time-Restricted Feeding

The goal or physiological purpose of any living system is the survival of the organism, and species as a whole. *Survival requires meeting the continually changing metabolic bioenergetic demands of the body. This demand is met by acquiring nutrients from the environment in order to provide the body with the energy required to perform biological functions and maintain homeostasis. Like most things in life, there is an optimal level of nutrient consumption, and both too little and too much can precipitate chronic diseases of aging.*

In addition to stress response, chronobiology, and microbiota issues Volume 2 of this book addresses the very important issue of calorie restriction, intermittent fasting, and time-restricted feeding with biological and clinical discussions including: 1) how the energy sensors and the fuel gauges of the body (AMPK and SIRT1) promote survival and slow the pace of aging in a way that increases longevity; 2) the notion of hormesis with regard to optimal metabolic balance between and among systems; 3) the importance of endogenous circadian biology (both synchronized and dyssynchronous) on insulin signaling and underlying molecular cascades; 4) the interwoven relationship between nutrient intake, energy sensing, insulin resistance, mitochondrial dysfunction, and chronic diseases of aging; 5) and finally, the effects of chronic overnutrition on metabolic signaling and accelerated aging.

Survival early in life is more metabolically demanding and requires a higher energy input. It is during these years that higher levels of physical fitness, aerobic capacity, and endurance are supported. During post-reproductive years, longevity is the main goal. It has been demonstrated that calorie restriction is beneficial in promoting longevity and preventing age-related chronic disease. Not only are there changes in optimal energy consumption based on the stage of life, but there are different genes that promote survival early or later in life. *The most favorable of these lifestyle parameters are fundamentally connected to the optimal activation of the energy sensors AMPK and SIRT1 to promote survival and slow the pace of aging, thereby increasing longevity.* These energy-sensing fuel gauges are coupled to gene programs of stress resistance, including DNA and cellular repair, antioxidant systems, autophagy, cell differentiation, and apoptosis. AMPK activity is upregulated during circumstances of calorie restriction, fasting, and exercise associated with accelerated ATP consumption. It is thought that the health benefits of calorie restriction are induced by AMPK-dependent inhibition of mTOR, acting to improve insulin resistance, promote mitochondrial biogenesis, prevent obesity and metabolic disease, and increase overall lifespan (Figure 5.28).

Optimal insulin sensitivity is known to coincide with calorie restriction, while excess nutrient intake has been associated with the development of insulin resistance. One characteristic of insulin resistance that makes it pathological is the loss of the circadian cyclicity of insulin secretory and sensitivity patterns in metabolic tissues. *A common pathogenic behavior is chronic dietary overconsumption, not just in terms of total daily caloric intake, but the quantity of intake relative to the time of the day.* Nuclear hormone receptors lie at the intersection between the endogenous clocks and metabolism, coupling metabolic processes to cyclical circadian output. In a negative feedback loop/bidirectional relationship, poor sleep results in insulin resistance due to increased stress response and disruption of endogenous clock synchronicity, while insulin resistance promotes poor quality sleep by inducing obstructive sleep apnea.

Chronic overnutrition can be defined as long-term nutrient intake that exceeds energy expenditure demands, which can then exceed the capacity for mitochondrial production of ATP, resulting in mitochondrial dysfunction. When there is mitochondrial dysfunction or overload, oxidation of nutrients is prevented and is instead diverted to storage within the cell. Lipid accumulation that exceeds the storage capacity of the hepatocyte eventually leads to insulin resistance and ultimately accelerated aging.

One underlying message of this chapter is the notion of hormesis, the perfect balance of stress. It is thought that stress at a lower level can be vitalizing, while stress at higher levels can be harmful or lethal. This balance is individual and depends on the stage of the life cycle and can change dynamically over the course of a lifetime. This concept of hormesis can be used for physical exercise. The beneficial effects of activity and diet, both intensity and quantity, should be considered relative to the stage in the life cycle. Hormesis, importantly, can also be applied to caloric intake. There is a level that is optimal; based on energy expenditure as well as the stage in the life cycle. Intake that is too low or too high can lead to oxidative stress and inflammation as well as insulin resistance and/or hyperinsulinemia. Many of the signaling pathways mentioned previously depend on a balance between nutrient depletion and nutrient overabundance. Allostatic load is "the wear and tear on the body;" chronic stress over time can lead to increased susceptibility to chronic disease.

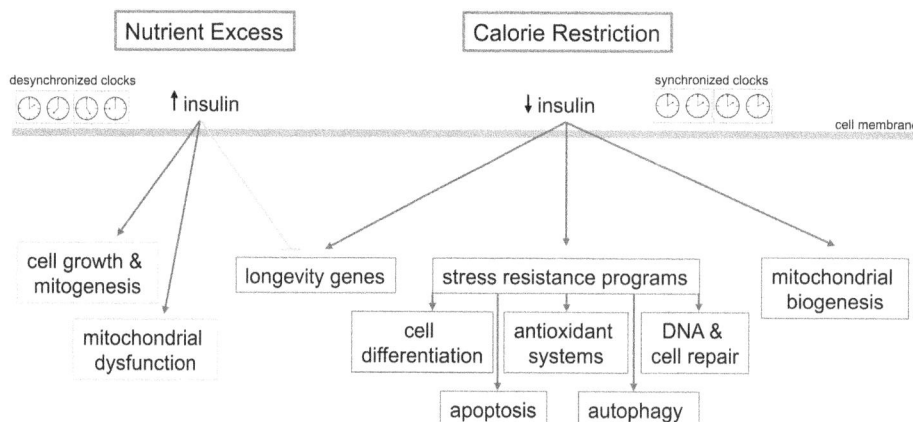

FIGURE 5.28 An illustrative calorie restriction summary. Excess nutrient consumption increases insulin levels and leads to desynchronization of biological clocks resulting in negative outcomes of cell growth, mitogenesis, mitochondrial dysfunction, and suppression of longevity genes. Alternately, calorie restriction maintains low insulin levels that preserve synchronized clock function and result in beneficial effects such as induction of longevity gene transcription, activation of stress resistance programs, and mitochondrial biogenesis.

5.5.5 The Microbiota

A greatly underappreciated aspect of human health is the crucial role played by the microbes with whom we share our bodies and upon whom we are unequivocally dependent. While the collection of microbes, or microbiota, is composed of bacteria, archaebacteria, eukaryotic microbes (including fungi), and viruses, this book focuses on the impacts of the bacterial members, the most thoroughly studied and understood. Roughly 95% of the resident bacteria—who outnumber us cell per cell—live in cooperative guilds (functional ecosystems) on the surface of the colon (the large intestine). It is here that we have some of our most critical interactions with the outside world and offer a hospitable, responsive environment for the microbiota.

Astoundingly, while bacteria populated the earth roughly 4.2 billion years and the first *Homo sapiens* emerged nearly four billion years later, we speak the same language, using shared signaling molecules and receptors, supporting one another in our mutual quests for life. *Humans have had no choice but to co-evolve with the bacteria, making our more complex human capabilities totally dependent on their presence. Indeed, the bacteria are not parasites or even visitors; they are inextricable components of the "supraorganism" that functions coherently and in an enhanced manner due to our synergistic mutualism,* as was so aptly realized by Nobel laureate Joshua Lederberg. Indeed, the components of the healthy supraorganism sing to the tune of the same circadian rhythms, share the same diet (though humans generally get the first pickings), and experience the same physical and psychogenic stresses.

While Hippocrates (460–370 BC) was convinced that "all diseases begin in the gut" and Antonie van Leewenhoek (in the 1680s) noted striking differences in microbes between oral and fecal samples as well as between healthy and diseased individuals, the importance of the gut microbiota regarding health was outweighed by the attention shifted toward getting rid of infectious agents (recognized as being microorganisms in the 20th century), knowing they could cause illness and even death. It has been a mere 10–20 years during which we are finally coming to appreciate that human health is critically dependent upon the health of our microbiota—for better and worse.

A major chapter in Volume 2 is devoted to this topic. Among the marvels of the microbiota that will be addressed therein are: 1) its necessity in activating and training our innate and adaptive immune systems (some 70% of which resides in our gut), enabling them to differentiate between symbiotic bacteria vs. pathogenic bacteria and healthy human cells vs. cancer cells; 2) its role as a so-called "second brain" associated with its own enteric nervous system that communicates back and forth with the same signaling molecules and neurotransmitters of the human brain; 3) its ability to elicit far-reaching transcriptional and epigenetic activity throughout the body; 4) its ability to turn dietary fibers indigestible to humans into indispensable vitamins, regulatory molecules, and other essential products; and (5) its balance of "healthy" versus "unhealthy" strains that has profound impacts on our health, mediating such modern—indeed, epidemic—ailments as obesity, type 2 diabetes, cardiovascular disease, a host of autoimmune diseases, a range of mental illnesses (e.g., depression, anxiety, ADD, ADHD, autism, and schizophrenia), Alzheimer's and Parkinson's diseases, irritable bowel syndrome, compromised liver and kidney function, and a variety of cancers. Like an endocrine organ, the impacts of the microbiota extend throughout the body and are unquestionably fundamental to our health (Figure 5.29).

For the modern-day medical practitioner, the time has come to embrace the potential of the microbiota to address health in new ways. With urbanized lifestyle-driven shifts in our microbiota leading to reductions in the diversity and fraction of

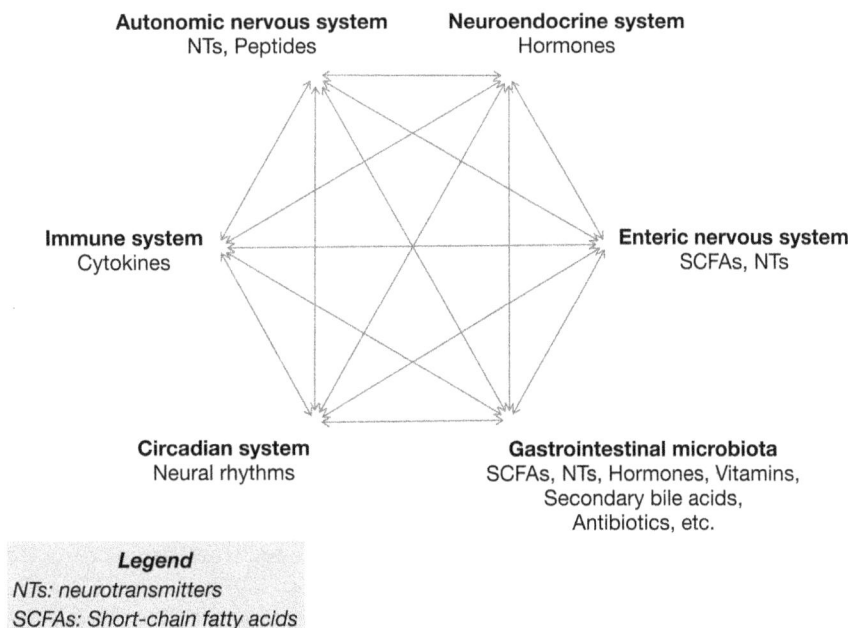

FIGURE 5.29 An illustrative microbiota summary. An interwoven, bidirectional, and self-amplifying web of parameters that affect microbial health.

healthy bacterial strains in the gut, the onus is upon our medical practitioners to examine the spectrum of bacteria and bacterial metabolites that might be foundational to the panoply of related illnesses and consider restoring the gut with critically missing microbes from a repository of such microbes. Patient-specific restorations could be more effective than the standard use of medicines that often address symptoms rather than causes, or hurt the body's immune system, for example. Metabolic or Physiological Fitness Landscapes could be employed to great effect to identify appropriate precision restoration treatments that have the potential to reverse a wide range of the modern westernized metabolic- and inflammatory-related health epidemics. We refer the reader to a dedicated chapter in Volume 2 which presents a comprehensive discussion of the role of microbiota in maintaining good health.

5.5.6 Insulin Resistance

Insulin resistance is a cardinal underlying mechanism in the pathogenesis of chronic disease states. In many cases, insulin resistance results from adaptive responses in efforts to promote survival under circumstances of scarce energy availability. Insulin signaling pathways typically induce mechanisms of cell survival and growth in the presence of required bioenergetic machinery such as healthy mitochondria. Although it is linked to metabolic disease, insulin resistance also plays a role in healthy physiology under conditions of periodic cyclicity such as the fasting/feeding circadian cycles seen in humans. Insulin resistance additionally results as an adaptation to cope with excess energy stores. Under these conditions it serves as a protective mechanism for metabolic tissue and prevents these tissues from being exposed to the threat of increased energy influx. There are currently two predominant theories of the exact etiology of insulin resistance that debate whether it precedes or follows a state of hyperinsulinemia. Another less widely discussed theory involves the notion of subclinical endotoxicosis promoting inflammation, which ultimately leads to hyperinsulinemia and secondarily to insulin resistance. Each of these arguments is discussed in greater detail in the chapter on insulin resistance.

Insulin secretion and signaling are each under circadian regulation. Moreover, insulin signaling governs the timing of biological clock activity. This bidirectional relationship highlights the significance of food as the strongest external cue for peripheral clocks. The loss of circadian rhythm of daytime insulin sensitivity and secretion alternating with nocturnal insulin resistance is a primary driver and component in the pace of aging and in the pathogenicity of chronic diseases of aging. Furthermore, non-circadian insulin resistance is integrally linked to mitochondrial dysfunction. *Consequently, there is a matrix of feed-forward self-amplifying loops of redox and inflammatory stress that are intrinsic to the most basic elements of metabolic disease and also include imbalances of energy and acid-base.* The relationship of redox, energy, and acid-base parameters of metabolic homeostasis is highlighted by a strikingly tandem correlation between the Nernst (redox), Gibbs free energy, and Henderson-Hasselbalch (acid-base) equations.

The energy sensors AMPK and SIRT1 are primary metabolic regulators with a strong circadian influence. They connect energy-consuming and producing pathways (in the case of AMPK) and redox stress resistance programs (in the case of SIRT1) in an inextricably coupled fashion to biological clocks for the maintenance of metabolic homeostasis.

Fundamental to the development of metabolic disease is the decoupling between the cytosolic glycolysis pathway that converts glucose to pyruvate and mitochondrial oxidative combustion (oxidative phosphorylation). This is especially important in tissues (such as skeletal and cardiac muscle, adipose tissue, and brain) where glucose metabolism is dependent on insulin signaling. This is specific to GLUT4-mediated translocation of glucose into the cell and the enzyme complex pyruvate dehydrogenase complex (PDC) catalyzed decarboxylation of pyruvate to acetyl-CoA in mitochondria prior to feeding into the TCA cycle.

The term metabolic flexibility is classically reserved for the circadian "metabolic switch" that occurs in skeletal muscle at the transition of the fasting-feeding (nocturnal-daytime) cycle. Accordingly, fatty acid oxidation occurs during the nocturnal hours while glucose oxidation occurs during the day. The loss of circadian rhythm is therefore tantamount to the conversion from cyclical (healthy) to noncyclical chronic (unhealthy) insulin resistance.

Metabolic flexibility is lost in the setting of mitochondrial dysfunction. Thus, there is a decoupling of mitochondria from cytosolic glucose bioenergetic metabolism. Accordingly, glucose combustion cannot be completed by the process of oxidative phosphorylation in the cells of any tissue where mitochondria are not functional. Thus, cells are inflexible in the sense of adaptability to changes in energy substrate availability, and the presence of ectopic lipid deposits ensures that fatty acid oxidation (FAO) outcompetes glucose oxidation. The significance of this inflexibility is rooted in the greater oxidative stress generated by FAO versus glucose oxidation. The exact mechanisms of this decoupling of these pathways are detailed in the insulin resistance chapter (in Volume 2 of this book).

The perspective of mitochondrial function in the clinical approach to diabetes has been a crucial missing piece. There are currently a few drugs that may be used to help restore mitochondrial function in the treatment of diabetes and other metabolic disorders. One newly approved drug, 6j, targets the enzyme complex PDC and thus helps repair the coupling of glucose metabolism in the cytosolic glycolysis pathway to pyruvate with mitochondrial oxidative metabolism. Another drug, Imeglimin, targets nuclear hormone receptor (NRH) coactivator PGC1a to promote mitochondrial biogenesis. It is also relevant in the connection of insulin signaling and secretion to circadian biology and NHR biology. Separate chapters in Volume 2 are dedicated to the Biology of Time and NHR metabolic regulation. Thus, these drugs are highly relevant to the focus of these discussions. Another chapter in Volume 2 is dedicated to the topics of insulin signaling and resistance. Insulin signaling is discussed from a genomics and molecular perspective to unveil relevant pathways critical to the pathogenesis of insulin resistance. On a more translational and clinical basis, insulin resistance is examined in Volume 2 in the contexts of cancer, type 2 diabetes, cardiovascular disease, obesity, neuroendocrinology, and Alzheimer's disease (Figure 5.30).

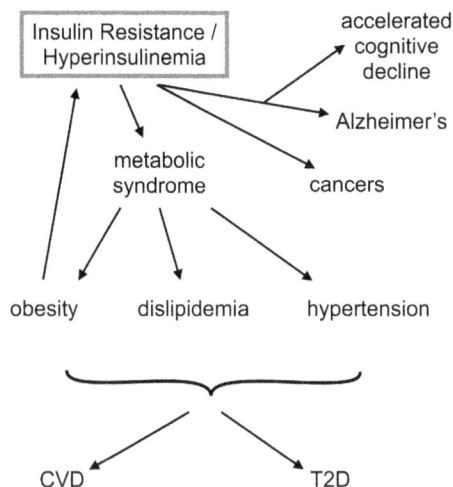

FIGURE 5.30 Schematic illustration of the role played by insulin resistance and hyperinsulinemia in the pathogenesis of chronic diseases such as obesity, cardiovascular disease, type 2 diabetes, Alzheimer's disease, cancer, and accelerated cognitive decline.

5.5.7 Mitochondrial Function and Dysfunction and Insulin Resistance

This chapter is where we connect all the dots regarding the bioenergetics of metabolism. We detail the crucial function of mitochondria in health and discuss how the inextricable and bidirectional link between dysfunction of mitochondria and resistance to insulin contributes to chronic diseases of aging. We consider underlying and intertwined factors such as the stress response, circadian rhythms, gut microbiota, immune system response, and bioenergetics.

The central ideas covered in this book are metabolic health, metabolic disease, the initiation of metabolic health and disease, and the therapeutics aimed at slowing down, stopping, or even reversing metabolic dysfunction. Metabolic health is directly linked to the energy producing organelles, mitochondria. Dysfunction of mitochondria is associated with insulin resistance and changes in VO_2 max and VO_2 submax. Skeletal muscle biopsy with cell mitochondrial stress testing can assess VO_2 max and VO_2 submax using rotenone and other electron transport chain inhibitors.

In the sense that mitochondrial function and insulin signaling are inseparably intertwined, mitochondrial dysfunction is interwoven in the fabric of insulin resistance. Whether the presence of insulin resistance and mitochondrial dysfunction overlaps completely remains unclear. Comparable to "the chicken or the egg", the question of which of these parameters came first in the pathogenic process is posed. *We would argue that this bidirectional, feed-forward relationship may be initiated by either mitochondrial dysfunction or insulin resistance, depending on patient-specific circumstances and genetic susceptibilities.* On the one hand, it may be that all patients either develop insulin resistance as a cause, and subsequently an effect, of mitochondrial dysfunction, or it may be initially caused by mitochondrial dysfunction. In either case, if one accepts the premise that all chronic diseases of aging are rooted in mitochondrial bioenergetic compromise, then all individuals who eventually develop a chronic disease would also in parallel manifest underlying pathologies of insulin resistance and mitochondrial dysfunction. For counterarguments exploring the other sides of insulin resistance and dysfunctional mitochondria, see Sidebar 12.

5.5.8 Chronic Diseases of Aging as Metabolic Disorders

Volume 2's primary aim is to provide a fresh new perspective on biological systems and human physiology, especially related to metabolism and metabolic diseases, drawing largely on modern concepts in physics and information sciences.

The final chapter is dedicated to the discussion of how the interwoven nature of mitochondrial dysfunction and insulin resistance contribute to the progression of chronic diseases of aging such as cancer, Alzheimer's disease, and cardiovascular disease. Taking into consideration other parameters such as the stress response, gut microbiota, circadian rhythms, immune responses, and bioenergetics, these diseases are discussed and presented as metabolic disorders. Finally, we propose a new model for personalized precision medicine utilizing the Physiological Fitness Landscape.

Cancer is approached as a metabolic disease and discussed in the context of insulin resistance, bioenergetics, and metabolic flexibility. Therapeutic approaches such as dietary alterations including calorie restriction, fasting, and ketogenic diets as well as the use of pharmaceuticals like metformin and NSAIDs are also detailed within this chapter. Alzheimer's can likewise be seen as a metabolic disorder due to the high energy demands of the brain and its resultant sensitivity to mitochondrial dysfunction. We outline how disruption in mitochondrial efficiency promotes insulin resistance and affects nerve cells leading to such neurodegenerative disorders and associated cognitive impairments. Cardiovascular disease and cardiomyopathy are also associated with insulin resistance, obesity, and metabolic inflexibility and are examined as metabolic disorders here.

This chapter concludes with an explanation of the proposal of a new model of medicine that is The Physiological Fitness Landscape. The significance of control and order parameters and the function of the PFL in creating a map to the restoration of health are discussed in detail. We emphasize the most important factors in maintaining a healthy life: vitalizing stress, diverse microbiota, proper chronophysiology, acid-base balance, redox homeostasis, and a healthy regimen of exercise and diet that keeps inflammation at bay. Altogether, we show how physical science can inform medicine.

REFERENCES

1. Nice-sugar Study Investigators. (2009). "Intensive versus Conventional Glucose Control in Critically Ill Patients". *N. Engl. J. Med.* 2009(360):1283–1297.
2. Krebs, H.A. and Kornberg, H.L. (1957). "Energy Transformations In Living Matter". In: *Energy Transformations in Living Matter*. Springer, Berlin, Heidelberg.

3. Falconer, K. (2003). *Fractal Geometry*. Wiley. Hoboken, NJ. ISBN 978-0-470-84862-3:308.

4. (a) Anderson, P.W. (1972). "More is Different". *Science* 177(4047):393–396. (b) Li, R. and Bowerman, B. (2010). "Symmetry Breaking in Biology". *Cold Spring Harb. Perspect. Biol.* 2(3):a003475. doi:10.1101/cshperspect.a003475.

5. Longo, G. and Montévil, M. (2014). "From Physics to Biology by Extending Criticality and Symmetry Breakings". *Perspectives on Organisms.* Springer, Berlin Heidelberg pp. 161–185.

6. (a) Selye, H. (1976). "The Stress Concept". *Can. Med. Assoc. J.* 115(8):718. PMID: 20312787. PMCID:PMC1878840. (b) Selye, H. (1979). "Correlating Stress and Cancer". *Am. J. Proctol. Gastroenterol. Colon Rect. Surg.* 30(4):18–20, 25–28.

7. Cannon, W.B. (1929). "Organization for Physiological Homeostasis". *Physiol. Rev.* 9(3):399–431.

8. Cannon, W. (1932). *Wisdom of the Body.* W.W. Norton & Company, United States. ISBN 978-0393002058. (b) Quick, J.C. and Spielberger, C.D. (1994). "Walter Bradford Cannon: Pioneer of Stress Research". *Int. J. Stress Manag.* 1(2):141–143. doi:10.1007/BF01857607.

9. McEwen, B.S. (2006). "Protective and Damaging Effects of Stress Mediators: Central Role of the Brain". *Dial. Clin. Neurosci.* 8(4):367.

10. Sterling, P. and Eyer, J. (1988). "Allostasis: A New Paradigm to Explain Arousal Pathology". In: Fisher, K. and Reason, J. (eds) *Handbook of Life Stress, Cognition and Health* (pp. 629–649). John Wiley and Sons, New York, NY.

11. Sterling, P. (2004). "Principles of Allostasis: Optimal Design, Predictive Regulation, Pathophysiology, and Rational Therapeutics". In: Schulkin, J. (ed) *Allostasis, Homeostasis, and the Costs of Physiological Adaptation* (pp. 17–64). Cambridge University Press, Cambridge. doi:10.1017/CBO9781316257081.004.

12. Friedman, J.M. and Halaas, J.L. (1998). "Leptin and the Regulation of Body Weight in Mammals". *Nature* 395(6704):763–770.

13. Saper, C.B., Chou, T.C. and Elmquist, J.K. (2002). "The Need to Feed: Homeostatic and Hedonic Control of Eating". *Neuron* 36(2):199–211.

14. McEwen, B.S. (2000). "The Neurobiology of Stress: From Serendipity to Clinical Relevance". *Brain Res.* 886(1):172–189.

15. Sapolsky, R.M. (2004). *Why Zebras Don't Get Ulcers: The Acclaimed Guide to Stress, Stress-Related Diseases, and Coping*, 3rd rev ed. W. H. Freeman Publisher .ISBN 978-0-7167-3210-5. New York, NY.

16. Patel, V.L., Yoskowitz, N.A., Arocha, J.F. and Shortliffe, E.H. (2009 February). "Cognitive and Learning Sciences in Biomedical and Health Instructional Design: A Review with Lessons for Biomedical Informatics Education". *J. Biomed. Inform.* 42(1):176–197. doi:10.1016/j.jbi.2008.12.002. Epub 2008 Dec 24. PMID:19135173.

17. Romano, C.A. and Pangaro, L.N. (2014). "What is a Doctor and What is a Nurse? A Perspective for Future Practice and Education". *Acad. Med.* 89(7):970–972. doi:10.1097/ACM.0000000000000277. PMID:24979165. (b) O'Malley, P.G. and Pangaro, L.N. (2016). "Research in Medical Education and Patient-Centered Outcomes: Shall Ever the Twain Meet?" *JAMA Intern. Med.* 176(2):167–168. doi:10.1001/jamainternmed.2015.6938. PMID:26641544.

18. (a) Johnson, N.E., Maas, M.B., Coleman, M., Jozefowicz, R. and Engstrom, J. (2012). "Education Research: Neurology Training Reassessed. The 2011 American Academy of Neurology Resident Survey results". *Neurology* 79(17):1831–1834. (b) Thompson Stone, R., Tollefson, T., Epstein, R., Jozefowicz, R.F. and Mink, J.W. (2017). "Education Research: Positive Effect of Scheduled Faculty Modeling on Clerkship Student Bedside Skills Exposure And Learning". *Neurology* 88(24):e236–e239.

19. (a)Ahmed, M.N., Toor, A.S., O'Neil, K. and Friedland, D. (2017 May-Jun) "Cognitive Computing and the Future of Health Care Cognitive Computing and the Future of Healthcare: The Cognitive Power of IBM Watson has the Potential to Transform Global Personalized Medicine". *IEEE Pulse* 8(3):4–9. doi:10.1109/MPUL.2017.2678098. PMID:28534755. (b) Hamilton, J.G., Genoff Garzon, M., Westerman, J.S., Shuk, E., Hay, J.L., Walters, C., Elkin, E., Bertelsen, C., Cho, J., Daly, B., Gucalp, A., Seidman, A.D., Zauderer, M.G., Epstein, A.S. and Kris, M.G. (2019). ""A Tool, Not a Crutch": Patient Perspectives About IBM Watson for Oncology Trained by Memorial Sloan Kettering". *J. Oncol. Pract.* 15(4):e277–e288. doi:10.1200/JOP.18.00417.

20. Tomas, A., Jones, B. and Leech, C. (2020). "New Insights into Beta-cell GLP-1 Receptor and cAMP Signaling". *J. Mol. Biol.* 432(5):1347–1366. doi:10.1016/j.jmb.2019.08.009. Epub 2019 Aug 22. PMID:31446075.

21. (a) Cornell, S. (2020). "A Review of GLP-1 Receptor Agonists in Type 2 Diabetes: A Focus on the Mechanism of Action of Once-Weekly Agents". *J. Clin. Pharm. Ther.* 45 (Suppl 1):17–27. doi:10.1111/jcpt.13230. PMID:32910490. PMCID:PMC7540167. (b) Nauck, M.A., Quast, D.R., Wefers, J. and Meier, J.J. (2020). "GLP-1 Receptor Agonists in the Treatment of Type 2 Diabetes—State-of-the-art". *Mol. Metab.* 2020 Oct 14:101102. doi:10.1016/j.molmet.2020.101102. Epub ahead of print. PMID:33068776. (c) Brunton, S.A. and Wysham, C.H. (2020). "GLP-1 Receptor Agonists in the Treatment of Type 2 Diabetes: Role and Clinical Experience to Date". *Postgrad. Med.* 132(Suppl 2):3–14. doi:10.1080/00325481.2020.1798099. Epub 2020 Sep 8. PMID:32815454.

22. Baggio, L.L. and Drucker, D.J. (2020 September 25) "Glucagon-like Peptide-1 Receptor Co-agonists for Treating Metabolic Disease". *Mol. Metab.*:101090. doi:10.1016/j.molmet.2020.101090. Epub ahead of print. PMID:32987188.

23. Hassan, A., Sharma Kandel, R., Mishra, R., Gautam, J., Alaref, A. and Jahan, N. (2020). "Diabetes Mellitus and Parkinson's Disease: Shared Pathophysiological Links and Possible Therapeutic Implications". *Cureus* 12(8):e9853.

24. Li, Y., Perry, T., Kindy, M.S., Harvey, B.K., Tweedie, D., Holloway, H.W., Powers, K., Shen, H., Egan, J.M., Sambamurti, K., Brossi, A., Lahiri, D.K., Mattson, M.P., Hoffer, B.J., Wang, Y. and Greig, N.H. (2009). "GLP-1 Receptor Stimulation Preserves Primary Cortical and Dopaminergic Neurons in Cellular and Rodent Models of Stroke and Parkinsonism". *Proc. Natl Acad. Sci. USA* 106(4):1285–1290.

25. Pittas, A.G., Harris, S.S., Eliades, M., Stark, P. and Dawson-Hughes, B. (2009). "Association between Serum Osteocalcin and Markers of Metabolic Phenotype". *J. Clin. Endocrinol. Metab.* 94(3):827–832.

26. (a) DeFronzo, R.A. and Abdul-Ghani, M. (2011). "Type 2 Diabetes Can Be Prevented with Early Pharmacological Intervention". *Diabetes Care* 34(Suppl 2):S202–S209. doi:10.2337/dc11-s221. PMID:21525456. PMCID:PMC3632162. (b) Gastaldelli, A., Gaggini, M. and DeFronzo, R. (2017). "Glucose Kinetics: An Update and Novel Insights into Its Regulation by Glucagon and GLP-1". *Curr. Opin. Clin. Nutr. Metab. Care* 20(4):300–309. doi:10.1097/MCO.0000000000000384. PMID:28463898.

27. Van den Berghe, G., Wouters, P., Weekers, F., Verwaest, C., Bruyninckx, F., Schetz, M., Vlasselaers, D., Ferdinande, P., Lauwers, P., Bouillon, R. (2001). Intensive insulin therapy in critically ill patients. *New England journal of medicine*, *345*(19), 1359–1367.

28. Schrödinger, E. (1992). *What Is Life?: With Mind and Matter and Autobiographical Sketches*. Cambridge University Press. Cambridge, England.

6

Science Seen Through the Lessons of Life

6.1 A Bird's-Eye Overview of the Book's Messages

While the two volumes cover a huge swath of scientific territory, being a practicing physician and caring deeply for my patients, my thoughts are never far from the human being as an emotional as much as a physical entity. This book's primary aim has been to provide a fresh new perspective on biological systems and especially on human physiology. The primary focus has been on metabolism and metabolic diseases, drawing to a large extent on modern concepts in physics and information science. In particular, metabolic processes have been described and elucidated through the lens of synchronization within the framework of quantum metabolism, which explains allometric scaling laws that relate metabolic rate to body weight.

The loss of organizational coherence associated with pathological changes results in accelerated aging and metabolic disease states. I have been particularly fascinated with the complexity and exquisite organizational perfection of the human body studied as a hierarchical coordinated system of systems. This evolutionary organizational achievement has directed me toward the newer areas of computational biology, namely systems biology, big data analytics, and bioinformatics. Systems biology has been inspired by the principles of engineering and hence it describes biological organisms as composed of highly integrated systems of wholes and systems within systems. Bioinformatics systematizes and analyzes huge amounts of biochemical and biological as well as clinical data using the computational power and algorithms developed specifically to handle unimaginably large reams of data points so they can be organized and eventually understood by the scientist. *Some key concepts that physics provided generously to biology include entropy, information, free energy, fitness landscape, and symmetry breaking. They allow for the introduction of organizing frameworks and inference of principles that govern biological processes.* These principles include entropy reduction, information gain, and free energy minimization, which biological organisms exploit in order to not only survive but thrive and successfully compete for the limited resources available in an environmental niche.

On the other hand, medicine is concerned with the issues of optimum health and disease. Linking these two concepts logically and evaluating these states quantitatively calls for a view of human physiology based on the theory of phase transitions and critical phenomena, which is ideally suited to describe a change of one stable state into another taking place in a complex dynamical system like the human body. This branch of physics, i.e. the physics of phase transitions, has been translated across various fields to applications in sociology, economics, finance, and ecology, to name but a few examples of its powerful reach. Surprisingly, medicine has not yet embraced the concept that a transformation from the state of health to disease can be understood as a phase transition occurring in a living system. This observation alone leads to the introduction of important implications, in particular the *identification of control parameters, order parameters, response functions, and susceptibilities.* These quantitative measures of any physical state and its stability status can be readily translated into physiology and medicine on a patient-specific basis. If properly implemented within clinical practice, this would be of great importance not only to diagnostics but also for therapy design and modification, if necessary, in order to improve it.

SIDEBAR 1: A NOVEL PROPOSAL FOR PERSONALIZED PRECISION MEDICINE

It is important to appreciate the fact that the state of health or disease is a dynamical notion and hence requires a quantitative assessment of its stability, i.e. response to an external perturbation. This is at variance with a highly limited assessment of the state of health that typically takes a snapshot in time by performing such tests as blood work or imaging. A snapshot corresponds to a point in phase space at an instant in time, but what is required is a trajectory, i.e. a time course from point A to point B, especially when the system is externally perturbed or stressed. A lack of response, or limited response, to perturbation indicates a robustly stable physiological state, but it could also be a pathological state, for example, drug resistance in a cancer patient. The currently used method of assessing a set of fixed parameters following tests may not be sufficient to address properly the issue of the acuteness of the symptoms of a disease. Rather, the time course of these parameters should be evaluated as a better insight into the disease progression and prognosis. Such quantitative approaches could in the future revolutionize both diagnostics and therapeutics within the precision medicine paradigm that is by definition patient-specific and evolvable over time. This is generally accepted but seldom, if ever, implemented algorithmically in the course of diagnostic and therapeutic procedures.

In order to better understand health and disease and hence improve clinical outcomes, I have proposed and promoted in this book the foundational notion of the physiological fitness landscape as applied to medicine and physiology. This powerful physics-inspired idea allows one to see disease states and disease-free states as parts of a multidimensional manifold spanned by control and order parameter axes and mapped into physiological measures of patient-specific fitness. This would enable physicians to design a therapy regimen that navigates within this complex multidimensional topography in a manner similar to traveling through highly undulated terrain. This is a metaphor for not only a compass and a sheet of paper with a map drawn on it, but also the knowledge of the local elevation, i.e. the location of peaks and valleys. With this type of information, we can find the easiest path from point A (disease) to point B (health) with minimized risk of failure (disease progression or even death). Conversely, one can use it for a more accurate prediction of disease progression, especially in the case of incurable or drug-resistant pathologies. The metaphor of landscape navigation applies to personalized therapy optimization and calls for a choice of treatment that favors "mountain passes" and valleys over risk-prone mountain ridges and peaks that cannot be conquered. It also informs the physician of insurmountable obstacles ahead and a most likely outcome of the natural progression of the disease within the patient-specific context.

In general, as can be readily appreciated from the contents of this book, I have been interested in promoting an interdisciplinary approach to medicine making full use of advanced concepts developed in physics, computer science, and information science as well as mathematics over the past few decades. Indeed, this book is an intellectual tour de force that challenges the reader to view biological systems and medicine as a very complex field sitting at the intersection of not only biology and physiology but also philosophy, physics, biochemistry, and material science. Yes, this is an ambitious undertaking that I have been pursuing for a number of years with substantial personal, financial, and professional sacrifice. As a practicing and committed physician, my main obligation is to the patients and their families, as well as to my own family. However, by writing this book and taking a lot of time away from my primary obligations I hope to advance the field of medicine and affect future generations of physicians and hence their patients in a positive if not transformative way. I also believe in the idea of life-long education and I have never been satisfied to rest on my laurels. While it may take a long time for these ideas to be accepted and implemented in the practice of medicine, I am optimistic that they will bear concrete fruit in terms of better clinical outcomes. As has been well documented in the work of Kuhn [1], every scientific and technological revolution starts with an idea that is usually at first rejected, then doubted, then gradually accepted, and finally, universally embraced as obvious.

This text has taken us through several basic science chapters that are of direct importance to the objective of drawing insights and inspiration from physics to better understand medicine and physiology. I have started this journey by presenting key aspects of thermodynamics such as work, heat, entropy, information, and finally, ideas found in the theory of phase transitions (order and control parameters) that directly impact our understanding of transitions from health to disease (and hopefully back). The chapter on the motor proteins and the machinery of life provides an overview of the molecular-level processes that are essential for sustaining life. They are all fueled by the universal currency of biological energy, ATP. One of the consequences of this seemingly simple observation is discussed at length in the chapter on quantum biology, a nascent but rapidly growing field at the intersection of biology and physics that promises a scientific revolution in biology and medicine by positing that some of the key processes in living systems are quantum mechanical in nature. I do realize that most people who chose the life sciences as their area of intellectual curiosity and professional development did so at least partly to avoid mathematics and physics. However, as truth-seekers, we cannot ignore an important and transformative perspective that shines the light on a major mystery, which is life itself. The conclusions stemming from this hypothesis are far-reaching, especially for explaining synchronization and coherence on an organism-wide scale.

While the use of quantum mechanics in biology is still being generally met with skepticism, the fact that quantum mechanics explained photosynthesis is of cardinal importance. If plants have evolved to efficiently capture electromagnetic energy by the use of quantum tunneling, it would make no sense to expect that mammalian organisms, which are higher on the evolutionary ladder, would "forget" such a major advance in adaptation. Darwinian evolution theory has taught us that competitive advantages are not only retained but improved upon. Therefore, it is more than reasonable to expect that animals, by the use of oxidative phosphorylation in their mitochondria, would do so with the help of quantum physics, especially in the electron transfer chain processes. I have described in detail how it's done and how quantum mechanics explains the allometric scaling laws of metabolism.

The following chapter in this book has delved into a number of modern physics topics applied to biology that resulted in the birth of a new field of science called systems biology. It includes topics such as abstract but powerful concepts of chaos, fractals, and nonlinear dynamics. The selection of these topics was not dictated by the ambition to impress but motivated to provide the reader with an easy-to-grasp overview of powerful modern advances in the physical sciences that offer major insights into the way living systems operate. *Numerous organs in the body have been shown to have a self-similar or fractal structure.* For example, the liver, the blood vessels in the lungs, or the placenta are all fractal objects. This geometrical feature of human anatomy results in a very profound physiological consequence such as the allometric scaling laws of metabolism or the effects of drug elimination by the liver in terms of fractal pharmacokinetics. Likewise, *chaotic dynamics have been found to be a critical aspect of both healthy brain and heart dynamics.* Therefore, ignoring such powerful mathematical paradigms imperils our understanding of human physiology and hence our ability to diagnose and treat a range of diseases from atrial fibrillation all the way to schizophrenia and epilepsy.

The laws of thermodynamics provide rules for the transformation of mass and energy from one form to another. The second law of thermodynamics is unique in physics because it explains the origin of the arrow of time, which is the basis for the gradual deterioration of all physical matter including biological systems. Biological cycles of time define the reversibility of biological processes, but over time they, too, degrade and are superimposed with the arrow of time that gives rise to aging and the inevitability of biological death. *Biology has evolved specifically to protect the entropy reduction achieved by the biochemical processes and subcellular structures forming the molecular machinery of the cell perfected by living systems.* Molecular motors are ubiquitous microscopic metaphors for machine engines. They are driven by ATP molecules using the first law of thermodynamics whereby the energy contained in their phosphate bonds is used to perform work, just like petroleum is converted into the mechanical work of an internal combustion engine.

In both cases, mass motion in the form of inter- and intracellular transport is generated by motors, with a remarkable design similarity between molecular motors and mechanical engines based on the same rotor and stator geometry. *The connection between electricity, magnetism, and energy generation is another parallel invoked in the electron transport chain of mitochondria that may be based on quantum electromagnetism* underpinning the phenomenon of quantum metabolism. It should not be overlooked that quantum biological processes are very likely the keys that can unlock the mysteries of cognition positioned at the center of quantum biological effects. Quantum biology offers us the best hope we have today for solving the mind–body problem. Other areas of quantum biology include magneto-reception in the retina of the eye of migratory birds and possibly humans, the mechanisms of olfaction, vision, and possibly the efficiency of protein folding processes.

A living system can be better understood, as we proposed in this book, by the extrapolation of the theory of special relativity with the following caveats: 1) the system of organized complexity is transient; 2) its peak complexity with organizational and functional perfection is ultimately lost; and 3) the loss of this organizational complexity accelerates, defining the loss of time and progressive speeding up of the aging process leading to a diminished healthspan and lifespan.

The most fundamental cycle of a living system is the one defining the metabolic rate, which is measured as joules or kilocalories in molecules of ATP produced per unit time. As stated throughout this book, metabolism is what distinguishes living from nonliving systems. In the quantum regime, metabolic rate is more efficient than in the classical regime in the sense that more ATP is produced per unit substrate per unit time. Critically, however, this *biological energy currency in the form of ATP is produced in a correlated fashion across the physiological system.* This is done in such a way that the coherent and spatially synchronized nature of the system is translated into the work of maintaining biological and physiological homeostasis. By spanning a large volume and hence a large mass of the system, this metabolic synchronization is consistent with the allometric scaling laws of physiology. As a result, larger and organizationally more complex organisms

such as humans have higher metabolic efficiency in the quantum regime of energy production than the constituent parts of the same organism such as tissues, organs, and cells. Therefore, the organizational complexity of the human body leads to greater efficiency than that which would be extrapolated from individual cells comprising the human body, hence fractional and not linear scaling laws emerge for metabolism. This means that *multicellular organisms are an improvement on unicellular organisms and multi-organ organisms are a further improvement on multi-cellular organisms.* This is not only due to the sheer use of the "economy of scales" for greater energy-transduction efficiency but also because of organ and tissue specialization. A parallel trend in achieving greater and greater organizational perfection can easily be drawn for human societies as they evolved from hunter-gatherer types to agrarian types to largely urban to nation-states and eventually to highly interconnected trading blocks, etc.

In physical systems, temperature variations strongly affect their properties. Biological systems mostly operate at a fixed and highly controlled (physiological) temperature. It has been argued by Lloyd Demetrius [2] that instead, time plays an analogous role to temperature in biological systems. However, biological or physiological time should be understood in a relative, not absolute sense. When time dilates, more biological energy in terms of the currency of ATP is produced relative to oxygen consumption compared to the case when it would be constrained. In the latter case, when time accelerates (as in aging), energy is lost to heat due to entropy production, and hence energy transformation is less efficient both in the sense of the percentage of ATP conversion from energy substrates and in terms of it being utilized in a synchronized and coherent manner to do the work of maintaining cellular homeostasis and therefore physiology. *The greater amount of work produced in a given duration of time, in a quantum sense, parallels the correlated nature of energy production over a larger volume of the system.* As demonstrated in the publications of Demetrius and his collaborators, this process follows a fractional power law that reflects the mechanical foundation of energy production in living systems. In terms of physiology, it is important to underscore a potential and even likely connection between the organism-wide coherence and the synchronicity of the quantum metabolism of energy production as the most basic biological cycle, which under normal conditions is correlated across the organism's metabolic physiology that, in turn, is driven by circadian molecular clocks on a macroscopic hierarchical scale.

Biological age is a relative quantity. Unfortunately, most of us do not live the maximum human lifespan but a shortened lifespan, which is biologically correlated to the "oldest" vital organ system, the most time-accelerated tissue. For example, somebody can be in generally good health with the exception of accelerated cognitive decline and dementia. Conversely, an individual can have absolute mental clarity and generally good health with the exception of advanced cardiovascular disease and, consequently, may have experienced multiple myocardial infarctions and have left ventricular dysfunction and a history of coronary artery bypass surgery. In these two examples, the biologically oldest physiological system is the brain and the cardiovascular system, respectively. They are the metaphorical weak links preventing the perfect synchronization of whole-body physiology including metabolism. *The biological age of many tissues may approximate relatively closely the chronological age to a much greater extent than diseased tissue, which has a far more divergent biological age compared to the chronological age of the individual. The faster the relative pace of aging, the greater the chronologic time.*

I proposed in the discussion that we should attempt to translate the two laws of thermodynamics to clinical healthcare. As a consequence, we can infer that we do not often live to the maximum human lifespan but instead our biological age approximates well below the limit of 120 years, with varying degrees of divergence from the actual chronological age. Why is it that very few people are fortunate enough to have such a long lifespan, and those who do, typically have a long healthspan until divergence between chronological and biological age accelerates as maximum human lifespan is approached? There are a number of factors affecting our aging rate and they are discussed at length in this book. *Chronic stress, poor nutrition, lack of exercise, poor sleep patterns, and lack of social connections, among other factors, strongly accelerate our rate of aging.* Related to this is systemic chronic inflammation, which exceeds the normal healthy physiological thermogenesis. Alternatively, people that have a shorter healthspan and lifespan have an earlier divergence between biological and chronological age, which may be particularly accentuated in any given tissue. *Genetic predisposition often determines the weakest link in our vulnerability to chronic disease and typically reflects a shortening of both the healthspan and lifespan. As biological aging accelerates, the fundamental nature of time accordingly is consistent with the shift from quantum to classical mode of metabolic energy production as the set of cycles that are fundamental to living systems drastically becomes compromised in terms of its synchronicity and connectivity with adaptive biological and physiological behavior.*

Foundationally, at a more microscopic hierarchical scale both temporally and spatially, the coherence of biological ATP energy production and the correlated synchronously coordinated endogenous biological clocks lose their organizational level of reduced entropy and symmetry. Accordingly, as the flow of energy declines between interacting elements of biological systems, the entropy production rate not only increases but accelerates and correlates with redox disturbances and aggressive modifications of cellular constituents that also correlate with disturbances to acid-base balance. Thus, movement away from these basic parameters of homeostasis defined by Gibbs free energy, redox, and acid-base balance signals the acceleration of the pace of aging. Ultimately this gives rise to the onset of clinical disease states and results in mortality, the state at which metabolism ceases to exist and the second law of thermodynamics takes over.

The notion of emergence refers to nonlinear interactions between the parts of a system that evolve as a bottom-up process to a new system whole, which is unpredictable by any linear extrapolation of the parts. Intriguingly, the modern disciplines of quantum physics and biophysics describe linear systems in the sense of there being an entanglement or superposition of the parts of a system as singular without separate interacting parts. That is, the linearity is rooted in the notion

that the "parts" of the system are inextricably part and parcel of the same entity. Quantum biology in this sense is rooted in quantum mechanics. Therefore, it represents an exception to the nonlinearity property of biological systems.

Furthermore, what is truly profound regarding human and other living systems is how the metabolic cycle in the quantum mode of energy production slows the passage of time by maintaining the exquisite biochemical organizational complexity of the body. *Quantum metabolism promotes a lower metabolic rate but a higher metabolic efficiency than classical metabolism,* which can be accounted for on the basis of three processes or properties, all of which have to take place. First, there is a greater quantity of ATP production per unit time per unit substrate as a result of the greater efficiency of oxidative phosphorylation versus glycolysis. Second, there is maximum efficiency in the quantum versus the classical range of oxidative phosphorylation. In this range, there is the lowest level of associated oxidative and inflammatory stress and therefore a reduced entropy production rate (see the concept of "take-over threshold" in Chapter 3). Third, quantum metabolism has a lower ATP requirement per unit of body mass because the correlated nature of energy production in maintaining biological and physiological homeostasis is more efficient. Because it requires less ATP production, and hence less mobilization of energy from adipose and glycogen stores, quantum metabolism requires less cognitive and physical effort for biological and physiological processes involved in housekeeping activities.

This explains the easy fatigability that occurs with aging coinciding with the quantum mode of energy production becoming increasingly compromised. Accordingly, physical and cognitive requirements of voluntary effort are increased in order to overcome the metaphorical friction that accompanies the higher entropy production rate. This may be thought of in terms of collisions between molecules that are intrinsic to the heat production and inflammatory processes. The acceleration of aging accompanies the impairment of redox and free energy homeostasis that results in the loss of biological complexity (see Chapter 3). Similarly, many metabolic diseases, including cancer, can be better understood in terms of the loss of quantum metabolism mode of energy production leading to inefficiency and heat generation (e.g. via inflammation). The slowing of time in terms of sensory perceptions and behavior as conscious experiences intuitively must link to the scientifically established notion of quantum metabolism. However, the fully mechanistic elucidation of these processes remains incomplete. While consciousness, or cognitions, as quantum phenomena remain controversial, the proposed scientific and intuitive foundations are compelling, and moreover, they are difficult to explain by currently available alternative theories. It is somewhat puzzling to this author that virtually no presently proposed theories of consciousness or cognition discuss metabolism as a foundational aspect of the functioning of the substrate for consciousness, the human brain, whose metabolic energy is unparalleled.

Even if this is not experimentally proven yet, in the author's opinion *biological clocks may be manifestations of quantum phenomena in biology. In particular, cryptochrome may be directly responsible for this due to its light sensitivity such that a single photon (a quantum of electromagnetic energy) can change the conformational state of this molecule*. This is very similar to bacteriorhodopsin, a proton pump, which operates by being activated by a single photon absorption event as long as the photon corresponds to blue light, i.e. has sufficiently high energy to trigger the conformational change required. Compelling evidence is being accumulated that many molecular biology processes, especially highly efficient ones, involve quantum mechanisms at least partially in gaining their evolutionary advantage.

The arrow of time defined by the unidirectional movement of heat and particles toward greater randomness (and lower temperature) correlated with dissipation or entropy generation, appears to be temporarily violated by the negative entropy generation (or information creation) of organized complexity within living systems as a result of work required to maintain their structure and function. This results in slowing down the arrow of time by capturing energy as usable heat in the chemical reactions and bonds that form the living system. *The energy that is captured to do the work of maintaining the far-from-equilibrium state of homeostasis of the cells, and hence the organism, effectively slows time by the generation of cycles.* Cyclical periodicity in theory returns the system to a previous point that is time-invariant, equating it to the time crystals mentioned above. Taken together, *living systems are defined by metabolic activity in the context of cycles and cycles within cycles whereby the metabolic cycle is the most fundamental temporal scale of the living system.* It gives rise to many higher-level hierarchical scales of cycles in time with the implicit purpose of resisting the arrow of time. This prolongs the time duration of the high-information negative entropy state of organized complexity and strengthens the living state. It is important to state that from the point of view of a biological organism time is defined by the duration of a cycle. The shortest cycle is a metabolic turnover time to produce a single molecule of ATP. Longer cycles are multiples of this unit just like a minute has 60 seconds and an hour 60 minutes. What biologists and biochemists prefer, instead of using these natural cycle times, is to express the number of products obtained in a unit of time that is natural to the measuring device, for example, moles per second. In this manner, we represent processes in terms of convenience to us rather than relevance to biological systems.

We have dedicated almost an entire chapter to advancing the *Physiological Fitness Model, which is an adaptation of similar concepts developed in the physics of phase transitions*. The simplicity and power of this methodology have led to Nobel Prizes in physics and the model has already shown to be applicable to physiology, although inexplicably has not yet been utilized for clinical medicine. This model has explosive potential for major implications in the field of clinical medicine. The concept of time and its application to biology is innately relevant to the pace of aging, to human health and metabolic disease, and by extension to all chronic diseases of aging. Accordingly, the topic of circadian rhythms, molecular clocks, and nuclear receptors occupies a prominent place in this book. It is worth noting that in 2017 the Nobel Prize in Medicine and Physiology was awarded to Hall, Young, and Rosbash for their pioneering discovery of molecular clocks, which gives a timely boost to the topic we discuss here. I also

hope that this could generate particular interest due to the timeliness of molecular clocks (pun not intended!).

A final necessary point of this discussion is that, although the emphasis in this book has been placed on the broadening of robust scientific data, the ultimate interface with patients in clinical medicine first and foremost requires a non-algorithmic intuitive interaction that understands an individual's fears, expectations, biases, belief systems, etc. Although nowadays physicians face grave and relatively imminent risk of losing our profession to computers, artificial intelligence algorithms, and technicians, it should be stressed that *personal interactions between patients and physicians have a fundamental and powerful therapeutic value in society*. It is quite possible to integrate what is missing from the optimal execution of the medical profession without disregard for, or losing sight of, the critical role of highly trained and expert clinicians inextricable from the relationships and therapeutic experience with patients.

In the following sections, I would like to share some anecdotes from my own life and professional medical experience that illustrate the various aspects presented in this book at a scientific level so the reader can relate the sometimes arcane topics discussed in this book to everyday experiences. My own life is an example of these intersecting spheres of human experience and I would like to share some glimpses into both my professional and personal life with you, the reader. Moreover, the purpose behind these relatable stories is to extend the application of ideas of science to broader scales of human physiology and behavior. These applications can be invoked as a guide to priorities, to succeeding in various endeavors, and to better experiencing the joy of living.

The following anecdotes also represent extrapolations from the lessons of metabolism that are more representative of the message from Volume 1. This message can be viewed as highlighting the human potential from a metabolic perspective, mainly characterizing the core or intrinsic years of youth, typically the late teens and 20s (for sports, music, health, and the notion of free will having the greatest capacity to self-actualize). These stories can be viewed as life lessons that resonate with the scientific insights showcased in this volume but also as a bridge to the biology and medicine focus provided in Volume 2 of this book.

6.2 Anecdotes and Their Morals

6.2.1 Anecdote 1: Football Teams

The lessons that can be learned from the Somerville High School football team that had an abysmal 0–10 season, not once, but during two consecutive seasons, are both tangible and almost cliché, as well as abstract and nothing short of astonishing. While some of the players moved on from high school and other players were new to the varsity team, there remained an entanglement of spirit that is profoundly illustrated in the fellowship off the field and a level of play on the field together. This experience was rooted in a single state of mind of this amazing team at that unique period of time. The outcome was a combined regular-season record of 19–1 in the subsequent two years and an improbable New Jersey state division championship victory! The tangible lessons learned by the players, their families, and the fans are that hard work pays off, that team sports teach athletes to work together, that mistakes are a powerful learning tool, and that the pain of having made those mistakes builds resilience that is part of the process of self-improvement. This carries over to connectedness, networking, and solidarity that bridges success in life and immensely enhances the enjoyment of the ride. This is because the vitalizing nature of goal-oriented behavior in sync with others and with one's own core values of free will and willpower attains those goals. The abstract lessons learned from this story further underscore mind–body phenomena of extraordinarily stable synchronized states of behavior. Each individual team player, and the composite formed by each player becoming an element of the greater system whole, represent single systems as superposition states, which are coupled in a phase-coherent manner forming a strongly entangled collective state.

The language used above is borrowed from quantum mechanics, as often these amazing examples of human achievement border on unexplainable by mechanistic rules in the sense of classical Newtonian mechanics, which is defined by nonlinear and sometimes non-local relationships. The single-system states are linear in the sense that there are no interactions and hence no potential conflicts, or trajectory changes, between the goal state and its attainment. The goal is superposed and focused among the component players of the team like a laser beam, the accomplishment predictable in the sense that it is entangled with the state of free will or goal state even before the goal itself is realized. What provides a powerful elucidation of this effect is the theory of special relativity, which holds that time as an absolute quantity does not exist, and accordingly, the past, present, and future occur simultaneously in a four-dimensional space–time continuum. This is the temporal component of entanglement, which is a property of the singularity among the team players and which enables their composite ambition to reach its collective fulfillment.

A more general idea is that reminiscence, referring, for example, to music we have heard or events in the past, could be a quantum phenomenon in the sense of time invariance. This means that the forward and backward propagation through time is occurring simultaneously in our minds, forming a quantum superposition of states. Perhaps this is also an effect taking place in the minds of performing artists and athletes who, when in the so-called "zone", create a coherent superposition of states in their minds that is composed of both past events and current situations to reinforce the desired effect in the (near) future. These champion athletes literally visualize their future success in their mind and they seem to force the desired outcome of the infinite range of possible scenarios for events. An appropriate analogy would be a coherent in-phase superposition of electromagnetic states in a laser, which leads to a powerful monochromatic beam as opposed to a random superposition of out-of-phase frequencies from a visible spectrum of electromagnetic radiation forming white light. White light can illuminate an object but does not have nearly the powerful energetic effect of laser light. White light is composed of all the possible colors in the spectrum while laser light selects only one color, just like a champion athlete is able to "snatch

victory out of the jaws of defeat", to invoke a well-known quote attributed to the US Representative James Seddon.

Linearity as a characteristic of the quantum state assures that there are no thought processes that stray from a single uniform and correlated perspective and the outcome is proportional to the effort put into it. A frequently employed strategy in football prior to a game-changing field goal is for the opposing team to call a timeout. This gives the kicker time to "think about it". In quantum physics, this "thought" event can be equated to an observation being made, also called measurement, that collapses the quantum state on a classical result. This measurement event is caused by the energy involved in the observation or thought process. On a macroscopic scale, the superposition of cognitions represents a single state of the mind–body system. The collapse of this superposition state results in the loss of coherence and a state of moving toward randomness and disorganization. This is manifested as conflicting states, many different and interfering perspectives resulting from the decoherence of a single state of the mind–body system, or collective mind–body (in the case of a sports team) state of spatiotemporal entanglement. The outcome is often a disconnection of the goal state from its realization. Thus, the outcome becomes unpredictable as the goal state and attainable state are no longer in sync, or in unity with each other. Here, uncontrolled and negative emotions dysregulate cognition rather than developing cognition in a top-down fashion regulating and controlling positive emotion. The latter is the case in a superposition-synchronized state of mind–body and behavior.

The singular collective state of mind–body of the 2017 Somerville High School football team may be aptly described as tantamount to the superposition state of quantum mechanics, such as a coherent state of a laser or a ferromagnet on a macroscopic scale in a physical system. Also tantamount is the manifestation of a macroscopic biological phenomenon as a temporally organized and spatially correlated synchronized state of metabolic circadian rhythms translating into healthy physiology and balanced behavior. The nearly miraculous transformation of the perennial loser team into an unbeatable championship team in the 2017 season is a real-life manifestation of a change from thermally induced decoherence to a self-generated quantum coherent state in a system of strongly interacting particles (football players). We can only speculate how major a role the coach played in this transformation but in either case, physical analogies can easily be found. After all, laser action requires an energy input and, similarly, every winning team receives such input from coaching staff, fans, family, and friends.

Another example of this effect involves a much more well-known football team, namely the New York Giants. The New York Giants were described as the most improbable winners in recent memory against the undefeated New England Patriots in the 2007 Super Bowl match (Figure 6.1). A documentary of the game was recently dedicated to the offensive front line. Each of the five players who protected the quarterback, Eli Manning, and opened holes for the running backs were interviewed throughout the documentary describing their thoughts about key plays in the course of the game. Analogous to the Somerville High School football team as state divisional

FIGURE 6.1 The New York Giants in the 2007 Super Bowl. The most improbable run by a wildcard team had just culminated in winning Super Bowl XLII in what has been widely considered the greatest upset in all of professional sports history. This herculean overachievement exemplifies a mind–body coherence and phase-locking involving the players of a team into a singular whole, capable of out-competing another group of more talented individuals even at the top of their game. This appears to be a macroscopic manifestation of similar interactions of metabolism at the molecular scale, which is responsible for the ultimate perfection of organism-wide physiology. Source: author-supplied photo, containing image used with permission from Getty Images/Win McNamee.

champions discussed above, the New York Giants exemplify at the professional sports level the power of synchronized performance between the individual units, the players, in a superposition coherent state as a team. This correlated synchronization between players is called phase-locking and displays arguably a quantum phenomenon at the macroscopic scale of team sports. The quantum nature of such astonishing displays of athleticism (also exemplified by other areas of human achievement, e.g. music, ballet, theater, etc.) is so extraordinary that it can be equated to the exquisite organizational perfection of living systems themselves.

This book has advocated the usefulness of physics concepts in biology and medicine beyond the traditional applications of mechanics to biomechanics, fluid dynamics to hemodynamics, thermodynamics to bioenergetics, or electrical conductivity to neurophysiology. While quantum biology is gaining recognition in the mainstream of science, Einstein's theory of relativity is very abstract and is commonly thought to be confined to within pure physics, especially astrophysics. The application of the theory of special relativity to biology is proposed in this book in the sense of the relative nature of time that carries explanatory power for the pace of aging. This is particularly pertinent in regard to the effect of relativistic time dilation that this book proposed to apply in the context of biology for the first time. Accordingly, optimal health and metabolic efficiency correlate to maximally dilated time. Moreover, the extension of this concept of time dilation to synchronized world-class qualities among players of a team is compelling.

In the documentary mentioned above, three of the linemen similarly recounted the astonishing perception that time slowed down when important events of the game were at a crescendo. The watershed moment in the game came in the fourth quarter with under one minute left in the game when the ball was somewhere close to mid-field. It was a third down and long with the score of 14–10 in favor of New England. On the play Shaun O'Hara, the team center, said that he looked back and saw Eli "about to be sacked before rising from the ashes, breaking the only tackle he ever broke in his life". Eli then launched the ball about 30 yards down the middle of the field to David Tyree, closely covered by three Patriot defensive backs. Two of the Giants linemen, O'Hara and David Diehl, both independently commented that the ball seemed like it was in the air a very long time. That is, from the relative perspective and perception of these players, time slowed down and they felt as if they were watching events in slow motion. These statements are uncannily reminiscent of the time dilation in the special theory of relativity. Two plays later, with only 30 seconds to go on the clock, Eli hit Plaxico Burress on a pass to the corner of the end zone for the game-winning touchdown (Figure 6.2). Rich Seibert commented that "the ball was in the air so long I felt I could have run down and caught it before Plaxico did". Still, O'Hara commented in a personal communication that 30 seconds for 80 yards was far too much time to put the ball back in the hands of Tom Brady, who is arguably the best quarterback in professional football history. Fortunately for the New York Giants and their fans, the players finished out the game executing it to perfection and sealed the deal for Super Bowl XLII. Both sides of the ball for the Giants had to play a perfect game because given the phenomenal talent assembled on the Patriot team, there was no room for mistakes or dissonance.

There are many intriguing applications of the notion of special relativity to biological systems here. For example, the pace of aging may be dilated/delayed by a phenomenon of quantum

FIGURE 6.2 The Helmet Catch. This stunning catch by David Tyree occurred at a watershed "do or die" moment with 59 seconds remaining in Super Bowl XLII. The NY Giants, considered a hopeless underdog, were down by a score of 14–10 and were on the disappointing precipice of falling short of overcoming almost insuperable odds of defeating the touted single best team in NFL history. A "hail Mary" 13-yard downfield pass by a scrambling Eli Manning on a third and five after breaking perhaps the only tackle of his career was caught at the Patriots' 24-yard line, keeping the Giants' drive alive. Tyree was intensely covered by multiple defenders in the middle of the field, but impossibly managed to maintain control of the ball by trapping it against his helmet while being tackled, virtually defying laws of physics. Source: used with permission from AP/Gene J. Puskar [3].

metabolism versus constrained/accelerated redox stress. Also, there are a host of broader examples spotlighted in this epilogue that hopefully improve the relatability of this mathematical and abstract concept. This particular example of the Helmet Catch captures the notion of free will, described in various places including section 9.6 below. While only demonstrated mathematically as a physical property, that time as an absolute quantity does not exist, consistent with Einstein's theory of special relativity, it is metaphorically intuitive to extend this concept to the biological domain. Thus, the past, present, and future are happening simultaneously across all dimensions of the physical universe. In optimal cases, to some extent, they occur in a special quantum entangled sense, the fabric of four dimensions of living systems within it. Accordingly, it is plausible that free will has the potential to connect the future to the present and past before it even happens.

It is the quantity of time that harnesses the potential and magic of life, and this 2007 Giants–Patriots game epitomized the notion. Time is a defining characteristic of mechanical,

electrical, or biochemical cycles, which are behind the functional organization of living systems. The unit of time describing the periodicity of a cyclical process represents its most important quantity. When the energy of the system over time is not lost, and the periodicity stays the same, time is maximally dilated and hence exists in a quantum sense with theoretically zero entropy production since reversible or reproducible processes generate no entropy within the system and dissipate no heat outside the system.

Moral 1: There is a clear connection between human physiology and human behavior.

6.2.2 Anecdote 2: Synchronization in Music

Special relativity, and especially time dilation, is consistent with the notion that time as an absolute quantity does not exist. This has deep and profound implications for the entanglement of human biology, physiology, and even inter-individual scales of societal interactions. In Chapter 3, examples, which included photosynthesis and mitochondrial electron transfer chain tunneling, were given of relativistic quantum effects at a molecular biology level. However, many macroscopic physiological effects can be also mentioned in this connection as unexplainable, such as the amazing reflex speed of elite athletes' reactions, for example, baseball catchers, tennis players, and basketball players, all of whom at the peak of their performance defy the empirical laws of physiology in terms of reaction time and spatiotemporal coordination. Many of us have also experienced premonitions or strange feelings of a strong conviction that something either very good or very bad was imminent to happen which later on did happen. These are examples of real-life phenomena that defy the laws of causality and the linearity of time, as if at these moments we had access to future events. This type of perception of major future developments can sometimes be seen at a societal scale with the emergence of fatalistic trends in the fine arts or poetry, as was the case in pre–World War II Europe. For example, the decadent trends of nihilism, Dadaism, or existentialism occurred as if these artists were sensing impending doom and catastrophe and had sensory access to future events.

On a personal level, as we grow older and gain a longer time perspective, the lines between distant past, recent past, and present become increasingly blurred and events that took place decades ago seem as fresh as if they happened just yesterday. It seems that we, or at least our minds, reside in a spatiotemporal continuum in which time has lost its linear, sequential character and our mind can travel freely through the multi-dimensional space–time continuum, even being able to have glimpses into the future. This is consistent with the notion that in the quantum world, time disappears and the past, present, and future co-exist. For example, this can have a very powerful effect on an individual as exemplified by the experience of music from a distant past that transports us mentally to that period, creating a state of hyper-reality by superimposing the emotional state of teenage years on those of late middle age. A similar experience can occur when reconnecting with old friends from 40 years ago after many years of not being in touch with one another. Other examples of time seemingly standing still or being able to access events in the past or in the future are discussed below and each of these examples can be better understood in light of the concepts of quantum phenomena and special relativity. In the context of the theme of this book, most relevant are the explosive implications of these seemingly abstract, yet very subjectively relatable concepts to the future of medicine.

Synchronization can be viewed as a hybrid of quantum physics with time dilation allowing an absolute synchronization and coherence and classical physics where frequencies and phases of oscillations are correlated. The latter can be exemplified by music orchestrated by members of a band. The quantum/classical hybrid synchronization is what is proposed in this book for metabolism as both quantum and classical manifestations of energy production. Moreover, extended to the realm of consciousness, synchronization of present sensory experiences with memories of the past generates the feeling of reminiscence. Memories acquired over a lifetime are attached to the associated emotional states, which can be stored in tubulin dimers of microtubules in neurons of the hippocampus as intimated by the Orch OR theory of Hameroff and Penrose (see Figure 6.3 for microtubule structure). New sensory experiences carried in microtubules of the brain's cortex could synchronize with stored memories and manifest as reminiscence, a quantum biophysical phenomenon of human cognition, which could then emerge as a component of the totality of consciousness (definable as an awareness of being aware).

Within the quantum picture of microtubules, referred to by Hameroff as "the quantum underground" [4], the past, present, and future have blurred boundaries and as long as decoherence and wavefunction collapse do not destroy this superposition state, the human mind may be able to freely transcend the artificial boundaries imposed by the linear concept of time. Perhaps this is what happens when mystical or out-of-body experiences are reported and often summarily dismissed by skeptics. This can also provide an explanatory consequence for pathologies such as Alzheimer's disease where memory is gradually lost, perhaps as a result of degradation of stored quantum/classical states in neuronal microtubules. Just like the loss of quantum coherence is linked in our metabolism to the increasing energetic inefficiency leading to insulin resistance and then to diabetes and related co-morbidities, an analogous deterioration in cognitive decline may be associated with the quantum decoherence taking place in an aging brain. A gradually increased frequency of "senior moments" and cognitive decline may then lead to dementia and Alzheimer's disease, which robs the person of more than health, the composite

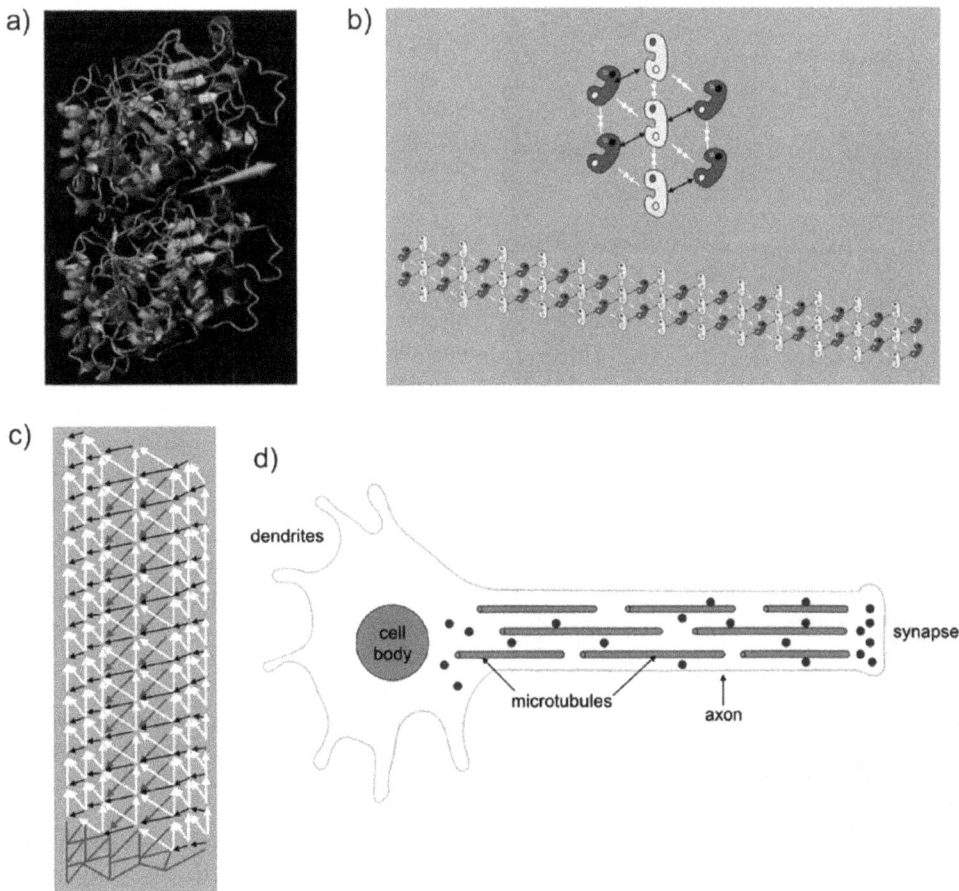

FIGURE 6.3 a) The structure of a tubulin dimer with its dipole moment, as shown by the arrow on the right. b) A hexagonal neighborhood of tubulin dimers in a microtubule lattice. c) The tubulin lattice comprises an entire microtubule. d) An array of microtubules as part of the axon in a neuronal cell.

record of one's life, past present, and future. It now appears that mitochondrial dysfunction plays a central role in all diseases of aging, both metabolic and neurodegenerative.

Like sports in the US and in many countries of the world, music is another pastime endemic in modern societies. Just as enrapturing world-class athletic talent reaches its peak at a young age, typically in the third decade of life, the same can be said of musicians. In fact, music imparts a therapeutic richness to the human body and soul. For example, although it can be argued that classical music or opera may be more sophisticated than the hard rock genre typified by the music of Led Zeppelin, the intoxicating impact of the synchronized performance of rock musicians rivals that of any genre of music. This British band, with its stardom reaching the zenith in the 1970s, was the incarnation of solidarity between the four band members allowing the product of their music to be greater than the sum of the individual contributions. That is, the band as an ensemble, in its effusively expressive totality, provided a rhapsodic listening experience for many people of that generation who loved their music. Although today, several decades later, time has taken its toll on the looks of these former rock stars, in its heyday Led Zeppelin, particularly Robert Plant and Jimmy Page, largely defined the essence of the youthful concept of "cool" of that era (Figure 6.4).

FIGURE 6.4 Led Zeppelin. The androgynous youthfulness coupled with the astonishing synchronization of a band's musical talent may be an example of the same laws that underpin the beautiful organizational perfection of human physiology in the state of optimum health.

The orchestra-like performance of these four amazing musicians was so rich and in sync that it had an ethereal and intangible quality that became entangled with the sensory cortex and consciousness of its listeners, I included. This can be said of all great musical bands who are capable of lifting the mood of their listeners and sometimes providing transformative power to the fans' lives. However, special relativity application to the rate of aging is also insightful in this case of aging of the rock stars mentioned above (and many other people, of course), which may be at least partly due to the lifestyle choices made. As elaborated earlier in the book, both the metabolic rate and efficiency in the state of optimum health exhibit the maximum level of the quantum mode of energy production. Metabolism is a hybrid composed of classical and quantum modes of energy production, but over a lifespan, there is a progressive loss of the quantum metabolism component and a deterioration in the quality of classical metabolism as mitochondrial dysfunction progresses. Chronically unhealthy psychogenic stressors (both in quantity and quality), circadian behaviors, and dietary consumption (quantity and quality) are the major contributors to altering the diversity and compositional pattern of microbiota, amplifying the magnitude and chronicity of the stress response and disturbing circadian synchronized physiology. The inextricable nature of this web of pathogenic parameters fundamentally underpins metabolic disease and the accelerated pace of aging.

The fundamental principle that I believe defines the impact of the listener's sensory experience of a performing musician is the notion of quantum entanglement. Entanglement in this sense has an uplifting and hence therapeutic effect by amplifying an aspect of the person's core belief or emotions. Whether this is only a metaphor of the quantum entanglement effect or a real macroscopic manifestation of the weirdness of quantum mechanics, only time will tell. In my personal opinion, the therapeutic nature of "musical entanglement" is rooted in the potential to synchronize emotion, which I propose to be a quantum effect that reaches deep down into the brain's structure and, usually in a very positive way, affects its function.

Within the only molecular-level theory of consciousness thus far, the direction of polarity of dipole moments of tubulin dimers comprising neuronal microtubules of the brain is at the basis of the mechanism that mediates the human sensory experience. Can this be directly connected to quantum entanglements between one person's brain and that of a perfectly synchronized player on a winning team? If so, perhaps the exalting experience of listening to mood-changing music that transports us in our mind to years past, or accidentally reconnecting with a friend from a bygone era accidentally, may generate a quantum superposition state in our mind, which feels perfectly real, with the past and present co-existing if only for a split second. At the next hierarchical level, we could contemplate creating such a quantum superposition state in the collective "mind" of a society where remains of the past zeitgeist co-exist with the present collective unconscious. For example, the British may still feel the past glory of the British Empire as a real persistent trait of their society while simultaneously experiencing themselves as a modern member of the European family of nations. This is not just a bit of nostalgia but a quantum superposition of two different collective states of social

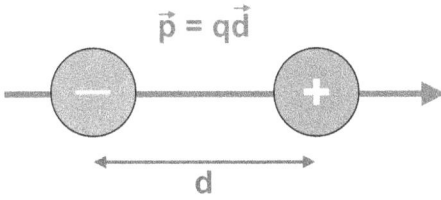

FIGURE 6.5   A dipole moment is the product of the magnitude of electric charge and the distance between a positive and negative charge forming it.

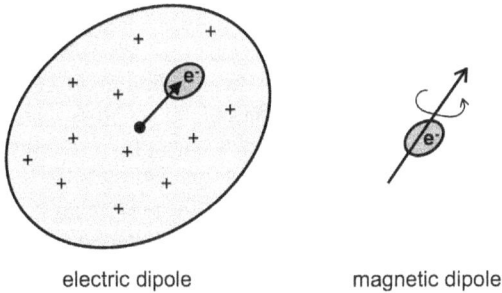

electric dipole   magnetic dipole

FIGURE 6.6   Electric and magnetic dipoles. An electric dipole occurs when a positive and a negative electric charge are separated by a distance in space. A magnetic dipole requires the movement of electric charges around an axis in space.

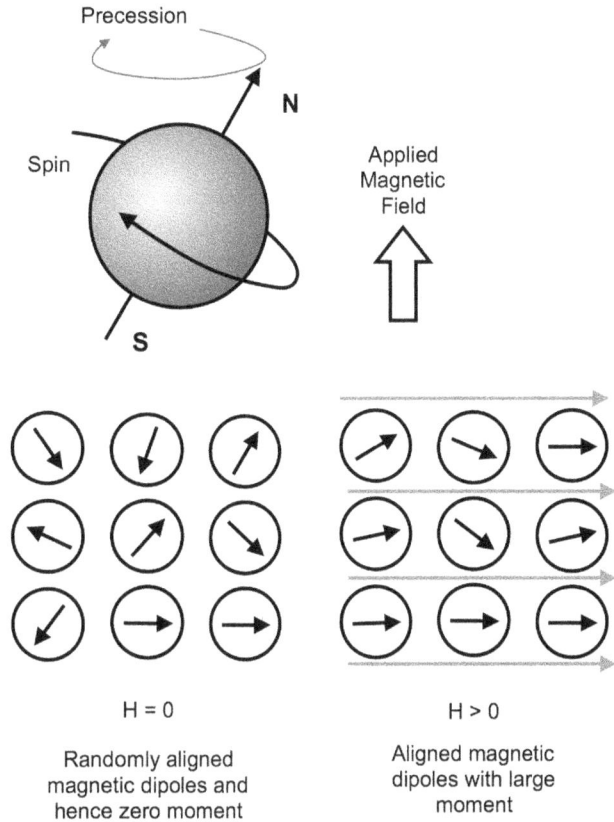

H = 0   H > 0

Randomly aligned   Aligned magnetic
magnetic dipoles and   dipoles with large
hence zero moment   moment

FIGURE 6.7   Illustration of the concept of a magnetic moment (spin) and its alignment with applied magnetic fields. In the panels below this is extended to ensembles of spins, which can be oriented randomly (left) or aligned with a field (right). In the latter case, this gives rise to the net magnetization of the substance.

"state of mind". Technical details behind this theory of consciousness can be found in the chapter on quantum biology and in the references provided therein.

Suffice it to say here that a dipole moment is the product of the magnitude of charge and the distance between a positive and negative charge forming it (Figure 6.5). The biophysical phenomenon of consciousness can be viewed as analogous to a macroscopic scale of quantum events that occur in physical systems such as the laser pointer (see Section 4.6 "Can Objections to Quantum Biology Be Overcome?") or a magnet (see Section 4.15 "Biological Motors"). The magneton is the basic building block of the larger magnet and the analogue to the electric dipole (Figure 6.6). The electric dipole, also represented in quantum mechanics, has been invoked in some theories of consciousness as the basic unit that is involved in conscious thought or perception, cognition. The Bohr magneton, on the other hand, is the basic unit of the overall magnetization responsible for the magnetic force and it consists of a single spin 1/2 produced by an unpaired electron in the outermost shell of an atom or a free electron. Protons and neutrons also have a spin 1/2 unit.

Magnetism is an example of a macroscopic quantum phenomenon. The force produced by the magneton's magnetic moment is analogous to a dipole moment of a pair of opposite electric charges (Figure 6.7). The magneton moment refers to the force created by the direction of spin of the electron whereas the dipole moment refers to the force generated by the location of the electron itself within the shell rotating around the central nucleus of an atom and is calculated with respect to the positive charge of the nucleus. Consider the direction of electron spin of all the tiny magnetons of a magnet in a paramagnetic state. They are randomly oriented in many directions

and, as a result, there is no net influence on surrounding magnetic fields or electric currents. In a ferromagnetic state, all (or a large number) the magnetons are aligned along the same direction creating a macroscopically observable magnetic state, which is sometimes referred to as a broken-symmetry state.

The broken symmetry here refers to a preferred direction of magnetization, which is present in a ferromagnetic state and absent in a paramagnetic state. Individual players on a winning team, due to their perfect synchronization, generate a powerful force much like laser light or a magnetization state, which is much more powerful than white light or a paramagnetic state, respectively. Within each player, there must also be perfect synchronization as much as synchronization between the players of the team, so there are two levels of organization in place to reach a championship state of mind within the team. This is similar to cells being organized into a tissue or an organ and tissues and organs forming a healthy and strong organism. The unity of purpose and organizational synchrony represents a quantum phenomenon of superposition, all contributing components of a larger system occupying the same quantum state at the same time. This is exactly what occurs with the electromagnetic waves generated in a laser pointer, namely a macroscopic coherent quantum state emerges (see Section 4.4 "Decoherence").

Can the conscious state of the human mind be analogous to the coherent state of a magnet or a laser? The human race has existed on this planet for a mere 100 or perhaps 200 thousand years and through its inventiveness produced a new state of matter, namely coherent light of the laser. Why would not Mother Nature, with its 4.5 billion years of experimentation on planet Earth with countless life forms and their replicas subjected to evolutionary pressures, arrive at an even more impressive solution, namely the quantum mind? The composite of all the correlated dipole moments of a microtubule in neurons in the brain could be that coherent quantum state that can generate cognitions, or consciousness, responsible for behavior on the scale of a human mind (and body to which the mind is entangled). Whether this hypothesis is correct or not, time will only tell. However, in the opinion of this author, it is not unreasonable to consider it. Better yet, experimental methods should be researched and tests carried out on the underlying assumptions of this theory, namely the presence of quantum states in biological systems such as neurons.

Every year on Memorial Day weekend, an i95 radio station takes votes from its listeners to rate the top 500 rock songs of all time. The 100 top songs are the best of the best and include songs composed and performed by the Beatles, the Rolling Stones, Billy Joel, Motown, Lynyrd Skynyrd, the Doors, Eric Clapton, Michael Jackson, Stevie Wonder, and so on. Virtually every single year almost without exception for almost 50 years since its release in 1971, "Stairway to Heaven" was voted number one. It was serendipitous or perhaps intentional that Jimmy Page said the song "Stairway to Heaven" "crystallized" [5] the essence of Led Zeppelin. A crystal lattice itself displays the mysterious behavior of a quantum system that is in a superposition state in both space and time. Accordingly, the atomic elements of a crystal lattice repeat themselves not only with perfect spatial symmetry but time symmetry as well. That is, the arrangement of atoms in a crystal may change over time, but they do so with a temporal periodicity such that they return to their starting positions with clock-work precision due to the vibrational waves known as phonons that propagate through the crystal. Therefore, in essence, these atoms do not change over the course of time. This is referred to as time invariance. A system with spatial and temporal symmetry looks the same independent of the observer's orientation of perspective in time or space. The entanglement of the musical product in the collective consciousness of its listeners underscores the quantum nature of its effect that does not change with time.

Time invariance of certain core values in human society may also exemplify a quantum entanglement of consciousness. If a living system were to display such fidelity with respect to quantum behavior, it would be tantamount to a mechanically frictionless perpetual motion machine whereby energy would not escape the system as unusable heat. Instead, it could then be utilized to do the work of maintaining the far-from-equilibrium state of biological homeostasis at a cellular level and hence the physiological homeostasis of the organism. The song "Stairway to Heaven" can thus be seen as a symbolic stairway to a state of permanence, to perfection, and to a singular state, the quantum state of coherence. This also manifests synchronization and an entangled superposition state, which was metaphorically described in the case of top athleticism

and musical mastery. In such cases, there are no conflicting perspectives, "when all is one and one is all". Robert Plant stated in the course of an interview he gave that "It feels like time gallops! Once upon a time we made Led Zeppelin 1 [the first Led Zeppelin album] in 36 hours. Now time just gallops, moves at the blink of an eye". The relatively small amount of energy that was once required to do the work of a given project now requires a much greater energetic input (and therefore effort) because such a large percentage of the energy is lost to heat that is unavailable to do work. The perception of time was so much slower back in the day because, in the quantum sense, energetic input was correlated and thus effort was not required to overcome the metaphorical friction at a bio-molecular level and hence there was little or no heat loss and entropy production. Entropy production rate defines the rate of energy lost from the system, which is unusable to do work and is consistent with the theory of special relativity. We proposed in this book that entropy production rate parallels the acceleration of time, the pace of aging, and the perception of the loss of time in conscious human beings. In the younger quantum state, the entropy production rate is only minimal, and therefore the passage of time that accompanies the production of work appears to be shorter. It is not surprising therefore that the 1970s was literally yesterday in the minds of people who grew up in that era.

Thus, aging per se erodes this potential biologically, both from the perspective of human health and disease, as well as performance in sports and arts. The notion of free will, an intangible force connecting the present with the past and the future, whereby the future can become the present and the past before it actually happens (by the notion of free will). Again, here is another application of special relativity from an intuitive sense. Thus, it often has the capacity to affect an outcome, which defines accomplishment reductionism as an established scientific methodology that analyzes a physical problem by breaking it up into constituent parts and investigating the smaller pieces independently, thereby simplifying the problem. In the physical sciences reductionism is the paragon methodology that can be traced all the way back to Sir Isaac Newton in the 1600s who single-handedly developed the principles of mechanics (as well as other fields of science). He postulated his laws of motion, the second of which states that force equals mass times acceleration ($F = ma$), whereby if two of these three parameters are known, the third can be predicted. Thus, reductionism may be ideally applied to linear systems by using equations like this to predict the evolution of the system or to extrapolate the system to a larger scale to make appropriate estimates by simple rescaling. Reductionism was the crown jewel of 20th-century science that has proven to be foundational for analytical deductive reasoning and methodology in studying physical, chemical, and then biological systems with applications to medicine. However, living systems are composed of component parts across many hierarchical scales, vastly separated by orders of magnitude on both the time and length scales, each with its own laws or rules of interaction. Hence, the trajectory of behavior across these disciplinary planes changes.

The scales range from the atomic level (or even smaller) in physics to molecular level in chemistry, macromolecular in biochemistry, molecular aggregate in molecular biology and cell biology, and ultimately macroscopic in physiology and behavioral sciences. Accordingly, the assimilation of these various scales of scientific inquiry in a living system is unpredictable because it involves many processes, under the general banner of "symmetry breaking", that represent the formation of new symmetries and the elimination of old ones. Crossing these boundaries via symmetry-breaking phenomena is a nonlinear phenomenon par excellence and hence it poses major challenges to our time-tested methodologies. Fundamentally, this underscores the limitations of modern-day scientific methodology and it has major consequences for not only medical research but also for clinical medicine. In the latter case, we do not even have our own laws, only empirical observations and inductive reasoning at our disposal. Therefore, effectual and safe clinical patient care must be precision-based, applicable to an individual's parameters rather than broadly applied to populations based on statistical averages as we currently do. The hallmarks of a living system such as a human being are complexity of interactions and inherent unpredictability over time.

Returning to metaphorical examples from everyday life, the astounding and engaging display of talented and synchronized entertainment by athletes, team sports, soloist musicians, and musical bands that approximate perfection may serve to exemplify the concept of singularity. This appears to occur within individuals and also between them at macroscopic scales of human behavior and may even highlight the quantum phenomena of superposition of states including "phase-locking" connecting individuals of the collective whole of a musical band or a competitive sports team as discussed earlier. Full development of the human brain takes at least two decades to occur. Between peak development and the onset of senescence, which typically begins to occur about the age of 30, there exists astonishing and exquisite organizational perfection in maximum dilation of time that accompanies the greatest potential for its translation into the gob-smacking entertainment served in the examples of Led Zeppelin and the New York Giants described above. Additionally, it can be posited that these can be seen as extensions of the temporally organized and spatially correlated synchronized states of metabolic circadian rhythms, which translate into physiology and behavior. However, physiology alone, without recourse to psychology and the power of the human mind is insufficient to explain some anomalous examples and experiences as will become evident in the next anecdote.

Moral 2: Synchrony, entanglement, and special relativity can be found in unexpected places.

6.2.3 Anecdote 3: The Power of Placebos

A clinical study in Europe reported the outcome of a treatment given to cancer patients who were told that they would likely experience alopecia due to the effects of the drug. However, the study was designed in such a way that only half of the enrolled patients received the drug while the other half of the patients were given a placebo. It turned out that approximately a third of the placebo-given patients developed alopecia, which of course is astonishing because they were not exposed to the drug, which causes alopecia.

Perhaps an even more interesting example involves a patient who presented with a type of blood cancer, a form of lymphoma characterized by a "softball-sized" conglomerative tumor. It was concluded on the diagnosis that the patient had only days left to live. The patient stated that his doctor told him that a certain drug could cure the disease. However, the doctor knew that there was nothing to disclaim and that the drug would be useless in treating this type of cancer. Nonetheless, having great empathy for the patient, his doctor decided to administer him the drug. Amazingly, within a few days, the patient experienced miraculous recovery with a rapid shrinking of the tumor. Years later the patient was determined to be completely free of clinical manifestations of cancer. However, he came across a report, which stated this drug was ineffective in treating his type of cancer. The patient then became depressed and dispirited, which led to a recurrence of the clinical cancer state a very short time thereafter and presented again with severe manifestations of the same type of cancer as before.

He then went back to see his doctor. His doctor, having witnessed the powerful impact of the placebo given to the patient, decided to deny this report's veracity and tell the patient that this report was completely false. Moreover, the doctor insisted that the drug had indeed been demonstrated to be exceptionally efficacious. In addition, the doctor told the patients that he had access to a new and improved formulation of the drug. He then proceeded to give his patient intravenous saline, which was of course a pharmacologically benign placebo. Again, the same scenario ensued and the patient once again experienced another remarkable remission of the cancer state. Years later the patient once more came across and read another article reporting how the drug clearly does not work, which was followed again by a recurrence of his cancer. The disease ultimately did take the patient's life. This story speaks volumes about the power of placebos and the power of the human mind over the human body.

One important theme here is the robustness of interconnections at all scales of living systems that determine heath via the physiological fitness function. This links the intracellular molecular scale to inter-cellular and inter-organ/tissue scales of an individual or an organism, all the way to inter-individual/organismic social scales or to even phenomena of entanglement that span ranges independent of spatial and temporal locality. The notion of superposition that defines unity in a quantum sense must phenomenologically underpin the exquisite organizational perfection of complexity in the infinite wisdom of nature. This book seeks to highlight this from a scientific perspective. However, humans are unique in the sense of their high level of consciousness whereby successful endeavors require connections between individuals quite analogous to what occurs at the intracellular scale. Furthermore, human social behavior involves networking needed to achieve a shared desired outcome.

When activities are properly orchestrated, these social networks resemble a phenomenon occurring in less-evolved species, which exhibit an increased level of collective intelligence as they interact. For example, a colony of ants forming a bridge, a school of fish swimming in unison, or a flock of birds flying together as one can achieve much more than the individual animals within the colony when their collective effort is needed for the common goal of survival. This may also be equated to team sports where team members coordinate their behavior both at a conscious and subconscious level. In humans, the element of consciousness is a layer that underscores the idea of the uniqueness of individuals. This is a powerful concept connecting human behavior to the definition of a successful life, one that fulfills the physiological meaning of purpose.

The notion of purpose is found at the intersection of philosophy and material science. Philosophy addresses the issue of the meaning of life while material science investigates the substrates that provide the physical context for living systems and their processes. In this sense, both emotional and physical health are inextricably linked to the notion of stress. Exposure to bearable stress in any living system builds resilience. The uniqueness of individuals includes the areas where handling challenges has variable degrees of resilience based on experiences and talents or affinities, or lack thereof. There is no greater feeling of empowerment than the capacity to control the outcome of a challenge that benefits oneself or others. The ability to help others elevates oneself at least in proportion to the extent that it helps the other individual(s). This is rooted in the sense of the value of being part of a greater whole and serves as a mechanism for building the connectedness of that whole at a conscious level. This in turn provides the expectations for reciprocity of help in the realm of challenges whereby others have greater resilience and capacity for meeting the demands of those challenges than us.

This type of connectedness engages the vitalizing potential of stress whereby the dose and quality of stressors can be dissipated among a network of individuals making it less difficult to bear by each individual. This type of support system is analogous to mechanical supports such as columns and pillars that allow an enormous force of gravity exerted by a massive skyscraper to be subdivided into these support structures preventing the building from collapsing onto itself. In the context of human interactions, this support network acts in reciprocal directions and can empower the individuals and further strengthen the network as a functioning whole. The meaning of conscious purpose can be better understood when one considers the notion of free will and the willpower to strive for those outcomes that are consistent with one's identity, core values, and inner sense of purpose. This is perhaps a fundamental basis for why we live and is the underpinning factor behind mind–body phenomena, that is, the link between mental and physical health.

Conversely, at all levels of isolation when interconnections are broken leading to a loss of complexity of living systems, humans and others, within and between organisms, devitalizes the physical nature of life. In these cases, the life force succumbs to the physical inevitability of the second law of thermodynamics. Invoking this perspective is critical to achieving and prolonging the fulfilling nature of the human living experience. It is proposed that the brain evolved for the purpose of the acquisition of food. Teleologically speaking, at the dawn of time we humans were awakened by a negative

energy balance, hungry and foraging for food. However, we were not the largest or the fastest animal in the jungle, nor the only ones foraging for food. To avoid being prey and becoming the next meal for a predator, the stress response must have evolved in a manner that is closely linked to the neuronal and neuroendocrine system responsible for energy balance, satiety, and hunger systems in the brain. Indeed, that is exactly how the hypothalamus is hardwired into the stress and appetite regulatory centers of the brain. Dealing with stress due to life-threatening danger lurking around the corner on a daily basis may have also stimulated the formation of social groups that provided protection and shared life-sustaining resources. The importance of human interactions and inter-connectedness is an important aspect of human health. We are social creatures and the state of our health strongly depends on how closely related we are to our fellow human beings. Moreover, it is not necessary to be either hugely popular or have hundreds of real or "Facebook" friends.

Sometimes a single individual can have an enormously positive influence on our lives, as is briefly illustrated in the following anecdote. This story can also be seen as relating to the role of another modern physics effect, namely quantum entanglement that was mentioned earlier in this chapter in a different context. In this case, it is the emotional entanglement in people's lives that makes a true difference to our well-being.

Moral 3: Mind can win over matter and hence humans are not bioreactors.

6.2.4 Anecdote 4: Human Interconnectedness

The first story I would like to share involves an unexpected encounter with Gregg Allman of the Allman Brothers Band, of whom I am a huge fan. I noticed him on a flight I was on from Baton Rouge to Newark. He was seated in the first-class cabin, while I was back in economy. Nevertheless, I decided to walk up to him from the back of the plane to introduce myself as a big fan. The seat next to him was empty and he told me to sit down. He was drinking shots of Seagram's Seven, which he offered me, although I unfortunately felt awkward and declined. We were talking about the classic song "Whipping Post", and a song from a new album at the time which he liked, "Queen of Hearts". It was an incredibly fun and unexpected experience! Before I left the seat, he pulled out a promotional picture of himself and signed it "To Brian … And to the Good Times!" (Figure 6.8). My impression of him was that he is one of the nicest, kindest men you could ever meet.

When I met Gregg Allman, it was at the end of another long day and week, in addition to the grind of traveling. I was tired and in an unexcited mood during my final year in fellowship training at Tulane University. I was 34 years old, still with plenty on my shoulders before I could "start my life". Both the Internal Medicine and Endocrinology Boards lay ahead as well as starting a practice, which I had no experience doing. Gregg Allman was the frontman of the Allman Brothers, who were a major part of the musical influence that began in the 1960s. In a strong sense, this band was part of the pantheon

GREGG ALLMAN

FIGURE 6.8 Signed photograph by Gregg Allman. The joys and successes in life that we all experience in our unique ways are part of an interwoven fabric of humanity. A central aspect of this social entanglement is not only to reach one's own maximum potential of achievement but to connect it to the enrichment or betterment of the lives of others. This results in an intrinsic sense of awe that a lone quest for stardom cannot give.

in the culture of the arguably most expressive generation in human history. This was the baby boomer generation, the most defined and revolutionary generation of any before and any since. I was in grade school when I began hearing the Allman Brothers, influenced by my brother, who is five years older than I am. By the time I was in high school, I was enthralled by their music. Gregg Allman was close to the beginning of this generation, and I was near the end of it. Gregg was larger than life. His voice and his band's music elevated people's moods all over the world and for many years. It was a phenomenon. An invitation by him to sit and join in personal conversation transformed my state of mind. Instantly, I went from feeling dull and apathetic to having great excitement that lasted the entire day. A grinding trip had become electrifying and fun. Where did that burst of energy come from? It came from a powerful and real connection with another person who happened to be one of my favorite musicians of all time.

Moreover, the above story speaks to something that may seem indirect, but that is foundational to the nature of the stress response and even essential to achieving the joys and successes out of life. The stress response across many perspectives and all scales of human biology is a major theme of both volumes of this book, with deep implications to the future of medicine. The experiences that build resilience and evolve personal skill sets, coupled with the heterogeneity of genetic talents, give each of us value, to ourselves and for others. This notion underscores an often-unconscious drive to better us; not only does it lift our potential and feeling of accomplishment, but nothing ennobles our own psyche more than elevating the opportunities or mood of others. This in my opinion puts a finger on the pulse of a reciprocally positive vitalizing experience of human connection. The feeling of awe meeting Gregg Allman represents perhaps a microcosm of what we all strive to do for others, albeit we all do it in our uniquely different contexts.

The next story is about the important positive influence a physical education aide had on my son, who has cerebral palsy. This person who could at first glance be viewed as only peripherally relevant from the perspective of my son, Matthew, played a major role in his sense of happiness and belonging at his school. This kind and unassuming gentleman unexpectedly died recently after he had just turned 54. My son had a great affection for him and vice versa. He helped Matthew from the age of five through eighth grade. He worked hard with my son to have him walk up part of the distance to receive his diploma in eighth grade. He set up props to lean on and worked on and practiced for months. This culminated with a tear-jerking five-minute standing ovation at the time Matthew received his diploma. They kept in touch over the years and recently Matthew told him he was graduating from Raritan Valley Community College this year. The gentleman replied how proud he was of Matthew's accomplishments. The gentleman's wife and daughter texted my son that he always had a big smile on his face whenever he spoke of him, that he was so inspired by Matthew. At the gentleman's wake, there were thousands of people showing their respects. This shows how small acts of kindness can build up and result in a powerful legacy affecting individuals and groups alike. When people in this world interact in a positive and encouraging way with others, their impact can be very profound and long-lasting. We become entangled with people by simple acts of kindness that we may underestimate at the time but those affected may keep warm memories for the rest of their lives and in turn affect others in a positive way. This is particularly true from the perspective of people like my son who love other people and only see the good in others in spite of or perhaps even because of having disadvantages in life. It is people like this gentleman who bring true happiness to strangers in mysterious ways.

The above story is also linked with the next short anecdote as it involves my son Matthew who is pictured below on his walker when he was still in middle school. In this picture we see him meeting a pitcher for the New York Mets, Al Leiter (Figure 6.9). Al was a starting pitcher in the Major Leagues for over 19 seasons, which included playing for the Yankees, Blue Jays, and Mets. This heart-warming photo is another example of an important connection between two strangers that vitalizes that human psyche, not just for the child, in this case, but also for the adult. Not surprisingly, Al Leiter, a true gentleman and a class act on the pitch and off the pitch, has won nearly every philanthropic award the MLB has to offer. There is something fundamentally human and indubitably superlative and self-elevating yourself by helping others. Not only does one strengthen the person on the receiving end of help and encouragement but this is returned in many mysterious ways. These imponderable explanations for the colossal subconscious reward and motivations for acts of human kindness hint at an underlying biophysical quantum nature to this realm of connectedness.

In school, we are exposed to learning not just in the classroom, but outside of it as well. Character is learned behavior through instruction, experience, and examples of role models, but it depends on the willingness to learn. Learning in the classroom is easier when there is a strong connection between the student and the teacher. This, like all forms of human

FIGURE 6.9 Matthew meeting Al Leiter and his wife. The power of connection is rooted in biophysical phenomena that range in scale from smaller than a human cell to beyond the boundaries of a single person. Few things ennoble the psyche more than lifting another person's mood. This picture speaks the proverbial million words about bringing joy to another person and expecting nothing in return but getting back the greatest gift of all: making another human happy. Who in this picture gives and who receives joy? Source: author-supplied photo.

connection, is bidirectionally motivated. Several years ago, my wife and I went to an especially memorable parent–teacher meeting at our son's school. We introduced ourselves to one of the teachers, and when we told her that Matthew was our son, she literally lit up! She recounted a story when she was having a particularly bad day with her students who were looking disinterested, nodding off, or even sleeping in class. When she looked at Matthew, she explained how he had a beaming smile and gave her a thumbs-up gesture. It instantly changed her mood, and a healthy and lasting "team" spirit, which was spontaneously formed at that moment. This is no more than metaphorical to what is seen at all hierarchical scales of interactions and signaling that underpins healthy human physiology. Moreover, it represents a deeply interwoven parameter of human health which synchronizes all organs and tissues of a living system.

On the field, the power of team sports, whether you are a player, a coach, staff member, or indeed just a fan, bonds are innately and subconsciously formed that sometimes last a lifetime. These attachments are predicated on sacrifice, common purpose, solidarity, and mutual respect. Moreover, sacrificial behaviors are typically instinctive and subliminal, impelled by personal traits, the joy of it, and the desire to be part of a "team".

Matthew's physical disability has never compromised his passion for sports. He had competed every year from the ages of 7 to 22 at the state and national level for the Paralympic Sports Club, Children's Lightning Wheels, sponsored by Children's Specialized Hospital. He always routes, loudly cheers for, and cares mightily for his teammates. Many of these kids were less fortunate, with greater physical disabilities than his. Matthew's acute awareness of performing at a personal best level has been more important than winning. Reaching his maximum potential was the paramount motivation to lifting his spirits and the spirits of his teammates, family, and friends.

When not competing himself, he lives vicariously through those who are. In high school, one of his greatest joys was being part of the football roster as the team's statistician. He shared awesome pleasure when the team won but felt a deep pain that would cause him to cry when they lost. His distress in fact encouraged him during sophomore year to ask the coach if it would be allowed for him to give the players a half-time speech. The coach supported his request and later praised Matthew for his "rousing pep rallies". At the football awards dinner at the end of the season that year, all graduating seniors were asked who in their lives inspired them the most; two of them stated Matthew Fertig. The following year, he was awarded the "12th man award", the most inspiring person on the roster. This award was newly conceived with the idea that it sets a precedent for other deserving students to be honored this way in future years.

Matthew's connection to his teachers and Lightning Wheels teammates, which was strikingly similar to that with his high school football team companions, followed the same pattern as at the highest level of team sports. An example that springs to mind is that of the players of the New York Giants football team. The magic of the connection in each case manifests its reciprocal and feed-forward self-amplifying nature. In the case of Matthew's friendly encounter with the NY Giants, his attraction was potentiated to some extent, but not entirely, by a celebrity presence. More significant for him has always been an innate desire to translate experiences in life to forming new "teams" wherever and whenever possible.

Sports provides both literal and metaphorical endogenous shots of adrenaline at many levels of connectedness, from the engagement as a competing athlete in the sport itself, to being a team member, to the fellowship and camaraderie amongst fans, and also to the less common but even more heart-warming examples of friendly, elegant and respectful behavior involving fans of opposing teams. Having a personal rapport with players on a professional sport's team is another "team" in and of itself that gives a sense of "skin in the game" and higher pulses of both adrenaline (excitement) and dopamine (reward). The biggest reward is when the endogenous changes in physiology are bidirectionally amplifying, as can be sensed and appreciated from some of the pictures below.

Finally, it should be underscored that the human connectedness that occurs between someone like Matthew and any professional athlete, like Shaun O'Hara and Tiki Barber, pictured in Figure 6.10, speaks volumes about the athlete's class and character. Their kindness is motivated by the intrinsic pleasure of lifting the soul of others. It is also worth saying that Matthew lived the first three months of his life fighting for his life on a mechanical ventilator. The upregulated autonomic and hormonal branches of his stress response during this early critical period created a lifelong "hair trigger alert" and reduced tolerance to stress. It has long been my opinion that Matthew's visceral sensitivities to others and his affinity for bonding with others are inherently biological. Social connectedness is interwoven into the physiological cloth that protects the stress response. The NY Giants went on to win two amazing Super Bowls in 2007 and 2011 when Matthew was 10 and 14 years old, respectively, but his connection to the team will last his lifetime. It is appropriate to quote in this connection the words of Sir Isaac Newton: "if I can see far, it's because I'm standing on the shoulders of Giants". While Newton was referring to different "Giants", the situation he described in this quote parallels the ones related above.

Moral 4: Nobody is an island; social networks can make us much stronger.

A team comes in many forms. There is an implicit beauty in the reciprocal sharing of organic self-amplifying circles of joy, sometimes sadness but always meaning and purpose. The hallmark of a team is the greater strength and resilience of the collective than the sum of its parts. Success takes many directions.

6.2.5 Anecdote 5: A 40-Year-Old[1] Professional Athlete

How can a 40-year-old professional athlete compete with 20- to 30-year-olds at their peak physical and mental development? As argued in this book, special relativity when applied to biology indicates a gradual slowing down of bio-energetic metabolism that underpins an acceleration of the androgynous aging process beyond age 25. This implies a marked difference in physiological performance strongly favoring the younger athletes. Furthermore, there may be psychological disadvantages in the older athletes due to the awareness on their part of the physiological deterioration. The more rapidly aging individual will be more acutely aware of the process than an individual whose experience is that of a still improving athlete. This invokes Einstein's theory of special relativity, which argues that time does not exist as an absolute quantity, and when the rate of passage of time is greater than that experienced by those around you, it feels that the opposite is true. Time and thus aging are captured in metabolic rate and hence the faster that rate, the slower the aging. The 40-year-old professional athlete who has not lost a step has an optimal stress response with the interwoven entire gestalt of extrinsic and intrinsic parameters of human health. These parameters include stress resilience from psychogenic to molecular levels of redox, microbiota composition and diversity, goal-oriented behaviors including diet, and synchronized circadian endogenous oscillating clocks that together underpin a metabolism that is the quantum manifestation of energy production. This 40-year-old athlete is entering his eighth Super Bowl appearance at the time of the writing of this epilogue. This is not even closely approximated by any other athlete in any type of sport in American history, surpassing such greats as Michael Jordan in basketball, Wayne Gretzky in hockey, and Babe Ruth in baseball. Tom Brady has already won five of the seven Super Bowls he played in, with the only losses, both of them, coming against Eli Manning and the NY Giants who were overwhelming underdogs each time. This almost superhuman performance of these great athletes exemplifies the quantum regime of metabolism that corresponds to an aging process that has not lost its internal synchronization of clocks within clocks but, equally importantly, is still firmly in step with the natural circadian rhythm of the Earth cycling around its own axis.

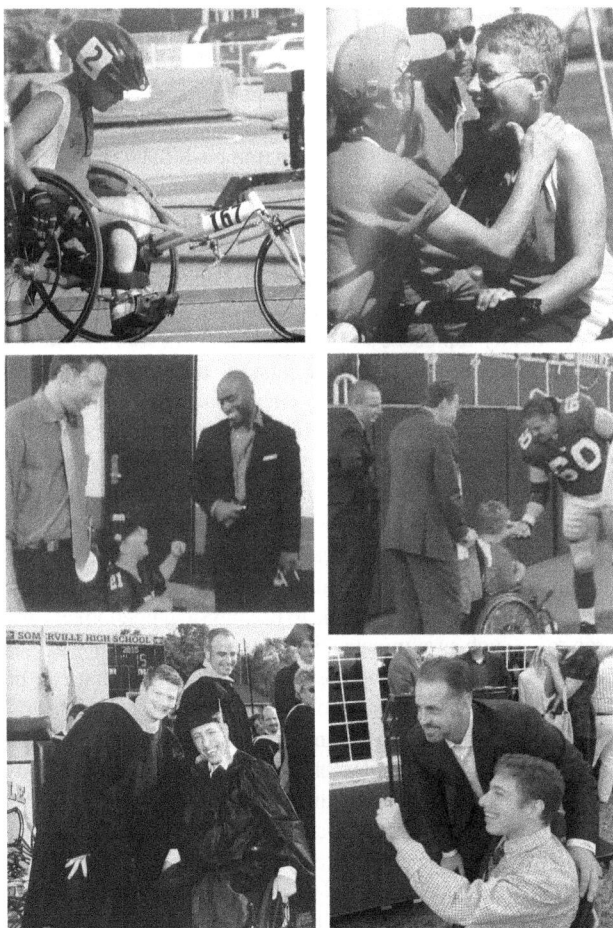

FIGURE 6.10 Matthew competing in Lightning Wheels, meeting members of the New York Giants (Tiki Barber behind the scenes as well as Shaun O'Hara and Amani Toomer on the field), graduating from high school, and wearing Shaun O'Hara's Super Bowl ring. These moments throughout Matthew's life illustrate the importance of human connectedness and its powerful impact on mental and physiological health. Source: author-supplied photos.

Moral 5: Mind can win over matter and information is in a constant struggle with entropy.

6.3 The Essence of This Book's Message

Synchronized energy production in the human body is described by quantum metabolism driven by coherent endogenous clock-controlled gene outputs. This mechanism and its malfunctions are linked to aging and metabolic diseases of aging by including the concepts of time/cycles, stress response, and diet/microbiota as control parameters that mediate disease. They crucially initiate pathology via molecular mechanisms that drive inflammation and oxidative modifications. Each control parameter perturbs the others in a vicious circle of cause-effect amplification due to the existence of interlocking feedback loops. Importantly, disturbances in these control parameters accelerate the pace of aging as a function of impaired metabolic rate and efficiency, and the inextricably entangled compromise of redox and free energy homeostasis. Prolonged stress response alters circadian timing, diet, and microbiota and is nonlinearly compounded to the detriment of health. The neuroendocrine and autonomic nervous systems mediate allostasis but if prolonged, cause an allostatic overload that disturbs homeostasis. metabolic disease is a continuum, which *is* the aging process. The chronic diseases of aging are essentially the unavoidable or ineluctable effects of disturbances in redox and energy forms of metabolic homeostasis. Thus, aging per se erodes this potential biologically, both from the perspective of human health and disease, as well as performance in sports and arts as briefly discussed in the anecdotes above. The notion of free will, an intangible force connecting the present with the past and the future, should not be overlooked since the future can become the present and the past before it actually happens as is one of the lessons of quantum mechanics and special relativity, which defies our intuition. Thus, it often has the capacity to affect the outcome of our efforts and enables us to accomplish the (seemingly) impossible.

In order to simplify the complexity of the problem, in this book, we have proposed the concept of physiological fitness landscape where control parameters are represented graphically as horizontal axes and the vertical axis maps the fitness function of the person measured in response to them. Valleys in the resultant stress response plots correspond to stability regions but peaks and ridges delineate boundaries between neighboring regions. These different areas of relative stability refer to optimal health, a pre-diabetic state, and advanced disease, respectively. Almost all chronic diseases of aging share the same control parameters, hence a common framework should be developed instead of a fragmented approach. Three critical aspects on the periphery of modern medical interventions: chronophysiology, microbiota, and prolonged stress, are entering the mainstream of medical research thanks to advances in understanding how they affect our health. Their inclusion within the physiological fitness landscape methodology offers a practical approach to optimal solutions for healthy aging and precision-medicine therapies for age-related diseases.

6.4 Understanding Biology and Medicine through the Lens of Physics

As exemplified above by the use of concepts borrowed from physics and other exact sciences, we can better understand the complexities of the human body, in particular, special relativity and time dilation. These concepts depend on the speed of the observer's frame of reference in physics and have been described in this book as having great relevance to human physiology and to the issue of aging in particular. This tells us that time is not an absolute quantity but depends on the frame of reference, and in the case of physiology, this frame of reference entangles human biology and physiology and connects the individual constituent systems across the scales of time (cycles) and space (sizes). Moreover, quantum effects, seen to be more and more important to cell biology with examples such as photosynthesis, mitochondrial electron transfer chain, and the molecular mechanisms of human sensory receptors, are profoundly illuminating both physiological and psychological phenomena such as exceptional feats of athleticism, intellectual brilliance, and artistic performance, to name but a few examples. Furthermore, inter-individual interactions as would be characteristic of championship football teams or musical ensembles, leading to amazing synchronization of action and purpose as described above can be seen as manifestations of synchronization, phase-locking, or perhaps even quantum entanglement and coherence. Each of these examples and scales of applications of physical concepts in biology can simplistically represent an independent application. However, it is the conviction of this writer that the concepts of quantum mechanics and special relativity cut across scales, indeed serve as interlocking connections between scales and introduce the missing ingredient of organizational synchronization in biology. It is expected that translating these concepts into the field of medicine can lead to explosive advances in the future.

The above praise of physical concepts is not meant to suggest that physics should be blindly implemented in the context of biology. What we propose is to use physical phenomena and properties such as symmetry breaking or susceptibility first as qualitative metaphors and then as characteristics that can be quantified when defined in biological and physiological terms. The ultimate advantage of using physics to understand biological systems in the complex environment of the human body is to improve predictability and outcome. One way of doing this is by using computerized algorithms that invoke principles of control and order parameters of the system being studied. In the former case, they can be either extrinsic (external) or intrinsic (internal) in relation to the system studied. The goals in this regard are six-fold as is briefly outlined in the following.

We need to acquire as much relevant biomedical data about an individual patient as possible, optimally to the extent that a computerized algorithm is capable of ingesting, assimilating, and processing this information. One should also be cautioned against information overload. Sometimes, too much data can lead to an overdetermined model that is not very useful at all.

In other cases, too much information can end up creating noise and obfuscating the problem. In still other situations, spurious patterns seen as correlation do not indicate causation.

Ideally, the computer algorithm developed on the basis of the physiological fitness landscape model must also be fed the totality of published and unpublished relevant knowledge worldwide and be updated regularly. This can lead to the refinement of the underlying model and its parameters. A first step in model development is to reconcile its results with the existing data, i.e. make it retrospectively correct. The second and decisive step in model development is to make predictions and validate them with empirical outcomes of future measurements, i.e. make the model prospectively correct.

An often ignored or discarded aspect of medical studies is the information that can be extracted from failed clinical trials or studies that came to wrong conclusions based on correct data. Such negligence of information can occur when patients are not properly stratified into cohorts with uniform criteria. These studies can still be useful and the data can be repurposed to provide valuable information.

The computer algorithm to be developed for the above purpose must be able to align the generic information from population-wide studies for personalized medicinal application to the case of a single patient. What this practically means is that the so-called normal blood test results obtained as average values within the standard deviation for a broad cross-section of the population are informative guideposts. However, for a particular patient, it is equally (or more important) to follow the trajectory of these data points over time as the disease develops or as pharmacological interventions occur.

This algorithm should be used both for diagnostic and therapeutic purposes, and in the latter case the multidimensional dynamic (time-dependent) physiological fitness landscape should be employed. The fitness landscape maps a trajectory of the multidimensional state of health of the individual, which ought to be recalibrated over time intervals following continuous data collection. Achieving a stable state of recovery can be then seen based on the trajectory traced in this multidimensional space.

The present-day, seemingly punitive, bedside medical records should one day have the capacity to meet these high standards of computerized big data collection and data analytics affordably and effectively. Our information and computer technology are advancing at a break-neck speed and the biomedical sector is ready for a revolution.

Having stated the goals for algorithmic approaches to medicine, we should balance this with a counterpoint. The other ultimate advantage to invoking physics to understand states of human health and disease involves not the computer brain but rather the human mind. The greater the number of perspectives for understanding a single concept, the deeper the insight and the higher the potential for problem solving. As a clinical physician, I and all of my colleagues in this noble profession are essentially engineers, applied scientists for whom all that is available are the perspectives we were taught. While these perspectives have a wonderful value, biological complexity is daunting and the clarity to see the trajectory in the course of the susceptibility state of a disease or the outcome of their clinical intervention is limited. Just like a person who

has greater empathy and ability to understand the viewpoints of others is capable of stronger relationships, a clinician with broader and more diverse perspectives, and hence robustness of understanding any given concept, will have a better ability to apply that concept innovatively and creatively to find solutions to problems related to health and disease.

6.5 Calming Words of Advice for the Patient

This book, while primarily aimed at my fellow clinical physicians and medical researchers, can also be found useful by the general public. The message for the educated public, stripped of all the scientific jargon, can be boiled down to the following simple advice. There are three primary extrinsic control parameters that affect either positively or negatively our state of health depending on their status. These are:

Diet: the quantity, quality, but also timing of our diet enormously affect our state of health and, when persistently neglected, can lead to serious disease states. As argued throughout this book metabolic rate and efficiency that in the state of optimum health has the maximum level of QM as a hybrid of classical and quantum metabolic modes of energy production, but over a lifespan, there is a progressive loss of the quantum metabolism component and a deterioration in the quality of classical metabolism as mitochondrial dysfunction progresses.

Chronobiology: circadian behavior, which means synchronizing our molecular clocks with the external rhythms of light and seasons. Most importantly, the quantity and quality of our sleep patterns as well as respecting the cyclicity of natural processes in our lifestyle ensures a solid base for maintaining a long health- and lifespan.

Stress management: our stress response involves not only dealing with toxic stress and vitalizing stress but also tolerable stress. In each of the three cases, oversimplification and inattention to nuance can lead to either over-reaction or trivialization. For example, shift work may at first glance appear to be unhealthy but it turns out that it is so only when the schedule alternates. Likewise, stress is an individual factor and when is perceived by the person as manageable, even seen from the outside as dangerously high, may in fact result in no detrimental effects on the person experiencing it.

Chronically unhealthy psychogenic stressors (in terms of both quantity and quality), circadian behaviors, and dietary consumption (quantity and quality) are the major contributors to altering the diversity and compositional pattern of microbiota, amplifying the magnitude and chronicity of the stress response and disturbing circadian synchronized physiology. The inextricable nature of this web of pathogenic parameters fundamentally underpins metabolic disease and the accelerated pace of aging.

Therefore, the key obligation of a family physician is to delve into these underlying root problems for many chronic diseases as well as susceptibility to acute disease states. Conversely, patients should be able to understand how lifestyle changes, especially in the major categories related to the three extrinsic parameters listed above, can lead to disease states, and hence the healing process should begin by mitigating these aspects first. This, of course, is easier said than done. What may stand in the way is discussed below and it relates to our free will and willpower.

6.6 A Few Words about Free Will

Free will is who we are. It is the essence of our individuality. It defines our potential for action, expression, and choice selection. We are free to determine our place in the world, in other words, we are not a deterministic epiphenomenon. No one tells us to stand up or to sit down. We ultimately only tell ourselves this or any measure of interacting with the world that dictates our buoyancy, serenity, or success. Self-doubt is self-defeating and causes a loss of connections to the core values that define our free will. This can be viewed as a quantum process, as invoked in various philosophical discourses on consciousness. It may perhaps involve a wave function of the brain generated from the tubulin states of neuronal microtubules. This wave function carries a potential that has many superimposed possible outcomes simultaneously translatable into physical outcomes. As discussed elsewhere in this book, one powerful result of such a hypothesis could involve an elaborately and coherently orchestrated manifestation of physiology that determines a behavioral outcome. Those superimposed wave functions eventually collapse into a single reality that may or may not be the desired outcome of free will but an unintended consequence of prior actions. As amply demonstrated by quantum physicists, the mere act of observing a quantum wave function causes its decoherence, or collapse. The observer may or may not be part of the system, hence a potential issue with the term "free" in free will. In terms of quantum computing technology, this causes practical problems. The massive size of a quantum computer is required to accommodate the cryogenic requirements for its operations. Even a vanishingly small amount of heat of the infrared rays generated by its observation alone is enough to collapse the quantum wave.

Intriguingly, a thought itself is a form of observation. The content of the thought reflects the biasing field that can affect the internal organization of dipolar excitations in the neuronal tubulin system. This can then cause a consequent propagation of action potential and an associated neuronal firing event. This is the idea that can explain the effect of "freezing a kicker". This takes place when the coach of the other football team calls a timeout to give the kicker time to think about the kick that he is about to make. If the kicker "thinks too much" he chokes, in effect defeating his own free will. Thus, the phenomenon of free will is exploited in competitive sports, which is a "play" on the quantum nature of free will mediated potential.

Human potential on an individual scale is impossible to define. It is highly variable but the effects are far-reaching, ranging from physiological health to success in sports, relationships, or a career. Most importantly, meeting the potential of free will connects us to our core values that allow finding personal joy and happiness in the journey of life.

6.7 On the Importance of Connections at All Levels

The break in connections of a living system means a loss of complexity. The robustness of interactions at every hierarchical scale from molecular to cellular to systems biology, the integration of organ systems to an organismic whole with maximum health is tantamount to maximum connectedness. The breakdown in the richness of these interconnections across hierarchical scales is in essence compartmentalization of systems biology, which results in disease. Metabolism drives this connectedness as perhaps best described in Chapter 3 when connections between vital organ system functions become reduced. The broader system complexity is lost and may be exemplified by the loss of heart rate variability as discussed above in this volume. In the most literal sense, this loss of complexity equates to disease morbidity. The progressive loss of complexity over time evolves the chronic diseases of aging and the vulnerability to mortality, which is in parallel with this progression. Metaphorically, a single metabolic tendril connecting vital organ system function is life support.

The loss of the rich entanglement of connectedness leads to ensuing isolation between systems of many hierarchical planes of systems within the system of systems, the organism whole, the individual, etc. This is typically induced and accelerated by toxic connections to or a loss of healthy connections from external parameters of the environment. This framework is represented by three major extrinsic control parameters discussed above in this chapter that connect ultimately to the intrinsic order parameters of the health or disease of mitochondrial function, and states of redox and free energy. The loss of free energy to heat coincides with and directly correlates with subclinical inflammation, redox disturbance, and mitochondrial dysfunction. This heat in the form of energy that cannot be utilized to do the work of biological maintenance and physiology is the result of the loss of connectedness.

The most under-recognized element of public health is the emotional stress response, which itself has unhealthy toxic connections to inflammatory mediators derived from the physical processes of disease. Social isolation from family, friends, co-workers, colleagues, and community represents a crucial extrinsic or external control parameter that may precipitate or prolong the hormonal and autonomic branches of the neural endocrine stress response that mediate immune dysfunction and inflammation. When chronic, this leads to premature human diseases of aging. The deterioration of connectedness at all levels of the entangled web, both within the boundaries of a human being and beyond, is a manifestation of disease. Empathy, compassion, social awareness, and intuition are critical personality traits that must be exercised to reduce the person's chronic stress response that ironically is often a function of elevating oneself by helping others with their stress response. This is fundamental to the health of the individual and of society.

Friction beyond manageable levels is what breaks connections, causing the liberation of heat. In the hypothetical perpetual motion machine, the absence of friction allows the machine, or object, once set in motion to keep moving without ever coming to a stop. In the process, no heat is liberated and the entropy production rate is zero. This hypothetical system is a purely quantum phenomenon, and as such is the counterpart to Newton's first law of classical mechanics. Again, this is a hypothetical model, because nothing, either man-made (yet) or in nature, is purely quantum as a whole system. A purely quantum living system would be age-less, and in fact, the evolutionary pressure to reproduce would be lost in such an idealized situation. Nonetheless, the transformation of energy from one form to another, for example, the quantum metabolism of ATP production, does exist at relevant degrees within the functioning of an overall individual of the healthy stages of his or her life. Other manifestations of quantum metabolism may occur in the realm of human consciousness and cognition, as well as potentially albeit not yet proven, underpinning the synchronized and coherent nature of circadian molecular clocks involving the core clock component, cryptochrome, and hence circadian metabolic physiology as described in Chapter 3.

There are other more established examples of quantum biology in humans such as in the process of olfaction. Since metabolism may be described by the first law of thermodynamics, in the simple sense of the process of transformation of energy from one form to another. All these examples are manifestations of metabolism in one form or another. Moreover, like the connectedness that determines the health of an individual, quantum connections are likely to extend beyond the cutaneous boundaries to involve entanglement between human beings (and even pets). Such phenomena, known as phase-locking, describe the synchronicity and coherence of the artistic examples in music and sports described above. Like a cancer cell that becomes detached from regulatory control, and spreads throughout the body as a system, a single player who causes friction within a sports team or musical band, it is the root of disease that easily and rapidly metastasizes throughout the system. Analogously, the political divisiveness in the United States currently is a signal of a feed-forward deteriorating instability, which we will briefly discuss next.

6.8 The Physiological Fitness Landscape and Society

The concept of the physiological fitness landscape is rooted in the physics and mathematics of systems undergoing changes in their structural and dynamical stability. These are very general considerations involving a tendency of any physical system to return to its energetic equilibrium. In mechanics, this implies a state of lowest energy, in thermodynamics a state that corresponds to a maximum entropy under external constraints, etc. Translating this to a living system, a tendency to maintain energetic equilibrium means homeostasis. When a living system is subjected to stress, its response to stress is called allostasis and its aim is also to maintain stability but in the presence of stress. Every physical and also biological system has a limit to its stability under an ever-increasing amount of external perturbation or stress. In the case of physical systems, this could lead to a phase transition into another state of matter. For example, a piece of metal subjected to an increase in temperature will eventually melt and transition from solid to liquid. A human being subjected to an external stressor such as a growing amount of peer pressure will eventually lead to a breaking point and may develop a psychosomatic disease, clinical depression, or worse. What these examples illustrate is a complex landscape in the space of parameter values that define various stressors and physiological states. Valleys correspond to equilibria and peaks to transitions between equilibrium states that are typically separated by potential barriers that allow us to return to equilibrium if stress is within a tolerable amount. Putting the system over the "hump" and into the next equilibrium state by external forces usually means a catastrophic transformation, a breaking point such as excess mechanical force out on a piece of solid will break it into two or applied to a sheet of glass will shatter it into pieces.

This metaphor can also be translated into the context of sociology and politics, which is very much a topic of current discourse within both the mainstream media (MSM) and social media. Any social network exhibits collective behavior, very much akin to that of an organism or a physical system composed of many elements that interact forming a coherent unit. This social network could be defined by citizenship (country), religious belief (faith), or linguistic group (language or dialect). Such networks tend to maintain stability when left to their own devices but are also subjected to external stresses and sometimes internally generated transformations. When external stresses or internal rearrangements reach critical points, these systems may face an existential crisis that corresponds to reaching a peak in their own fitness landscape. At this point, the system is extremely sensitive to even the smallest perturbation and this may result in its transition to a neighboring equilibrium state with virtually no possibility of returning to the old equilibrium when the potential barrier is large enough.

In the case of societies or countries, such instability points highlight dramatic transformations such as revolutions, civil wars, or changes in the economic systems. Well-known examples include the French Revolution, the Civil War in the United States, or the overthrow of communism in Poland in 1989. These are truly fundamental changes in the stability of the system and in the above examples, they occurred virtually from within following a period of gradual increase of the societal "temperature", which is meant to represent internal conflict or divisions. This could also result from external causes such as invasion, aggression, trade war, etc. In the period preceding such a dramatic transition, one can observe an increased social dynamics, fluctuating opinions, a greater amount of internal conflict and division, and a heightened state of general anxiety. There is also a possibility of a phase separation where parts of the whole become autonomous or independent due to the major differences in their aims and goals. This can be illustrated by the split between India and Pakistan along religious lines after achieving independence from British colonial rule. During the Civil War or the Russian revolution, this situation manifested itself for a time but both parts coalesced into a single stable unit in the end when one of the two opposing sides lost and the other won. Czechoslovakia, on the other hand, underwent a peaceful separation into two independent countries of the Czech Republic and Slovakia along linguistic lines.

The election of Donald Trump in 2016, and the vote for Brexit in the UK earlier that year, were events that can be now seen as internally generated stressors that moved both the US and UK societies, respectively, on a trajectory far away from the traditional stable equilibrium (homeostasis) and toward allostatic overload. At the time of writing this book, in 2020, it is too early to tell whether the ensuing phase separation between the strongly polarized and divided parts of each of these two societies will cause a transition to a new and potentially dramatically different equilibrium state or lead to a stronger and more unified country at the end of the period of upheaval. In the opinion of this writer, whether the outcome of this process will be positive or negative will entirely depend on the strict adherence to the core values that made both the US and UK such strong bastions of civility, namely free speech, individual liberties, democracy, equal rights for all, and a free-market economy. These fundamental principles used to lie at the core of societal equilibrium and hence the political parties in both countries, as in more or less all of the West, always tended to compete for the center, sometimes having lots in common and very little that differentiated between their platforms.

Today, the differences are striking and the common ground almost non-existent. In the UK, the main difference between Labour and the Conservative Party was literally a continental divide, i.e. whether or not the UK should belong to the EU. In the US, the divide between the Republicans and Democrats appears to be along ideological lines that demarcate preferences between capitalist and socialist economies, nation-state versus global village perspectives, one common language or a multilingual society, traditional American melting pot or a multicultural framework to be used for nation-building, etc. The further this polarization progresses, the more it is analogous to reaching an allostatic overload point of no return and the more worried we should be about this country staying together as a coherent unit irrespective of the fact that it is the most powerful state that ever existed. It may still lose its stability due to the internally generated stress and either collapse or splinter into fragments. As a physician, I see this as a manifestation of a pathological state and hope for a cure that may bring the state of homeostasis back into the political fitness landscape of the US. As argued above, politics or social life in general can be viewed through the same lens as the life of an organism, i.e. as a struggle to find balance facing challenges and opposing forces.

6.9 Striving for Balance Amongst Complexity

The overarching purpose of the two volumes of this book is to invoke physics and other disciplines to better understand biological systems and human diseases. The more perspectives we have, the greater our capacity to problem solve, and the more successful the future of medicine will be. The notion of systems of parts and wholes, and of perspectives, should be applied to all aspects of society including politics. Heterogeneity is part of nature's design. The interlocking of strengths and weaknesses of people results in developing essential building blocks toward the greater health of individuals and society as a whole. Movement to the far right or to the far left isolates elements from the common good. Perspectives become increasingly diminished, myopic and polarized. Heat is generated as connections are broken as a result of the friction.

Cognitions degenerate into emotions of fear, hostility, anger, and aggression on the societal scale. This self-amplifying and self-destructive circuit is a classical sign of metastatic disease. This is very reminiscent of standing on the mountain ridge in a fitness landscape and being at a tipping point in a highly susceptible state. A small perturbation leads to either the danger of falling off the cliff to another attractor state or falling back to the familiar valley of stability. The choice is between internal conflict or civil war versus civic discourse and a healthy and mutually respectful debate. Hoping that cool heads will prevail, this author is optimistic that American society will find a way back to civility and a tradition of robust democracy and belief in a common good. An even more optimistic attitude should be associated with our crystal ball gazing into the future of medicine and human health. We are entering into an era of unparalleled medical progress translating into long and healthy lives for more and more people on our planet.

6.10 A New Perspective

Why is this book relevant and what can I do with this information? Let us start with the following observations. As described in the section entitled "Application of Molecular Biology to Clinical Enigmas", if as physicians we did everything right in accordance with standards of care, we are only correct less than half of the time. We do the best that we can with the privilege of serving our patients using the knowledge available to us at the time. This book provides a new and hopefully useful perspective that recognizes proper characterization of metabolism as providing the core information necessary for problem solving in clinical medicine. One of the novelties we propose is the identification and quantification of control and order parameters that pertain to both health and disease and lead to the development of fitness landscapes that can inform the physician and the patient how to best maintain the state of health or return to it from a pathological state. Expertise in understanding metabolic function is critical to not only the development of a robust multidimensional set of order parameters and their measurements, but also to providing physicians with clinically intuitive and significantly non-algorithmic skills. This is based on the fundamental understanding of extrinsic and intrinsic control parameters of health that are characteristic properties of each individual. These parameter values change over time as we age and succumb to disease, Hence, as an example, periodic updating through the performance of stress tests would be needed.

The purpose of this book is not to serve as a "cookbook-style how-to guide", marketed as a popular book that can be monetized. Rather, it aims to provide a broad perspective

informed by current scientific advances explaining what regulates health and how best to maintain it. Also, this book is an expression of appreciation for the complexity of biological systems and the inchoate nature of clinical medicine. For the reader with innate intellectual curiosity, the author hopes that this book offers a survey of a colorful and textured tapestry of medicine, biology, and physics indicating directions toward a deeper exploration of the laws of physics as a guide to a better understanding of biological behavior. Further, the fitness landscape provides a model that can be employed for practical problem solving that prevents, manages, and even potentially reverses disease and accelerated aging. In the author's opinion, this promises to play not only a useful but a central role in the future of medicine.

6.11 The Bridge from Physiology to Spirituality

Clinical physicians are not basic scientists. We are applied scientists, engineers of human health. We use both logic and experience to best treat our patients. However, the mysteries of the human body often make a rational argument elusive. Instead of strictly following logic and the scientific method we often defer to both individual experiences as physicians and the collective experience of our profession in terms of the standard of care. Accordingly, we collectively possess an enormous potential for the integration of the sciences in order to find a deeper meaning and truth in the therapeutic experience.

Reductionism is based on the linearity of scientific political methodology. It was the crown jewel of the 20th-century philosophy of science. However, it is fraught with profound limitations, especially relevant in explaining the complexity of biological systems, and its application to medicine. An uncritical application of reductionist thinking underscores a very dangerous game that we are playing due to the contextual nature of biology and the inability to predict outcomes.

Intriguing but profound, we have to imagine how linearity brings us full circle. The notion of perspective is an inherent conflict when there are multiple points of view or representations of the truth. A single perspective defines both ultimate reality and maximum efficiency. A single perspective is the linearity of quantum physical phenomena of entanglement and superposition of states into a single state. Superposition of states representing a single state involves linearity in the sense that there are no interactions with outside states because the states of concern superimpose as a single fabric of reality. Examples of entanglement and superposition of states on a macroscopic scale include quantum metabolism in biology, among other examples described in Chapter 3 (From Quantum Biology to Quantum Medicine) in this book. An important example also includes the phenomenon of consciousness or conscious cognitions that is a non-biological perception but only recognizable from the perspective of biological experience. Specific examples of this type include mind-body phenomena of extraordinary athletic, entertainment, or intellectual accomplishments. Moreover, the therapeutic effects of the placebo are far more powerful than traditionally appreciated. These phenomena require a superposition state of the mind-body system. Even the broader context of the entanglement invokes out-of-body experiences of a warm, welcoming world often including deceased relatives. Thus, it is compelling that the superposition of states that explains the beautiful complexity and exquisite organizational perfection of the state of human health, manifested in the phenomenon of quantum metabolism as an example, is only a localized component part of the greater whole in the superposition of consciousness.

In addition to the superposition state of quantum metabolism as a macroscopic biological phenomenon, an even larger more macroscopic hierarchical scale in biology involves the temporally organized correlated state of circadian synchronized metabolic physiology and behavior-coordinated scale of biological molecular clocks. This major insight into biological molecular clocks was recognized in 2017 by the award of the Nobel Prize of Medicine and Physiology. These molecular clocks may be understood as a level of integration translating a quantum manifestation of energy production into the synchronous scale of metabolic physiology.

Finding joy in the conscious experience is an opportunity to live life to the fullest in synchrony with the fundamentals of our biological organization. On the other hand, stressful and painful experiences build resilience just like building muscle through exercise. This resilience is based on neuronal connections of the deep cortical layers of the brain responsible for consciousness, or cognitions. Building the strength of these synaptic connections accompanies an increase of the animated nature of the deeper levels of awareness, and awareness of awareness, in the living experience. This is tantamount to escaping narcissism and self-preoccupation with a greater appreciation for perspectives that lay outside of the self. Furthermore, it promotes the human qualities of the prefrontal cortex, namely empathy, compassion, social awareness, and intuition, with greater interpersonal connectedness. Indeed, the interest and capacity for understanding other people's perceptions, feelings, and perspectives, and helping them to find solutions and to resolve personal anguish or pain, offers the greatest opportunity to elevate and vitalize our own personal sense of self. The inability to do this defines the state of uncontrollable and devitalizing stress rooted in the top-down regulation of cognition by emotion. We as physicians have both an opportunity and a responsibility to transcend our own self-centeredness and help others overcome their health-related limitations.

NOTE

1. This book was written over the course of ten years, with this passage being written three years ago. Tom Brady is now 43 years old.

REFERENCES

1. Kuhn, T.S. (2012). *The Structure of Scientific Revolutions.* University of Chicago Press. Chicago, Illinois.
2. Demetrius, L. (1997). Directionality principles in thermodynamics and evolution. *Proceedings of the National Academy of Sciences*, 94(8), 3491–3498.
3. Ref for Figure 6.2 Photograph of the "helmet catch" from the 2007 Super Bowl. www.bostonherald.com/sports/patriots/2016/01/super_bowl_xlii_giants_derail_patriots_bid_for_perfection
4. Craddock, T.J., Hameroff, S.R. and Tuszynski, J.A. (2017). The "Quantum Underground": Where Life and Consciousness Originate. In: *Biophysics of Consciousness: A Foundational Approach* Poznanski, R. R., Tuszynski, J. A., & Feinberg, T. E. (Eds.), New Jersey, NJ, (pp. 459–515).
5. Crowe, C. (1975). "The Durable Led Zeppelin". *Rolling Stone.* https://www.rollingstone.com/music/music-news/the-durable-led-zeppelin-36209/

Index

For Product Safety Concerns and Information please contact our EU
representative GPSR@taylorandfrancis.com
Taylor & Francis Verlag GmbH, Kaufingerstraße 24, 80331 München, Germany

www.ingramcontent.com/pod-product-compliance
Lightning Source LLC
Chambersburg PA
CBHW080715220326
41598CB00033B/5430

9 780367 712259